T0201615

Data Analysis Techniques for Physical Scientists

Data Analysis Techniques for Physical Scientists is a comprehensive guide to data analysis techniques for physical scientists, providing a valuable resource for advanced undergraduate and graduate students, as well as seasoned researchers. The book begins with an extensive discussion of the foundational concepts and methods of probability and statistics under both the frequentist and Bayesian interpretations of probability. It next presents basic concepts and techniques used for measurements of particle production cross sections, correlation functions, and particle identification. Much attention is devoted to notions of statistical and systematic errors, beginning with intuitive discussions and progressively introducing the more formal concepts of confidence intervals, credible range, and hypothesis testing. The book also includes an in-depth discussion of the methods used to unfold or correct data for instrumental effects associated with measurement and process noise as well as particle and event losses, before ending with a presentation of elementary Monte Carlo techniques.

Claude A. Pruneau is a Professor of Physics at Wayne State University, from where he received the 2006 Excellence in Teaching Presidential award. He is also a member of the ALICE collaboration, and conducts an active research program in the study of the Quark Gluon Plasma produced in relativistic heavy ion collisions at the CERN Large Hadron Collider. He has worked as a Research Fellow at both Atomic Energy for Canada Limited and McGill University, and is a member of the American Physical Society, Canadian Association of Physicists and the Union of Concerned Scientists.

Data Analysis Techniques for Physical Scientists

CLAUDE A. PRUNEAU
Wayne State University, Michigan

CAMBRIDGE
UNIVERSITY PRESS

University Printing House, Cambridge CB2 8BS, United Kingdom

One Liberty Plaza, 20th Floor, New York, NY 10006, USA

477 Williamstown Road, Port Melbourne, VIC 3207, Australia

4843/24, 2nd Floor, Ansari Road, Daryaganj, Delhi - 110002, India

79 Anson Road, #06-04/06, Singapore 079906

Cambridge University Press is part of the University of Cambridge.

It furthers the University's mission by disseminating knowledge in the pursuit of education, learning and research at the highest international levels of excellence.

www.cambridge.org
Information on this title: www.cambridge.org/9781108416788
DOI: 10.1017/9781108241922

First published 2017

Printed in the United Kingdom by TJ International Ltd. Padstow Cornwall

A catalogue record for this publication is available from the British Library

ISBN 978-1-108-41678-8 Hardback

To my son Blake

Contents

Preface

Physics students typically take a wide range of advanced classes in mechanics, electromagnetism, quantum mechanics, thermodynamics, and statistical mechanics, but sadly, receive only limited formal training in data analysis techniques. Most students in experimental physics indeed end up gleaning the required material by reading parts of a plurality of books and scientific articles. They typically end up knowing a lot about one particular analysis technique but relatively little about others. Paradoxically, modern experiments in particle and nuclear physics enable an amazingly wide range of very sophisticated measurements based on diverse analytical techniques. The end result is that beginning students may have a rather limited understanding of the many papers they become coauthors of by virtue of being members of a large scientific collaboration. After twenty years of teaching "physics" and carrying out research in heavy-ion physics, I figured I should make an effort to remedy this situation by creating a book that covers all the basic tools required in the data analysis of experiments at RHIC, the LHC, and other large experimental facilities.

This was a fairly ambitious project given that the range of techniques employed in today's experiments is actually quite large and rather sophisticated. In the interest of full disclosure, I should state that the scope of the project changed several times, at times growing and at others shrinking. Eventually, I decided for a book in three parts covering (I) foundational concepts in probability and statistics, (II) basic and commonly used advanced measurement techniques, and (III) introductory techniques in Monte Carlo simulations targeted, mostly, toward the analysis and interpretation of experimental data. As such, it became impossible to present detailed descriptions of detector technologies or the physical principles they are based on. But as it turns out, high-quality data analyses are possible even if one is not familiar with the many technical details involved in the design or construction of detectors. Detector attributes relevant for data analyses can in general be reduced to a statement of a few essential properties, and it is thus possible to carry out quality analyses without a full knowledge of all aspects of a detector's design and operation. I have thus opted to leave out detailed descriptions of detector technologies as well as particle interactions with matter and focus the discussion on some representative and illustrative examples of data calibration and analyses. Detailed discussions of detector technologies used in high-energy nuclear and particle physics may, however, be found in a plurality of graduate textbooks and technical texts. Additionally, I have also omitted few big and important topics such as interferometry (HBT), jet reconstruction, and neutral networks, for which very nice and comprehensive books or scientific reviews already exist.

Overall, this book essentially covers all basic techniques necessary for sound analyses and interpretation of experimental data. And, although it cannot cover all analysis techniques used by modern physicists, it lays a solid foundation in probability and statistics, simulation techniques, and basic measurement methods, which should equip conscientious and dedicated students with the skill set they require for a successful career in experimental nuclear or particle physics, and such that they can explore more advanced techniques on their own.

I should note, in closing, that although this book targets primarily students in nuclear and particle physicists, it should, I believe, prove to be a useful introduction to data analysis for students working in other fields, including astronomy and basically all other areas of the physical sciences. It should also, I hope, provide a useful reference for more advanced and seasoned scientists.

I would like to express my sincere acknowledgments to the many people who, through discussions and advices, have helped shape this book. These include Monika Sharma, Rosie Reed, Robert Harr, Paul Karchin, and Sergei Voloshin, who through questions and comments have helped me plan or contributed various improvements to the book. I also wish to acknowledge the important contributions of several undergraduate and graduate students, most particularly Nick Elsey, Derek Everett, Derek Hazard, Ed Kramkowski, Jinjin Pan, Jon Troyer, and Chris Zin, who served as guinea-pigs for some fractions of the material. I am grateful to my colleagues Giovanni Bonvicini, from the CLEO Collaboration, for providing a Dalitz plot; Yuri Fisyak and Zhangbu Xu, from the STAR (Solenoidal Detector at Relativistic Heavy Ion Collider [RHIC]) Collaboration, for their contribution of a dE/dx plot; and my former postdoctoral student, Sidharth Prasad, for producing exemplars of unfolding. I also acknowledge use of several sets of results from the STAR collaboration, publicly available from the collaboration's website, for the generation of figures presenting examples of flow measurements and correlation functions. I am particularly indebted to colleagues Drs. Jean Barrette, Ron Belmont, Jana Bielcikova, Panos Christakoglou, Kolja Kauder, William Llope, Prabhat Pujahari, Sidharth Prasad, Joern Putschke, and William Zajc for their detailed reading and feedback on various sections of the book corresponding to their respective areas of expertise and interest. I also wish to acknowledge Ms. Heidi Kenaga and Ms. Theresa Kornak for their meticulous proofreading of the manuscript and for being so nice in correcting my Frenglish.

Finally, I wish to acknowledge that a large fraction of the graphs and figures featured in this book were created with ROOT, Keynote, and Graphic Converter. Several of the ROOT macros I wrote for the generation of figures will be made available at the book website.

Claude A. Pruneau

How to Read This Book

Not all students, instructors, and practitioners of the field of experimental physics may have the inclination, the time, or the need to study this book in its entirety. Indeed, only a selected few may have the opportunity to read the book from cover to cover. This should not be a problem, however, because the material is organized in large blocks that are reasonably self-sufficient, and ample references to earlier or upcoming chapters, as the case may be, are included in the narrative. The book also includes a number of specialized or in-depth topics that may be skipped in a first reading. Such topics include, for instance, the formal definition of probability in §2.2, the notion of Fisher information discussed in §4.7, the technique of Kalman filtering introduced in §5.6 and for which a detailed example of application is presented in §9.2.3, as well as discussions of track and vertex reconstruction presented in §§ 9.2 and 9.3. This said, the book is designed to progressively develop and approach topics, and it should then be possible to study the material in a variety of ways, adapting the depth and breadth of coverage. The following are recommended lists of chapters and sections that should be covered given specific and targeted needs.

- Introductory course in probability and statistics:
  Chapters 2 (§§2.1, 2.3–2.11), 3 (§§3.1–3.13), 4 (§§4.1–4.6), 5 (§§5.1–5.5), 6 (§§6.1–6.6), 7 (§§7.1, 7.2, 7.4, 7.7), 13
- Advanced course in probability and statistics:
  Chapters 1, 2, 3, 4, 5, 6, 7, 13
- Introductory course in data analysis techniques (one semester):
  Chapters 1, 2 (§§2.1, 2.3–2.11), 3 (§§3.1–3.13), 4 (§§4.1–4.6), 5 (§§5.1–5.5), 6 (§§6.1–6.6), 8 (§§8.1–8.6), 9 (§§9.1, 9.2), 13
- Advanced course in data analysis techniques (two semesters):
  Chapters 1, 2 (§§2.1, 2.3–2.11), 3 (§§3.1–3.13), 4 (§§4.1–4.6), 5 (§§5.1–5.6), 6 (§§6.1–6.7), 7, 8 (§§81–8.6), 9 (§§9.1, 9.2), 12, 13, 14
- Course on correlation functions (one semester):
  Chapters 2 (§§2.5–2.13), 4 (§§4.3, 4.5, 4.6), 10, 11, 12, 13

Of course, instructors using this book should feel free to select and change the order of topics to suit their specific needs. For instance, Monte Carlo methods are formally introduced in Chapter 13 but it is often useful and inconvenient to use and discuss some of these concepts along with materials of the early chapters (e.g., Chapters 2–7).

1 The Scientific Method

1.1 What the Brain Does

From the moment a child is born, her brain is flooded with signals generated by her senses, including smells, sounds, images, touches, and tastes, as well as various other messages produced by her internal organs. Her brain readily engages in processes of pattern recognition and classification to "organize" all these data. This pattern recognition forms the basic elements of perception or views of reality the child will gather from this moment on. Slowly but surely, her brain will start to organize the data into mental models of the world that surrounds her: *If I cry, somebody comes, feeds me, holds me, and it feels good.*

At first, her perceptions are crude and simple, but with time, they progressively become more and more sophisticated. And soon, with the emergence of language abilities, her modeling of the world acquires tags and labels. She communicates the things she likes and dislikes, her wants, and what she doesn't want. And, as her senses and motor skills develop, she becomes an avid explorer of her immediate environment, discovering and adding to the categories, that is, the models already partially in place. The process is slow, sometimes laborious, but ineluctably, her brain engages in increasingly sophisticated pattern recognition and progressively forms complex and subtle representations of the world. This process eventually explodes in a rich and intense search for meaning, through her adolescence and early adulthood, as her brain nears full maturity. In time, her brain becomes a sophisticated pattern recognition engine that excels at detecting, sometimes inventing, patterns of all kinds, and making connections between the various entities that inhabit her world. Quite naturally, and with perhaps relatively limited awareness of her mental processes, she becomes an intuitive model builder, arguably some form of prescientific state of mind endowed with curiosity, a vast capacity for inquiry, and extensive intellectual resources that enable her not only to witness the world, but also become one of its actors, to experiment, and even to shape reality.

Though the storyline of the development of a human from infant to adulthood is fascinating, the reader might ask why it is relevant as an introduction to a book on data analysis techniques?

The answer is rather simple: much like a growing child, modeling the world and experimenting is in fact what scientists do. Indeed, the raison d'être of scientists is to discover, observe, and formulate models of the world and reality. Could it be then that we humans are scientists by our very nature, that is, by simple virtue of the inner workings of our minds? Surely, the evolution of a child's brain and mental processes just briefly depicted seems a rather universal process, and this does suggest that all humans are endowed with a natural and innate ability to become scientists. The fact that only relatively few among us end up making science their primary occupation and profession may not diminish this capacity in the least, but we need to acknowledge that as each of us emerges into adulthood, we develop varied interests and skill sets and engage in diverse activities, each accompanied by distinct and at times seemingly incompatible forms of discourse. Our capacity to detect or make up patterns does not vanish, however, and we continue through adulthood to seek understanding and meaning in the people and events that surround us. Evidently, given that our models of the world are based on our individual experiences and circumstances (including formal and informal education), we collectively end up having a plurality of views and interpretations of reality, some of which may clash drastically, or even violently.

Can all these views and forms of discourse be true simultaneously, or are there models that constitute a closer representation of reality? And if so, is there a privileged form of discourse and method that can enable us to reach, progressively perhaps, a more robust and truthful model of reality? Is there, in fact, such a thing as reality? These are both powerful and complex questions that many great thinkers have reflected upon through time. Indeed, ideas on such matters abound and many philosophers, through history, have claimed ownership of the truth. Amid all these ideas, one particular form of discourse and inquiry has risen and developed. It is both humble and powerful: it admits its innate incapacity to reach a perfect truth, but provides the means to progressively and systematically identify better views of reality. We call it the **scientific method**.

1.2 Critics of the Scientific Method

Science, as an empirical form of inquiry, finds its roots in the work of Copernicus, Galileo, and Newton. In Europe, before and around their times, the dominant philosophical view was that human perceptions and reasonings are intrinsically fallible and cannot be trusted, and that we should consequently rely on the word of God embodied in the Bible to guide our views and interpretation of reality. Copernicus, who had an interest in the motion of heavenly bodies, most particularly the planets, came to the realization that the Ptolemaic tables that had been used to describe the motion of the planets for several centuries were quite inaccurate and developed the notion that better observations of their positions through time would reveal the geocentric Ptolemaic model is wrong and that the planets, Earth included, all revolve around the Sun. Galileo and Newton, much like Copernicus, would champion the notion that the heavens are not perfect, and that careful observations can reveal much about the nature of things.

Arguably, it was Galileo who made the greatest breakthrough. Equipped with the telescopes he had built, he proceeded to discover mountains on the Moon, spots on the surface of the Sun, the Sun's rotation, phases of Venus, and satellites orbiting Jupiter. He would then conclude that the heavens are not immutable or perfect, and demonstrably show that empirical observations are not only possible but also powerful in their capacity to reveal new phenomena and new worlds.

Newton would later use precise observations of Mars, in particular, to demonstrate that the planets follow elliptical orbits with the Sun at one foci, and formulate a theory postulating the existence of a force acting at a distance between the Sun and the planets: gravity.

Together, the work of Galileo and of Newton would exemplify the notion that empirical knowledge is possible, particularly when the senses are enhanced (e.g., with a telescope) and reasonings framed into a powerful mathematical language (e.g., calculus). Together, these works would provide the impetus for a new view of the world and a new approach to scholarly works: empirical observations coupled with detailed mathematical representations of world entities and their relations can lead to great advances in our ability to understand and shape the world. Science, or the Scientific Method, as we now think of it, was born.

In spite of these great early successes, several philosophers would argue that while empirical methods have merits and do enable the formulation of models or theories of the world that work, such models could never be proven absolutely correct. David Hume (1711–1776), in particular, argued that it is factually impossible to deduce universal generalizations (i.e., models that always apply) from series of finitely many observations, and consequently, that inductive reasoning, and therefore causality, cannot, ultimately, be justified rationally. The notion has obvious merits. For instance, to use a rather trivial example, consider whether the fact that all zebras *you* might have so far seen in your life featured a black and white striped mane implies that all zebras are necessarily striped black and white? Obviously not: you have not seen all zebras in existence on our planet and thus cannot conclude all zebras feature black and white stripes. In fact, golden zebras, featuring a pigmentation abnormality characterized by the lack of melanin color pigments, do exist.

A less trivial example of Hume's point involves Newton's three laws of mechanics and his Law of Universal Gravitation. Following the initial successes of the theory in explaining the observations of Mars' orbit, Newton's laws of mechanics and gravity were tested repeatedly throughout the eighteenth and nineteenth centuries and found to be exquisitely accurate. Universal Gravitation also featured great predictive power best exemplified by the discovery of the planet Neptune based on calculations by the French astronomer Urbain Le Verrier, that accounted for observed anomalies in the orbit of Uranus. It seemed fitting, indeed, to qualify the law as universal. Yet, neither the three laws of mechanics nor the law of gravity would indefinitely survive the test of time. Indeed, with the publication of a paper, in 1907, by Albert Einstein, it emerged that the laws of Newton were in fact incorrect, or as many physicists prefer to say, incomplete. Evidently, demonstrating the inadequacy of Newton's laws (or more properly stated, the underlying principle of Galilean transformations) would require observations involving light and objects moving at large velocities. A few years later, in 1915, with the publication of a paper on general relativity, Einstein would also put into question Newton's amazingly successful theory of gravity. General relativity would eventually find confirmation in observations of the precession of the orbit of Mercury and the deflection of light by the Sun's gravity measured by Sir Arthur Eddington in 1919 [78]. No matter how many observations could be successfully explained by Newton's Law of Universal Gravitation, its failure to successfully explain the magnitude of Mercury's precession and the proper deflection of light by the Sun would provide tangible evidence of its inaccuracy as a model of reality. Hume was certainly right. Yet, he also obviously overstated the case. Though not perfect, Newton's laws proved to be immensely useful in explaining the world as well as building devices and artifacts that would ease humans' lives. Indeed, effective empirical knowledge is possible, even though it is bound to forever be tentative.

1.3 Falsifiability and Predictive Power

The notion of empirical knowledge as tentative was properly clarified by Sir Karl Popper, who argued that although a scientific theory cannot be proven correct by any finite number of experiments, it can be **falsified**; that is, it may be proven false (wrong) in the appropriate context. Indeed, it suffices to observe a single non-black and white striped zebra to conclude that the theory that all zebras are black and white striped is incorrect. Likewise, measurements of the precession of Mercury and the deflection of light would demonstrate the inadequacy of Newton's Law of Universal Gravitation. The theory could not be proven right, but it could be proven wrong. Limited empirical knowledge is thus possible, insofar as it can be falsified, that is, demonstrated false in an appropriate context. We will see in Chapter 6 that **falsifiability** has led to the notion of **null hypothesis** and (scientific) **hypothesis testing**. A null hypothesis typically champions the accepted theory (e.g., Law of Universal Gravitation). Experiments are then conducted to test the null hypothesis and the theory remains unchallenged as long as it is deemed acceptable based on measured data. An accumulation of observations that support the null hypothesis increases its plausibility (degree of belief) without ever proving it is universally correct. However, if (reliably) measured data are observed to deviate sizably from the null hypothesis (i.e., data values predicted based on the null hypothesis), the null hypothesis is considered rejected (false) and the theory challenged. And if the number of observations that challenge the null hypotheses becomes large, or the observed deviations can be considered irreconcilably large, the theory is eventually abandoned.

Alas, nothing is that simple. An experiment can go wrong and produce results that improperly reject the null hypothesis. The rejection is then considered an **error of the first kind (type I)**. Alternatively, experimental results can also falsely support the null hypothesis, and the unwarranted acceptance of the hypothesis is known as an **error of the second kind (type II)**. Regrettably, pushing this argument to its absurd limit, postmodern philosophers have formulated the notion that even falsification is impossible. Is the golden zebra actually a zebra or something else? Can one be sure of anything at all; can one reliably say anything at all? This absurd line of argument has led to the postmodern notion that all theories are equally valid; that all forms of discourse and inquiries are equally valid; and science, most particularly the scientific method, does not constitute a privileged vehicle to acquire and validate (or falsify) knowledge. This view, however, is considered far too extreme by most modern scientists given that it blindly neglects the tangible and significant advances made by science in the last three centuries. The post-postmodernistic view, as one might call it, accepted by most modern scientists, is that Popper was in fact essentially correct but the notion of **predictive power** must augment basic falsification. In other words, scientific models must be capable of making predictions that can be tested against careful and reproducible observations or measurements.

Indeed, much like a child's brain evolves to become a model builder and learn to learn, philosophers and scientists have, over time, examined and pondered how to learn from Nature and eventually formulated what is now known as the **scientific method**. This said, the scientific method is not fundamentally different today than it was, say, fifty or one hundred years ago. Although certainly more articulated, debated, and written about, it remains rather simple at its very core: observe a phenomenon of interest, formulate a mathematical model, verify how the model accounts for past observations, and use it to predict variants of the phenomenon and future observations. And, following Popper, models shall be readily abandoned if falsified by observed data. However, falsifiability is often largely insufficient: models based on distinct and perhaps incompatible assumptions may often be concomitantly supported by a specific dataset. It is thus necessary to identify extensions to existing measurements and examine where conflicting models may deviate appreciably from each other, and thereby provide grounds for additional testing and falsification.

1.4 A Flawed but Effective Process!

At this point, it is important to stress that perceptions and experimental measurements do not in fact need to be perfect for science to progress. As we will discuss at length in Chapter 12, no perception or measurement is in fact ever perfect: there are invariably resolution effects that smear the values of measured (physical) quantities; there are signal recognition or reconstruction issues that lead to signal losses and biases; and there may be background signals that may interfere with a measurement. But it is in the very nature of the scientific method to base measurements on techniques and processes that are well established and can be modeled with a high degree of reliability, and as such, can be corrected for experimental or instrumental effects. Indeed, a detailed understanding of the measurement process enables the formulation of (measurement) models that enable precise and reliable corrections of measured data. Although our sensory perceptions may not provide an objective view of reality, our technology-based measurements can. And if and when corrections applied to data are deemed questionable, the correction procedures can usually be studied and improved through better design of the measurement apparatus and protocol.

1.5 Science as an Empirical Study of Nature

The first model assumption to be made is perhaps that a particular phenomenon might be interesting or useful to study. Although such a statement may seem trivial, particularly in modern scientific cultures, it amounts to a relatively new idea, which dates back to the times of Copernicus and Galileo. Before the Copernican revolution,[1] the prevalent

[1] And sadly still today in some conservative religious cultures.

attitude, among learned individuals and scholars, was that wisdom cannot be gained by observations of the world but must rather be based on sacred texts, such as the Bible, and the writings of ancient Greek philosophers, most particularly those of Aristotle and Plato. But the tremendous discoveries Galileo achieved with his small telescope made it abundantly clear that new knowledge and wisdom can in fact be gained by observing Nature with our own senses, or with "machines" that enhance them.

Early scientific observations mostly involved direct sensory inputs. But in time it became clear that our senses suffer from several limitations. Indeed, there is great variance as to what might be considered a loud sound or a bright light intensity by different individuals. There is also great variability or lack of reproducibility in the observations of a single observer. Our eyes and ears, in particular, dynamically adapt to the environment and thus do not provide reliable measurements of luminosity and loudness, among several other observables. Our senses are also limited in their sensitivity. For instance, our eyes and ears cannot perceive very weak light and sound signals, and they may easily be damaged by excessively bright or loud sources. They also have rather crude abilities to detect the difference between two distinct sensory inputs (e.g., just noticeable loudness difference, pitch difference, etc.). In stark contrast, technologies based on previously acquired scientific knowledge alleviate most, if not all, of our senses' shortcomings.

In the course of time, scientists and engineers have learned to design and build devices that vastly surpass human sensory capabilities in sensitivity, precision, and accuracy, as well as dynamic range. Consider that Galileo's first telescope, with an aperture of 26 mm, could collect roughly four times the light his unaided eyes could and provided modest magnification (14×). So equipped, Galileo was able to "see" phenomena and features of Nature that were otherwise impossible to detect with human eyes alone, such as the phases of Venus, mountains on the Moon, and so forth. But modern telescopes have far surpassed the reach and prowess of Galileo's first telescope. The Hubble Space Telescope (HST) in Earth's orbit and the Keck telescopes atop Mauna Kea, Hawaii, have effective apertures 90,000 and 1,600,000 times larger than those of the human eye, respectively, thereby enabling astronomers to "see" objects at distances that Galileo himself could perhaps not even comprehend. Other orbital telescopes, such as the Wilkinson Microwave Anisotropy Probe, the Spitzer Space Telescope, the Chandra X-ray Observatory, and the Fermi Gamma-ray Space Telescope have extended the range of the human eye so it is now essentially possible to exploit the entire electromagnetic spectrum in our study of the Universe. Recent technological advances have also made it possible to detect and study gravitational waves produced by large objects in rapid motion, thereby also extending humans' very crude and primitive ability to sense gravity.

Technology not only amplifies, extends, or improves human perceptions; it also enables scientists to select, prepare, and **repeat** the conditions of particular observations. It is then possible to bring specific phenomena into focus while eliminating others, or at the very least suppress uninteresting and spurious effects. This is perhaps best epitomized by experiments at the Large Hadron Collider (LHC), such as ATLAS, CMS, and ALICE, that study collisions of specific beams of well-defined energy with vast arrays of high-granularity and high-sensitivity sensors that enable precise detection of particles produced by collisions. Much like the Keck and HST telescopes, the LHC detectors provide observational capabilities that far surpass any individual human ability, and thereby enable detailed exploration of the structure of elementary particles and the forces that govern them.

Obtaining such fantastic capabilities evidently involves many challenges. The cost of these facilities is extremely large, and their complexity is commensurate with their cost. The design, construction, and operation of these very complex machines require large international collaborations with scientists and engineers of varied and advanced skill sets. Complexity also brings challenges in the areas of student training, detector maintenance and operation, as well as data analysis, and thus often necessitates narrow training and specialization. It also brings about the need for elaborate detector calibration and data correction procedures. Fortunately, these large experimental facilities stand on the shoulders of prior facilities and experiments. They were indeed not designed totally from scratch and scientists involved in these organizations have inherited and perfected clear and precise protocols to handle all matters of data calibration and reconstruction, some of which are briefly discussed in Chapters 12 and 14.

1.6 Defining Scientific Models

While the notion that humans are natural-born modelers is enticing, it tells us very little about the requirements for scientific modeling, that is, the elaboration of scientific models or theories of the world that are falsifiable and endowed with predictive power. Surely, the plurality of religions and philosophies humans adopt or inherit from their parents should be a clear sign that reaching an objective view of reality is anything but a simple process. Yet, the fact that we all share a common predicament is a good indicator that we partake in the same reality and that although it may be difficult to reach an objective and comprehensive view of this reality, the task nonetheless remains feasible, if only by small increments. What then should be the defining elements of a scientific model of reality?

Broadly speaking, a model may consist of any constructs used to represent, describe, or predict observations. In this context, quantum mechanics, the theory of evolution, and creationism may then qualify as models of reality, at least to the extent that they provide a means to represent and interpret reality. These models, however, differ greatly in their capacity for falsification as well as in their predictive power. To be scientific, and thus used reliably or in a credible authoritative fashion, a model must satisfy a few minimal requirements that, alas, are not met by all models elaborated, shared, or inherited by humans in their daily lives. To be scientific, a model must reasonably circumscribe its range or scope of applicability; it must identify and clearly define observables or quantities of interest to the model; it must be internally consistent and logical; and it must be capable, following Popper's argument, of falsification as well as genuine and meaningful predictive power.

To be clear, Popper's notion of models as tentative implies models are not reality but at best fragmentary representations or images of reality. But three hundred years of modern science have taught us that although all models remain tentative, one can nonetheless have very high expectations from scientific models, and that in fact it is possible to reach increasingly sophisticated and powerful insights into the inner workings of our universe.

Parenthetically, it should be stressed that a person or organization presenting a model or set of ideas as the ultimate and final view of reality (i.e., reality itself) should most likely not be trusted...

Scientific models are built on the basis of well-defined entities and designed to describe these and the relations between them. Although a mathematical formulation of these relations is not absolutely required, it is typically useful because it enables, in most cases, a clear path to falsifiability and model predictability.

Roadmaps constitute an interesting basic example of a model of reality. Typically presented on a flat surface, they depict the position of entities of interest (e.g., roads, landmarks, and various artifacts) relative to a reference position, and are designed to provide guidance to travelers. Representation as flat surfaces is of course not meant to imply the world is flat but merely a necessity imposed by the media used to present maps. Roads, landmarks, and artifacts are labeled and coded to help users find their location and a path to their destination. Although somewhat simplistic as a model of reality, a map provides a good example of a basic scientific model. It is based on well-defined entities, has a clear purpose (representation of the human environment), provides predictive power (i.e., how to reach one's destination), and is falsifiable. The reliability of a map is particularly important. If roads are improperly represented or labeled, users of the map may experience delays and frustrations in reaching their destination. The map is thus easily falsified: if you reach an intersection and the road names posted at the section do not match those shown on the map, suggests that the map may not be accurate. There might be a typo on the map. It could be dated and not representative of recent changes in the road structure, and so forth. But it is also conceivable that postings are missing or improperly placed. The user could also be confused about his or her actual location. In essence, both the data (observed street names) and the model (road names shown on the map) can be wrong. The map remains nonetheless useful insofar as the number of typos or mistakes is relatively modest. A practical user would thus not dispose of the map simply because it features a few mistakes or inaccuracies. In fact, a savvy road traveler would figure out how to use her location to update and correct the map. The model could then be salvaged and thus reusable in the future without a repeat of the same frustrations. But then again, if whole regions of a city have been remodeled, an old map has lost its utility and should be disposed of, unless perhaps its uniqueness justifies keeping it as a museum piece.

While somewhat trivial, the roadmap example illustrates many of the facets and properties of a scientific model. It models well-defined entities of the real world, it indicates their relations (e.g., the Empire State building is south of the Chrysler building), it is falsifiable, and has some predictive power (i.e., how to get to one's destination). But as a model of reality, it is rather primitive and limited. It does not tell us why the streets were built the way they were, who named them, why the given names were chosen, and so forth. It is descriptive and useful but features no dynamics, evolution, or causal relationships.

Scientific models formulated in the physical and biological sciences typically seek to provide not only a descriptive account of reality but also the dynamics, that is, the causal and evolutionary interrelationships connecting the entities of a model. Classical mechanics (Newton's laws) is a prime example of a scientific model featuring descriptive components (kinematics: representation of motion) and causal components (dynamics: forces and causes of the motion). Likewise, the theory of evolution involves a descriptive component, the taxonomy of species, and a dynamic component that describes how species are connected through time, how they evolve. In both cases, it is the dynamical component of the theory that is of greatest interest because it tells us how systems change, and indeed how they evolve. Dynamical models feature great predictive power and capacity for falsifiability. For instance, not only does classical mechanics provide for a description of the motion of objects, called **kinematics**; it also features **dynamics**, which enables predictions of where objects will be in the future based on their current location and models of the forces through which they interact. If these force models are wrong, so will be the predictions. Likewise, the theory of evolution, first formulated by Darwin, empowers us to understand the relations and connections between species and how environmental constraints shaped them over long periods of time to become what they are today.

By stark contrast, creationism, as well as intelligent design, are models of reality that are totally devoid of content, falsifiability, and predictive power. To be sure, one is obviously at a liberty to posit that God created the Universe on a Sunday, exactly 6,000 years ago. That includes, of course, photons traveling through space as if they came from galaxies located hundreds of millions of light years away. And given there are billions of galaxies in the visible universe (e.g., a simple extrapolation from the Ultra Deep Field survey completed by the Hubble Space Telescope), that makes God an amazing being indeed. But why are some stars red and others blue? Why are some galaxies shaped like spirals and others like footballs? Why does Earth have an atmosphere rich in oxygen and capable of sustaining life, while the other planets do not? Because God made it that way? But why did God make these things that way? The faithful respond: Don't ask. It is the mystery of the creation. But how does this inform us about the world we live in? Are you sure the Universe was created on a Monday, not a Tuesday, or was it a Saturday? Did the wise men who wrote the Bible know about other galaxies, dinosaurs, and stellar nucleo-synthesis? Does the Bible provide a path to such discoveries and for a falsifiable representation of reality? Sadly, it does not. For sure, it tells an evocative story. But the story has no reliable markers, no real capacity for cross-checks and thus no falsifiability. And more importantly, it has no factual predictive power. It is thus of little use as a basis to model reality. Change the religion, change the book (e.g., the Koran, the Torah, the Bhagavad Gita, the Vedas, the Avestan, etc.), and the actors change names, the narrative changes, the commandments also change a little, but the conclusion remains the same: sadly, as models of reality these books have no trustworthy content, no falsifiability, no predictive power, and thus no real usefulness.

No reliable model of reality means no trustworthy path to knowledge, no tangible and reliable source of meaning and ethics. It means chaos. And chaos it is across our beautiful blue planet. Witness cultural and religious factions claiming they own their lands as well as the truth, and worst, readily conducting genocides, in the name of God, to eliminate whatever groups disagree with them and stand in their path. But it does not have to be that way. The scientific method does work. It is slow but robust and the models (knowledge) of reality it provides, while innately tentative and incomplete, are steadily bringing our scientific civilization to a greater and clearer vision of our Universe, our origins, and our nature. To quip, I would suggest that those who wish to get a true spiritual experience should pick up a physics or biology textbook, because there is sure no better way to embrace reality than science. But brace yourself, it takes work!

1.7 Challenges of the Scientific Method: Paradigm Shifts and Occam's Razor

Though science works, scientific modeling of reality is not without challenges of its own. One such challenge was clearly put to light by Thomas Kuhn in his book, *The Structure of Scientific Revolutions* [132]. To understand the issue, let us briefly consider the transition between the geocentric and heliocentric views of the world that occurred at the eve of the scientific revolution following the works and writings of Nicolaus Copernicus.

Born in 1473 in the town of Torun, Poland, Copernicus was orphaned at an early age and taken under the tutelage of his maternal uncle, Lucas Watzenrode the Younger (1447–1512), a very influential bishop of Poland, who provided for his education. Copernicus studied law and medicine, but his true passion was astronomy. Noticing that the positions of the planets were considerably off compared to predictions provided by the Alfonsine tables,[2] he became convinced that Ptolemy's geocentric view of the world was incorrect and he proposed a heliocentric model in which all the planets revolve around the Sun, except the Moon, which revolves around the Earth. Much time would pass before Copernicus's heliocentric model became widely accepted, but it eventually did, thanks in part to the work of Kepler, Galileo, and Newton. In time, astronomers would successively also dethrone the Sun and our galaxy as the center of the Universe.

The central point of this story is the geocentric model with its deferents and epicycles. The fact of the matter is that an arbitrary number of nested epicycles could be added to the geocentric model to fix it and provide a very accurate model of the apparent motion of the planets. With sufficiently many epicycles, the model could be made reliable for several decades, perhaps centuries. Kuhn's point is that based on observations of the apparent motion of the planets alone, it would not have been possible to readily falsify the geocentric model augmented with an arbitrary number of nested epicycles. Thankfully, several other observations, including the fact that Venus has phases incompatible with Ptolemy's geocentric model, the aberration of light, and Foucault's pendulum, would provide incontrovertible falsification power to reject the geocentric model in favor of the heliocentric model.

The capacity to mathematically represent the apparent motion of the planets with the wrong model, however, remains a serious issue. Thomas Kuhn realized the same type of issue could arise within models discussing various other aspects of reality. In essence, Kuhn understood that mathematical models describing a portion of reality (e.g., geocentric motion of the planets, classical description of particle motion, etc.) may be artificially

[2] Astronomical tables prepared on the request of thirteenth century King Alfonso X of Castile, based on Claudius Ptolemy's geocentric model of the motion of planets.

augmented with fixes and artifices until drastically different types of observations render a model untenable and requires what he called a **paradigm shift**.

Figuratively speaking, the need for the addition of epicycles may in part arise because of the limited quality of the data or poorly understood features of the measurement process. It could also amount to an attempt to temporarily fix the model while waiting for additional and better quality data, or a new model capable of providing a more encompassing view of the phenomenon or system of interest, and so on. One could obviously expound further on this topic, but the main point of interest, as far as this book is concerned, is that mathematical models used for the representation of natural phenomena can always be made more complicated to account for unusual features of the data. How then can scientists judge whether a scientific model and its mathematical realization provide a proper and sufficient representation of a phenomenon of interest (i.e., reality)? Why is Newton's model of orbital motion better than Ptolemy's? Why was it necessary to invent quantum mechanics?

An answer to such questions is often provided in the form of Occam's razor,[3] a principle stating that among competing hypotheses, the simplest, that is, the one with the fewest assumptions, should be selected. In mathematical terms, this translates into selecting the (fit) model with the least number of free parameters that is consistent with the data. Indeed, why use a cubic or quartic polynomial to fit a set of data if the precision of the data does not warrant it? A straight line or parabola might be sufficient, unless perhaps one has prior reasons to believe a higher degree polynomial must be used. An important aspect of data analysis then involves the evaluation of errors and of the techniques to assess or test whether a model or hypothesis constitutes an appropriate and sufficient representation of the data.

Understanding experimental errors and how they propagate to model properties derived from the data is thus a central aspect of the scientific method. We thus devote several sections, throughout the book, on this very important topic. An intuitive notion of error and techniques of error propagation are first discussed in Chapter 2 after the introduction of the concept of probability. A more precise definition of the concept of errors is introduced in Chapter 4 on the basis of estimators and statistics. A full characterization of errors, however, requires the notions of confidence level and confidence intervals discussed in Chapter 6. The notion of confidence interval is slightly modified in Bayesian inference and renamed credible interval in Chapter 7. Equipped with the notions of confidence intervals and data probability models, it then becomes possible, in Chapter 6, to fully address the notion of (scientific) hypothesis tests. A mathematical implementation of Occam's razor, based on the Bayesian interpretation is finally discussed in Chapter 7.

Occam's razor alone, however, is usually insufficient to abandon a particular model. A more convincing line of arguments is generally needed. Indeed, and to get back to our example, it is not Occam's razor that made scientists reject Ptolemy's model, but the observation of phenomena completely inconsistent with a geocentric universe, including the existence of Venus phases, the aberration of light, and Foucault's pendulum, and just as importantly the immense predictive power of Newton's mechanics and Law of Universal Gravitation. Newton's model is not better than Ptolemy's merely because it has fewer

[3] A problem-solving principle attributed to William of Ockham (c. 1287–1347).

parameters (in fact, it has quite a few as well, the mass of the planets, the size and eccentricity of their orbits, etc.) but because its higher level of abstraction provides a unifying principle (e.g., the force of gravity) that enables falsifiable predictions of the motion of objects on Earth as well as in the heavens. Likewise, the nonclassical concepts of wave function and quantization enabled accurate quantitative descriptions of large classes of phenomena that were otherwise intractable within the classical physics paradigm.

Kuhn argued that competing paradigms are frequently incommensurable, as competing and irreconcilable accounts of reality, and that scientists cannot rely on objectivity alone. Accordingly, the necessity for paradigm shifts may then involve a certain degree of subjectivity. It remains nonetheless that it would have been completely impossible to realize a flyby of Pluto (NASA Horizon spacecraft, summer 2015) based on Ptolemy's geocentric model but the exploit was readily achievable based on Newton mechanics. Likewise, designing microtransistors and computer chips would be inconceivable within the framework of classical mechanics but became a tremendously successful outcome of the development of quantum mechanics. Similar conclusions may also be stated for the tremendous progress achieved in biology, most particularly in genetics. It is rather clear that such advances could not have been possible without the guiding principles of Darwin's theory of evolution. There is little or no subjectivity associated with these facts. Falsifiability may be temporarily compromised by the addition of epicycles but the overarching predictive power and technical prowess of scientific theories developed after successive paradigm shifts are objective indications that although our models of reality remain tentative, science as a whole has made tremendous progress in formulating meaningful and powerful theories of reality. There is indeed little doubt that science is closer to the true nature of reality today than it was 50, 100, or 2,000 years ago.

1.8 To Err Is Human, to Control Errors, Science!

One of the very first things science students learn in the laboratory is that making mistakes in the execution of an experimental procedure and the acquisition of data is terribly easy. Committed students usually figure out how to improve their lab work and reduce or even eliminate the number of mistakes made in their execution of experimental protocols. But students must also learn that although the capacity to conduct reliable experiments is a skill one can hone through training and repetition, the risk of errors, and the need to understand them, never goes away. There are many types and sources of errors, of course, but at the end of the day, even if an experimental procedure is executed flawlessly, there remain irreducible sources of errors and uncertainties determined by the measurement process itself, known as **process noise**, and the instrument read-out, known as **measurement noise**. And although improvements in the design of an experimental apparatus or procedure may reduce these errors, they can never eliminate them completely.

But, as we have argued earlier, the true power of the scientific method resides in its capacity to falsify models and its ability to make accurate predictions. This implies it is absolutely necessary to understand the errors of a measurement, identify their sources and types, and make the best efforts to reduce their amplitude. Indeed, the capacity to reject a model (falsification of the null hypothesis), and adopt another one (adoption of an alternative hypothesis), based on a specific measurement is chiefly determined by the measurement's precision.

Evidently, having an estimate of the error involved in a measurement does not mean one knows the value of the error (unless perhaps, the value being measured is already known). If one did, it would suffice to subtract the error from the measured value and one would achieve an error-free measurement. Having an estimate of the error of measurement instead means that one can assess the likelihood of deviations of any given size from the true value of the observable. In this context, measured values are viewed as random variables, that is, observables that may deviate uncontrollably from their true value due either to their intrinsic nature (e.g., the exact decay time of a radioactive nuclei cannot be predicted, only the average lifetime is known) or variability associated with a macroscopic process involving a large number of elementary subprocesses (e.g., fluctuations of the number of collisions experienced by a high-energy particle traversing a large chunk of material). One is thus constrained to model measurements according to the language of probability and statistics. Assessing experimental errors thus requires a probabilistic model of the measurement process, and extraction of meaningful scientific results necessitates statistical analysis and inference. This is largely what this book is about.

1.9 Goals, Structure, and Layout of This Book

Traditionally, scientists, most particularly physicists, have made use of probability and statistical techniques rooted in the so-called frequentist interpretation of probabilities, which basically regards probabilities as relative frequencies observed in the limit of an infinite number of trials or measurements. However, a growing number of scientists, including physicists, now make use of techniques based on the Bayesian interpretation of probabilities, which assigns hypotheses a certain degree of belief or plausibility. This rapid increase in the adoption of Baysian probabilities and inference techniques stems in part from the relatively recent developments of the theory of probability as logic by Jaynes and others, the elegance and power of the interpretation, as well as the development and articulation of numerous tools to deal with practical scientific problems. The majority of works and scientific publications, however, are still based on the traditional methods of the frequentist interpretation. This book thus attempts to cover data analysis techniques commonly used in astro-, nuclear, and particle physics based on both approaches.

This book targets graduate students and young scientists. It is not intended as a text for mathematicians or statisticians. So, while we spend a fair amount of time introducing the foundations of the theory of probability, and discuss, among other topics, properties of estimators and statistical inference in great detail, we leave out detailed proofs of many of the theorems and results presented in the book, and rather concentrate on the interpretation and applications of the concepts. Note that at variance with more elementary books on probability and statistics, we assume the reader to be reasonably proficient in elementary mathematical and computational techniques, and thus also leave out many of the detailed numerical calculations and derivations found in such texts.

The book is divided into three parts. The first part provides a foundation in probability and statistics beginning with a formal definition of the concepts of probability and detailed discussions of notions of statistics, estimators, fitting methods, confidence intervals, and statistical tests. These concepts are first explored from a classical or frequentist perspective in Chapters 4–6, but also discussed within the Bayesian paradigm in Chapter 7.

The second part of the book introduces a variety of measurements techniques commonly utilized in nuclear and particle physics. It begins, in Chapter 8, with the introduction of basic observables of interest in particle/nuclear physics and exemplars of basic measurement methods. Selected advanced topics are presented in Chapters 9–11. Chapter 9 introduces the notion of event reconstruction, and presents examples of track reconstruction and fitting, as well as primary and secondary vertex reconstruction techniques, while Chapters 10 and 11 present extensive descriptions of correlation function measurement methods. These topics then form grounds for a detailed discussion of error assessment and data correction methods, including efficiency corrections, spectrum unfolding techniques, and various other techniques involved in the correction of correlation functions and flow measurements, presented in Chapter 12.

Part III introduces basic Monte Carlo techniques, in Chapter 13, and elementary examples of their use for simulations of real-world phenomena, as well as for detailed evaluations of the performance and response of complex detection apparatuses, in Chapter 14.

There is much to discuss. Let us begin!

PART I

FOUNDATION IN PROBABILITY AND STATISTICS

2 Probability

2.1 Modeling Measurements: A Need for Probability

Gambling aficionados are accustomed to the notion that when they roll a pair of dice repeatedly, they obtain, each time, very different number combinations, and unless the dice have been heavily tampered with, it is impossible to control or predict which faces will roll up. The same may be said of the balls in a lottery machine: it is not possible to actually predict which balls will be drawn. This seems rather obvious. Is it not? Well, actually no! The classical mechanics that govern these processes is wholly deterministic. Given enough information about the initial kinematic conditions of a roll as well as the properties of the dice and the table they are rolled on, it should be possible to calculate, at least in principle, which face the dice will actually land on. This is perhaps less obvious, but it is true. Phenomena such as a roll of dice or the breakdown of a window by an impact are ruled by deterministic laws of physics and should thus be predictable. The problem, of course, is that they are immensely complicated phenomena involving a large succession of events. For instance, a proper calculation of the trajectory of a set of dice would require knowledge of their exact speed and orientation when they leave the player's hand. One would also need to account for friction, the air pressure and temperature, the exact elasticity of the dice and all components of the table on which they roll, etc. And because dice can bounce several times against each other and on the table, one should have to follow their complete trajectory, accounting for whatever imperfections the dice or the table might feature. This is a rather formidable task that is unlikely to ever be accomplished, even with modern supercomputers. Effectively, if dice are thrown with enough vigor, they should bounce and roll so many times as to make the calculation practically impossible. For all intents and purposes, the roll of fair dice is truly a random phenomenon: its outcome is unpredictable and all faces have an equal probability of rolling up.

Scientific experiments are obviously not a form of gambling, but the many physicochemical processes involved in measurements have much in common with a dice roll. Typical measurements involve a large succession of macroscopic and microscopic processes that randomly alter their outcome. Effectively, repeated measurements of a given physical quantity (e.g., the position, momentum, or energy of an object) also yield different values that seemingly fluctuate and adopt a random pattern, which, at best, clusters near the actual value of the observable of interest.

The fluctuations stem in part from the technique used to carry out the measurements. For instance, ten people who measure the length of, say, a table with a tape measure, will report slightly different values, even though they might use the same tape measure. The differences have to do with the way these different people position the tape measure next to the table, how they read the tape measure, whether they keep it extended without stretching it, and so on. Measurement accuracy may also be limited by the physical process used to carry out the measurement as well as the observable measured. For example, scatterings and energy losses in a detector randomly affect the momentum of the particles measured in a magnetic spectrometer. If a given particle is kicked this or that way, its direction and energy are slightly altered, and the momentum determined based on the particle's trajectory is slightly off from the actual value.

The bottom line is that measured values of physical observables are likely to vary from measurement to measurement. Additionally, inspection of the values obtained in a sequence of measurements will reveal that the specific value obtained in a given measurement cannot be predicted. The measured values appear as a sequence of seemingly random numbers. Yet a table has a length, and measurements of this length yield values that cluster around a typical value that one expects should be representative of the actual length of the table. Likewise, repeated measurements of particle positions, momenta, and so on shall yield values that vary from measurement to measurement but should cluster around the actual values of these observables.

Throughout this book, we adopt the view that the outcome of measurements of scientific observables is, for all intents and purposes, a random process. The outcomes of a given measurement, or a succession of measurements, shall be considered random variables. But randomness does not imply all values of an observable are equally probable: a reasonably well designed and carefully carried out experiment shall yield values that cluster near and about the actual value of the observable. Whoever has carried out a succession of measurements of a well-defined observable according to a specific measurement protocol can in fact attest that such clustering occurs. We will assume that it is possible, at least in principle, to formulate a probabilistic model of the measurement process and the clustering of its outcomes. In other words, although it is impossible to predict the value of a specific measurement with absolute certainty, it should be possible to formulate a model that provides the probability of any given outcome or the distribution of values obtained after several measurements. Although each specific outcome is random, the probability model should describe, overall, the probability of any given outcome according to a specific mathematical function called the probability distribution.

Evidently, a model of the measurement process shall be as good as the efforts put into understanding it as well as the prior knowledge available about the process and the apparatus used to carry out the experiment. The job of a statistician shall thus be to find and apply statistical methods that enable a trustworthy characterization or **inference** of measured observables and scientific models used to describe phenomena of interest, even though the measurement model may not be completely accurate. A difficulty arises that statistical inference is contingent on the notion of probability and how this notion is applied to the description and characterization of experimental results.

Two main paradigms, known as frequentist and Bayesian, are commonly used to define and interpret the notion of probability. The frequentist paradigm assumes measurement fluctuations are determined by a parent distribution representing the relative frequency of values or ranges of values of an observable. The parent distribution is a priori unknown but can be determined, at least in principle, in the limit of an infinite number of measurements. The Bayesian paradigm dispenses with the need for an infinite number of measurements by shifting the discussion into a hypothesis space in which all components of the measurement process, including the probability model of the measurement and its parameters, are considered as hypotheses, each endowed with a degree of plausibility, that is, a probability.

It is the purpose of §2.2 of this chapter to motivate and introduce the two interpretations of probability these paradigms are based on. A discussion of the pros and cons of the two interpretations is initiated in §2.3 but will continue throughout the first part of this book. The remainder of this chapter discuss mathematical concepts pertaining to and used in both paradigms. Section 2.4 introduces Bayes' theorem, the notions of inference, as well as the concepts of sample and hypothesis space, while §2.5 defines discrete and continuous random variables, probability distribution, probability density distributions, as well as cumulative distributions and densities. Sections 2.6 and 2.7 next introduce functions of random variables and techniques for the characterization of distributions. Multivariate distributions and their moments are discussed in §2.8 and §2.9, respectively. Section 2.10 introduces the notions of characteristic function, moment-generating function, and examples of their application, including a proof of the very important central limit theorem. With these tools in hand, we proceed to introduces notions of measurement errors and random walk processes in §2.11 and §2.12, respectively. The chapter ends, in §2.13, with the definition of cumulants, which have broad utility in probability and statistics, and constitute, in particular, an essential component of correlation and flow analyses conducted in the field of high-energy heavy-ion collisions.

2.2 Foundation and Definitions

The notion of probability can be intuitively introduced on the basis of the relative frequency of the phenomena of interest. For instance, to establish that a cubic die has been tampered with, it suffices, in principle, to roll it a very large number of times, and count how many times each face rolls up. The relative frequency of each face, f_i, with $i = 1, \ldots, 6$ is the number of times, N_i, each face rolls up divided by the total number of rolls, N_{Tot}.

$$f_i = \frac{N_i}{N_{\text{Tot}}} \tag{2.1}$$

The frequencies f_i provide an indication of the likelihood, that is, the **probability**, of rolling any face i in subsequent rolls of this same die. A die is considered fair if all faces have the same probability, that is, the same frequency.

The notion of frequency as a probability also applies in virtually all forms of experimental measurement. For example, repeated measurements of the universal gravitational constant, G (Big G), are expected to yield values in the neighborhood of the actual value of the constant. The number of times values are observed in a specific ranges $[G, G + dG]$, relative to the total number of observations, can then be used to estimate the probability of observing G in that range. It is thus natural to establish a connection between notions of probability and relative frequency. This leads to the **frequentist interpretation** of the notion of probability.

The frequentist interpretation of probability is most concerned with the outcome of measurements or observations. The observations, either discrete or continuous numbers, are elements of sets, known as **sample space**, defining the domain of observable values. One can thus naturally and formally introduce the notion of probability based on values associated with elements or subsets of the sample space. Such a definition, introduced in 1933 by Kolmogorov, is presented in §2.2.1. The frequentist interpretation becomes problematic if and when dealing with systems involving phenomena that cannot be observed more than once (e.g., the collapse of the Tacoma Narrows bridge, a star exploding in a supernova, or the Big Bang), or when trying to express the plausibility or truthfulness of statements about the world. How indeed does one quantify the probability that a statement about dark matter or the Higgs boson is correct if it does not involve numerical values, whether discrete or continuous (e.g., dark matter does not consist of known baryons, or there exists only one Higgs boson)? Obviously such statements are the result of inference based on prior knowledge of the world and the outcome of experimental measurements with specific measured data. One then needs a robust and well-defined method to translate one's certainty about prior knowledge and experimental results into statements about the world. One must also be able to combine such statements according to the rules of logic. This then creates the need to extend logic (or the calculus of predicates) to include the notion of plausibility or likelihood. Such an extension was developed largely by E. T. Jaynes and his collaborators in the 1970s. We briefly present the basic tenets and rules of the foundation of probability as logic in §§ 2.2.3 and 2.2.4.

2.2.1 Probability Based on Set Theory

The concept of probability embodies the notion of randomness, in other words, the fact that one cannot predict with complete certainty the outcome, or value, of a particular measurement. It may be formally defined in the context of **set theory**. Indeed, measurement outcomes can be viewed as members of a set. We call **sample space**, denoted **S**, the set consisting of elements corresponding to actual and possible outcomes of a measurement. The set **S** can be divided into subsets **A**, **B**, **C**, and so on. A subset **A** of **S**, noted $\mathbf{A} \subseteq \mathbf{S}$, may be empty and contain no elements. It may alternatively contain one, few, or all elements of **S**. It is then possible to assign **A** with a real number, noted $p(A)$. This number is called the **probability of A** provided it obeys the following three axioms:

$$p(A) \geq 0 \text{ for } \mathbf{A} \subseteq \mathbf{S}, \tag{2.2}$$

$$p(A \cup B) = p(A) + p(B) \text{ for } \mathbf{A} \cap \mathbf{B} = 0, \tag{2.3}$$

$$p(S) = 1. \tag{2.4}$$

The first axiom stipulates that every subset **A** belonging to **S** has a probability $p(A)$ larger than or equal to zero. The second axiom states that the probability assigned to the union

$\mathbf{A} \cup \mathbf{B}$ of two disjoint subsets **A** and **B** (i.e., $\mathbf{A} \cap \mathbf{B} = 0$) is the sum of their probabilities $p(A)$ and $p(B)$. The third axiom states that the probability assigned to **S** is unity.

Introducing the notation $\overline{\mathbf{A}}$ to mean the complement of **A** in **S**, such that $\mathbf{A} \cup \overline{\mathbf{A}} = \mathbf{S}$, one can demonstrate the following properties (see Problem 2.1):

$$p(\bar{A}) = 1 - p(A), \tag{2.5}$$

$$p(\bar{A} \cup A) = p(S) = 1, \tag{2.6}$$

$$0 \leq p(A) \leq 1, \tag{2.7}$$

$$\text{if } \mathbf{A} \subset \mathbf{B}, \text{ then } p(A) \leq p(B), \tag{2.8}$$

$$\text{if } \mathbf{A} \text{ is a null subset, then } p(A) = 0. \tag{2.9}$$

The first expression indicates that the probability of the complement of **A** is equal to 1 minus the probability of **A** itself. The second expression indicates that the probability of the union of **A** and its complement, which can be identified as **S**, has a probability equal to one. The third expression states that the probability of **A** can take any value in the range [0, 1], including the values 0 and 1. The last expression states that the probability of an empty subset is null.

Consider a "measurement" of an observable (variable) X that can take any specific value or group of values corresponding to elements of the set **S**. Further consider that a measurement of X might yield any values in the set. This variable is then considered a **random variable**, and the axioms (2.2–2.4) define the probability of measuring X within a given subset **A**. Note that X may represent a single value, a range, or combinations of values, as we will illustrate in the text that follows. Additionally, if values of X are restricted to the set of integers, such as in counting experiments (e.g., number of people recovering from an illness thanks to a medication or the number of particles produced in a nucleus–nucleus collision), X is said to be a **discrete random variable**, whereas if X belongs to a subset or the entire set of real numbers **R**, it constitutes a **continuous random variable**.

2.2.2 Conditional Probability and Statistical Independence

Let us now consider a sample space **S**, with subsets **A** and **B** such that $p(B) \neq 0$. One then defines the **conditional probability**, noted $p(A|B)$, as the probability of **A** given **B**:

$$p(A|B) = \frac{p(A \cap B)}{p(B)}. \tag{2.10}$$

This corresponds to the probability of observing the random variable X within **A** when it is also within **B**. It is relatively straightforward to show that the notion of conditional probability satisfies the axioms of probability introduced previously (see Problem 2.2). In fact, the probability $p(A)$ can itself be viewed as a conditional probability $p(A) = p(A|S)$ since $p(S) = 1$ by construction.

Two subsets **A** and **B**, and the measurement outcomes they represent, are said to be **independent** if they satisfy the condition

$$p(A \cap B) = p(A)p(B). \tag{2.11}$$

This condition means that the probability that X is a member of **A** and **B** **simultaneously** is equal to the product of the probabilities of X being in **A** and **B** independently. This enables the evaluation of the conditional probability $p(A|B)$:

$$p(A|B) = \frac{p(A \cap B)}{p(B)} = \frac{p(A)p(B)}{p(B)} = p(A) \quad \text{(statistical independence).} \tag{2.12}$$

Consequently, when **A** and **B** are statistically independent, the conditional probability of **A** given **B** is equal to the probability of **A** itself, in other words, the probability of **A** does not depend on **B**. Likewise, the conditional probability of **B** given **A** is equal to the probability of **B** itself:

$$p(B|A) = \frac{p(A \cap B)}{p(A)} = \frac{p(A)p(B)}{p(A)} = p(B) \quad \text{(statistical independence).} \tag{2.13}$$

It is quite important to realize that the notion of **statistical independence** or **independent subsets**, $p(A \cap B) = p(A)p(B)$, differs from that of **disjoint subsets,** $\mathbf{A} \cap \mathbf{B} = 0$. Indeed, if the intersection $\mathbf{A} \cap \mathbf{B}$ is empty, it is not possible for an element x (a measurement outcome) to be simultaneously part of **A** and **B**; the probability of observing such an element (measurement value) is thus null, and therefore different in general from $p(A)p(B)$.

Clearly, the conditional probabilities $p(A|B)$ and $p(B|A)$ are not independent. To establish their relation, first consider that by definition of the conditional probability, one has

$$p(A|B) = \frac{p(A \cap B)}{p(B)}, \tag{2.14}$$

and similarly,

$$p(B|A) = \frac{p(B \cap A)}{p(A)}. \tag{2.15}$$

Given the commutativity of the intersection of two sets, $\mathbf{B} \cap \mathbf{A} = \mathbf{A} \cap \mathbf{B}$, one finds that the two conditional probabilities are related as follows:

$$p(A \cap B) = p(B|A)p(A) = p(A|B)p(B), \tag{2.16}$$

provided neither $p(A)$ nor $p(B)$ is null. This implies it is possible to calculate the conditional probability $p(B|A)$ as follows:

$$p(B|A) = \frac{p(A|B)p(B)}{p(A)}. \tag{2.17}$$

This expression, known as **Bayes' theorem,**[1] is a fundamental relationship in probability theory and finds use in a wide range of practical applications, as we shall discuss in this and subsequent chapters of this book.

It is useful to further explore the properties of conditional probabilities by partitioning the set **S** into finitely many subsets, $\mathbf{A}_i$ with $i = 1, \ldots, n$. Consider, for instance, n subsets $\mathbf{A}_i$, whose union is by construction equal to **S**, in other words, such that $\mathbf{S} = \bigcup_i \mathbf{A}_i$. Assume

[1] Thomas Bayes (1702–1761) was an English mathematician and Presbyterian minister known for the development of the mathematical theorem that bears his name.

further that none of these subsets are null, $p(A_i) \neq 0$, and that they are all disjoint, $\mathbf{A}_i \cap \mathbf{A}_j = 0$ for $i \neq j$. Then it is possible to show (see Problem 2.3) that

$$p(B) = \sum_i p(B|A_i)p(A_i). \tag{2.18}$$

This result is known as the **law of total probability**.

Combining the law of total probability with Bayes' theorem, one finds

$$p(A|B) = \frac{p(B|A)p(A)}{\sum_i p(B|A_i)p(A_i)}, \tag{2.19}$$

which is particularly useful in the estimation of the probability of a specific hypothesis given measured results. We present an example of an application of Bayes' theorem and the law of total probability in the following subsection.

Example: A Problem about Problematic Parts

Statement of the Problem

10P100Bad, a famous auto-parts supplier, conducted a study of the parts it produces in a year. The study revealed 90% of the produced parts are within specifications while the remainder 10% are not. Stricken with revenue losses, the company decided to hide this fact and sell both the good and defective parts. A client, the struggling car company Don'tFoolMe, suspected there was a problem with the parts and designed its own test to determine whether the components it bought from 10P100Bad were within tolerances. Unfortunately, their test was not very precise. They estimated that their test had a probability of 1% to identify a component as nondefective even though it was, and a probability of 0.5% to reject a nondefective part. Calculate the probability a bad part would not be rejected by their test and used in the construction of a car.

Solution

Let us establish what we know from the statement of the problem. First, we know that the prior probability, $p(D)$, a part is defective is 10%:

$$p(D) = 0.1, \tag{2.20}$$

$$p(\bar{D}) = 0.9. \tag{2.21}$$

We also know the probability, $p(A|D)$, of accepting a defective part is 1%,

$$p(A|D) = 0.01, \tag{2.22}$$

$$p(\bar{A}|D) = 0.99, \tag{2.23}$$

and the probability, $p(\bar{A}|\bar{D})$, of rejecting a nondefective part is 0.5%. One thus writes

$$p(A|\bar{D}) = 0.995, \tag{2.24}$$

$$p(\bar{A}|\bar{D}) = 0.005. \tag{2.25}$$

We now seek the probability, $p(D|A)$, that a part accepted in the fabrication of the cars is defective. By virtue of the total probability theorem, one writes

$$\begin{aligned} p(D|A) &= \frac{p(A|D)p(D)}{p(A|D)p(D) + p(A|\bar{D})p(\bar{D})}, \\ &= \frac{0.01 \times 0.1}{0.01 \times 0.1 + 0.995 \times 0.9}, \\ &= \frac{0.001}{0.001 + 0.8955} = 0.0011. \end{aligned} \tag{2.26}$$

One then concludes the car maker Don'tFoolMe has a probability of 0.1% of integrating defective parts in its fleet. That sounds like a financial disaster in the making...

2.2.3 A Need for Probability as an Extension of Logic

The notion of probability based on a sample space as a set of numbers representing the outcome of measurements is rather restrictive. Indeed, as already stated, one would also like to quantify the plausibility of general statements about the world or specific phenomena. For instance, modern cosmologists are concerned with the notion that the expansion of the universe (e.g., the fact that all observable galaxies recede from one another at speeds that grow proportionally to the distance that separate them) might accelerate over time. One would then like to use existing data to quantify the plausibility of statements such as the expansion of the universe has been constant through times, or the expansion has accelerated during the last n billion years, and so forth. Indeed, one would like to express a probability for either statements given the state or prior knowledge about the universe and measured data (e.g., based on type Ia supernovae). In all scientific endeavors, one is interested in stating/quantifying the plausibility of statements made about the world or specific phenomena. The problem, of course, is that one typically deals with limited data and that all data involve finite errors.

The scientific process, applied to a specific inquiry of a specific system, involves predictive statements about measurements based on one or several models of the system (e.g., space is warped by the presence of massive objects such as stars). Models can then be used to make predictions concerning phenomena occurring in the realm of the system (e.g., the warp of space around a star was used by A. Einstein to predict the deflection of star light by the Sun observable during an eclipse of the Sun and a sizable contribution to the precession of planet Mercury in its orbit around the Sun). Measurements of observables of interest (e.g., a shift in the apparent position of stars near the line of sight of the eclipse, and the measured rate of precession of Mercury's orbit) can then be used to infer, after due statistical analysis of the measured values and their errors, whether the model predictions are correct. It then becomes natural to enlarge (or replace) the notion of sample space with the notion of hypothesis space consisting of (logical) statements about the phenomenon in question (or the world in general). Probability may then be viewed as an extension of logic where propositions (predicates) are ascribed a plausibility, likelihood, or degree of belief.

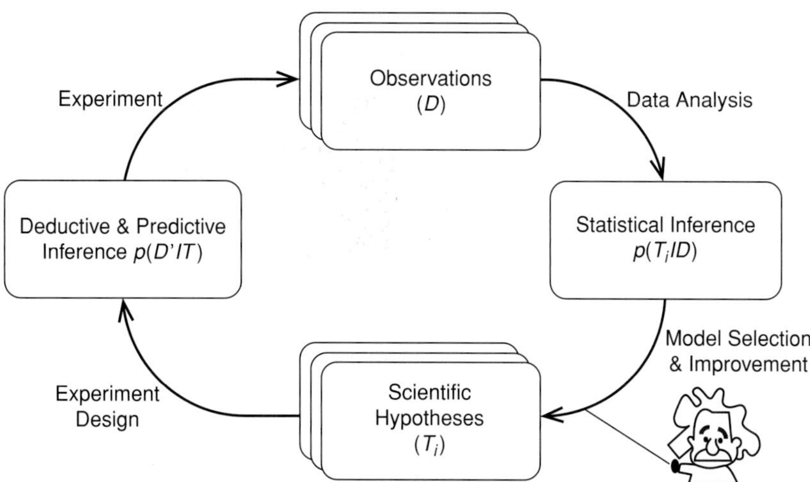

Fig. 2.1 Schematic summary of the scientific method as a cyclical process involving stages of modeling, deductive (predictive) inference, measurements, statistical analyses, and inference.

A proper foundation of this approach was progressively developed by several authors including C. Shannon [172], H. Jeffreys [118, 119], and R. T. Cox [69], arguably culminating with the work of Jaynes published posthumously in 2003 [117]. Several articles and books have extended Jaynes' work.

An extension of the notion of probability to include logic is attractive because one can use the formally well-established rules of logic to combine logic statements and reason based on these predicates in a sound and robust manner. Of course, the point of associating probabilities to logical statements is that one can then express the probability of some statements being true if logically derived from (or entailed by) prior statements when the veracity of such statements is itself not perfectly well established. The probability of a statement derived from several others thus depends on the individual probabilities of these statements.

Probability as logic is also particularly useful because it enables a scientific discourse based on logic while accounting formally and robustly for the limited amount of information one may have about a specific phenomenon (or perhaps the universe as a whole). The scientific method, which nominally involves the formulation of hypotheses and the realization of measurements designed to test these hypotheses, may then be formulated as a cycle, illustrated in Figure 2.1, involving predictive inference based on models, deployment of one or several experiments whose results are analyzed statistically (statistics) and used to carry out formal tests of the hypotheses or models (statistical inference), and eventually to decide which of the models has the highest probability, that is, is most consistent with the measured data. This may then be followed by theoretical work leading to additional or tweaked hypotheses and models, which then form the basis for new experiments or measurements. The cycle repeats, eventually leading to significant scientific advances.

Physical models of nature are well covered in courses on classical mechanics, electrodynamics, quantum field theory, and so forth. But such courses typically ignore the details of the predictive process leading to the formulation of experiments, the statistical analysis of the data, and the statistical inference involved in transforming observations (measurements) into conclusions about the veracity or appropriateness of the models. One of the purposes of this book is to fill this gap by providing a relatively detailed presentation of the methods of predictive inference, data analysis techniques (including techniques used to correct for instrumental effects, e.g., biases and defects), parameter estimation, hypothesis testing, and statistical inference.

2.2.4 Probability as an Extension of Logic

Extending logic to include the notions of plausibility and probability is not a trivial matter. Much like the generalization of geometry to non-Euclidean space, much freedom appears to exist, at least at the outset, in the manner in which this can be done. Cox [69] and Jaynes [117] formulated such an extension starting from three **desiderata**, that is, three sets of properties or attributes a theory of probability based on logic should satisfy. The desiderata and the foundational theoretical structure are here stated without much of a discussion or proof. Readers interested in digging into the foundations of the theory should consult the book by Jaynes [117] or the more recent book by Gregory [97].

Cox and Jaynes have reasoned that an extension of logic including probabilities should satisfy the following three desiderata:

1. The degree of plausibility of statements is represented by real numbers.
2. The measure of plausibility must behave rationally: as new information supporting the truth of a particular statement (predicate) is obtained, its plausibility must increase continuously and monotonically.
3. A theory of probability as logic must be consistent:
   a. There must be structural consistency: if a conclusion can be reasoned along many paths, all paths (based on the same information) must yield the same result, that is, the same value of plausibility.
   b. It should be possible to account for all information relevant to a particular problem in a quantitative manner. In other words, it should be possible to assign a degree of plausibility to prior information and account for it in the reasoning process.
   c. Equivalent states of knowledge must be represented by equivalent values of plausibility.

Jaynes demonstrated that these desiderata entail two rules, a sum rule and a product rule, that provide a foundation for a probability theory based on logic. We here state these two rules without demonstration. Derivations of these rules may be found in refs. [97, 117].

The sum rule is written

$$\text{Sum Rule}: \quad p(A|B) + p(\overline{A}|B) = 1 \tag{2.27}$$

The notation $p(X)$, stated p of X, represents the probability (plausibility) of the predicate X, which may consist of any properly constructed combinations of simpler predicates.

The symbols A and B represent two such logic predicates (rather than subsets of a sample space), that is, statements about the world or a particular phenomenon. The vertical bar is used to indicate that one considers the probability of the predicate on the left, given or assuming that the predicate on the right is true. For instance, the notation $p(A|B)$ corresponds to the probability that the statement A is true when the statement B is known to be true. A short horizontal line over a letter is here used to indicate the negation of the statement: $\overline{A}$ (commonly stated non-A) represents the logical negation of the statement A. The sum rule encodes the rather obvious and sensible notion that the sum of the probability of A being true given B and the probability of its negation $\overline{A}$ (also given B) is equal to unity and therefore exhausts all possibilities.

The product rule concerns conditional probabilities and is written

$$\text{Product Rule}: \quad p(A, B|C) = p(B|A, C)p(A|C) \tag{2.28}$$

The comma notation A, B expresses a logical conjunction, that is, a logic "AND" between the propositions A and B. The statement A, B can be true if and only if both A and B are true. The notation $p(A, B|C)$ thus expresses the probability of A, B being true when C is known to be true, whereas $p(B|A, C)$ corresponds to the probability of B being true when A, C is true, that is, when it is known that both A and C are true. The product rule tells us that the probability of A, B being true given C is known to be true is equal to the product of the probability of B being true when A and C are known to be true, and the probability of A being true when C is known to be true.

Clearly, since the conjunction operation commutes, that is, $A, B = B, A$, the product rule may also be written

$$p(A, B|C) = p(B, A|C) = p(A|B, C)p(B|C) \tag{2.29}$$

We will see in the text below that the product rules entails Bayes' theorem but let us first consider the relation between the sum and product rules to the axioms of probability stated in §2.10.

An attentive reader will have noted that the two rules are not mere accidents but were essentially designed to be consistent with the axioms of probability based on set theory. Consider, for illustrative purposes, two statements A and B as follows:

- A: Random variable X is found in the subset $\mathbf{A}$ of the sample space $\mathbf{S}$.
- B: Random variable X is found in the subset $\mathbf{B}$ of the sample space $\mathbf{S}$.

Let us first consider a case where $\mathbf{B}$ is the complement of $\mathbf{A}$ in $\mathbf{S}$, that is, $\mathbf{B} = \overline{\mathbf{A}}$. The sum rule applied to the statement A tells us

$$p(A|S) + p(\overline{A}|S) = 1 \tag{2.30}$$

Substituting B in lieu of $\overline{A}$, one has

$$p(A|S) + p(B|S) = 1 \tag{2.31}$$

In the language of set theory (§2.2.1), this may be written

$$p(A \cup B) = p(A) + p(B) = 1 \tag{2.32}$$

given $\mathbf{A} \cap \mathbf{B} = 0$ by virtue of the definition $\mathbf{B} = \overline{\mathbf{A}}$. The sum rule, Eq. (2.27), thus embodies the axioms given by Eqs. (2.2–2.4).

The product rule similarly includes and extends the conditional probability definition given by Eq. (2.10): let C state that the random variable X is found within $\mathbf{S}$. Since $\mathbf{S}$ corresponds, by construction, to the entire sample space spanned by X, the statement C is always true by construction. A logical AND between C and some arbitrary statement D is thus equal to D ($D, C = D$). The proposition C can thus be omitted. The product rule

$$p(A, B|C) = p(A|B, C)p(B|C) \tag{2.33}$$

may then be written

$$p(A, B) = p(A|B)p(B). \tag{2.34}$$

For the statement A, B to be true, the variable X must be found simultaneously in A and B. This is possible only if the intersection of these two subsets, $\mathbf{A} \cap \mathbf{B}$, is nonempty. One then finds that the conditional probability $p(A|B)$ is given by

$$p(A|B) = \frac{p(A \cap B)}{p(B)}, \tag{2.35}$$

which is precisely the definition of conditional probability given by Eq. (2.10).

Let us now return to the expression of the product rule given by Eq. (2.28). As stated earlier, thanks to commutativity of the AND operation, one may write

$$p(A, B|C) = p(B, A|C). \tag{2.36}$$

By virtue of the product rule, this expression may then be written

$$p(A|B, C)p(B|C) = p(B|A, C)p(A|C), \tag{2.37}$$

which corresponds to Bayes' theorem:

$$p(A|B, C) = \frac{p(B|A, C)p(A|C)}{p(B|C)} \tag{2.38}$$

Indeed, following the same line of arguments as earlier in this paragraph, that is, identifying C as stating that the variable is found in the sample space $\mathbf{S}$, one recovers Eq. (2.17):

$$p(A|B) = \frac{p(B|A)p(A)}{p(B)}. \tag{2.39}$$

We note in closing this section that the sum rule may also be written

$$\text{Extended Sum Rule}: \; p(A + B|C) = p(A|C) + p(B|C) - p(A, B)|C). \tag{2.40}$$

The plus sign is here used to represent a logical disjunction, that is, a logical OR operation. The expression $p(A + B|C)$ thus corresponds to the probability of A or B being true, given that C is known to be true. It is relatively straightforward to show that this expression is consistent with the simpler sum rule (Eq. 2.27) as well as the axioms (2.2–2.4).

2.3 Frequentist and Bayesian Interpretations of Probabilities

The definitions based on set theory and probability as logic discussed in the previous sections naturally lead to two distinct interpretations of the notion of probability. In one, the so-called **frequentist interpretation**, a probability is regarded as a limiting value of the relative frequency of an outcome (either a discrete number or a range of values) when the number of trials becomes infinite. In the other, often referred to as **subjective probability**, the notion of probability expresses the plausibility of specific statements, which can be interpreted as degree of belief said statements are true. It is also commonly referred as **Bayesian interpretation** of probability.

Scientists, particularly physicists, have long made use of the frequentist interpretation and developed a great many tools to assign errors to measurements, and conduct statistical tests of scientific hypotheses. Studies of phenomena where the notion of limiting frequencies does not readily apply have prompted many scientists to pay more attention and embrace the Bayesian interpretation in their experimental studies and toward the inference of conclusions derived on their experiments. The Bayesian interpretation is now used in a growing number of scientific applications.

2.3.1 Frequentist Interpretation

The **frequentist interpretation** derives its name from the fact that a probability can often be regarded as a **limiting relative frequency**. In this context, the elements of a set **S** correspond to the possible outcomes of a measurement considered to be repeatable an **arbitrary large** number of times. A subset **A** of **S**, commonly referred to as an **event**, amounts to a set of possible outcomes of the measurement or observation. A particular event is said to occur whenever a measurement yields a member of a given subset (e.g., the subset **A**). A subset consisting of one element denotes a single elementary outcome. The probability of **A** thus corresponds to the fraction of all events yielding this particular outcome in the limit when measurements are hypothetically repeated infinitely many times:

$$p(\mathrm{A}) = \lim_{n\to\infty} \frac{\text{number of occurrences of } \mathbf{A}}{n}. \tag{2.41}$$

The probability of the occurrence of nonelementary outcomes may be determined from the probability of individual outcomes, consistent with the axioms expressed in Eqs. (2.2–2.4) because, by construction, the fraction of occurrences is always greater than zero and less than or equal to unity. The frequentist interpretation of probability forms the basis of the branch of mathematics known as **classical statistics**, also known as **classical inference** and **frequentist statistics**. All tasks, techniques, and methods based on the frequentist interpretation of probability are also said to form or be part of the **frequentist inference paradigm**.

Critics of the frequentist approach argue that given an infinite number of measurements is clearly not possible, the limit $n \to \infty$ cannot be achieved or verified in practice. The probability $p(A)$ of a set of outcomes, A, consequently cannot be determined with perfect precision. Effectively, one must assume that a particular measurement may be represented by a specific parent probability distribution. A frequentist statistician must then establish, on the basis of a finite number of measurements, whether a particular distribution or model properly describes the measurement(s) at hand. This may fortunately be accomplished on the basis of statistical tests discussed in §6.4. One may then compare several models and tests, based on finite data samples, determine which model is best compatible with the measured data.

In spite of the aforementioned conceptual difficulty, the frequentist interpretation is used routinely in science texts on probability and statistics, and by scientists in their analyses. It is typically considered appropriate and sufficient whenever one deals with scientific observations that can be repeated many times. It is, however, somewhat problematic whenever a measurement or event (e.g., the Big Bang, a supernova, a volcanic eruption, or the collapse of a bridge) cannot be repeated. It also makes it impossible to directly and explicitly integrate scientific hypotheses in the probabilistic discourse.

2.3.2 Subjective Interpretation

The **subjective interpretation of probability**, also called **Bayesian interpretation of probability**, is used increasingly in the physical sciences and many other scientific fields. It can be formulated based on both set theory and probability as logic, but we will argue, throughout this text, that a formulation based on probability as logic is far more interesting, convenient, and powerful.

Within the foundation of probability based on set theory, an event is regarded as a statement that the observable X is an element of the subset **A**. The quantity $p(A)$ may then be interpreted as the degree of belief the statement **A** might be true:

$$p(A) = \text{degree of belief that the statement } \mathbf{A} \text{ is true.} \tag{2.42}$$

In this context, the Bayesian interpretation assumes it is possible to construct the sample space **S** in terms of elementary hypotheses that are mutually exclusive, in other words, implying that only one statement is actually true. A set consisting of multiple such disjoint subsets is therefore true if any one of the subsets it contains is true. And one then has $p(S) = 1$.

Jaynes' definition of probability as an extension of logic readily extends the realm of the probability discourse. Indeed, dealing with predicates rather than sets and subsets, it becomes naturally possible to discuss the probability of statements about any world entities that can be expressed within the calculus of predicate. This implies, in particular, that the probabilistic discourse is no longer confined to the outcome of measurements and observations (being members of sets) but can be augmented to include statements about models, scientific hypotheses, and so forth. In this context, the quantity $p(A)$ may then be interpreted as the degree of belief the proposition A might be true:

$$p(A) = \text{degree of belief that the proposition } A \text{ is true,} \tag{2.43}$$

where the proposition A is not restricted to statements about measurement outcomes but can include statements about models, scientific hypotheses, and so forth.

The concept of subjective probability is closely related to Bayes' theorem and forms the basis of **Bayesian statistics** and **Bayesian inference**. It also forms the basis of the **Bayesian inference paradigm**.

In this context, one can then consider probabilities of the form $p(B|A)$, where A expresses a specific scientific hypothesis (e.g., a statement about a model or a model parameter), while B might represent the hypothesis that a specific experiment will yield a specific outcome (i.e., a specific discrete value or a continuous value in specific range). The conditional probability $p(B|A)$ then represents the degree of belief that B is observed given a hypothesis A is true. As such, the Bayesian inference paradigm provides a convenient framework, discussed in great detail in Chapter 7, to gauge the merits of one or competing models (or theories) relating to a specific measurement or set of measurements.

Given a certain theory, T, one might assign a certain **prior probability**, $p(T)$, that this theory is a valid model of the world (or set of experimental results). The probability, $p(D|T)$, called the **likelihood**, then provides an estimate of the degree to which measured data, D, can be expected based on the theory T. The conditional probability $p(T|D)$ thus provides a **posterior probability** that the theory, T, is true, conditioned by the data (measurements). According to Bayes' theorem, this posterior may be written

$$p(T|D) \propto p(D|T) \times p(T). \tag{2.44}$$

In this context, data are considered as facts and thus taken as true.[2] Bayes' theorem then enables the evaluation of the probability $p(T|D)$ that the theory T might be true, given the data. In other words, the merits of the theoretical hypothesis T can be gauged and evaluated based on the available data. This leads to the notion of hypothesis testing, which is first discussed in the context of the frequentist paradigm in Chapter 6 and more directly and naturally within the Bayesian paradigms in Chapter 7.

Given a dataset D, the merits of different theories or hypotheses, $T_1, T_2, \ldots$ can in principle be compared. Ideally, a particular theory T_i might emerge to have a posterior probability much larger than the others, $p(T_i|D) \gg p(T_j|D)$ for $j \neq i$, and would then become the favored theory. In practice, it is often the case that several competing models or hypotheses yield relatively weak and similar posterior probabilities. The available data are then considered insufficient to discriminate between the models.

It is fair to note that Bayesian statistics does not provide, ab initio, any particular method to determine the prior probability, $p(T)$. In the absence of prior inferences based on other theories, models, or data, it might be set to unity. The likelihood probability, $p(D|T)$, then provides the sole basis for the evaluation of $p(T|D)$. In other situations, there could be

[2] One should bear in mind, however, that measured values might need substantial corrections to be fully representative of the observable of interest. This implies that the probability model accounting for a specific measurement should include a proper probabilistic description of relevant instrumental effects or that raw measurement values can be "corrected" to account for such instrumental effects. This important topic is discussed in Chapter 12.

older data that enables an evaluation of the prior $p(T)$ before the experiment is conducted. The "new" data can then be seen as improving the knowledge about T. Quite obviously, the value of $p(T|D)$ is subject to the prior hypothesis as well as the data. This then leads to a framework that is subjective, hence the notion of subjective interpretation of probability. Although this might be seen as a weakness, one should stress that once a prior $p(T)$ and the likelihood $p(D|T)$ are determined, Bayes' theorem unambiguously provides an estimate of the posterior probability $p(T|D)$. In a scientific context, this provides for a mechanism to submit models and theories to strict and constraining tests of validity. Examples of such tests are presented in Chapter 7.

Unfortunately, Bayesian statistics is often considered in opposition to classical (frequentist) statistics. In fact, some problems discussed within the frequentist and Bayesian paradigm yield contrasting and incompatible solutions. Hard-core frequentists argue that the notion of degree of belief in a prior hypothesis leads to arbitrary posteriors and thus reject the Bayesian paradigm altogether. Some Bayesian statisticians argue that the definition of probability in terms of a limit, Eq. (2.41), is itself artificial or arbitrary, and thus reject the frequentist paradigm. Can there be a common ground?

It may be argued that Bayesian statistics in fact includes the frequentist interpretation as a special case, and as such provides a broader and more comprehensive context for data analysis. The formulation of probability as logic discussed in §2.2.4 naturally embodies the Bayesian interpretation of probability. Indeed, probability defined as an extension of logic deals with predicates or statements about the world (or a particular phenomenon), and assigns a certain degree of plausibility to these predicates. Predicates are thus viewed as elements of a hypothesis space rather than a simple set of numerical values. It is then possible, as we already argued, to consider more general and elaborate problems of inference. We will come back to this idea in more detail in Chapter 7. This said, it should also be clear that predicates considered in a particular analysis may also be formulated solely on the basis of sets of values, and the corresponding probabilities of these values can then be viewed as limiting frequencies, that is, frequencies that would be observed should an infinite number of observations be made. For instance, it is reasonable to consider that the outcome of a measurement will yield a certain element of **S** a certain fraction of the time. A prior, $p(A)$, may thus be regarded as the degree of belief that a certain probability distribution dictates the outcome of a measurement. The conditional probability $p(B|A)$ then provides the degree of belief that the given probability distribution yields an outcome B within **S**. The subjective interpretation thus effectively encompasses the relative frequentist interpretation if one admits the implicit proviso that $p(A) = 1$. A subjective interpretation may, however, also be associated with cases in which the concept of frequency is not readily or meaningfully applicable. For instance, while the notion that a certain quantity X lies within a specific interval can be determined in both interpretations, the determination of confidence intervals with the frequentist interpretation assumes it is reasonable to use a specific parent probability distribution to carry out the calculation of the interval. In effect, this assumes one has a reasonably high degree of belief that a specific probability distribution is a proper representation of the outcome of an experiment. Effectively, the prior, which corresponds to the probability that a specific probability determines the outcome of a measurement, is assumed to have maximal probability. The frequentist interpretation can thus indeed be viewed as "special case" of the subjective interpretation.

2.4 Bayes' Theorem and Inference

Whether working within the frequentist interpretation or the Bayesian interpretation of probability, Bayes' theorem is ideally suited toward statistical inference analyses, that is, analyses where one wishes to establish the optimal value of model parameters, estimates of their errors, or which of many competing hypotheses has the highest probability of asserting the truth about a particular system or phenomenon. Although frequentist inference can and will be considered in this context, it is far more convenient to introduce the concept of inference within the Bayesian paradigm using the notion of probability as an extension of logic because generic statements about scientific hypotheses (i.e., models, model parameters, etc.) can be evaluated in a single formal and robust mathematical setting where the prior plausibility of hypotheses as well as data are considered.

2.4.1 Basic Concepts of Bayesian Inference

Let H_i, with $i = 1, \ldots, n$, represent a set of n propositions asserting the truth of competing hypotheses. Given the very nature of the scientific process, these hypotheses are formulated out of a particular context. Let us represent relevant statements from this context (also known as prior information) as I. We will additionally represent measured data in terms of a proposition D. For inference purposes, Bayes' theorem may then be written

$$p(H_i|D, I) = \frac{p(D|H_i, I)p(H_i|I)}{p(D|I)}. \tag{2.45}$$

The quantity $p(D|H_i, I)$ represents the probability of observing the data D if both H_i and I are true. It is commonly called likelihood of the data D based on the hypothesis H_i, or simply likelihood function, and noted $\mathbf{L}(H_i)$. The quantities $p(H_i|I)$ and $p(H_i|D, I)$ represent the prior and posterior probability of the hypothesis H_i. The probability $p(H_i|I)$ is based solely on prior knowledge whereas $p(H_i|D, I)$ includes both the prior knowledge and the new knowledge provided by the measurement D. The denominator, $p(D|I)$, is seemingly more cryptic but it corresponds to the probability of obtaining the data D given the prior information available on the system or phenomenon. Although it may be difficult to assess this probability directly, note that it can be computed in terms of the law of total probability

$$p(D|I) = \sum_i p(D|H_i, I)p(H_i|I) \tag{2.46}$$

where the sum is taken over all hypotheses that can be formulated about the system. All in all, this factor provides a normalization factor that ensures that the sum over the probability of all hypotheses, given the data and prior information, is equal to unity:

$$\sum_i p(H_i|D, I) = 1. \tag{2.47}$$

2.4.2 Hypothesis vs. Sample Space

Within the frequentist approach, one is focused on the outcome of measurements and techniques mostly to utilize these measurements to extract information about a phenomenon or system. Measured observables may be discrete (e.g., number of particles observed in a specific proton–proton collision) or continuous (e.g., the momenta of produced particles). Observed values, collectively called **sample**, may then be viewed as elements of either a discrete set (i.e., a subset of **Z**, the set of integers) or a continuous set (i.e., a subset of **R**, the set of real numbers). The measured values are thus considered random outcomes, or random variables, either discrete or continuous, from a **parent population** known as a **sample space**.

The Bayesian approach shifts the focus toward statements about the data and hypotheses or models used to described the data. The goal is indeed to use the measured data to establish the plausibility (or degree of belief) of various hypotheses or statements formulated about a system (phenomenon) and the data it produces. Hypotheses may concern various characterizations of the data, model parameters, or even a model as a whole. They may be formulated either in terms of discrete statements (e.g., dark matter exists; there is only one Higgs boson; etc.) or in the form of continuous statements (e.g., the Hubble constant lies in the range $[H_0, H_0 + dH_0]$; the mass of the Higgs boson is in the range $[M, M + dM]$, etc.). The Bayesian approach thus enlarges, so to speak, the sample space associated with the outcome of measurement observables to include a space of hypotheses or model statements that can be made about a system both before and after the measurement is conducted. It is then concerned with assigning degrees of belief, or plausibility, to each of these hypotheses or model statements.

Strictly speaking, hypotheses are not random variables, but specific statements about a phenomenon or reality at large. Indeed, dark matter either exists or does not, but it is not a random phenomenon. Likewise, physical quantities such as the speed of light or Planck constant have specific values and thus cannot be legitimately regarded as random variables. The true (and precise) values of the observables factually remain unknown, however, so the Bayesian notion of degree of belief that the value of a physical quantity might lie in a given interval thus makes good sense. Yet, measurements involve a number of effects that may effectively smear or seemingly randomize observed values. A Bayesian statistician must then account for the measurement outcomes with a probability model of the measurement process. One concludes that while an observable of interest might not be random, measured instances of the observable will invariably appear random. Consequently, insofar as probabilities are regarded as degrees of belief, there is no philosophical difficulty or contradiction in considering prior probabilities that an observable X might lie within a range $[x_0, x_0 + dx]$ while measurement instances have a probability $p(x|x_0)$, determined by the measurement process, to be found in the range $[x, x + dx]$. Effectively, we conclude that both x_0 and x can be treated as random variables.

2.5 Definition of Probability Distribution and Probability Density

Whether one adheres to the frequentist or Bayesian interpretation of probabilities, one is faced with either discrete or continuous variables. With finitely many discrete values, it is obviously possible to assign a (finite) probability to each value separately but if the number of discrete values is infinite, or if the variables are continuous, one must introduce the notion of probability density. We discuss basic features and properties of discrete variable first, in §2.5.1, and consider continuous variables next, in §2.5.2.

2.5.1 Discrete Observables and Probability Distribution Functions

Consider, for instance, a game of dice in which players throw two cubic dice at a time on a mat. The faces of the dice are labeled with numbers ranging from 1 to 6. The game may then involve betting on the sum of the dice values rolled in a given throw. Clearly, the sum of dice rolled takes only a finite number of discrete values from 2 to 12 and is thus a **discrete outcome**. There is only one way to get a sum of 2 or 12, but several ways to roll a 6 or 7. The outcome of the bet should then be decided based on the probability of a given roll.

One should remark, once again, that a roll of dice is nominally a phenomenon that can be described in terms of deterministic laws of physics. Indeed, given specific initial conditions (i.e., the position, orientation, translational and rotational speed of the dice), one could in principle predict the outcome of a roll provided the elastic properties of the dice and the table on which they roll are well known. In practice, the properties of the dice and table are not so well known, and measuring the initial conditions of a roll with sufficient precision is rather tricky. In essence, not enough is known about the system (the two dice and mat) to enable an accurate calculation of the outcome of a roll, that is, on which face the dice will stop rolling. The outcome of a dice roll thus appears unpredictable and the sum of the top faces may then be regarded as a random variable. Given the geometrical symmetry of a die, it is natural to assume all faces are equally likely. Within the frequentist interpretation, and for a fair die, one expects that all six faces should have the same frequency after rolls have been repeated a very large number of times, while in the Bayesian interpretation, one may ab initio express the belief (or plausibility) that the perfect symmetry of a die implies each face has a probability of $1/6$ of rolling up.

More generally, one may be concerned with the determination of the probabilities of values taken by one or several discrete random variables. For example, a marketer might be concerned with the number of people showing up at a special public event based on ads published in newspapers or played on radio stations, whereas an astronomer might be interested in counting how many supernovae explosions were detected in a specific night with a powerful telescope. In these and other discrete systems, as for a roll of dice, one assumes the systems are not sufficiently well known (either by virtue of their macroscopic complexity or their inherent nondeterministic character) to predict a specific outcome with certainty, and one must then assess either a frequency (frequentist approach) or degree of plausibility (belief) that specific values might be observed.

In the case of a measurement of one discrete random variable, n, the sample space consists of all (integer) values, or combinations of values, the variable can take. Assuming the number of such values is finite, one is then concerned with the probability, $p(n)$, of each element individually. In a roll of a pair of dice, for instance, one might want to know the probability of rolling a 7. The quantity $p(n)$ is thus a function that represents how the probability of values of n is distributed across the sample space $\mathbf{S}$ and is known as the **Probability Distribution Function**, or PDF.

By virtue of the third axiom, Eq. (2.4), the sum of the probabilities of all outcomes must be unity. The sum of the probabilities $p(n)$ must thus satisfy the condition:

$$\sum_{n \in \mathbf{S}} p(n) = 1. \tag{2.48}$$

2.5.2 Continuous Observables and Probability Density Functions

Obviously, there are also cases in which measured observables can take continuous values. Examples of continuous variables include the temperature of the atmosphere at sunset, the barometric pressure during a storm, the strength of the electric and magnetic fields produced by an antenna, or the momentum of particles produced by nuclear collisions, and so on. One may be interested in studying how such quantities vary with time, position, or other variables. Alternatively, one might be interested in the very precise determination of "constants" of nature, such as the speed of light in vacuum, the lifetime of the ^{14}C radioisotope, or the cross section of a particular nuclear reaction. Within the context of the Bayesian approach, one may also be interested in considering continuous hypotheses. This is the case, for instance, when a particular model parameter or observable cannot be observed directly but must be inferred from one or several other measurements. One can then formulate continuous hypotheses stating that a continuous observable O lies with a given range $[O, O + \Delta O]$.[3]

While a physical quantity is known (or assumed) to have a single and unique value, repeated measurements would yield continuous values that fluctuate, seemingly arbitrarily, from measurement to measurement. One is thus faced with **continuous random variables**, which are either elements of a continuous sample space (frequentist approach) or a continuous hypothesis space (Bayesian approach). Either way, a space of continuous (random) variables, or combination of random variables, is obviously infinite. It is thus not meaningful to talk about the probability of a specific value. One is instead concerned with the probability of measuring values in specific finite intervals (e.g., $[O, O + \Delta O]$). However,

[3] Note that the same letter is here used to represent both a logical proposition and the observable it is concerned with.

in the limit of vanishingly small intervals, $\Delta O \to 0$, one can introduce the notion of **probability density**.

Let us consider an experiment whose outcome consists of a single continuous observable X. The sample space **S** associated with this measurement may thus consists of a subset of **R**, the set of real numbers. Given continuous subsets of **R** have infinite cardinality, it is not meaningful to consider the probability of a single value, x. One can, however, consider the probability that such an observed value x will be found within the infinitesimal interval $[x, x + dx]$. We shall here assume this probability exists and can be evaluated with a function $f(x)$ known as **Probability Density Function**, hereafter noted, PDF:

$$\text{Probability to observe } X \text{ in } [x, x + dx] = f(x)\, dx. \tag{2.49}$$

Note 1: In general, statisticians use capital letters (e.g., X, Y, W, etc.) to denote or identify the name of observables (observable quantities, variables), while lowercase letters (e.g., x, y, w) are used to label specific instances of these variables. For instance, the variable identifying the position of a particle might be defined as X while a specific measurement (i.e., a specific instance) of the observable would be typically written as x. In physics, however, lowercase and uppercase letters are typically used to denote different quantities or observables (although perhaps related). The uppercase/lowercase convention used by statisticians may then become difficult to apply. We thus (mostly) adhere to the convention in Part I when introducing generic and foundational concepts and relinquish the convention in Parts II and III of the book when discussing physics concepts.

In the frequentist interpretation, $f(x)\, dx$ corresponds to the fraction of times the value x is found in the interval $[x, x + dx]$ in the limit that the total number of measurements is infinitely large, while in the Bayesian interpretation, this quantity provides the degree of belief an observed value x might be in that range, without any particular assumption as to whether the experiment can actually be repeated. Additionally, note that in this context, a continuous variable X could also represent a statement or hypothesis about a parameter of a model used to describe the system or phenomenon. It is thus legitimate to consider probability densities in hypothesis space as well as in sample space.

By virtue of the third axiom (Eq. 2.4), a PDF must be normalized such that the probability of any outcome is unity. The sum of the probabilities of all outcomes is thus the integral of the function $f(x)$ over the entire sample space **S** (or hypothesis space as the case may be) spanned by the observable X:

$$\int_{\mathbf{S}} f(x)\, dx = 1. \tag{2.50}$$

It is important to reemphasize that the notion of probability density function applies to both measurement outcomes and continuous model hypotheses. In fact, much of the discussion that follows in this and following chapters about probability densities is applicable to continuous variables in both sample and hypothesis spaces without much regard as to whether they are discussed in the context of either interpretation.

Note 2: Throughout this text, we use the abbreviation PDF and the notations $f(x)$ and $p(x)$ for both probability distribution functions and probability density functions. In general, it shall be clear from the context whether one is considering discrete or continuous variables and, correspondingly, probability distributions or probability densities.

2.5.3 Cumulative Distribution and Density Functions

It is useful to introduce the notions of **Cumulative Distribution Function** and **Cumulative Density Function**, both hereafter denoted CDF. Given a PDF $f(x)$, the CDF $F(x)$, may be formally defined as a cumulative (sometimes called **running**) sum or integral of the function $f(x)$. For a discrete probability distribution, one has

$$F_n = \sum_{i=0}^{n} f(x_i). \tag{2.51}$$

whereas for a density one has

$$F(x) = \int_{-\infty}^{x} f(x')\,dx'. \tag{2.52}$$

Obviously, the function $F(x)$ amounts to the probability a random variable X takes a value smaller or equal to x. One can thus alternatively first define $F(x)$ as the probability of obtaining an outcome less than or equal to x and evaluate the PDF, $f(x)$, as a derivative of F:

$$f(x) = \frac{dF(x)}{dx}. \tag{2.53}$$

The two definitions are equivalent if the distribution $F(x)$ is well behaved, that is, if it is everywhere differentiable. This can be formalized mathematically, but the notion of "everywhere differentiable" will be sufficient for the remainder of this text.

The concept of cumulative distribution is illustrated in Figure 2.2, which displays the cumulative integral, Eq. (2.52), of the uniform distribution (top panel) and the **standard normal distribution** (bottom panel). The definitions of these distributions and their properties are formally introduced in §§3.4 and 3.9. Note that since both distributions are symmetric relative to their **median**,[4] indicated with a vertical dash line, $F(x)$ takes a value of 0.5 at that point. This means there is a probability of 50% that the value of x will be smaller than $x = 0$. Additionally, note that in the limit of $x \to \infty$, the cumulative function $F(x)$ converges to unity; in other words, the probability of observing any value of x is unity. This is indeed guaranteed by the normalization of $f(x)$ defined by Eq. (2.50).

It is useful to introduce an alternative notation for PDFs. Given a PDF $f(x)$ expresses the probability of occurrences, or events, being observed in the interval $[x, x + dx]$, one may also write

$$f(x) \equiv \frac{1}{N}\frac{dN}{dx}, \tag{2.54}$$

[4] The notion of median is formally defined in §2.7.

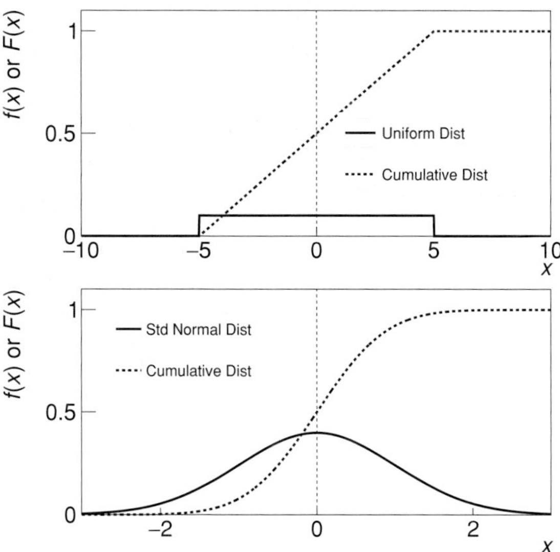

Fig. 2.2 Illustration of the concept of cumulative distribution for a uniform distribution (top) and (bottom) a standard normal (Gaussian) distribution.

where N represents the integral of the function dN/dx over its domain which corresponds to the sample (or hypothesis) space of x. One gets

$$\int_{-\infty}^{\infty} \frac{1}{N}\frac{dN}{dx}dx = \frac{1}{N}\int dN = \frac{1}{N}N = 1, \tag{2.55}$$

which satisfies the normalization condition given by Eq. (2.50). The function dN/dx thus represents a density of counts or events in the interval $[x, x + dx]$, and dividing this density by its integral yields the probability density $f(x)$ with proper normalization.

The function notation, $f(x)$, has the advantage of being compact and is used predominantly throughout this book. However, the differential notation $N^{-1}dN/dx$ explicitly presents $f(x)$ for what it is: a density. We thus use it whenever it is important to emphasize the notion of density, most particularly when discussing differential cross sections and correlation functions in Chapters 8 and 9, and toward the definition of histograms introduced in §4.6.

2.6 Functions of Random Variables

Consider a measurement of the momentum, p, of particles produced in proton–proton collisions at some fixed energy. Given the production of particles is a stochastic (random) phenomenon, the momentum of the measured particles can be regarded as a random quantity. Assuming all produced particles are pions, one can determine their energy based on the relativistic relation, $E = \sqrt{p^2c^2 + m^2c^4}$, where m is the mass of the pion and c the speed of light. Since p is a random variable, E is consequently also a random variable. This is true in general: functions of random variables are themselves random variables and it is often of interest to determine or characterize the probability density of such random variables.

Let us first introduce a continuous function, $q(x)$, of a single continuous random variable X. Since X is random, the application of $q(x)$ on observed values x yields random values, and it is thus legitimate to consider these as instances of a random variable Q. Our goal shall be to determine the PDF of Q given a PDF for X.

Let $f(x)$ be the PDF of X. By definition, it represents the probability of observing X in the range $[x, x+dx]$.[5] An observation is an event, and the specific variable used to represent this event should thus be immaterial. This means the probability of observing that event is **conserved** or **invariant** when the variable representing the event is changed or transformed. Let us then consider the same event from the point of view of the variables X and Q. Within the sample space of X, the probability of the event may be expressed

$$f(x)\,dx = \text{Probability an event takes place in } [x, x+dx], \tag{2.56}$$

whereas from within the sample space of Q, one has

$$g(q)\,dq = \text{Probability an event takes place in } [q(x), q(x+dx)], \tag{2.57}$$

where we introduced $g(q)$ as the PDF of the random variable Q. These two expressions represent the same subset of events and must then be equal. In order to find a relation between $g(q)$ and $f(x)$, let us write the function $q(x+dx)$ as a truncated Taylor series:

$$q(x+dx) = q(x) + \left.\frac{dq}{dx}\right|_x dx + O(2). \tag{2.58}$$

In the limit $dx \to 0$, the quantity dq may then be written

$$dq = q(x+dx) - q(x) = \left.\frac{dq}{dx}\right|_x dx. \tag{2.59}$$

Now, given the expressions (2.56) and (2.57) are equal, one can write

$$g(q) = f(x(q))\left|\frac{dq}{dx}\right|^{-1}. \tag{2.60}$$

Since the function $g(q)$ represents a probability density, one ensures it is positive definite by using the absolute value $|dq/dx|$ in Eq. (2.60). Additionally, note that if $x(q)$ is multivalued, one must include all values of x that map onto a specific value of q.

As a simple example of a multivalued problem, consider the function $q(x) = x^2$ with inverse $x = \pm\sqrt{q}$, and

$$\left|\frac{dx}{dq}\right| = \frac{1}{2\sqrt{q}}.$$

[5] In this context, the interpretation of densities in terms of limiting frequency or degree of belief is somewhat immaterial and we carry out the discussion in terms of the frequentist interpretation for convenience, but the reasonings and results are identical within the Bayesian interpretation.

Given there are two roots with equal contributions, one must include a factor of 2 in Eq. (2.60). One thus gets

$$g(q) = \frac{f(x(q))}{\sqrt{q}}. \tag{2.61}$$

The concept of probability density is readily extended to functions of multiple random variables in §2.8, but first it is useful and convenient to introduce commonly used properties of PDFs.

2.7 PDF Characterization

While a PDF carries the maximum amount of information on the behavior of a random variable and the phenomenon or physical quantity it represents, it is often desirable, convenient, and at times sufficient to reduce or transform this information into a set of appropriately chosen properties or functions. For a known PDF, these properties are uniquely defined and can usually be calculated exactly. If the PDF is unknown or partially known, the properties must be estimated based on functions of sampled (measured) data known as **statistics** and **estimators**.[6] Several types of properties are of interest toward the characterization of PDFs and for the modeling of data. We first introduce and discuss, in this section, the notions of **α-point**, **mode**, **expectation value**, **moments**, **centered moments**, and **standardized moments**. The notions of **characteristic function** and **moment-generating functions** are introduced in §2.10 whereas **cumulants** are defined in §2.13. The notions of **covariance** and **factorial moments** are discussed in §2.9 and §10.2 respectively, after the introduction of multivariate random functions in §2.8.

2.7.1 α-Point and Median

The concept of **alpha-point**, also called **quantile of order** α, and noted α-point, is introduced by defining a quantity x_α that demarcates the probability of x being smaller than α:

$$F(x_\alpha) = \alpha \text{ with } 0 \leq \alpha \leq 1. \tag{2.62}$$

The corresponding value x_α is thus the inverse:

$$x_\alpha \equiv F^{-1}(\alpha). \tag{2.63}$$

The value $x_{1/2}$, called the **median**, is a special case of the α-point commonly used to estimate the typical value of a random variable, given there is a 50% probability that measured values of x are smaller. The uniform and Gaussian PDFs shown in Figure 2.2 are symmetric about $x = 0$. Their 1/2-point is consequently $x_{1/2} = 0$ and the probabilities of x being smaller or larger than 0 are equal.

[6] A formal definition of the notion of a statistics is presented in §4.3.

For a discrete random variable x_i, with probability $p(x_i)$, the cumulative distribution

$$F(x) = \sum_{x_i \leq x} p(x_i) \tag{2.64}$$

spans discrete values only. The α-point determined by Eq. (2.63) is thus strictly exact only for discrete values x_α.

2.7.2 Mode

The **mode** of a PDF $f(x)$ is defined as the value of the random variable x for which the PDF has a maximum.

A PDF with a single maximum is called a **unimodal distribution**. The position of the mode may be obtained either by inspection or by finding the extremum of the distribution, in other words, by finding the value x where the PDF has a null first derivative with respect to x. Given that

$$f'(x) \equiv \frac{df(x)}{dx}, \tag{2.65}$$

the mode of the distribution is thus

$$x_{\text{mode}} = f'^{-1}(0). \tag{2.66}$$

PDFs with two or more maxima are known as **bimodal** and **multimodal**, respectively. As for unimodal distributions, their modes can be obtained either by direct inspection or by finding the zeros of the function $f'(x)$.

Figure 2.3 illustrates the concepts of **median**, **mode**, and **mean** of a unimodal PDF as well as the notion of a bimodal PDF.

2.7.3 Expectation Value (Mean)

The expectation value, noted E[x], of a continuous random variable distributed according to a PDF $f(x)$ is defined as

$$\text{E}[x] \equiv \int_{-\infty}^{\infty} x f(x)\, dx \equiv \mu. \tag{2.67}$$

This expression is commonly called **population mean**, **mean**, or **average value** of x. In this book, depending on the context, it will also be denoted as μ, μ_x, or $\langle x \rangle$, where the brackets $\langle\ \rangle$ refer to a population or domain average:

$$\text{E}[x] \equiv \mu \equiv \mu_x \equiv \langle x \rangle. \tag{2.68}$$

The expression E[x] is also commonly referred to as first moment, noted $\mu_1{}'$, of the PDF $f(x)$ for reasons that will become obvious in the following discussion. Brackets [x] rather than parentheses (x) are used to indicate that E[x] is not a function of x but rather represents a certain value of x determined by the shape of the distribution $f(x)$. Indeed, it is a property of the PDF itself and thus not a function of a particular value of x. If $f(x)$ is a strongly peaked function, then the mean is likely to be near the mode of the function. However,

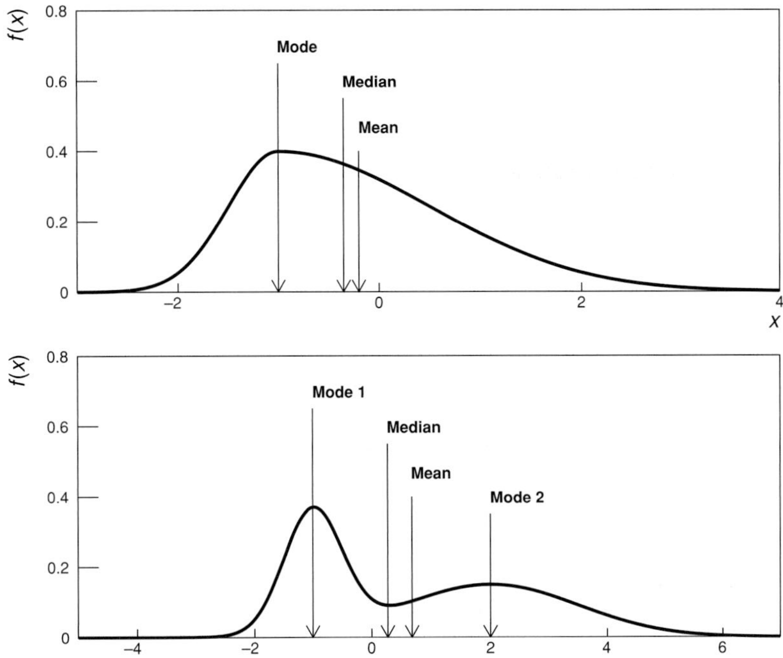

Fig. 2.3 (Top) Illustration of the concepts of median, mode, and mean for a single mode distribution. (Bottom) Example of a bimodal distribution.

for bimodal or multimodal functions, the mean is typically not representative of a specific peak, unless perhaps one peak strongly dominates the others.

As an example of calculation of the expectation value of a function, let us evaluate the mean of the uniform distribution $p_u(x)$ defined in the range $a \leq x \leq b$ according to

$$p_u(x|a, b) = \begin{cases} \frac{1}{b-a} & \text{for } a \leqslant x \leqslant b, \\ 0 & \text{elsewhere,} \end{cases} \tag{2.69}$$

where the denominator $b - a$ ensures the proper normalization of the distribution when $p_u(x|a, b)$ is integrated over $\mathbf{R}$. The mean of x is defined according to

$$\langle x \rangle \equiv \mu_x \equiv \int_{-\infty}^{\infty} p_u(x|a, b) x \, dx. \tag{2.70}$$

Substituting the definition, Eq. (2.69), for $p_u(x|a, b)$, and proceeding with the integration, one finds

$$\langle x \rangle = \frac{1}{b-a} \int_a^b x \, dx, \tag{2.71}$$

$$= \frac{1}{b-a} \left. \frac{x^2}{2} \right|_a^b, \tag{2.72}$$

$$= \frac{a+b}{2}, \tag{2.73}$$

which corresponds to the middle of the interval $[a, b]$, and is thus indeed representative of typical values of x determined by $p_u(x|a, b)$.

The notion of expectation value is quite general. In fact, instead of the expectation value of x, one can calculate the expectation value of any functions of x. Let us here denote such a function as $q(x)$. Clearly, since x is a random variable, so shall be $q(x)$. As in §2.6, let $g(q)$ denote the PDF of q. By definition, the expectation value of q shall be

$$\mathrm{E}[q] \equiv \int_{-\infty}^{\infty} q g(q)\, dq. \tag{2.74}$$

Recall from §2.6 that the probability of x being in the interval from x to $x + dx$ must be equal to the probability of q being in the interval from $q(x)$ to $q(x + dx)$. The preceding expectation value may then be written

$$\mathrm{E}[q] = \int_{-\infty}^{\infty} q(x) f(x) \frac{dx}{dq}\, dq = \int_{-\infty}^{\infty} q(x) f(x)\, dx. \tag{2.75}$$

The expectation value of the function $q(x)$, given the PDF $f(x)$, is thus equal to the inner product of $q(x)$ by $f(x)$.

2.7.4 Moments, Centered Moments, and Standardized Moments

A special and important case of the function $q(x)$ involves powers of x. One defines the nth **algebraic moment** (also simply called the nth moment) of the PDF $f(x)$, denoted μ'_n, as

$$\mu'_n \equiv \mathrm{E}[x^n] = \int_{-\infty}^{\infty} x^n f(x)\, dx. \tag{2.76}$$

Obviously, the mean, μ, is a special case of Eq. (2.76) and corresponds to the first ($n = 1$) moment μ'_1.

It is also convenient to consider moments relative to the mean μ. These are defined according to

$$\mu_n \equiv \mathrm{E}[(x - \mathrm{E}[x])^n] = \int_{-\infty}^{\infty} (x - \mu)^n f(x)\, dx, \tag{2.77}$$

and are named **centered moments** or **moments about the mean**. Hereafter, we use the expression **centered moment** exclusively.

A first special and important case to consider is the second centered moment μ_2. It corresponds to the variance, noted Var$[x]$, of the PDF,

$$\mu_2 \equiv \mathrm{Var}[x] \equiv \mathrm{E}[(x - \mathrm{E}[x])^2] = \int_{-\infty}^{\infty} (x - \mu)^2 f(x)\, dx = \sigma^2, \tag{2.78}$$

where σ corresponds to what is commonly known as the **standard deviation** of the PDF. Note that the notation for the variance varies across texts: many authors use the notations $V[x]$, $V(x)$, or Var(x). In this text, we use the [ ] notation to emphasize that the variance is a functional of the PDF, that is, a property of the PDF rather than, strictly speaking, a function of x. Additionally, depending on the context of our discussions, we shall use several alternative notations for the variance of x as follows:

$$\mu_2 \equiv \text{Var}[x] \equiv \sigma^2 \equiv \sigma_x^2 \equiv \langle (x-\mu)^2 \rangle \equiv \langle \Delta x^2 \rangle, \tag{2.79}$$

where $\Delta x \equiv x - \mu$, and the brackets $\langle \, \rangle$ here again refer to a population or domain average.

One can then show (see Problem 2.4) that the variance may be expressed in terms of the second and first moments:

$$\mu_2 \equiv \text{E}[x^2] - \mu^2 = \mu'_2 - \mu^2. \tag{2.80}$$

One also readily verifies that the variance Var[x] measures the spread of x about its mean value μ, as illustrated in Figure 2.4 with four selected Gaussian distributions. The Gaussian distribution $p_G(x|\mu, \sigma)$, formally introduced in §3.9, is defined according to

$$p_G(x|\mu, \sigma^2) = \frac{1}{\sqrt{2\pi}\sigma} \exp\left[-\frac{(x-\mu)^2}{2\sigma^2} \right], \tag{2.81}$$

where the factor $\sqrt{2\pi}\sigma$ enables proper normalization of the distribution when it is integrated over **R**. Distributions shown in Figure 2.4 are symmetric about the origin and thus have a mean $\langle x \rangle$ equal to zero. Setting $\mu = 0$ in Eq. (2.81), one proceeds to calculate the variance of x according to

$$\text{Var}[x] = \text{E}[x^2] - \mu^2 = \text{E}[x^2], \tag{2.82}$$

$$= \frac{1}{\sqrt{2\pi}\sigma} \int_{-\infty}^{\infty} \exp\left(-\frac{x^2}{2\sigma^2} \right) x^2 \, dx. \tag{2.83}$$

Substituting $z = x/\sigma$, the preceding expression becomes

$$\text{Var}[x] = \frac{\sigma^2}{\sqrt{2\pi}} \int_{-\infty}^{\infty} \exp\left(-\frac{z^2}{2} \right) z^2 \, dz. \tag{2.84}$$

The integral is readily determined to equal $\sqrt{2\pi}$ from basic definite integral tables. The variance of the Gaussian distribution is then

$$\text{Var}[x] = \sigma^2. \tag{2.85}$$

Comparing the distributions plotted in Figure 2.4 for selected values of σ, one finds, indeed, that a Gaussian with a large spread in x features a large variance, whereas a narrow Gaussian has a small variance. This result can be readily extended to distributions of arbitrary shapes: broadly distributed PDFs have a large variance whereas narrowly distributions have a small variance.

It is often useful to compare the higher moments, $n > 2$, of a distribution to the standard deviation. This may be accomplished with **standardized moments**, μ_k^{std}, defined as ratios of the kth centered moments and the kth power of the standard deviation:

$$\mu_k^{\text{std}} = \frac{\mu_k}{\sigma^k}. \tag{2.86}$$

The standard moments hence correspond to kth moments normalized with respect to the standard deviation. The power of k is required in the normalization given moments scale as x^k. This implies the standardized moments are scale invariant; in other words, rescaling

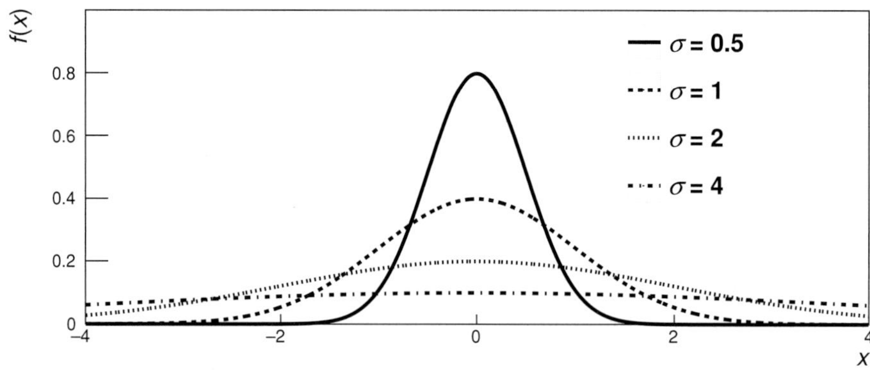

Fig. 2.4 Four Gaussian PDFs (see §3.9) with mean, $\mu = 0$, and standard deviations σ of 0.5, 1.0, 2.0, and 4.0. The variance measures the breadth of a distribution. The distribution shown as a solid line is the narrowest and has the smallest variance. The other distributions are wider and thus have larger variances.

of the variable x by an arbitrary factor α leads to changes in the kth moment and the standard deviation by a factor α^k, but their ratio is invariant, that is, independent of α. The standardized moments μ_k^{std} are dimensionless numbers for the same reason. Also note that the first standardized moment vanishes because the first moment about the mean is null, while the second standardized moment equals unity because the second moment about the mean is the variance. The third and fourth standardized moments are called skewness and kurtosis, respectively. These are discussed in the next two sections.

2.7.5 Skewness

The **skewness** of a distribution is commonly denoted γ_1 or Skew[x] in the literature. It is formally defined as the third standardized moment of a distribution:

$$\gamma_1 \equiv \text{Skew}[x] \equiv \mu_3^{\text{std}} = \frac{\mu_3}{\sigma^3}. \tag{2.87}$$

Skewness is essentially a measure of the asymmetry of a distribution. Consider, for instance, the distributions shown in Figure 2.5. The solid line curve displays a Gaussian distribution (defined in §3.9) with mean, $\mu = 0$, and width, $\sigma = 0.6$. It is by construction symmetric about its mean and therefore has null skewness. The dash and dotted curves are constructed as two juxtaposed half Gaussian distributions. Their peaks are both at $x = 0$ (same as for the black curve) but the left and right widths of the dash (dotted) curve are 0.6 (2.0) and 1.5 (0.6), respectively. Focusing on the dash curve, one finds the right side of the distribution tapers differently than the left side. These tapering sides, called low and high side tails, provide a visual means for determining the sign of the skewness of a distribution. The skewness is typically negative if the low side (left) tail is longer than the high side (right) tail. It is then said to be left-skewed (dotted curve in Figure 2.5). If the high side (right) tail is longer, the distribution is said to be right-skewed (dashed curve). In a skewed (unbalanced, lopsided) distribution, the mean is farther out in the long tail than is the median. Distributions with zero skewness, such as normal distributions, are symmetric

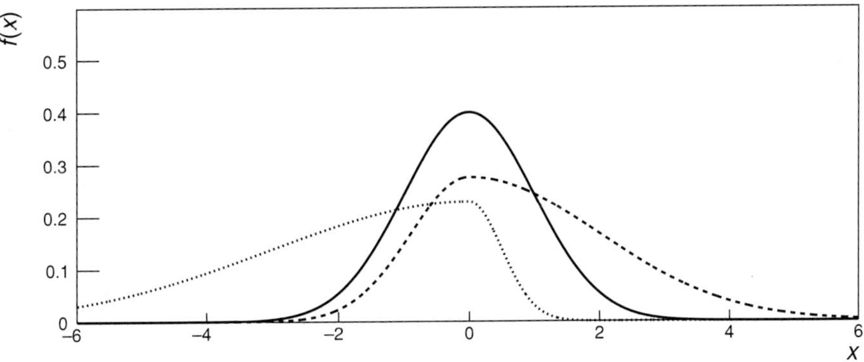

Fig. 2.5 Illustration of the notion of skewness. The solid line curve is by construction symmetric and has, as such, zero skewness. The dashed and dotted curves have longer high side and low side tails and consequently have positive and negative skewness, respectively.

about their mean: in effect, their mean equals their median and mode. Note, however, that it is not sufficient for the mean of a distribution to be right (left) of the median to conclude it is right (left) skew. Multimodal and discrete distributions, in particular, may violate this simple expectation.

Karl Pearson[7] (1857–1936) suggested several alternative measures of skewness:

$$\text{Pearson mode skewness: } \frac{\mu - \text{mode}}{\sigma}, \tag{2.88}$$

$$\text{Pearson's 1st skewness coefficient: } 3\frac{\mu - \text{mode}}{\sigma}, \tag{2.89}$$

$$\text{Pearson's 2nd skewness coefficient: } 3\frac{\mu - \text{median}}{\sigma}. \tag{2.90}$$

These are, however, not frequently used to characterize data in nuclear and particle physics.

2.7.6 Kurtosis

Two definitions of **kurtosis** are commonly used in modern statistical literature. The first corresponds to the old kurtosis and is often noted Kurt[x]. It is defined as the fourth standardized moment of a distribution. As such, it corresponds to the ratio of the fourth centered moment by the fourth power of the standard deviation:

$$\text{Kurt}_{\text{old}}[\text{x}] \equiv \frac{\mu_4}{\sigma^4}. \tag{2.91}$$

Modern kurtosis is usually called **excess kurtosis**. Denoted γ_2, it is defined as the ratio of the fourth cumulant by the square of the second cumulant (see §2.13.1 for the definition

[7] Influential English mathematician generally credited for establishing the discipline of mathematical statistics.

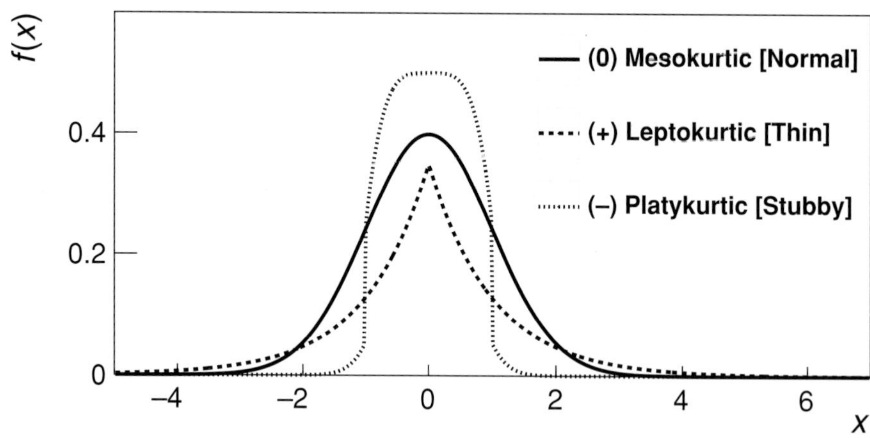

Fig. 2.6 Illustration of the notion of kurtosis. The solid line is a Gaussian distribution and has zero excess kurtosis. The dashed curve has a peaked distribution with long tails and thus has a positive kurtosis while the dotted curved has a flat top with short tails and hence is characterized by a negative kurtosis.

of cumulants):

$$\gamma_2 = \frac{\kappa_4}{\kappa_2^2} = \frac{\mu_4}{\sigma^4} - 3. \tag{2.92}$$

The "minus 3" conveniently makes kurtosis of the Gaussian distribution equal to zero. More importantly, the definition in terms of cumulants implies the (excess) kurtosis of a sum of n independent random variables x_i is equal to the sum of the kurtosis of these n variables divided by n^2. This can be written (see Problem 2.6):

$$\gamma_2 \left[\sum_{i=1}^{n} x_i \right] = \frac{1}{n^2} \sum_{i=1}^{n} \gamma_2 \left[x_i \right]. \tag{2.93}$$

Such a simple scaling by n^2 does not arise with the "old" kurtosis definition.

Figure 2.6 schematically illustrates the notion of kurtosis. A high kurtosis distribution has a sharper peak and longer, fatter tails than a Gaussian distribution, while a low kurtosis distribution has a more rounded peak and shorter, thinner tails. Distributions with zero excess kurtosis are called **mesokurtic**, or **mesokurtotic**. The most obvious example of a mesokurtic distribution is the Gaussian distribution (see §3.9). A few other well-known distributions can be mesokurtic, depending on their parameter values. For example, the binomial distribution (see §3.1) is mesokurtic for $p = 1/2 \pm \sqrt{1/12}$. A distribution with positive excess kurtosis is called **leptokurtic**, or **leptokurtotic**. A leptokurtic distribution has a more acute peak around its mean and fatter tails than a Gaussian distribution (narrower peak and more probable extreme values). Examples of leptokurtic distributions include the Laplace distribution and the logistic distribution; such distributions are sometimes called super-Gaussian. A distribution with negative excess kurtosis is called **platykurtic**, or **platykurtotic**. The shape of a platykurtic distribution features a lower, wider peak around the mean (i.e., a lower probability than a normally distributed variable of values near the mean) and thinner tails; in other words, extreme values (both smaller and larger than the mean) have a larger probability than a normally distributed variable. The uniform distribution (see §3.4) is a prime example of a platykurtic distribution. Another example involves the Bernoulli distribution (see §3.1), with $p = 1/2$, obtained, for example, for the number of times one obtains "heads" when flipping a coin (i.e., a coin toss), for which kurtosis is -2. Such distributions are sometimes termed sub-Gaussian.

The notion of kurtosis is not as frequently used as the notion of variance but nonetheless remains of interest in general to characterize the shape of a distribution. It finds specific applications in nuclear physics with the study of net charge fluctuations and the determination of the charge susceptibility of the quark gluon plasma (see §11.3.3).

2.7.7 Credible Range

Perhaps the most basic way to characterize a set of data is to describe the **range** it covers. The range of the data, as the word suggests, is simply the difference between the highest and lowest observed values of a random variable. It is consequently straightforward to determine, although it may be somewhat misleading for distributions with long low or high side tails (with low probability). In such cases, the range is subject to large fluctuations when dealing with small samples and is thus rather unreliable in characterizing the bulk of a distribution. It may then be preferable to use the notion of **interquartile range** instead, which is evaluated as the difference between the higher and lower quartiles. The lower and higher quartiles correspond to 1/4-point and 3/4-point, respectively.

The notions of **deciles** and **percentiles** are also commonly used to report placements within a distribution as fractions of a population, sample, or distribution with values smaller or equal to a given decile or percentile. For instance, students receiving a 99 percentile score on a physics exam have good reasons to be proud because they were among the top 1% of all test-takers.

The dispersion or spread of a distribution may also be reported by quoting its full width at half maximum (or FWHM), as illustrated in Figure 2.7. The FWHM presents the advantage, relative to the standard deviation, of being fairly immune to the effects associated with low or high side tails. As such, it is useful to characterize the width of the main body of unimodal distributions. It is easy to verify (see Problem 2.9) that the FWHM of a Gaussian distribution is

$$\text{FWHM} = 2.35\sigma \qquad \text{Gaussian distribution.} \tag{2.94}$$

2.8 Multivariate Systems

Practical scientific problems are rarely limited to measurements of a single (random) variable. Especially in particle and nuclear physics, modern experiments involve measurements of large numbers of physical variables simultaneously. At the detector level, measured quantities amount to voltages produced by sensors, which can eventually be interpreted

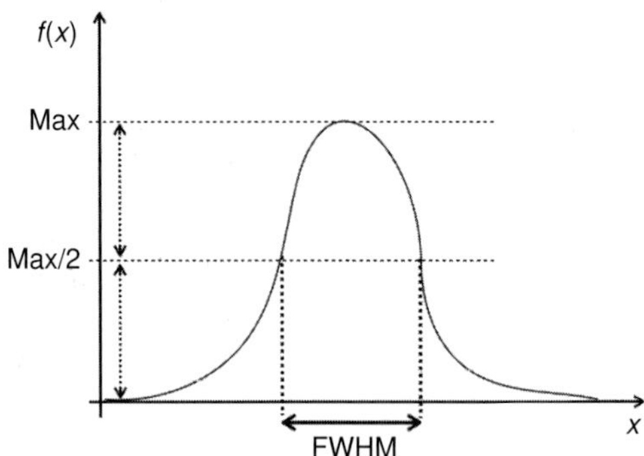

Fig. 2.7 Illustration of the notion of full width at half maximum (FWHM).

as particle positions, momenta, or energies, and possibly a host of other physical quantities. Nuclear collisions produce varying number of particles with random values of momentum, energy, or even particle species. The number of physical variables involved in nucleus–nucleus interactions may vary from a handful in soft proton–proton interactions to thousands in head-on Pb on Pb collisions at the Large Hadron Collider. A fraction of these variables may be correlated, while others may be completely independent. However, it is usually not known a priori which variables are statistically independent and which others are correlated. One is then compelled, at least conceptually, to formulate the notion of multivariate (i.e., multiple variables) probability densities that encompass all measured variables. However, for simplicity's sake and without loss of generality, we will first examine the notion of multivariate probability densities using two variables only.

This discussion can be carried out equivalently in terms of the frequentist and Bayes approaches to probability. Here, we will adopt the Bayesian perspective and use the language of probability as logic. See [67] for an introduction of the same concepts based on sets and the frequentist approach.

2.8.1 Joint Probability Density Functions

Let us consider a system involving two continuous random observables (or model parameters). Let X represent the hypothesis (or statement) that the first observable lies in the range $[x, x + dx]$ and Y that the second lies in the range $[y, y + dy]$ given some prior information about the system, I. The conjunction (AND) of the two hypotheses, X, Y, is true if both hypotheses are true. The quantity $p(X, Y|I)$ then expresses the degree of belief that the hypotheses X and Y might be true jointly (i.e., simultaneously). Given X and Y are continuous hypotheses, the conjunction X, Y is also a continuous hypothesis, and $p(X, Y|I)$ thus corresponds to a **joint probability density function**

$$p(X, Y|I) = \lim_{\Delta x, \Delta y \to 0} \frac{p(x \le X < x + \Delta x, y \le Y < y + \Delta y|I)}{\Delta x \Delta y}, \tag{2.95}$$

that expresses the degree of belief the variables X and Y be found jointly in the intervals $[x, x + dx]$ and $[y, y + dy]$, respectively.

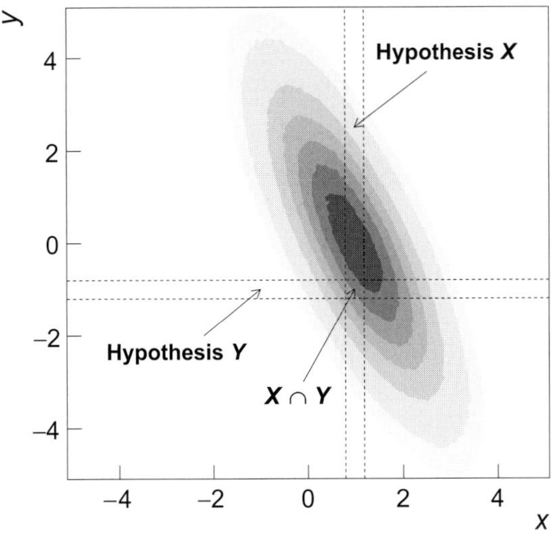

Fig. 2.8 Contour plot of the joint probability $f(x, y)$ of two variables X and Y generated according to a 2D Gaussian model. The horizontal dashed lines delimit hypothesis Y while the vertical dashed lines represent the range of hypothesis X. The probability of a pair (x, y) to lie within the square given by the conjunction of X and Y is equal to $f(x, y)\Delta x \Delta y$, with $\Delta x \Delta y$ being the surface area of the square.

Evidently, one could also consider alternative hypotheses that the observables X and Y are found in other intervals. Summing probabilities of hypotheses spanning the entire hypothesis space, **H**, yields unity. One thus gets the normalization condition

$$\iint_{\mathbf{H}} p(X, Y|I)\, dX\, dY = 1. \tag{2.96}$$

The notion of joint probability density is illustrated in Figure 2.8 which shows a probability density function of two variables, x and y, in the form of an iso-contour plot (the boundaries between different shades of gray delineate loci of equal probability density). While x and y are both random variables with Gaussian distributions, their values are not independent of one another: large values of x tend to be accompanied by small (i.e., negative) values of y and small (i.e., negative) values of x tend to be observed in conjunction with large values of y. The two variables are then said to be **correlated**. A measure of the degree of correlation between two variables can be obtained with the notion of covariance discussed in §2.9.3.

2.8.2 Marginal Probability Density Functions

Let us now assume that the PDF $p(X, Y|I)$ is given. It is obviously also of interest to determine the probability density $p(X|I)$ corresponding to the probability that the hypothesis X is true irrespective of other hypotheses. We next show $p(X|I)$ is readily obtained by integration of $p(X, Y|I)$ over all hypotheses Y. This operation is referred to as **marginalization** in the statistics literature.

To demonstrate this result, let us first assume the hypotheses Y are discrete and may be labeled Y_i, with $i = 1, \dots, n$. Let us further assume the Y_i are collectively exhaustive, $\sum_{i=1}^{n} Y_i = 1$, and mutually exclusive, $Y_i, Y_j = 0$ for $i \neq j$. One can then write

$$p\left(\sum_{i=1}^{n} Y_i|I\right) = 1. \tag{2.97}$$

Let us then consider the probability $p(X, \sum_{i=1}^{n} Y_i|I)$ asserting the degree of belief that X and $\sum_{i=1}^{n} Y_i$ are jointly true. Application of the product yields

$$p\left(X, \sum_{i=1}^{n} Y_i|I\right) = p\left(\sum_{i=1}^{n} Y_i|I\right) p\left(X|\sum_{i=1}^{n} Y_i, I\right) \tag{2.98}$$

$$= 1 \times p\left(X|\sum_{i=1}^{n} Y_i, I\right). \tag{2.99}$$

But since $\sum_{i=1}^{n} Y_i = 1$, the conjunction of this and the prior I is simply I, and one gets

$$p\left(X, \sum_{i=1}^{n} Y_i|I\right) = p(X|I), \tag{2.100}$$

which is the result we are looking for. We must now express the left-hand side of this expression in terms of probabilities $p(X, Y_i|I)$. This is readily accomplished by noting that a conjunction (AND) can be distributed onto a disjunction (OR), that is,

$$X, \sum_{i=1}^{n} Y_i = \sum_{i=1}^{n} X, Y_i. \tag{2.101}$$

Since the Y_i are mutually exclusive, the propositions X, Y_i are also mutually exclusive, and one can use the extended sum rule, Eq. (2.40), to obtain

$$p(X|I) = p\left(X, \sum_{i=1}^{n} Y_i|I\right) = \sum_{i=1}^{n} p(X, Y_i|I). \tag{2.102}$$

Our derivation was based on discrete statements Y_i. Let us now assume there is an infinite number of such statements (collectively exhaustive and mutually exclusive); we must then replace the sum by an integral and the probabilities by densities, and one gets the sought for result:

$$p(X|I)\,dx = \left(\int_{\mathbf{H}} p(X, Y|I)\,dy\right) dx \tag{2.103}$$

or simply

$$p(X|I) = \int_{\mathbf{H}} p(X, Y|I)\,dy. \tag{2.104}$$

The probability density function $p(X|I)$ is said to be the marginal probability density of $p(X, Y|I)$, or alternatively, one can say that $p(X, Y|I)$ has been marginalized or that the "uninteresting" parameter Y has been eliminated by **marginalization**. Quite obviously, given conjunctions commute, that is, $A, B = B, A$, one can achieve the marginalization of X in the same fashion:

$$p(Y|I) = \int_{\mathbf{H}} p(X, Y|I)\, dx. \tag{2.105}$$

The use of probability notation $p(X|I)$, $p(Y|I)$, $p(X, Y|I)$, and so forth, may become rapidly tedious. It is thus convenient to introduce an alternative notation based on more traditional function notation (common in the frequentist interpretation). For instance, representing the density $p(X, Y|I)$ by a function $f(x, y)$, it is common to denote marginal probabilities $p(X|I)$ as $f_x(x)$ and one writes

$$f_x(x) = \int f(x, y), dy, \tag{2.106}$$

$$f_y(y) = \int f(x, y), dx, \tag{2.107}$$

where the integrals are taken over the domains of the integrated variables.

Experimentally, the PDF $f(x, y)$ may be estimated using a two-dimensional histogram involving a very large number of measurements of pairs (x, y). The marginal PDFs $f_x(x)$ and $f_y(y)$ may then be estimated from projections of the two-dimensional histogram onto axes x and y, respectively, as illustrated in Figure 2.9. Two-dimensional and multidimensional histograms and their projections are formally discussed in §4.6.

2.8.3 Conditional Probability Density Functions

Given a known joint PDF $p(X, Y|I)$ and the marginal PDFs $p(X|I)$ and $p(Y|I)$ (or in more traditional notation: $f(x, y)$, $f_x(x)$, and $f_y(y)$), it is also of interest to evaluate the probability density, $p(X|Y, I)$, corresponding to the probability density that X is true when Y is known to be true. Since X and Y are continuous hypotheses, this amounts to the conditional probability density for X to be in the interval $[x, x + dx]$ given that Y is known to be in $[y, y + yx]$. Applying the product rule onto $p(X, Y|I)$, we readily get

$$p(X, Y|I) = p(Y|I)p(X|Y, I). \tag{2.108}$$

Rearranging, and substituting the expression obtained in Eq. (2.105) for $p(Y|I)$, one gets

$$p(X|Y, I) = \frac{p(X, Y|I)}{p(Y|I)} \tag{2.109}$$

$$= \frac{p(X, Y|I)}{\int_{\mathbf{H}} p(X, Y|I)\, dx}, \tag{2.110}$$

which is the **Conditional Probability Density Function** of X given Y. Evidently, the commutativity of the conjunction operation implies one can also write

$$p(Y|X, I) = \frac{p(X, Y|I)}{p(X|I)} \tag{2.111}$$

$$= \frac{p(X, Y|I)}{\int_{\mathbf{H}} p(X, Y|I)\, dy}, \tag{2.112}$$

which is the **Conditional Probability Density Function** of Y given X.

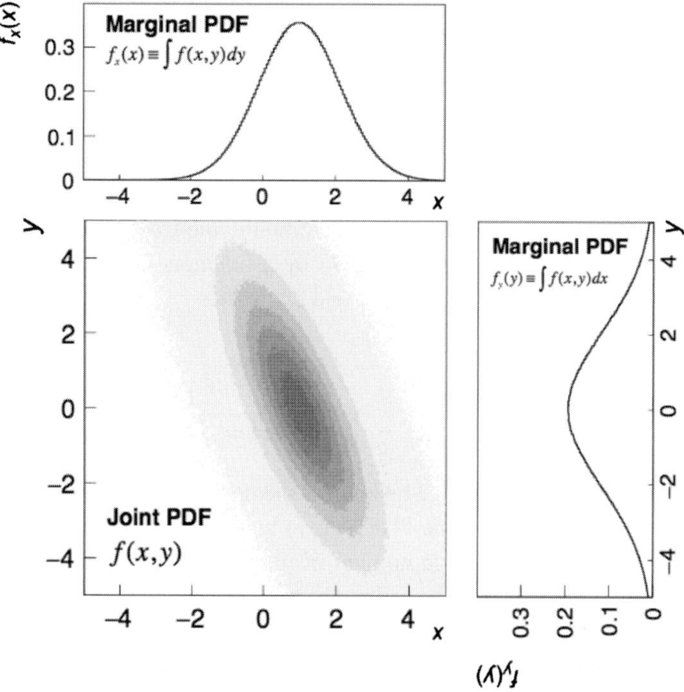

Fig. 2.9 Illustration of the notion of marginal probability. The contour plot presents the joint probability $f(x, y)$ of two variables x and y defined according to a 2D Gaussian model. The right and top panels show the marginal probabilities $f_y(y)$ and $f_x(x)$ obtained by integrating $f(x, y)$ over x and y, respectively.

Here again, it also convenient to introduce a somewhat more traditional function notation, and one writes

$$h_x(x|y) \equiv \frac{f(x,y)}{f_y(y)} = \frac{f(x,y)}{\int f(x',y)\,dx'} \tag{2.113}$$

$$h_y(y|x) \equiv \frac{f(x,y)}{f_x(x)} = \frac{f(x,y)}{\int f(x,y'),\,dy'}, \tag{2.114}$$

which defines $h_x(x|y)$ as a conditional probability density of x given y and $h_y(y|x)$ as a conditional probability density of y given x.

We stress that the conditional PDF $p(X|Y, I)$, or equivalently $h_x(x|y)$, must be regarded as a function of a single variable x in which y is treated as a constant value. It expresses the probability (density) of getting a certain value of x given a specific value of y has already been observed, and conversely for $p(Y|X, I)$.

We discuss in §4.6.3 how to obtain estimates of $h_y(y|x)$, experimentally, from a two-dimensional histogram by projection onto the x-axis of a slice taken at a specific value of y. In general, different choices of the constant y, for instance y_1 and y_2, lead to different conditional probabilities noted $h_x(x|y_1) \neq h_x(x|y_2)$, as illustrated in Figure 2.10. Note, however, that given both functions are PDFs, they both satisfy the normalization

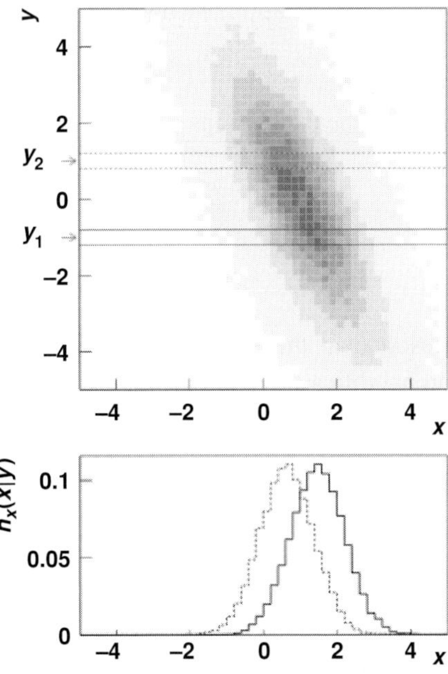

Fig. 2.10 Illustration of the notion of Conditional Probability Density Function. The scatterplot presents the joint probability density $f(x, y)$ of two variables x and y randomly generated according to a 2D Gaussian model. The bottom panel show the conditional probabilities $h_x(x|y_1)$ (solid line) and $h_x(x|y_2)$ (dashed line) obtained by integrating $f(x, y)$ along y in the ranges $[y_1 - \epsilon, y_1 + \epsilon]$ and $[y_2 - \epsilon, y_2 + \epsilon]$, respectively (with $\epsilon = 0.2$).

condition:

$$\int_{-\infty}^{-\infty} h_x(x|y_1)\, dx = 1, \tag{2.115}$$

$$\int_{-\infty}^{-\infty} h_x(x|y_2)\, dx = 1. \tag{2.116}$$

2.8.4 Bayes' Theorem and Probability Densities

Combining Eqs. 2.109 and 2.111 (or equivalently Eqs. 2.113 and 2.114), one arrives at an expression of Bayes' theorem in terms of the marginal and conditional PDFs of continuous variables x and y:

$$p(X|Y, I) = \frac{p(Y|X, I)p(X|I)}{p(Y|I)} \tag{2.117}$$

where the functions $p(X|Y, I)$, $p(Y|X, I)$, $p(X|I)$, and $p(Y|I)$ are probability density functions. Using the alternative function notation introduced in the preceding text, this may be written:

$$h_x(x|y) = \frac{h_y(y|x) f_x(x)}{f_y(y)}. \tag{2.118}$$

Continuing with this notation, we note that Eqs. (2.113, 2.114) can also be written

$$f(x, y) = h_y(y|x)f_x(x) = h_x(x|y)f_y(y), \tag{2.119}$$

One then obtains the expressions

$$f_x(x) = \int_{-\infty}^{-\infty} h_x(x|y)f_y(y)\,dy, \tag{2.120}$$

$$f_y(y) = \int_{-\infty}^{-\infty} h_y(y|x)f_x(x)\,dx, \tag{2.121}$$

which correspond to the law of total probability applied towards the determination of marginal probabilities $f_x(x)$ and $f_y(y)$.

Indeed, given $f_y(y)$ and $h_x(x|y)$, one can use Eq. (2.120) to derive the density $f_x(x)$. As we shall discuss in §12.3, Eq. (2.120) can be used, in particular, to fold and unfold smearing and efficiency effects associated with instrumental artifacts provided a model of the detector performance is available. One can also use Eq. (2.120), or Eq. (2.121), to account for physical effects. For instance, the function $f_y(y)$ might represent the momentum spectrum of a full particle jet (composed of neutral and charged particles) produced in elementary nuclear interactions, and $h_x(x|y)$ could model the probability of measuring charged jet momenta x given a full jet momentum y. The function $f_x(x)$, calculated with Eq. (2.120), would then represent the momentum distribution of charged jets.

Next, recall from Eq. (2.11) that if two hypotheses A and B are independent, the probability of their conjunction must satisfy $p(A, B|I) = p(A|I)p(B|I)$. Two continuous variables x and y can thus be considered **statistically independent** if their joint PDF factorizes as follows:

$$f(x, y) = f_x(x)f_y(y) \quad \text{(statistical independence)}. \tag{2.122}$$

This, in turn, implies that the conditional PDFs $h_y(y|x)$ and $h_x(x|y)$ are the same for all values of x and y. Indeed, substituting the preceding expression for $f(x, y)$ in Eqs. (2.113) and (2.114), one gets

$$h_y(y|x) = \frac{f_x(x)f_y(y)}{f_x(x)} = f_y(y) \quad \text{(statistical independence)}, \tag{2.123}$$

$$h_x(x|y) = \frac{f_x(x)f_y(y)}{f_y(y)} = f_x(x) \quad \text{(statistical independence)}, \tag{2.124}$$

from which we conclude that if two variables x and y are statistically independent, their conditional probability densities equal their marginal densities. This implies that having knowledge of one variable does not influence the probability of the other in any way. Conversely, finding that conditional densities $h_y(y|x)$ and $h_x(x|y)$ depend on x and y, respectively, would be a sure indication that the two variables are not statistically independent.

2.8.5 Extension to $m > 2$ random variables

The preceding discussion can be readily extended to measurements involving any number m of random variables, x_i, $i = 1, \ldots, m$. The probability of measuring the m variables in ranges $[x_i, x_i + dx_i]$ defines the joint PDF $f(x_1, x_2, \ldots, x_m)$. Given this PDF is a function of multiple variables, one can define several marginal probabilities $f_{x_i}(x_i)$:

$$f_{x_i}(x_i) = \int \cdots \int f(x_1, x_2, \ldots, x_m)\, dx_1 dx_2 \ldots dx_{i-1} dx_{i+1} \ldots dx_m. \tag{2.125}$$

One can also define marginal PDFs that are functions of several variables. For two variables, e.g., x_1 and x_2, one gets

$$f_{x_1,x_2}(x_1, x_2) = \int \cdots \int f(x_1, x_2, \ldots, x_m)\, dx_3 dx_4 \ldots dx_m, \tag{2.126}$$

which can easily be generalized to any two (or more) variables.

The extension of conditional probabilities to multiple variables proceeds similarly. For instance, the conditional probability density of getting x_1 given values $x_2, \ldots, x_m$ may be written

$$h_{x_1}(x_1 | x_2, \ldots x_m) = \frac{f(x_1, x_2, \ldots, x_m)}{\int f(x_1', x_2, \ldots, x_m)\, dx_1'}. \tag{2.127}$$

This expression can be generalized to obtain the conditional PDF of finding several variables. For instance, the conditional PDF of x_1, x_2, given $x_3, \ldots, x_m$ is given by

$$h_{x_1,x_2}(x_1, x_2 | x_3, \ldots x_m) = \frac{f(x_1, x_2, \ldots, x_m)}{\int f(x_1', x_2', \ldots, x_m)\, dx_1' dx_2'}. \tag{2.128}$$

The methods based on Bayes' theorem and the law of total probability presented earlier for functions of two variables can be readily extended to calculate marginal and conditional PDFs of several variables (see Problem 2.11).

2.8.6 Multivariate Functions of Random Variables

Equipped with the notion of multivariate probability densities introduced earlier in §2.8, we proceed to discuss multivariate functions of random variables.

Let $q(x_1, \ldots, x_n)$ represent a function of multiple random variables $x_1, x_2, \ldots, x_n$.[8] Obviously, given the x_i are continuous random variables, the value $q(x_1, \ldots, x_n)$ may also be regarded as a random variable characterized by a probability density $g(q)$. The probability (density) of specific values q is determined in part by the function itself and in part by the likelihood of getting combinations of $x_1, x_2, \ldots, x_n$ that yield that given value. In turn, this likelihood is determined by the joint probability density $f(x_1, x_2, \ldots, x_n)$ of the variables. The probability of observing a value q in the range $[q, q + dq]$ may then be obtained by summing all relevant combinations of values of $x_1, \ldots, x_n$, that is, all such values that yield a value q in that range. This may be written

$$g(q)\, dq = \int_{d\Omega} f(x_1, x_2, \ldots, x_n)\, dx_1 dx_2 \cdots dx_n, \tag{2.129}$$

[8] Again here, the variables x_i may represent continuous hypotheses (Bayesian interpretation) or the outcome of some series of measurements (frequentist interpretation).

where the volume element $d\Omega$ encloses the region in $x_1, x_2, \ldots, x_n$ space between the two hypersurfaces $q(x_1, x_2, \ldots, x_n) = q$ and $q(x_1, x_2, \ldots, x_n) = q + dq$. The size and shape of $d\Omega$ are obviously determined by the function $q(x_1, \ldots, x_n)$ itself. The preceding integral is thus generally nontrivial and its evaluation may require numerical methods, including Monte Carlo methods discussed in §13.2. However, there are several interesting cases that can be handled analytically, a few of which we examine in the following because they are frequently encountered in data analysis problems.

First, consider a case where two independent random variables X and Y are distributed according to PDFs $f_x(x)$ and $f_y(y)$, respectively. Suppose we wish to calculate the PDF $f_z(z)$ corresponding to a function $z(x, y)$ of the two variables. Because X and Y are independent variables, the joint PDF $f(x, y)$ is simply the product of the functions $f_x(x)$ and $f_y(y)$. The determination of $f_z(z)$ thus reduces to the relatively simple integral

$$f_z(z)\,dz = \int_{d\Omega} f_x(x) f_y(y)\,dxdy, \tag{2.130}$$

where the domain of integration $d\Omega$ includes all combinations of x and y satisfying $z \equiv z(x, y)$.

Let us proceed with three specific examples, starting with the integration for $z = x \pm y$. One writes

$$f_z(z)\,dz = \int_{-\infty}^{\infty} f_x(x)\,dx \int_{z\pm x}^{(z+dz)\pm x} f_y(y)\,dy. \tag{2.131}$$

The inner integral is carried over an infinitesimal range dz across which the function f_y does not change. One gets

$$f_z(z)\,dz = dz \int_{-\infty}^{\infty} f_x(x) f_y(z \pm x)\,dx, \tag{2.132}$$

which implies the density $f_z(z)$ may be written

$$f_z(z) = \int_{-\infty}^{\infty} f_x(x) f_y(z \pm x)\,dx. \tag{2.133}$$

Alternatively, reversing the order of integrations, one finds

$$f_z(z) = \int_{-\infty}^{\infty} f_x(z \pm y) f_y(y)\,dy. \tag{2.134}$$

This result can also be obtained using δ-functions. We can enforce the requirement $z = x \pm y$ by inserting a δ-function $\delta\left(z - (x \pm y)\right)$ into Eq. (2.130) while carrying out the integration over all possible values of both x and y. The integration over one of the variables then becomes trivial and one gets

$$f_z(z) = \int_{-\infty}^{\infty} f_x(x)\,dx \int_{-\infty}^{\infty} f_y(y)\delta(z - (x \pm y))\,dy, \tag{2.135}$$

$$= \int_{-\infty}^{\infty} f_x(x) f_y(z \pm x)\,dx. \tag{2.136}$$

This expression is commonly written $f_z = f_x \otimes f_y$ and is called the **Fourier convolution** of f_x and f_y. Note that in practical situations, complications may arise because measurements of the function f_x and f_y may be limited to ranges $x_{\min} \leq x \leq x_{\max}$ and $y_{\min} \leq y \leq y_{\max}$, beyond which the functions do not necessarily vanish but cannot be measured. The evaluation of the integral must consequently be limited to the boundaries of the measurement exclusively. Such cases are encountered, for instance, in measurements of correlation functions discussed in Chapter 10. Fourier convolutions are also commonly encountered in smearing or resolution modeling of the response of detectors (see, e.g., §12.2.3).

A generalization of the preceding result for a function $z(x_1, x_2, \ldots, x_n)$ which is a linear combination of the variables x_i

$$z = \sum_{i=1}^{n} c_i x_i, \tag{2.137}$$

yields

$$f_z(z) = \int_{-\infty}^{\infty} dx_1 \cdots \int_{-\infty}^{\infty} dx_{n-1} f\left(x_1, \ldots, x_{n-1}, \left[z - \frac{1}{c_n}\sum_{i=1}^{n-1} c_i x_i\right]\right). \tag{2.138}$$

Next, consider the integration of Eq. (2.130) for $z = xy$, which we carry out by inserting the δ-function $\delta\,(z - xy)$ into the convolution integral:

$$f_z(z) = \int_{-\infty}^{\infty} dx \int_{-\infty}^{\infty} dy f_x(x) f_y(y) \delta\,(z - xy)\,. \tag{2.139}$$

The δ-function $\delta\,(g(x))$ may be written

$$\delta\,(g(x)) = \sum_i \frac{\delta\,(x - x_i)}{|g'(x_i)|}, \tag{2.140}$$

where $g'(x_i)$ is the derivative of $g(x)$ with respect to x evaluated at the roots x_i of $g(x)$. In the case under consideration, we have $g(x) = z - xy$. There is a single root $x_o = z/y$ and $|g'(x_o)| = |y|$. One then gets

$$f_z(z) = \int_{-\infty}^{\infty} f_x(z/y) \frac{f_y(y)}{|y|} dy. \tag{2.141}$$

This function f_z is known as the **Mellin convolution** of f_x and f_y. Mellin convolutions are very useful in physics. They provide, in particular, a convenient technique toward the calculation of jet fragmentation functions and DGLAP evolution (see, e.g., [150] § 20).

As a third case, we consider the function $z = x/y$. The convolution integral may then be written

$$f_z(z) = \int_{-\infty}^{\infty} dx \int_{-\infty}^{\infty} dy f_x(x) f_y(y) \delta\,(z - x/y)\,, \tag{2.142}$$

which yields after integration over x:

$$f_z(z) = \int_{-\infty}^{\infty} |y| f_x(yz) f_y(y)\, dy. \tag{2.143}$$

2.9 Moments of Multivariate PDFs

We proceed to extend the notion of expectation value introduced in §2.7.3 to include functions of several variables. For instance, the mean or expectation values of a function $q(x_1, x_2, \ldots, x_n)$ may be written

$$\mu_q \equiv \mathrm{E}[q(\vec{x})] = \int_{-\infty}^{\infty} \cdots \int_{-\infty}^{\infty} q(\vec{x}) f(\vec{x})\, dx_1 \cdots dx_n, \tag{2.144}$$

where for the sake of simplicity, we have introduced a vector $\vec{x} = (x_1, \ldots, x_n)$ to denote the dependence over the variables $x_1, x_2, \ldots, x_n$ and the function $f(\vec{x})$ corresponds to the joint PDF $f(x_1, x_2, \ldots, x_n)$ of the variables $x_1, x_2, \ldots, x_n$. Let us examine the calculation of the expectation value, Eq. (2.144), for selected and particularly relevant cases of the function $q(\vec{x})$.

2.9.1 First-Order Moments of x_i

In general, the function $q(\vec{x})$ may consist of linear or nonlinear functions of the variables x_i. However, as a first and simple case, it is useful to choose any of the n random variables x_i and calculate their first moments μ_i according to

$$\mu_i \equiv \mathrm{E}[x_i] = \int_{-\infty}^{\infty} \cdots \int_{-\infty}^{\infty} x_i f(\vec{x})\, dx_1 \cdots dx_n. \tag{2.145}$$

In this context, the symbols μ_i correspond to the means of each of the random variables x_i and should not be confused with the higher moments of a single variable introduced earlier in this chapter. It is convenient to represent the means μ_i as a vector of n elements

$$\vec{\mu} = (\mu_1, \mu_2, \ldots, \mu_n), \tag{2.146}$$

which one can interpret as the mean of the PDF in the full n-dimension space spanned by the random variables x_i.

2.9.2 Variance of x_i

Next, consider the variance of the function $q(\vec{x})$. By definition, one has

$$\begin{aligned} \sigma_q^2 \equiv \mathrm{Var}[q] &= \mathrm{E}[(q(\vec{x}) - \mu_q)^2] \\ &= \int_{-\infty}^{\infty} \cdots \int_{-\infty}^{\infty} (q(\vec{x}) - \mu_q)^2 f(\vec{x})\, dx_1 \cdots dx_n. \end{aligned} \tag{2.147}$$

Choosing once again $q(\vec{x}) = x_i$, one obtains the variance of the multivariate PDF relative to each of the n random variables x_i:

$$\begin{aligned} \sigma_i^2 \equiv \mathrm{Var}[x_i] &= \mathrm{E}[(x_i - \mu_i)^2] \\ &= \int_{-\infty}^{\infty} \cdots \int_{-\infty}^{\infty} (x_i - \mu_i)^2 f(\vec{x})\, dx_1 \cdots dx_n. \end{aligned} \tag{2.148}$$

2.9.3 Covariance of Two Variables x_i and x_j

It is also useful to consider the expectation value of products such as $x_i x_j$ for $i, j = 1, \ldots, n$, and $i \neq j$. We are more specifically interested in centered moments of two variables x_i and x_j relative to their respective means and define the **covariance**, $\text{Cov}[x_i, x_j]$, of variables x_i and x_j, with $i \neq j$ as:

$$\text{Cov}[x_i, x_j] \equiv \text{E}[(x_i - \mu_i)(x_j - \mu_j)], \tag{2.149}$$

$$= \int_{-\infty}^{\infty} \cdots \int_{-\infty}^{\infty} (x_i - \mu_i)(x_j - \mu_j) f(\vec{x})\, dx_1 \cdots dx_n. \tag{2.150}$$

The covariance of two random variables measures the degree to which the variables are **correlated**, or **covarying**. Consider, for instance, the covariance of two variables x_1 and x_2 for a density $f(x_1, x_2)$. The preceding expression becomes

$$\text{Cov}[x_1, x_2] = \int (x_1 x_2 - x_1\mu_2 - \mu_1 x_2 + \mu_1\mu_2) f(x_1, x_2)\, dx_1 dx_2, \tag{2.151}$$

where μ_1 and μ_2 are the mean values of variables x_1 and x_2, respectively. Splitting the four terms of the integrand, we get

$$\begin{aligned}\text{Cov}[x_1, x_2] = &\int x_1 x_2 f(x_1, x_2)\, dx_1 dx_2 - \mu_2 \int x_1 f(x_1, x_2)\, dx_1 dx_2 \\ &- \mu_1 \int x_2 f(x_1, x_2)\, dx_1 dx_2 + \mu_1\mu_2 \int f(x_1, x_2)\, dx_1 dx_2,\end{aligned} \tag{2.152}$$

and noting that the integrals of the second and third terms are simply μ_1 and μ_2, respectively, while the integral of the last term is unity by virtue of the normalization of $f(x_1, x_2)$, Eq. (2.152) reduces to

$$\text{Cov}[x_1, x_2] = \int x_1 x_2 f(x_1, x_2)\, dx_1 dx_2 - \mu_1\mu_2 \tag{2.153}$$

A null covariance indicates the variables x_1 and x_2 may be statistically independent. Indeed, $\text{Cov}[x_1, x_2] = 0$ means one can write

$$\int x_1 x_2 f(x_1, x_2)\, dx_1 dx_2 = \mu_1\mu_2 = \int x_1 f_{x_1}(x_1)\, dx \int x_2 f_{x_2}(x_2)\, dx_2. \tag{2.154}$$

which in turn suggests $f(x_1, x_2) = f_{x_1}(x_1) f_{x_2}(x_2)$ as expected, if the variables x_1 and x_2 are statistically independent. However, we will see later in this section that the factorization (and statistical independence) is not strictly guaranteed by a null covariance, $\text{Cov}[x_1, x_2] = 0$.

The interpretation of the notion of covariance is best illustrated with the practical examples of joint probability densities of two random variables X and Y presented in Figure 2.11. The joint distributions shown in panels (a–c), are defined as product of two

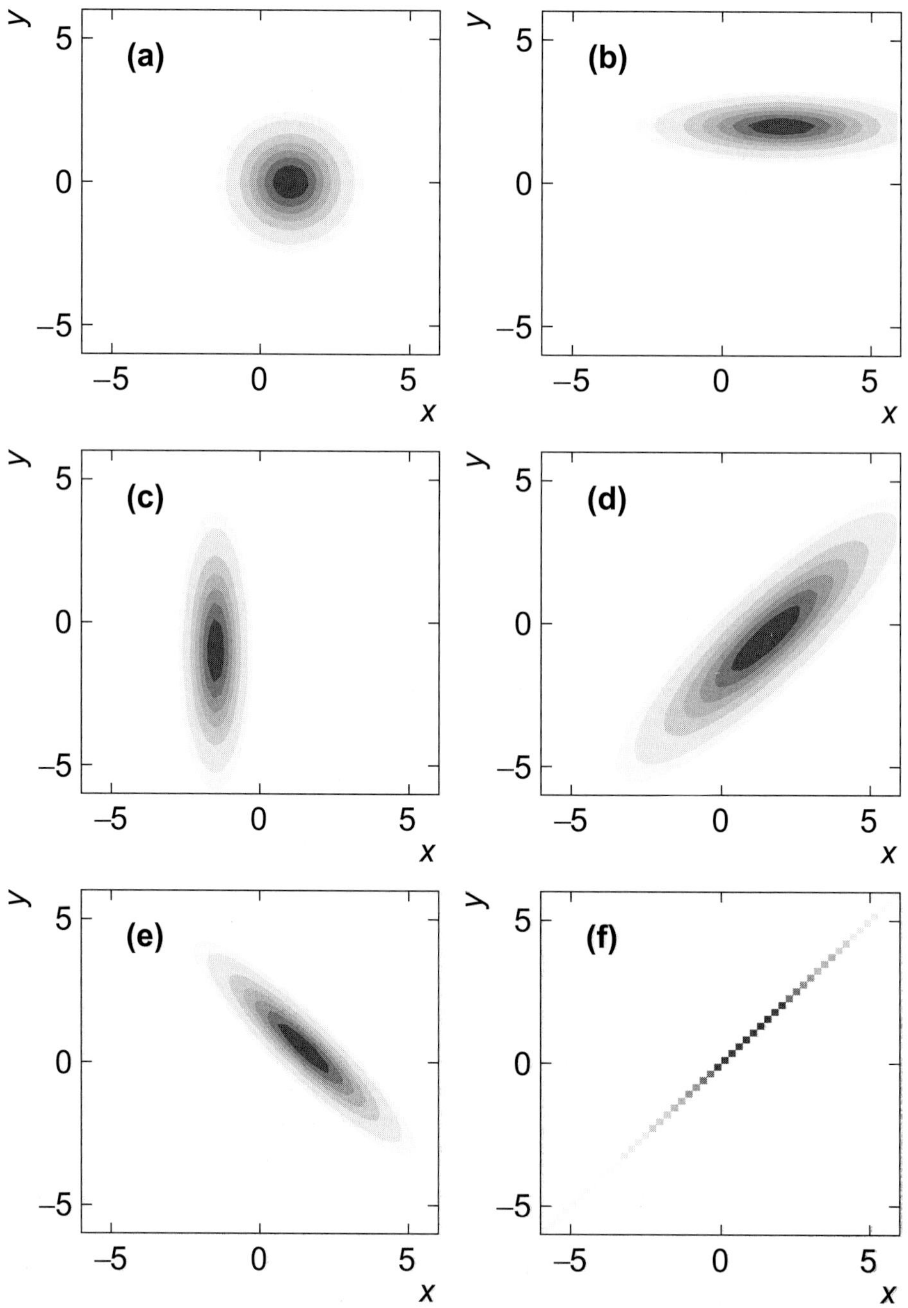

Fig. 2.11 Illustration of the notion of covariance of two random continuous variables. Panels (a), (b), and (c) present examples of uncorrelated variables whereas panels (d) and (e) show examples of fluctuations with positive and negative covariance, respectively. Panel (f) presents a special case where the covariance is maximal (Pearson coefficient is unity). See text for details.

independent Gaussians according to

$$p_{a-c}(x, y) = \frac{1}{\sqrt{2\pi}} \exp\left[-\frac{(x-\mu_x)^2}{2\sigma_x^2}\right] \times \frac{1}{\sqrt{2\pi}} \exp\left[-\frac{\left(y-\mu_y\right)^2}{2\sigma_y^2}\right] \tag{2.155}$$

with means μ_x, μ_y and standard deviations σ_x, σ_y, for variables x and y, respectively. The joint PDF shown in (a) features equal standard deviations $\sigma_x = \sigma_y$ whereas those shown in (b) and (c) are defined with $\sigma_x > \sigma_y$ and $\sigma_x < \sigma_y$, respectively. Given their definition as a product of independent Gaussian distributions, one readily verifies that the covariance of x and y for these distributions may be written

$$\begin{aligned}\text{Cov}[x, y] = & \frac{1}{\sqrt{2\pi}} \int \exp\left[-\frac{(x-\mu_x)^2}{2\sigma_x^2}\right](x-\mu_x)\, dx \\ & \times \frac{1}{\sqrt{2\pi}} \int \exp\left[-\frac{\left(y-\mu_y\right)^2}{2\sigma_y^2}\right]\left(y-\mu_y\right)\, dy\end{aligned}$$

and is thus null because the expectation values of x and y equal μ_x and μ_u, respectively, by definition of the Gaussian distribution.

The distributions shown in panels (d–f) represent correlated joint distributions of the variables X and Y defined according to

$$\begin{aligned} x &= \mu_x + r_1 + \alpha r_2, \\ y &= \mu_y + r_1 - \alpha r_2, \end{aligned} \tag{2.156}$$

where r_1 and r_2 represent independent Gaussian distributed random variables

$$\begin{aligned} p_1(r_1) &= \tfrac{1}{\sqrt{2\pi}} \exp\left[-\tfrac{r_1^2}{2\sigma_1^2}\right], \\ p_2(r_2) &= \tfrac{1}{\sqrt{2\pi}} \exp\left[-\tfrac{r_2^2}{2\sigma_2^2}\right], \end{aligned} \tag{2.157}$$

with null expectation values, $\text{E}[r_1] = \text{E}[r_1] = 0$, and standard deviations σ_1 and σ_2, respectively. The parameter α is set to unity in panels (d-e) and to zero in panel (f). Panel (d) illustrates a case with $\sigma_1 \gg \sigma_2$ implying fluctuations of r_1 are much larger than those of r_2 and thus dominate the values of both x and y, whereas panel (d) features a case with $\sigma_1 \ll \sigma_2$ in which fluctuations of r_2 are much larger than those of r_1. Dominant fluctuations of r_1 (over those of r_2) lead to covarying values of x and y: when x increases, so does y on average as seen in panel (d). In contrast, dominant fluctuations of r_2 lead to anti-correlated values of x and y: as x increases, y tends to decrease (on average).

In order to calculate the covariance of x and y for these three distributions, we first express r_1 and r_2 in terms of x and y by inversion of Eq. (2.156) with $\alpha = 1$:

$$\begin{aligned} r_1 &= \tfrac{1}{2}\left(x + y - \mu_x - \mu_y\right), \\ r_2 &= \tfrac{1}{2}\left(x - y - \mu_x + \mu_y\right). \end{aligned} \tag{2.158}$$

On can then calculate the joint PDFs of x and y in terms of the joint PDF of r_1 and r_2 as follows:

$$p_{d-f}(x,y) = \frac{d^2N}{dxdy} = \frac{d^2N}{dr_1dr_2}\left|\frac{\partial(r_1,r_2)}{\partial(x,y)}\right|. \tag{2.159}$$

The Jacobian $|\partial(r_1,r_2)/\partial(x,y)| = 1/2$ is readily calculated from Eq. (2.158). One then obtains

$$p_{d-f}(x,y) \equiv \frac{1}{2}\frac{1}{\sqrt{2\pi}\sigma_1}\exp\left(-\frac{r_1^2}{2\sigma_1^2}\right)\frac{1}{\sqrt{2\pi}\sigma_2}\exp\left(-\frac{r_2^2}{2\sigma_2^2}\right). \tag{2.160}$$

Substituting values for r_1 and r_2 from Eq. (2.158), one gets

$$\begin{aligned} p_{d-f}(x,y) = {} & \frac{1}{2}\frac{1}{\sqrt{2\pi}\sigma_1}\exp\left[-\frac{\left(x+y-\mu_x-\mu_y\right)^2}{8\sigma_1^2}\right] \\ & \times\frac{1}{\sqrt{2\pi}\sigma_2}\exp\left[-\frac{\left(x-y-\mu_x+\mu_y\right)^2}{8\sigma_2^2}\right], \end{aligned} \tag{2.161}$$

which cannot be factorized in terms of functions of x and y independently, owing to the fact that the two variables are correlated. Calculation of the covariance of x and y for these distributions is best accomplished in terms of the variables r_1 and r_2 as follows:

$$\text{Cov}[x,y] = \int p_{d-f}\,(x-\mu_x)\left(y-\mu_y\right)dxdy, \tag{2.162}$$

$$= \int \frac{d^2N}{dr_1dr_2}(r_1+r_2)(r_1-r_2)\,dr_1dr_2, \tag{2.163}$$

$$= \int p_1(r_1)p_2(r_2)\left(r_1^2-r_2^2\right)dr_1dr_2, \tag{2.164}$$

$$= \int p_1(r_1)r_1^2dr_1 - \int p_1(r_2)r_2^2dr_2, \tag{2.165}$$

$$= \sigma_1^2-\sigma_2^2, \tag{2.166}$$

where in the third line we used the fact that d^2N/dr_1dr_2 factorizes, in the fourth line the normalization to unity of the Gaussian distribution, and in the last line, we substituted the variance σ_1^2 and σ_2^2 of r_1 and r_2, respectively. One finds that for $\sigma_1 > \sigma_2$, when the fluctuations of r_1 are larger than those of r_2, that the covariance of x and y is positive, reflecting that an increase of x is on average accompanied by an increase of y, while for $\sigma_1 < \sigma_2$, the covariance is negative, corresponding to the reverse behavior, that is, an increase of x is on average accompanied by a decrease of y. Lastly, we remark that in the case of panel (f), the variables x and y are perfectly correlated ($\alpha = 0$), thereby implying that the covariance of x and y equals the variance of x.

2.9.4 Covariance Matrix V_{ij}

For a multivariate distributions, $f(\vec{x})$, with $\vec{x} = (x_1, x_2, \ldots, x_n)$, it is useful to extend the notion of covariance to also encompass the notion of variance. This enables the definition of a covariance matrix, V_{ij}, as follows:

$$V_{ij} \equiv \mathrm{E}[(x_i - \mu_i)\left(x_j - \mu_j\right)], \tag{2.167}$$

$$= \int_{-\infty}^{\infty} \cdots \int_{-\infty}^{\infty} (x_i - \mu_i)\left(x_j - \mu_j\right) f(\vec{x})\, dx_1 \cdots dx_n, \tag{2.168}$$

where all values $i, j = 1, \ldots, n$ are allowed, including $i = j$.

More generally, given two functions $q_1(\vec{x})$ and $q_2(\vec{x})$ of n variables $\vec{x} = (x_1, x_2, \ldots, x_n)$, the covariance $\mathrm{Cov}[q_1, q_2]$ is calculated as follows:

$$\begin{aligned}
\mathrm{Cov}[q_1, q_2] &\equiv \mathrm{E}[\left(q_1 - \mu_{q_1}\right)\left(q_2 - \mu_{q_2}\right)], \\
&= \mathrm{E}[q_1 q_2] - \mu_{q_1}\mu_{q_2}, \\
&= \int_{-\infty}^{\infty}\int_{-\infty}^{\infty} q_1 q_2 g(q_1, q_2)\, dq_1 dq_2 - \mu_{q_1}\mu_{q_2}, \qquad (2.169) \\
&= \int_{-\infty}^{\infty} \cdots \int_{-\infty}^{\infty} q_1(\vec{x}) q_2(\vec{x}) f(\vec{x})\, dx_1 \cdots dx_n - \mu_{q_1}\mu_{q_2},
\end{aligned}$$

where $g(q_1, q_2)$ is the joint probability density for q_1 and q_2 while $f(\vec{x})$ is the joint PDF for $\vec{x} = (x_1, x_2, \ldots, x_n)$. Obviously, $\mathrm{Cov}[q_1, q_2]$ is by construction invariant under permutation of the variables q_1 and q_2:

$$\mathrm{Cov}[q_1, q_2] = \mathrm{Cov}[q_2, q_1]. \tag{2.170}$$

This implies the matrix V_{ij} defined earlier is *symmetric* by construction and its elements satisfy

$$V_{ij} = V_{ji}. \tag{2.171}$$

2.9.5 Correlation Coefficients ρ_{ij}

The off-diagonal matrix elements V_{ij}, $i \neq j$, measure the degree of correlation between the random variables x_i and x_j. Given each of the variables in general exhibits a finite variance, $V_{ii} > 0$, it is useful to quantify the covariances of two variables x_i and x_j relative to their respective variances by introducing correlation coefficients ρ_{ij} defined as

$$\rho_{ij} = \frac{V_{ij}}{\sqrt{V_{ii}V_{jj}}} = \frac{V_{ij}}{\sigma_i \sigma_j}. \tag{2.172}$$

These coefficients are commonly referred to as Pearson correlation coefficients in the literature. It can be shown they are bound in the range $-1 \leq \rho_{ij} \leq 1$ (see Problem 2.12).

2.9.6 Interpretation and Special Cases

The covariance $\mathrm{Cov}[x, y]$ characterizes the degree of correlation between the random variables X and Y. There are phenomena such that when x is greater than its mean μ_x, y is also likely (on average) to be larger than its mean μ_y, and conversely, when $x < \mu_x$ so does $y < \mu_y$. The differences $x - \mu_x$ and $y - \mu_y$ are together positive or negative, which implies the two quantities are positively correlated, $\text{Cov}[x, y] > 0$. Other phenomena show a negative correlation, $\text{Cov}[x, y] < 0$. In this case, when $x < \mu_x$, one is more likely to observe $y > \mu_y$, and conversely when $x > \mu_x$, one observes $y < \mu_y$. In still other phenomena, there is no preference $y < \mu_y$ or $y > \mu_y$ when x is smaller or larger than its mean; the two variables are thus uncorrelated. This occurs when the joint PDF $f(x, y)$ can be factorized, $f(x, y) = f(x)f(y)$. Then clearly one gets $\text{Cov}[x, y] = 0$:

$$f(x, y) = f(x)f(y) \quad \text{implies } \text{E}[x, y] = \text{E}[x]\text{E}[y]. \tag{2.173}$$

However, the converse is not necessarily true. A null covariance does not always guarantee the joint PDF factorizes. Indeed, there can be cases where the integral, Eq. (2.149), is null even though the joint PDF does not factorize. For instance, consider the case $y = x^2$ for a uniform PDF, $f(x)$, defined in the range $[-1, 1]$. Obviously, y and x are then perfectly correlated by construction. Yet one finds

$$\begin{aligned}
\text{Cov}\,[y, x] &= \text{Cov}\left[x^2, x\right], \\
&= E\left[x^3\right] - E\left[x^2\right] E\left[x\right], \\
&= 0 - 0 \times E\left[x^2\right], \\
&= 0.
\end{aligned} \tag{2.174}$$

This implies that although a null covariance, $\text{Cov}\,[y, x] = 0$, in general suggests the variables x and y are statistically independent, one must be cautious to reach this conclusion too hastily and verify against pathological cases such as the foregoing one.

2.10 Characteristic and Moment-Generating Functions

The characteristic function (CF), noted $\phi_x(t)$, of a real valued random variable is defined, in the complex plane, as the inverse Fourier transform of the probability density function (PDF) of this variable while the Moment-Generating Function (MGF), denoted $M_x(t)$, is defined on the set of real numbers as the Laplace transform of the PDF. By construction, a CF (or MGF) is uniquely defined by its PDF, and conversely, a PDF is uniquely defined by a CF (or MGF). CF and MGF may then be used as alternative definitions and characterizations of the behavior of a random variable. CF and MGF are particularly useful toward the calculation of the moments of PDFs. They also are useful in the calculation of other PDF properties, such as their limiting behavior, and in the demonstration of several important theorems of probability and statistics. The CF and MGF of a PDF have very similar definitions based on Fourier and Laplace transforms and can often be used interchangeably. It is important to note, however, that the MGF of a PDF does not always exist. This is particularly the case when one or more of the expectation values $\text{E}[x^k]$ of a PDF diverge (e.g., the Breit–Wigner PDF). By contrast, one can show that the characteristic function of a real-valued PDF always exists, although it does not entail the existence of its moments.

2.10.1 Definitions of $\phi_x(t)$ and $M_x(t)$

The characteristic functions $\phi_x(t)$ of a random variable x with PDF $f(x)$ is defined as the expectation value of the function e^{itx}

$$\phi_x(t) = E\left[e^{itx}\right] = \int_{-\infty}^{\infty} e^{itx} f(x)\,dx, \tag{2.175}$$

and as such essentially corresponds to the inverse Fourier transform of the function $f(x)$. The MGF, $M_x(t)$, is defined as the Laplace transform of $f(x)$ for values t limited to the set of real numbers:

$$M_x(t) = E\left[e^{tx}\right] = \int_{-\infty}^{\infty} e^{tx} f(x)\,dx, \qquad t \in \mathrm{R} \tag{2.176}$$

There appears to be very little difference between the two definitions. We will, however, see that the expectation value $E\left[e^{ts}\right]$ of some PDFs diverges and $M_x(t)$ consequently does not always exist. At the same time, one can show that $\phi_x(t)$ always exists although it may be of limited use in practice.

Equation (2.175) establishes a one-to-one relationship between the PDF $f(x)$ of a variable x and its characteristic function $\phi_x(t)$. This implies that if $f(x)$ is not known a priori, it may be determined on the basis of the Fourier transform of its characteristic function $\phi(x)$:

$$f(x) = \frac{1}{2\pi}\int_{-\infty}^{\infty} e^{-itx}\phi_x(t)\,dt. \tag{2.177}$$

This property is useful, as we shall see in Section 2.10.3, to determine the PDF of sums of variables and to derive several important results in probability and statistics.

2.10.2 Calculation of the Moments μ'_k of a PDF

Let us first show that both $\phi_x(t)$ and $M_x(t)$ can be used to calculate the moments of a PDF.

Expansion of the exponential e^{tx} in series yields

$$e^{tx} = 1 + tx + \frac{t^2x^2}{2!} + \frac{t^3x^3}{3!} + \cdots + \frac{t^nx^n}{n!} + \cdots \tag{2.178}$$

The expectation value $E\left[e^{tx}\right]$ may then be written

$$E\left[e^{tx}\right] = 1 + tE\left[x\right] + \frac{t^2E\left[x^2\right]}{2!} + \frac{t^3E\left[x^3\right]}{3!} + \cdots + \frac{t^nE\left[x^n\right]}{n!} + \cdots \tag{2.179}$$

given t is here considered a "parameter" of the function. Since the expectation values $E\left[x^k\right]$ correspond to the moments μ'_k of the PDF, one may then write

$$E\left[e^{tx}\right] = 1 + t\mu' + \frac{t^2\mu'_2}{2!} + \frac{t^3\mu'_3}{3!} + \cdots + \frac{t^n\mu'_n}{n!} + \cdots \tag{2.180}$$

Derivatives of this expression with respect to t evaluated at $t = 0$ yield the moments μ'_k of the PDF:

$$\mu'_k = \left.\frac{d^k}{dt^k}E\left[e^{tx}\right]\right|_{t=0} = \left.\frac{d^k}{dt^k}M_x(t)\right|_{t=0}. \tag{2.181}$$

The same reasoning for $\phi_x(t)$ yields (see Problem 2.18)

$$\mu'_k = i^{-k} \frac{d^k}{dt^k} \phi_x(t) \bigg|_{t=0} . \tag{2.182}$$

We demonstrate the use of this expression for the calculation of the moments of the Gaussian distribution. The characteristic function of the Gaussian distribution is obtained by calculating the integral

$$\phi(t) = E\left[e^{itx}\right] = \int_{-\infty}^{\infty} \frac{e^{-\frac{(x-\mu)^2}{2\sigma^2}}}{\sqrt{2\pi}\sigma} e^{itx}\, dx. \tag{2.183}$$

Using integration tables, and after some basic algebra, one gets

$$\phi(t) = \exp\left(i\mu t - \tfrac{1}{2}\sigma^2 t^2\right). \tag{2.184}$$

The first moment (i.e., the mean) of the Gaussian PDF is obtained by evaluating the first derivative of Eq. (2.184) at $t = 0$, with respect to t. One gets

$$\mu'_1 = i^{-1} \frac{d}{dt} \phi(t) \bigg|_{t=0} = i^{-1} \left(i\mu - \sigma^2 t\right) \exp\left(i\mu t - \tfrac{1}{2}\sigma^2 t^2\right)\Big|_{t=0} = \mu, \tag{2.185}$$

which is the anticipated result. Similarly, calculation of the second moment μ'_2 requires a second derivative of the characteristic function, also evaluated at $t = 0$. One finds

$$\begin{aligned} \mu'_2 &= i^{-2} \frac{d^2}{dt^2} \phi(t) \bigg|_{t=0} \\ &= i^{-2} \left\{ -\sigma^2 \exp\left(i\mu t - \tfrac{1}{2}\sigma^2 t^2\right) + \left(i\mu - \sigma^2 t\right)^2 \exp\left(i\mu t - \tfrac{1}{2}\sigma^2 t^2\right) \right\}\Big|_{t=0} \\ &= \sigma^2 + \mu^2, \end{aligned} \tag{2.186}$$

which yields

$$\text{Var}[x] = \mu'_2 - (\mu'_1)^2 = \sigma^2, \tag{2.187}$$

also as anticipated. The calculation of higher moments of the Gaussian distribution proceeds in a similar fashion.

The CFs and MGFs of the probability distributions used in this textbook are provided in Tables 3.1 to 3.8, as are moments of these distributions.

Note that although the characteristic function of a PDF always exist, its derivatives, Eq. (2.182), may not. This is, for instance, the case of the Cauchy PDF and, by extension, the Breit–Wigner PDF (see §3.15). One can show that the characteristic function of the Cauchy distribution is $\phi(t) = \exp(-|t|)$. This function is not differentiable at $t = 0$. Its moments consequently do not exist (i.e., they diverge).

2.10.3 Sum of Random Deviates

Imagine one has n independent random variables $x_1, \ldots, x_n$, with respective PDFs $f_1(x_1), \ldots,$ $f_n(x_n)$ and their corresponding characteristic functions $\phi_1(t), \ldots, \phi_n(t)$. Let us construct a new random variable z as the sum of the variables $x_1,\ldots, x_n$:

$$z = \sum_{i=1}^{n} x_i. \tag{2.188}$$

We wish to determine the probability density function $f(z)$ on the basis of the PDFs $f_1(x_1), \ldots,$ and $f_n(x_n)$. Rather than explicitly integrating the product of the functions $f_i(x_i)$ as in Section 2.8.6, it turns out to be simpler to first calculate the characteristic function $\phi_z(t)$.

$$\phi_z(t) = E\left[e^{itz}\right], \tag{2.189}$$

$$= \int \cdots \int \exp\left(it \sum_{i=1}^{n} x_i\right) f_1(x_1)\cdots f_n(x_n)\, dx_1 \cdots dx_n, \tag{2.190}$$

which may also be written

$$\phi_z(t) = \int e^{itx_i} f_1(x_1)\, dx_1 \cdots \int e^{itx_n} f_n(x_n)\, dx_n, \tag{2.191}$$

$$= \phi_1(t)\cdots\phi_n(t). \tag{2.192}$$

The characteristic function of a random variable z, which is a sum of random variables x_i, is thus simply the product of the characteristic functions of these variables. The PDF $f(z)$ can then be obtained by inverse Fourier transform:

$$f(z) = \frac{1}{2\pi} \int_{-\infty}^{\infty} \phi_z(t) e^{-itz}\, dt, \tag{2.193}$$

$$= \frac{1}{2\pi} \int_{-\infty}^{\infty} \phi_1(t)\cdots\phi_n(t) e^{-itz}\, dt. \tag{2.194}$$

As a first example of application of this theorem, let us determine the probability density function, $f(z)$, of a variable z defined as the sum of two Gaussian variables x_1 and x_2 with means μ_1 and μ_2 and widths σ_1 and σ_2, respectively. Using (2.191), we write the characteristic function of the sum of Gaussian variables as

$$\phi_z(t) = \phi_1(t)\phi_2(t), \tag{2.195}$$

where $\phi_1(t)$ and $\phi_2(t)$ are characteristic functions of the Gaussian variables x_1 and x_2, respectively. Substituting the expression of the characteristic function for a Gaussian variable, Eq. (2.184), we find

$$\phi_z(t) = \exp\left[i\mu_1 t - \tfrac{1}{2}\sigma_1^2 t^2\right] \exp\left[i\mu_2 t - \tfrac{1}{2}\sigma_2^2 t^2\right], \tag{2.196}$$

$$= \exp\left[i(\mu_1 + \mu_2)t - \tfrac{1}{2}\left(\sigma_1^2 + \sigma_2^2\right)t^2\right]. \tag{2.197}$$

Introducing the variables $\mu_z = \mu_1 + \mu_2$ and $\sigma_z = \sqrt{\sigma_1^2 + \sigma_2^2}$, we find that the preceding expression may be rewritten

$$\phi_z(t) = \exp\left[i\mu_z t - \tfrac{1}{2}\sigma_z^2 t^2\right], \tag{2.198}$$

which is itself the expression of the characteristic function of a Gaussian PDF with mean μ_z and σ_z. We have thus demonstrated the notion that the sum of two Gaussian variables yields another Gaussian variable with a mean equal to the sum of the means of the individual variables and a width equal to the sum, in quadrature, of their respective widths.

The preceding calculation can easily be applied to the difference between two Gaussian variables $\Delta x = x_1 - x_2$ (see Problem 2.20). One finds the difference has a mean $\mu_{\Delta x} = \mu_1 - \mu_2$ and a variance $\sigma^2_{\Delta x} = \sqrt{\sigma_1^2 + \sigma_2^2}$. One may also extend the preceding calculation (see Problem 2.21) to a sum of an arbitrary number n of Gaussian deviates x_i with mean μ_i and width σ_i. Note that a similar result applies for a sum of n Poisson variables (see Problem 2.22).

2.10.4 Central Limit Theorem

The **Central Limit Theorem** (CLT) stipulates that the sum X of n independent continuous random variables, x_i, $i = 1, \ldots, n$, taken from distributions of mean μ_i and variance V_i (or σ_i^2), respectively, has a probability density function $f(X)$ with the following properties:

1. The expectation value of X is equal to the sum of the means μ_i,

$$\langle X \rangle \equiv \mathrm{E}[X] = \sum_{i=1}^{n} \mu_i. \tag{2.199}$$

2. The variance of $f(X)$ is given by the sum of the variances σ_i^2,

$$\langle \Delta X^2 \rangle \equiv \mathrm{Var}[X] = \sum_{i=1}^{n} V_i = \sum_{i=1}^{n} \sigma_i^2. \tag{2.200}$$

3. The function $f(X)$ becomes a Gaussian in the limit $n \to \infty$.

The CLT finds applications in essentially all fields of scientific study. Whether discussing the behavior of complex systems, or scrutinizing the details of scientific measurements, one finds that measured variables are in general influenced by a large number of distinct effects and processes. The more complex they are, the larger the number of effects and processes. This means that fluctuations, and consequently measurement errors tend to have a Gaussian distribution. The CLT is thus truly important, and that is why so much emphasis is also given to discussions of Gaussian distributions.

The proof of items 1 and 2 of the CLT is straightforward and is here presented first. The third item is less obvious and discussed next at greater length.

Let us assume the variables x_i follow PDFs $f_i(x_i)$ of mean μ_i and variance V_i. Given X is defined as a sum

$$X = \sum_{i=1}^{n} x_i, \tag{2.201}$$

one can write

$$\mathrm{E}[X] = \mathrm{E}\left[\sum_{i=1}^{n} x_i\right] = \sum_{i=1}^{n} \mathrm{E}[x_i] \tag{2.202}$$

since sum operations commute. Thus, insofar as the mean μ_i are defined, one gets item 1 of the theorem

$$\mathrm{E}[X] = \sum_{i=1}^{n} \mu_i. \tag{2.203}$$

The calculation of the variance of $f(X)$ proceeds similarly:

$$\mathrm{Var}[X] = \mathrm{E}\left[(X - \mathrm{E}[X])^2\right], \tag{2.204}$$

$$= \mathrm{E}\left[\left(\sum_{i=1}^{n} x_i - \sum_{i=1}^{n} \mu_i\right)^2\right], \tag{2.205}$$

$$= \mathrm{E}\left[\left(\sum_{i=1}^{n} (x_i - \mu_i)\right)^2\right]. \tag{2.206}$$

To calculate the square within the expectation value, note that one can write $\left(\sum_{i=1}^{n} a_i\right)^2 = \sum_{i=1}^{n} a_i^2 + \sum_{i\neq j=1}^{n} a_i a_j$, where diagonal terms are separated from nondiagonal terms. The variance thence becomes

$$\mathrm{Var}[X] = \sum_{i=1}^{n} \mathrm{E}\left[(x_i - \mu_i)^2\right] + \sum_{i\neq j=1}^{n} \mathrm{E}\left[(x_i - \mu_i)(x_j - \mu_j)\right]. \tag{2.207}$$

The expectation value in the first term corresponds to the variance of the variables x_i, whereas the expectation value in the second term yields the covariances of variables x_i and x_j. Since the variables x_i and x_j are assumed to be independent, these covariances vanish, and one is left with the anticipated result

$$\mathrm{Var}[X] = \sum_{i=1}^{n} \mathrm{Var}[x_i]. \tag{2.208}$$

If the measurements are not independent, Eq. (2.199) still holds but Eqs. (2.200, 2.208) do not and Eq. (2.207) must be used instead.

We next turn to the derivation of the third item of the central limit theorem, i.e., the notion that in the large n limit, the sum of n variables follows a Gaussian distribution. It is already clear from the addition theorem presented in §2.10.3 that the CLT applies for n variables x_i that are Gaussian distributed. Our task is now to demonstrate that the theorem holds also for deviates with arbitrary PDFs.

Equation (2.208) tells us that the variance of the sum z is equal to the sum of the variances σ_i^2. Introducing the average variance $\sigma^2 = (\sum_{i=1}^{n} \sigma_i^2)/n$, one thus expects the variance of the sum z to scale as $n\sigma^2$. It is then convenient to introduce

$$y_i = \frac{x_i - \mu_i}{\sqrt{n}}, \tag{2.209}$$

which satisfies $\mathrm{E}[y_i] = 0$ and $\mathrm{Var}[y_i] = \sigma_i^2/n$. We may thus recast the problem of finding the PDF of $z = \sum x_i$ in terms of $z' = \sum y_i$, which by construction shall have a null mean and a variance equal to σ^2. Our task is thus reduced to the calculation of the characteristic function $\phi_{z'}(t)$ given by

$$\phi_{z'}(t) = \prod_{i=1}^{n} \phi_i(t), \tag{2.210}$$

where $\phi_i(t)$ are the characteristic functions of variables y_i. Rather than using specific expressions for the functions $\phi_i(t)$, let us use a generic expansion of the function e^{ity_i} and write

$$\phi_i(t) = \mathrm{E}[e^{ity_i}] = \sum_{m=0}^{\infty} \frac{1}{m!}(it)^m \mathrm{E}[y_i^m]. \tag{2.211}$$

Insertion of this expression in Eq. (2.210) with values from Eq. (2.209) for y_i yields

$$\phi_{z'}(t) = \prod_{i=1}^{n} \left(\sum_{m=0}^{\infty} \frac{i^m t^m}{m! n^{m/2}} \mathrm{E}[(x_i - \mu_i)^m] \right). \tag{2.212}$$

Let us introduce the shorthand notation $V_i^m = \mathrm{E}[(x_i - \mu_i)^m]$ and recall that by construction $V_i^1 = 0$. The calculation of the foregoing product of sums, although tedious, is simple if terms are grouped in powers of t. One gets at the lowest order in the following expression:

$$\begin{aligned} \phi_{z'}(t) = 1 &- \frac{t^2}{2n} \sum_{j=1}^{n} V_j^2 - \frac{it^3}{3! n^{3/2}} \sum_{j=1}^{n} V_j^3 \\ &+ \frac{t^4}{4! n^2} \sum_{j=1}^{n} V_j^4 + \frac{t^4}{2!2! n^2} \sum_{j_1=1}^{n} V_{j_1}^2 \sum_{j_2=1}^{n} V_{j_1}^2 + O(5). \end{aligned} \tag{2.213}$$

For the sake of convenience, let us further introduce the average of the moments E_j^m:

$$\overline{V^m} = \frac{1}{n} \sum_{j=1}^{n} E_j^m. \tag{2.214}$$

Note the special case $\sigma^2 = \overline{V^2}$. In Eq. (2.213), we replace the sums by factors $n\overline{V^m}$. The characteristic function then reduces to

$$\phi_{z'}(t) = 1 - \frac{t^2}{2}\overline{V^2} - \frac{it^3}{3! n^{1/2}}\overline{V^3} + \frac{t^4}{4! n}\overline{V^4} + \frac{t^4}{2!2!}\overline{V^2}\,\overline{V^2} + O(5). \tag{2.215}$$

Terms in $1/n$ vanish in the large n limit. After substitution of $\overline{V^2}$ by σ^2, this expression thus further reduces to

$$\phi_{z'}(t) = 1 - \frac{t^2\sigma^2}{2} + \frac{t^4\sigma^4}{4} + O(6) = \exp(-\sigma^2 t^2/2), \tag{2.216}$$

which corresponds to the characteristic function of a Gaussian with mean zero and variance σ^2. Transforming back to the variable $\sum_i x_i$, one obtains a Gaussian with mean equal to $\sum_i \mu_i$ and variance $n\sigma^2 = \sum_i \sigma_i^2$. This completes the proof of the CLT.

A few words of caution are in order. First, the preceding discussion is strictly valid only when moments of the PDFs $f(x_i)$ exist. Distributions with very long tails, such as the

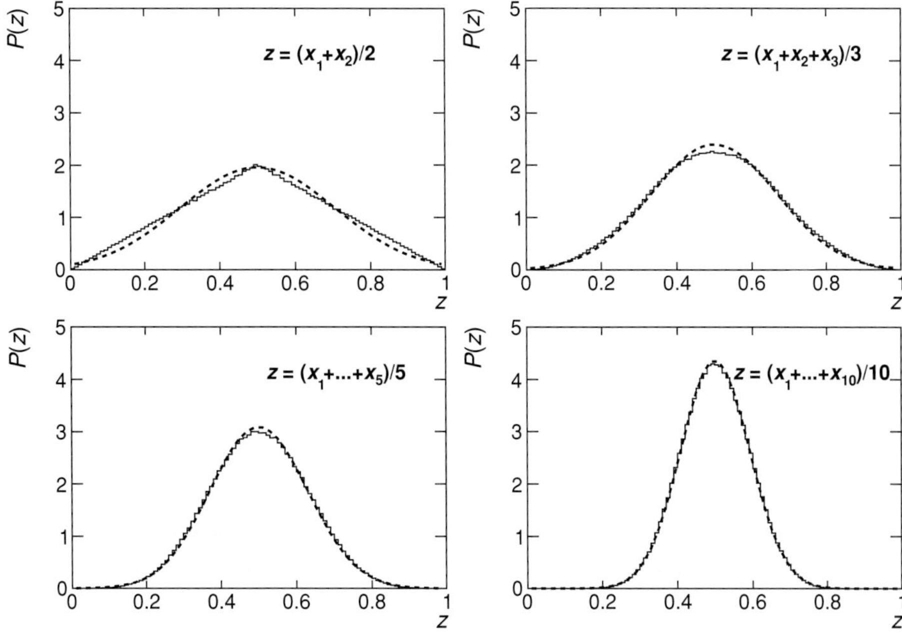

Fig. 2.12 Illustration of the Central Limit Theorem: Histograms of scaled sums ($z = \frac{1}{n}\sum_{i=1}^{n} x_i$) of $n = 2, 3, 5$, and 10 uniformly distributed random deviates in the range [0, 1] obtained with 1 million entries. Dashed curves: Gaussian distribution with standard deviations $\sigma = 1/\sqrt{12n}$.

Breit–Wigner distribution or the Landau distribution,[9] which have divergent expectation values $E[x^k]$, will thus elude the CLT. In practice, in specific analyses one may be more concerned with the rate at which the PDF of a finite sum of many random variables converges to a Gaussian distribution. Obviously, as per Eq. (2.198), a sum of Gaussian deviates itself follows a Gaussian distribution. It is, however, difficult to quantify the rate at which the sum of variables with arbitrary PDFs might converge to a Gaussian distribution without a sophisticated and detailed analysis. Still, it is usually relatively easy to test whether sufficient convergence is achieved at given n by using simple Monte Carlo simulations such as those illustrated in Figure 2.12 in which scaled sums of n uniformly distributed deviates, $z = \frac{1}{n}\sum_{i=1}^{n} x_i$, are compared with a normal distribution of mean $\mu = 0.5$ and standard deviation $\sigma = 1/\sqrt{12n}$.

2.11 Measurement Errors

Consider repeated measurements of a set of m observables $\vec{X} = (X_1, X_2, \ldots, X_m)$ yielding a set of n random variables $\{\vec{x}_1, \vec{x}_2, \ldots, \vec{x}_n\}$. If the experimental conditions of the n

[9] A probability distribution named after Soviet physicist Lev Landau and commonly used to model the energy of particles traversing a medium of finite thickness.

measurements are identical, the dispersion of values is a result of the measurement process and it is reasonable to assume that the measured values are drawn from a common parent population with a PDF $f(\vec{x})$. The PDF $f(\vec{x})$ is evidently not completely known but one can use the measured sample to obtain estimates of the means $\vec{\mu} = (\mu_1, \mu_2, \ldots, \mu_m)$ and covariance matrix V_{ij} of the variables x_i. Techniques to obtain such estimates will be discussed in detail in Chapter 4. In this section, let us simply assume these estimates are somehow available.

2.11.1 Error Propagation

Suppose we wish to evaluate a certain observable Y based on a function $y(\vec{x})$ of the variables X_i. If the PDF $f(\vec{x})$ was known, one could proceed, as in §2.8.6, and obtain the PDF $g(y)$ corresponding to the function $y(\vec{x})$. With this PDF in hand, it would then be possible to evaluate the mean value μ_y, which one could then adopt as the known value of the observable Y. The standard deviation σ_y would then characterize the uncertainty on μ_y. But, since $f(\vec{x})$ is unknown, it is not possible to determine $g(y)$ ab initio. What can then be done?

It turns out that although it is not possible to fully determine the PDF $g(y)$, one can nonetheless estimate its mean μ_y and variance σ_y based solely on the means $\vec{\mu}$ and the covariance matrix V_{ij}. This can be accomplished by calculating the Taylor expansion of the function $f(\vec{x})$ in the vicinity of $\vec{x} = \vec{\mu}$ truncated to first order:

$$y(\vec{x}) = y(\vec{\mu}) + \sum_{i=1}^{n} \left[\frac{\partial y}{\partial x_i}\right]_{\vec{x}=\vec{\mu}} (x_i - \mu_i) + O(2). \tag{2.217}$$

By definition, one has $\mathrm{E}[x_i - \mu_i] = 0$, and the expectation value of y is thus to first order:

$$\mathrm{E}[y(\vec{x})] \approx y(\vec{\mu}). \tag{2.218}$$

Let us next calculate the expectation value of y^2 to estimate the variance of the preceding estimate:

$$\begin{aligned} \mathrm{E}[y^2(\vec{x})] &= y^2(\vec{\mu}) + 2y(\vec{\mu}) \sum_{i=1}^{n} \left[\frac{\partial y}{\partial x_i}\right]_{\vec{\mu}} E\left[x_i - \mu_i\right] \\ &+ E\left[\left(\sum_{i=1}^{n} \left[\frac{\partial y}{\partial x_i}\right]_{\vec{\mu}} (x_i - \mu_i)\right)\left(\sum_{j=1}^{n} \left[\frac{\partial y}{\partial x_j}\right]_{\vec{\mu}} (x_j - \mu_j)\right)\right] + O(3). \end{aligned} \tag{2.219}$$

The second term cancels out because $\mathrm{E}[x_i - \mu_i] = 0$, and we must therefore keep the next order in the Taylor expansion. Given the derivatives $\partial y/\partial x_i|_{\vec{x}=\vec{\mu}}$ are constants, the third term can be rewritten

$$\sum_{i=1}^{n}\sum_{j=1}^{n} \left(\left[\frac{\partial y}{\partial x_j}\right]_{\vec{x}=\vec{\mu}} \left[\frac{\partial y}{\partial x_i}\right]_{\vec{x}=\vec{\mu}} E\left[(x_j - \mu_j)(x_i - \mu_i)\right]\right). \tag{2.220}$$

The expectation value $E\left[\left(x_j - \mu_j\right)(x_i - \mu_i)\right]$ is the covariance of the variables x_i and x_j, which we denote V_{ij}. The expectation value $\text{E}[y^2(\vec{x})]$ may thus be written

$$\text{E}[y^2(\vec{x})] \approx y^2(\vec{\mu}) + \sum_{i,j=1}^{n} \left[\frac{\partial y}{\partial x_i}\right]_{\vec{x}=\vec{\mu}} \left[\frac{\partial y}{\partial x_j}\right]_{\vec{x}=\vec{\mu}} V_{ij}, \tag{2.221}$$

from which we conclude that the variance σ_y^2 is given to first order by

$$\sigma_y^2 \approx \sum_{i,j=1}^{n} \left[\frac{\partial y}{\partial x_i}\frac{\partial y}{\partial x_j}\right]_{\vec{x}=\vec{\mu}} V_{ij}. \tag{2.222}$$

More generally, given a set of function of m functions $y_1(\vec{x}), \cdots, y_m(\vec{x})$, the covariance matrix of these variables may be calculated as follows (see Problem 2.24):

$$U_{kl} = \text{Cov}[y_k, y_l] \approx \sum_{i,j=1}^{n} \left[\frac{\partial y_k}{\partial x_i}\frac{\partial y_l}{\partial x_j}\right]_{\vec{x}=\vec{\mu}} V_{ij}. \tag{2.223}$$

Given that the quantities involved have finitely many indices, $k, l = 1, \ldots, m$ and $i, j = 1, \ldots, n$, it is convenient to represent Eq. (2.223) in matrix notation as follows:

$$\mathbf{U} = \mathbf{AVA}^T, \tag{2.224}$$

where we introduced a matrix of derivatives $\mathbf{A}$ defined as

$$A_{ij} = \left.\frac{\partial y_i}{\partial x_j}\right|_{\vec{x}=\vec{\mu}}, \tag{2.225}$$

and the notation $\mathbf{A}^T$ stands for the transpose of matrix $\mathbf{A}$:

$$\left(\mathbf{A}^T\right)_{ij} = (\mathbf{A})_{ji}. \tag{2.226}$$

Equations (2.222, 2.223) provide the basis for error propagation used in multivariate scientific problems. They are also used in signal processing, for example, with Kalman filtering discussed in §5.6.

2.11.2 Uncorrelated Error Propagation

Error propagation readily simplifies when the n variables x_i are mutually independent, that is, uncorrelated. In such cases, one has $V_{ii} = \sigma_i^2$ and $V_{ij} = 0$ for $i \neq j$. Eqs. (2.222, 2.223) thus reduce to

$$\sigma_y^2 = \sum_{i=1}^{n} \left[\frac{\partial y}{\partial x_i}\right]^2_{\vec{x}=\vec{\mu}} \sigma_i^2, \tag{2.227}$$

and

$$U_{kl} = \sum_{i=1}^{n} \left[\frac{\partial y_k}{\partial x_i}\frac{\partial y_l}{\partial x_i}\right]_{\vec{x}=\vec{\mu}} \sigma_i^2. \tag{2.228}$$

It is important to note that the partial derivatives $\partial y_k/\partial x_i$ and $\partial y_l/\partial x_i$ are generally nonzero, the off-diagonal matrix elements U_{kl}, $k \neq l$ are thus nonvanishing, and the matrix $\mathbf{U}$ is nondiagonal even though the covariance matrix $\mathbf{V}$ is.

2.11.3 Basic Examples of Error Propagation

Equations (2.222–2.228) may be used, for instance, to estimate the errors on sums and products of random variables. For a sum $y = x_1 + x_2$, one finds the error is (see Problem 2.25)

$$\sigma_y^2 \approx \sigma_1^2 + \sigma_2^2 + 2V_{12} \quad \text{(error on } \; y = x_1 + x_2\text{)}, \tag{2.229}$$

whereas for a product $y = x_1 \times x_2$, one gets (see Problem 2.27)

$$\frac{\sigma_y^2}{y^2} \approx \frac{\sigma_1^2}{x_1^2} + \frac{\sigma_2^2}{x_2^2} + \frac{2V_{12}}{x_1 x_2} \quad \text{(error on } \; y = x_1 \times x_2\text{)}. \tag{2.230}$$

The preceding expressions simplify to the *usual* equations introduced in elementary physics courses when variables x and y are independent, i.e., when $V_{12} = 0$.

Note that, as per our discussion of the sum of Gaussian deviates in §2.10.3, the estimate given by Eq. (2.229) of the variance of a sum $y = x_1 + x_2$ is exact if the variables x_1 and x_2 are Gaussian deviates.

2.11.4 Covariance Matrix Diagonalization

Consider once again a large set of joint measurements of m random variables $x_1, \ldots, x_m$ yielding mean values $\mu_1, \ldots, \mu_m$ and a covariance matrix $V_{ij} = \text{Cov}[x_i, x_j]$. Nonvanishing off-diagonal elements of this matrix shall indicate the random variables are correlated and thus not independent of one another. Characterizing errors on such a system of variables is nontrivial in general. It should indeed be clear that it is factually incorrect to use only the variances V_{ii} to state uncertainty estimates for the variables x_i. Indeed, since the x_i are mutually correlated, so are their errors. We will see in §6.1.5 that statements about the uncertainties of correlated variables require correlated error regions, which may be quite cumbersome to represent for $m > 2$. However, should there be a way to obtain a system of m variables $y_1, \ldots, y_m$ by linear transformation of the random variables x_i

$$y_i = \sum_{j=1}^{m} R_{ij} x_j, \tag{2.231}$$

such that the covariance matrix $U_{ij} = \text{Cov}[y_i, y_j]$ is diagonal, the characterization and modeling of the system could be carried out in terms of m independent variables and would thus be far simpler.

Let us verify that the preceding linear transformation exists by inserting Eq. (2.231) for y_i in the formula of the covariance matrix elements U_{ij}:

$$U_{ij} = \sum_{k,k'=1}^{m} R_{ik} R_{jk'} \text{Cov}\left[x_k, x_{k'}\right]. \tag{2.232}$$

Substituting $V_{kk'}$ for Cov $[x_k, x_{k'}]$, we get

$$U_{ij} = \sum_{k,k'=1}^{m} R_{ik} V_{kk'} R_{jk'} \tag{2.233}$$

$$= \sum_{k,k'=1}^{m} R_{ik} V_{kk'} R^T_{k'j}, \tag{2.234}$$

which in matrix notation may be written

$$\mathbf{U} = \mathbf{R}\mathbf{V}\mathbf{R}^T. \tag{2.235}$$

In order to identify the elements of the matrix **R**, let us consider the linear equation

$$V\vec{r}^{(i)} = \alpha_i \vec{r}^{(i)}, \tag{2.236}$$

where α_i and $\vec{r}^{(i)}$ are eigenvalues and eigenvectors of **V**, respectively. Techniques to determine eigenvalues are presented in various textbooks on linear algebra and are thus not discussed here. By construction, the eigenvectors are required to obey the orthogonality condition

$$\vec{r}^{(i)} \cdot \vec{r}^{(j)} = \sum_{k=1}^{m} r_k^{(i)} r_k^{(j)} = \delta_{ij}. \tag{2.237}$$

In cases where two or more eigenvalues are equal, the direction of the corresponding vectors are not uniquely defined but can be chosen arbitrarily to meet the foregoing condition. The n rows of the transformation matrix **R** may then be written as the n eigenvectors $\vec{r}^{(i)}$ of the matrix **V**, and one can verify by simple substitution that **U** has the required property, that is, it consists of a diagonal matrix:

$$U_{ij} = \sum_{k,l=1}^{m} R_{ik} V_{kl} R^T_{lj} = \sum_{k,l=1}^{m} r_k^{(i)} V_{kl} r_l^{(j)}, \tag{2.238}$$

$$= \alpha_j \sum_{k=1}^{m} r_k^{(i)} r_k^{(j)}, \tag{2.239}$$

$$= \alpha_j \delta_{ij}. \tag{2.240}$$

We conclude that the variances of the transformed variables $y_1, \ldots, y_n$ are given by the eigenvalues α_i of the original covariance matrix V, and indeed that all off-diagonal elements of the covariance matrix **U** are null. It is thus possible to model or characterize the system of variables $\{x_i\}$ on the basis of the n independent random variables y_i.

Note in closing that by virtue of Eq. (2.237), one finds that the transpose of matrix **R** is equal to its inverse, $\mathbf{R}^{-1} = \mathbf{R}^T$ (see Problem 2.32). The matrix **R** consequently corresponds to an orthogonal transformation carrying a rotation of the vector x that leaves its norm invariant.

Analytical solutions of Eq. (2.236) are simple for 2×2 and 3×3 matrices but rapidly become cumbersome or even intractable for larger matrices. Problem 2.33 discusses a case involving the diagonalization of a 2×2 covariance matrix analytically. Diagonalization of larger matrices are usually carried out numerically using popular software packages such as MATLAB®, Mathematica®, and ROOT [59].

2.12 Random Walk Processes

A random walk is a process consisting of a succession of random steps. It could be the path followed by a molecule as it travels in a liquid or a gas, the behavior of financial markets, or the production of particles with collective behavior (e.g., flow). As it turns out, the notion of random walk is central to the description of many stochastic phenomena, most notably diffusion processes and the description of collective flow phenomena observed in heavy-ion collisions. Diffusion and scattering processes amounts to a succession (i.e., a sum) of many "small" individual scattering processes. They thus constitute a natural application of the central limit theorem.

We here discuss random walk processes of increasing complexity starting, in §2.12.1, with random motion in one dimension. The discussion is then extended to two dimensions in §2.12.2. Extensions to three or more dimensions are readily possible and addressed in the problem section.

2.12.1 Random Walks in One Dimension

Let us first consider a random walk in one dimension with fixed step size along the x-axis. Starting from the origin, a walker flips a fair coin to decide in which direction to go next. Heads, he moves along $+x$ by one unit; tails, he moves along $-x$ by one unit. After he lands at the new position, he flips the coin, and repeats the process for several steps denoted by an index i. Each move is represented by $x_i = \pm 1$ and the two values have equal probability: $P(+1) = P(-1) = 0.5$.

To determine the probability of finding the walker at certain position x after a large number of steps n, let us consider the sum of all steps from the walker's initial position, $S_n = \sum_{j=1}^{n} x_j$. Our task is thus to determine the probability density of finding the walker at certain position S_n. The series $\{S_n\}$ is called simple random walk on the set of integers $\mathbf{Z}$.

Let us first evaluate the mean and variance of S_n. Since the values $x = +1$ and $x = -1$ are equally probable, the expectation value of x (a single step) is $\langle x \rangle = \mathrm{E}[x] = 0$ and its variance equal $\sigma_1 = \mathrm{Var}[x] = 1$. The mean and variance of S_n are thus, respectively,

$$\langle S_n \rangle = \mathrm{E}[S_n] = \sum_{j=1}^{n} \mathrm{E}[x_j] = 0, \tag{2.241}$$

$$\sigma_n^2 = \mathrm{Var}[S_n] = \sum_{j=1}^{n} \mathrm{E}[x_j^2] + \sum_{i \neq k=1}^{n} \mathrm{E}[x_i x_j] = n, \tag{2.242}$$

since $\mathrm{E}[x_i x_j] = \mathrm{E}[x_i]\mathrm{E}[x_j] = 0$ for all $i \neq j$, and $\mathrm{E}[x_j^2] = 1$.

To find the PDF of S_n, we invoke the central limit theorem, and obtain in the large n limit,

$$\frac{1}{N}\frac{dN}{dS_n} = \frac{1}{\sqrt{2\pi}\sigma_n} \exp\left(-\frac{S_n^2}{2\sigma_n^2}\right). \tag{2.243}$$

The foregoing reasoning can be repeated for a one-dimensional random walk with variable step size. Assuming for instance that the step size is uniformly distributed in the range $[-L/2, L/2]$, the expectation value of a single step remains null, and its variance is $\sigma_1^2 = L^2/12$, as per Eq. (3.89). The expectation value of the sum S_n thus also remains null in the large n limit, and its variance becomes $\sigma_n^2 = n\sigma_1^2$, while the form of the PDF remains the same.

It is also interesting to consider the introduction of a biased random step. Returning for instance to the case of a fixed size step random walk, let us assume that the probability of a step in the positive direction is p while that of a step in the negative direction is $1 - p$. The expectation value of a single step then becomes $\mathrm{E}[x] = 2p - 1$ and its variance $\mathrm{Var}[x] = 2p(1 - p) = \sigma_1^2$. The mean of the sum is consequently shifted, $\langle S_n\rangle = n(2p - 1)$, and the variance becomes $\sigma_n^2 = 2np(1 - p) = n\sigma_1^2$. The PDF of S_n is then

$$\frac{1}{N}\frac{dN}{dS_n} = \frac{1}{\sqrt{2\pi}\sigma_n} \exp\left(-\frac{(S_n - \langle S_n\rangle)^2}{2\sigma_n^2}\right). \tag{2.244}$$

Alternatively, one may also consider a forward biased Gaussian distributed random walk. Let the PDF of one step, x, be given by

$$\frac{1}{N}\frac{dN}{dx} = \frac{1}{\sqrt{2\pi}\sigma_1} \exp\left(-\frac{(x - \langle x\rangle)^2}{2\sigma_1^2}\right), \tag{2.245}$$

where $\langle x\rangle$ and σ_1 are respectively the mean and RMS of the step size. By virtue of the addition theorem (2.193), the PDF of the sum S_n of n random steps thus has the same form as Eq. (2.244) but with mean $\langle S_n\rangle = n\langle x\rangle$ and variance $\sigma_n^2 = n\sigma_1^2$.

2.12.2 Random Walks in 2D

We next proceed to determine the characteristics of a two-dimensional random walk with fixed step but arbitrary direction. Let $x_i = \cos\phi_i$ and $y_i = \sin\phi_i$ be the projections of the unit step size in the the x–y plane along the x- and y-axes, respectively. All directions ϕ_i being equally probable, we write $P(\phi_i) = (2\pi)^{-1}$ in the range $[0, 2\pi]$. Sums of projections along the x- and y-axes are denoted $S_{x,n} = \sum_{j=1}^{n} x_i$ and $S_{y,n} = \sum_{j=1}^{n} y_i$, respectively. They define a displacement vector $\vec{S}_n$, with modulus $S_n = \sqrt{S_{x,n}^2 + S_{y,n}^2}$, and direction relative to the x-axis given by $\psi_n = \tan^{-1}(S_{y,n}/S_{x,n})$.

By construction, $\mathrm{E}[x_i] = \mathrm{E}[y_i] = 0$, $\mathrm{E}[x_i^2] = \mathrm{E}[y_i^2] = \mathrm{E}[\cos^2\phi_i] = 1/2$, which entails $\mathrm{E}[S_{x,n}] = \mathrm{E}[S_{y,n}] = 0$ and $\sigma_n^2 = \mathrm{Var}[S_{x,n}] = \mathrm{Var}[S_{y,n}] = n/2$. In the large n limit, both $S_{x,n}$ and $S_{y,n}$ are Gaussian distributed:

$$\frac{1}{N}\frac{d^2N}{dS_{x,n}dS_{y,n}} = \frac{1}{2\pi\sigma_n^2} \exp\left(-\frac{S_{x,n}^2}{2\sigma_n^2}\right) \exp\left(-\frac{S_{y,n}^2}{2\sigma_n^2}\right), \tag{2.246}$$

which simplifies to

$$\frac{1}{N}\frac{d^2N}{S_n dS_n d\psi} = \frac{1}{2\pi\sigma_n^2}\exp\left(-\frac{S_n^2}{2\sigma_n^2}\right) \tag{2.247}$$

after substitution of the modulus of $\vec{S_n}$ for the sum of the square of its components. Note that the PDF is independent of ψ. One thus concludes, as expected, that all directions are equally probable. Integration over the angle ψ yields

$$\frac{1}{N}\frac{dN}{dS_n} = \frac{S_n}{\sigma_n^2}\exp\left(-\frac{S_n^2}{2\sigma_n^2}\right). \tag{2.248}$$

The variance $\sigma_n^2 = n\sigma_1$ implies the "typical" length of S_n scales as the square root of the number of steps, that is, $\sqrt{n}$.

The foregoing two-dimensional random walk can be modified to describe motion biased toward an arbitrary direction ψ. This type of biased random walk can be used to model collective flow, that is, the concerted motion of produced particles, observed in the study of heavy-ion collisions at medium to high energy. To this end, let $P(\phi_i) = (2\pi)^{-1}[1 + 2v_m \cos(m(\phi_i - \psi))]$, where the coefficient v_m, commonly called **flow coefficient of order** m is usually much smaller than unity. For a fixed value of ψ, and a fixed step size, r, the expectation values of $x_i = r\cos(m\phi_i)$ and $y_i = r\sin(m\phi_i)$ are (see Problem 2.35):

$$\mathrm{E}[x_i] = v_m r\cos(m\psi), \tag{2.249}$$

$$\mathrm{E}[y_i] = v_m r\sin(m\psi), \tag{2.250}$$

while the second moments are

$$\mathrm{E}[x_i^2] = \mathrm{E}[y_i^2] = \frac{r^2}{4\pi}. \tag{2.251}$$

The variances of x_i and y_i are thus respectively

$$\sigma_{x,1} \equiv \mathrm{Var}[x_i] = \frac{r^2}{4\pi} - r^2 v_m^2\cos^2(m\psi) \approx \frac{r^2}{4\pi}, \tag{2.252}$$

$$\sigma_{y,1} \equiv \mathrm{Var}[y_i] = \frac{r^2}{4\pi} - r^2 v_m^2\sin^2(m\psi) \approx \frac{r^2}{4\pi}, \tag{2.253}$$

where the right-hand side approximations hold if the coefficients are small ($v_m << 1$). The expectation values of the sums $S_{x,n}$ and $S_{y,n}$ are consequently

$$\langle S_{x,n}\rangle = \mathrm{E}[S_{x,n}] = nrv_m\cos(m\psi), \tag{2.254}$$

$$\langle S_{y,n}\rangle = \mathrm{E}[S_{y,n}] = nrv_m\sin(m\psi), \tag{2.255}$$

while their standard deviations are

$$\sigma_{x,n} = \sqrt{n}\sigma_{x,1}, \tag{2.256}$$

$$\sigma_{y,n} = \sqrt{n}\sigma_{y,1}. \tag{2.257}$$

In the large n limit, both $S_{x,n}$ and $S_{y,n}$ are Gaussians

$$\frac{1}{N}\frac{dN}{dS_{x,n}} = \frac{1}{\sqrt{2\pi}\sigma_{x,n}} \exp\left[-\frac{(S_{x,n} - \langle S_{x,n}\rangle)^2}{2\sigma_{x,n}^2}\right], \tag{2.258}$$

$$\frac{1}{N}\frac{dN}{dS_{y,n}} = \frac{1}{\sqrt{2\pi}\sigma_{y,n}} \exp\left[-\frac{\left(S_{y,n} - \langle S_{y,n}\rangle\right)^2}{2\sigma_{y,n}^2}\right]. \tag{2.259}$$

For small values of v_m, one has $\sigma_{x,n}^2 \approx \sigma_{y,n}^2 = \sigma_n^2 = n\frac{r^2}{4\pi}$, and one can thus write (see Problem 2.36)

$$\frac{1}{N}\frac{dN}{S_n dS_n d\theta} = \frac{1}{2\pi\sigma_n^2} \exp\left[-\frac{\left(\vec{S}_n - \langle\vec{S}_n\rangle\right)^2}{2\sigma_n^2}\right], \tag{2.260}$$

where we introduced $\vec{S}_n = (S_{x,n}, S_{y,n})$, its average

$$\langle\vec{S}_n\rangle = (nrv_m \cos(m\psi), nrv_m \sin(m\psi)), \tag{2.261}$$

and θ, the angle between the vectors $\vec{S}_n$ and $\langle\vec{S}_n\rangle$.

The vectors $\langle\vec{S}_n\rangle$ and $\vec{S}_n$ represent the expectation value and a particular realization of the random walk with parameter v_m, respectively. A single realization of the random walk, $\vec{S}_n$, can be regarded as a measurement of $\langle\vec{S}_n\rangle$. It is thus interesting to consider how precise the "measurements" of the modulus S_n and angle ψ are relative to the expectation values $\langle S_n\rangle$ and $\tan^{-1}(\langle S_{y,n}\rangle/\langle S_{x,n}\rangle)$. Integration of (2.260) over S_n yields (see Problem 2.37)

$$\frac{1}{N}\frac{dN}{d\theta} = \frac{1}{\pi}\exp(-\chi^2)\left\{1 + z\sqrt{\pi}\left[1 + \text{erf}(z)\right]\exp(z^2)\right\} \tag{2.262}$$

where $z = \chi\cos(\theta)$, erf(x) is the error function, and $\chi \equiv \langle S_n\rangle/\sigma_n$.

Integrations over θ gives (see Problem 2.37)

$$\frac{1}{N}\frac{dN}{dS_n} = \frac{1}{\sigma_n^2}\exp\left(-\frac{(S_n^2 + \langle S_n\rangle^2)}{2\sigma_n^2}\right) I_0\left(\frac{\langle S_n\rangle |S_n|}{\sigma_n^2}\right), \tag{2.263}$$

where $I_0(z)$ represents the modified Bessel function of the first kind and of order 0. Note that the expressions (2.262) and (2.263) are independent of the step size r but depend on the magnitude of the coefficient v_m.

2.13 Cumulants

Cumulants provide a powerful tool for the study of multiple variable correlations and are the basis of several analysis techniques used in nuclear and particle physics, many of which are presented in Chapters 10 and 11.

2.13.1 Cumulant Definition

The cumulants κ_n of a random variable x, relative to a specific probability distribution, are formally defined in terms of the cumulant-generating function $g(t)$, which is the logarithm of the moment-generating function of that probability distribution:

$$g(t) = \ln\left(\mathrm{E}\left[e^{tx}\right]\right). \tag{2.264}$$

The cumulants κ_n are obtained as coefficients of the power series expansion of $g(t)$:

$$g(t) = \sum_{n=1}^{\infty} \kappa_n \frac{t^n}{n!}. \tag{2.265}$$

If $g(t)$ is available in close analytical form, the cumulants can then be calculated according to

$$\kappa_n = \left.\frac{d^n g(t)}{dt^n}\right|_{t=0} \tag{2.266}$$

As an example, consider the calculation of the cumulants of the binomial distribution with moment-generating function (from Table 3.1)

$$M(t) = \left(1 - p + pe^t\right)^n. \tag{2.267}$$

The function $g(t)$ becomes

$$g(t) = \ln(M(t)) = n\ln\left(1 - p + pe^t\right). \tag{2.268}$$

Derivatives of order n of $g(t)$ yield the cumulants

$$\kappa_1 = \left.\frac{d}{dt}g(t)\right|_{t=0} = np, \tag{2.269}$$

$$\kappa_2 = \left.\frac{d^2}{dt^2}g(t)\right|_{t=0} = np(1-p), \tag{2.270}$$

$$\kappa_3 = \left.\frac{d^3}{dt^3}g(t)\right|_{t=0} = n(2p^3 - 3p^2 + p), \tag{2.271}$$

etc.

The use of cumulants is convenient in the analysis of independent random variables consisting of a sum of two or more independent variables. For instance, let us define a random variable $z = x + y$, where x and y are two statistically independent variables with cumulant-generating functions $g_x(t)$ and $g_y(t)$, respectively. Let us calculate the cumulant-generating function $g_z(t)$ and show that cumulants of z of all orders n are equal to the sum of the n order cumulants of x and y (additivity property).

$$g_z(t) = \ln\left(\mathrm{E}\left[e^{t(x+y)}\right]\right), \tag{2.272}$$

$$= \ln\left(\mathrm{E}\left[e^{tx}\right]\mathrm{E}\left[e^{ty}\right]\right), \tag{2.273}$$

$$= \ln\left(\mathrm{E}\left[e^{tx}\right]\right) + \ln\left(\mathrm{E}\left[e^{ty}\right]\right), \tag{2.274}$$

$$= g_x(t) + g_y(t), \tag{2.275}$$

where in the second line we have use the statistical independence of x and y to factors their expectation values. Since the derivative (at any order) of a sum of two functions is equal to the sum of the derivatives of the functions (also at all orders), one finds that the cumulants of z equal the sum of the cumulants of x and y as follows:

$$\kappa_n^{(z)} = \left.\frac{d^n}{dt^n} g_z(t)\right|_{t=0}, \tag{2.276}$$

$$= \left.\frac{d^n}{dt^n} g_x(t)\right|_{t=0} + \left.\frac{d^n}{dt^n} g_y(t)\right|_{t=0}, \tag{2.277}$$

$$= \kappa_n^{(x)} + \kappa_n^{(y)}. \tag{2.278}$$

Let $c \in \mathbf{R}$, a constant. Cumulants are easily verified to have the following basic properties:

$$\kappa_1(x+c) = \kappa_1(x) + c \quad \text{shift} - \text{equivariance}, \tag{2.279}$$

$$\kappa_n(x+c) = \kappa_n(x) \quad \text{for } n \geq 2, \text{ shift invariance}, \tag{2.280}$$

$$\kappa_n(cx) = c^n \kappa_n(x) \quad \text{homogeneity}. \tag{2.281}$$

Additionally, note that by definition (2.264) of the cumulant-generating function $g(t)$, one can write

$$M(t) = \exp\left(g(t)\right), \tag{2.282}$$

$$1 + \sum_{n=1}^{\infty} \frac{\mu_n{}' t^n}{n!} = \exp\left(\sum_{k=1}^{\infty} \frac{\kappa_n t^n}{n!}\right). \tag{2.283}$$

Using this expression, one can derive the following recursion formula between the moments $\mu_n{}'$ and the cumulants κ_n,

$$\kappa_n = \mu_n{}' - \sum_{m=1}^{n-1} \binom{n-1}{m-1} \kappa_m \mu_{n-m}{}' \tag{2.284}$$

One finds that the first moment equals the first cumulant while the second and third central moments equal the second and third central cumulants, respectively. The fourth cumulant is related to the excess kurtosis, $\kappa_4 = \mu_4 - 3\mu_2^2$. Higher cumulants are more complicated polynomial functions of the moments.

2.13.2 Joint Cumulants

As for moments, one can also define joint cumulants. The joint cumulants of random variables $x_1, x_2, \ldots x_n$ are defined as derivatives of the joint cumulant-generating function

$$g(t_1, t_2, \ldots, t_n) = \ln\left(\mathrm{E}\left[e^{\sum_{j=1}^{n} t_j x_j}\right]\right). \tag{2.285}$$

The first-order cumulant of n random variables may be written

$$\kappa[x_1, x_2, \ldots, x_n] = \sum_{P} (|P|-1)!\,(-1)^{|P|-1} \prod_{B \in P} \mathrm{E}\left[\prod_{i \in B} x_i\right], \tag{2.286}$$

where P stands for partitions of $\{1, 2, \ldots, n\}$, $|P|$ is the number of parts in such partitions, B runs through all the blocks of the partition, while i enumerates elements of any given partition.[10] With two variables x and y, one gets the covariance

$$\kappa[x, y] = \mathrm{E}[xy] - \mathrm{E}[x]\,\mathrm{E}[y], \tag{2.287}$$

and with three variables x, y, and z, one obtains

$$\kappa[x, y, z] = \mathrm{E}[xyz] - \mathrm{E}[xy]\,\mathrm{E}[z] - \mathrm{E}[xz]\,\mathrm{E}[y] - \mathrm{E}[yz]\,\mathrm{E}[x] + 2\mathrm{E}[x]\,\mathrm{E}[y]\,\mathrm{E}[z], \tag{2.288}$$

which for two identical variables, $x = y$, reduces to

$$\kappa[x, x, z] = \mathrm{E}\left[x^2 z\right] - \mathrm{E}[xz]\,\mathrm{E}[x] - \mathrm{E}\left[x^2\right]\mathrm{E}[z] + 2\mathrm{E}[x]^2\,\mathrm{E}[z]. \tag{2.289}$$

Given the expectation value of a product of statistically independent variables is the product of their expectation values, one easily verifies that any cumulant involving two or more independent variables is null by definition. And if all n random variables are identical, the joint cumulant is simply the nth ordinary cumulant.

The formula (2.286) expressing cumulants in terms of moments can be inverted. One finds

$$\mathrm{E}[x_1, x_2, \ldots, x_n] = \sum_P \prod_{B \in P} \kappa[x_i;\, i \in B]. \tag{2.290}$$

For instance, the moment of the product xyz is given by

$$\mathrm{E}[xyz] = \kappa[x, y, z] + \kappa[x, y]\kappa[z] + \kappa[x, z]\kappa[y] + \kappa[y, z]\kappa[x] + \kappa[x]\kappa[y]\kappa[z]. \tag{2.291}$$

Joint cumulants are quite important in the analysis of particle correlations in high-energy physics, most particularly in the study of multiparticle correlation functions and collective motion (flow) discussed in Chapters 10 and 11.

Exercises

2.1 Demonstrate the properties listed under Eq. (2.5).

2.2 Verify that the notion of conditional probability satisfies the three axioms defining the notion of probability.

2.3 Verify the expression Eq. (2.18) known as the law of total probability.

2.4 Derive the expression for the variance given by Eq. (2.80).

2.5 Show that the skewness, γ_1, can be equivalently defined as the ratio of the third cumulant κ_3 and the third power of the square root of the second cumulant $\kappa_2^{3/2}$.

2.6 Show that the excess kurtosis of a sum of n independent random variables x_i is equal to the sum of the kurtosis of these n variables divided by n^2 (Eq. 2.93).

[10] Examples of applications of the notions of cumulant, partitions, and blocks are presented in Chapter 11.

2.7 Derive the expression Eq. 2.60 for the density $g(q)$ obtained for a function $q(x)$ of continuous random variable x distributed according to a PDF, $f(x)$.

2.8 Calculate expressions for the density $g(a)\,da$ given functions $a(x) = x$ and $a(x) = x^4$, assuming the continuous variable x has a PDF, $f(x)$.

2.9 Verify that the full width at half maximum (FWHM) of a normal distribution is 2.35σ.

2.10 Derive a method to estimate the values $x_{i,o}$ defined by Eq. (4.64) for PDFs $f(x) \propto \exp(-x/\lambda)$ and $f(x) \propto (k + x)^{-\beta}$, where k and β are two unknown constants. Hint: Use interpolation of the yields in bins $i - 1$ and $i + 1$ to estimate the constants λ and β bin by bin.

2.11 Derive an expression similar to Eq. (2.118) for $h_{x_1,x_2}(x_1, x_2|x_3, \dots x_m)$, where $m > 2$.

2.12 Show that the Pearson coefficients defined by Eq. (2.172) are bound in the range $-1 \le \rho_{ij} \le 1$ by construction.

2.13 Calculate the first, second, third, and fourth moments of the uniform distribution defined as follows:

$$p_U(x; \alpha, \beta) = \begin{cases} \frac{1}{\beta-\alpha} & \alpha \le x \le \beta \\ 0 & \text{otherwise} \end{cases} \tag{2.292}$$

2.14 Calculate the first, second, third, and fourth moments of the triangular distribution defined as follows:

$$p_T(x; \alpha, \beta) = \begin{cases} \frac{2(x-\alpha)}{(\beta-\alpha)^2} & \alpha \le x \le \beta \\ 0 & \text{otherwise} \end{cases} \tag{2.293}$$

2.15 Show that given a partition of a set S into n mutually disjoint subsets A_i, with $i = 1, n$, the probability $P(B)$, of a set $B \subset S$ may be written as follows:

$$P(B) = \sum_i P(B|A_i)P(A_i). \tag{2.294}$$

Hint: Disjoints subsets have a null intersection, $A_i \cap A_j = 0$ for $i \ne j$.

2.16 Combine Bayes' theorem with the law of total probability to obtain the following expression:

$$P(A|B) = \frac{P(B|A)P(A)}{\sum_i P(B|A_i)P(A_i)}. \tag{2.295}$$

2.17 A neural network is designed and trained to classify particles entering an electromagnetic calorimeter as photon (P), electron (E), or hadron (H) based on the longitudinal and transverse patterns of energy deposition they produce in the calorimeter. A Monte Carlo simulation is used to estimate the neural net performance summarized in Table 2.1. The notations P_a, E_a, and H_a are used to indicate that the energy deposition pattern is due to a photon, an electron, or a hadron. Data analyzed with the neural network provide for fractions 0.05, 0.15, and 0.8 of photons, electrons, and hadrons, respectively. Calculate the relative rates produced by the nuclear reaction under study.

Table 2.1 Particle Identification Probabilities Used in Problem 2.17

$P(P_a\|P) = 0.98$	$P(P_a\|E) = 0.06$	$P(P_a\|H) = 0.05$
$P(E_a\|P) = 0.01$	$P(E_a\|E) = 0.90$	$P(E_a\|H) = 0.15$
$P(H_a\|P) = 0.01$	$P(P_a\|E) = 0.04$	$P(H_a\|H) = 0.80$

2.18 Derive the expression (3.62) giving the moments of a PDF in terms of derivatives of its characteristic function.

2.19 Use Eq. (3.62) to calculate the moments of (a) the uniform distributions, (b) the exponential distribution, (c) the t-distributions, and (d) the χ^2-distribution.

2.20 Show that the difference between two independent Gaussian deviates $\Delta x = x_1 - x_2$ has a mean $\mu_{\Delta x} = \mu_1 - \mu_2$ and a variance $\sigma^2_{\Delta x} = \sqrt{\sigma_1^2 + \sigma_2^2}$.

2.21 Extend Eq. (2.229) and find an expression for the error of a sum of several independent Gaussian deviates $x_1, x_2, \ldots, x_n$.

2.22 Extend Eq. (2.229) and find an expression for the error of a sum of several independent Poisson deviates $x_1, x_2, \ldots, x_n$.

2.23 Show that the characteristic function of $f(w) = \frac{1}{\sqrt{2\pi w}} e^{-w/2}$ is given by $\phi_w(t) = (1 - 2it)^{-1/2}$.

2.24 Derive the expression (2.223) for the covariance of functions $y_i(\vec{x})$ and $y_j(\vec{x})$.

2.25 Demonstrate the expression $\sigma_y^2 \approx \sigma_1^2 + \sigma_2^2 + 2V_{12}$, given by Eq. (2.229), corresponding to the error on the sum of random variables x_1 and x_2.

2.26 Demonstrate the expression $\sigma_y^2 \approx \sigma_1^2 + \sigma_2^2 + 2V_{12}$, given by Eq. (2.229), corresponding to the error on the difference of random variables x_1 and x_2.

2.27 Demonstrate Eq. (2.230) corresponding to the error on the product of correlated random variables x_1 and x_2.

2.28 Calculate the error on a ratio $y = x_1/x_2$ obtained by dividing correlated random variables x_1 and x_2.

2.29 Show that $\lambda_\pm = \frac{1}{2}\left[\sigma_1^2 + \sigma_2^2 \pm \sqrt{\sigma_1^2 + \sigma_2^2 - 4(1-\rho^2)\sigma_1^2\sigma_2^2}\right]$ indeed satisfies Eq. (5.66), and find the eigenvectors r_+ and r_-.

2.30 Show that if the intersection of two subsets A and B is null (i.e., for $A \cap B = 0$), the subsets cannot be independent, and determine the value of $P(A \cap B)$.

2.31 Given PDFs $g(x)$ and $h(y)$ for random variables x and y respectively, calculate the PDF of variable z defined as

- $z^2 = x^2 + y^2$.
- $\tan^{-1}(x/y)$.

2.32 Verify by direct substitution of the eigenvectors $\vec{r}$ into $\mathbf{A}$ defined by Eq. (2.231) that $\mathbf{A}$ satisfies $\mathbf{A}^{-1} = \mathbf{A}^T$ (i.e., its inverse is equal to its transpose), and that it is consequently an orthogonal transformation of the vector x that leaves its norm invariant.

2.33 Consider a two-dimensional covariance matrix $\mathbf{V}$ defined as follows:

$$\mathbf{V} = \begin{pmatrix} \sigma_1^2 & \rho\sigma_1\sigma_2 \\ \rho\sigma_1\sigma_2 & \sigma_2^2 \end{pmatrix}. \tag{2.296}$$

Solve the eigenvalue equation $(\mathbf{V}-\lambda)\vec{r}=0$ and show the eigenvalues may be written

$$\lambda_{\pm}=\frac{1}{2}\left[\sigma_1^2+\sigma_2^2\pm\sqrt{\sigma_1^2+\sigma_2^2-4\left(1-\rho^2\right)\sigma_1^2\sigma_2^2}\right] \tag{2.297}$$

with eigenvectors given by

$$r_+=\begin{pmatrix}\cos\theta\\ \sin\theta\end{pmatrix}\; r_-=\begin{pmatrix}-\sin\theta\\ \cos\theta\end{pmatrix} \tag{2.298}$$

such that

$$\theta=\frac{1}{2}\tan^{-1}\left(\frac{2\rho\sigma_1\sigma_2}{\sigma_1^2-\sigma_2^2}\right)$$

and

$$\mathbf{A}=\begin{pmatrix}\cos\theta & \sin\theta\\ -\sin\theta & \cos\theta\end{pmatrix}.$$

2.34 Consider a system with two random (or fluctuating) observables x and y. Explore the types of correlation that might arise between these two observables by using a linear combination of two randomly generated numbers r_1 and r_2:

$$\begin{aligned} x&=a+b_1r_1+b_2r_2\\ y&=c+d_1r_1-d_2r_2 \end{aligned} \tag{2.299}$$

where a, b_1, b_2, c, d_1, and d_2 are arbitrary constants. Calculate the mean and variance of the observable x and y as well as the covariance of x and y. Discuss conditions under which x and y might yield (a) independent variables, (b) correlated variables with a positive covariance, (c) correlated variables with a negative covariance, and (d) correlated variables with a Pearson coefficient equal to unity. Assume the random variables r_1 and r_2 have null expectation values, $E[r_1]=E[r_2]=0$, and nonvanishing variances σ_1^2 and σ_2^2, respectively.

2.35 Verify that the first-order and second-order moments associated with a two-dimensional random walk are given by Eqs. (2.249) and (2.251).

2.36 Derive Eq. (2.260) from Eq. (2.258) for the expression of the probability density $\frac{1}{N}\frac{dN}{S_n dS_n d\theta}$ of the sum vector $\vec{S}_n$.

2.37 Verify the expressions (2.262) and (2.263) by explicitly integrating Eq. (2.260). Hints:

$$\begin{aligned} I_o(z)&=\frac{1}{\pi}\int_o^{\pi}e^{\pm z\cos(\theta)}d\theta\\ I_n(z)&=\frac{1}{\pi}\int_o^{\pi}e^{\pm z\cos(n\theta)}d\theta \end{aligned} \tag{2.300}$$

2.38 Show that the expression $\mathbf{U}=\mathbf{AVA}^T$ is equivalent to Eq. (2.223).

3 Probability Models

As we discussed in Chapter 2, Bayes' theorem and the Bayesian approach to probability constitute central components of the scientific method. Recall that Bayes' theorem may be used, in particular, to infer the posterior probability of a scientific hypothesis, H_i based on measured data D according to $p(H_i|D, I) = p(D|H_i, I)p(H_i|I)/p(D|I)$. Additional hypotheses, H_j, with $j \neq i$, may also be tested in the same way. The posterior probabilities $p(H_i|D, I)$ associated with these hypotheses may then be compared to establish which is best supported by the measured data D. Calculations of posterior probabilities, however, require one has techniques to calculate both the likelihoods $p(D|H_i, I)$ and the priors $p(H_i|I)$. The determination of prior probabilities is a topic all of its own that we address in Chapter 7. In this chapter, we discuss, through several concrete examples, how prior knowledge about a phenomenon or system, denoted I, may be formulated to obtain **data probability models**, that is, probability distributions and probability densities that model phenomena of interest and can thus be used to calculate, based on specific hypotheses, the likelihood of data. We begin in §3.1, §3.2, and §3.3 with basic models including the binomial distribution, the multinomial distribution, and the Poisson distribution. Several continuous distributions, including the uniform distribution, the exponential distribution, and the normal distribution, are introduced in later sections.

Professional scientists make use of a vast array of functions and probability distributions that cannot all be covered in a single text. The functions discussed in this chapter were selected for their foundational value, their broad range of application, or because they are needed in later chapters. Short examples of application are included in this chapter whenever possible and more extensive or specialized uses are discussed in later chapters of the book. Clearly, this chapter may be read both as an introduction to practical probability distributions/densities as well as a reference manual for such distributions.

3.1 Bernoulli and Binomial Distributions

3.1.1 Definition of the Bernoulli Distribution

Named after Swiss mathematician Jacob Bernoulli[1] (1654–1705), the **Bernoulli distribution** is a discrete probability distribution that takes exclusively two values: 1 with success

[1] Contributed works in calculus, particularly the calculus of variations, and in the field of probability.

probability ε, and 0 with "failure" probability $1-\varepsilon$:

$$p_{\mathrm{Ber}}(n|\varepsilon) = \begin{cases} 1-\varepsilon & \text{probability of failure, } n=0, \\ \varepsilon & \text{probability of success, } n=1. \end{cases} \tag{3.1}$$

It may alternatively be written

$$p_{\mathrm{Ber}}(n|\varepsilon) = (1-\varepsilon)^{1-n}\varepsilon^{n}, \tag{3.2}$$

where $n = 1$ stands for success and $n = 0$ for failure.

Arguably the simplest example of the Bernoulli distribution is the single toss of a fair coin, which has a probability $\varepsilon = 0.5$ to come up heads or tails. The particle detection process may also be regarded as a "Bernoulli experiment." In this case, ε represents the probability of detecting a single particle with a given apparatus.

3.1.2 Definition of the Binomial Distribution

In the context of the Bayesian approach to probability (i.e., probability as logic), one can formulate prior information about the system, denoted I, by stating that the single toss of a coin (or an attempt to detect a particle) is an event, E, with two possible outcomes that we denote S, "*the event was a success*," and $\overline{S}$, "*the event was NOT a success*." The event E may then be written

$$E = S + \overline{S}, \tag{3.3}$$

where the + operator corresponds to a logical OR. Since S and $\overline{S}$ are mutually exclusive (i.e., $S, \overline{S}$ = false, or $S \cap \overline{S} = 0$), the probability of the event E may be written

$$p(E|I) = p(S|I) + p(\overline{S}|I) = \varepsilon + (1-\varepsilon) = 1. \tag{3.4}$$

A series of n tosses may then be regarded as a succession of Bernoulli events, each with two possible outcomes, but a unit probability. This then leads us naturally to the introduction of the binomial distribution.

The **binomial distribution** arises in the context of identical, independent measurements, tosses, or samplings, repeated a finite number of times. As such, it is an extension of the Bernoulli distribution for $n > 1$. Each measurement, observation, or sample is assumed to be an independent trial with a probability of success denoted by ε. A set of N measurements can also be regarded as a single measurement, or event, yielding n successful trials and $N-n$ "failures." If one were to carry out the N trial measurements several times, the resulting value of n would occur with a probability given by the binomial distribution, $p_B(n|N,p)$:

$$p_{\mathrm{B}}(n|N,p) = \frac{N!}{n!(N-n)!}\varepsilon^{N}(1-\varepsilon)^{N-n}. \tag{3.5}$$

The binomial distribution may evidently be derived using the frequentist approach to probability. It is more interesting to discover, however, how it naturally arises also in the context of the Bayesian approach, and in particular, how the calculus of predicates of probability as logic provides a programmatic and robust recipe to derive this result.

Prior information about the series of tosses, I, may be summarized in two statements: each toss is a Bernoulli events with two mutually exclusive outcomes, and the n tosses are mutually independent, i.e., the outcome of one does not affect the outcome of the others.

As stated earlier, a single toss (or draw) may be regarded as an event, $E = S + \overline{S}$, with unit probability. A series of n tosses may also be regarded as an event, which we shall denote B_n. This event consists of n independent Bernoulli tosses we note E_i. Since the occurrence of these n mutually independent events is known, the aggregate event B_n may then be viewed as an AND between all these individual elementary events: the first event happened and then the second, and then the third, and so forth. One may then write

$$B_n \equiv E_1, E_2, \ldots, E_n \tag{3.6}$$

Additionally, since each event E_i has two possible outcomes, S_i and $\overline{S}_i$, one can also write

$$B_n = (S_1 + \overline{S}_1), (S_2 + \overline{S}_2), \ldots, (S_n + \overline{S}_n). \tag{3.7}$$

Recall that the commas represent logical ANDs between the different events. The AND may then be carried out as a multiplication of the n operands of the expression. For $n = 2$, one gets

$$\begin{aligned} B_2 &= (S_1 + \overline{S}_1), (S_2 + \overline{S}_2) && (3.8)\\ &= S_1, S_2 + S_1, \overline{S}_2 + \overline{S}_1, S_2 + \overline{S}_1, \overline{S}_2 && (3.9) \end{aligned}$$

while for $n = 3$ one gets

$$\begin{aligned} B_3 = {} & S_1, S_2, S_3 + S_1, S_2, \overline{S}_3 + S_1, \overline{S}_2, S_3 + \overline{S}_1, S_2, S_3 \\ & + S_1, \overline{S}_2, \overline{S}_3 + \overline{S}_1, \overline{S}_2, S_3 + \overline{S}_1, S_2, \overline{S}_3 + \overline{S}_1, \overline{S}_2, \overline{S}_3 \end{aligned} \tag{3.10}$$

and similarly for larger values of n. The probability of a particular sequence, for instance, $S_1, S_2, \overline{S}_3$, is calculated with the product rule:

$$\begin{aligned} p(S_1, S_2, \overline{S}_3|I) &= p(S_1|I)p(S_2, \overline{S}_3|S_1, I) \\ &= p(S_1|I)p(S_2|S_1, I)p(\overline{S}_3|S_1, S_2, I) \end{aligned} \tag{3.11}$$

But recall that the prior information includes the notion that the outcomes of distinct tosses or draws are independent of one another. This means that $p(S_2|S_1, I) = p(S_2|I)$ and $p(\overline{S}_3|S_1, S_2, I) = p(\overline{S}_3|I)$. The probability of the event $S_1, S_2, \overline{S}_3$ is thus simply

$$p(S_1, S_2, \overline{S}_3) = p(S_1|I)p(S_2|I)p(\overline{S}_3|I). \tag{3.12}$$

The prior information I also tells us that the probabilities of success and failure are constant and independent of the draw. We can thus write

$$p(S_1, S_2, \overline{S}_3) = p(S|I)p(S|I)p(\overline{S}|I) = \varepsilon^2(1 - \varepsilon). \tag{3.13}$$

Similar conclusions are obtained for other combinations of S and $\overline{S}$ and for larger values of N. Indeed, we find the probability of a specific outcome depends on the number of successes and failures. For an experiment consisting of N trials, the calculation thus simply becomes a matter of keeping track of the number of ways a given number of S and $\overline{S}$ can occur. This is of course given by the binomial coefficient.

$$
{}^{N}C_n = \frac{N!}{n!(N-n)!} \equiv \binom{N}{n}. \tag{3.14}
$$

The probability of a sequence containing n times a success S and $N-n$ times a failure $\overline{S}$ is thus indeed given by the binomial distribution

$$
\begin{aligned}
p(n|N, I) &= \frac{N!}{n!(N-n)!} p(S|I)^n p(\overline{S}|I)^{(N-n)} &\quad (3.15)\\
&= \frac{N!}{n!(N-n)!} \varepsilon^n (1-\varepsilon)^{(N-n)}. &\quad (3.16)
\end{aligned}
$$

Although the foregoing derivation is notably lengthier than a proof based on the frequentist approach, it has the important advantage of being transparent and programmatic: once the prior information I is given, the probability of any given sequence is readily written and calculated. There is no need for particular intuition or tricks.

3.1.3 Properties of the Binomial Distribution

It is relatively simple to verify that the sum of the probabilities of values of n from zero to N equals unity. Summing these different probabilities amounts to summing the probabilities of all terms of an event B_n. But an event B_n has, by construction, a probability equal to the product of the events it contains. Since all the events E_i have unit probability, the sequence event B_n thus itself also has unit probability. Alternatively, one can also use Pascal's binomial theorem, which may be written

$$
(x+y)^N = \sum_{n=0}^{N} \binom{N}{n} x^{N-n} y^n. \tag{3.17}
$$

Setting $x = \varepsilon$ and $y = 1 - \varepsilon$, one finds the left-hand side of Eq. (3.17) yields $(x+y)^N = (\varepsilon + 1 - \varepsilon)^N = 1$, and the right-hand side becomes the sum of the probabilities sought after.

Pascal's binomial theorem may also be used to obtain the moments of the binomial distribution. The mean, μ, of the binomial distribution is obtained by calculating the expectation value $\mathrm{E}[n]$. One gets

$$
\mu \equiv \mathrm{E}[n] = \sum_{n=0}^{N} n \frac{N!}{n!(N-n)!} p^N (1-p)^{N-n} = Np. \tag{3.18}
$$

Similarly, the variance, σ^2, of the binomial distribution is given by

$$
\sigma^2 = \mathrm{Var}[n] = \mathrm{E}[n^2] - (\mathrm{E}[n])^2 = Np(1-p). \tag{3.19}
$$

These two results can be verified (see Problem 3.2) using a binomial decomposition of $(1-p)^{N-n}$ in series. It is, however, simpler to calculate these, and higher moments, using the characteristic function (defined in §2.10) of the binomial distribution (see Problem 3.2).

Table 3.1 Properties of the binomial distribution

Function	$p_B(n\|N,\varepsilon) = \frac{N!}{n!(N-n)!}\varepsilon^N(1-\varepsilon)^{N-n}$
Characteristic Fct	$\phi(t) = \left(1-\varepsilon+\varepsilon e^{it}\right)^N$
Moment Gen Fct	$M(t) = (1-\varepsilon+\varepsilon e^{t})^N$
Moments	
1st	$\mu = \mu' = N\varepsilon$
2nd	$\mu_2' = N\varepsilon(1-\varepsilon+N\varepsilon)$
3rd	$\mu_3' = Np(1-3\varepsilon+3N\varepsilon+2\varepsilon^2-3N\varepsilon^2+N^2\varepsilon^2)$
4th	$\mu_4' = N\varepsilon(1-7\varepsilon+7N\varepsilon+12\varepsilon^2-18N\varepsilon^2$ $+6N^2\varepsilon^2-6\varepsilon^3+11N\varepsilon^3+N^3\varepsilon^3)$
Centered moments	
2nd	$\sigma^2 \equiv \mu_2 = N\varepsilon(1-\varepsilon)$
3rd	$\mu_3 = N\varepsilon(1-\varepsilon)(1-2\varepsilon)$
4th	$\mu_4 = N\varepsilon(1-\varepsilon)\left(3\varepsilon^2(2-N)+3\varepsilon(N-2)+1\right)$
Skewness excess	$\gamma_1 = \frac{1-2\varepsilon}{\sqrt{N\varepsilon(1-\varepsilon)}}$
Kurtosis excess	$\gamma_2 = \frac{6\varepsilon^2-6\varepsilon+1}{N\varepsilon(1-\varepsilon)}$
1st cumulant	$\kappa_1 = N\varepsilon$
Other cumulants	$\kappa_{i+1} = \varepsilon(1-\varepsilon)\frac{d\kappa_i}{d\varepsilon}$

Figure 3.1 displays examples of the binomial distribution obtained for various values of ε. The characteristic function, skewness, kurtosis, and other properties of the binomial are presented in Table 3.1.

It is interesting to consider the shape of the binomial distribution for arbitrarily large values of N, as illustrated in Figure 3.2, where the distribution is plotted as a function of

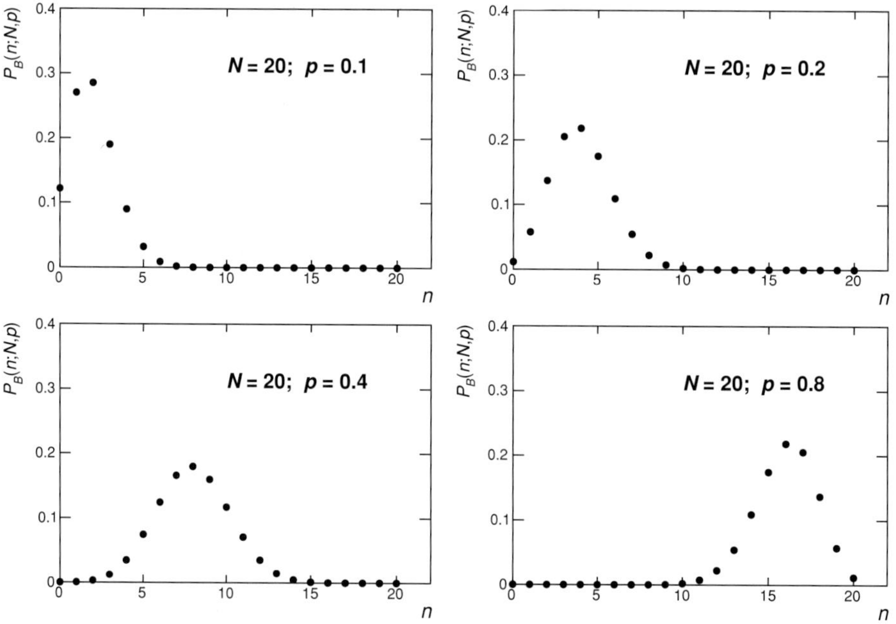

Fig. 3.1 Examples of the binomial distribution for $N = 20$ and selected values of p.

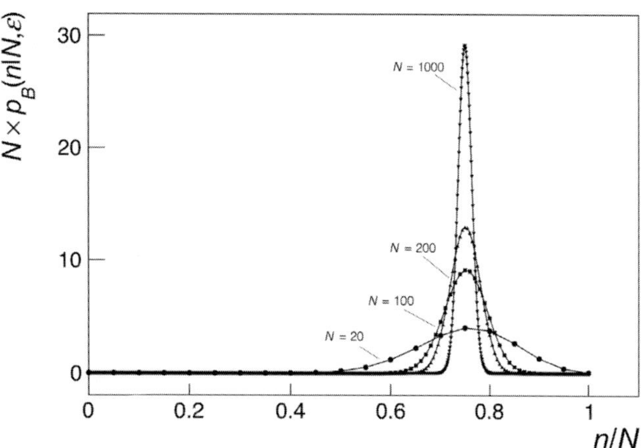

Fig. 3.2 Evolution of the binomial distribution with increasing N, plotted for $\varepsilon = 0.75$.

n/N. One finds that for increasing values of N, the distribution (plotted as a function of n/N) becomes progressively narrower, and in the limit $N \to \infty$ becomes a delta function $\delta(\varepsilon)$, that is, the distribution peaks at $n/N = \varepsilon$.

The binomial distribution finds a wide variety of applications ranging from games of chance (toss of a coin), opinion polls, and particle physics, where it plays a central role in modeling detection efficiencies, as we discuss briefly in the example provided in the next section and in greater detail in §12.2.2.

3.1.4 Example: Measurements of Particle Multiplicity Variance

Consider a measurement of particle multiplicity production in proton–proton or nucleus–nucleus collisions at relativistic energy. Assume a large acceptance detector is used to measure the particle production. For various technical reasons, it is usually impossible to measure all particles emitted within the detector acceptance. One characterizes a detector's response by stating the detector efficiency, ε, defined as the probability of measuring any given particle. Assuming the detection of a particle is not affected by the presence of other particles, we will show, in Chapter 12, that the efficiency can be estimated as the ratio of the average number of detected particles, $\langle n \rangle$, by the number of particles produced within the acceptance of the detector, $\langle N \rangle$:

$$\varepsilon = \frac{\langle n \rangle}{\langle N \rangle}. \tag{3.20}$$

The bracket notation $\langle \rangle$ is hereafter used to represent the average of the variables n and N over all measured collisions.

To obtain this result, consider that in essence, each particle has a probability ε of being detected. This implies that for a given N, the number of measured particles n should fluctuate. Given a particle is either detected or not detected (success or failure), the probability distribution function (PDF) of n is a binomial distribution with parameters N and probability ε:

$$p_B(n|N,\varepsilon) = \frac{N!}{n!(N-n)!}\varepsilon^N(1-\varepsilon)^{N-n}. \tag{3.21}$$

Particle production in elementary collisions is intrinsically a stochastic phenomenon. We assume the number of produced particles, N, can be described by some a priori unknown PDF, $p_P(N)$. Its two lowest moments and variance are in principle calculated as follows:

$$\langle N\rangle = \mathrm{E}[N] = \sum p_P(N)N, \tag{3.22}$$

$$\langle N^2\rangle = \mathrm{E}[N^2] = \sum p_P(N)n^2, \tag{3.23}$$

$$\mathrm{Var}[N] = \mathrm{E}[N^2] - \mathrm{E}[N]^2. \tag{3.24}$$

But given that the PDF itself is not known a priori, these calculations are not readily possible, and one must actually *measure* these moments. The problem, of course, is that the detection efficiency is smaller than unity, and fluctuations will take place in the detection process as well as in the particle production process. For a given N, measured values n will fluctuate according to the binomial distribution. But since N itself fluctuates, one must fold the probability of production and detection. The probability of detecting n particles, $p_m(n)$, may then be expressed as

$$p_m(n) = \sum_N p_B(n|N,\varepsilon)p_P(N), \tag{3.25}$$

where the sum is taken on all possible values of N. Figure 3.3 presents plots of the distribution $p_m(n)$ for the trivial case of $p_P(N) = \delta(N)$ and illustrates how measured multiplicities can widely fluctuate for small values of efficiency. Given our interest is to estimate the moments of particle production, we proceed to calculate the moments of the measured distribution. The average measured multiplicity n is given by

$$\begin{aligned}
\langle n\rangle = \mathrm{E}[n] &= \sum_n p_m(n)n, \\
&= \sum_n\sum_N p_B(n|N,\varepsilon)p_P(N)n, \\
&= \sum_N p_P(N)\sum_n p_B(n|N,\varepsilon)n, \\
&= \varepsilon\sum_N p_P(N)N, \\
&= \varepsilon\langle N\rangle,
\end{aligned} \tag{3.26}$$

where in the third line we commuted the sums, and in the fourth, we used the mean (Eq. 3.18) of the binomial distribution. The average measured multiplicity, $\langle n\rangle$, is thus, as expected, the product of the average produced multiplicity $\langle N\rangle$ by the efficiency ε. The second moment can be calculated in a similar fashion (see Problem 3.8):

$$\left\langle n^2\right\rangle = \varepsilon\,(1-\varepsilon)\,\langle N\rangle + \varepsilon^2\left\langle N^2\right\rangle. \tag{3.27}$$

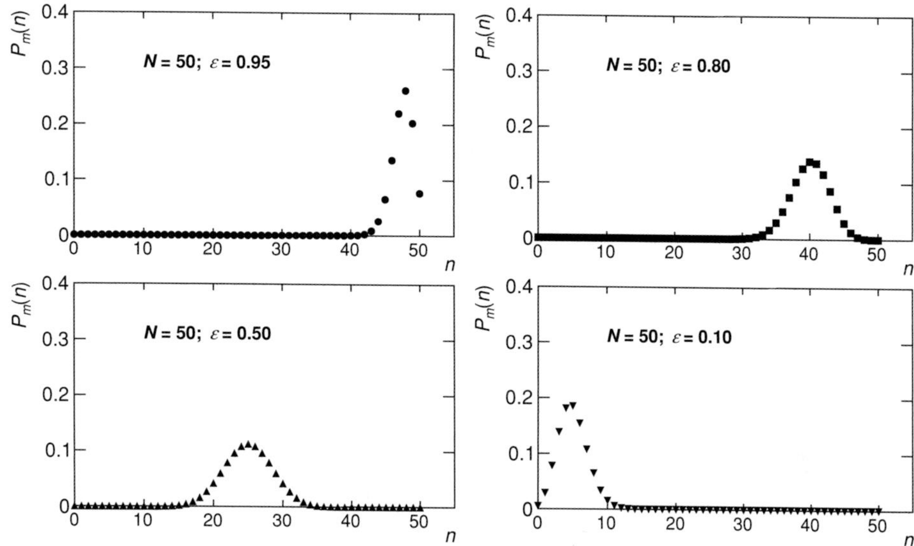

Fig. 3.3 Examples of multiplicity distributions obtained with efficiencies of 95%, 80%, 50%, and 10% for a produced multiplicity consisting of a single value of 50 particles.

The variance of the measured multiplicity is thus

$$\text{Var}[n] = \langle n^2 \rangle - \langle n \rangle^2 \tag{3.28}$$

$$= \varepsilon^2 \text{Var}[N] + \varepsilon (1 - \varepsilon) \langle N \rangle . \tag{3.29}$$

While the average measured multiplicity is proportional to the produced multiplicity, the relationship between the variances of the measured and produced multiplicities is not as straightforward, given that it involves a term that is proportional to the multiplicity and a nontrivial dependence on the efficiency, $\varepsilon (1 - \varepsilon)$. As we will discuss in further detail in Chapter 12, this implies that observables based on the variance of the multiplicity are not *robust* and cannot be trivially corrected for efficiency effects by taking simple ratios. However, as we shall see in that same chapter, factorial moments lend themselves naturally to such corrections.

3.1.5 Negative Binomial Distribution

The negative binomial distribution, as its name suggests, is intimately related to the binomial distribution. Here again, one considers successive trials that may yield successes S or failures $\overline{S}$. However, instead of considering the number of successes given n trials, one studies the number of trials completed before a prescribed number of success, k, is encountered. In this context, the last trial performed is by definition a success, and prior to this success, there must have been $k - 1$ successes in the previous $n - 1$ trials. The probability of k successes, with the last one being a success, thus yields the **negative binomial**

distribution, which may be written

$$p(n|k, \varepsilon) = \binom{n-1}{k-1} \varepsilon^k (1-\varepsilon)^{n-k}, \tag{3.30}$$

where ε denotes the probability of success S, as in the case of the regular binomial distribution.

The negative binomial distribution is particularly useful toward the description of produced particle multiplicity distributions measured in proton–proton collisions at high beam energy (see, e.g., [197] and references therein) or the number of photons produced by a blackbody at finite temperature [110].

3.1.6 A Technical Note

Computation of the binomial, multinomial (§3.2), and Poisson (§3.3) distributions requires the evaluation of the factorial of numbers that can be arbitrarily large. Although binomial and multinomial coefficients are finite, their evaluation can run into severe numerical instabilities, or floating point overflows, even with modern computers. Given that one is typically interested in ratios of factorial multiplied by powers of "p," it is often convenient to calculate the logarithms of the coefficients rather than their actual values. This can easily be accomplished using Stirling's[2] factorial formula:

$$n! \approx \sqrt{2\pi n}\left(\frac{n}{e}\right)^n. \tag{3.31}$$

3.2 Multinomial Distribution

3.2.1 Definition

The **multinomial distribution** is an extension of the binomial distribution applicable when a measurement (random selection, or trial) can lead to many different outcomes of definite probability.

Consider, for instance, an "experiment" consisting of picking balls at random from a vat containing 25 red balls, 50 green balls, 100 yellow balls, and 25 black balls. The total number of balls is 200. Balls are replaced in the vat after they have been picked. The probability of picking one red ball in one trial equals the ratio of the number of red balls to the total number of balls in the vat, $p_{\text{red}} = 0.125$. Likewise, the probabilities to randomly pick a green, yellow, or black ball are $p_{\text{green}} = 0.25$, $p_{\text{yellow}} = 0.5$, and $p_{\text{black}} = 0.125$, respectively; and the sum of the four probabilities is unity. Next, consider the probability of picking two red balls and one green ball in one trial (consisting of a random selection of three balls). The probability of this outcome is obviously proportional to the product

[2] James Stirling, Scottish mathematician, 1692–1770.

$p_{\text{red}}^2 p_{\text{green}}$, i.e., it is proportional to the product of probabilities of the different colors to a power equal to the number of balls of the given color being picked out.

Let us next consider, more generally, that a measurement could yield m types of objects or outcomes (e.g., four distinct colors, five types of particle species, etc.). Let us identify these outcomes with an index i spanning values $1, 2, \ldots, m$, and denote their respective probabilities p_i. By definition, the sum of these m probabilities equals unity:

$$\sum_{i=1}^{m} p_i = 1. \tag{3.32}$$

Next, consider the probability of getting $n_1, n_2, \ldots, n_m$ instances of type $i = 1, \ldots, m$ within a sample of N objects. The probability of a specific combination of $n_1, n_2, \ldots, n_m$ instances of each type is proportional to the product of their respective probabilities to the power n_i:

$$\text{Probability} \propto p_1^{n_1} \times p_2^{n_2} \times \cdots \times p_m^{n_m}. \tag{3.33}$$

There are many ways to generate a certain combination of $n_1, n_2, \ldots, n_m$ objects of types $i = 1, \ldots, m$. One must thus sum the probabilities of all these different sequences, while accounting for the fact that the order of the objects in a particular sequence is irrelevant. There are $N!$ possible sequences of N objects (outcomes) but since objects of a given type i are all equivalent (indistinguishable), one must divide the total number of sequences by the number of equivalent sequences of each type, $n_1! \ldots n_m!$. One then obtains the multinomial distribution

$$p_M(n_1, \ldots, n_m|N, p_1, \ldots, p_m) = \frac{N!}{n_1! n_2! \cdots n_m!} p_1^{n_1} \times p_2^{n_2} \times \cdots \times p_m^{n_m}, \tag{3.34}$$

where by construction $\sum_{i=1}^{m} n_i = N$. It is often convenient to write the multinominal distribution using multinomial coefficients defined as follows:

$$\binom{N}{n_1, n_2, \ldots, n_m} \equiv \frac{N!}{n_1! n_2! \cdots n_m!}. \tag{3.35}$$

And using the product notation $\prod_{i=1}^{m} x_i = x_1 \times \cdots x_m$, one gets

$$p_M(n_1, \ldots, n_m|N, p_1, \ldots, p_m) = \binom{N}{n_1, \ldots, n_m} \prod_{i=1}^{m} p_i^{n_i}. \tag{3.36}$$

A derivation of the multinomial distribution may also be obtained with the Bayesian technique used in §3.1 for the derivation of the binomial distribution. In this case, the prior information I may be stated as follows: (1) there are m mutually exclusive and logically independent outcomes O_i, with $i = 1, \ldots, m$; (2) the outcome of a given draw is independent of the outcome of other draws. A draw, or event E, may thus be represented as a disjunction (i.e., an OR) of the m different outcomes, which in this context is equivalent to their sum:

$$E = O_1 + O_2 + \cdots + O_m. \tag{3.37}$$

A series of N draws, considered as an event M_N, may then be written

$$M_N = (O_1 + O_2 + \cdots + O_m)^N \tag{3.38}$$

Table 3.2 Properties of the multinomial distribution

Function	$p_M(n_1,\ldots,n_m\|N;p_1,\ldots,p_m)=\frac{N!}{n_1!n_2!\cdots n_m!}p_1^{n_1}\times p_2^{n_2}\times\cdots\times p_m^{n_m}$
Characteristic Fct	$\phi(t_1,t_2,\ldots,t_m)=\left(\sum_{j=1}^{m}p_je^{it_j}\right)^N$
Moment Gen Fct	$M(t_1,t_2,\ldots,t_m)=\left(\sum_{j=1}^{m}p_je^{t_j}\right)^N$
Moments	
1st	$\mu_i=\mu_i'=Np_i$
2nd	$\mu_{2,i}'=Np_i(1-p_i+Np_i)$
Covariance	$\text{Cov}[n_i,n_j]=-Np_ip_j$

Repeating the reasoning carried for the binomial distribution, we find that the probability of a specific draw depends on the number of occurrences of each type. An event may then be described as consisting of n_1 items of type O_1, n_2 items of type O_2, and so on, with $\sum_i n_i = n$. The number of ways such outcomes may occur is given by multinomial coefficients (3.35). The probability of a specific outcome with $(n_1, n_2, \ldots, n_m)$ is thus indeed given by Eq. (3.36).

3.2.2 Properties of the Multinomial Distribution

Moments of the multinomial distribution are easily calculated based on its characteristic function

$$\phi(t_1,t_2,\ldots,t_m)=\left(\sum_{j=1}^{m}p_je^{it_j}\right)^N. \tag{3.39}$$

One readily verifies that the mean and variance of variables n_i have expressions similar to those found for the binomial distribution:

$$\mu_i = \text{E}[n_i] = Np_i, \tag{3.40}$$

$$\sigma_i^2 = \text{Var}[n_i] = Np_i(1-p_i). \tag{3.41}$$

One must, however, also consider the correlation between different variables. This can be quantified by calculating the covariance, noted $\text{Cov}[n_i, n_j]$, of two variables n_i and n_j such that $i \neq j$. The calculation of these covariances can also be carried out based on the characteristic function of the multinomial distribution (see Problem 3.7). One gets

$$\text{Cov}[n_i,n_j] = -Np_ip_j. \tag{3.42}$$

The covariance is negative because the total number of objects sampled is fixed. If one value n_i increases, the others must decrease on average, and they therefore have a negative correlation.

The binomial and multinomial distributions are known as **exchangeable distributions** because they satisfy

$$p(O_j|O_{j-1},I) = p(O_j|O_{j+1},I), \tag{3.43}$$

where O_j stands for the outcome of the jth draw. This equation may seem rather peculiar since it indicates that the probability of an outcome O_j, given a previous outcome O_{j-1}, is known to have occurred is equal to the probability of the same outcome given a subsequent outcome O_{j+1} is known to occur. How can this be causally possible? The answer is that it is not: the connection between these probabilities is purely logical, not causal. This peculiar property of Bayesian probabilities is discussed at length in §4.5 of ref. [97] . The exchangeable property is also shared by the **hypergeometric distribution**

$$p(r|N, M, n) = \frac{\binom{M}{r}\binom{N-M}{n-r}}{\binom{N}{n}}, \tag{3.44}$$

which expresses the probability of an event (a draw) involving r outcomes S in n tries from a pool of objects containing N objects, M of which are of type S and the remainder $N - M$ are of type $\overline{S}$ [117].

3.3 Poisson Distribution

3.3.1 Definition

The **Poisson distribution** is a discrete probability distribution that expresses the probability of a number of occurrences or events in a fixed interval of space or time. It is defined by the following prior information I or model: the probability of an event to occur in a given time interval $[t, t + dt]$ (or alternatively spatial interval $[x, x + dx]$) is defined as $p\,dt$, where p is a constant value independent of t (or x). Furthermore, the occurrence of an event at time t has no influence on the occurrence of events taking place in any other time intervals before or after the time t.

Let us define a predicate E as stating that no event occurred in the time interval $[0, t + dt]$ and let a function $q(t)$ represent the probability that no event occurred during the interval $[0, t]$. We then write

$$p(E|I) \equiv q(t + dt). \tag{3.45}$$

The statement E can be also formulated as the conjunction (i.e., an AND) between two propositions E_1 and E_2 such that E_1 states that no event occurred in the interval $[0, t]$, and E_2 states that no event occurred in the interval $[t, t + dt]$:

$$E = E_1, E_2 \tag{3.46}$$

We apply the product rule to express the probability of E in terms of the probabilities of E_1 and E_2:

$$p(E|I) = p(E_1, E_2|I) = p(E_1|I)p(E_2|E_1, I). \tag{3.47}$$

Since the two time intervals are mutually exclusive, and by virtue of the model hypothesis that the occurrence of an event at time t has no influence on the occurrence of events taking place in any other time intervals, we get

$$p(E|I) = p(E_1|I)p(E_2|I). \tag{3.48}$$

The probabilities $p(E|I)$ and $p(E_1|I)$ are quite obviously equal to $q(t + dt)$ and $q(t)$, respectively, while the probability $p(E_2|I)$ is equal to $1 - pdt$ by virtue of our first model hypothesis. The preceding expression thus becomes

$$q(t + dt) = q(t)\,(1 - pdt)\,. \tag{3.49}$$

Defining the difference $dq = q(t + dt) - q(t)$, we obtain the differential equation

$$\frac{dq}{dt} = -pq(t), \tag{3.50}$$

which holds solutions of the form

$$q(t) = q(0)\exp(-pt), \tag{3.51}$$

with $q(0) = 1$ as an obvious initial condition. This gives the exponential distribution (revisited in §3.5) expressing the probability that no events occurred in the time interval $[0, t]$ under the condition of the model (prior information I) defined previously.

Let us next consider a subsidiary proposition, noted C_n, that exactly n counts (or events) occurred at ordered times $t_1 < t_2 \ldots < t_n$ during the time interval $[0, t]$, and each within infinitesimal intervals $dt_1, dt_2, \ldots, dt_n$. The proposition may then be regarded as a conjunction of $2n + 1$ simpler propositions, that there were no events in the interval $[0, t_1]$, one event during $[t_1, t_1 + dt_1]$, no events during $[t_1, t_2]$, one event during $[t_2, t_2 + dt_2]$, and so on. Applying once again the product rule and the model hypothesis that distinct time intervals are independent, one can write

$$\begin{aligned} p(C_n|p, I) &= \exp\left(-pt_1\right) \times pdt_1 \times \exp\left(-p(t_2 - t_1\right) \times pdt_2 \times \ldots \\ &\quad \times \exp\left(-p(t_n - t_{n-1}\right) \times pdt_n \times \exp\left(-p(t - t_n\right) \\ &= \exp\left(-pt\right) p^n dt_1 \ldots dt_n. \end{aligned} \tag{3.52}$$

In order to obtain the probability of observing exactly n events in the time interval $[0, t]$, irrespective of their exact times, we must sum the probabilities $p(C_n|p, I)$ by accounting for the fact that each of the n events may happen at any time in the interval $[0, t]$ while respecting the order $t_1 < t_2 \ldots < t_n$, that is, such that t_1 can occur anytime in $[0, t_2]$, t_2 can occur anytime in $[0, t_3]$, and so on.

$$p(n|p, t, I) = \exp\left(-pt\right) p^n \int_0^t dt_n \ldots \int_0^{t_3} dt_2 \int_0^{t_2} dt_1. \tag{3.53}$$

The time ordered integral yields $t^n/n!$ and we thus obtain the Poisson distribution

$$p(n|p, t, I) = \frac{\exp\left(-pt\right)(pt)^n}{n!}. \tag{3.54}$$

It is common practice to express the distribution in terms of the average number of counts in the finite time interval $[0, t]$, denoted $\lambda = pt$:

$$p_P(n|\lambda) = \frac{\exp(-\lambda)\,\lambda^n}{n!} \tag{3.55}$$

3.3.2 Properties of the Poisson Distribution

Let us first verify that the sum of the probabilities of all values of n yields unity:

$$\sum_{n=0}^{\infty} p_P(n|\lambda) = \sum_{n=0}^{\infty} \frac{\lambda^n}{n!} e^{-\lambda}, \tag{3.56}$$

$$= e^{-\lambda} \sum_{n=0}^{\infty} \frac{\lambda^n}{n!}. \tag{3.57}$$

In the second line, the sum corresponds to a Taylor expansion of e^{λ}. Substituting this expression for the sum, we get the anticipated result:

$$\sum_{n=0}^{\infty} p_P(n|\lambda) = e^{-\lambda} e^{\lambda} = 1. \tag{3.58}$$

The distribution is thus properly normalized and qualifies for use as a PDF.

The moments of the Poisson distribution, $\mu'_k = \mathrm{E}[n^k]$, may be obtained by direct sum of Eq. (3.55), but are more conveniently calculated as derivatives of the characteristic function of the distribution which we proceed to evaluate:

$$\phi_n(t) = \mathrm{E}\left[e^{itn}\right] = \sum_{n=0}^{\infty} e^{itn} \frac{e^{-\lambda}\lambda^n}{n!}, \tag{3.59}$$

$$= e^{-\lambda} \sum_{n=0}^{\infty} \frac{\left(\lambda e^{it}\right)^n}{n!}. \tag{3.60}$$

The sum corresponds to the Taylor expansion of $e^{\lambda e^{it}}$. We thus obtain

$$\phi_n(t) = e^{\lambda(e^{it}-1)} \tag{3.61}$$

as the characteristic function of the Poisson distribution. The moments μ'_k of the Poisson distribution are then obtained from kth order derivatives of this expression

$$\mu'_k = i^{-k} \left.\frac{d^k}{dt^k}\phi_n(t)\right|_{t=0}. \tag{3.62}$$

The first and second moments are

$$\mu'_1 = \mathrm{E}[n] = i^{-1} \left.\frac{d}{dt} e^{\lambda(e^{it}-1)}\right|_{t=0} = \lambda, \tag{3.63}$$

$$\mu'_2 = \mathrm{E}[n^2] = i^{-2} \left.\frac{d^2}{dt^2} e^{\lambda(e^{it}-1)}\right|_{t=0} = \lambda^2 + \lambda \tag{3.64}$$

Table 3.3 Properties of the Poisson distribution

Function	$p_P(n\|\lambda) = \frac{\lambda^n}{n!}e^{-\lambda}$
Characteristic Fct	$\phi(t) = e^{\lambda(e^{it}-1)}$
Moment Gen Fct	$M(t) = e^{\lambda(e^{t}-1)}$
Moments	
1st	$\mu = \mu_i' = \lambda$
2nd	$\mu_2' = \lambda(1+\lambda)$
3rd	$\mu_3' = \lambda\left(1+3\lambda+\lambda^2\right)$
4th	$\mu_4' = \lambda\left(1+7\lambda+6\lambda^2+\lambda^3\right)$
Central moments	
2nd	$\mu_2 = \lambda$
3rd	$\mu_3 = \lambda$
4th	$\mu_4 = \lambda(1+3\lambda)$
Skewness excess	$\gamma_1 = \lambda^{-1/2}$
Kurtosis excess	$\gamma_2 = \lambda^{-1}$

from which we obtained the variance (2nd centered moment) of the Poisson distribution:

$$\text{Var}[n] = \mu_2' - (\mu_1')^2 = \lambda. \tag{3.65}$$

The fact that the variance and the mean are equal is the hallmark of the Poisson distribution. It may be used as a rough diagnostic of measured distributions to establish whether they exhibit a **Poissonian behavior**. Higher moments, shown in Table 3.3, are easily obtained by calculating higher derivatives of Eq. (3.61).

The Poisson distribution is illustrated in Figure 3.4 for selected values of the parameter λ. Examples of application of the distribution are discussed in §§3.3.3 and 3.3.4. A derivation of the Poisson distribution as a limiting case of the binomial distribution is presented in §3.3.5 and its behavior in the large λ limit discussed in §3.3.6. The gamma distribution, presented in §3.6, may be regarded as an extension of the Poisson distribution for continuous values of n (i.e., $n \in \mathbb{R}$).

3.3.3 Example 1: Radioactive Decay Rates

The Poisson distribution is particularly useful in the context of processes (or systems) in which the probability of a certain phenomenon is small. For instance, consider repeated measurements of the number of decays of a radioactive source in a specific time interval Δt. The decay rate (activity) of the source is defined as the number of (nucleus) decays per time interval. It is proportional to the number N of nuclei present in the source and inversely proportional to the nucleus lifetime τ:

$$\frac{dN}{dt} = \frac{N}{\tau}. \tag{3.66}$$

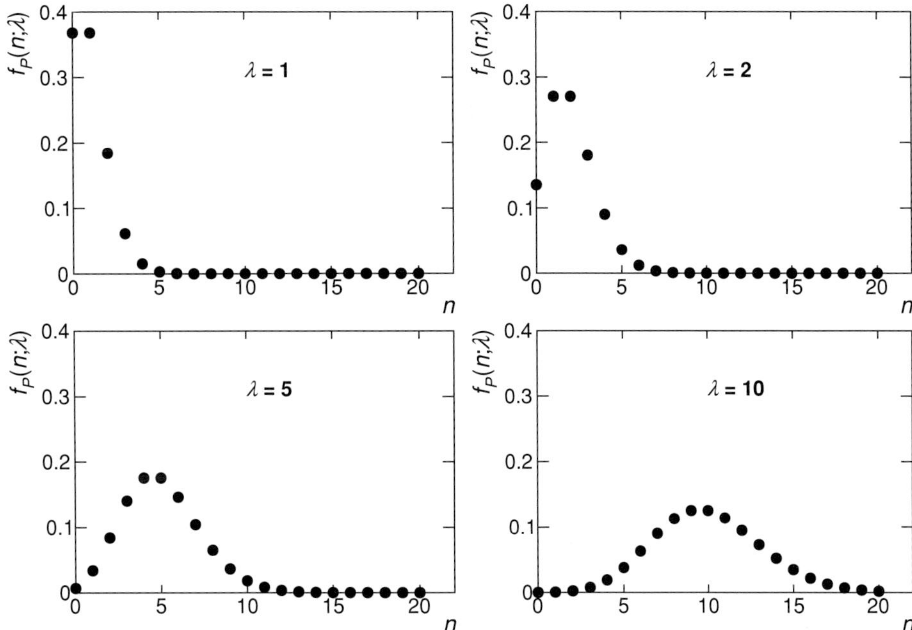

Fig. 3.4 Examples of the Poisson distribution with four different values of the parameter λ.

The expected (i.e., average) number of decays, $\langle n \rangle$, in a finite time interval Δt is thus

$$\langle n \rangle = \frac{dN}{dt}\Delta t = \frac{N}{\tau}\Delta t. \tag{3.67}$$

This result is strictly valid if N does not significantly vary during the time interval Δt and is thus applicable for sufficiently large N and small Δt. The decay of radioactive nuclei is, however, a stochastic process. While one can determine the average number of decays in a time interval of a specific duration, the number of decays in successive measurements of the number of decays in Δt will yield random numbers, n, that fluctuate about the mean $\langle n \rangle$. Given that decays are independent of one another, and for large sources (i.e., large N), one can model the probability of n decays during the time interval Δt with the Poisson distribution:

$$p_P(n; \langle n \rangle) = \frac{\langle n \rangle^n}{n!} e^{-\langle n \rangle}. \tag{3.68}$$

This implies it is possible to use the average of the number of decays, $\langle n \rangle$, observed over repeated measurements, to determine the lifetime τ of the nucleus even though it may be quite large relative to the duration of the experiment, or even the typical human lifespan. Uncorrelated fluctuations of the number n from measurement to measurement are commonly referred to as **Poisson fluctuations** or **Poisson noise**.

Although the Poisson distribution is defined for discrete values of the variable n, it can nonetheless be treated as a continuous function of a variable x, provided it is integrated over sufficiently large ranges $\Delta x \gg 1$. In fact, one finds that in the very large n limit, the

Poisson distribution reduces to a Gaussian distribution (see §3.9 for a demonstration of this result).

3.3.4 Example 2: Detector Occupancy

Imagine you are instructed to design a time-of-flight hodoscope for a large particle physics experiment. You are told that one can expect an average of 100 particles in the acceptance of the hodoscope. You are then required to design the detector in such way that detector units of the hodoscope should have a hit occupancy, in other words, a rate of particle crossing each units, that does not exceed 5%. How many units or segments should the hodoscope then be made of?

Let us assume that, to first order, there exists no correlation between the particles entering the acceptance of the detector. The average occupancy of a detector segment is thus equal to the ratio, λ, of the average number of particles, $\langle n\rangle$, to the number of segments, N:

$$\lambda = \frac{\langle n\rangle}{N}. \tag{3.69}$$

One can use Poisson statistics to calculate the probability that a particular segment will be traversed by n particles in a given event:

$$P(n) = \frac{\lambda^n e^{-\lambda}}{n!}. \tag{3.70}$$

We wish to limit the probability of multiple occupancy of a segment to less than 5%.

$$P(n \geq 2) = 1 - P(0) - P(1) \leq 0.05. \tag{3.71}$$

Substituting the expression of the Poisson distribution, one obtains

$$\begin{aligned} P(\text{multiple occupancy}) &= 1 - e^{-\lambda} - \lambda e^{-\lambda}, \\ &\approx 1 - 1 + \lambda - \lambda + \lambda^2 + O(2), \\ &= \lambda^2 \leq 0.05. \end{aligned} \tag{3.72}$$

We then conclude the hodoscope should comprise at least $N = 447$ segments to limit the multiple hit occupancy to no more than 5%.

3.3.5 Poisson Distribution as a Limit of the Binomial Distribution

We derived the Poisson distribution as the probability of a given number of occurrences in a fixed interval of time, but it may also be obtained as a limit of the binomial distribution for $\varepsilon \to 0$, $N \to \infty$ with $\lambda = \varepsilon N$. This may be readily demonstrated based on the limiting behavior of the characteristic function of the binomial distribution. Recall from §2.10 that the characteristic function of a PDF can be considered as an alternative definition of the distribution. A limit of the binomial distribution may then be used to define a "new" probability distribution. We will quickly show this new distribution is in fact the Poisson distribution.

One finds, by simple integration, that the characteristic function of the binomial distribution, Eq. (3.5), is given by

$$\phi(t) = \left[\varepsilon\left(e^{it} - 1\right) + 1\right]^N. \tag{3.73}$$

Let $z = \lambda\left(e^{it} - 1\right)$. Using the binomial theorem, Eq. (3.73) may then be written

$$\phi(t) = \left(1 + \frac{z}{N}\right)^N = \sum_{k=0}^{N} \frac{N!}{(N-k)!k!}\frac{z^k}{N^k}. \tag{3.74}$$

In the limit $N \to \infty$, this yields $\exp(z)$. One thus obtains

$$\phi(t) = \exp\left[\lambda\left(e^{it} - 1\right)\right], \tag{3.75}$$

which happens to be the characteristic function of the Poisson distribution. The Poisson distribution thus constitutes a limiting case of the binomial distribution for $\varepsilon \to 0, N \to \infty$ with $\lambda = \varepsilon N$.

$$\lim_{\substack{N\to\infty \\ \varepsilon \to 0}} p_B(n|N,p) = p_P(n|\lambda) = \frac{\lambda^n}{n!}e^{-\lambda}. \tag{3.76}$$

3.3.6 Large λ Limit of the Poisson Distribution

In the previous section, we saw that the Poisson distribution is a limiting case of the binomial distribution for $\varepsilon \to 0$, $N \to \infty$ and finite values of $\lambda = \varepsilon N$. It is also interesting to consider the behavior of the Poisson distribution $p_P(n|\lambda)$ in the limit $\lambda \to \infty$. Although the Poisson distribution is formally defined for discrete (integer) values of n, we can treat n as a continuous variable x in the limit of large λ as long as we integrate the function over an interval much larger than unity. Given the standard deviation of the Poisson is equal to the square root of λ, it is convenient and meaningful to define

$$x = \frac{n - \lambda}{\sqrt{\lambda}}, \tag{3.77}$$

which has, by construction, a mean of zero and a variance equal to unity. The characteristic function (3.75) of the PDF of x may then be written

$$\phi_x(t) = E\left[e^{itx}\right], \tag{3.78}$$

$$= E\left[e^{itn/\sqrt{\lambda}}e^{-it\sqrt{\lambda}}\right], \tag{3.79}$$

$$= e^{-it\sqrt{\lambda}}E\left[e^{itn/\sqrt{\lambda}}\right], \tag{3.80}$$

$$= e^{-it\sqrt{\lambda}}\ \phi_n\left(\frac{t}{\sqrt{\lambda}}\right), \tag{3.81}$$

where by definition ϕ_n is the characteristic function of the Poisson distribution. We substitute the expression of the characteristic function of the Poisson distribution in the preceding and get

$$\phi_x(t) = \exp\left[\lambda\left(e^{it/\sqrt{\lambda}} - 1\right) - it\sqrt{\lambda}\right]. \tag{3.82}$$

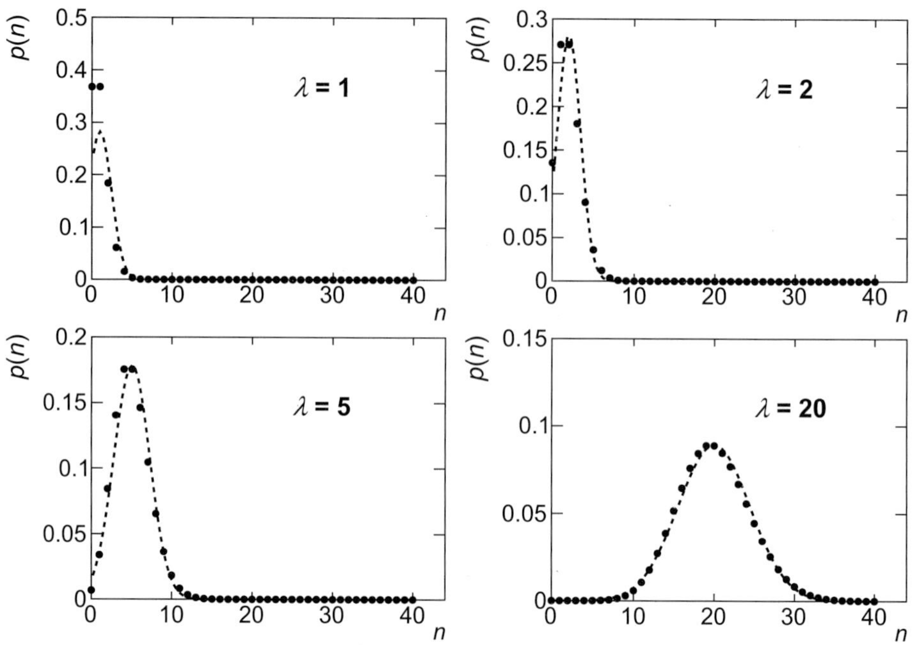

Fig. 3.5 Illustration of the convergence of the Poisson distribution (solid circle) with the normal distribution (dashed line) in the "large" λ limit. Normal distributions are plotted for $\mu = \lambda$ and $\sigma = \sqrt{\lambda}$.

Expansion of the innermost exponential yields $1 + \frac{it}{\sqrt{\lambda}} + \frac{1}{2}(\frac{it}{\sqrt{\lambda}})^2 + O(3)$ in the limit $\lambda \to \infty$. Equation (3.82) thus reduces to

$$\phi_x(t) = \exp\left(-\tfrac{1}{2}t^2\right), \tag{3.83}$$

which is the expression of the characteristic function of the standard normal distribution already introduced in §2.10.2 and further discussed in §3.9. One thus concludes that for large values of λ, the variable n follows a Gaussian distribution. In other words, for large values of λ, the Poisson distribution may be approximated by a Gaussian distribution with mean and variance both equal to λ. Figure 3.5 illustrates how the Poisson distribution quickly converges toward the Gaussian distribution with mean $\mu = \lambda$ and standard deviation $\sigma = \sqrt{\lambda}$ for increasing values of the parameter λ.

3.4 Uniform Distribution

3.4.1 Definition

The **uniform distribution**, as its name suggests, implies all values of a continuous variable, x, are equally probable in a given range $[\alpha, \beta]$. It is defined in the range $[-\infty, \infty]$ as

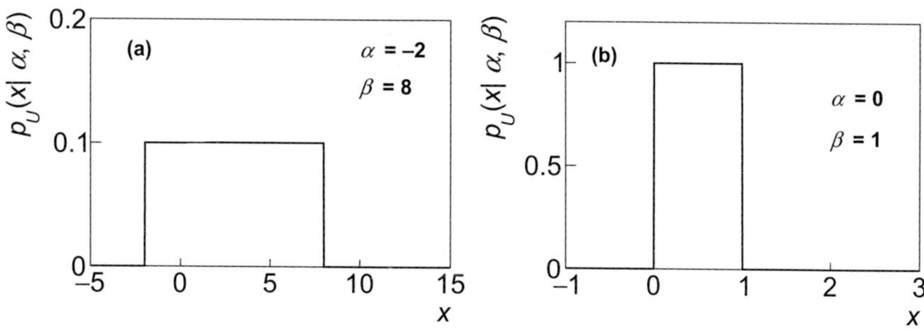

Fig. 3.6 (a) Example of the uniform distribution, $p_u(x|\alpha, \beta)$ with $\alpha = -2.0, \beta = 8.0$. (b) Standard uniform distribution.

follows:

$$p_u(x|\alpha, \beta) = \begin{cases} \frac{1}{\beta - \alpha} & \alpha \leq x \leq \beta, \\ 0 & \text{otherwise.} \end{cases} \tag{3.84}$$

Figure 3.6 displays an example of the uniform distribution with values $\alpha = -2$ and $\beta = 8$. The probability density has a constant value of 0.1 in the range $-2 \leq x \leq 8$, and is null otherwise. The area under the curve equals unity and corresponds to the probability of observing x in the range $-2 \leq x \leq 8$: all values of x in that range are equally probable, and those outside of the range are *impossible*.

The cumulative density function (CDF) of p_u is given by

$$P_u(x) = \int_{-\infty}^{x} p_u(x'|\alpha, \beta)\, dx', \tag{3.85}$$

$$= \begin{cases} 0 & \text{for } x < \alpha, \\ (x - \alpha)/(\beta - \alpha) & \text{for } \alpha \leq x \leq \beta, \\ 1 & \text{for } x > \beta. \end{cases} \tag{3.86}$$

Setting $\alpha = 0$, and $\beta = 1$, one gets the **Standard Uniform Distribution**, $p_u(x|0, 1)$.

3.4.2 Properties of the Uniform Distribution

The mean, μ, of the uniform distribution is obtained by calculating the expectation value of x for this PDF.

$$\mu \equiv \mathrm{E}[x] = \int_{-\infty}^{\infty} p_u(x|\alpha, \beta) x\, dx, \tag{3.87}$$

$$= \frac{1}{\beta - \alpha} \int_{\alpha}^{\beta} x\, dx = (\beta + \alpha)/2. \tag{3.88}$$

Table 3.4 Properties of the uniform distribution

Function	$p_u(x\|\alpha,\beta) = \begin{cases} \frac{1}{\beta-\alpha} & \alpha \le x \le \beta \\ 0 & \text{otherwise} \end{cases}$
Characteristic Fct	$\phi(t) = E\left[e^{itx}\right] = \left(e^{it\beta} - e^{it\alpha}\right)/(it\,(\beta-\alpha))$
Moment Gen Fct	$M(t) = E\left[e^{tx}\right] = \frac{e^{t\beta}-e^{t\alpha}}{t(\beta-\alpha)}$
Mode	Anywhere in the range $\alpha \le x \le \beta$
Median	$(\alpha+\beta)/2$
Moments	
1st	$\mu = \mu'_i = \frac{\beta+\alpha}{2}$
2nd	$\mu'_2 = \frac{\alpha^2+\alpha\beta+\beta^2}{3}$
kth	$\mu'_k = \frac{1}{k+1}\sum_{i=0}^{k} \alpha^i \beta^{k-i}$
Central moments	
2nd	$\sigma^2 \equiv \mu_2 = \frac{(\beta-\alpha)^2}{12}$
Skewness excess	$\gamma_1 = 0$
Kurtosis excess	$\gamma_2 = 6/5$

The variance, σ^2, is readily obtained by calculating $\text{E}[x^2] - \text{E}[x]^2$ as follows (see Problem 3.13):

$$\sigma^2 \equiv \text{Var}[x] = \frac{1}{\beta-\alpha}\int_{-\infty}^{\infty} (x-\mu)^2 p_u(x|\alpha,\beta)\,dx, \tag{3.89}$$

$$= \frac{(\beta-\alpha)^2}{12}. \tag{3.90}$$

From this expression, we conclude that a flat distribution of width L has a standard deviation $L/\sqrt{12}$. The skewness, the kurtosis excess, and other higher moments may be obtained similarly and are listed in Table 3.4.

3.4.3 Random Number Generation

The uniform and standard uniform distributions find use in a variety of applications, and most particularly, random number generation.

Given a set of random numbers $\{r_i\}$, with $i = 1, \ldots, n$ distributed according to the standard uniform distribution $p_u(x|0,1)$, one can trivially obtain random numbers following a uniform (but nonstandard) distribution $p_u(x|\alpha,\beta)$ using a simple linear transformation as follows:

$$r'_i = \alpha + (\beta-\alpha)\,r_i. \tag{3.91}$$

In fact, pseudo-random number generators (defined in Chapter 11) provided with popular mathematical software libraries (e.g., Mathematica®, MATLAB®, ROOT) typically feature a "core" random number generator that produces a standard uniform distribution, which may then be transformed, according to Eq. (3.91), to obtain arbitrary uniform distributions.

The standard uniform distribution can also be transformed to generate random numbers according to a wide variety of less trivial PDFs. For instance, consider the function $f(x)$ and its cumulative distribution, $y = F(x)$:

$$y = F(x) = \int_{-\infty}^{x} f(x')\,dx'. \tag{3.92}$$

By construction, one has

$$\frac{dy}{dx} = \frac{d}{dx}\int_{-\infty}^{x} f(x')\,dx' = f(x). \tag{3.93}$$

As we saw in §2.6, the probability of an event is invariant under a change of variable $x \to y$. Let $g(y)$ represent the PDF of y, that is, the probability of observing y in the range $[y, y + dy]$. One can then use Eq. (2.60) to express $g(y)$ in terms of $f(x)$. One gets

$$g(y) = f(x)\left|\frac{dx}{dy}\right| = f(x)\left|\frac{dy}{dx}\right|^{-1} = 1, \tag{3.94}$$

and we conclude that $g(y)$ is a uniform PDF. This implies that any PDF $f(x)$ of a continuous variable may in principle be expressed (and calculated) in terms of a uniform distribution. This property is particularly important for the generation of random numbers discussed in §13.3.3.

3.4.4 Example: Position Resolution of a Basic Particle Hit Detector

Magnetic spectrometers are used to measure the momentum of charged particles produced by nuclear collisions. In the bending plane, charged particles follow circular trajectories. The radius of these trajectories, or tracks, is proportional to the momentum of the particles and inversely related to the magnetic field (B). A precise knowledge of the B field therefore enables a measurement of the particle momenta. This requires a determination of the radius of curvature of the tracks, which is often accomplished by measuring "hit" positions along the track, in other words, positions where the particles traverse specific detector planes. In the most basic spectrometers, these detector planes are equipped with a collection of contiguous segments that can each detect the passage of a single particle. As a particle passes through a given segment or unit, it deposits energy and generates a signal. Assuming a particle can generate a signal or "hit" only in one segment, one finds that the hit position resolution is intrinsically limited by the width of the unit. The hit position is typically assumed to be the center of the segment. That means there is an uncertainty on the position proportional to the width of the segment. Let us estimate this error on the position by assuming that the particle has an a priori equal probability of crossing the unit anywhere along its width:

$$P(x) = \frac{1}{\Delta L}, \tag{3.95}$$

where ΔL is the actual width of the unit. By construction, the average position, $\langle x \rangle$, shall be the center of the unit. We can then use the standard deviation of the uniform distribution

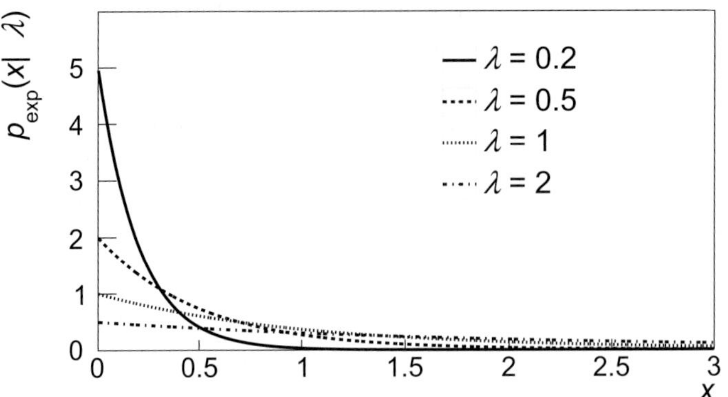

Fig. 3.7 Exponential distribution $p_{exp}(x|\lambda)$ for selected values of λ.

as an estimate of the error, σ_x, on the position:

$$\sigma_x = \text{Var}[x] = \int_0^{\Delta L} (x - \langle x \rangle)^2 P(x)\, dx = \frac{\Delta L}{\sqrt{12}}. \tag{3.96}$$

This simple estimate of the error can be used in a wide variety of contexts to estimate errors on measurements where the probability density is (approximately) uniform.

3.5 Exponential Distribution

3.5.1 Definition

The **exponential probability distribution** is defined as follows:

$$p_{\text{exp}}(x|\lambda) = \begin{cases} \lambda^{-1} \exp(-x/\lambda) & \text{for } x \geq 0, \\ 0 & \text{for } x < 0. \end{cases} \tag{3.97}$$

It is determined by a single parameter, λ, which sets the falloff rate of the exponential, as illustrated in Figure 3.7.

The factor λ^{-1} ensures the proper normalization of $p_{\text{exp}}(x|\lambda)$ as a PDF. Indeed, calculating the cumulative distribution function of $p_{\text{exp}}(x|\lambda)$, one finds

$$P_{\text{exp}}(x|\lambda) = \int_{-\infty}^{x} p_{\text{exp}}(x'|\lambda)\, dx', \tag{3.98}$$

$$= \begin{cases} 1 - \exp(-x/\lambda) & \text{for } x \geq 0, \\ 0 & \text{for } x < 0, \end{cases} \tag{3.99}$$

which converges to unity in the limit $x \to \infty$.

3.5.2 Interesting Features of the Exponential Distribution

As we saw in §3.3, the exponential distribution naturally arises in the context of stochastic phenomena where the probability of an event to occur in a given time interval $[t, t+dt]$ has a fixed value pdt where p is a constant value independent of t, and such that the occurrence of an event at time t has no influence on the occurrence of events taking place in any other time intervals before or after the time t. The event rate may then be described in terms of a constant τ^{-1} that specifies the average rate of events per unit of time. The Poisson distribution then corresponds to the probability of observing N events in a given time interval Δt:

$$p(N|\Delta t) = \frac{1}{N!} (t/\tau)^N e^{-t/\tau}, \tag{3.100}$$

and the probability of having no events, that is, $N = 0$, during an interval Δt is given by the exponential distribution $e^{-t/\tau}$.

The exponential PDF is useful in a variety of contexts. It can, for instance, describe the probability of observing the decay of a specific nuclear isotope as a function of time, t. The quantity λ then corresponds to the lifetime of the isotope and is usually noted τ instead. The exponential PDF may also be used to represent the probability of weakly decaying particles (such as D mesons) to decay at a certain distance from their point of production.

It is interesting to consider the time interval T between two events relative to the time interval Δt. Since no events occur between the two events for $T > \Delta t$, we can write

$$p(T > \Delta t) = e^{-\Delta t/\tau}. \tag{3.101}$$

Additionally, given the expression of the CDF, Eq. (3.98), one can also conclude that the probability of observing two events separated by a time T smaller than Δt is given by

$$p(T \le \Delta t) = 1 - e^{-\Delta t/\tau}, \tag{3.102}$$

since the right-hand side of Eq. (3.102) is the complement of the probability $p(T > \Delta t)$.

The exponential distribution is often said to have "no memory." For instance, if no event occurred until a time t_1, the probability of having no event in a subsequent time t_2 is independent of t_1. For a fixed value t_1, this may be written

$$p(T > t_1 + t_2 | T > t_1) = \frac{e^{-(t_1+t_2)/\tau}}{e^{-t_1/\tau}} = e^{-t_2/\tau} = p(T > t_2), \tag{3.103}$$

which indeed implies that the probability of $T > t_1 + t_2$ does not depend on t_1. Since the probability of an event taking place in a certain time interval does not depend on what happened during prior time intervals, the exponential can hence be said to have no memory. This should come as no surprise, however, as per the derivation of the distribution presented in §3.3.1 in terms of independent events.

Table 3.5 Properties of the exponential distribution

Function	$p_{\exp}(x\vert\lambda) = \lambda^{-1}\exp(-x/\lambda)$
Characteristic Fct	$\phi(t) = (1 - i\cdot\lambda t)^{-1}$
Moment Gen Fct	$M(t) = (1 - \lambda t)^{-1}$
Mode	0
Median	$\ln(2)\lambda$
Moments	
1st	$\mu = \mu' = \lambda$
2nd	$\mu'_2 = 2\lambda^2$
nth	$\mu'_n = \lambda^n n!$
Central moments	
2nd	$\sigma^2 \equiv \mu_2 = \lambda^2$
nth	$\mu_n = \Gamma(n+1,-1)\lambda^n/e$
Skewness excess	$\gamma_1 = 2$
Kurtosis excess	$\gamma_2 = 6$

3.5.3 Moments of the Exponential Distribution

The expectation value of x is (see Problem 3.14)

$$\mu = \mathrm{E}[x] = \frac{1}{\lambda}\int_0^\infty x e^{-x/\lambda}\,dx = \lambda, \tag{3.104}$$

and the variance is given by

$$\mathrm{Var}[x] = \frac{1}{\lambda}\int_0^\infty (x-\lambda)^2 e^{-x/\lambda}\,dx = \lambda^2. \tag{3.105}$$

Other moments of the distribution are listed in Table 3.5. More generally, one can show (see Problem 3.14) that the moments of the exponential distribution are given by

$$\mu'_n = \lambda^n n! \tag{3.106}$$

whereas the central moments are given by

$$\mu_n = \frac{\Gamma(n+1,-1)}{e\lambda^n}, \tag{3.107}$$

where $\Gamma(a, b)$ is an incomplete gamma function.

3.5.4 Example: Radiocarbon Decay and Half-Life

A convenient and useful way to visualize the exponential PDF is by means of the concept of half-life. Consider carbon-14 (^{14}C), an unstable carbon isotope consisting of six protons and eight neutrons. Commonly referred to as radiocarbon, ^{14}C decays spontaneously into a nucleus of nitrogen-14 (^{14}N, which consists of seven protons and seven neutrons) by emission of an electron and an anti-neutrino. This decay is governed by the weak nuclear force. It involves the transmutation of one of the eight neutrons of the ^{14}C nucleus into a proton, the production of an electron (which conserves the total electric charge), and the emission of a neutrino (which conserves energy and momentum):

$$^{14}_{6}\text{C} \rightarrow ^{14}_{7}\text{N} + e^- + \overline{\nu}_e \tag{3.108}$$

The isotope ^{14}C is naturally present in all organic materials. It forms the basis of the radio-carbon dating method used to determine the age of archaeological, geological, and hydro-geological artifacts. The isotope ^{14}C has a half-life ($t_{1/2}$) of 5,730 years. This means that given a sample of n ^{14}C nuclei, it should take (on average) 5,730 years for half of these nuclei ($n/2$) to be decayed. After an additional 5,730 years, one-half of the remaining half would have decayed, thereby leaving us with one-fourth ($n/4$) of the original amount. After three half-lives, one-eighth remains; after four half-lives, one-sixteenth is left, and so on. This can be written

$$n_{\text{undecayed}}(t) \sim \left(\frac{1}{2}\right)^{t/t_{1/2}}, \tag{3.109}$$

or equivalently

$$n_{\text{undecayed}}(t) \sim \exp\left(\frac{-t}{t_{1/2}/\ln 2}\right). \tag{3.110}$$

The quantity $t_{1/2}/\ln 2$ is readily identified as the mean life, τ, of the isotope:

$$\tau = t_{1/2}/\ln 2 = t_{1/2}/0.693147. \tag{3.111}$$

The half-life of ^{14}C, 5,730 years, makes it ideal for "dating" archeological samples that are a few to several thousand of years old. In practice, the technique appears to be limited to about 10 half-lives, corresponding to roughly 60,000 years.

3.6 Gamma Distribution

3.6.1 Definition

The **gamma distribution** (not to be confused with the gamma function, $\Gamma(x)$, see later) is encountered in a wide variety of situations and fields including econometrics, life sciences, as well as physical sciences. As such, it is commonly represented in terms of (at least) three distinct parameterizations in the scientific literature. In this book, we will use the notation most popular in Bayesian statistics, which involves a shape parameter, α, and a rate parameter β:

$$p_\gamma(x|\alpha,\beta) = \frac{\beta^\alpha x^{\alpha-1} e^{-x\beta}}{\Gamma(\alpha)} \quad \text{for } x \geq 0 \;\; \alpha,\beta > 0, \tag{3.112}$$

where $\Gamma(x) = \int_0^\infty z^{n-1}\exp(-z)dz$, is the **gamma function** for $x > 0$. One readily verifies that the exponential distribution (§3.5) and the χ^2-distribution (§3.9) are two special cases

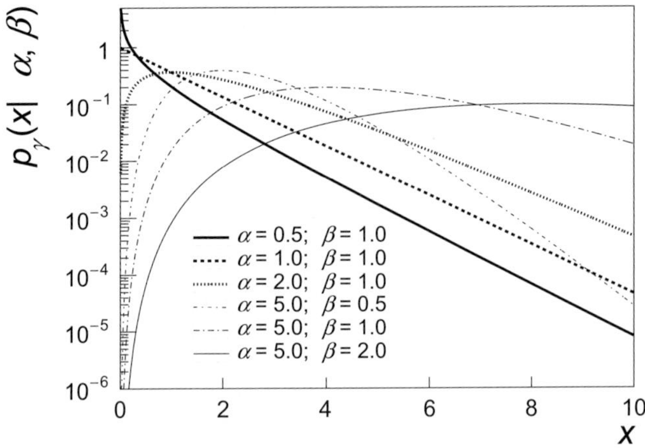

Fig. 3.8 Gamma distributions $p_\gamma(x|\alpha, \beta)$ for selected values of α and β.

of the gamma distribution, while integer values for the shape parameter α yield **Erlang distributions**, which corresponds to sums of α exponentially distributed random deviates, each with a mean $1/\beta$.

As illustrated in Figure 3.8, the gamma distribution spans a wide range of shapes and forms determined by the parameters α and β. It is thus quite convenient for modeling the PDF of random variables and is used in a wide variety of applications and systems.

Integration of p_γ in the interval $[0, x]$ yields the CDF denoted:

$$P_\gamma(x|\alpha, \beta) = \int_0^x p_\gamma(z|\alpha, \beta)\, dz = \frac{\gamma(\alpha, \beta x)}{\Gamma(\alpha)} \tag{3.113}$$

where $\gamma(\alpha, \beta x)$ is known as **incomplete gamma function**. The CDF function $P_\gamma(x|\alpha, \beta)$ is commonly called **regularized gamma function**. It can be expressed in closed form if α is a positive integer. Indeed, since p_γ vanishes, by construction, both at $x = 0$ and for $x \to \infty$, integration by part yields:

$$P_\gamma(x|\alpha, \beta) = 1 - \sum_{k=0}^{\alpha-1} \frac{(\beta x)^k}{k!} \exp(-\beta x) \tag{3.114}$$

$$= \exp(-\beta x) \sum_{k=\alpha}^{\infty} \frac{(\beta x)^k}{k!}, \tag{3.115}$$

where in the second line, we substituted the complement of the partial Taylor series of $\exp(\beta x)$ found in the first line.

Table 3.6 Properties of the gamma distribution

Function	$p_\gamma(x\|\alpha,\beta) = \frac{\beta^\alpha x^{\alpha-1}e^{-x\beta}}{\Gamma(\alpha)}$
Characteristic Fct	$\phi(t) = (1 - it/\beta)^{-\alpha}$
Moment Gen Fct	$M(t) = (1 - t/\beta)^{-\alpha}\ t < \beta$
Mode	$(\alpha - 1)/\beta$ for $\alpha \geq 1$
1st Moment	$\mu = \mu' = \alpha/\beta$
2nd Central moment	$\sigma^2 \equiv \mu_2 = \alpha/\beta^2$
Skewness excess	$\gamma_1 = 2/\sqrt{\alpha}$
Kurtosis excess	$\gamma_2 = 6/\alpha$

3.6.2 Moments of the Gamma Distribution

Moments and other basic properties of the gamma distribution are listed in Table 3.6.

3.7 Beta Distribution

3.7.1 Definition

The **beta distribution** is defined as

$$p_\beta(x|\alpha,\beta) = \frac{\Gamma(\alpha+\beta)}{\Gamma(\alpha)\Gamma(\beta)}x^{\alpha-1}(1-x)^{\beta-1}, \tag{3.116}$$

over the domain [0, 1] for real exponent values $\alpha, \beta > 0$ and as such may be regarded as an extension of the binomial distribution.

3.7.2 Properties of the Beta Distribution

Figure 3.9 illustrates how judicious choices of the **shape parameters** α and β produce functions that span a rather wide range of functional shapes in the domain $0 < x < 1$. As such, the beta distribution is extremely convenient and often used toward the expression of prior probabilities in Bayesian statistics (§7.2.2).

Remarkably, one finds that if a likelihood function is a binomial distribution, multiplying the prior by the likelihood yields another beta distribution. The beta distribution is then said to be a **conjugate prior** of the binomial distribution. In general, in cases in which the prior and posterior are members of the same family of distributions, with a given type of likelihood function, ones considers the prior and the likelihood to be **conjugate distributions**. Other cases of conjugate distributions include the normal distribution when used both as prior and likelihood functions, as well as the gamma distribution when used a prior for a Poisson likelihood. A more in-depth discussion of conjugate priors is presented in §7.3.3.

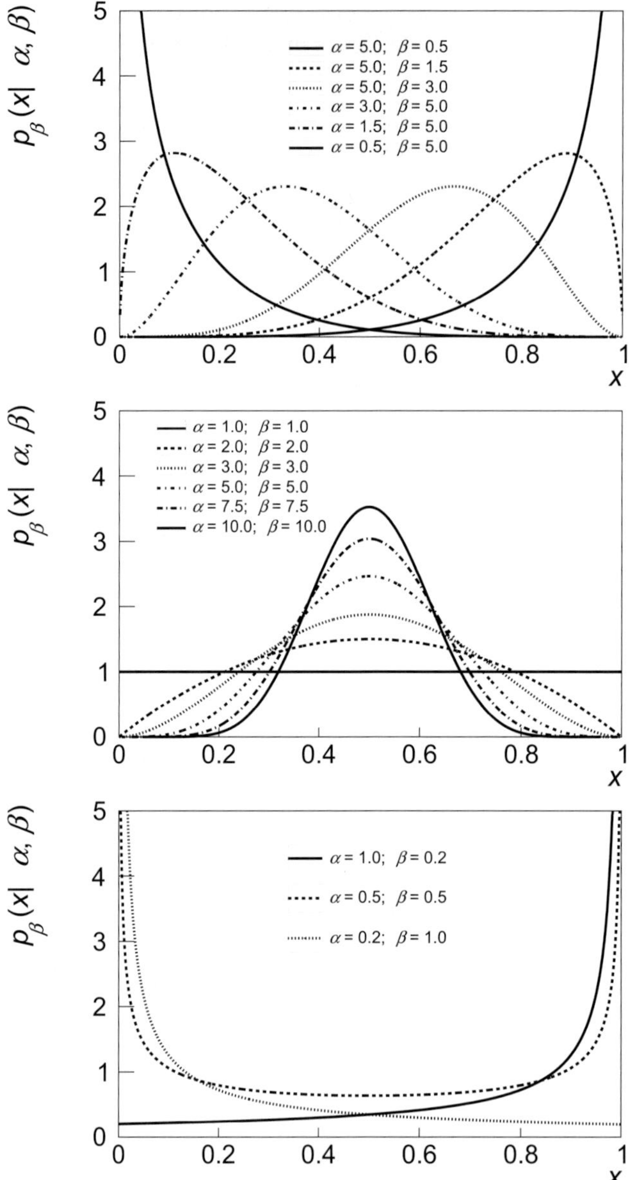

Fig. 3.9 Examples of the beta distribution $p_\beta(x|\alpha, \beta)$ for selected values of α and β.

Basic properties of the beta distribution are listed in Table 3.7. The moment-generating function of the beta distribution is calculated in terms of the expectation value of e^{tx}

$$M_X(\alpha, \beta, t) = \mathrm{E}[e^{tx}] = \int_0^1 e^{tx} p_\beta(x|\alpha, \beta)\, dx. \tag{3.117}$$

This expression corresponds to the confluent hypergeometric function

$$M_X(\alpha, \beta, t) = {}_1F_1(\alpha, \alpha + \beta, t), \tag{3.118}$$

Table 3.7 Properties of the beta distribution

Function	$p_\beta(x) = \frac{\Gamma(\alpha+\beta)}{\Gamma(\alpha)\Gamma(\beta)} x^{\alpha-1}(1-x)^{\beta-1}$
Characteristic Fct	${}_1F_1(\alpha, \alpha+\beta, it)$
Moment Gen Fct	${}_1F_1(\alpha, \alpha+\beta, t)$
Mode	$(\alpha-1)/(\alpha+\beta-2)$ for $\alpha, \beta > 1$
Median	$\approx (\alpha-1/3)/(\alpha+\beta-2/3)$ for $\alpha, \beta > 1$
1st Moment	$\mu = \alpha/(\alpha+\beta)$
2nd Central moment	$\text{Var}[x] = \alpha\beta/(\alpha+\beta)^2(\alpha+\beta+1)$
Skewness	$\gamma_1 = 2/(\beta-\alpha)\sqrt{\alpha+\beta+1}/(\alpha+\beta+2)\sqrt{\alpha\beta}$
Kurtosis excess	$\gamma_2 = 6\left[(\alpha-\beta)^2(\alpha+\beta+1) - \alpha\beta(\alpha+\beta+2)\right]/$ $\alpha\beta(\alpha+\beta+2)(\alpha+\beta+3)$

which may be written as an infinite series [5]

$$M_X(\alpha, \beta, t) = 1 + \sum_{k=1}^{\infty} \left(\prod_{j=0}^{k-1} \frac{\alpha + j}{\alpha + \beta + j} \right) \frac{t^k}{k!}. \tag{3.119}$$

Similarly, the characteristic function of the beta distribution may be shown to be equal to

$$\phi_X(\alpha, \beta, t) = {}_1F_1(\alpha, \alpha+\beta, it) \tag{3.120}$$

$$= 1 + \sum_{k=1}^{\infty} \left(\prod_{j=0}^{k-1} \frac{\alpha + j}{\alpha + \beta + j} \right) \frac{(it)^k}{k!}. \tag{3.121}$$

3.8 Dirichlet Distributions

3.8.1 Definition

Much like the beta distribution may be regarded as an extension of the binomial distribution, the **Dirichlet distribution** may be viewed as an extension of the multinomial distribution. It is defined as a function of k continuous variables x_i and α_i according to

$$p_D(\vec{x}|\vec{\alpha}) = \frac{\Gamma(\alpha_1 + \alpha_2 + \cdots + \alpha_k)}{\Gamma(\alpha_1)\Gamma(\alpha_2)\cdots\Gamma(\alpha_k)} x_1^{\alpha_1 - 1} x_2^{\alpha_2 - 1} \cdots x_k^{\alpha_k - 1}, \tag{3.122}$$

with the following constraints:

$$\vec{x} = (x_1, x_2, \ldots, x_k) \qquad 0 < x_i < 1; \qquad \sum_{i=1}^{k} x_i = 1,$$

$$\vec{\alpha} = (\alpha_1, \alpha_2, \ldots, \alpha_k) \qquad \alpha_i > 0; \qquad \sum_{i=1}^{k} \alpha_i \equiv \alpha,$$

Table 3.8 Properties of the Dirichlet distribution

Function	$f(\vec{x}\vert\vec{\alpha}) = \frac{\Gamma(\alpha_1+\alpha_2+\cdots+\alpha_k)}{\Gamma(\alpha_1)\Gamma(\alpha_2)\cdots\Gamma(\alpha_k)} x_1^{\alpha_1-1} x_2^{\alpha_2-1}\cdots x_k^{\alpha_k-1}$
Mode	$x_i^{Mode} = (\alpha_i - 1)/(\sum_{j=1}^k \alpha_j - k)$ for $\alpha_i > 1$
Mean	$E[x_i] = \alpha_i / \sum_{j=1}^k \alpha_j$
Variance	$\text{Var}[x_i] = \alpha_i(\alpha - \alpha_i)/\alpha^2(\alpha+1)$
Covariance	$\text{Cov}[x_i, x_j] = -\alpha_i\alpha_j/\alpha^2(\alpha+1)$ for $i \neq j$
Note:	$\alpha = \sum_{i=1}^k \alpha_i$

where $\Gamma(z)$ is the regular gamma function. The Dirichlet distribution simplifies to a beta distribution for $k = 2$. Also note that the constraint $\sum_{i=1}^k x_i = 1$ implies the distribution is effectively $(k-1)$ dimensional.

The Dirichlet distribution find applications, much like the beta distribution, in the context of Bayesian inference with finite statistics (see, e.g., §12.3.8).

Marginalization of the Dirichlet distribution, that is, reduction to one variable x_i, yields the beta distribution with parameters $r = \alpha_i$ and $s = \alpha - \alpha_i$.

$$p_\beta(x_i|r = \alpha_i, s = \alpha - \alpha_i) = \frac{x_i^{\alpha_i-1}(1-x_i)^{\alpha-\alpha_i-1}}{\beta(\alpha_i, \alpha - \alpha_i)}. \tag{3.123}$$

3.8.2 Properties of the Dirichlet Distribution

The properties of the Dirichlet distribution are summarized in Table 3.8.

3.9 Gaussian Distribution

3.9.1 Definition

The **Gaussian distribution** is defined in terms of a continuous variable x in the domain $[-\infty, \infty]$:

$$p_G(x|\mu, \sigma^2) = \frac{1}{\sqrt{2\pi}\sigma} \exp\left[-\frac{(x-\mu)^2}{2\sigma^2}\right]. \tag{3.124}$$

The distribution is given many names: statisticians and mathematicians use the term **normal distribution** while social scientists commonly refer to it as the **bell curve** because of its curved flaring shape. Physicists typically prefer the appellation **Gaussian distribution** in honor of Carl Friedrich Gauss[3] (1777–1855). Indeed, although the distribution was likely "discovered" by De Moivre as an approximation of the binomial distribution, and

[3] German mathematician and scientist who made major contributions in number theory, statistics, analysis, differential geometry, geodesy, geophysics, and physics. Gauss is considered by many as the greatest mathematician since antiquity.

subsequently used by Laplace in 1783 to study measurement errors, its first extensive application in a scientific context is commonly attributed to Gauss, who, in 1809, used the distribution toward the analysis of astronomical data.

The name and qualifier **normal** is perhaps best suited for the distribution. Indeed, one finds that many distributions tend to the normal distribution in the large N limit. One also finds that the sum of a very large number of random variables has a normal distribution. This fact is embodied in the central limit theorem, which we discussed in §2.10.4. The normal distribution thus, indeed, describes that which is common or "normal." But given this text is targeted primarily toward physical scientists, we will hereafter mostly use the Gaussian distribution appellation commonly in vogue in physics.

Note that statisticians often use the notation $X \sim \mathcal{N}(\mu, \sigma^2)$ to indicate that a variable X is distributed according to a normal distribution of mean μ and variance σ^2. The distribution may also be defined in terms of a **precision parameter** ξ defined as the multiplicative inverse of the variance

$$\xi = \frac{1}{\sigma^2}, \tag{3.125}$$

and is then written

$$p_G(x|\mu, \xi) = \sqrt{\frac{\xi}{2\pi}} \exp\left[-\frac{1}{2}\xi\,(x-\mu)^2\right]. \tag{3.126}$$

3.9.2 Moments of the Gaussian Distribution

We first note that while the function $p_G(x|\mu, \sigma^2)$ does not have an analytical primitive (indefinite integral), its definite integral over the domain $[-\infty, \infty]$ nonetheless exists and can be shown to equal unity. The function is thus properly defined and usable as a probability density. This stems from the inclusion of the normalization factor $\sqrt{2\pi}\sigma$ (see Problem 3.15):

$$\int_{-\infty}^{\infty} \exp\left[-\frac{(x-\mu)^2}{2\sigma^2}\right] dx = \sqrt{2\pi}\sigma. \tag{3.127}$$

One verifies, through integration by parts, that the two parameters μ and σ correspond to the mean and the variance of the distribution, respectively:

$$\mathrm{E}[x] = \frac{1}{\sqrt{2\pi}\sigma}\int_{-\infty}^{\infty} \exp\left[-\frac{(x-\mu)^2}{2\sigma^2}\right] x\,dx = \mu, \tag{3.128}$$

and

$$\mathrm{Var}[x] = \frac{1}{\sqrt{2\pi}\sigma}\int_{-\infty}^{\infty} \exp\left[-\frac{(x-\mu)^2}{2\sigma^2}\right] (x-\mu)^2\,dx = \sigma^2. \tag{3.129}$$

It is immaterial whether one considers σ, σ^2, or ξ, as the parameter determining the width of the distribution. However, given the Gaussian distribution is commonly used to

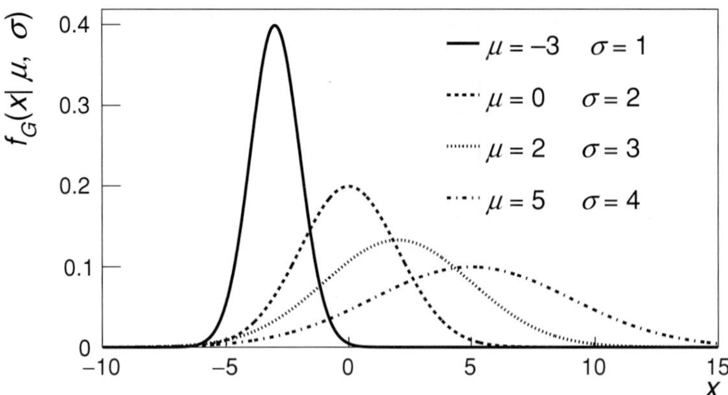

Fig. 3.10 Gaussian distribution $p_G(x|\mu, \sigma^2)$ for selected values of μ and σ^2.

describe fluctuations in the outcome of measurements, the precision label given to ξ is both practical and advantageous: indeed, as ξ increases, the width of the distribution decreases and thus reflects a measurement of greater precision. Evidently, σ^2 and σ are also both useful given they respectively represent the variance and standard deviation of the distribution.

Figure 3.10 displays examples of the Gaussian distribution for various combinations of μ and σ. Higher moments of the distribution are best calculated using the characteristic function (or moment-generating function) of the distribution listed in Table 3.9 (see Problem 3.16).

A special case of the Gaussian distribution, known as **standard normal distribution**, is obtained by setting $\mu = 0$ and $\sigma = 1$:

$$\varphi(x) \equiv p_G(x|\mu = 0, \sigma^2 = 1) = \frac{1}{\sqrt{2\pi}} \exp\left(-\frac{x^2}{2}\right). \tag{3.130}$$

The cumulative distribution of $\varphi(x)$, noted $\Phi(x)$, is

$$\Phi(x) = \frac{1}{\sqrt{2\pi}} \int_{-\infty}^{x} \exp\left(-x'^2/2\right) dx' = \frac{1}{2}\left[1 + \mathrm{erf}\left(x/\sqrt{2}\right)\right], \tag{3.131}$$

where $\mathrm{erf}(x)$ is the **error function**, defined as

$$\mathrm{erf}(x) = \frac{2}{\sqrt{\pi}} \int_0^x e^{-t^2} dt. \tag{3.132}$$

It is also useful to be familiar with the **complementary error function** $\mathrm{erfc}(x)$, defined as

$$\mathrm{erfc}(x) = 1 - \mathrm{erf}(x) = \frac{2}{\sqrt{\pi}} \int_x^{\infty} e^{-t^2} dt. \tag{3.133}$$

Table 3.9 Properties of the Gaussian (normal) distribution

Function	$p_G(x;\mu,\sigma^2) = (2\pi)^{-1/2}\sigma^{-1}\exp\left(-(x-\mu)^2/2\sigma^2\right)$
Characteristic Fct	$\phi(t) = \exp\left(i\mu t - \sigma^2 t^2/2\right)$
Moment Gen Fct	$M(t) = \exp\left(\mu t + \sigma^2 t^2/2\right)$
Mode	μ
Median	μ
Moments	
1st (Mean)	$\mu = \mu'$
2nd	$\mu_2' = \mu^2 + \sigma^2$
3rd	$\mu_3' = \mu^3 + 3\mu\sigma^2$
4th	$\mu_4' = \mu^4 + 6\mu^2\sigma^2 + 3\sigma^4$
Centered moments	
2nd (Variance)	σ^2
3rd	0
4th	σ^4
Skewness excess	$\gamma_1 = 0$
Kurtosis excess	$\gamma_2 = 1$

Additionally, note that the error function is odd in x:

$$\text{erf}(-x) = -\text{erf}(x) \tag{3.134}$$

The error function, erf(x), and the complementary error function, erfc(x), cannot be calculated analytically but are tabulated in many reference books. They are also provided in most mathematical software packages (e.g., Mathematica®, MATLAB®, and ROOT).

An arbitrary Gaussian distribution, $p_G(y;\mu,\sigma^2)$, function off a variable y with a mean μ and variance σ^2 can be transformed into $\varphi(x)$ through a simple change of variable $x = (y-\mu)/\sigma$. Accordingly, the cumulative integral,

$$P_G(y|\mu,\sigma^2) = \int_{-\infty}^{y} p_G(y'|\mu,\sigma^2)\,dy', \tag{3.135}$$

$$= \frac{1}{2}\left[1 + \text{erf}\left(\frac{x-\mu}{\sqrt{2}\sigma}\right)\right], \tag{3.136}$$

is equal to $\Phi(x)$.

Returning to the standard normal distributions, $\varphi(x)$, it is useful to consider the probability $p_{DS}(\Delta)$ that the variable x will be found within the double-sided (DS) interval $-\Delta \le x \le \Delta$. It is calculated according to

$$p_{DS}(\Delta) \equiv \frac{1}{\sqrt{2\pi}}\int_{-\Delta}^{\Delta}\exp(-t^2/2)\,dt, \tag{3.137}$$

$$= \Phi(\Delta) - \Phi(-\Delta), \tag{3.138}$$

$$= \text{erf}(\Delta/\sqrt{2}), \tag{3.139}$$

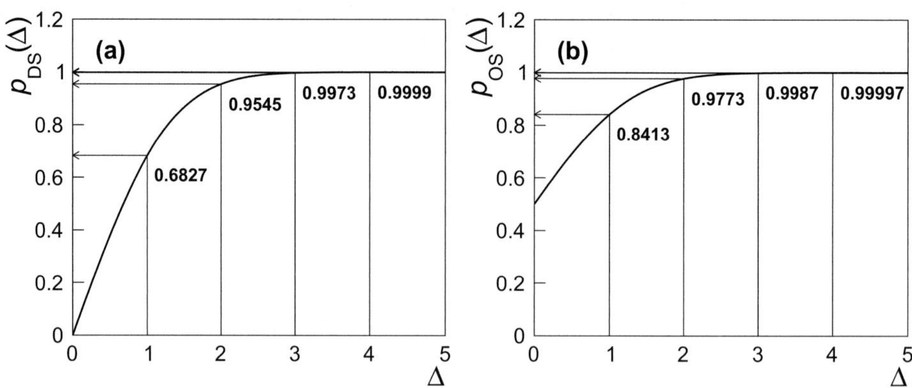

Fig. 3.11 Probability that a random variable x distributed according to a standard normal distribution is found within (a) $\pm\Delta$ of the origin and (b) in the interval $x \leq \Delta$.

shown in Figure 3.11a. The probability that x falls within one "sigma" ($\pm\sigma$) of the mean is 0.6827 and amounts to 0.9545, 0.9973, and 0.999937 for $\Delta = 2\sigma$, 3σ, and 4σ, respectively. One-sided intervals are also of interest when a variable is found at or near a physical limit and one wishes to know the probability of $x \leq \Delta$. One must then consider the one-sided probability $p_{\text{OS}}(\Delta)$ defined by

$$p_{\text{OS}}(\Delta) \equiv \Phi(\Delta) = \frac{1}{\sqrt{2\pi}} \int_{-\infty}^{\Delta} \exp(-t^2/2)\, dt \tag{3.140}$$

$$= \frac{1}{2} + \frac{1}{2}\text{erf}\left(\Delta/\sqrt{2}\right), \tag{3.141}$$

shown in Figure 3.11b.

The Gaussian distribution has applications in very many practical situations. This stems largely from the central limit theorem (discussed in §2.10.4), which states that the sum of n independent random numbers x_i, $i = 1, \ldots, n$ of means μ_i and variances σ_i^2 becomes a Gaussian distribution with mean $\mu = \sum_{i=1}^{n} \mu_i$ and variance $\sigma^2 = \sum_{i=1}^{n} \sigma_i^2$ in the large n limit. The theorem applies under general conditions regardless of the individuals PDFs of the variables x_i. We will discuss several applications of the central limit theorem and the Gaussian distribution throughout this book.

3.10 Multidimensional Gaussian Distribution

The **multidimensional Gaussian distribution** finds applications in systems involving simultaneously several variables with Gaussian distributions. For instance, consider a system involving m variables, $\vec{x} = (x_1, \ldots, x_m)$ of mean value $\vec{\mu} = (\mu_1, \ldots, \mu_i)$. The multidimensional Gaussian distribution may be written in its most general form:

$$p_{\text{MG}}(\vec{x}) = K \exp\left[-\frac{1}{2}(\vec{x} - \vec{\mu})^T \mathbf{Q} (\vec{x} - \vec{\mu})\right], \tag{3.142}$$

where K is a normalization constant, which we evaluate in the following, and $\mathbf{Q}$ is an $m \times m$ symmetric matrix. There exists a transformation

$$\vec{y} = \mathbf{A}\vec{x} \tag{3.143}$$

such that $\mathbf{Q}$ can be diagonalized. The matrix $\mathbf{A}$ is orthogonal, satisfies $\mathbf{A}^{-1} = \mathbf{A}^T$, and has a determinant $|\mathbf{A}| = 1$. It is then possible to reduce Eq. (3.142) to a simpler form involving a diagonal matrix $\mathbf{D} = \mathbf{AQA}^T$. Defining $\vec{\nu} = \mathbf{A}\vec{\mu}$, Eq. (3.142) may then be written

$$p_{\text{MG}}(\vec{x}) = K \exp\left[-\frac{1}{2}(\vec{y} - \vec{\nu})^T \mathbf{AQA}^T (\vec{y} - \vec{\nu})\right], \tag{3.144}$$

$$= K \exp\left[-\frac{1}{2}\left(\sum_{i=1}^{m} D_{ii}(y_i - \nu_i)^2\right)\right]. \tag{3.145}$$

The elements D_{ii} can readily be identified as the multiplicative inverse of the variances, $\sigma_{y,i}^{-2}$, of the variables y_i. The normalization constant K is thus equal to $(2\pi)^{-m/2} \prod \sigma_{y,i}^{-1} = (2\pi)^{-m/2}|D|^{-1/2}$. By extension, one also concludes that the matrix $\mathbf{Q}$ corresponds to the inverse, $\mathbf{V}^{-1}$, of the covariance matrix of variables x_i. By construction, one can thus write

$$\mathbf{V} = \mathbf{A}^T \mathbf{D}^{-1} \mathbf{A}, \tag{3.146}$$

and the multi-Gaussian distribution becomes

$$p_{\text{MG}}(\vec{x}) = \frac{1}{(2\pi)^{m/2}|V|^{1/2}} \exp\left[-\frac{1}{2}(\vec{x} - \vec{\mu})^T \mathbf{V}^{-1} (\vec{x} - \vec{\mu})\right], \tag{3.147}$$

where we also expressed the constant K in terms of the determinant of the covariance matrix $\mathbf{V}$.

As an example of application of the multidimensional Gaussian distribution, consider a Gaussian two-particle correlation function in pseudorapidity.[4] Let η_1 and η_2 represent the pseudorapidity of the two particles. Their relative and average rapidities are written $\Delta\eta = \eta_1 - \eta_2$ and $\bar{\eta} = 0.5(\eta_1 + \eta_2)$, respectively. Let us model the correlation strength between the two particles according to their relative and average rapidities using Gaussian distributions:

$$C(\Delta\eta, \bar{\eta}) \propto \exp\left(-\frac{\Delta\eta^2}{2\sigma_{\Delta\eta}^2}\right) \exp\left(-\frac{\bar{\eta}^2}{2\sigma_{\bar{\eta}}^2}\right), \tag{3.148}$$

where $\sigma_{\Delta\eta}$ and $\sigma_{\bar{\eta}}$ are the widths of the correlations in $\Delta\eta$ and $\bar{\eta}$, respectively. Let us define a matrix D as follows:

$$D = \begin{pmatrix} \sigma_{\Delta\eta}^{-2} & 0 \\ 0 & \sigma_{\bar{\eta}}^{-2} \end{pmatrix}. \tag{3.149}$$

[4] The notion of correlation function will be defined in Chapter 10. However, all that matters here is that the function considered depends on two Gaussian random variables.

Let us also define a matrix A to transform coordinates η_1 and η_2 into $\Delta\eta$ and $\bar{\eta}$ according to

$$\begin{pmatrix} \Delta\eta \\ \bar{\eta} \end{pmatrix} = \begin{pmatrix} 1 & -1 \\ \frac{1}{2} & \frac{1}{2} \end{pmatrix} \begin{pmatrix} \eta_1 \\ \eta_2 \end{pmatrix}. \tag{3.150}$$

The preceding correlation function may then be written in terms of η_1 and η_2 as follows:

$$C(\eta_1, \eta_2) = \frac{1}{2\pi\sigma_{\Delta\eta}\sigma_{\bar{\eta}}} \exp\left(-\frac{1}{2}\vec{\eta}^T Q\vec{\eta} \right), \tag{3.151}$$

with

$$\vec{\eta} = \begin{pmatrix} \eta_1 \\ \eta_2 \end{pmatrix}, \tag{3.152}$$

and

$$Q = A^T DA = \begin{pmatrix} \sigma_{\Delta\eta}^{-2} + \frac{1}{4}\sigma_{\bar{\eta}}^{-2} & -\sigma_{\Delta\eta}^{-2} + \frac{1}{4}\sigma_{\bar{\eta}}^{-2} \\ -\sigma_{\Delta\eta}^{-2} + \frac{1}{4}\sigma_{\bar{\eta}}^{-2} & \sigma_{\Delta\eta}^{-2} + \frac{1}{4}\sigma_{\bar{\eta}}^{-2} \end{pmatrix}, \tag{3.153}$$

which is the inverse of the covariance matrix of η_1 and η_2.

3.11 Log-Normal Distribution

3.11.1 Definition

The **log-normal distribution** is defined as

$$p_{\ln}(x|\mu, \sigma) = \frac{1}{\sqrt{2\pi}\sigma} \frac{1}{x} \exp\left(-\frac{(\ln x - \mu)^2}{2\sigma^2} \right) \qquad x > 0, \tag{3.154}$$

where μ and σ^2 are the mean and variance, respectively, of the distribution of the random variable $y = \ln x$ which, by construction, has a normal distribution. The cumulative distribution function is obtained by integration:

$$P_{\ln}(x|\mu, \sigma) = \frac{1}{2}\text{erfc}\left(-\frac{\ln x - \mu}{\sqrt{2}\sigma} \right) = \Phi\left(-\frac{\ln x - \mu}{\sigma} \right), \tag{3.155}$$

where erfc and Φ are the complementary error function, and the standard normal CDF, respectively.

3.11.2 Properties of the Log-Normal Distribution

The variables μ and σ^2 are clearly not the mean and variance of the log-normal distribution. Indeed, direct calculation shows (see Problem 3.23) the mean and variance of $p_{\ln}(x|\mu, \sigma^2)$

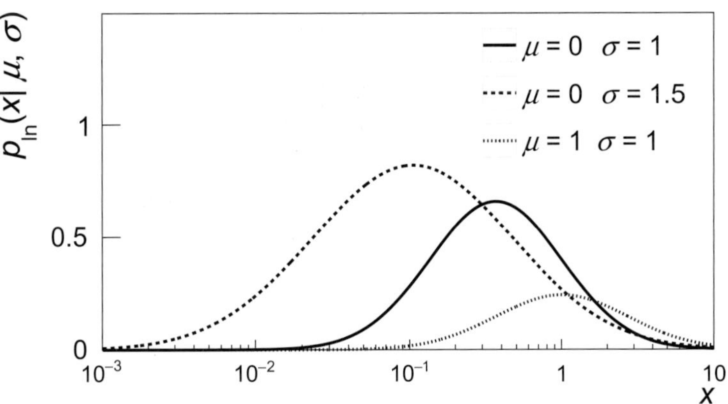

Fig. 3.12 Log-normal distribution $p_{\ln}(x; \mu, \sigma)$ with selected values of μ and σ.

are as follows:

$$\mathrm{E}[x] = \exp\left(\mu + \frac{1}{2}\sigma^2\right), \tag{3.156}$$

$$\mathrm{Var}[x] = \left[\exp(\sigma^2) - 1\right]\exp\left(2\mu + \sigma^2\right). \tag{3.157}$$

Other moments and properties of the log-normal distribution are presented in Table 3.10. Examples of the log-normal distribution are shown in Figure 3.12.

3.11.3 Generalized Central Limit Theorem

The log-normal distribution is of particular interest with regards to random variables that are the product of many random variables. For instance, consider n random variables z_1, $z_2, \ldots, z_n$ and their product:

$$x \equiv \prod_{i=1}^{n} z_i. \tag{3.158}$$

We introduce variables w_i defined such that $z_i = e^{w_i}$, and $y \equiv \sum_{i=1}^{n} w_i$. By construction, one has $x = e^y$, and the probability $p(x)\,dx$ equals $p(y)\,dy$. This can also be written

$$p(x) \equiv p(y)\left|\frac{dy}{dx}\right|. \tag{3.159}$$

For a reasonably large n, one can apply the central limit theorem and conclude y is Gaussian distributed with mean $\bar{y} \equiv \sum_{i=1}^{n} \bar{w}_i$ and variance $\sigma^2 \equiv \sum_{i=1}^{n} \sigma_{w,i}^2$. The variable x consequently has a log-normal distribution. Indeed, given $\frac{dy}{dx} = \frac{1}{x}$, one obtains:

$$p(x) \equiv \frac{1}{\sqrt{2\pi}\sigma}\frac{1}{x}\exp\left(-\frac{(\ln x - \mu)^2}{2\sigma^2}\right). \tag{3.160}$$

Table 3.10 Properties of the log-normal distribution

Function	$p_{\ln}(x; \mu, \sigma) = (2\pi)^{-1/2}(\sigma x)^{-1} \exp\left(-(\ln x - \mu)^2 / 2\sigma^2\right)$
Characteristic Fct	$\phi(t) = \frac{1}{2} + \frac{1}{2}\text{erf}((\ln x - \mu)/\sqrt{2\sigma^2})$
Mode	$\exp(\mu - \sigma^2)$
Median	$\exp(\mu)$
Mean	$\mu = \exp\left(\mu + \frac{1}{2}\sigma^2\right)$
Variance	$\sigma^2 = (\exp(\sigma^2) - 1)\exp(2\mu + \sigma^2)$
Skewness excess	$\gamma_1 = \left(\exp(\sigma^2) + 2\right)\sqrt{\exp\sigma^2 - 1}$
Kurtosis excess	$\gamma_2 = \exp(4\sigma^2) + 2\exp(3\sigma^2) + 3\exp(2\sigma^2) - 6$

The central limit theorem (§2.10.4) can then be "generalized" as follows:

1. The sum of a large number of random variables has a normal distribution.
2. The product of a large number of random variables has a log-normal distribution.

The generalized central limit theorem is applicable, in particular, to a succession of n gain (or attenuation) devices with gains (attenuation factors) $z_i = e^{a_i}$, where the coefficients a_i, which determine the gains, are approximately normally distributed.

The characteristic function and lowest moments of the log-normal distribution are presented in Table 3.10.

3.12 Student's *t*-Distribution

3.12.1 Definition

Student's *t*-distribution, commonly known as the *t-distribution*, involves a family of continuous probability distributions that arise when estimating the average of populations known to have a normal distribution.

Suppose a random variable, X, is known to be normally distributed but with unknown mean and variance. One can then proceed to estimate the mean of the distribution based on a sample consisting of n measurements. As we shall see in Chapter 4, the mean of this sample provides an unbiased estimator of the mean of the distribution. However, the sample mean is expected to deviate, sample by sample, from the actual mean of the distribution. We will show in §6.4 that the probability distribution of the sample mean is given by the t-distribution:

$$p_t(t|\nu) = \frac{\Gamma\left(\frac{\nu+1}{2}\right)}{\sqrt{\nu\pi}\,\Gamma\left(\frac{\nu}{2}\right)}\left(1 + \frac{t^2}{\nu}\right)^{-\frac{\nu+1}{2}}, \tag{3.161}$$

where $\nu = n - 1$ is the number of degrees of freedom being sampled.

Table 3.11 Properties of Student's t-distribution

Function	$p_t(\nu) = \frac{\Gamma(\frac{\nu+1}{2})}{\sqrt{\nu\pi}\Gamma(\frac{\nu}{2})}\left(1+\frac{t^2}{\nu}\right)^{-\frac{\nu+1}{2}}$
Characteristic Fct	$\phi(t) = \frac{K_{\nu/2}(\sqrt{\nu}\|t\|)(1+\nu/2)}{\Gamma(\nu/2)2(\nu/2-1)}$ for $\nu > 0$
	$K_{\nu/2}(x)$ is a Bessel function
Moment Gen Fct	Undefined
Mode	0
Median	0
Mean	$\mu = 0$
Variance	$\sigma^2 = \frac{\nu}{\nu-2}$ for $\nu > 2$
Skewness excess	$\gamma_1 = 0$
Kurtosis excess	$\gamma_2 = \frac{6}{\nu-4}$ for $\nu > 4$

The t-distribution is particularly useful in the context of Student's t-test, which quantifies the likelihood that a measured value lies within a given confidence interval of a predicted value or the results of other measurements (discussed in §6.4). It is symmetric about the origin, $t = 0$, and its moments, relatively simple to calculate, are listed in Table 3.11.

Figure 3.13 presents examples of the t-distribution for selected values of the number of degrees of freedom, ν, compared to a standard normal distribution (thin solid line). For small values of ν, the t-distribution has tails that slowly taper off to zero at infinity. The distribution, however, eventually converges to the normal distribution for a large (infinite) number of degrees of freedom, that is, when the mean is evaluated from very large samples.

3.12.2 Properties of Student's t-Distribution

Properties of the Student's t-distribution are presented in Table 3.11.

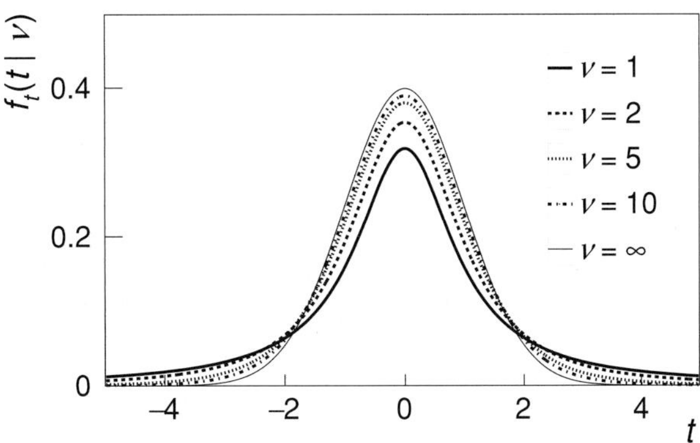

Fig. 3.13 Student's t-distribution for selected values of the number of degrees of freedom, ν, compared to the standard normal distribution (thin solid line).

3.13 Chi-Square Distribution

3.13.1 Definition

The χ^2 **distribution** arises in the context of data modeling with the maximum likelihood and least-squares methods discussed in §5.2. We will see that given a dataset consisting of n points (x_i, y_i), $i = 1, \ldots, n$, and a model $f(x|\vec{\theta})$, dependent on m unknown parameters $\vec{\theta} = \{\theta_1, \theta_2, \ldots, \theta_m\}$, one can evaluate the goodness of a fit based on the χ^2 function defined as

$$\chi^2 = \sum_{i=1}^{n} \frac{\left[y_i - f(x_i|\vec{\theta})\right]^2}{\sigma_i^2}, \tag{3.162}$$

where σ_i^2 are the errors on the n measurements y_i. For measurements governed by Gaussian statistics, that is, when repeated measurements at a given value of x yield values of y distributed according to a normal distribution, the PDF of the χ^2 values is given by the χ^2-distribution defined as

$$p_{\chi^2}(z|n) \equiv \frac{1}{2^{n/2}\Gamma(n/2)} z^{n/2-1} e^{-z/2} \qquad \text{for } n = 1, 2, \ldots. \tag{3.163}$$

We will derive this distribution in detail in §3.13.3. In this section, let us focus on its properties and interpretation. We first note that the χ^2 distribution of the continuous variable z evaluated in the domain $0 \le z < \infty$ is a special case of the gamma distribution (§3.6) with $\beta = 1/2$ and $\alpha = n/2$. In this context, the variable n is an integer called **number of degrees of freedom**. The evaluation of the gamma function, $\Gamma(x)$, is thus restricted to x taking integer and half-integer values only. For integers $n \ge 1$, one has $\Gamma(n) = (n-1)!$ whereas for odd values of n, the evaluation of $\Gamma(n/2)$ is achieved with the recursive formula

$$\Gamma(x+1) = x\Gamma(x) \qquad \text{with } \Gamma(1/2) = \sqrt{\pi}. \tag{3.164}$$

The number of degrees of freedom n is a measure of the number of independent variables in a system or model. By definition, an unconstrained dataset containing n independent random variables has n degrees of freedom. However, imagine fixing the mean of the n values y_i prior to their measurements. If the $n-1$ first values are obtained randomly and independently, the nth value becomes automatically constrained by the a priori fixed value of the mean: it cannot be considered random. The dataset has consequently only $n-1$ degrees of freedom. Now, imagine fixing the second moment of the distribution as well. It is easy to verify that once the $n-2$ first measured values have been obtained, the last two are automatically determined by the first $n-2$ values and the specified values of the first and second moments of the distribution. Specifying additional moments (or other linearly independent parameters of the model) thus reduces the number of degrees of freedom: each additional model parameter reduces the number of degrees of freedom by one unit. The number of degrees of freedom of a dataset consisting of n random values fitted to a model involving m parameters is thus $N - m$.

Table 3.12 Properties of the χ^2 distribution

Function	$p_{\chi^2}(z; n) \equiv \frac{1}{2^{n/2}\Gamma(n/2)} z^{n/2-1} e^{-z/2}$
Characteristic function	$(1-2it)^{n/2}$
Mode	$\max(n-2, 0)$
Median	$\approx n(1-2/9n)^3$
Mean	$\mu = n$
Variance	$\sigma^2 = 2n$
Skewness excess	$\gamma_1 = \sqrt{8/n}$
Kurtosis excess	$\gamma_2 = 12/n$

3.13.2 Properties of the Chi-square Distribution

The mean and variance of the χ^2 distribution may be calculated by integration by parts (see Problem 3.18). One gets

$$\mathrm{E}[z] = \int_0^\infty z \frac{1}{2^{n/2}\Gamma(n/2)} z^{n/2-1} e^{-z/2}\, dz = n, \tag{3.165}$$

and

$$\mathrm{Var}[z] = \int_0^\infty (z-n)^2 \frac{1}{2^{n/2}\Gamma(n/2)} z^{n/2-1} e^{-z/2}\, dz = 2n. \tag{3.166}$$

Other properties of the χ^2 distribution are listed in Table 3.12.

The χ^2 distribution is illustrated in Figure 3.14a for various values of the number of degrees of freedom n. One observes that all distributions exhibit a peak near n. As shown in Figure 3.14b, this translates in having a value of the χ^2 per degrees of freedom,

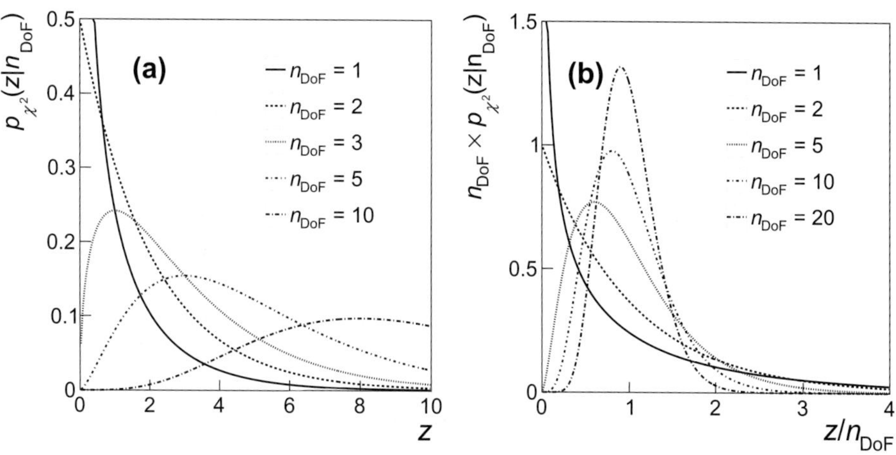

Fig. 3.14 (a) χ^2 probability distribution for selected numbers of degrees of freedom n_{DoF}; (b) scaled χ^2 probability distribution plotted as a function of the χ^2 per degrees of freedom, χ^2/n_{DoF}, for selected values of n_{DoF}.

$\chi^2/\text{DoF} = \chi^2/n$, peaked near unity. Indeed, note that while the mean of the χ^2/DoF distribution is unity for all values of n, the mode of the distribution differs from unity, with the largest deviations for a small number of degrees of freedom, n.

It is useful to consider the addition of two independent, χ^2-distributed, random variables Z_1 and Z_2 with n_1 and n_2 degrees of freedom, respectively. We will show in §3.13.4 that the sum $Z = Z_1 + Z_2$ is characterized by a χ^2 distribution with $n_1 + n_2$ degrees of freedom. This implies that a merged dataset consisting of two independent datasets behaves as one would intuitively expect, that is, the variable Z shall have a χ^2 distribution determined by the total number of degrees of freedom of the two sets. However, this conclusion does not apply if the two datasets are not mutually independent. This is the case, in particular, if a subset of the first dataset is also included in the second dataset.

3.13.3 Derivation of the χ^2 Distribution

In order to derive the expression of the PDF $p_{\chi^2}(z|n)$ of a χ^2 variable with n degrees of freedom, we appeal to the addition theorem, Eqs. (2.191, 2.193), discussed in §3.13. We define z as the sum of the n error weighted deviations of the measurements y_i, with error σ_i, from the expected values μ_i:

$$z = \sum_{i=1}^{n} \frac{(y_i - \mu_i)^2}{\sigma_i^2} = \sum_{i=1}^{n} w_i, \tag{3.167}$$

with $w_i = (y_i - \mu_i)^2/\sigma_i^2$. By virtue of Eq. (2.191), we can write the characteristic function of z as a product of the characteristic functions of the variables w_i.

$$\phi_z(t) = \prod_{i=1}^{n} \phi_{w,i}(t). \tag{3.168}$$

We thus need to determine the function $\phi_{w,i}(t)$. To accomplish this, we first define variables $y'_i = (y_i - \mu_i)/\sigma_i$ and note that since we assumed the variables y_i have normal distributions with mean μ_i and width σ_i, then the variables y'_i follow the standard normal distribution

$$f(y'_i) = \frac{1}{\sqrt{2\pi}} e^{-y_i'^2}. \tag{3.169}$$

The PDFs of variables w_i may then be written

$$p_w(w_i) = 2f(y'_i)\left|\frac{dy'_i}{dw_i}\right| = \frac{1}{\sqrt{2\pi w_i}} e^{-w_i/2}, \tag{3.170}$$

where the factor of 2 accounts for the fact that two values of y'_i yield the same value w_i. One can show (see Problem 2.23) that the characteristic function of this expression is

$$\phi_{w_i}(t) = (1 - 2it)^{-1/2}. \tag{3.171}$$

According to Eq. (3.168), the characteristic function of z is then given by

$$\phi_z(t) = (1 - 2it)^{-n/2}. \tag{3.172}$$

Finally, the inverse Fourier transform of this expression yields the χ^2 distribution with n degrees of freedom (see Problem 3.19)

$$p_{\chi^2}(z|n) = \frac{1}{2^{n/2}\Gamma(n/2)} z^{n/2-1} e^{-z/2}. \tag{3.173}$$

3.13.4 Combining the χ^2 of Datasets

The χ^2 distribution is used in a variety of analyses including the evaluation of the goodness-of-fit (§5.3) and tests of compatibility of two independent data samples (§6.6.5). Additionally, since it is natural to expect that a greater amount of data should lead to better constraints on model parameters, it is important to consider how the combination of two independent datasets characterized by χ^2 distributed random variables Z_1 and Z_2 with n_1 and n_2 degrees of freedom, respectively, should behave.

Since Z_1 and Z_2 are assumed to originate from independent datasets, they can be considered independent variables and the characteristic function of their sum Z may then be written

$$\begin{aligned} \phi_Z(t) &= \phi_{Z_1}(t) \times \phi_{Z_1}(t) \\ &= (1-2it)^{-n_1/2} \times (1-2it)^{-n_2/2} \\ &= (1-2it)^{-(n_1+n_2)/2}, \end{aligned} \tag{3.174}$$

where the last line corresponds to the characteristic function of a χ^2 distributed variable with $n_1 + n_2$ degrees of freedom. We thus conclude that the sum Z of two independent χ^2 distributed random variables Z_1 and Z_2 is indeed also a χ^2 distributed random variable with $n_1 + n_2$ degrees of freedom.

3.14 *F*-Distribution

The F-distribution, also known as Snedecor's F-distribution or the Fisher–Snedecor distribution,[5] arises in analyses in which one wishes to establish whether two (or more) datasets have significantly different variances. Such situations commonly occur in manufacturing and quality control: all manufacturing processes lead to finite variance of the quality (specifications) of products. If and when manufacturers change components of their manufacturing process, they may worry that the variance of some particular product specifications (e.g., concentration of medicine in manufactured pills, strength of steel beams used in construction, etc.) may have increased, thereby leading to more product recalls and loss of profits. Similarly, particle detector components built and operated by different agencies or

[5] So named after Ronald Fisher (1890–1962), English statistician, mathematician, and biologist, and George W. Snedecor (1881–1974), American statistician and mathematician.

technicians may also have different performance variance and consequently impact resolution and detection efficiency differently (e.g., energy resolution of electromagnetic or hadronic calorimeters, §12.2). The F-test is specifically designed for this purpose.

The F-test is formulated on the basis of two independent, χ^2 distributed, random variables Y_1 and Y_2 with n_1 and n_2 degrees of freedom, respectively. One defines the variable f as the ratio of the χ^2 per degree of freedom of two samples:

$$f = \frac{Y_1/n_1}{Y_2/n_2}. \tag{3.175}$$

It can be shown ([5] and references therein) that the random variable f is distributed according to the PDF

$$p(f|n_1, n_2) = \begin{cases} \frac{\Gamma((n_1+n_2)/2)}{\Gamma(n_1/2)\Gamma(n_2/2)} \left(\frac{n_1}{n_2}\right)^{\frac{n_1}{2}} \frac{f^{(n_1-2)/2}}{[1+fn_1/n_2]^{(n_1+n_2)/2}} & f > 0 \\ 0 & \text{elsewhere,} \end{cases} \tag{3.176}$$

known as the F-distribution. One additionally finds that the expectation value of f is

$$\langle f \rangle = \frac{n_2}{n_2 - 2}, \tag{3.177}$$

for $n_2 > 2$. Remarkably, this expression is independent of n_1. The variance depends on both n_1 and n_2, however, and one has [5]:

$$\text{Var}[f] = \frac{n_2^2(2n_2 + 2n_1 - 4)}{n_1(n_2 - 1)^2(n_2 - 4)} \tag{3.178}$$

for $n_2 > 4$.

3.15 Breit–Wigner (Cauchy) Distribution

3.15.1 Definition

The **Breit–Wigner** (BW) distribution is named after Gregory Breit[6] (1899–1981) and Eugene Paul Wigner[7] (1902–1995), who contributed to its development and the elucidation of its properties. It is a continuous distribution defined over the domain $-\infty < x < \infty$ as follows:

$$p_{\text{BW}}(x|x_0, \Gamma) = \frac{1}{\pi\Gamma/2} \left[\frac{\Gamma^2/4}{\Gamma^2/4 + (x - x_0)^2} \right]. \tag{3.179}$$

The parameters x_0 and Γ characterize the position (centroid) and width of the distribution, respectively. A special case of the BW distribution, with $x_0 = 0$ and $\Gamma = 2$, is the Cauchy

[6] Russian-born American physicist famous for his work on particle resonant states carried out in collaboration with Wigner, and research with Edward Condon on proton–proton dispersion.

[7] Hungarian-America physicist and mathematician who shared the Nobel Prize in Physics in 1963 for his work on the theory of the atomic nucleus and elementary particles.

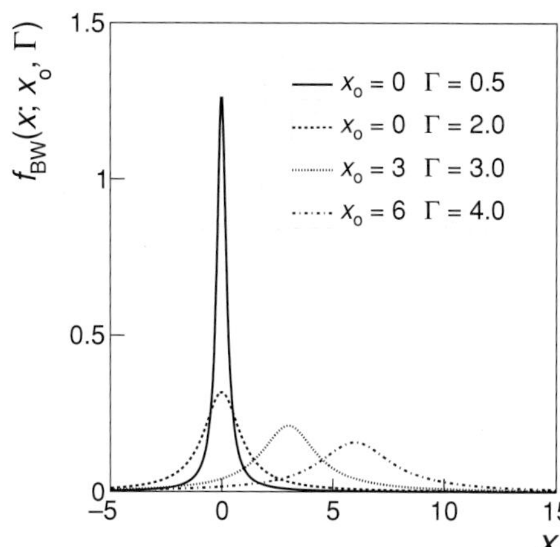

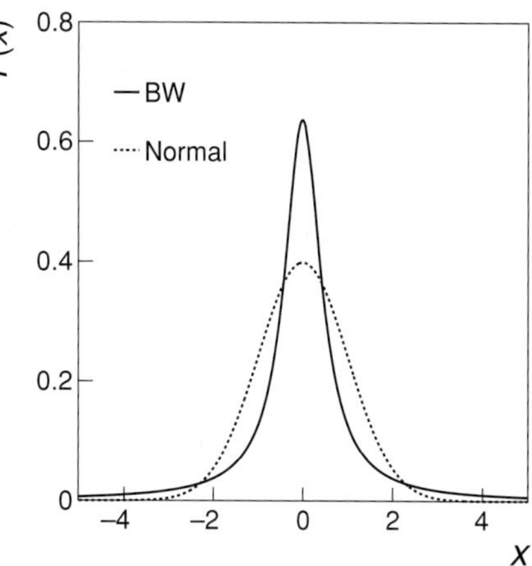

Fig. 3.15 (Left) Breit–Wigner distribution, $p_{BW}(x|x_0, \Gamma)$, for selected values of x_0 and Γ. (Right) comparison of the Breit–Wigner (solid line) and normal (dashed line) distributions for $\Gamma = \sigma = 1.0$.

distribution:

$$p_C(x) = \frac{1}{\pi\,(1 + x^2)}. \tag{3.180}$$

The BW has a particular importance in the calculation of resonant nuclear scattering cross-sections, in which the parameters x_0 and Γ correspond to the mass and width of the resonance, respectively. As such, it finds many applications in experimental data analysis of short-lived particles (e.g., see §8.5.1).

Figure 3.15 (left) illustrates the BW distribution for selected values of the parameters x_o and Γ. The BW distribution may at first sight appear similar in shape to the Gaussian distribution, even though its mathematical expression is obviously quite different. It has, however, rather different properties. Figure 3.15 (right) displays BW and Gaussian distributions centered at the origin ($x_0 = \mu = 0$), and with equal widths ($\Gamma = \sigma$). Indeed, one finds that the two distributions are similar, as they both peak at the origin and taper off asymptotically to zero at infinity. Note, however, that the BW distribution exhibits a much narrower peak and tapers off more slowly than the Gaussian distribution.

In fact, the BW distribution decreases toward zero so slowly that the integrals $\int_0^\infty x f(x)\,dx$ and $\int_{-\infty}^0 x f(x)\,dx$ are individually divergent (see Problem 3.25). This implies that the expectation value E[x] and all higher moments of the distributions are also divergent. That said, by virtue of the symmetry of the distribution about x_0, this value obviously corresponds to the median and can be used to characterize the centroid of the distribution. Likewise, while the parameter Γ cannot be identified as a moment of the distribution, it is nonetheless useful to characterize its width.

3.15.2 Relativistic Breit–Wigner Distribution

In the context of high-energy collision studies where the energy and momenta of particles can be rather large, one modifies the foregoing Breit–Wigner distribution into a Relativistic

Breit–Wigner (RBW) distribution written

$$p_{\text{RBW}}(E|M,\Gamma) = \frac{k}{(E^2 - M^2)^2 + M^2\Gamma^2}, \tag{3.181}$$

where E, M, Γ are the center-of-mass energy that produces the resonance, the mass of the resonance, and its decay width, respectively. The constant k is given by

$$k = \frac{2\sqrt{2}M\Gamma\gamma}{\pi\sqrt{M^2 + \gamma}}, \tag{3.182}$$

with

$$\gamma = \sqrt{M^2\left(M^2 + \Gamma^2\right)}, \tag{3.183}$$

written in natural units, $\hbar = c = 1$. In general, the width Γ may be a function of the energy E when it constitutes a large fraction of the mass M of the resonance (e.g., $\rho^0 \to \pi + \pi$).

3.16 Maxwell–Boltzmann Distribution

3.16.1 Definition

The **Maxwell–Boltzmann distribution**, often called the Maxwell speed distribution, is a probability density describing the velocity distribution of molecules of an idealized gas in thermodynamic equilibrium at a specific temperature T. It can also be used to describe the velocity or energy distribution of an idealized hadron (or parton) gas produced in heavy ion collisions, provided these can be considered equilibrated at a specific temperature. First derived by Maxwell in the late nineteenth century, the distribution may be written

$$p_{\text{MB}}(v) = 4\pi\sqrt{\left(\frac{m}{2\pi kT}\right)^3} v^2 \exp\left(-\frac{mv^2}{2kT}\right), \tag{3.184}$$

where m stands for the mass of the molecules (particles), v their velocity, T the temperature of the gas, and k the Boltzmann constant. The distribution is illustrated in Figure 3.16 for several values of $\sqrt{kT/m}$.

3.16.2 Properties of the MB Distribution

The most probable velocity, v_{mp}, corresponding to the mode of the distribution, is obtained by setting the derivative of the distribution with respect to v equal to zero. One finds

$$v_{\text{mp}} = \sqrt{\frac{2kT}{m}}. \tag{3.185}$$

The mean speed is given by the expectation value $\text{E}[v]$ of the distribution:

$$\langle v \rangle \equiv \text{E}[v] = \int_0^\infty v p_{\text{MB}}(v)dv = \sqrt{\frac{8kT}{\pi m}} = \frac{2}{\pi}v_{\text{mp}} \tag{3.186}$$

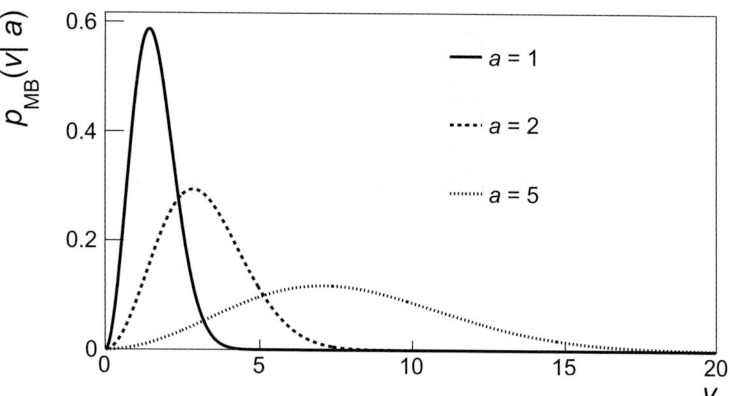

Fig. 3.16 Examples of the Maxwell–Boltzmann distribution for selected values of the parameter $a = \sqrt{kT/m}$.

Usually of greater interest, because it enables the evaluation of the average kinetic energy of the particles, is the root mean square speed, v_{rms}:

$$v_{\text{rms}} \equiv \sqrt{\langle v^2 \rangle} = \left\{ \int_0^\infty v^2 p_{\text{MB}}(v) dv \right\}^{1/2} = \sqrt{\frac{3kT}{m}} = \sqrt{\frac{3}{2}} v_{\text{mp}}. \tag{3.187}$$

3.16.3 Derivation of the MB Distribution

The Maxwell–Boltzmann distribution can be derived using the kinetic theory of gases. In the nonrelativistic limit, the kinetic energy of molecules of type i is

$$E_i = \frac{p_i^2}{2m_i}, \tag{3.188}$$

where p_i and m_i represent the momentum and mass of the molecules (particles), respectively. The number of particles in state i, noted N_i is given by

$$\frac{N_i}{N} = \frac{1}{Z} g_i \exp\left(-\frac{E_i}{kT} \right), \tag{3.189}$$

where T is the equilibrium temperature of the gas, g_i is a degeneracy factor corresponding to the number of microstates that have the same energy levels (or mass), k is the Boltzmann constant, and N is the total number of particles in the system. This can be calculated with the partition function of the system,

$$Z = \sum_j g_j \exp\left(-\frac{E_j}{kT} \right). \tag{3.190}$$

The probability density function, p_{p}, for finding a molecule (particle) at a specific momentum $\vec{p} = (p_x, p_y, p_z)$ can thus be written

$$p_{\text{p}}(p_x, p_y, p_z) = \left(\frac{1}{2\pi mkT} \right)^{3/2} \exp\left(-\frac{p_x^2 + p_y^2 + p_z^2}{2mkT} \right), \tag{3.191}$$

Table 3.13 Properties of the Maxwell–Boltzmann distribution

PDF	$p_{\mathrm{MB}}(x) = \sqrt{\frac{2}{\pi}\frac{x^2}{a^3}} \exp[-x^2/2a^2]$
CDF	$\mathrm{erf}(\frac{x}{\sqrt{2}a}) - \sqrt{\frac{2}{\pi}} \frac{x\exp(-x^2/(2a^2))}{a}$
Mode	$\sqrt{2}a$
Mean	$\mu = 2^{3/2}\pi^{-1/2}a$
Variance	$\sigma^2 = a^2(3\pi - 8)/\pi$
Skewness	$\gamma_1 = 2^{3/2}(16 - 5\pi)/(3\pi - 8)^{3/2}$
Kurtosis	$\gamma_2 = 4(-96 + 40\pi - 3\pi^2)/(3\pi - 8)^2$

which implies that the momenta p_x, p_y, and p_z are each normally distributed about the origin and with a standard deviation $\sqrt{mkT}$.

The Maxwell–Boltzmann distribution can be expressed in terms of the energy of the particles by a simple variable transformation:

$$p_{\mathrm{E}}(E)\, dE = p_{\mathrm{p}}(\vec{p})\, d^3p = p_{\mathrm{p}}(\vec{p}) 4\pi p^2 dp. \tag{3.192}$$

Given $dE/dp = p/m$, one gets

$$p_{\mathrm{E}}(E)\, dE = 2\sqrt{\frac{E}{\pi}} \left(\frac{1}{kT}\right)^{3/2} \exp\left(\frac{-E}{kT}\right) dE. \tag{3.193}$$

The properties of the distribution are summarized in Table 3.13.

Exercises

3.1 A vat is filled with 5 yellow, 10 red, 15 green, and 20 blue balls. Calculate the probability of drawing 5 balls of the same color if (a) the balls are put back in the vat after each draw, and (b) if the balls are kept out.

3.2 Calculate the mean, variance, excess skewness, and excess kurtosis of the binomial distribution using a decomposition of $(1 - p)^{N-n}$. Repeat these calculations using the moment-generating function of the binomial distribution.

3.3 The designers of a system of interception of ballistic missiles have estimated that their system has an efficiency of 99.8% to intercept any given missile directed at a country. Determine the probability that the system would intercept all of 50 missiles fired against it. How many missiles would an aggressor need to have a probability greater than 0.5 of having one, two, and five missiles hit their target?

3.4 At their peak, typically during August 9–14, the Perseids meteor showers have a rate of 60 meteors per hour, visible from the Northern Hemisphere. Calculate the probability of observing fewer than 10 meteors in 30 minutes, and the probability of observing more than 40 in 10 minutes.

3.5 Use the Stirling formula, Eq. (3.31), to derive the large n limit of the Poisson distribution. Hint: Define $x = \frac{n-\lambda}{\sqrt{\lambda}}$ and use the Stirling formula to express the Poisson distribution in terms of a standard distribution.

3.6 Measurements at the Intersecting Storage Rings (ISR) and at Fermi National Accelerator Laboratory (FNAL) have shown that the multiplicity of charged particles produced in proton–proton collisions can be described with relatively high precision with a negative binomial distribution. Assuming that all produced particles are pions and neglecting charge conservation effects, derive an expression for the conditional probability of the number of positive pions given fixed numbers of negative and neutral pions. Additionally find an expression for the joint probability of the multiplicity of positive and negative charge particles. Hint: See ref. [197].

3.7 Calculate the mean, variance, and covariance of the multinomial distribution.

3.8 Verify the derivation of Eq. (3.26) and calculate the variance of the multiplicity distribution discussed in §3.1.4.

3.9 The STAR Time Projection Chamber (TPC) detector features 45 sensor rows, thereby enabling reconstruction of charged particles consisting of up to 45 "hits." There is limited efficiency of finding a "hit" on each of the pad rows. Make a plot of the distribution of the number of hits on measured tracks for hit finding efficiencies of 90%, 75%, and 50%. Assume for simplicity's sake that all charged particles cross all 45 pad rows.

3.10 Numerically determine the mean, variance, skewness, and kurtosis of the distributions evaluated in the previous problem. Compare these estimates to those expected from the relationships found in Table 3.1.

3.11 Discuss the trivial distribution of net charge that can be observed for total charged particle multiplicities of 50, 100, 200, and 500 particles. Assume the mean of these net-charge distributions is null. Calculate the variance of each of these distributions.

3.12 Calculate the mean, variance, excess skewness, and excess kurtosis of the Poisson distribution.

3.13 Calculate the mean, variance, excess skewness, and excess kurtosis of the Uniform distribution.

3.14 Calculate the mean, variance, excess skewness, and excess kurtosis of the Exponential distribution. Derive a general expression for all moments and central moments of the Exponential distribution.

3.15 Verify the normalization, Eq. (3.127), of the normal distribution.

3.16 Calculate the mean, variance, excess skewness, and excess kurtosis of the Gaussian distribution. Derive a general expression for all moments and central moments of the Gaussian distribution.

3.17 Calculate the first three cumulants, κ_n, $n = 1, 2, 3$ of the Poisson and Gaussian distributions.

3.18 Calculate the mean, and variance of the χ^2 distribution.

3.19 Calculate the inverse Fourier transform of $\phi_z(t) = (1 - 2it)^{-n/2}$ and derive the expression (3.173) of the χ^2 distribution.

3.20 Calculate the probability that an unstable particle, for example, a lambda baryon, decays within a distance x of its site of production. Assume the particle has a half-life $t_{1/2}$ and neglect relativistic effects.

3.21 Derive a relationship between the mean lifetime of a particle and its half-life.

3.22 Calculate the mean and variance of the χ^2 distribution using integration by parts.

3.23 Derive the expressions Eq. (3.156) for the mean and variance of the log-normal distribution.

3.24 Calculate the probability the χ^2 of a fit (where the χ^2-distribution applies) is larger than the number of degrees of freedom, n, for values of n equal to 2, 4, 8, 16, and 32.

3.25 Show that the integral $\int_0^\infty E^n f_{\text{BW}}(E)\, dE$, where $f_{\text{BW}}(E; x_o, \Gamma)$ is the Breit–Wigner distribution, does not converge for n equal to or larger than 1.

3.26 Implement a computer program to carry out numerical calculations of the cumulative integral of the Gaussian and Breit–Wigner distributions. Use this program to calculate the probability that values lie beyond 1.5 σ, 2.5 σ, and 3.5 σ from the mode of these distributions (Use $\Gamma = \sigma$ for the Breit–Wigner distribution).

3.27 Derive the expression (3.191) for the Maxwell–Boltzmann distribution in terms of the momenta p_x, p_y, and p_z.

3.28 Use Eq. (3.191) to derive the expression of the Maxwell–Boltzmann given by Eq. (3.184).

3.29 Show that in thermodynamic equilibrium, the energy per degree of freedom, ϵ, of the molecules of a gas (of identical particles) is distributed as a chi-squared distribution with one degree of freedom according to

$$f_\epsilon(\epsilon)\, d\epsilon = \sqrt{\frac{1}{\epsilon \pi k T}} \exp\left[\frac{-\epsilon}{kT}\right] d\epsilon.$$

4 Classical Inference I: Estimators

Statistical inference is a mathematical activity, or process, intent on extracting meaningful information from measurements in order to characterize systems, establish best model parameters, or determine the plausibility (truthfulness) of hypotheses and models. Indeed, once experimental data have been acquired, one wishes either to characterize the data by some ad hoc functional form, or compare the measured data distribution to some model predictions. In either case, the functional form may involve "free" parameters not constrained by theory and to be determined by the data. There are also cases where no specific functional form is known a priori and as simple a characterization of the data as possible is sufficient or required. It may be sufficient, for instance, to evaluate the mean, variance, and selected higher moments of the data. Inference also includes the notion of statistical test, that is, the use of statistical techniques to assess the adequacy of scientific models and hypotheses.

Traditionally, statistical inference is a subfield of probability and statistics concerned with the determination of the properties of the underlying distribution of datasets based on statistical analysis techniques largely based on the frequentist interpretation of probability. However, the field of inference has greatly expanded in recent decades, and inference may now be approached from a plurality of paradigms,[1] or schools of thought, corresponding to different interpretations of the notion of probability and approaches to the inference process. We will restrict our discussion to the frequentist and Bayesian paradigms, the former still most commonly used in the physical sciences, and the latter receiving a fast growing level of interest in physical and life sciences, engineering, as well as economics and financial fields. Other commonly less adopted paradigms, including the likelihood paradigm, the total survey error paradigm, and the survey sampling paradigm, are discussed in the statistics literature [15, 96, 167].

Similarities and distinctions between the goals and methods of the frequentist and Bayesian inference paradigms are briefly outlined in §4.1. The remainder of this chapter as well as Chapters 5 and 6 present detailed discussions of classical inference concepts and techniques while Bayesian inference methods are explored in Chapter 7.

[1] The term paradigm is here used in the sense defined by Kuhn [132]. Restricted to statistical inference, this amounts to classifying statisticians and scientists according to the inference framework they use in the conduct of their statistical analyses.

4.1 Goals of Frequentist Inference

Inference problems are traditionally classified according to the type of statements or conclusions they produce. These include:

1. Point estimates, which involve the determination of a particular observable, properties of a distribution, or model parameters designed to characterize a particular population
2. Interval estimates, most commonly known as confidence intervals within which a particular observable, property of a distribution, or model parameter would be found with a definite and specific probability if the experiment or sampling could be repeated indefinitely
3. Rejection of a hypothesis based on a statistical test predicated on a probabilistic model of the data
4. Clustering and classification of data points

By contrast, in the Bayesian paradigm the inference process essentially involves only two categories of tasks known as **model selection** and **parameter estimation** from which a plurality of subsidiary statements can be derived or summarized.

Data analyses in astronomy, particle physics, and nuclear physics rely increasingly on the Bayesian paradigm but nonetheless still make extensive use of the traditional frequentist methods. We thus endeavor to discuss a wide range of inference problems under both paradigms, and present, in this and the following chapters, discussions of inference techniques developed in both paradigms.

In traditional statistical inference, the characterization and modeling of datasets relies on mathematical functions or procedures known as **estimators**, which we formally define in §4.3. Essential properties of estimators are discussed in §4.4, while examples of implementation for basic notions of mean, variance, and covariance are presented in §4.5. Data modeling (commonly called **fitting**) based on the method of **Maximum Likelihood** ($\mathbb{ML}$) and the relatively simpler method of **Least Squares** ($\mathbb{LS}$) are introduced in §5.1 and §5.2, as are a number of associated technical problems, including interpolation/extrapolation, and weighted means of results of distinct experiments. A brief discussion of the notion of goodness-of-fit follows in §5.3. However, for a more in-depth discussion of the notion of statistical test applied to functional fits, see §6.6.5.

4.2 Population, Sample, and Sampling

Before proceeding with the introduction of the concepts of statistics and estimators, it is important to discuss the underlying notions of sample and population. Definitions of these concepts are presented in §4.2.1 while their characterization and representation are briefly discussed in §§4.2.2 and 4.2.3, respectively.

4.2.1 Population and Statistical Sample

In statistical analysis, a **population** is defined as a complete set of items or entities that share one or several properties. For instance, the population of Americans living in the continental United States shares a geographical location, a language, a political system, and so on. The concept of population is not restricted to humans; one may of course consider populations of animals, plants, or any other types of entities of interest. For instance, one may wish to study and characterize the fleet (population) of cars used in America, and contrast it to the fleets of cars used in Europe or Asia; or one might be interested in studying the population of stars composing the Milky Way. The notion of population, applied in statistics, is thus as broad and generic as human interests in studying the world around us.

By contrast, a **statistical sample** is a subset drawn from the population and used for the purpose of a statistical analysis. If the sample is sufficiently large and obtained properly, it is assumed that traits and characteristics of the entire population can be adequately inferred from corresponding characteristics of the sample.

It is useful to introduce the notion of **subpopulation** consisting of a specific subset of a population and defined based on one or more additional properties shared by the elements or items of this subset. For instance, the United States car fleet may be subdivided into several subpopulations according to the nationality (origin) of the manufacturers: Americans, Europeans, Asians, and so on. Items of a sample are in general heterogeneous and may not all share additional properties of interest. A sample must then be distinguished from a subpopulation.

Analyses in terms of subpopulations are useful, for instance, in pharmacology because different subpopulations (e.g., based on ethnic or hereditary history) may have different responses to medicines or treatments. In nuclear physics, one may distinguish the properties (e.g., momentum spectrum) of distinct particle subpopulations (e.g., pions, kaons, or protons) produced by a given type of elementary collision at a specific incident energy. In some cases, a great difference (variance) may exist between subpopulations. Modeling and understanding the population dynamics is then best accomplished by examining subpopulations individually. Distinctions between populations and subpopulations are somewhat arbitrary. A specific analysis may indeed focus on a particular subpopulation, which effectively becomes the "population of interest." The larger population, of which the given population is a subset, may then be referred to as the **parent population**. It is also common to refer the population from which a sample is drawn as the **parent population of the sample**.

4.2.2 Sample Characterization

It is typically not possible to acquire information about all elements of a population. The population may be too large or it may not be possible, ab initio, to access or enumerate all its elements. Population **sampling** is thus necessary. Sampling consists in selecting individuals or acquiring information about finitely many individuals or items of the population. The acquired subsets, called **samples**, may vary in size, but it is a general rule that the larger they are, the better the inference about properties of the whole population will be.

The sample space of a random variable X, defined in §2.2, corresponds to the entire space spanned by X. A population and measured samples of a population thus typically correspond to subsets of the sample space.

A **complete sample** is a set of items that includes all elements of a population.[2] As stated earlier, various practical or technical issues may limit the size of a sample. Statistical analyses involving complete samples are thus rare, in general. Since finite samples are typically the rule, statisticians must ensure that acquired samples are **unbiased**. Biases may occur, for instance, if subpopulations of a population feature distinct traits or behaviors in regard to an observable of interest. In this context, a sample consisting of elements drawn mostly from one particular subpopulation cannot be representative of the other subpopulations and thus the population as a whole.

An **unbiased sample** is a set of objects chosen from a complete sample using a drawing or sampling process that is not influenced by the properties of the elements of the population. Particular subpopulations are thus not favored or neglected and the ensuing statistical analysis and inference is thus representative of the whole population.

Truth be told, **sampling bias** occurs in all sciences and is often unavoidable. For instance, early studies of stars of the Milky Way greatly suffered from a brightness bias. Stars appear dimmer to an observer the farther away they are. Studies of stellar properties were thus initially biased, that is, influenced predominantly by classes of brighter stars. Similarly, studies of human populations may be biased based on geographical location, gender, religion, ethnicity, and so forth.

The problem arises, obviously, that it is generally not possible to draw complete samples, and one must devise techniques to avoid biasing collected samples. At the outset, it might appear that the collection of a **random sample**, also known as a **probability sample**, might solve this problem. In practice, observational biases are often introduced by the techniques or instruments used to acquire the data sample.

In astronomy, for instance, studies of galactic properties are somewhat biased by the observational wavelength window of the telescope: distant galaxies that recede from us at high velocity have large red shifts and may be difficult or impossible to measure with an optical (visible light) telescope.

In particle and nuclear physics, measurements of unbiased samples of events and produced particles are de facto impossible. Indeed, the triggering methods and devices used in the detection of collisions (events) and measurements of particle properties invariably feature technical limitations. Particles are measured within finite kinematic ranges, known as **detection acceptance**, and events may be missed if they fail to activate the devices used as trigger detectors. Collisional events acquired with detectors and procedures that minimally influence or bias the properties of measured events and the particles they contain are said to feature a **minimal bias**. By extension, triggering devices and procedures enabling the collection of such data are said to be **minimum bias triggers**. Interestingly, the volume of data produced in high-energy physics experiments is often so large that it is not possible to detect, digitize, and record all collisions. Special triggering detectors and procedures are then devised to bias and select collisions and particles of particular interest. It is then

[2] However, a complete sample is not required to cover the entire sample space of a random variable.

possible to selectively increase the relative abundance of specific types of particles or events (e.g., events susceptible of containing a top quark or a Higgs boson).

More generally, it is important to acknowledge that measurements can be influenced by various instrumental effects (discussed in Chapter 12) that cause loss or neglect of individual items of a population, or that smear the properties of these items. The loss or neglect of items is typically characterized by **sampling efficiency** and/or **detection efficiency** (commonly just called efficiency). Efficiency losses, if not uniform across all values of a particular observable, may severely bias the measured values and the conclusions reached about the population. In experimental physics, most particularly in nuclear and particle physics, particle and collision (event) losses are unavoidable and typically depend on observables of interest such as the momentum of particles, or the multiplicity produced in nuclear collisions. Fortunately, techniques exist to assess detection efficiencies and their dependencies on observables of interest, and it is then possible to correct sampled data for efficiency losses. Similarly, smearing or measurement resolution can also introduce measurement biases, but techniques to evaluate such effects are also available and may be used to correct the data. These topics are discussed at length in §12.3.

From the preceding discussion, it should be clear that a statistical analysis of a data sample is but a part of the measurement process. Obviously, measurement methods and the observables of interest vary greatly across scientific disciplines, but it remains that the sampling of a population and the statistical analyses based on collected samples are only a small part of the scientific process. Accordingly, techniques used to gather data and the types of data collected thus vary considerably across disciplines. However, statistical techniques to extract information about the parent population are essentially universal, that is, more or less independent of the specificities of the data or their origin.

4.2.3 Sample Representation and Random Variables

Statistical samples may consist of discrete data, continuous data, or mixtures of the two. They may be as simple as a collection of values $\{o_i\}$, with $i = 1, \ldots, n$, of a particular observable O (e.g., the momentum of particles produced by a certain type of nuclear collisions) representing a particular trait or characteristic of individuals of a population. They may also be arbitrarily complex and gather jointly (simultaneously) multiple observables representing traits or properties of individuals of a population. Computationally, such collections of observables may be represented as vectors $\vec{x}$ or n-tuples. An n-tuple is an ordered list of n elements that may be heterogeneous in nature or type. For instance, in nuclear collision studies, one may represent the information reconstructed about measured particles in terms of tuples $(q, p_T, \eta, \phi, \text{PID})$, where q is the charge of the particle, p_T its transverse momentum, η its pseudorapidity, ϕ the azimuthal angle of production, and PID a variable identifying the species of the particle.

In statistical analyses, one may be interested in characterizing a population based on one or more components of the n-tuples. More specifically, one may be interested in determining the moments, covariances (cumulants), and (probability) distributions of these components, either jointly or individually. Computational techniques required to obtain such quantities are the domain of statistical inference. We begin our discussion of statistical inference techniques with the introduction of basic notions of statistics and estimators in §4.3. Properties of estimators are discussed in §4.4, and basic estimators presented in §4.5. Histograms, which provide graphical representations of the distribution of numerical data, are introduced and defined in §4.6. Classical inference techniques involving composite hypotheses and most particularly the optimization of model parameters toward the description of measured data are addressed in §§5.1 and 5.2.

4.3 Statistics and Estimators

4.3.1 Problem Statement

Let us consider an experiment consisting of a series of n distinct measurements of some phenomenon and yielding a dataset D. We will first assume that though distinct, the measurements are carried in the same conditions and from an unchanging system.[3] Let us consider that the dataset D consists of n measurements yielding random continuous vectors $\vec{x}_i$ of the same dimensionality. In the simplest case, each measurement would yield a single value x_i. In a slightly more sophisticated situation, the vectors could consists of pairs of numbers (x_i, y_i) considered as independent and dependent variables, respectively. More generally still, the vector $\vec{x}$ could represent an arbitrarily long list of values (i.e., an n-tuple) obtained from each measurement (i.e., each individual item of the sample). The sample space of each measurement would thus consist of all possible values of the multidimensional variable, $\vec{X}$, in other words, all values $\vec{x}_i$ a measurement of $\vec{X}$ can produce.

In a general scientific context, various states of prior knowledge may exist about the phenomenon being measured. If the phenomenon is newly observed, it is possible that very little prior information is available about it and one may then wish to characterize the sample (and the populations it is meant to represent) by its moments or other similar quantities. Alternatively, a great deal of information may already exist about the phenomenon, and one may wish to characterize the sample according to a specific mathematical model, or hypothesis, M, stated in terms of a function, $f(\vec{X}|\vec{\theta}, M)$, which expresses some relation between the elements of the vector $\vec{X}$. In this context, $\vec{\theta}$ could represent a finite set of parameters of the model M believed to properly describe the phenomenon, or being tested. For instance, in an analysis of the speed of receding galaxies, the model M could state the hypothesis that the relation between the recession speed of galaxies is strictly proportional to their distance. The model would then involve a single parameter consisting of the slope of the relation between the speed and the distance. It might then be the purpose of the analysis to infer which value or range of values of the slope, H_0, known as the Hubble constant, best fits the data. So, whether one wishes to characterize the population of interest according to moments or some more sophisticated model of the phenomenon, one is faced with a similar problem, which involves the determination of parameters that represent the whole

[3] We will see later in this chapter how these conditions may be lifted.

population based on a specific data sample D. This is known as a problem of **parameter estimation** and alternatively referred to as a **single point estimate**.

Parameter estimation may be discussed in both the frequentist and Bayesian paradigms. We first introduce the more traditional frequentist treatment in §4.3.2 and then show, in §4.3.3, that the Bayesian approach yields a more general and comprehensive solution to parameter estimation problems.

4.3.2 Parameter Estimation within the Frequentist Paradigm

For simplicity, let us first restrict our discussion by assuming that n successive measurements of a random variable X yield values, x_i, with $i = 1, \ldots, n$. The dataset, D, may then be represented as a vector $\vec{x} = (x_1, x_2, \ldots, x_n)$. Within the frequentist inference paradigm, one assumes there exists a probability density, $f(x)$, describing the likelihood (frequency) of observing specific values x_i, with $i = 1, 2, \ldots, n$, of the random variable X in n distinct measurements. The sample space of the experiment thus consists of a space of n-dimensions, with each dimension corresponding to the sample space of X. Successive deployments of this experiment are expected to yield fluctuating values of the vector $\vec{x}$. In order to evaluate the probability density function (PDF), $f_n(\vec{x})$, of the vector $\vec{x}$, one must first determine the probability density of a given set of n values measured in one experiment. We will assume the measurements are carried out independently but with identical conditions. Remember from Chapter 2 that the probability of a subset A of the sample space $\mathbf{S}$ consisting of the union of disjoint subsets A_i (with $A_i \cup A_j$ for $i \neq j$) is simply the product of the probabilities of the subsets. The PDF of $\vec{x}$ thus consists of the product of the probability densities of each of the values x_i. Since the measurement conditions are deemed identical, each measurement x_i is governed by the same PDF $f(x)$. One thus gets:

$$f_n(\vec{x}) = f(x_1)f(x_2)\cdots f(x_n) = \prod_{i=1}^{n} f(x_i). \tag{4.1}$$

The problem, of course, is that the function $f(x)$, or some of its parameters, are not known a priori. It thus becomes our goal to infer what this function might be or, at the very least, determine some of its properties (e.g., its moments) based on the measurements x_i.

Let us assume that the PDF $f_n(\vec{x})$ might depend on m a priori unknown parameters $\vec{\theta} \equiv \{\theta_1, \theta_2, \ldots, \theta_m\}$. We wish to construct a function of the n measured values x_i, denoted $\vec{x} = (x_1, x_2, \ldots, x_n)$, to estimate the parameters $\vec{\theta}$. In the literature, a function of the measurements $\vec{x}$ with no unknown parameters is called a **statistic**, while a statistic used to estimate one or several properties of a PDF is known as an **estimator**.

An experiment yields random values $\vec{x}$. A function of these $\vec{x}$, that is, an estimator consequently yields values $\vec{\theta}$, known as **estimates**, that are randomly distributed. Let us denote these estimates $\hat{\theta}$ to distinguish them from the actual value $\vec{\theta}$ (i.e., the true value of the parameters). $\hat{\theta}$ is commonly called **estimator** of $\vec{\theta}$. The term **estimator** usually refers to the mathematical function of the sample, whereas the term **estimate** is used for a particular value of the estimator obtained within a specific experiment, that is, with a specific sample $\{x_i\}, i = 1, \ldots, n$.

By its very nature and construction, an estimator $\hat{\theta}$ is expected to yield estimates that fluctuate dataset by dataset and randomly deviate from the true value $\vec{\theta}$ when the experiment is repeated. One hopes, however, that given sufficiently many values x_n, the estimator might **converge** to the actual value $\vec{\theta}$. In this context, it is interesting to consider the convergence properties (vs. the size of a sample) and errors associated with particular estimator definitions: how close an estimator $\hat{\theta}$ comes to the true value $\vec{\theta}$ depends on the quality of the measurements, the size of the sample, as well as the actual formulation or definition of the estimator. We, however, delay a discussion of these considerations until §4.4 after a discussion of Bayesian parameter estimation in the next section.

4.3.3 Parameter Estimation within the Bayesian Paradigm

Parameter estimation within the Bayesian paradigm has the same basic goal as in classical inference: one wishes to find parameter values $\vec{\theta}$ that best fit or characterize the observed data, D. However, the Bayesian approach also accounts for prior information that might be available about the system, the model, M, and the parameters themselves, $\vec{\theta}$, before the data are measured. Prior information about the parameters of the models, in particular, is encoded as a **prior probability density** $p(\vec{\theta}|I)$, where I represents relevant facts known about the system prior to the measurement(s). The data D thus provides a complement of information, which through Bayes' theorem provides a **posterior probability distribution** (density) of the parameters $\vec{\theta}$:

$$p(\vec{\theta}|D, I) = \frac{p(\vec{\theta}|I)p(D|\vec{\theta}, I)}{p(D|I)} \tag{4.2}$$

The probability $p(D|I)$, known as global likelihood of the data, can be calculated using the law of total probability:

$$p(D|I) = \int p(\vec{\theta}|I)p(D|\vec{\theta}, I)\prod_i d\theta_i \tag{4.3}$$

In effect, the probability density function $p(\vec{\theta}|D, I)$, known as posterior PDF, plays the same role as the PDF $f_n(\vec{x})$ in the frequentist paradigm and provides the probability of measuring specific values of the model parameters $\vec{\theta}$ based on the measured data D. The posterior $p(\vec{\theta}|D, I)$ is then said to be **conditioned by** the data D.

For a model with a single continuous parameter, θ, this corresponds to the density of probability of finding the parameter in the range $[\theta, \theta + d\theta]$, while for a model with multiple parameters, it amounts to the joint density of probability for finding the parameters θ_i simultaneously in the ranges $[\theta_i, \theta_i + d\theta_i]$. An estimator of the model parameters, denoted $\hat{\theta}$, may then be determined on the basis of the mode of the distribution, that is, the value of $\vec{\theta}$ yielding the maximum value of the PDF $p(\vec{\theta}|D, I)$. While it might be tempting to use the expectation value of the parameters $\vec{\theta}$

$$\langle\vec{\theta}\rangle \equiv \int \vec{\theta}p(\vec{\theta}|D, I)\prod d\theta_i. \tag{4.4}$$

in lieu of the mode, one should note that, in general, the distribution $p(\vec{\theta}|D, I)$ may not be symmetric about its mode. Indeed, if $p(\vec{\theta}|D, I)$ is asymmetric, skewed, and features long tails, the expectation value $\langle\vec{\theta}\rangle$ shall not have the highest probability and thus would not yield the most probable value of $\vec{\theta}$. It is thus essential to identify the true maximum position of $p(\vec{\theta}|D, I)$. In effect, whether using a frequentist or Bayesian approach, finding the parameters of a model that best match the data becomes a problem of optimization. Optimization of $f_n(\vec{x})$ and $p(\vec{\theta}|D, I)$ in the presence of a single free model parameter may be relatively simple, particularly for linear models, but it can be quite arduous for models involving several (free) parameters. Fortunately, many techniques exist to tackle this optimization task. Several of these techniques, relevant for linear and nonlinear data models, are discussed in Chapters 5 and 7.

We shall consider examples of estimators evaluated in the frequentist and Bayesian paradigms in §§4.5 and 7.4, respectively. But first, let us examine the properties of estimators as mathematical functions.

4.4 Properties of Estimators

Given estimators are arbitrary functions of the measured data values x_i, one should expect they provide estimates of observables of interest that may depend on their actual mathematical formulation, the size of the data sample used to compute them, as well as, perhaps, the probability distributions that govern the production and measurement of the observable X. It is thus sensible to wonder how precise or accurate estimates of an observable may be given a specific dataset, its size, and the actual mathematical formulation of the statistics used to estimate the observable. This implies statistics and estimators are not all born equal. It is thus of interest to briefly consider, in the following paragraphs, their properties or behavior under changes in the sample size or the probability distribution that governs an observable X of interest. The discussion presented here is at a similar level to that found in Cowan [67], but a more in-depth treatment of these concepts may be found in James [116].

4.4.1 Estimator Consistency

An estimator $\hat{\theta}$ is said to be **consistent** when estimates converge toward the true value of the parameter(s) as the number of observations x_i increases. A number of convergence criteria can be applied but it is common to require **consistency in probability** defined based on the convergence of the probability $p\big(|\hat{\theta}_n - \theta| > \varepsilon\big)$, where $\hat{\theta}_n$ is an estimator of θ based on n observations. An estimator is then said to be consistent if a value N exists such that

$$p(|\hat{\theta}_n - \theta| > \varepsilon) < \eta \tag{4.5}$$

for all $n > N$ and arbitrarily small (positive definite) numbers ε and η. This expression indicates $\hat{\theta}_n$ converges toward θ for increasing values of n, as schematically illustrated in Figure 4.1.

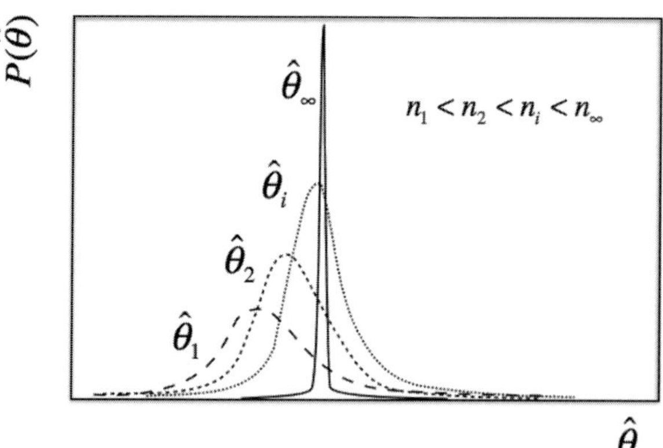

Fig. 4.1 An estimator $\hat{\theta}$ is said to be consistent if its probability density narrows and progressively converges to the true value (of an observable) as the number of observations n_i increases.

Consistency implies asymptotic convergence but does not guarantee that the precision of an estimator improves monotonically with the number of observations n. It is thus a necessary but insufficient requirement for an estimator to be useful. In other words, adding observations does not necessarily guarantee an increase in precision and smaller errors. Additional criteria are thus warranted.

4.4.2 Expectation Value of an Estimator

Statistics and estimators are multivariate functions of random variables. We can then apply the rules and techniques introduced in §§2.8–2.8.6 for such functions to determine their probability density functions and associated properties.

Let $\hat{\xi} \equiv \hat{\xi}(\vec{x})$ represent an estimator of interest calculated based on a sample of n measured values x_i that we shall represent, for convenience, in vector notation as $\vec{x} = (x_1, \ldots, x_n)$. Let us denote the PDF of the estimator $\hat{\xi}$ as $g(\hat{\xi}\,|\,\xi)$, where it is assumed that the estimator has some "true" value ξ. Calculation of this function, commonly known as **sampling distribution** of ξ, requires one makes assumptions concerning the probability distributions of each of the measured data x_i. In general, one requires a PDF $p(x_i|\vec{\theta}, H_i, I)$ based on a **probability model** or hypothesis H_i and prior information I for each of the observables x_i obtained in the data sample, where $\vec{\theta}$ represents one or several parameters of the models H_i which may be unknown a priori. In practice, if the x_i are independently sampled from the same population, it is reasonable to assume they may be described by the same model H and PDF $p(x|\vec{\theta}, H, I)$, that is, with the same model parameter(s) $\vec{\theta}$.

As we saw in §2.8.6, the probability distribution of a particular set of such values $\vec{x} = (x_1, \ldots, x_n)$, called the likelihood (of the data D) for parameter $\vec{\theta}$, and denoted $L(D|\vec{\theta}, H, I)$, is readily determined by multiplying the probability densities of each of the x_i. We thus write

$$L(D|\vec{\theta}, H, I) \equiv L(\vec{x}|\vec{\theta}, H, I) = \prod_{i=1}^{n} p(x_i|\vec{\theta}, H, I). \tag{4.6}$$

Indeed, the **likelihood function** $L(\vec{x}|\vec{\theta}, H, I)$ represents the probability of the data $\vec{x}$ given a hypothesis or model H with parameters $\vec{\theta}$. Consequently, regarding the estimator $\hat{\xi}$ as a function of the continuous random variable $\vec{X}$, its expectation value may be calculated according to Eq. (2.144), and one gets

$$\mathrm{E}[\hat{\xi}(\vec{x})] = \int \hat{\xi} g(\hat{\xi}|\xi)\, d\hat{\xi}, \tag{4.7}$$

$$= \int \cdots \int \hat{\xi}(\vec{x}) L(\vec{x}|\vec{\theta}, H, I)\, dx_1 \cdots dx_n, \tag{4.8}$$

$$= \int \cdots \int \hat{\xi}(\vec{x}) p(x_1|\vec{\theta}, H, I) \cdots p(x_n|\vec{\theta}, H, I)\, dx_1 \cdots dx_n. \tag{4.9}$$

Within the frequentist paradigm, this expectation value represents the mean value of $\hat{\xi}$ produced by an infinitely large (or very large) number of similar experiments yielding samples $\{x_i\}, i = 1, \ldots, n$, while in the Bayesian paradigm, it is simply the expectation value of $\hat{\xi}$ implied by the likelihood function L, the model H, and the parameter value $\vec{\theta}$ without regard as to whether the measurement can be repeated.

A case of particular interest arises when $\hat{\xi} \equiv \hat{\theta}_i$ is an estimator of one of the model parameters θ_i, that is, a function of the variables x_i chosen to yield an estimate of a particular model parameter θ_i of the distribution $p(x|\vec{\theta}, H, I)$. Two situations are typically encountered in this context. In one, the estimator consists of a simple statistic, that is, a formula involving no free parameters and whose calculation yields an estimate of the model parameter directly. In others, the statistics formula depends on the value of the parameter itself which is then considered unknown. An optimization procedure is then required to determine which parameter value is best compatible with the measured data.

Whether an estimator ξ is designed to determine some generic property of a distribution (e.g., one of its moments or cumulants), or a particular parameter θ_i of the model H that defines the distribution, it is legitimate to wonder how precise and accurate an estimate obtained from a finite dataset may be. This requires a discussion of the notions of **estimator bias** and **estimator precision**.

4.4.3 Estimator Bias

The **bias**, b_n, of an estimator is defined as the difference between its expectation value $\mathrm{E}[\hat{\xi}(\vec{x})]$ and the true value ξ for an "experiment" consisting of n observations. For an elementary (scalar) estimator, this is written

$$b_n[\hat{\xi}] \equiv \mathrm{E}[\hat{\xi}] - \xi = \mathrm{E}[\hat{\xi} - \xi]. \tag{4.10}$$

The bias b_n is an intrinsic property of the estimator $\hat{\xi}$, which depends on its formulation (i.e., its functional form), the sample size n, and the properties of the PDF being evaluated, but not on the sampled values. An estimator is said to be **unbiased** for all values of n and ξ if

$$b_n[\hat{\xi}] = 0, \tag{4.11}$$

or

$$\mathrm{E}[\hat{\xi}] = \xi. \tag{4.12}$$

If the bias is finite for finite values of n but vanishes in the limit $n \to \infty$, then the estimator is considered to be **asymptotically unbiased**. We present examples of unbiased and asymptotically unbiased estimators in §4.5.

It is worth noticing that the notion of bias could also be based on the median or the mode of the distribution $g(\hat{\xi}|\xi)$. The choice of $\mathrm{E}[\hat{\xi}]$ is indeed somewhat arbitrary, but its linear property as an operator greatly simplifies calculations of bias and thus makes for a convenient and practical choice. However, this choice implies that an estimator $\hat{\xi}$ is not necessarily the median or the mode of the PDF $g(\hat{\xi}|\vec{\xi})$. It might indeed be more or less probable to obtain estimates with values below (or above) the expectation value $\mathrm{E}[\hat{\xi}]$. It is also important to realize that an estimator may be biased even if it is consistent or unbiased even if inconsistent. For a more in depth discussion of this notion, see, for instance refs. [67, 116].

The linear nature of the expectation value operator also implies that while an estimator $\hat{\xi}$ might be unbiased, functions of this estimator are not necessarily unbiased. For instance, if η is an unbiased estimator of ξ, the square η^2 or the inverse $1/\eta$ are not unbiased because linear operators are not invariant under nonlinear variable transformations. A condition based on the median would guarantee such invariance but is often considered far less practical. Be that as it may, if η^2 is a consistent estimator of ξ^2, one can show that it will tend to be asymptotically unbiased. Consistency is thus usually considered, by statisticians, a more important property than the lack of bias. In practice, however, physicists typically do not have to worry too much about such subtleties.

4.4.4 Estimator Mean Square Error

The bias of an estimator determines its **accuracy**, that is, whether it tends toward the correct value (i.e., the true value) on average. One must also consider the estimator's **precision** and characterize the dispersion of values that might be obtained with successive measurements and the estimates computed from them. This is accomplished via a calculation of the **mean square error** (MSE), defined as

$$\mathrm{MSE} = E[(\hat{\theta} - \theta)^2]. \tag{4.13}$$

It is convenient to rewrite this expression by inserting $-E[\hat{\theta}] + E[\hat{\theta}]$ in the argument of the expectation value. One gets

$$\mathrm{MSE} = E[(\hat{\theta} - E[\hat{\theta}] + E[\hat{\theta}] - \theta)^2], \tag{4.14}$$

$$= E[(\hat{\theta} - E[\hat{\theta}])^2] + (E[\hat{\theta} - \theta])^2. \tag{4.15}$$

The first term of the second line corresponds to the variance $\mathrm{Var}[\hat{\theta}]$ of the estimator, whereas the second term is the square of its bias b. One thus finds that the MSE of an estimator equals the sum of its variance and the square of its bias:

$$\mathrm{MSE} = \mathrm{Var}[\hat{\theta}] + b^2. \tag{4.16}$$

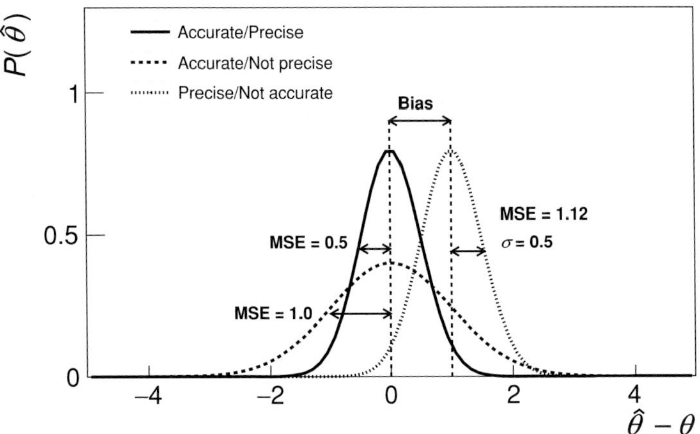

Fig. 4.2 Illustration of the concepts of **bias** and **Mean Square Error** (MSE).

The notions of bias and estimator mean square error were here introduced in terms of simple estimators or functions of the measured values, $\xi(\vec{x})$. These concepts are, however, readily extended to cases where model parameters (alternatively viewed as continuous model hypotheses) are obtained by maximization of a likelihood function or minimization of a χ^2 function. They in fact also apply to any type of measurement that extracts estimates of model parameters, whether based on a simple or composite hypotheses [116].

4.4.5 Epilog: Accuracy and Precision of Estimators

There is no unique method to construct estimators. Indeed, we will show later in this chapter that there exist several distinct criteria to define and build estimators, for a given model parameter θ, which produce distinct functional forms and associated PDF $g(\hat{\theta}|\theta)$. But, given a particular estimator, one can usually establish whether it has the desired properties, that is, whether it is consistent, whether it features a bias, has a small MSE, and so on. The notions of bias, estimator variance, and MSE are illustrated in Figure 4.2, which displays the PDFs of three hypothetical estimators $\hat{\theta}$, plotted as functions of the deviation $\hat{\theta} - \theta$.

An estimator and, more generally, a measurement method, are said to be unbiased if repeated measurements yield estimates that cluster around the actual value of the model parameter (null deviation on average). However, an estimator is said to be **biased** if the estimates it produces tend to systematically deviate from the actual value of the parameter. An experimental procedure and an estimator are said to be **accurate** if they yield results (e.g., estimates of model parameters or physical quantities) with no bias.

The variance of an estimator determines its precision. Obviously, smaller values of variance imply a more **precise** estimator and experimental technique. An estimator may, however, be very precise but biased and consequently not accurate. An experimental method and the estimator used to obtain parameter estimates should thus ideally be designed to have both high precision and accuracy. An estimator is deemed **optimal** if it has zero bias and minimum variance. Likewise, a measurement procedure is considered **valid** if it is both accurate and precise.

It is in the very nature of stochastic phenomena that by construction, all estimators have a limited precision, that is, a finite MSE. We show in §4.7.6 that the minimal variance of an estimator $\hat{\theta}$, called the **minimal variance bound** (MVB), is given by the following expression:

$$\text{MVB}(\hat{\theta}) = \left\langle \left(\frac{d}{d\theta} \ln L \right)^2 \right\rangle^{-1}, \tag{4.17}$$

where L is the likelihood function defined in Eq. (4.6). We will make use of this bound on several occasions later in this book.

We consider the formulation of basic estimators and examine some of their properties in the next section. Maximum likelihood and least square estimators are introduced in §5.1 and §5.2, respectively.

4.5 Basic Estimators

Consider a random sample $D = \{x_i\}$ consisting of n values x_i distributed according to a common but unknown PDF $p(x)$. Although it may not be possible to fully determine $p(x)$ based on the sample $\{x_i\}$, it is possible and may suffice, in many practical applications, to determine its lowest moments.

4.5.1 First Moment: Sample Mean

We first seek to obtain a function of the n values x_i that provides an estimator $\hat{\mu}$ of the expectation value $\mu = \text{E}[x]$. An obvious candidate for such an estimator is the **arithmetic mean**, $\bar{x}$, defined as

$$\hat{\mu} \equiv \bar{x} \equiv \frac{1}{n} \sum_{i=1}^{n} x_i. \tag{4.18}$$

It is commonly also called **sample mean** or just **mean**, and is often also noted $\langle x \rangle$.

By construction as a sum of the random variables x_i, $\bar{x}$ is a statistic and we can thus apply the techniques introduced in the previous paragraphs to calculate its value and properties. One expects $\bar{x}$ to deviate randomly from the expectation value $\mu = \text{E}[x]$ of the PDF $p(x)$. Indeed, one expects that repeated measurements of n values $\{x_i\}$ drawn from the same population, characterized by the PDF $p(x)$, should yield values of $\bar{x}$ that fluctuate, but nonetheless cluster about μ. For finite samples, deviations from μ are by construction finite and the variance of $\bar{x}$ is consequently also finite. One can show that the weak law of large numbers guarantees that $\bar{x}$ is a consistent estimator of the population mean μ [116].

It is also relatively straightforward to show that the estimator $\bar{x}$ converges to μ with zero bias in the limit $n \to \infty$.

The proof of unbiasedness is indeed relatively simple and also instructional. It proceeds on the basis of a calculation of the expectation value of $\bar{x}$ as follows:

$$\mathrm{E}[\bar{x}] = E\left[\frac{1}{n}\sum_{i=1}^{n} x_i\right]. \tag{4.19}$$

The evaluation of the expectation values requires an integral. However, the order of evaluation of this integral and the sum is inconsequential. One can thus write

$$\mathrm{E}[\bar{x}] = \frac{1}{n}\sum_{i=1}^{n} \mathrm{E}[x_i]. \tag{4.20}$$

The expectation value $\mathrm{E}[x_i]$ is by definition equal to

$$E\left[x_i\right] = \int \cdots \int x_i p(x_1)p(x_2)\cdots p(x_n)\, dx_1 dx_2 \cdots dx_n, \tag{4.21}$$

$$= \left(\int p(x_1)\, dx_1\right) \times \cdots \times \left(\int x_i p(x_i)\, dx_i\right) \times \cdots \times \left(\int p(x_n)\, dx_n\right), \tag{4.22}$$

$$= \int x_i p(x_i)\, dx_i = \mu, \tag{4.23}$$

in which we used the normalization of the PDF, $\int p(x_i)\, dx_i = 1$, and the definition of the first moment μ. Inserting this result in Eq. (4.20), we find

$$\hat{\mu} = \mathrm{E}[\bar{x}] = \frac{1}{n}\sum_{i=1}^{n} \mu = \mu. \tag{4.24}$$

We then conclude that the sample mean, $\bar{x}$, is indeed an unbiased estimator of the mean μ of the PDF $p(x)$.

4.5.2 Mean Square Error of the Sample Mean

It is useful to consider the variance $\mathrm{Var}[\bar{x}]$, or **Mean Square Error** (MSE), as an indicator of typical deviations of the sample mean, $\bar{x}$, from the actual value μ. $\mathrm{Var}[\bar{x}]$ gauges the variation of $\bar{x}$ for repeated experiments, that is, different sets $\{x_i\}$. As such, it can be used to estimate the error, or uncertainty, of $\bar{x}$ relative to the true value μ.

We evaluate the variance $\mathrm{Var}[\bar{x}]$ by inserting the expression of the sample mean into the definition of the variance:

$$\mathrm{Var}[\bar{x}] = \mathrm{E}[\bar{x}^2] - (\mathrm{E}[\bar{x}])^2, \tag{4.25}$$

$$= E\left[\left(\frac{1}{n}\sum_{i=1}^{n} x_i\right)^2\right] - \mathrm{E}\left[\frac{1}{n}\sum_{i=1}^{n} x_i\right]^2. \tag{4.26}$$

The calculation proceeds similarly as for the expectation value E[$\bar{x}$] (see Problem 4.5). One gets

$$\text{Var}[\bar{x}] = \frac{1}{n^2}\sum_{i,j=1}^{n} \text{E}[x_i x_j] - \frac{1}{n^2}\left(\sum_{i,=1}^{n} \text{E}[x_i]\right)^2, \tag{4.27}$$

$$= \frac{1}{n^2}\left(n\mu_2' + n(n-1)\mu^2\right) - \mu^2, \tag{4.28}$$

$$= \frac{1}{n}\left(\mu_2' - \mu^2\right), \tag{4.29}$$

$$= \frac{\sigma^2}{n}, \tag{4.30}$$

where μ_2' is the second moment of the PDF $p(x)$, and σ its (unknown) variance. We find that the variance of the sample mean $\bar{x}$ equals the variance of the PDF $p(x)$ representing the fluctuations of the random variable X divided by the number of measurements n. In other words, the standard error on the sample mean $\bar{x}$ is equal to the standard deviation of the PDF divided by the square root of the number of measurements carried out, that is, the size of the sample $\{x_i\}$:

$$\sigma_{\bar{x}} = (\text{MSE}\,[\bar{x}])^{1/2} = \frac{\sigma}{\sqrt{n}}. \tag{4.31}$$

The error on the sample mean thus depends on the PDF itself through σ, as well as the size of the sample used to estimate the mean. It can be arbitrarily reduced by increasing the sample size n. However, improvements are slow: to reduce the error by a factor of 2, one needs to quadruple the size of the data sample.

4.5.3 Estimator of the Variance

We next seek an estimator $\widehat{\sigma^2}$ of the variance σ^2 of the unknown distribution $p(x)$. Based on the expression of the sample mean, Eq. (4.18), it would seem reasonable to write

$$S^2 = \frac{1}{n}\sum_{i=1}^{n}(x_i - \bar{x})^2. \tag{4.32}$$

However, it turns out that this expression does not provide an unbiased estimator of σ^2. Indeed, let us evaluate its expectation value:

$$E\left[S^2\right] = E\left[\frac{1}{n}\sum_{i=1}^{n}(x_i - \bar{x})^2\right], \tag{4.33}$$

$$= \frac{1}{n}\sum_{i=1}^{n} E\left[\left(x_i - \frac{1}{n}\sum_{j=1}^{n} x_j\right)^2\right]. \tag{4.34}$$

We expand the square within the expectation value to get

$$\frac{1}{n}\sum_{i=1}^{n} E\left[x_i^2\right] - \frac{2}{n^2}\sum_{i=1}^{n} E\left[x_i \sum_{j=1}^{n} x_j\right] + \frac{1}{n^3}\sum_{i=1}^{n} E\left[\sum_{j=1}^{n} x_j \sum_{k=1}^{n} x_k\right]. \tag{4.35}$$

Since sum and expectation value operations commute, one can write

$$\frac{1}{n}\sum_{i=1}^{n} E\left[x_i^2\right] - \frac{2}{n^2}\sum_{i,j=1}^{n} E\left[x_i x_j\right] + \frac{1}{n^3}\sum_{i=1}^{n}\sum_{j,k=1}^{n} E\left[x_j x_k\right]. \tag{4.36}$$

One gets $E\left[x_i x_j\right] = E\left[x_i^2\right] = \mu_2'$ for $i = j$ and $E\left[x_i x_j\right] = E\left[x_i\right] E\left[x_j\right] = \mu^2$ for $i \neq j$ since the measurements are assumed to be uncorrelated. Equation (4.36) becomes

$$\tfrac{1}{n}\sum_{i=1}^{n}\mu_2' - \tfrac{2}{n^2}\sum_{i=1}^{n}\mu_2' - \tfrac{2}{n^2}\sum_{i\neq j=1}^{n}\mu^2 + \tfrac{1}{n^3}\sum_{i=1}^{n}\sum_{j=1}^{n}\mu_2' + \tfrac{1}{n^3}\sum_{i=1}^{n}\sum_{i\neq j=1}^{n}\mu^2. \tag{4.37}$$

The sums with $i \neq j$ involve $n(n-1)$ identical terms. Equation (4.37) thus simplifies to

$$\mathrm{E}[S^2] = E\left[\frac{1}{n}\sum_{i=1}^{n}(x_i - \bar{x})^2\right] = \frac{n-1}{n}\sigma^2, \tag{4.38}$$

from which we conclude that S^2 is indeed a biased estimator of the variance σ^2. Obviously, given that $\lim_{n\to\infty}(n-1)/n = 1$, we conclude S^2 is asymptotically unbiased. Note, however, that if the mean μ is known, then the estimator $\frac{1}{n}\sum_{i=1}^{n}(x_i - \mu)^2$ is unbiased (see Problem 4.6). In general, since the mean μ is a priori unknown, it remains desirable to find a nonbiased estimator of the variance. It is then possible to show (see Problem 4.7) that the estimator s^2, defined as

$$s^2 = \frac{1}{n-1}\sum_{i=1}^{n}(x_i - \bar{x})^2, \tag{4.39}$$

is in fact unbiased.

4.5.4 Mean Square Error of the Variance Estimator s^2

Following the procedure used in §4.5.2 to calculate the MSE of the sample mean, it is relatively simple, although somewhat tedious (see Problem 4.14), to show that the MSE of the estimator s^2 is given by

$$\mathrm{MSE}[s^2] = \mathrm{Var}[s^2] = \frac{1}{n}\left(\hat{\mu}_4 - \frac{n-3}{n-1}\hat{\mu}_2\right), \tag{4.40}$$

where the $\hat{\mu}_k$ are estimates of central moments (see Problem 4.9) of order k given by

$$\hat{\mu}_k = \frac{1}{n-1}\sum_{i=1}^{n}(x_i - \bar{x})^k. \tag{4.41}$$

Equation (4.40) reduces to the following expression for a Gaussian PDF:

$$\mathrm{MSE}[s^2] = \frac{2\sigma^4}{n^2}(n-1) \approx \frac{2\sigma^4}{n-1} \quad \text{Gaussian PDF.} \tag{4.42}$$

One thus concludes that the uncertainty on the estimate of the standard deviation, σ_s, is proportional to the actual variance of the PDF and scales inversely with the square root of the sample size n. This implies that for broadly distributed PDFs, a rather "large" sample must be acquired in order to obtain a reasonably precise estimate of the standard deviation.

4.5.5 Covariance and Pearson Coefficients

The result obtained in the previous section for the variance also applies to the covariance of joint measurements of two variables X and Y. Let us indeed consider a sample of n pairs of measurements $\{x_i, y_i\}$. One can show (see Problem 4.8) that the expression

$$\hat{V}_{xy} = \frac{1}{n-1}\sum_{i=1}^{n}(x_i - \bar{x})(y_i - \bar{y}) \tag{4.43}$$

provides an unbiased estimator of the covariance V_{xy} of the two variables X and Y.

It is also useful to consider the ratio, r_{xy}, of the covariance $\hat{V}_{xy}$ by the square root of the product of the variances of X and Y:

$$\hat{r}_{xy} = \frac{\hat{V}_{xy}}{\sqrt{s_x^2 s_y^2}} = \frac{\overline{xy} - \bar{x}\bar{y}}{\sqrt{\left(\overline{x^2} - \bar{x}^2\right)\left(\overline{y^2} - \bar{y}^2\right)}}. \tag{4.44}$$

This ratio is known as Pearson correlation coefficient.[4] It measures the strength of the correlation between the variables X and Y relative to their respective variances. When calculated for an entire population, the coefficient is typically designated by the Greek letter ρ (see §2.9.5) whereas the Latin letter r is used for a calculation based on a sample. It can be shown that $\hat{r}_{xy}$ is biased but converges to the Pearson coefficient ρ_{xy} in the large n limit (see Problem 4.15). One can also show (see Problem 4.16) that the coefficient lies in the range $[-1, 1]$. A value of 1 indicates that the variables Y and X are maximally correlated, and that a strict linear relation exists between the two variables:

$$y = |a|x. \tag{4.45}$$

Similarly, a value of -1 indicates that all data points lie on a single line $y = -|a|x$. A null value suggests there is no correlation between the variables.

4.5.6 Robust Moments

We saw in previous sections that the estimators of the moments of a distribution based on a sample of n values boil down to expressions of the form

$$\hat{\mu}'_k \equiv \langle x^k \rangle_n \equiv \frac{1}{n}\sum_{i=1}^{n} x_i^k \tag{4.46}$$

[4] First introduced by Francis Galton in the 1880s, and named after English mathematician Karl Pearson (1857–1936) generally credited with establishing the discipline of mathematical statistics.

for regular moments and

$$\hat{\mu}_k \equiv \langle (x - \hat{\mu}_1)^k \rangle_n \equiv \frac{1}{n} \sum_{i=1}^{n} (x_i - \langle x \rangle)^k . \tag{4.47}$$

for centered moments. While straightforward to implement in a computer program, these expressions feature important drawbacks. First, given the sums may proceed over a very large number of values, numerical errors are likely to occur, particularly for higher moments, due to rounding errors imposed by the finite number of bits used in floating point operations. Second, in the case of centered moments, the preceding formula requires one runs the analysis twice: once to obtain the mean $\langle x \rangle$, and once again to obtain the centered moments. Additionally, the two preceding formulas also present the disadvantage that the calculation must be run over all items x_i of the data sample before one obtains an estimate of the moments. Fortunately, these technical difficulties are easily circumvented by a simple reformulation of the preceding expressions.

Let us first consider the mean of a sample of n values and write

$$\begin{aligned} \langle x \rangle_n &= \frac{1}{n} \sum_{i=1}^{n} x_i, \\ &= \frac{1}{n} \left(x_n + \sum_{i=1}^{n-1} x_i \right), \qquad (4.48) \\ &= \frac{1}{n} x_n + \frac{(n-1)}{n} \langle x \rangle_{n-1}, \qquad (4.49) \end{aligned}$$

where in the third line, we have inserted $\langle x \rangle_{n-1}$ corresponding to the mean of the first $n-1$ values of the sample. Rearranging, we get

$$\langle x \rangle_n = \langle x \rangle_{n-1} + \frac{x_n - \langle x \rangle_{n-1}}{n}, \tag{4.50}$$

which says that the mean $\langle x \rangle_n$ obtained after n values is equal to the mean $\langle x \rangle_{n-1}$ obtained with $n-1$ values **plus** a correction Δ_n given by

$$\Delta_n = \frac{x_n - \langle x \rangle_{n-1}}{n}. \tag{4.51}$$

The same derivation technique can be applied to centered moments of order $k \geq 2$ (see Problem 4.17), and one obtains

$$\begin{aligned} \mu_k &= \langle (x - \langle x \rangle)^k \rangle \\ &= \tfrac{n-1}{n} \left\{ \left[1 - (1-n)^{k-1} \right] \Delta_n^k + \textstyle\sum_{l=2}^{k} \binom{l}{k} \langle (x - \langle x \rangle_n)^l \rangle_n \Delta_n^{k-l} \right\}, \qquad (4.52) \end{aligned}$$

which for $k = 2, 3$ may be written

$$\langle (x - \langle x \rangle)^2 \rangle_n = \frac{n-1}{n} \left\{ n\Delta_n^2 + \langle (x - \langle x \rangle)^2 \rangle_{n-1} \right\} \tag{4.53}$$

$$\langle (x - \langle x \rangle)^3 \rangle_n = \frac{n-1}{n} \left\{ (2n - n^2)\Delta_n^3 + 3 \langle (x - \langle x \rangle)^2 \rangle_n \Delta_n + \langle (x - \langle x \rangle_n)^3 \rangle_n \right\}. \tag{4.54}$$

The preceding expressions for centered moments are said to be **robust** because they involve numerically stable operations. Calculations proceed iteratively and do not involve addition or subtraction of huge numbers. Indeed, the correction Δ_n vanishes asymptotically for growing values of n, the calculated moments thus naturally tend to their "true" value as n iteratively increases. Values involved in the calculation all remain finite and rounding errors are thus minimized. The preceding expressions also involve the practical advantage that it is not necessary to proceed to an analysis of the full dataset in order to obtain the moments. The calculation of moments of all orders may be carried out iteratively and in parallel: one gets an estimate of all the required moments at each step n of the calculation. This enables "on the fly" analyses in which estimates of the moments may be obtained during the data acquisition process. It may then be possible to monitor the data collection, in real time, for shifts and varying trends that could indicate changes in the performance of detector components.

4.5.7 Other Estimators Commonly Used

Geometric Mean

Whenever the magnitude of a random variable is proportional to a product of n elementary random variables, x_i, it is convenient to replace the concept of arithmetic mean, which involves a sum of variables, with the **geometric mean**, defined as follows:

$$\bar{x}_{\text{geom}} = \sqrt[n]{x_1 x_2 \ldots x_n}. \tag{4.55}$$

The geometric mean finds applications in physics and engineering. A specific example involves the notion of average attenuation. Suppose a beam of light is directed through a series of n attenuators, each contributing a random attenuation, q_i, with $i = 1, \ldots, n$. The overall attenuation, Q, of the "signal" is thus equal to the product of all n attenuation factors, $Q = q_1 \times q_2 \times \ldots \times q_n$. The average attenuation, in other words, the attenuation per element, is thus $\bar{q} = \sqrt[n]{q_1 \times q_2 \times \ldots \times q_n}$, which is the geometric mean of the attenuations q_i. This means one could replace all n attenuators, q_i, by n identical attenuators, each yielding an attenuation of $\bar{q}$, and get the exact same result. A similar reasoning can be done with gain devices, such as amplifiers or photomultipliers.

Harmonic Mean

Another distribution characterization seldom encountered is the **harmonic mean**, $\bar{x}_{\text{harm}}$, of n values x_i defined as

$$\bar{x}_{\text{harm}} = \frac{n}{1/x_1 + 1/x_2 + \ldots + 1/x_n}. \tag{4.56}$$

Although less frequently used, this type of mean also has various applications. For instance, notes produced by the different strings of a guitar or violin sound pleasing (or harmonic) to the human ear[5] if their effective lengths are in the ratio 1: $\frac{1}{2}$: $\frac{1}{3}$:...The

[5] A fact originally noted by the ancient Greek Pythagoras.

harmonic mean of two notes is thus conveniently defined as twice the inverse of the sum of the reciprocal of their lengths. This stems from the fact that the human ear is sensitive to the frequency of the sound (the pitch), while the wavelength is determined by the length of the strings. The arithmetic mean of several frequencies is consequently equivalent to the harmonic mean of several string lengths.

4.6 Histograms

4.6.1 Basic Concept and Definition

Experimentally, various constraints limit the number of times a measurement of a particular observable X might be carried out. It is nonetheless desirable and of interest to generate simple visual representations of the acquired data. Indeed, whether one has acquired a sample of 10, 1,000, 1,000,000, or several billions data points, it is clearly useful to visualize these data in a graph. Different scenarios arise. It could be that the observable X of interest has a definite value, that is, is expected to take a single value. Measurement uncertainties may however lead to deviations from the actual (i.e., true) value of the observable. One may then be interested in visualizing how finitely many measured values $\{x_i\}$, $i = 1, \ldots, n$, are distributed. Alternatively, values of a continuous observable X could instead be expected to follow some model or probability distribution. It is then of interest to compare the density of measured values vs. x with the model or theoretical distribution. Either way, it is clear that plotting a graph of the number of instances or "events" as a function of the value of x is of interest. Indeed, one would like to compare the measured density distribution to the model. The problem, of course, is that we are dealing with a continuous variable and only dispose of a finite sample. One must then invoke a statistic, or estimator, that enables a sensible representation of the distribution of the continuous variable X. This is readily achieved by representing a set of measured values $\{x_i\}$, $i = 1, \ldots, n$, according to a construction consisting of the numbers of observed values, n_j, with $j = 1, \ldots, m$, in m finite width elements, called bins, of a partition of the range of allowed values of X. This construction is called a **histogram**.

In order to better understand the need for histograms, consider that many measurements can be essentially reduced to an experimental determination of the PDF of specific processes. In particle and nuclear physics, for instance, it is common to measure the momentum spectrum and angular distributions of particles produced in elementary interactions such as proton–proton, proton–nucleus, or nucleus–nucleus collisions at high energy. While one can model, at least in principle, the overall momentum and angular distributions, it is not possible to predict on a collision-by-collision, and particle-by-particle basis, what momenta and angle the particles will have. The momenta and angles are consequently random variables. However, a very large (infinite) number of measurements of momenta or production angles is expected to yield distributions of these variables that are representative of the nuclear interactions under consideration, that is, their respective PDFs. In practice, it is obviously impossible to carry out an infinite number of measurements. In fact, in many cases, the number of occurrences of a specific measurement outcome may be severely limited by small cross section, finite beam intensity (luminosity), experimental acceptance, or detection efficiency (see Chapter 12 for definitions of these concepts). Additionally, even if many occurrences are measured for a continuous variable X, it is in fact impossible to measure the probability of a specific value. It is, however, conceptually and practically meaningful to determine the number of occurrences in a finite interval of width Δx. In principle, one would prefer to carry out measurements with as small a value of Δx as possible. In practice, the measurement is meaningful only if a finite number of occurrences is observed in a given interval. One can therefore study histograms, in other words, the number of occurrences in segments of finite width Δx.

Histograms may be defined in one, two, or any number of dimensions and are used for the study of a single random variable as well as the simultaneous (joint) study of two or more variables. We begin our discussion of histograms with the definition of one-dimensional histograms in §4.6.2 and extend to two- and multidimensional histograms in §4.6.3.

4.6.2 One-Dimensional Histograms

A one-dimensional (1D) histogram of a single continuous variable X is defined over a finite range $x_{\min} \leq x < x_{\max}$. This range is divided or **binned** in a specific (but arbitrary) number, m, of contiguous segments, called **bins**. The bins are numbered from $j = 1$ to m. They may all be of equal width, $\Delta x = (x_{\max} - x_{\min})/m$, or defined to cover subranges of unequal widths with lower and upper bin boundaries noted $x_{\min,j}$ and $x_{\max,j}$, respectively. In the latter case, bin boundaries are usually defined to leave no gaps between contiguous bins, that is, such that $x_{\max,j} = x_{\min,j+1}$ for $j = 1, \ldots, m-1$.

The analysis of a dataset $\{x_i\}$ and the filling of the histogram proceed as follows. The bin contents, H_j, $j = 1, \ldots, m$ are initially set to zero. The values $\{x_i\}$, $i = 1, \ldots, n$, are then sequentially considered. Values x_i satisfying $x_{\min} \leq x_i < x_{\max}$ are set to generate an entry in the bin j such that $x_{\min,j} \leq x_i < x_{\max,j}$. Depending on the specific circumstances of an analysis, the entry may involve incrementing the bin content H_j by one unit (i.e., +1) or by some weight w_i determined by the value x_i. Once all entries have been carried out, the shape of the histogram is expected to be representative of the PDF of variable X. The histogram may finally be divided by a suitable normalization constant consisting, for instance, of the number of events analyzed, the total number of entries in the histogram, or the total weight of the entries. For unit increments (+1) per entry, division by the total number of entries yields normalized bin contents

$$h_i = H_i/n, \tag{4.57}$$

corresponding to the relative frequency of each bin. In the limit of an infinite number of measurements, $n \to \infty$, h_i is then expected to represent the probability, p_i, of occurrences of the random variable X in the range $x_{\min,i} \leq x < x_{\max,i}$,

$$\lim_{n\to\infty} h_i = p_i \equiv \int_{x_{\min,i}}^{x_{\max,i}} f(x)\,dx, \tag{4.58}$$

with the normalization[6]

$$\sum_{i=1}^{m} p_i = \int_{x_{\min}}^{x_{\max}} f(x)\,dx = 1. \tag{4.59}$$

The normalized bin content h_i is thus an estimator of the probability of obtaining a value of X in the range $x_{\min,i} \leq x < x_{\max,i}$. If resolution smearing and efficiency effects can be ignored (or at the very least neglected), the definition of h_i in terms of an arithmetic mean of the number of entries in the bin constitutes an unbiased estimator of the probability p_i of observing entries in that bin. Altogether, the bin contents h_i, $i = 1, \ldots, n$, thus provide an estimator of the probability distribution of the observable X.

When weights w_i are used to increment a histogram, H_i corresponds to the sum of all weights used to increment bin i. If the weights are themselves random variables, $h_i = H_i/n$ no longer represents a number of occurrences in bin i but is instead the mean weight in that bin:

$$\lim_{n\to\infty} h_i = \frac{1}{n}\langle \Sigma w_i \rangle = \int_{x_{\min,i}}^{x_{\max,i}} w f(x)\,dx. \tag{4.60}$$

The mean weight in bin i, denoted $\langle w_i \rangle$, is given by the ratio of $\langle \Sigma w_i \rangle$, and the number of entries in that same bin:

$$\langle w_i \rangle = \frac{\langle \Sigma w_i \rangle}{\langle \Sigma 1_i \rangle} = \frac{\int_{x_{\min,i}}^{x_{\max,i}} w f(x)\,dx}{\int_{x_{\min,i}}^{x_{\max,i}} f(x)\,dx}. \tag{4.61}$$

With weights equal to unity, and in the limit $n \to \infty$, we find that a histogram provides the probabilities p_i that the variable X occurs in the range $x_{\min,i} \leq x < x_{\max,i}$. Evidently, it is not possible to carry out an experiment with an infinite number of measurements. The values h_i one obtains experimentally thus provide estimates of the probabilities p_i, as defined in the foregoing. Globally, a normalized histogram h_i, $i = 1, \ldots, m$ thus provides an estimator of the relative probabilities p_i, $i = 1, \ldots, m$. Uncertainties associated with the estimates h_i are discussed later in this section.

Various plotting techniques are available to represent histograms graphically. Histograms are often represented as bar graphs such as those displayed in Figure 4.3. These histograms were created with a Monte Carlo (random) number generator (§13.3) according to a Gaussian distribution (§3.9) of mean 0.5, and standard deviation equal to 0.75. The histograms shown in panels (a), (b), (c), and (d) were obtained with sample sizes of 100, 1,000, 10,000, and 1,000,000 entries, respectively. While it is difficult to discern the exact shape of the distribution based on histogram (a), the shape becomes progressively more obvious with increased statistics, and with one million entries in panel (d), one can identify a Gaussian distribution. Panels (e) and (f) present histograms obtained with 100 distinct entries. The

[6] Obviously, this normalization applies only if there are no gaps between bins.

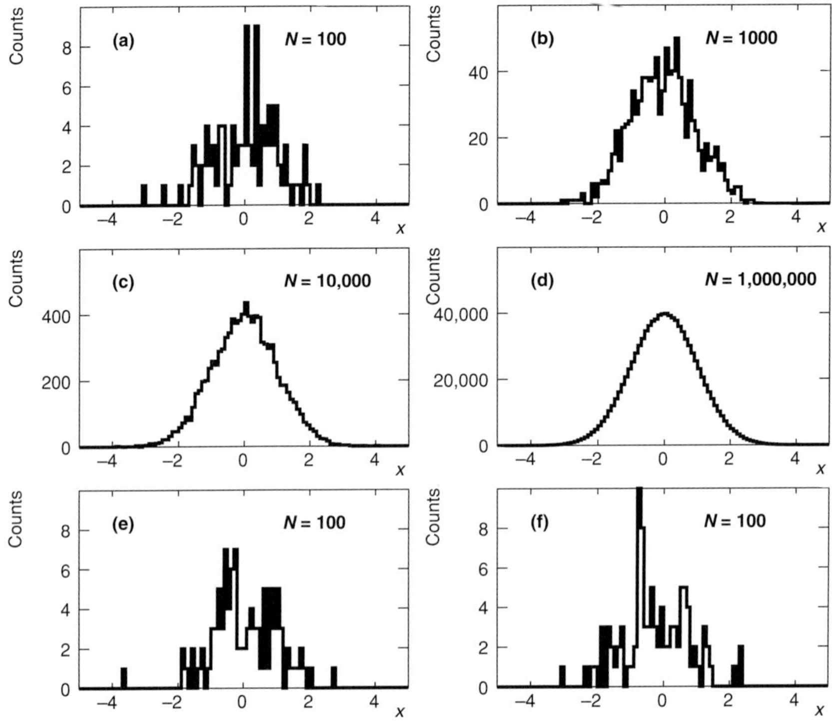

Fig. 4.3 Histograms obtained with a Monte Carlo generator designed to produce a standard normal distribution ($\mu = 0$, $\sigma = 1$). The size of the generated sample is shown in each panel. Histograms (a), (e), and (f) present distinct samples of size equal $n = 100$ which exemplify that distributions measured with fewer statistics (i.e., small sample) may considerably deviate from the actual PDF.

entries are mutually exclusive and different from those shown in panel (a). Comparing the three distributions (a), (e), and (f), one observes much variance in the shape, width, and height of the distributions. Obviously, with little statistics, it is difficult to precisely characterize the **parent** PDF, that is, the PDF that determines the measured distribution. But as the number of samples increases, one progressively becomes able to better discern and identify the shape of the distribution. This is typical of all measurements relying of finitely many samples or events.

While of evident interest, the quantities p_i defined earlier do not exactly correspond to the PDF, $f(x)$, one seeks to determine, but instead correspond to integrals of the PDF in bins $i = 1, n$. A theorem of elementary calculus, however, enables the determination of an estimate of the PDF. The theorem (used here without demonstration) stipulates that for each bin i, there exists a value $x_{i,0}$ such that

$$\int_{x_{\min,i}}^{x_{\max,i}} f(x)\,dx = f(x_{i,0})\,(x_{\max,i} - x_{\min,i}) = f(x_{i,0})\Delta x_i, \tag{4.62}$$

where $\Delta x_i = x_{\max,i} - x_{\min,i}$. It is important to realize that in general, the values $x_{i,0}$ do not correspond to the center of bins i. In the limit $n \to \infty$, given the value p_i, the value $f(x_{i,0})$ of the PDF at $x_{i,0}$ is then

$$f(x_{i,0}) = \frac{p_i}{\Delta x_i}. \tag{4.63}$$

For finite n, the value p_i is replaced by h_i, and one gets estimates $\hat{f}_i$ of the PDF value in each bin:

$$\hat{f}_i \equiv f_{\text{estimate}}(x_{i,0}) = \frac{h_i}{\Delta x_i}. \tag{4.64}$$

In histogram plots, the estimate $\hat{f}_i$ is usually represented as a continuous horizontal line across the bin width, as shown in Figure 4.3. If instead one wishes to plot $f(x)$ as an actual function, one must determine the abscissa x_i corresponding to the ordinate $\hat{f}_i$ in order to plot the data as a set of points $(x_i, \hat{f}_i)$. This is further discussed in Problem 2.10.

Given the finite size of the sample used in a measurement, one expects the measured estimates $\hat{f}_i$ might randomly deviate from the actual value of the function $f(x_i)$. It is thus also important to estimate the errors on each of these values in order to evaluate the precision of the measurement, and enable meaningful comparisons with other measurements or with models (see §6.4). If there are no bin-to-bin correlations, for experiments where the successive values of x measured are uncorrelated, the number of entries H_i in each bin is a random number determined by a Poisson distribution (§3.3) of mean μ_i given by

$$\mu_i = n \int_{x_{\min,i}}^{x_{\max,i}} f(x)\, dx = np_i, \tag{4.65}$$

and variance $\text{Var}[H_i] = \mu_i \approx H_i$. One therefore concludes that the values H_i shall converge to np_i in the large n limit. It also implies one can estimate the error (standard deviation) on the bin content as $\sqrt{H_i}$. The standard deviation of h_i is thus of the order of

$$\delta h_i \approx \text{Var}[h_i]^{1/2} = \sqrt{H_i}/n, \tag{4.66}$$

and since H_i equals nh_i, this amounts to the commonly known result $\delta h_i = \sqrt{h_i/n}$. The relative error of the estimate $\hat{f}_i = h_i/\Delta x_i$ is then

$$\frac{\delta \hat{f}_i}{\hat{f}_i} = \frac{\sqrt{h_i/n}}{h_i} = \frac{1}{\sqrt{n}} \frac{1}{\sqrt{h_i}} \to \frac{1}{\sqrt{n}} \frac{1}{\sqrt{p_i}}. \tag{4.67}$$

One concludes, as expected, that the relative error, $\delta \hat{f}_i/\hat{f}_i$, on an estimate of the PDF improves as the square root of the number of measurements n carried to determine the PDF. This is costly. Indeed, to get a reduction of the errors by a factor of 2, one must quadruple the statistics, whereas to achieve an improvement by a factor of 10, one must increase the dataset by a factor of 100, and so on. Consequently, one finds that significantly improving known results requires considerable increase of the size of existing data samples. In nuclear and particle physics experiments, this translates into demands for substantial increases in the beam luminosity (intensity), the duration of data acquisition periods, as well as detector acceptance and efficiency.

4.6.3 Multidimensional Histograms

Two-dimensional (2D) and multidimensional histograms are used to represent data with dependencies on two or more variables. They are defined and filled in much the same way as 1D histograms. The following discussion is limited to 2D histograms but can be readily extended to multidimensional histograms.

To construct a 2D histogram, one must first define the ranges $x_{\min} \le x < x_{\max}$, $y_{\min} \le y < y_{\max}$ and number of bins m_x and m_y used to plot the two random variables (here labeled x and y) of interest. Modern plotting software packages typically enable the creation of 2D histograms with different types of partitions: one can use bins of fixed or variable width on both axes, or mix fixed and variable widths on the two axes. It is customary to identify the bins using a pair of indices (i, j) where i and j range from 1 to m_x and m_y, respectively. A histogram is filled using a data sample consisting of n pairs $(x, y)_k$, with $k = 1, \ldots, n$. Given the variables are random observables, one expects the content of the bins, H_{ij}, to also be random variables. The content of each histogram is readily converted into an estimate of the joint probability, h_{ij}, of observing pairs (x, y) in bin (i, j) by dividing H_{ij} by the size, n, of the event sample. Given a joint PDF, $f(x, y)$, and in the large n limit, the probability, p_{ij}, of observing counts in bin (i, j) is given by

$$p_{ij} \equiv \lim_{n\to\infty} h_{ij} = \int_{x_{\min,i}}^{x_{\max,i}} \int_{y_{\min,j}}^{y_{\max,j}} f(x, y)\, dxdy. \tag{4.68}$$

As for 1D histograms, one notes that although the values p_{ij} enable a representation of $f(x, y)$, they are not equal to the PDF $f(x, y)$ but rather constitute, as per Equation (4.68), integrals of the probability density. However, values $(x_{o,i}, y_{o,j})$ exist for each bin such that

$$p_{ij} = f(x_{o,i}, y_{o,j})\Delta x_i \Delta y_j \tag{4.69}$$

where $\Delta x_i = x_{\max,i} - x_{\min,i}$ and $\Delta y_i = y_{\max,i} - y_{\min,i}$. Note that $(x_{o,i}, y_{o,j})$ does not usually coincide with the middle of the bin. Given a measured sample, h_{ij}, one can therefore obtain an approximation $\hat{f}$ of the joint PDF using

$$\hat{f}(x_{o,i}, y_{o,j}) = \frac{h_{ij}}{\Delta x_i \Delta y_j}. \tag{4.70}$$

As for 1D distributions, the relative errors $\delta \hat{f}_{ij}/\hat{f}_{ij}$ are given by

$$\frac{\delta \hat{f}_{ij}}{\hat{f}_{ij}} = \frac{\sqrt{h_{ij}/N}}{h_{ij}} = \frac{1}{\sqrt{n}}\frac{1}{\sqrt{h_{ij}}} \to \frac{1}{\sqrt{n}}\frac{1}{\sqrt{p_{ij}}}. \tag{4.71}$$

There are several techniques to graphically represent 2D histograms. Figure 4.4 illustrates various popular representations: (a) a scatterplot where the density of points is proportional to the bin content, (b) z-color where a range of colors (or shades of gray) is used

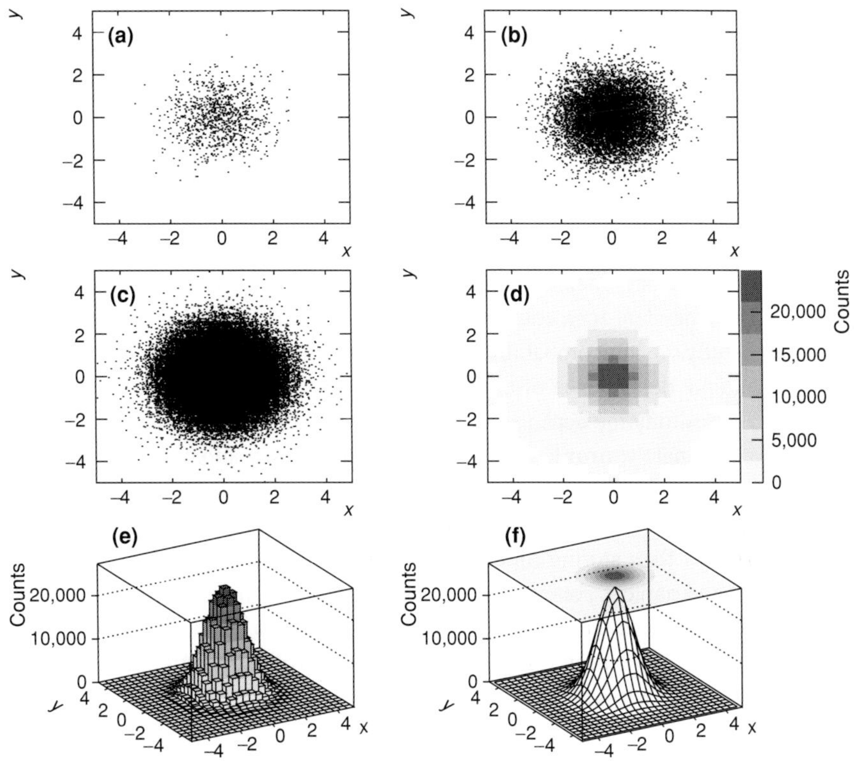

Fig. 4.4 Illustration of the notion of 2D histograms based on a 2D standard normal distribution ($\mu_x = \mu_y = 0$, $\sigma_x = \sigma_y = 1$): (a–c) scatterplots with 1,000, 10,000, and 1,000,000 entries. Panels (d, f) present alternative plotting methods (contour, lego, hybrid) for a histograms with 1,000,000 entries.

to represent the amplitude of the signal, (c) contour plot where lines show values of equal amplitude, as well as (d) lego plot, (e) surface plot, and (f) mixed z-color + surface where the amplitude is shown explicitly in a three-dimensional plot. Various other graphing techniques are available in plotting packages such as ROOT, MATLAB®, or Mathematica™. In principle, projections of the estimate of joint probability, h_{ij}, onto the x- and y-axes yield estimates of the marginal probabilities $\hat{f}_x(x)$ and $\hat{f}_y(y)$, respectively. Note, however, that since 2D histograms are typically filled using specific boundaries (both along the x- and y-axes), the projections are likely to be incomplete and not yield the full integrals given by Eqs. (2.106, 2.107) unless the probabilities p_{ij} are negligible outside the boundaries of the histogram. This means that the projections are estimates of the conditional probability densities $h_x(x|y_{\min} \le y < y_{\max})$ and $h_y(y|x_{\min} \le x < x_{\max})$. They provide PDFs of x and y given y and x, respectively, are in the ranges $y_{\min} \le y < y_{\max}$ and $x_{\min} \le x < x_{\max}$ defined by the boundaries of the histogram.

Slices of the histogram h_{ij} in y and x projected onto the x- and y-axis, respectively, may be used to obtain estimates of the conditional probabilities $h_x(x|y)$ and $h_y(y|x)$. Note, however, that the finite bin width used to create and fill histograms implies these estimates are valid for ranges of y and x, respectively, corresponding to the widths of the bins along these two axes. Slices consisting of several rows or columns can of course be added together to improve the statistical significance of the data. But this brings about a loss in the resolution of the conditions on y and x used to define the slices.

4.6.4 Profile Histograms

In the study of multivariate systems, one often encounter situations in which a variable Y exhibits significant event-to-event fluctuations as well as an explicit dependence on some independent (control) variable X. A two-dimensional histogram can evidently be used to study the joint probability of the two variables, but in many cases one is interested primarily in the dependence of the average $E[Y]$ on the variable X. Creating a 2D histogram Y vs. X to study the dependence of Y on X may then be fastidious and somewhat of an overkill. Fortunately, **profile histograms** provide an easy and convenient alternative.

Profile histograms are based on the same notion of partition of the domain of one of several control variables X, and can be drawn in one, two, or multiple dimensions. However, rather than storing multiple content bins for the dependent variable (Y), for each bin of the variable X, one reduces the information to three quantities for each X bin: the number of entries in the bin i, hereafter noted n_i; an estimate of the mean of Y, $\langle y_i \rangle$; and the variance, $\langle \Delta y_i^2 \rangle$, of the signal Y in the bin. A profile histogram is "filled" similarly as a regular histogram by calling a "fill" function

$$\text{profileHisto.fill}(x, y) \tag{4.72}$$

for a pair of values (x, y), where x is an instance value of the independent variable X and y a jointly measured value of the dependent variable Y. However, rather than incrementing a bin content, a profile histogram implements Eqs. (4.50, 4.53) to obtain robust estimates of the mean, $\langle y_i \rangle$, and the variance, $\langle \Delta y_i^2 \rangle$, of the signal Y in each bin i based on successive entries y. The mean and variance of Y in a given bin i (of X) are thus calculated for successive entries in that bin according to

$$\Delta_i = \frac{y - \langle y_i \rangle}{n_i + 1}, \tag{4.73}$$

$$\langle y_i \rangle' = \langle y_i \rangle + \Delta_i, \tag{4.74}$$

$$\langle \Delta y_i^2 \rangle' = \frac{n_i}{n_i + 1} \left[(n_i + 1)\Delta_i^2 + \langle \Delta y_i^2 \rangle \right], \tag{4.75}$$

$$n_i' = n_i + 1, \tag{4.76}$$

where primed quantities stand for updated values obtained after the inclusion of the $(n_i + 1)$th entry. The profile histogram may then be examined at any stage of an analysis to obtain estimates of the mean, $\langle y_i \rangle$, and centered moment, $\langle \Delta y_i^2 \rangle$, in each bin i. Additionally, for sufficiently large number of entries, n_i, the variance $\langle \Delta y_i^2 \rangle$ may be used to estimate the error on the mean in each bin according to

$$\delta \langle y_i \rangle = \sqrt{\frac{\langle \Delta y_i^2 \rangle}{n_i}}. \tag{4.77}$$

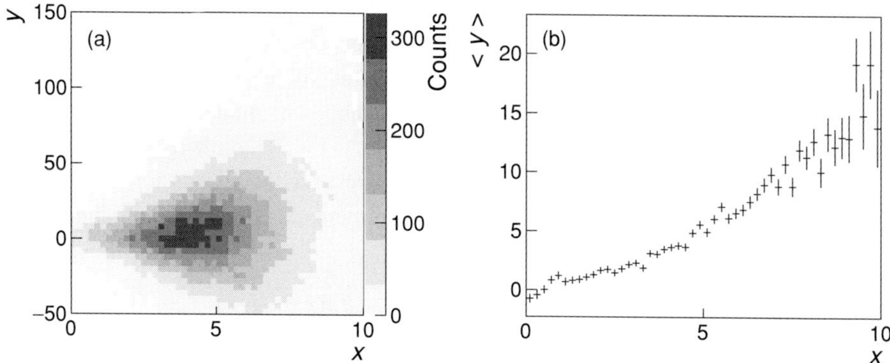

Fig. 4.5 Illustration of the concept of profile histogram. (a) 2D histogram showing the joint distribution of variables X and Y. (b) Profile histogram displaying the dependence of the average of Y on X. Error bars displayed indicate uncertainties on the mean of the signal Y.

This provides the additional advantage that it is readily possible to produce plots of the average $\langle Y\rangle(x)$ with errors bars corresponding to the error on the mean, or alternatively, the standard deviation of the signal.

Figure 4.5a presents an example of a joint distribution (2D histogram) of two variables X and Y, where the dependent variable Y exhibits large fluctuations at any given value of X. The dependence of Y on X, difficult to discern based on this 2D histogram, is readily identified in the profile histogram shown in panel (b) of the same figure.

4.7 Fisher Information

Given a probability model $p(x|\vec{\theta}, H, I)$ for an observable X and model H, it is legitimate to ask how measurements of X may constrain the values of the model parameters $\vec{\theta}$, that is, how much information is contained in the data and the functional form of $p(x|\vec{\theta}, H, I)$ that can constrain the value of the parameters $\vec{\theta}$. Two distinct quantities, the entropy S and the Fisher information $\mathbb{I}(\theta)$, are commonly used to characterize this information. The Fisher information and examples of calculation for specific PDFs are presented in this section, whereas the notion of entropy S and its connection to the Fisher information are discussed in §7.3.1. A technique to obtain a minimal bound on the variance of model parameters based on the Fisher information is presented in §4.7.6.

This section may be skipped in a first reading of this chapter.

4.7.1 A Basic Derivation of the Fisher Information

The function $p(x|\vec{\theta}, H, I)$ amounts to the probability (density) of observing values of x given model parameters $\vec{\theta}$. Changing the values of the model parameters modifies the probability and the likelihood of x being observed in any given range: the parameters determine the distribution of observable values x. One can thus say that the PDF $p(x|\vec{\theta}, H, I)$ contains some kind of "information" that dictates what values x may be observed given specific parameters $\vec{\theta}$. Conversely, given a measured value of x, the functional form of a PDF should also provide information about the (unknown) model parameters $\vec{\theta}$. Let us then examine how one can quantify this information.

For simplicity's sake, let us first restrict the discussion to a single parameter PDF $p(x|\theta, H, I)$. Changing θ by an increment $\Delta\theta$ produces a change in probability Δp approximately equal to

$$\Delta p \approx \frac{\partial p(x|\theta, H, I)}{\partial \theta} \Delta\theta. \tag{4.78}$$

The derivative $\partial p/\partial\theta$ determines how a change in θ affects the probability distribution. It also influences how a change in the measured distribution (i.e., values of x being more or less probable) impacts the most likely values of the parameter θ. The slope function $\partial p/\partial\theta$ thus contains much of the information one needs to establish a connection between the data and the model parameter. The magnitude of the probability (density) also matters, however. A given increment Δp should be far more significant if it effects a comparatively large change on the probability density. It is thus more relevant to consider relative changes of the density, that is, how the ratio $\Delta p/p$ is affected by a change $\Delta\theta$ of the parameter θ. To this end, one introduces the notion of **score**, hereafter noted ξ, and defined as

$$\xi(x|\theta) = \frac{\partial}{\partial \theta} \ln p(x|\theta, H, I)\bigg|_{\theta} \tag{4.79}$$

$$= \frac{1}{p}\frac{\partial p}{\partial \theta}\bigg|_{\theta}. \tag{4.80}$$

A high score means a given increment $\Delta\theta$ has a large impact on the probability density whereas a small score produces only a modest change. Evidently, given the density is a function of x, the score is itself also a function of x. One should thus consider the average score of the distribution; that is, one should calculate its expectation value $E[\xi]$. It is straightforward to verify, however, that this expectation value vanishes for all values of θ. Indeed, for "well-behaved" distributions, one has

$$\mathrm{E}[\xi] = \int \frac{1}{p(x|\theta)} \frac{\partial p(x|\theta)}{\partial \theta} p(x|\theta)\, dx, \tag{4.81}$$

$$= \int \frac{\partial p(x|\theta)}{\partial \theta} dx = \frac{\partial}{\partial \theta} \int p(x|\theta)\, dx, \tag{4.82}$$

$$= \frac{\partial}{\partial \theta} 1 = 0, \tag{4.83}$$

where in the second line, we assumed the density $p(x|\theta)$ is a regular function such that the order of the integral vs. x and the derivative with respect to θ may be interchanged, while in the third line, we used the normalization of the function. Clearly, the foregoing null expectation value is of little interest because it does not provide any means to use the score to extract information about the probability density $p(x|\theta)$. The second moment of the score, called Fisher information $\mathbb{I}(\theta)$, does, however.

$$\mathbb{I}(\theta) = \int \left[\frac{1}{p(x|\theta)} \frac{\partial p(x|\theta)}{\partial \theta} \right]^2 p(x|\theta)\, dx, \tag{4.84}$$

Since the integrant is by construction nonnegative, the Fisher information $\mathbb{I}(\theta)$ is positive definite for all values of θ, and its magnitude constitutes an indicator of the amount of information contained in the PDF. Probability densities with high scores in regions of high probability yield large integral values whereas densities with small scores yield small values of $\mathbb{I}(\theta)$. The Fisher information $\mathbb{I}(\theta)$ indeed constitutes an indicator of the information contained in a PDF $p(x|\theta)$.

It is interesting to note that given the expectation value of the score is null, the Fisher information amounts to the variance of the score. Densities with large Fisher information have a large score variance and provide better capabilities to determine and constrain model parameters.

The Fisher information may also be calculated in terms of the expectation value of the second derivative of the score. To demonstrate this statement, first compute the second derivative of the score relative to θ

$$\frac{\partial^2}{\partial \theta^2} \ln p(x|\theta) = \frac{\partial}{\partial \theta} \left(\frac{1}{p} \frac{\partial p}{\partial \theta} \right), \tag{4.85}$$

$$= \frac{1}{p} \frac{\partial^2 p}{\partial \theta^2} - \frac{1}{p^2} \left(\frac{\partial p}{\partial \theta} \right)^2, \tag{4.86}$$

and calculate its expectation value across the domain of x. Since

$$\mathrm{E}\left[\frac{1}{p} \frac{\partial p}{\partial \theta} \right] = \int \frac{\partial p}{\partial \theta}\, dx = \frac{\partial}{\partial \theta} \int p\, dx = 0, \tag{4.87}$$

$$\mathrm{E}\left[\frac{1}{p} \frac{\partial^2 p}{\partial \theta^2} \right] = \int \frac{\partial^2 p}{\partial \theta^2}\, dx = \frac{\partial^2}{\partial \theta^2} \int p\, dx = 0, \tag{4.88}$$

it follows, using Eq. (4.84), that

$$\mathbb{I}(\theta) = -\mathrm{E}\left[\frac{\partial^2}{\partial \theta^2} \ln p(x|\theta) \right], \tag{4.89}$$

where the expectation value is evaluated at a fixed value of θ.

4.7.2 Fisher Information Matrix

For a probability model involving m parameters $\vec{\theta} = (\theta_1, \theta_2, \ldots, \theta_m)$, the Fisher information becomes a covariance matrix with elements

$$\mathbb{I}_{ij}(\vec{\theta}) = \mathrm{Cov}\left[\frac{\partial}{\partial \theta_i} \ln p(x|\vec{\theta}), \frac{\partial}{\partial \theta_j} \ln p(x|\vec{\theta}) \right], \tag{4.90}$$

which for regular functions may be written as (Problem 4.19)

$$\mathbb{I}_{ij}(\vec{\theta}) = -\mathrm{E}\left[\frac{\partial^2}{\partial \theta_i \partial \theta_j} \ln p(x|\vec{\theta}), \right]. \tag{4.91}$$

4.7.3 Fisher Information Properties

The Fisher information features several important properties.

Additivity

Given two observables X and Y, the information obtained from independent measurements of these observables may be written

$$\mathbb{I}_{x,y}(\theta) = \mathbb{I}_x(\theta) + \mathbb{I}_y(\theta), \tag{4.92}$$

where $\mathbb{I}_x(\theta)$ and $\mathbb{I}_y(\theta)$ are the information obtained from measurements of X and Y, respectively. The additivity results from the fact that the variance of a sum of two independent variables is equal to the the sum of their respective variances.

Scaling

The property of additivity is readily extended to an "experiment" consisting of n elementary measurements of an observable X sampled from the same distribution $p(x|\theta)$. The Fisher information of the likelihood $p(\vec{x}|\theta)$ of the measured data $\vec{x} = (x_1, x_2, \ldots, x_n)$ is equal to n times the Fisher information of an individual measurement.

$$\mathbb{I}_{\vec{x}}(\theta) = n\mathbb{I}_x(\theta). \tag{4.93}$$

In this context, the Fisher information obtained from a single measurement of an observable X is commonly called **unit Fisher information** to distinguish it from that obtained from a sample of n measurements.

Transformation

The Fisher information of a probability model parameter θ depends on the derivative of the distribution relative to this parameter. But the parameterization of a model can be changed without changing its meaning. For instance, with a Gaussian PDF, specification of the width of the distribution may be done in terms of the standard deviation σ, the variance σ^2, or the precision $\xi = \sigma^{-2}$. These three parameters have different meanings, physical units, and numerical values, but the Gaussian PDF they define remains the same. In some fashion, the information associated to these different parameters cannot be too different. Indeed, their information must be related to one another. More generally, it is interesting to consider how the Fisher information $\mathbb{I}(\vec{\theta})$ transforms under a **reparametrization** $\vec{\theta} \rightarrow \vec{\lambda}(\vec{\theta})$.

Let us first consider a probability model involving a single parameter θ and calculate the information in terms of λ based on $\theta \equiv \theta(\lambda)$

$$\mathbb{I}_\lambda(\lambda) = \int \left[\frac{1}{p(x|\theta(\lambda))} \frac{\partial p(x|\theta(\lambda))}{\partial \lambda} \right]^2 p(x|\theta(\lambda))\, dx, \tag{4.94}$$

where we included the dependency of the parameter θ on λ explicitly to emphasize that the information is considered a function of λ rather than θ. Using a chain rule for the partial derivative, we get

$$\mathbb{I}_\lambda(\lambda) = \int \left[\frac{1}{p(x|\theta)} \frac{\partial p(x|\theta)}{\partial \theta} \frac{\partial \theta}{\partial \lambda} \right]^2 p(x|\theta)\, dx, \tag{4.95}$$

$$= \left(\frac{\partial \theta}{\partial \lambda} \right)^2 \int \left[\frac{1}{p(x|\theta)} \frac{\partial p(x|\theta)}{\partial \theta} \right]^2 p(x|\theta)\, dx, \tag{4.96}$$

where the probability is now considered a function of θ rather than λ and the square of the derivative $\partial\theta/\partial\lambda$ was factored out of the integral because it is independent of the observable x. We thus conclude that under a reparameterization $\theta \equiv \theta(\lambda)$, the Fisher information transforms according to

$$\mathbb{I}_\lambda(\lambda) = \mathbb{I}_\theta(\theta(\lambda)) \left(\frac{\partial \theta}{\partial \lambda} \right)^2. \tag{4.97}$$

For a multiparametric model involving m parameters, $\vec{\theta} = (\theta_1, \theta_2, \ldots, \theta_m)$, we saw in §4.7.2 that the information takes the form of a matrix. We may then consider how the information changes under a transformation $\vec{\theta} \to \vec{\lambda}$ defined by the $m \times m$ Jacobian matrix

$$\mathbf{J}_{ij} \equiv \frac{\partial \theta_i}{\partial \lambda_j}, \tag{4.98}$$

where the indices i and j span the m model parameters. Repeating the chain rule derivation, Eq. (4.95), for each pair of parameters $i, i' \to j, j'$, one finds

$$\mathbb{I}_{\lambda_j,\lambda_{j'}}(\vec{\lambda}) = \mathbb{I}_{\theta_i,\theta_{i'}}(\vec{\theta}) \left(\frac{\partial \theta_i}{\partial \lambda_j} \right) \left(\frac{\partial \theta_i'}{\partial \lambda_j'} \right). \tag{4.99}$$

The transformation may also be written in a convenient matrix form

$$\mathbb{I}_{\vec{\lambda}}(\lambda) = \mathbf{J}^T \mathbb{I}_{\vec{\theta}}(\vec{\theta}(\vec{\lambda}))\mathbf{J}. \tag{4.100}$$

We will use the transformation property of the Fisher information when discussing the notion of information invariance forming the basis of Jeffreys' priors introduced in §7.3.2.

4.7.4 Example 1: Fisher Information of a Binomial PDF

Consider a nuclear physics experiment involving a measurement of the rate of production of a certain particle species in a specific nuclear collision at a specific beam energy. The measurement is carried out with a given detector and within a specific kinematic range. Nominally, produced particles by the collisions are detected and counted within the apparatus. However, particle losses associated with various instrumental causes invariably occur. The number of particles detected, n, thus represents only a fraction of the number produced N and one defines the particle detection efficiency as the ratio of the number of observed particles by the number of produced particles:

$$\varepsilon \equiv \frac{n}{N}. \tag{4.101}$$

The efficiency is largely determined by the design of the apparatus but may be influenced by a host of factors internal and external to the apparatus. In practice, it is typically difficult

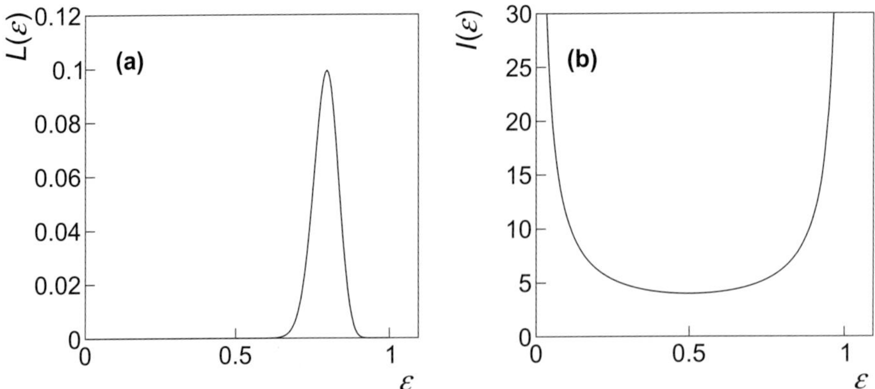

Fig. 4.6 (a) Plot of the likelihood function of an efficiency parameter ε given the successful detection of $n = 80$ particles and the prior knowledge that $N = 100$ particles actually passed through the detection apparatus. (b) Fisher information $\mathbb{I}(\varepsilon)$ contained in the binomial distribution plotted as a function of the efficiency (success rate) parameter ε.

to carry out, ab initio, a precise calculation of the efficiency of a specific device and one is often required to carry out a controlled measurement where both n and N are explicitly determined to calculate the efficiency. The difficulty arises that a measurement is intrinsically a stochastic process, akin to the toss of a coin. For a given particle passing through the detector, there is a probability ε that the particle will be detected and a probability $1 - \varepsilon$ that it will not. Assuming, as in §3.1.1, that the detection of any given particle is independent to that of others, the detection process can be modeled with a Bernoulli probability model

$$p(x|\varepsilon) = \varepsilon^x (1 - \varepsilon)^{1-x}, \tag{4.102}$$

where a value $x = 1$ represents a success, that is, a particle being detected, and a value $x = 0$ corresponds to a "failure," that is, a particle remaining undetected. Clearly, just a single toss of a coin is insufficient to determine whether a coin is fair, one needs to attempt the detection of a large number of particles to obtain an estimate of the efficiency. Let us thus consider a succession of N independent particles passing through the detector. As for coin tosses, the total number of success n (i.e., particle detected) relative to the number of trials N shall be given a binomial distribution

$$p(n|N, \varepsilon) = \frac{N!}{n!(N-n)!} \varepsilon^n (1 - \varepsilon)^{N-n}. \tag{4.103}$$

The function $p(n|N, \varepsilon)$ determines the likelihood of n particles being observed given N are produced when the detection efficiency is ε. But ε is unknown, and it is the purpose of the experiment to determine its value. One can then invoke the principle of maximum likelihood, discussed in §5.1, and obtain the $\mathbb{ML}$ estimate of ε as n/N, in which n is the number of actually detected particles, and N is the number of particles known to pass through the detector.

The likelihood function $p(n|N, \varepsilon)$ is plotted as a function of ε in Figure 4.6a for $N = 100$ and $n = 80$. One finds that the function has a maximum at $\varepsilon = 0.8$ corresponding indeed to the $\mathbb{ML}$ estimator value $80/100$. We evaluate the (unit) Fisher information of the binomial

PDF using Eq. (4.84):

$$\mathbb{I}(\varepsilon) = -\mathrm{E}\left[\frac{\partial^2}{\partial \varepsilon^2} \ln p(x|\varepsilon)\right] \tag{4.104}$$

$$= -\mathrm{E}\left[\frac{\partial^2}{\partial \varepsilon^2} \ln\left(\frac{N!}{n!(N-n)!}\varepsilon^n (1-\varepsilon)^{N-n}\right)\right] \tag{4.105}$$

$$= \mathrm{E}\left[\frac{n}{\varepsilon^2} + \frac{N-n}{(1-\varepsilon)^2}\right], \tag{4.106}$$

where ε is considered a constant for the purpose of the evaluation of the expectation value. Given the expectation value of n is equal to $N\varepsilon$, we find the Fisher information for a binomial distribution is

$$\mathbb{I}(\theta) = \frac{1}{\varepsilon^2}\mathrm{E}[n] + \frac{N - E[n]}{(1-\varepsilon)^2}, \tag{4.107}$$

$$= \frac{N\varepsilon}{\varepsilon^2} + \frac{N - N\varepsilon}{(1-\varepsilon)^2}, \tag{4.108}$$

$$= \frac{N}{\varepsilon(1-\varepsilon)}. \tag{4.109}$$

Equation (4.109) is plotted in Figure 4.6b. One observes that for values $\varepsilon = 1$ and $\varepsilon = 0$, the information diverges thereby reflecting the fact that the outcome of N trials is absolutely certain, that is, all or none of the particles are detected. The information reaches a minimum at $\varepsilon = 0.5$ corresponding to the value of ε for which outcomes are least certain. Indeed, for $\varepsilon = 0.5$, the odds of getting a success (particle detected) or failure (not detected) are equal and the outcome of a trial (measurement) is thus maximally uncertain.

4.7.5 Example 2: Fisher Information of a Gaussian Distribution

Let us calculate the Fisher information matrix of a Gaussian PDF:

$$p(x|\mu,\sigma^2) = \frac{1}{\sqrt{2\pi}\sigma} \exp\left[-\frac{(x-\mu)^2}{2\sigma^2}\right], \tag{4.110}$$

with unknown mean μ and variance σ^2. As per the definition, Eq. (4.91),

$$\mathbb{I}(\mu,\sigma^2) = -E\left[\begin{pmatrix} \frac{\partial^2}{\partial\mu^2} \ln p(x|\mu,\sigma^2) & \frac{\partial^2}{\partial\mu\partial\sigma^2} \ln p(x|\mu,\sigma^2) \\ \frac{\partial^2}{\partial\sigma^2\partial\mu} \ln p(x|\mu,\sigma^2) & \frac{\partial^2}{\partial\sigma^2\partial\sigma^2} \ln p(x|\mu,\sigma^2) \end{pmatrix}\right], \tag{4.111}$$

this requires calculation of derivatives of the log of the PDF with respect to μ and σ^2. First taking the log of $p(x|\mu,\sigma^2)$, we get

$$\ln p = -\frac{1}{2}\ln 2\pi - \frac{1}{2}\ln\sigma^2 - \frac{(x-\mu)^2}{2\sigma^2}. \tag{4.112}$$

Second derivatives of ln p yield

$$\frac{\partial^2}{\partial\mu^2}\ln p = -\frac{1}{\sigma^2}, \tag{4.113}$$

$$\frac{\partial^2}{\partial\sigma^2\partial\mu}\ln p = -\frac{x-\mu}{\sigma^4}, \tag{4.114}$$

$$\frac{\partial^2}{\partial\sigma^2\partial\sigma^2}\ln p = \frac{1}{2\sigma^4} - \frac{(x-\mu)}{\sigma^6}. \tag{4.115}$$

Inserting the preceding expressions in Eq. (4.111), and remembering that the expectation value of x is the mean μ, and that of $(x-\mu)^2$ is σ^2, the Fisher information matrix of a Gaussian PDF may then be written

$$\mathbb{I}(\mu,\sigma^2) = \begin{pmatrix} \frac{1}{\sigma^2} & 0 \\ 0 & \frac{1}{2\sigma^4} \end{pmatrix}. \tag{4.116}$$

We find that the information one may gain on the mean by measurements of x is inversely proportional to the variance σ^2. For a small or vanishing variance, the information diverges and the precision achievable in the determination of μ is thus very large. Conversely, for a large variance, the information is small, and the achievable precision rather limited. As for the variance, one finds that the information contained in $p(x|\mu,\sigma^2)$ is inversely proportional to the square of its value. Again in this case, one concludes that if the variance is small, it shall be relatively straightforward to obtain precise estimates of its value but estimates shall remain rather imprecise if σ^2 is large. It is important to note that the inverse square dependence implies that the variance may be more difficult to determine precisely than the mean. Additionally note that the diagonal terms are strictly null, implying that the information on μ and σ^2 are not correlated (although, technically speaking, the information on μ is entirely determined by σ^2).

In general, the off-diagonal elements of the Fisher information matrix of a PDF are nonvanishing. However, whenever $\mathbb{I}_{ij}(\vec{\theta}) = 0$ for $i \neq j$, the parameters θ_i and θ_j are said to be orthogonal. In the case of the Gaussian, we found the off-diagonal elements are null, thereby implying that μ and σ are orthogonal; that is, the determination of μ does not influence the determination of σ^2, and conversely. This largely stems from the symmetric nature of the Gaussian PDF. By contrast, PDFs that are not symmetric about their mean shall tend to have nonvanishing nondiagonal elements, and the determination of the corresponding parameters will be mutually correlated.

As we discuss next, the Fisher information of a given parameter determines a minimal variance bound achievable in an experimental measurement of its value. This means that the very shape and definition of a PDF determine the precision achievable in the experimental determination of its parameters.

4.7.6 Minimal Variance Bound

By definition, an unbiased estimator satisfies:

$$\langle\hat{\theta}\rangle = E[\hat{\theta}] = \int \hat{\theta} L\, dX = \theta, \tag{4.117}$$

where L is the likelihood of the measured data $\vec{x} = (x_1, x_2, \ldots, x_n)$ and dX stands for $dx_1 dx_2 \cdots dx_n$. We differentiate this expression with respect to θ. Since the estimator depends on the x_i alone, and assuming the limits of integration are independent of θ, one gets

$$\int \hat{\theta} \frac{dL}{d\theta} dX = 1. \tag{4.118}$$

Next, consider the normalization integral of L:

$$\int L \, dX = 1. \tag{4.119}$$

Differentiating with respect to θ produces

$$\int \frac{dL}{d\theta} dX = 0. \tag{4.120}$$

We then rewrite Eqs. (4.118, 4.120) as follows:

$$\int \hat{\theta} \frac{d \ln L}{d\theta} L \, dX \equiv \left\langle \hat{\theta} \frac{d \ln L}{d\theta} \right\rangle = 1, \tag{4.121}$$

$$\int \frac{d \ln L}{d\theta} L \, dX \equiv \left\langle \frac{d \ln L}{d\theta} \right\rangle = 0. \tag{4.122}$$

Next, multiply the second line by θ and subtract from the first line to get

$$\int \left(\hat{\theta} - \theta\right) \frac{d \ln L}{d\theta} L \, dX = \left\langle \left(\hat{\theta} - \theta\right) \frac{d \ln L}{d\theta} \right\rangle = 1. \tag{4.123}$$

We can now use the Schwarz inequality

$$\int u^2 \, dX \int v^2 \, dX \geq \left(\int uv \, dX \right)^2 \tag{4.124}$$

with $u = (\hat{\theta} - \theta)\sqrt{L}$ and $v = \frac{d \ln L}{d\theta} \sqrt{L}$ to obtain

$$\left(\int (\hat{\theta} - \theta)^2 L \, dX \right) \left(\int \left(\frac{d \ln L}{d\theta} \right)^2 L \, dX \right) \geq 1. \tag{4.125}$$

The first integral corresponds to the variance $\langle (\hat{\theta} - \theta)^2 \rangle$ of the estimator while the second integral is the average of the derivative of the logarithm of L with respect to θ. One then arrives at the lower bound:

$$V\left[\hat{\theta}\right] \geq \frac{1}{\left\langle (d \ln L / d\theta)^2 \right\rangle}. \tag{4.126}$$

The quantity $\left\langle (d \ln L / d\theta)^2 \right\rangle$ is the **Fisher information**, $\mathbb{I}(\theta)$, contained in the likelihood function L. Based on Eq. (4.89), we find that it may also be expressed (see also Problem 4.2)

$$I(\theta) = \left\langle d^2 \ln L / d\theta^2 \right\rangle. \tag{4.127}$$

Alternatively, one can also insert the expression for L in the preceding equation, and obtain (for a single parameter):

$$I(\theta) = -n \int \frac{d^2 \ln p(x|\theta)}{d\theta^2} p(x|\theta)\, dx. \tag{4.128}$$

This expression and Eq. (4.126) can be used to obtain the minimum sample size n required to achieve a given measurement precision of the parameter θ.

Exercises

4.1 Derive the expression given by Eq. (4.16) for the Mean Square Error (MSE).

4.2 Derive the expression Eq. (4.127) for $I(\theta)$.

4.3 Derive the expression Eq. (4.128) for $I(\theta)$.

4.4 Show that for a biased estimator, one must replace the numerator of Eq. (4.17) with $1 + db/d\theta$.

4.5 Verify that the standard error on the sample mean estimator is given by $\sigma/\sqrt{n}$ as in Eq. (4.30).

4.6 Show that S^2, defined by Eq. (4.32), is an unbiased estimator of the variance σ^2 when the population mean, μ, is known.

4.7 Show that s^2, defined by Eq. (4.39), is an unbiased estimator of the variance σ^2 when the population mean, μ, is not known.

4.8 Show that $\hat{V}_{xy}$, defined by Eq. (4.43), is an unbiased estimator of the covariance V_{xy}.

4.9 Show that the expression $\hat{m}_k = \frac{1}{n-1}\sum_{i=1}^{n}(x_i - \bar{x})^k$ provides an unbiased estimator of the kth central moment, μ_k.

4.10 Find unbiased estimators for the moments μ'_k and central moments μ_k.

4.11 Show that the variance of estimators $\widehat{\mu}'_k$ of the moments μ'_k is given by

$$V[\widehat{\mu}'_k] = \frac{1}{n}\left(\mu'_{2k} - \left(\mu'_k\right)^2\right).$$

4.12 Show that the covariance of estimators of the moments μ'_r and μ'_q is given by

$$\text{Cov}[\hat{\mu}'_r, \hat{\mu}'_q] = \frac{1}{n}\left(\mu'_{r+q} - \mu'_r\mu'_q\right).$$

4.13 Show that the variance of the estimator of the standard deviation $\hat{\sigma}$ is given by the following expression in the large n limit:

$$V[\hat{\sigma}] = \frac{\mu_4 - (\mu_2)^2}{4n\sigma^2}.$$

Additionally, show that for a Gaussian PDF, this expression reduces to

$$V[\hat{\sigma}] = \frac{\sigma^2}{2(N-1)}.$$

4.14 Show that the standard error of the estimator $s^2 = \frac{1}{n-1}\sum_{i=1}^{n}(x_i - \bar{x})^2$ is given by

$$V[s^2] = \frac{1}{n}\left(\mu_4 - \frac{n-3}{n-1}\mu_2\right),$$

where $\hat{\mu}_k$ are central moments of order k given by

$$\hat{\mu}_k = \frac{1}{n-1}\sum_{i=1}^{n}(x_i - \bar{x})^k .$$

4.15 Show that the expectation value and variance of the Pearson coefficient estimator $\hat{r}_{xy}$, defined by Eq. (4.44), are given by the following expressions:

$$E[\hat{r}_{xy}] = \rho - \frac{\rho(1-\rho^2)}{2n} + O(n^{-2})$$

$$V[\hat{r}_{xy}] = \frac{1}{n}(1-\rho^2)^2 + O(n^{-2}).$$

4.16 Prove the assertion that the estimator $\hat{r}_{xy}$ defined by Eq. (4.44) is bound in the range $[-1, 1]$.

4.17 Derive the expression Eq. (4.17) for robust centered moments of order k, and verify the expressions given by Eq. (4.53) and Eq. (4.54) for centered moments of second and third order, respectively.

4.18 Show the following expression yields a biased but asymptotically unbiased estimator $\hat{\lambda}$ of the decay constant $\lambda = 1/\tau$:

$$\hat{\lambda} = \frac{1}{\hat{\tau}} = n\left(\sum_{i=1}^{n} t_i\right)^{-1} . \tag{4.129}$$

4.19 Show that the Fisher information matrix defined by Eq. (4.90) may be written as

$$\mathbb{I}_{ij}(\vec{\theta}) = -\mathrm{E}\left[\frac{\partial^2}{\partial\theta_i\partial\theta_j}\ln p(x|\vec{\theta}),\right].$$

5 Classical Inference II: Optimization

In Chapter 4, we introduced the notions of statistics and estimators, discussed their basic properties, and examined specific basic estimators, most particularly those of moments and centered moments, used in classical inference. Although of obvious interest, moments are often insufficient to fully characterize the probability distribution governing a particular phenomenon or dataset. It is often the case that the general functional form of a distribution is known, but not fully specified. Indeed, while the functional dependence on a random variable X may be known, the function might have dependencies on finitely many model parameters $\vec{\theta} = (\theta_1, \theta_2, \ldots, \theta_m)$, which are a priori unknown or unspecified. For instance, it might be known that a particular dataset behaves according to a log-normal distribution, but the parameters μ and σ that determine the specific shape of the distribution could be unknown. Alternatively, one might be interested in determining the values of model parameters governing the dependence between dependent and independent variables. One is thus in need of methods to determine the parameters $\vec{\theta}$ of a distribution that best describe or match a set of measured data. Several such "fitting" methods exist, including the method of maximum likelihood, the extended method of maximum likelihood, various variants of least-squares methods, and Kalman filter methods. These are presented in §5.1, §5.1.7, §5.2, and §5.6, respectively. An excellent discussion of the relatively less used method of moments is presented in ref. [67].

The maximum likelihood and the least-squares methods are optimization problems: both involve the optimization of a **goal function**, often called **objective function** or **merit function**. The former involves *maximization* of a **likelihood function** while the latter requires *minimization* of a χ^2 **function**. Both techniques involve a search in (model) parameter space for an optimum value, that is, an extremum of an objective function. While such searches are relatively simple when the model involves only a few parameters, they may become particularly challenging when models involve a very large number of parameters. Several techniques exist to handle searches in multiparameter space and it is clearly not possible to cover them all in this introductory text. We thus focus our discussion on the general principles of the maximum likelihood and the least-squares methods in following sections of this chapter, and present a selection of optimization techniques pertaining to both methods in §7.6 after the introduction of Bayesian inference methods in §§7.2–7.4.

Once an optimum is achieved, one wishes to establish how good the fit really is. One thus requires a measure of the goodness-of-fit. Such a measure is discussed in §5.3 on the basis of the likelihood and chi-square functions. One is then particularly interested in evaluating errors on the parameters obtained in the optimization. This and related matters are discussed in §5.3. With parameter values and error estimates in hand, it becomes possible to extend or extrapolate the results predicted by the model. Techniques to evaluate the errors on such extrapolations are presented in §5.4, whereas §5.5 presents a technique to average the results (i.e., parameter values) obtained by two or more experimental studies.

5.1 The Method of Maximum Likelihood

Let $p(x|\vec{\theta})\,dx$ determine the probability a random variable X be found in the interval $[x, x + dx]$ given m parameters $\vec{\theta}$. Let us assume that the functional form of $p(x|\vec{\theta})$ is known but not the values of the parameters $\vec{\theta}$. Our goal is thus to obtain estimators of these parameters $\vec{\theta}$ based on the likelihood of a set $\{x_i\}$, $i = 1, \ldots, n$, of measured data.

5.1.1 Basic Principle of the $\mathbb{ML}$ Method

Let us assume the measurement of an observable X is repeated n times, thereby yielding a set of values $\{x_i\}$, $i = 1, \ldots, n$. If the parameters $\vec{\theta}$ were known, the data probability model embodied in the probability density function (PDF) $p(x|\vec{\theta})$ would give us the probability to obtain the value x_1 in the interval $[x_1, x_1 + dx]$, the value x_2 in the interval $[x_2, x_2 + dx]$, and so on. Assuming all n measurements are independent and yield uncorrelated results, the probability of measuring specifically the values $\vec{x} = (x_1, x_2, \ldots, x_n)$ is then simply the product of their respective probabilities:

$$p(x_1|\vec{\theta})\,dx_1 \times p(x_2|\vec{\theta})\,dx_2 \times \cdots \times p(x_n|\vec{\theta})\,dx_n = \prod_{i=1}^{n} p(x_i|\vec{\theta})\,dx_i. \tag{5.1}$$

If the PDF $p(x|\vec{\theta})$ is a good model of the data, one would expect the foregoing probability to be relatively large. Conversely, a poor model of the data should yield a low probability. Obviously, since $p(x|\vec{\theta})$ is a function of $\vec{\theta} = (\theta_1, \theta_2, \ldots, \theta_m)$, the value of the above probability must explicitly depend on the values of these m parameters. Well-chosen values of $\vec{\theta}$ should lead to a high probability, whereas a poor choice should result in a low probability. Since the intervals dx_i feature no dependency on the parameters $\vec{\theta}$, it is then sensible to seek an extremum of the product $\prod_{i=1}^{n} p(x_i|\vec{\theta})$ which corresponds to the **likelihood function** $L(\vec{x}|n, \vec{\theta})$ already introduced in Eq. (4.6):

$$L(\vec{x}|\vec{\theta}) \equiv \prod_{i=1}^{n} f(x_i|\vec{\theta}). \tag{5.2}$$

Nominally, $L(\vec{x}|\vec{\theta})$ corresponds to the joint PDF of the measured x_i given the parameters $\vec{\theta}$, but in this context, the data points are considered fixed (i.e., constant) and one seeks the values $\vec{\theta}$ that maximize the likelihood given the data points; in other words, one seeks the parameter values $\vec{\theta}$ such that the measured data points are the most probable.

The $\mathbb{ML}$ method specifically consists in seeking an extremum (a maximum actually) of $L(\vec{x}|\vec{\theta})$ relative to a variation of the m parameters $\vec{\theta}$:

$$\frac{\partial L(\vec{\theta})}{\partial \theta_i} = 0 \quad \text{for} \quad i = 1, \ldots, m, \tag{5.3}$$

Simultaneous solution of these equations yields the $\mathbb{ML}$ estimators $\hat{\theta}_i$ of the parameters θ_i. A valid solution exists provided the function L is differentiable with respect to the parameters θ_i, and the extremum is not on a boundary of the parameters' range. The $\mathbb{ML}$ estimator $\hat{\theta}$ thus corresponds to the most likely value of $\vec{\theta}$ based on the n data points x_i.

Conceivably, depending on the dataset, the form of the functional, and the number of parameters, several solutions may exist that correspond to **local maxima**. Great care must then be taken to fully explore the parameter space in order to find the parameters $\vec{\theta}$ with the largest likelihood L.

We emphasize that given that the solution of $\partial L(\vec{\theta})/\partial\theta_i = 0$ is obtained for a specific set of data points, the parameter values θ_i corresponding to the extremum thus constitute estimators of the actual values. As such, it is convenient to write the solution(s) with a hat, $\hat{\theta}_i$, which indicate they are estimators to be distinguished from the true values $\vec{\theta}_i$.

Proof that the $\mathbb{ML}$ method produces consistent and unbiased estimators may be found, for instance, in ref. [116].

The $\mathbb{ML}$ method is relatively easy to use and does not require data to be binned or histogrammed. We consider a few examples of application of the method in the following sections.

5.1.2 Example 1: $\mathbb{ML}$ Estimator of the Rate Parameter of the Poisson Distribution

Very massive stars end their existence in spectacular explosions known as supernovae. Supernovae are a relatively rare phenomenon taking place randomly in this and other galaxies. Imagine disposing of a large aperture telescope (several meters) equipped with a high-efficiency and high-resolution camera. You might then be interested in characterizing the rate of supernovae explosions according to the type of galaxy where they take place. Being a rare phenomenon, the number of observations n per time period (e.g., per night) may be modeled according to a Poisson distribution

$$p(n|\mu) = \frac{e^{-\mu}\mu^n}{n!}, \tag{5.4}$$

where μ represents the average rate of explosions (per night). Several nights of observation will be required to carry out the measurement. For the sake of simplicity, let us assume it is possible to observe the same region of the sky and for the same exact duration during $N = 100$ nights, with equal observational conditions. The number of observations made nightly are labeled n_i, $i = 1, \ldots, N$.

Our goal is to determine the rate μ using an $\mathbb{ML}$ estimator. We thus need an expression for the likelihood of the data $\{n_i\}$, $i = 1, \ldots, N$, given a specific value of μ. Equation (5.4) provides the probability of observing n explosions in one night given a mean μ. The likelihood of the data amounts to the probability of a sequence $n_1, n_2, \ldots, n_N$ and is thus simply the product of the probabilities $p(n_i|\mu)$ of observing n_i explosions during nights $i = 1, \ldots, N$:

$$L(\mu) = \prod_{i=1}^{N} \frac{e^{-\mu}\mu^{n_i}}{n_i!}, \tag{5.5}$$

$$= \left(\prod_{i=1}^{N} n_i!\right)^{-1} \exp(-N\mu)\,\mu^{(\sum_{i=1}^{N} n_i)}. \tag{5.6}$$

Given the exponential factors, it seems simpler to maximize the log of the likelihood, $\ln L$, rather than L. Indeed, since $\ln L$ is a monotonically increasing function of its argument L, an extremum of this argument shall also correspond to an extremum of the log, and conversely. The technique is then referred to as log-maximum-likelihood, or simply $\mathbb{LML}$ method. We proceed to seek an extremum of

$$\ln L(\mu) = -N\mu + \left(\sum_{i=1}^{N} n_i\right)\ln\mu - \sum_{i=1}^{N} \ln n_i, \tag{5.7}$$

by taking a derivative relative to μ

$$0 = \frac{\partial \ln L}{\partial \mu} = -N + \frac{\sum_{i=1}^{N} n_i}{\mu}, \tag{5.8}$$

which yields

$$\hat{\mu} = \frac{1}{N}\sum_{i=1}^{N} n_i. \tag{5.9}$$

We conclude that the $\mathbb{LML}$ estimator for the rate parameter μ of the Poisson distribution is equal to the arithmetic mean of the observations $\{n_i\}$, $i = 1, \ldots, N$.

As a concrete example of the method, we consider a simulation of a measurement of the number of supernova over a span of 100 nights. The number n_i of explosions observed nightly are generated with a Poisson random number generator, with a rate parameter $\mu = 2.5$, and shown in Figure 5.1 as a histogram. The rate parameter is assumed unknown in the remainder of the analysis. We then proceed to calculate the likelihood of the simulated data, $L(\mu)$, as function of the nightly rate μ in the range $[0, 4]$ according to Eq. (5.5). The likelihood $L(\mu)$ is vanishingly small for most values of μ but clearly peaks near $\mu = 2.5$. The estimate obtained with the $\mathbb{ML}$ method thus corresponds to an extremum (mode) of the likelihood function $L(\mu)$, which is indeed very close to the actual value of the parameter used in the simulation. We will see, later in this chapter, that the width of the likelihood function may be used to assess an error on the estimate.

5.1.3 Example 2: $\mathbb{LML}$ Estimator of the Decay Constant of the Exponential PDF

Consider a radiological experiment reporting n values $t_1, t_2, \ldots, t_n$ corresponding to decay times of some radioactive isotope X. If the production of this isotope involves no feed down from heavier isotopes, it is then legitimate to assume the data may be represented by

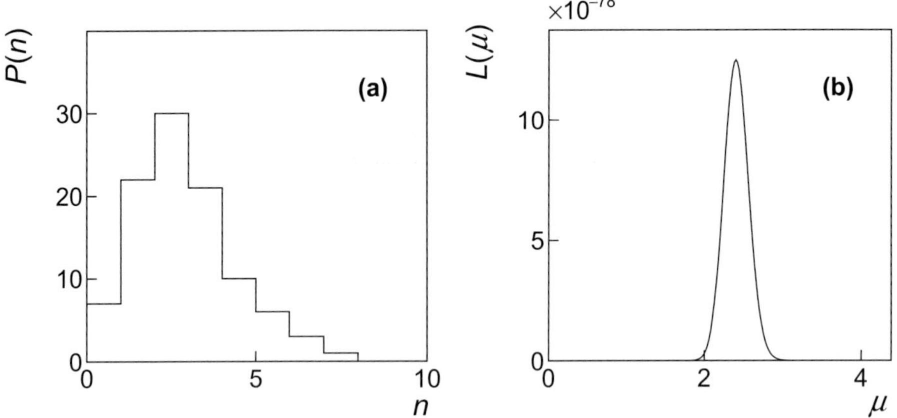

Fig. 5.1 (a) Distribution of the number of supernovae explosion per night in a simulated measurement spanning 100 nights for a rate parameter of $\mu = 2.5$. (b) Likelihood of the simulated data plotted as a function of μ (assumed unknown).

an exponential PDF:

$$p(t|\tau) = \frac{1}{\tau} e^{-t/\tau}. \tag{5.10}$$

Our goal is to determine the mean lifetime of the isotope, that is, find the parameter value τ that best represents the collected data and can then be used to characterize the isotope X.

The likelihood function is here the product of n exponential factors

$$L(\tau) = \prod_{i=1}^{n} p(t_i|\tau) = \prod_{i=1}^{n} \frac{1}{\tau} e^{-t_i/\tau}, \tag{5.11}$$

which readily transforms into the exponential of a sum

$$L(\tau) = \tau^{-n} \exp\left(-\sum_{i=1}^{n} t_i/\tau\right). \tag{5.12}$$

Here again, it is simpler to seek an extremum of $\ln L$ rather than L. We find

$$0 = \frac{d \ln L}{d\tau} = \left[-n\tau^{-1} + \tau^{-2}\sum_{i=1}^{n} t_i\right] \tau^{-n} \exp\left(-\sum_{i=1}^{n} t_i/\tau\right). \tag{5.13}$$

The lifetime τ and exponential are nonzero. The factor in square brackets must consequently be null. Solving for τ thus yields the $\mathbb{LML}$ estimator $\hat{\tau}$

$$\hat{\tau} = \frac{1}{n} \sum_{i=1}^{n} t_i, \tag{5.14}$$

as the arithmetic mean of the measured decay times. It is easily verified that $\hat{\tau}$ is an unbiased estimator of the mean lifetime τ (see Problem 5.1).

As a specific example of application of the estimator (5.14), imagine an experiment involving the measurement of decays of the unstable isotope ^{11}C, with a mean lifetime

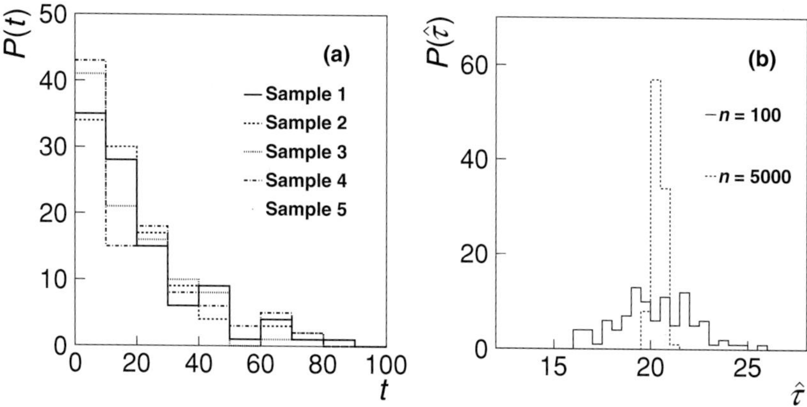

Fig. 5.2 (a) Histograms of decay times of ^{11}C measured in five different simulated experiments (samples). (b) Histogram of measured estimates $\hat{\tau}$ obtained from 100 samples based on measurements of 100 (solid line) and 5000 (dashed line) decay times.

of 20.334 s. We simulated such an experiment using the exponential random generator introduced in §13.3.3 and produced a set of hundred sequences of 100 decay time values. Five of these sequences were used to fill the histograms plotted in Figure 5.2a. We used Eq. (5.14) to calculate hundred estimates of the ^{11}C lifetime on the basis of these sequences. The values were used to fill a histogram of the estimators, shown in Figure 5.2b. We find that the estimates are broadly distributed about the value of 20.334 s used to generate the random numbers. We then repeated the simulations, but this time with datasets consisting of 5,000 points each. Estimates are once again computed on the basis of Eq. (5.14) and plotted in Figure 5.2b as the dotted histogram. As expected, we find these estimates are more narrowly concentrated about the mean. The estimates obtained with a sample size of 100 average to 20.19 s, whereas the estimates obtained with the sample size of 5,000 average to 20.35 s, which is closer to the value of 20.334 s used in their generation.

5.1.4 Example 3: $\mathbb{ML}$ Estimators of the Mean and Variance of a Gaussian PDF

As a third example of application of the $\mathbb{ML}$ method, we determine estimators of the mean and variance of a Gaussian PDF. Let us assume that n measured values x_i are distributed according to a Gaussian PDF with unknown mean μ and standard deviation σ. Given the exponential nature of the Gaussian PDF, it is once again convenient to use the logarithm of the likelihood function L:

$$\ln L(\mu,\sigma^2) = \ln\left(\prod_{i=1}^{n} p(x_i;\mu,\sigma^2)\right), \tag{5.15}$$

$$= \ln\left(\prod_{i=1}^{n} \frac{1}{\sqrt{2\pi}\sigma}\exp\left(-\frac{(x_i-\mu)^2}{2\sigma^2}\right)\right). \tag{5.16}$$

Transforming the log of the product as a sum of logs, and rearranging the terms, one gets

$$\ln L(\mu, \sigma^2) = \sum_{i=1}^{n} \left(-\ln\sqrt{2\pi} - \frac{1}{2}\ln\sigma^2 - \frac{(x_i - \mu)^2}{2\sigma^2} \right), \tag{5.17}$$

$$= -n\ln\sqrt{2\pi} - \frac{n}{2}\ln\sigma^2 - \sum_{i=1}^{n} \frac{(x_i - \mu)^2}{2\sigma^2}. \tag{5.18}$$

The extremum of $\ln(L)$ is found in the usual way by equating its derivatives with respect to μ and σ to zero. Let us first consider an estimator of the mean:

$$0 = \frac{\partial \ln L(\mu, \sigma^2)}{\partial \mu} = -\sum_{i=1}^{n} \frac{\partial}{\partial \mu} \left(\frac{(x_i - \mu)^2}{2\sigma^2} \right). \tag{5.19}$$

Solving for μ yields the $\mathbb{ML}$ estimator $\hat{\mu}$:

$$\hat{\mu} = \frac{1}{n} \sum_{i=1}^{n} x_i, \tag{5.20}$$

which, again in this case, is the arithmetic mean of the sampled values. We already showed, in §4.5.1, that the arithmetic mean of a sample is in general an unbiased estimator of true mean μ. It is thus simple to verify that this conclusion applies specifically to the Gaussian distribution also (see Problem 5.7).

We next proceed to find an estimator for the variance σ^2 of the distribution:

$$0 = \frac{\partial \ln L(\mu, \sigma^2)}{\partial \sigma^2} = -\frac{n}{2}\frac{1}{\sigma^2} + \frac{1}{2\sigma^4} \sum_{i=1}^{n} (x_i - \mu)^2. \tag{5.21}$$

Solution for σ^2 yields the estimator

$$\widehat{\sigma^2} = \frac{1}{n} \sum_{i=1}^{n} (x_i - \hat{\mu})^2, \tag{5.22}$$

which based on our generic discussion of the variance of estimators, in §4.5.3, is known to be an *asymptotically unbiased* estimator of the variance (also see Problem 5.2) with the expectation value

$$\mathrm{E}[\widehat{\sigma^2}] = (n-1)\sigma^2/n. \tag{5.23}$$

Also recall, from §4.5.3, that the estimator s^2 given by Eq. (4.39) is an unbiased estimator of the variance of distributions. We then expect that it also constitutes an unbiased estimator of the variance of a Gaussian PDF (see Problem 5.9). However, s^2 is *not* the $\mathbb{ML}$ estimator of σ^2 for a Gaussian PDF.

5.1.5 Errors

There are limited instances of $\mathbb{ML}$ estimators whose variance can be calculated analytically. For instance, the variance of the estimator (5.14) of the mean of an exponential decay,

$\hat{\tau} = \frac{1}{n}\sum_{i=1}^{n}$, can be computed by direct substitution in the expression of the variance. One gets

$$\text{Var}[\hat{\tau}] = \text{E}[\hat{\tau}^2] - \text{E}[\hat{\tau}]^2 = \frac{\tau^2}{n}, \tag{5.24}$$

which is identical in form to the expression (4.27) of the variance of the sample mean. The variance of $\hat{\tau}$ is a function of τ, the true mean of the PDF considered. This might seem rather problematic, since τ is a priori unknown: how indeed does one report the standard error based on an unknown quantity? However, we have found that $\hat{\tau}$ is an unbiased estimator of τ obtained by finding an extremum of the likelihood function L. One can thus write

$$\frac{\partial L}{\partial \tau} = \frac{\partial L}{\partial \hat{\tau}}\frac{\partial \hat{\tau}}{\partial \tau} = 0. \tag{5.25}$$

Since $\partial\hat{\tau}/\partial\tau \neq 0$, one concludes that an extremum for τ yields an extremum for $\hat{\tau}$, and conversely. It is thus legitimate to use $\hat{\tau}$ in lieu of τ to get an estimate of the variance. One can then report a measurement of τ as $\hat{\tau} \pm \hat{\tau}/\sqrt{n}$. This implies that if an experiment was repeated several times, with the same number of measurements n, one would expect the standard deviations of the results (i.e., estimates) to be $\tau/\sqrt{n}$. Note, however, that in cases where the distribution of estimates significantly deviates from a Gaussian distribution, it is more meaningful (and common) to report an error corresponding to the 68.3% confidence interval (see §6.1.2).

Analytical computation of the variance $\text{Var}[\hat{\theta}]$ of estimators of "complicated" observables may be tedious, difficult, or even impossible. Fortunately, a number of alternative computation techniques exist, few of which we briefly examine in the following. See also ref. [67] for an in-depth discussion of this technical topic.

The most basic technique, commonly applied, to estimate the variance of an estimator consists in carrying out several experiments yielding the quantity of interest. One repeats the experimental procedure several times and obtains several sets of measurements, $\{x_i\}_k$, where k is an index used to identify the different datasets. Estimators $\hat{\theta}_k$ are computed for each dataset k. One can then calculate the expectation values and variance of these estimators. Obviously, it may not always be possible to repeat a given experiment because of cost, lack of time, or because the observed phenomena might be unique by its very nature (e.g., observation of neutrinos from a particular supernova explosion). It may, however, be possible to split the dataset into several subsamples, each of which can be used to obtain distinct $\mathbb{ML}$ estimates. The variance of these estimates relative to the $\mathbb{ML}$ estimate of the full sample can thus be used to evaluate the variance of the estimator.

Whenever repetition or splitting of the data sample produced by an experiment is not an option, one may resort to a detailed simulation of the experiment to artificially create repeated evaluations of the estimator of interest. The idea is to replicate the conditions and procedure of a measurement in a Monte Carlo simulation. While such simulations of experiments will be discussed in detail in Chapter 14, the technique can be summarized as follows. Suppose one wishes to measure a variable X distributed according to a certain PDF $f(x|\theta)$. Once an estimate $\hat{\theta}$ of the parameter θ is obtained experimentally, one carries out repeated simulations of the experiment by generating instances of x based on $f(x|\hat{\theta})$, making sure experimental effects and correction procedures are properly taken into account. The simulated experiments yield estimates $\hat{\theta}_i$ of the parameter θ. Given a sufficient number of replications of the experiment, it is then possible to compute the variance of the values $\hat{\theta}_i$ and consequently obtain an estimate of the true variance of the estimator.

Yet another technique commonly applied to determine the variance of $\mathbb{ML}$ estimators is to use the minimal variance bound based on the Rao–Cramer–Frechet (RCF) inequality given by Eq. (4.17), which for several parameters $\vec{\theta} = (\theta_1, \ldots, \theta_m)$, may be written

$$(V^{-1})_{ij} = -\left\langle \frac{\partial^2 \ln L}{\partial\theta_i \partial\theta_j} \right\rangle = -\left. \frac{\partial^2 \ln L}{\partial\theta_i \partial\theta_j} \right|_{\vec{\theta}=\hat{\theta}}. \tag{5.26}$$

As an example, consider the calculation of the variance of estimators $\hat{\mu}$ and $\hat{\sigma}^2$ of the mean and variance of the Gaussian PDF. Using the log of the likelihood function (5.17), it is easy to calculate (see Problem 5.10) the second-order derivatives with respect to μ and σ^2:

$$\left\langle \frac{\partial^2 \ln L}{\partial \mu^2} \right\rangle = -\frac{n}{\sigma^2}, \tag{5.27}$$

$$\left\langle \frac{\partial^2 \ln L}{\partial \sigma^2} \right\rangle = -\frac{2n}{\sigma^2}, \tag{5.28}$$

$$\left\langle \frac{\partial^2 \ln L}{\partial \mu \partial \sigma} \right\rangle = 0. \tag{5.29}$$

The matrix V^{-1} is diagonal and is trivially inverted to yield the variances

$$\mathrm{Var}[\mu] = \left\langle \frac{\partial^2 \ln L}{\partial \mu^2} \right\rangle^{-1} = \frac{\sigma^2}{n}, \tag{5.30}$$

$$\mathrm{Var}[\sigma] = \left\langle \frac{\partial^2 \ln L}{\partial \sigma^2} \right\rangle^{-1} = \frac{\sigma^2}{2n}. \tag{5.31}$$

One finds that the variances Var[μ] and Var[σ] are proportional to the variance σ^2 of the Gaussian PDF and inversely proportional to the size n of the data sample used to carry out the estimate. This is a rather general property that holds in the large n limit for most estimators. It expresses the well-known result that statistical errors are inversely proportional to the square root of n, the sample size.

The variance of $\mathbb{ML}$ estimators may also be determined using a graphical technique. The technique is based on the RCF bound discussed earlier and the fact that, near an extremum of the likelihood function L, one can expand the function in a Taylor series about the $\mathbb{ML}$ estimate $\hat{\theta}$:

$$\ln L(\theta) = \ln L(\hat{\theta}) + \frac{1}{2} \left[\frac{\partial^2 \ln L}{\partial \theta^2} \right]_{\theta=\hat{\theta}} (\theta - \hat{\theta})^2 + O(3), \tag{5.32}$$

where we omitted the first-order term in $\partial \ln L/\partial\theta$, which by construction, vanishes at the extremum. Based on Eq. (5.26), this can be written

$$\ln L(\theta) \approx \ln L_{\max} - \frac{(\theta - \hat{\theta})^2}{2\hat{\sigma}_{\theta}^2}, \tag{5.33}$$

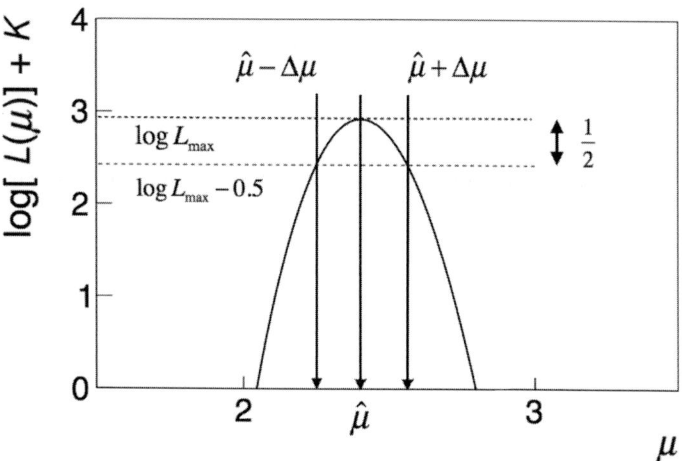

Fig. 5.3 Determination of a parameter error with the log-likelihood graphical method: The solid curve is the log of the likelihood function of the simulated data presented in Figure 5.1. An arbitrary constant K was added to $\ln L$ for convenience of presentation. The log-likelihood function is approximately parabolic in the vicinity of the extremum located at $\hat{\mu} = 2.4$ and the value $\log L - 1/2$ is found at 2.56. The standard error $\Delta\mu$ thus approximately amounts to $2.56 - 2.4 = 0.16$.

which means the error can be estimated graphically on the basis of

$$\ln L(\hat{\theta} \pm \sigma_\theta) = \ln L_{\max} - \frac{1}{2}. \tag{5.34}$$

The principle of the graphical method is illustrated in Figure 5.3, which presents a graph of the logarithm of the likelihood function of the rate parameter μ corresponding to simulated measurements, discussed in §5.1.2, of the number of supernovae observations detected nightly over a span of 100 days. The standard error on the parameter μ is here obtained by graphically finding the values $\hat{\mu} - \Delta\mu$ and $\hat{\mu} + \Delta\mu$ corresponding to $log[L_{\max}] - 1/2$, where $L_{\max}$ is the maximum likelihood observed.

5.1.6 Maximum Likelihood Fit of Binned Data

The $\mathbb{LML}$ method enables a relatively straightforward estimation of parameters for PDFs such as exponential or Gaussian distributions. However, it becomes impractical when the number of observations of random variable x becomes excessively large. This could be the case, for instance, if there is insufficient memory to store all the measured values. There are also issues of numerical accuracy for very large data samples. It is then often desirable, or more convenient, to carry out a *fit* based on a histogram of the data (§4.6). The range of interest $[x_{\min}, x_{\max}]$ is partitioned into m bins, chosen to permit identification of the relevant PDF features while accounting for the finite resolution of the measurements, the size of the sample, and the memory available. The bins are not required to be of equal size; one can use arbitrary bin boundaries $[x_{\min,i}, x_{\max,i}]$ for bins $i = 1, \ldots, m$. Consider a sample consisting of N measured values histogrammed into m bins. Each bin will contain a number of entries n_i (with $i = 1, \ldots, m$) representative of the PDF that characterizes the measured data. The number of entries in the whole histogram may then be expressed as a vector $\vec{n} = (n_1, n_2, \ldots, n_m)$. Obviously, each value n_i is subject to statistical fluctuations. However, the expectation value of the number of entries in bin i, noted ν_i, can be expressed as a function of the unknown parameters $\vec{\theta}$ of the PDF, given by the following expression:

$$\nu_i(\vec{\theta}) = N \int_{x_{\min,i}}^{x_{\max,i}} p(x|\vec{\theta})\, dx. \tag{5.35}$$

It is convenient to denote the expectation values ν_i in a vector form $\vec{\nu} = (\nu_1, \nu_2, \ldots, \nu_m)$ also. It is the purpose of the fit to determine the parameter(s) $\vec{\theta}$ most consistent with the measured values, in other words, the values that yield an extremum of the likelihood function $L(\vec{\theta})$.

The vector (histogram) $\vec{n}$ may be viewed as a single measurement of the m-dimensional vector $\vec{x}$. If all measurements of x are independent of one another, the values are uncorrelated. It implies that the number of entries in the m bins are uncorrelated (although they are obviously determined by the PDF). In this context, the measurement of the vector $\vec{n}$ with a total number of entries N amounts to a random partition of N draws into m bins, each with expectation $\nu_i(\vec{\theta})$. This corresponds to a multinomial PDF. The joint probability to measure the vector $\vec{n}$, given a total number of entries N and expectations $\vec{\nu}$, is then given by

$$p_{\text{joint}}(\vec{n}|\vec{\nu}) = \frac{N!}{n_1!n_2!\cdots n_m!}\left(\frac{\nu_1}{N}\right)^{n_1}\left(\frac{\nu_2}{N}\right)^{n_2}\cdots\left(\frac{\nu_m}{N}\right)^{n_m}, \tag{5.36}$$

where the ratio ν_i/N expresses the probability of getting entries in bin i. The logarithm of this expression yields the log-likelihood function

$$\ln L(\theta) = \ln\left(\frac{N!}{n_1!n_2!\cdots n_m!}\right) + \sum_{i=1}^{m} n_i \ln\left(\frac{\nu_i}{N}\right). \tag{5.37}$$

Clearly, the first term of the right-hand side involves variables that are independent of the PDF parameter(s) $\vec{\theta}$. Since we are seeking an extremum of L, it is unnecessary to keep track of these constants and the search for values $\vec{\theta}$ that maximize L can thus proceed on the basis of the second term alone by whatever optimization method is available or practical (see §7.6 for examples of such methods). When the number of bins is very large, $m \gg N$, then the binned method becomes equivalent to the standard ML method. This method is thus insensitive to the presence of null bins in the histograms, in stark contrast to the least-squares method discussed in §5.2.

5.1.7 Extended Maximum Likelihood Method

We saw in previous sections that the ML method is useful to determine unknown parameters $\vec{\theta} = (\theta_1, \theta_2, \ldots, \theta_n)$ of a PDF $p(\vec{x}|\vec{\theta})$ given a set of n measured values $\vec{x} = (x_1, x_2, \ldots, x_n)$. But what if the number of observations n is itself a random variable determined by the process or system under observation. This is the case, for instance, in measurements of scattering cross section where the number of produced particles is itself a random variable, or in observations and counting of radioactive decays of an unknown substance over a fixed period of time.

Let us consider processes in which the fluctuations of n are determined by a Poisson distribution with mean λ. The likelihood of observing n values $\vec{x}$ may then be described by an **extended likelihood function** as follows:

$$L(\lambda;\vec{\theta}) = \frac{\lambda^n e^{-\lambda}}{n!}\prod_{i=1}^{n} p(x_i|\vec{\theta}). \tag{5.38}$$

In general, λ may be regarded as a function of $\vec{\theta}$, $\lambda \equiv \lambda(\vec{\theta})$, or conversely $\vec{\theta} \equiv \vec{\theta}(\lambda)$. The log of the likelihood function may then be written

$$\ln L = n \ln \lambda(\vec{\theta}) - \lambda(\vec{\theta}) - \ln(n!) + \sum_{i=1}^{n} \ln p(x_i|\vec{\theta}), \tag{5.39}$$

$$= -\lambda(\vec{\theta}) + \sum_{i=1}^{n} \ln(\lambda(\theta) p(x_i|\vec{\theta})), \tag{5.40}$$

where in the second line, we have dropped unnecessary constants and use the fact that the log of a product of distinct factors equals the sum of their logs. Maximization of $\ln L$ is thus, in general, dependent on both the observed value of n as well as the measured values $\vec{x}$. The observed value n thus constrain the model parameters $\vec{\theta}$ and conversely, the observed values x_i constrain λ.

The maximization of $\ln L$ greatly simplifies, of course, if λ and $\vec{\theta}$ are independent. Derivatives of $\ln L$ relative to these variables yield independent conditions:

$$\frac{\partial \ln L}{\partial \lambda} = \frac{\partial}{\partial \lambda}(n \ln \lambda - \lambda) = 0, \tag{5.41}$$

$$\frac{\partial \ln L}{\partial \theta} = \frac{\partial}{\partial \theta}\sum_{i=1}^{n} \ln p(x_i|\vec{\theta}) = 0. \tag{5.42}$$

The first line yields $\hat{\lambda} = n$ whereas the second line amounts to the regular likelihood method, which is independent of λ. Use of the extended maximum likelihood thus appears to provide little gain over the regular method in this case.

The extended method remains of interest, nonetheless, for cases where the function $p(x_i|\vec{\theta})$ may be expressed as a linear combination of several (linearly independent) elementary functions

$$p(x|\vec{\theta}) = \sum_{k=1}^{m} \theta_k p_k(x), \tag{5.43}$$

with

$$\int p_k(x)\, dx = 1. \tag{5.44}$$

Such a situation arises when an observable can be expressed as a combination of finitely many signals, each with their distinct PDF $p_k(x)$, and unknown relative probability describable with parameters θ_k. A specific example of this situation involves the energy deposition of charged particles in a detector volume. One finds that the energy loss is a stochastic process that depends on particle types. The energy loss profile of particles at a given momentum (observed in a specific scattering experiment) thus depends on the relative probabilities of the different species (e.g., electron, pion, kaon, and so forth). The functions $p_k(x)$ could thus represent the energy loss profiles of different particle types while the parameters θ_k would represent their relative probabilities constrained by

$$\sum_{k=1}^{m} \theta_k = 1. \tag{5.45}$$

Evidently, one could treat this case as a regular application of the maximum likelihood method with $m-1$ independent parameters (i.e., with $\theta_m = 1 - \sum_{k=1}^{m-1} \theta_k$) but it is advantageous to carry out the search for an extremum of the likelihood function using all parameters θ_k and λ simultaneously.

Substituting the linear combination (5.43) for $p(x_i|\vec{\theta})$ in Eq. (5.40) yields

$$\ln L = -\lambda + \sum_{i=1}^{n} \ln \left(\sum_{k=1}^{m} \lambda \theta_k p_k(x_i) \right). \tag{5.46}$$

Let $\mu_k = \lambda\theta_k$ represent the mean value of the number of instances of type k. Using Eq. (5.45), $\ln L$ may then be written

$$\ln L(\vec{\mu}) = -\sum_{k=1}^{m} \mu_k + \sum_{i=1}^{n} \ln \left(\sum_{k=1}^{m} \mu_k p_k(x_i) \right), \tag{5.47}$$

where the parameters μ_k are not subjected to any constraints. The total number of events n may then be treated as a sum of independent Poisson variables μ_i and optimization of $\ln L$ thus yields estimates $\hat{\mu}_k$ of the mean of each of the types k. While mathematically equivalent to the independent optimization of λ and θ_k, this approach involves the advantage that all parameters are treated equally and one obtains the contributions of each type k directly.

5.2 The Method of Least-Squares

The $\mathbb{ML}$ and $\mathbb{LML}$ methods enable the determination of parameters that best characterize a data set given a specific PDF assumption. Although these methods are powerful, they may become fastidious, inconvenient, or impractical in many cases. However, an alternative, called the **Least-Squares** ($\mathbb{LS}$) method, is available and applicable in a wide range of situations.

We show in §5.2.1 how the $\mathbb{LS}$ method can be formally derived from the $\mathbb{ML}$ method in cases where the measured values can be considered Gaussian random variables. The $\mathbb{LS}$ method is, however, commonly applicable to problems of parameter estimation in which the Gaussian variable hypothesis is not strictly valid.[1]

[1] The $\mathbb{LS}$ method typically yields reasonable results whether or not fluctuations of the dependent variable y are Gaussian. However, one must be careful with the interpretation of error estimates obtained with non-Gaussian deviates.

The formal derivation and definition of the $\mathbb{LS}$ method presented in §5.2.1 may be skipped in a first reading. Section 5.2.2 discusses the simple case of straight line fit and linear regression, which may be skipped by readers already familiar with these basic notions. Polynomial fits and progressively more complex minimization problems are presented in §§5.2.3–5.2.7.

5.2.1 Derivation of the $\mathbb{LS}$ Method

Consider a set of observations where a variable y is measured as a function of a variable x. Let us assume there exists, at least in principle, some relation between the two quantities. A basic measurement shall consist of n pairs (x_i, y_i) where the values x_i are assumed to be the control variable (whether explicitly controlled or not) and the values y_i are assumed to be functions of x_i. In this context, the variables y and x are also commonly called **dependent** and **independent** variables, respectively. For simplicity's sake, we will here assume the x_i are known without error. The $\mathbb{LS}$ method can, however, be generalized for cases where both x and y carry measurement errors. We will further assume the y_i are Gaussian random variables, that is, variables with a Gaussian PDF, $p_G(y_i|\mu_i, \sigma_i)$, defined by Eq. (3.124). This is meant to imply that if it were possible to repeat the measurement several times at the same value x_i, the measured values y_i would be distributed according to a Gaussian PDF of definite mean, μ_i and width σ_i. The values μ_i are assumed to depend on x_i in some manner we model with a function $f(x|\vec{\theta})$ that depends on one or more parameters, $\vec{\theta} = (\theta_1, \theta_2, \ldots \theta_m)$, of a priori unknown value:

$$\mu_i \equiv f(x_i|\vec{\theta}). \tag{5.48}$$

This type of measurement is rather general. Consider as a simple example, a measurement of the position, x, vs. time, t, of a car subjected to some unknown but constant acceleration (see example in §5.2.2). An $\mathbb{LS}$ fit of the data might then yield the value of this acceleration. Alternatively, one could measure the temperature (dependent variable) along a bar of metal (position) when the extremities of the bar are submitted to finite temperature differences, and a model of heat conduction could be used to describe the temperature profile along the bar. The possibilities are endless. One can envision measuring any physical quantity as a function of some other variable, be it time, space, currents, or electric and magnetic fields, and seek to model the relation between them. All one needs is a function, $y = f(x|\vec{\theta})$, modeling the relationship between y and x based on some "free" parameters, that is, model parameters $\vec{\theta}$ of unknown or unspecified value. We will introduce the $\mathbb{LS}$ method for problems involving a single independent variable, x, and one dependent variable, y, but it can be readily extended to an arbitrary number of independent and dependent variables.[2]

The goal of the $\mathbb{LS}$ method is to find the value(s) $\vec{\theta}$ that maximize the probability of getting the measured values y_i. But since the n variables y_i are by assumption distributed according to Gaussian PDFs $p_G(y_i|\mu_i, \sigma_i)$ of mean μ_i and width σ_i, we will apply the $\mathbb{LML}$ method to determine the value(s) of the parameters $\vec{\theta}$ that maximize the probability

[2] In this context, the phrase *independent variable* implies that x is a control variable, i.e., its values can be selected, or controlled, while the variable y adopts values possibly determined by x and is, as such, dependent on x.

of measuring the values y_i at the given x_i. We use the vector notations $\vec{x}$ and $\vec{y}$ to denote all values x_i and y_i, respectively. The likelihood L of the measured values $\vec{y}$ depends on the joint probability, $g_{\text{joint}}(\vec{y}, \vec{x}|\vec{\theta})$. Assuming the n points (x_i, y_i) are measured independently and are thus uncorrelated, g_{joint} reduces to the product of the probabilities of all pairs (x_i, y_i):

$$g_{\text{joint}}(\vec{y}|\vec{\mu}, \vec{\sigma}) = \prod_{i=1}^{N} p_G(y_i|x_i; \mu_i, \sigma_i), \tag{5.49}$$

$$= \prod_{i=1}^{N} \frac{1}{\sqrt{2\pi\sigma_i^2}} \exp\left(-\frac{(y_i - \mu_i(x_i))^2}{2\sigma_i^2}\right), \tag{5.50}$$

where the values $\mu_i(x_i)$ are determined by the model function, $f(x|\vec{\theta})$, which is dependent on the unknown or unspecified parameter(s) $\vec{\theta}$[3]. We will assume there exists an estimate for the widths σ_i, that is, that values σ_i can be inferred either from the data directly, or on the basis of some theoretical considerations. The likelihood function of the values y_i may then be written:

$$L(\vec{y}|\vec{x}, \vec{\sigma}, \vec{\theta}) = \prod_{i=1}^{N} \frac{1}{\sqrt{2\pi\sigma_i^2}} \exp\left(-\frac{(y_i - f(x_i|\vec{\theta}))^2}{2\sigma_i^2}\right). \tag{5.51}$$

Note that this expression of the likelihood function assumes the data points are uncorrelated. Extension to a case in which the data points are correlated is relatively simple and will be discussed in §5.2.4.

Given its formulation as a product of exponentials, it is convenient to consider the logarithm of the likelihood function. Since the logarithm, $\ln(x)$, grows monotonically with x, a search for an extremum of the log of the likelihood function shall yield parameter values that maximize the likelihood function itself:

$$\ln L = -\frac{1}{2}\sum_{i=1}^{N} \ln\left(2\pi\sigma_i^2\right) - \frac{1}{2}\sum_{i=1}^{N} \frac{\left(y_i - f(x_i|\vec{\theta})\right)^2}{\sigma_i^2}. \tag{5.52}$$

The first term is not a function of the parameter(s) $\vec{\theta}$ and can be ignored in the search for an extremum of the log of the likelihood function. Consequently, maximization of $\ln(L)$ only involves the second term of Eq. (5.52) and it is thus convenient to define a **chi-square function** as follows:

$$\chi^2(\vec{\theta}) = \sum_{i=1}^{N} \frac{\left[y_i - f(x_i|\vec{\theta})\right]^2}{\sigma_i^2}. \tag{5.53}$$

Given the negative sign in front of the sum in Eq. (5.52), maximization of $\ln L$ then amounts to a minimization of the χ^2 function. This forms the basis of the $\mathbb{LS}$ method.

Proof that the method of least-squares produces consistent and unbiased estimators may be found for instance in ref. [116].

[3] The values x_i are taken as given, i.e., selected a priori. One consequently does not consider the probability of having such values. Only the y_i are considered random variables and assigned a probability.

The fact that the minimization of the $\chi^2(\vec{\theta})$ function is equivalent to the maximization of the log-likelihood is based on the assumption that the random variables y_i are Gaussian distributed. Although this is not always true in practice, we note that by virtue of the central limit theorem, the multitude of random phenomena that produce the random character of the y_i, implies their distributions are nearly Gaussian in general. The $\mathbb{LS}$ method thus usually constitutes a reasonable approximation of the $\mathbb{ML}$ method. There are nonetheless cases where this approximation is not valid, and one must use the $\mathbb{ML}$ method, rather than the $\mathbb{LS}$ method, to carry out a search for optimal parameter(s) $\vec{\theta}$.

The parameters that minimize the χ^2 function are called $\mathbb{LS}$ estimators and are noted $\hat{\theta}_1, \hat{\theta}_2, \ldots, \hat{\theta}_m$ or simply $\hat{\theta}$ with the understanding that there are m such parameters. Additionally, the function is commonly called "chi-square" even in cases where the individual measurements y_i do not have Gaussian PDFs.

The χ^2 function, Eq. (5.53), was obtained based on the additional assumption that the N variables y_i are uncorrelated. When this assumption is not valid, and there are significant correlations among the variables y_i, one must transform Eq. (5.53) in terms of variables that are independent. In §5.2.4, we will introduce a technique to accomplish this in the context of the $\mathbb{LS}$ method. However, we first consider two cases of χ^2 minimization that do not involve such correlations: straight-line fits are presented in §5.2.2 while more general polynomial fits of order n are discussed in §5.2.3.

5.2.2 Straight-Line Fit and Linear Regression

Arguably the simplest and most common case of application of the $\mathbb{LS}$ method is for the determination of the parameters of a **straight line**, applicable when a variable y is known, or believed, to depend linearly on an independent variable x:

$$y = f(x|a_0, a_1) = a_0 + a_1 x. \tag{5.54}$$

Given a set of n measured points (x_i, y_i), with $i = 1, \ldots, n$, our goal is to find the values of the slope a_1 and the ordinate at the origin a_0 that minimize the chi-square function, χ^2, defined by Eq. (5.53). By virtue of our choice of model, the χ^2 is now a function of the parameters a_0 and a_1, which can be written

$$\chi^2 = \sum_{i=1}^{N} \frac{[y_i - f(x_i|a_0, a_1)]^2}{\sigma_i^2}. \tag{5.55}$$

We substitute the expression (5.54) for $f(x|a_0, a_1)$ in Eq. (5.55) to obtain a χ^2 function that explicitly depends on a_0 and a_1:

$$\chi^2(a_0, a_1) = \sum_{i=1}^{N} \frac{(y_i - a_0 - a_1 x_i)^2}{\sigma_i^2}. \tag{5.56}$$

We seek an extremum of this function (a minimum, actually) relative to variations of a_0 and a_1. This is accomplished by finding values of these two parameters for which derivatives with respect to a_0 and a_1 are null simultaneously:

$$\frac{\partial \chi^2(a_0, a_1)}{\partial a_0} = 0, \tag{5.57}$$

$$\frac{\partial \chi^2(a_0, a_1)}{\partial a_1} = 0. \tag{5.58}$$

Computation of these two derivatives yields

$$0 = \frac{\partial \chi^2(a_0, a_1)}{\partial a_0} = -2\sum_{i=1}^{N} \frac{(y_i - a_0 - a_1 x_i)}{\sigma_i^2}, \tag{5.59}$$

$$0 = \frac{\partial \chi^2(a_0, a_1)}{\partial a_1} = -2\sum_{i=1}^{N} \frac{(y_i - a_0 - a_1 x_i)x_i}{\sigma_i^2}. \tag{5.60}$$

Dropping the common multiplicative factors and rearranging, we get

$$a_0 \sum_{i=1}^{N} \frac{1}{\sigma_i^2} + a_1 \sum_{i=1}^{N} \frac{x_i}{\sigma_i^2} = \sum_{i=1}^{N} \frac{y_i}{\sigma_i^2}, \tag{5.61}$$

$$a_0 \sum_{i=1}^{N} \frac{x_i}{\sigma_i^2} + a_1 \sum_{i=1}^{N} \frac{x_i^2}{\sigma_i^2} = \sum_{i=1}^{N} \frac{x_i y_i}{\sigma_i^2}. \tag{5.62}$$

It is convenient to define the following quantities:

$$\begin{aligned} S &\equiv \sum_{i=1}^{N} \frac{1}{\sigma_i^2} & S_x &\equiv \sum_{i=1}^{N} \frac{x_i}{\sigma_i^2} \\ S_{xx} &\equiv \sum_{i=1}^{N} \frac{x_i^2}{\sigma_i^2} & S_y &\equiv \sum_{i=1}^{N} \frac{y_i}{\sigma_i^2} \\ S_{xy} &\equiv \sum_{i=1}^{N} \frac{x_i y_i}{\sigma_i^2} & \Delta &\equiv S_{xx} S - S_x^2 \end{aligned} \tag{5.63}$$

Eq. (5.61) may then be rewritten

$$a_0 S + a_1 S_x = S_y, \tag{5.64}$$

$$a_0 S_x + a_1 S_{xx} = S_{xy}, \tag{5.65}$$

or equivalently in matrix form:

$$\alpha \vec{a} = \vec{b}, \tag{5.66}$$

where we introduced the matrix α as well as the vectors $\vec{a}$ and $\vec{b}$ defined as follows:

$$\alpha = \begin{pmatrix} S & S_x \\ S_x & S_{xx} \end{pmatrix} \vec{a} = \begin{pmatrix} a_0 \\ a_1 \end{pmatrix} \vec{b} = \begin{pmatrix} S_y \\ S_{xy} \end{pmatrix}. \tag{5.67}$$

To solve for $\vec{a}$, we multiply both sides of Eq. (5.66) by the inverse α^{-1}:

$$\vec{a} = \alpha^{-1} \vec{b}. \tag{5.68}$$

The inverse of α is

$$\alpha^{-1} = \frac{1}{\Delta}\begin{pmatrix} S_{xx} & -S_x \\ -S_x & S \end{pmatrix}. \tag{5.69}$$

The estimators $\hat{a}_0$ and $\hat{a}_1$, which minimize the χ^2 function, may thus be written

$$\hat{a}_0 = \frac{1}{\Delta}\left(S_y S_{xx} - S_x S_{xy}\right), \tag{5.70}$$

$$\hat{a}_1 = \frac{1}{\Delta}\left(S S_{xy} - S_x S_y\right). \tag{5.71}$$

The calculation of the coefficients S, S_x, S_{xx}, and so on relies exclusively on the points (x_i, y_i). Equations (5.70 and 5.71) then yield estimators $\hat{a}_0$ and $\hat{a}_1$ of the straight-line parameters that best fit the measured data points. Note that if the values y_i are computed without estimates of their standard deviations, σ_i, it suffices to set all values σ_i equal to unity in the foregoing calculations to obtain estimates of $\hat{a}_0$ and $\hat{a}_1$. However, the interpretation of the χ^2 of the fit in terms of a χ^2-distribution is not strictly possible in this case, nor are meaningful estimates of the errors on $\hat{a}_0$ and $\hat{a}_1$.

The same mathematical procedure applies whether one considers a straight-line fit or a linear regression. The term **fit** is, however, usually reserved for problems (or systems) where a model is used to infer a linear relationship between the dependent and independent variables. For instance, Hubble's law ($V = Hz$, with $z = \Delta\lambda/\lambda$) states there is a linear relation between the receding velocity, V, and the redshift, z, of galaxies. A linear fit carried out on a set of measured points, (z_i, V_i), consequently yields an estimate of the Hubble constant H. By contrast, the term **linear regression** is typically used for cases where no model is known a priori, or whenever large variances characterize both variables. The foregoing procedure thus yields an estimate of the trend between the variables, akin to an estimate of correlation. A linear regression can, for instance, be used to characterize the relationship between the height and weight of humans in a given population (see Figure 5.4).

The fact that the measurements y_i each carry an error σ_i implies the model parameters a_0 and a_1 are known with limited precision only. We can estimate their respective errors using the error propagation technique introduced in §2.11, Eqs. (2.222, 2.223). Given that only the coefficients S_y and S_{xy} are functions of y_i, we can write

$$\frac{\partial a_0}{\partial y_j} = \frac{1}{\Delta}\left(S_{xx}\frac{\partial S_y}{\partial y_j} - S_x\frac{\partial S_{xy}}{\partial y_j}\right), \tag{5.72}$$

$$\frac{\partial a_1}{\partial y_j} = \frac{1}{\Delta}\left(S\frac{\partial S_{xy}}{\partial y_j} - S_x\frac{\partial S_y}{\partial y_j}\right). \tag{5.73}$$

The derivatives of S_y and S_{xy} with respect to y_j yield $1/\sigma_j^2$ and x_j/σ_j^2, respectively. Inserting these values in the preceding expressions, we get

$$\frac{\partial a_0}{\partial y_j} = \frac{1}{\Delta}\left(S_{xx}\frac{1}{\sigma_j^2} - S_x\frac{x_j}{\sigma_j^2}\right), \tag{5.74}$$

$$\frac{\partial a_1}{\partial y_j} = \frac{1}{\Delta}\left(S\frac{x_j}{\sigma_j^2} - S_x\frac{1}{\sigma_j^2}\right). \tag{5.75}$$

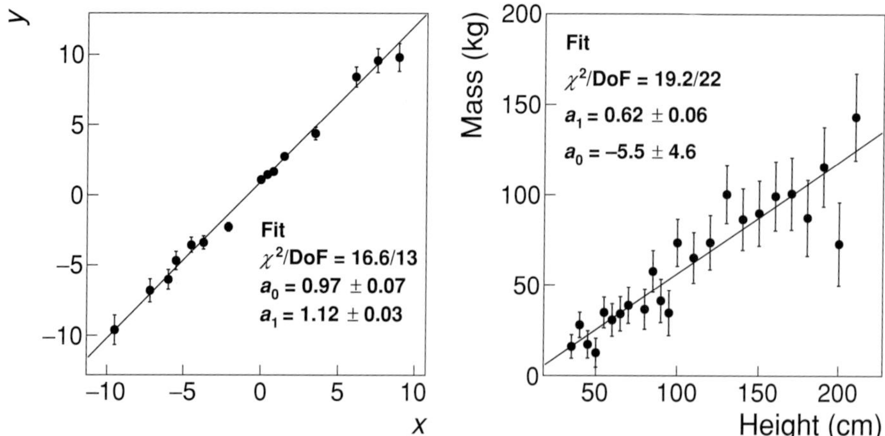

Fig. 5.4 (a) Straight-line fit on simulated data. (b) Example of a linear regression between the weight and height of children in an arbitrary population sample.

Assuming the y_i are independent, the variances $\sigma^2_{a_0}$ and $\sigma^2_{a_1}$ may now be estimated using Eq. (2.222), which we rewrite here in terms of a_0, a_1 as function of y_i:

$$\sigma^2_{a_0} = \sum_{j=1}^{N} \left[\frac{\partial a_0}{\partial y_j} \right]^2 \sigma_j^2, \tag{5.76}$$

$$\sigma^2_{a_1} = \sum_{j=1}^{N} \left[\frac{\partial a_1}{\partial y_j} \right]^2 \sigma_j^2. \tag{5.77}$$

Substituting the derivatives (5.74) in the preceding expressions, we get after simplification (see Problem 5.11):

$$\sigma^2_{a_0} = \frac{S_{xx}}{\Delta}, \tag{5.78}$$

$$\sigma^2_{a_1} = \frac{S}{\Delta}. \tag{5.79}$$

The variances $\sigma^2_{a_0}$ and $\sigma^2_{a_1}$ correspond respectively to the elements $(\alpha^{-1})_{11}$ and $(\alpha^{-1})_{22}$ of the inverse of matrix α. This is no mere accident and in fact derives from a general result we will discuss in §5.2.5.

5.2.3 $\mathbb{LS}$ Fit of a Polynomial

The $\mathbb{LS}$ method introduced in the previous section for linear fits is readily extended to polynomial fits of any order. For instance, let us assume the data may be represented by a polynomial of order m:

$$f(x) = a_o + a_1 x + \ldots + a_m x^m = \sum_{j=0}^{m} a_j x^j. \tag{5.80}$$

For notational convenience, we represent the $m+1$ parameters a_i as a vector $\vec{a} = (a_0, a_1, \ldots, a_n)$. Once again, we assume the measurements y_i are independent. The χ^2 function may then be written

$$\chi^2(\vec{a}) = \sum_{i=1}^{N} \frac{\left(y_i - \sum_{k=0}^{m} a_k x_i^k\right)^2}{\sigma_i^2}. \tag{5.81}$$

We seek the values of the parameters a_j, $j = 0, \ldots, m$, that yield a minimum χ^2. This is accomplished by setting all derivatives of χ^2 with respect to the parameters a_j equal to zero simultaneously:

$$\frac{\partial \chi^2}{\partial a_j} = -2 \sum_{i=1}^{N} \frac{\left(y_i - \sum_{k=0}^{m} a_k x_i^k\right)}{\sigma_i^2} \frac{\partial}{\partial a_j} \left(\sum_{k=0}^{m} a_k x_i^k\right) = 0. \tag{5.82}$$

The derivative of $\sum_{k=0}^{m} a_k x_i^k$ with respect to a_j yields x^j. Equation (5.82) thus simplifies to

$$\frac{\partial \chi^2}{\partial a_j} = -2 \sum_{i=1}^{N} \frac{\left(y_i - \sum_{k=0}^{m} a_k x_i^k\right) x_i^j}{\sigma_i^2} = 0. \tag{5.83}$$

We rearrange and separate the terms to get

$$\sum_{i=1}^{N} \frac{y_i x_i^j}{\sigma_i^2} = \sum_{i=1}^{N} \sum_{k=0}^{m} \frac{a_k x_i^k x_i^j}{\sigma_i^2} = \sum_{k=0}^{m} a_k \sum_{i=1}^{N} \frac{x_i^k x_i^j}{\sigma_i^2}. \tag{5.84}$$

The index j takes values from 0 to m, and the index i runs from 1 to N. Equation (5.84) hence corresponds to $m+1$ equations that must be solved simultaneously. It is convenient to define a matrix α and a column vector $\vec{b}$ with the elements

$$\alpha_{kj} = \sum_{i=1}^{N} \frac{x_i^k x_i^j}{\sigma_i^2}, \tag{5.85}$$

$$b_j = \sum_{i=1}^{N} \frac{y_i x_i^j}{\sigma_i^2}. \tag{5.86}$$

Equation (5.84) may then be written as a linear equation:

$$\alpha \vec{a} = \vec{b}, \tag{5.87}$$

which yields the solution

$$\hat{a} = \alpha^{-1} \vec{b}, \tag{5.88}$$

in which we use the notation $\hat{a}$ to emphasize that the preceding expression yields estimators of the model parameters a_i, $i = 1, \ldots, m$.

A polynomial fit may thus be accomplished with the following three steps: (1) calculation of the matrix α, (2) calculation of the column vector $\vec{b}$, and (3) inversion of the matrix

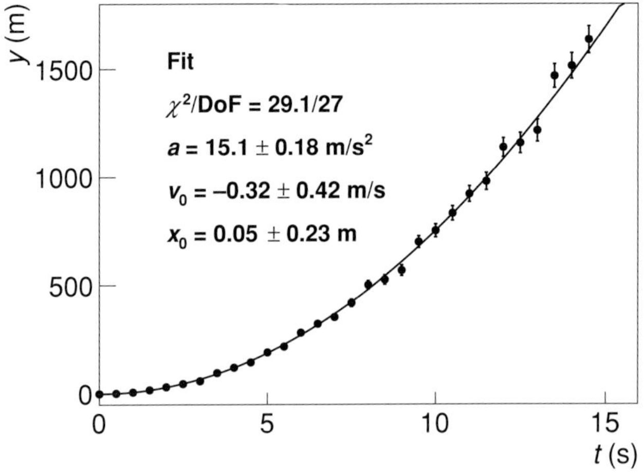

Fig. 5.5 Quadratic fit of simulated data (t_i, y_i) where the y_i are measured altitudes vs. times t_i of a rocket subjected to a constant acceleration a.

α and multiplication by $\vec{b}$ according to Eq. (5.88). Techniques and programs to invert matrices and solve linear equations are described in various textbooks and are available in various software packages.

Figure 5.5 displays a quadratic fit of simulated data describing the altitude y of a rocket as a function of time. The data were generated with the constant acceleration model $y(t) = 0.5at^2 + v_o t + y_o$, with values $a = 15$ m/s^2, $v_o = 0$, and $y_o = 0$. Measurement errors were simulated with Gaussian deviates with widths $\sigma_i = 5.00 + 0.03 * y$. The fit was carried out using the same quadratic model and assumed knowledge of the measurement errors. Trajectory parameters obtained from the fit are within statistical errors of the values used for the generation of the simulated trajectory.

We will show in §5.2.5 that the variances of the estimators $\hat{a}$ are given by the diagonal elements of the inverse matrix α. However, for polynomials of high-order m, the matrix α is prone to become **ill conditioned**, and its inversion may become numerically unstable. It is possible to partly remedy this problem by using **orthogonal polynomials**. Indeed, orthogonal polynomials, or any other complete basis of orthogonal functions, enable a straightforward and unique decomposition of arbitrary (continuous) functions. As such, they typically produce fit coefficients, for each element of the basis, that are nearly independent of one another.

5.2.4 $\mathbb{LS}$ Fit for Correlated Variables y_i

In §5.2.1, we showed that minimization of the χ^2 function, defined by Eq. (5.53), is equivalent to the maximum of the likelihood function L of measuring the values y_i. The derivation assumed the values y_i are mutually independent. There are, however, several classes of measurements that yield correlated variables y_i, that is, with nonvanishing covariances,

$\text{Cov}[y_i, y_j] \neq 0$ for $i \neq j$. Given the existence of such correlations, one expects the minimization of Eq. (5.53) to yield incorrect results because the values will be given inappropriate weights. Proper weights may be restored if one can transform the variables y_i, with nonzero covariances $\text{Cov}[y_i, y_j]$, into a set of variables z_i with $\text{Cov}[z_i, z_j] = 0$ for $i \neq j$. This can be readily accomplished by using the inverse of the covariance matrix of the variables y_i. Given $V_{ij} = \text{Cov}[y_i, y_j] = \text{E}[y_i y_j] - \text{E}[y_i]\text{E}[y_j]$, the χ^2 function may then be written

$$\chi^2(\vec{\theta}) = \sum_{i,j=1}^{N} (y_i - f(x_i|\vec{\theta}))(V^{-1})_{ij}(y_j - f(x_j|\vec{\theta})), \tag{5.89}$$

in which expectation values $\mu_i(x_i|\vec{\theta})$ were replaced by model functions $f(x_i|\vec{\theta})$. Note that if the variables y_i are uncorrelated, then the covariance matrix V_{ij} is diagonal, its inverse is a diagonal matrix with coefficients $\left(2\sigma_i^2\right)^{-1}$, and Eq. (5.53) is thus recovered. The parameters that minimize the function χ^2 are called $\mathbb{LS}$ estimators and are noted $\hat{\theta}_1, \hat{\theta}_2, \ldots, \hat{\theta}_m$ or simply $\hat{\theta}$. As in all other cases, minimization of χ^2 proceeds by equating the derivatives of Eq. (5.89) with respect to θ_i to zero. We discuss general implementations of the method for linear and nonlinear models in §§5.2.5 and 5.2.7 respectively, and an implementation for binned data in §5.2.6.

5.2.5 Generalized Linear Least-Squares Fit

The $\mathbb{LS}$ fit method is applicable for fits of any function $f(x|\vec{\theta})$ but is particularly well suited and considerably simplifies when f is a linear function of its parameters a_j, $j = 1, \ldots, m$

$$f(x_i|\vec{a}) = \sum_{j=1}^{m} a_j f_j(x_i), \tag{5.90}$$

where the coefficients $f_j(x_i)$ may be arbitrary functions of x, not just powers of x, as in the case of simple polynomial fits discussed in §5.2.3. However, the functions $f_j(x_i)$ must be linearly independent and may not depend on the model parameters a_j. **Orthogonal functions**, in particular, present the advantage that the coefficients a_j are not correlated. Commonly used functions, beside powers of x, include Fourier decompositions, orthogonal polynomials, and Legendre polynomials.

We repeat the steps carried out for fits with polynomials to obtain the estimates $\hat{a}_j$. However, we include the possibility, discussed in the previous section, that the data y_i might be correlated. For notational convenience, we introduce coefficients F_{ij} defined as follows:

$$F_{ij} = f_j(x_i). \tag{5.91}$$

The χ^2 function becomes

$$\chi^2 = \sum_{i,k=1}^{N} \left(y_i - \sum_{j=1}^{m} F_{ij} a_j \right) (V^{-1})_{ik} \left(y_k - \sum_{j'=1}^{m} F_{kj'} a_{j'} \right), \tag{5.92}$$

which may also be formulated in a convenient matrix form

$$\chi^2 = (\vec{y} - \mathbf{F}\vec{a})^T V^{-1} (\vec{y} - \mathbf{F}\vec{a}), \tag{5.93}$$

in which $\vec{y} - \mathbf{F}\vec{a}$ is considered an $N \times 1$ column vector, $\vec{y}$ being a column vector representing all N entries y_i. $\mathbf{F}$ is an $N \times m$ matrix with elements equal to the coefficients F_{ij}, and $\vec{a}$ is an $m \times 1$ column vector containing the parameters a_j. The notation $\mathbf{O}^T$ is used to denote the transpose of matrix $\mathbf{O}$.

We find the minimum of χ^2 by differentiating with respect to a_p, with $p = 0, \ldots, m$:

$$0 = \frac{\partial \chi^2}{\partial a_p}, \tag{5.94}$$

$$= -\sum_{i,k=1}^{N} \left(\sum_{j=1}^{m} F_{ij}\delta_{jp} \right) (V^{-1})_{ik} \left(y_k - \sum_{j'=1}^{m} F_{kj'}a_{j'} \right)$$
$$- \sum_{i,k=1}^{N} \left(y_k - \sum_{j=1}^{m} F_{kj}a_j \right) (V^{-1})_{ik} \left(\sum_{j'=1}^{m} F_{ij'}\delta_{j'p} \right), \tag{5.95}$$

$$= -2 \sum_{i,k=1}^{N} F_{ip}(V^{-1})_{ik} \left(y_k - \sum_{j'=1}^{m} F_{kj'}a_{j'} \right). \tag{5.96}$$

On the second line, we used $\partial a_j / \partial a_p = \delta_{jp}$, in which δ_{jp} is the Kroenecker symbol:

$$\delta_{ij} = \begin{matrix} 1 & \text{for i = j}, \\ 0 & \text{for i} \neq \text{j}. \end{matrix}$$

We next took the sum $\sum_{j=1}^{m} F_{ij}\delta_{jp} = F_{ip}$ and made use of the fact that the inverse of matrix $\mathbf{V}$ is symmetric. The preceding expression is succinctly expressed in matrix form:

$$0 = \mathbf{F}^T V^{-1} (\vec{y} - \mathbf{F}\vec{a}), \tag{5.97}$$

which we rewrite as

$$\mathbf{F}^T V^{-1} \mathbf{F}\vec{a} = \mathbf{F}^T V^{-1} \vec{y}. \tag{5.98}$$

As in prior sections, it is convenient to introduce the matrix α and column vector $\vec{b}$, defined as

$$\alpha = \mathbf{F}^T V^{-1} \mathbf{F}, \tag{5.99}$$

$$\vec{b} = \mathbf{F}^T V^{-1} \vec{y}. \tag{5.100}$$

Equation (5.98) becomes

$$\alpha \vec{a} = \vec{b}. \tag{5.101}$$

Solving for $\vec{a}$, we get

$$\hat{a} = \alpha^{-1} \vec{b}. \tag{5.102}$$

This expression provides estimates $\hat{a} = (\hat{a}_0, \ldots, \hat{a}_m)$ that are linear functions of the measurements $\vec{y}$. It can thus be computed analytically. The inversion of large matrices, however, becomes rather tedious for large values of N or m and is thus best carried out numerically on a computer. In practice, it is often most efficient or simply convenient to use numerical algorithms, such as the one described in §5.2.7. It can be shown that the foregoing estimates $\hat{a}_j$ have zero bias and minimum variance.

Errors (variances) on the parameters may be obtained using the error propagation technique introduced in Eqs. (2.222, 2.223) and used in §5.2.2. The covariance matrix U_{ij} of the fit estimators $\hat{a}_i$ and $\hat{a}_j$may be written

$$U_{ij} = \sum_{k,k'=1}^{N} \frac{\partial a_i}{\partial y_k} V_{kk'} \frac{\partial a_j}{\partial y_{k'}}. \tag{5.103}$$

In order to compute the derivatives $\partial a_i / \partial y_k$, we note that Eqs. (5.100) and (5.102) may be combined to obtain

$$\vec{a} = \left(\mathbf{F}^T V^{-1} \mathbf{F}\right)^{-1} \mathbf{F}^T V^{-1} \vec{y}. \tag{5.104}$$

We thus get

$$\frac{\partial \vec{a}}{\partial \vec{y}} = \left(\mathbf{F}^T V^{-1} \mathbf{F}\right)^{-1} \mathbf{F}^T V^{-1}. \tag{5.105}$$

The covariance matrix $\mathbf{U}$ of the estimators $\hat{a}_i$ may then be written

$$\mathbf{U} = \left(\mathbf{F}^T V^{-1} \mathbf{F}\right)^{-1} \mathbf{F}^T V^{-1} V \left[\left(\mathbf{F}^T V^{-1} \mathbf{F}\right)^{-1} \mathbf{F}^T V^{-1} \right]^T. \tag{5.106}$$

The matrix $\left(\mathbf{F}^T V^{-1} \mathbf{F}\right)$ and its inverse are by construction symmetric. Equation (5.106) thus simplifies to

$$\mathbf{U} = \left(\mathbf{F}^T V^{-1} \mathbf{F}\right)^{-1}. \tag{5.107}$$

By construction, $\mathbf{U}$ is an $m \times m$ symmetric matrix. Its diagonal elements U_{jj} correspond to the variances, $\text{Var}[\hat{a}_j]$, of the estimators a_j and as such should provide estimates of the errors on each of the fit parameters. However, the nondiagonal elements U_{ij}, corresponding to covariances $\text{Cov}[\hat{a}_i, \hat{a}_j]$ of the estimators a_i and a_i are in general non-null, even if the matrix $\mathbf{V}$ is itself diagonal. The errors on the parameters a_j are correlated and thus cannot be specified independently, that is, for each parameter individually.

It is instructive to consider the covariance matrix $\mathbf{U}$ in terms of second-order derivatives of the χ^2 function. Toward this end, we will calculate the second-order derivatives of the χ^2 function based on Eq. (5.96).

$$\left.\frac{\partial^2 \chi^2}{\partial a_r \partial a_s}\right|_{\vec{a}=\hat{a}} = -2 \frac{\partial}{\partial a_r} \sum_{i,i'=1}^{N} F_{is} \left(V^{-1}\right)_{ii'} \left(y_{i'} - \sum_{j'=0}^{m} F_{i'j} a_j \right), \tag{5.108}$$

$$= 2 \sum_{i,i'=1}^{N} F_{is} \left(V^{-1}\right)_{ii'} F_{i'r}, \tag{5.109}$$

$$= 2 \left(F^T V^{-1} F\right)_{sr}. \tag{5.110}$$

The order of the derivatives is inconsequential and the matrix $F^T V^{-1} F$ is symmetric, that is, equal to its transpose:

$$\left(F^T V^{-1} F\right)^T = F^T \left(V^{-1}\right)^T F = F^T V^{-1} F, \tag{5.111}$$

since V and V^{-1} are symmetric matrices. But as per Eq. (5.107), we found that the covariance matrix $\mathbf{U}$ is equal to $\left(F^T V^{-1} F\right)^{-1}$. We thus obtain the interesting and useful result:

$$\left.\frac{\partial^2 \chi^2}{\partial a_r \partial a_s}\right|_{\vec{a}=\hat{a}} = 2\left(U^{-1}\right)_{sr} \tag{5.112}$$

In the vicinity of the solution $\hat{a}$ (i.e., near the minimum of the χ^2-function), it is legitimate to write

$$\chi^2(\vec{a}) = \chi^2(\hat{a}) + \frac{1}{2}\sum_{i,j=0}^{m} \left.\frac{\partial^2 \chi^2}{\partial a_i \partial a_j}\right|_{\vec{a}=\hat{a}} (a_i - \hat{a}_i)\left(a_j - \hat{a}_j\right) + O(3) \tag{5.113}$$

Note that the absence of first derivatives stems from the fact that the series expansion is carried out at the minimum in which first-order derivatives vanish implicitly. Substituting the expression (5.112) for the second-order derivative, we obtain

$$\chi^2(\vec{a}) = \chi^2(\hat{a}) + \sum_{i,j=0}^{m} \left(U^{-1}\right)_{ij} \delta a_i \delta a_j, \tag{5.114}$$

in which $\delta a_i = a_i - \hat{a}_i$. In order to interpret this expression, consider for illustrative purposes a case in which $\mathbf{U}$ is a diagonal with elements σ_i^2. The inverse $\mathbf{U}^{-1}$ thus has diagonal elements $1/\sigma_i^2$ and Eq. (5.114) therefore implies that the χ^2 shall increase by one unit when deviations $\delta a_i = \sigma_i$ from $\hat{a}$ are considered. This is a rather generic result. It tells us that the χ^2 increases by one unit when a fit parameter is varied away from its optimal value by one standard deviation (while all other coefficients are kept constant and equal to their value at the χ^2 minimum).

Equation (5.114) provides numerical and graphical techniques to estimate and visualize the errors on the estimators $\hat{a}_i$ as illustrated in Figure 5.6 for cases involving one- and two-parameter fits. Panel (a) displays 15 simulated measurements of a constant but noisy signal of amplitude $y = 10$ fitted with a constant polynomial $y(x) = a_0$. The fit yields a value $a_0 = 9.67 \pm 0.81$ with a minimum $\chi^2 = 19.6$ for 14 degrees of freedom. Panel (c) shows the dependence of the fit χ^2 on a_0 and displays how the errors on a_0 may be obtained by increasing the χ^2 by one unit. Panel (b) displays a noisy linear signal fitted with a first order polynomial, $y(x) = a_0 + a_1 x$. Panel (d) displays iso-contours of the fit χ^2 plotted as a function of a_0 and a_1 for values $\chi^2 = \chi^2_{\min} + 1$ and $\chi^2 = \chi^2_{\min} + 2$. The symmetric and circular aspects of the contour indicate the parameters a_0 and a_1 are essentially uncorrelated. Their errors are thus independent and shown as one standard deviation errors in panel (b) based on the $\chi^2 = \chi^2_{\min} + 1$ contour.

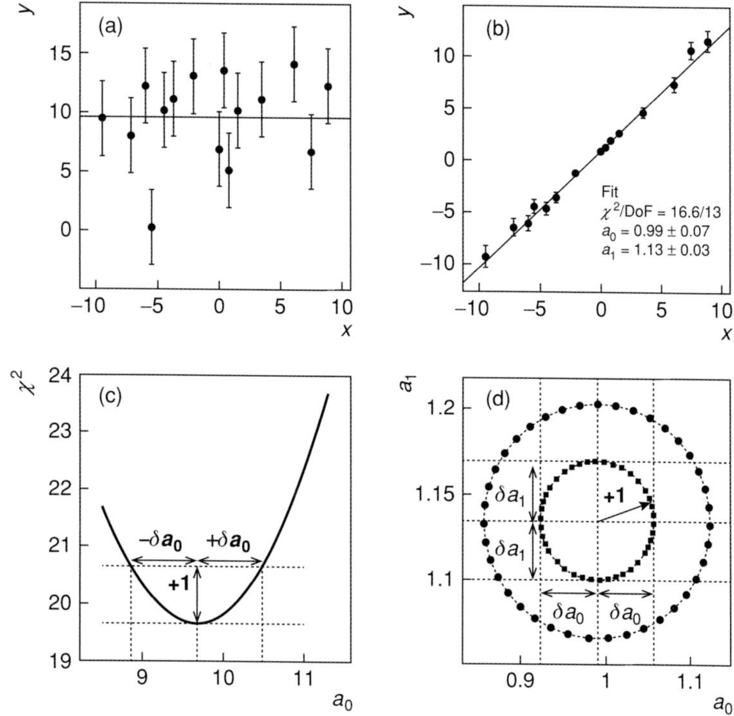

Fig. 5.6 Graphical interpretation of fit χ^2 for one- and two-parameter fits. (a) Noisy constant signal; (b) linear signal dependence; (c) χ^2 of the constant signal fit vs. the fit parameter a_0; (d) χ^2 iso-contours at $\chi^2_{\min} + 1$ and $\chi^2_{\min} + 2$ as a function of the fit parameters a_0 and a_1.

5.2.6 $\mathbb{LS}$ Fit with Binned Data

The $\mathbb{LS}$ method discussed in prior sections enables fits with arbitrary functions (models) for datasets containing an arbitrary number N of data points (x_i, y_i). However, it is not always possible or practical to handle all data points (x_i, y_i) individually. Often, one wishes to fit a model to data that have been binned into a histogram. We saw in §5.1.6 that the $\mathbb{ML}$ method readily enables model parameter estimation with histograms. But the $\mathbb{ML}$ method can be cumbersome and one consequently wishes to extend the $\mathbb{LS}$ method for fits of binned data also.

Consider a measurement in which data have been binned into a histogram consisting of n bins. Let H_i represent the number of entries in bin i, with $i = 1, \ldots, n$. We wish to fit the data with some function $f(x|\vec{\theta})$ determined by $m \geq 1$ parameter(s) $\vec{\theta}$, which have yet to be estimated. Let N be the total number of entries in the histogram. The function $f(x|\vec{\theta})$ is used as a model of the data. One thus expects the number of entries in bin i to be given by

$$\mu_i(\vec{\theta}) = N p_i(\vec{\theta}) = N \int_{x_{i,\min}}^{x_{i,\max}} f(x|\vec{\theta})\, dx, \tag{5.115}$$

in which we have defined the probability $p_i(\vec{\theta})$ that there will be entries in bin i. The parameters $\vec{\theta}$ are determined by minimization of the χ^2 function:

$$\chi^2(\vec{\theta}) = \sum_{i=1}^{N} \frac{\left(H_i - \mu_i(\vec{\theta})\right)^2}{\sigma_i^2}, \tag{5.116}$$

in which the variance σ_i^2 of the number of entries H_i must be estimated either from the data or from the model. There are thus few options as to how to proceed with the minimization of χ^2.

The $\mathbb{LS}$ fit can be somewhat simplified if the number of bins n is very large and such that there are just a few entries in each bin. In this case, the content of each bin may be reasonably well described by Poisson distributions and the variance of the number of entries in bin i is equal to the mean number of entries "predicted" by the model, $\mu_i(\vec{\theta})$. The χ^2 function may then be written as

$$\chi^2(\vec{\theta}) = \sum_{i=1}^{N} \frac{\left(H_i - \mu_i(\vec{\theta})\right)^2}{\mu_i(\vec{\theta})} = \sum_{i=1}^{N} \frac{\left(H_i - Np_i(\vec{\theta})\right)^2}{Np_i(\vec{\theta})}. \tag{5.117}$$

The functions $p_i(\vec{\theta})$ are integrals of $f(x|\vec{\theta})$ dependent on the unknown parameters $\vec{\theta}$ and the boundaries of each bins. Minimization of the χ^2 by analytical methods, described in prior sections, thus become intractable and one must then use numerical techniques such as those presented in §5.2.7.

An alternative approach consists in approximating the variances σ_i^2 by the number of entries H_i in each bin. Such a substitution is reliable if the bin contents H_i are uncorrelated and sufficiently large to provide a reliable estimate of the fluctuations. This leads to the **Modified Least-Squares** ($\mathbb{MLS}$) fit method based on the minimization of the χ^2 function defined as

$$\chi^2(\vec{\theta}) = \sum_{i=1}^{n} \frac{\left(H_i - Np_i(\vec{\theta})\right)^2}{H_i}. \tag{5.118}$$

Again in this case, the minimization of the χ^2 involves integrals $p_i(\vec{\theta})$ and as such is best handled by numerical methods. However, note that since the denominator includes the bin content H_i, the method will fail whenever the number of entries in one or more bin is null. It may be possible to remedy this problem by rebinning, that is, grouping bins together, or by using variable size bins.

Recall that in the case of the $\mathbb{ML}$ method, a multinomial function is used to estimate the expectation value of the number of entries per bin μ_i rather than a Poisson PDF. One can show that the variance of the $\mathbb{ML}$ estimate converges faster to the minimum variance bound than the $\mathbb{LS}$ or $\mathbb{MLS}$ estimates. This implies that for fits of binned data, it is preferable to use $\mathbb{ML}$ estimators whenever feasible.

5.2.7 Numerical $\mathbb{LS}$ Methods

We have so far focused on applications of the $\mathbb{LS}$ method for fits of linear models of the form given by Eq. (5.90). However, scientific analyses and data parameterization commonly involve nonlinear models that cannot be readily linearized. Examples of nonlinear functions commonly used include the Gaussian and Breit–Wigner distributions. It is also often desirable to add or mix elementary functions. For instance, one might wish to represent a signal and its background as a sum of a Gaussian distribution and a polynomial of order m as follows:

$$f(x|a_i, N, \mu, \sigma) = \sum_{i=0}^{m} a_i x^m + N \exp\left(-\frac{(x-\mu)^2}{2\sigma^2}\right). \qquad (5.119)$$

Clearly, this function's dependence on μ and σ is nonlinear, and given the addition of the "background" terms $\sum a_i x^m$ cannot be linearized by taking a logarithm of the function. Nonlinear models abound in physics, and in science in general. It is thus particularly important to consider model parameter estimation in the context of such models.

Numerical χ^2 minimization methods are a particular case of the more general case of optimization (or extremum finding) encountered both in classical and Bayesian inference problems. Rather than discussing numerical approaches piecemeal in this and following chapters, we present a systematic and comprehensive discussion of numerical techniques and algorithms in §7.6.

5.3 Determination of the Goodness-of-Fit

As we saw in §5.2.1, the $\mathbb{LS}$ method is derived from and therefore strictly equivalent to the $\mathbb{ML}$ method whenever the measurements y_i are Gaussian random variables. When this condition is met, the $\mathbb{LS}$ estimators $\vec{\theta}$ obtained by χ^2 minimization consequently coincide with those obtained by the $\mathbb{ML}$ method. Once estimators are known, and for a given set of data points, one can then seek the probability of getting a certain χ^2 value.

It is convenient to introduce **normalized deviates**, denoted $z_i(\vec{\theta})$, and defined as

$$z_i(\vec{\theta}) = \big(y_i - \mu_i(\hat{\theta})\big)/\sigma_i, \qquad (5.120)$$

in which $\hat{\theta}$ are the estimator values obtained in the fit. By construction, a normalized deviate $z_i(\vec{\theta})$ measures the deviation between an observed value y_i and the value $\mu_i(\hat{\theta})$ predicted by the model for x_i, according to parameters $\hat{\theta}$, and relative to the standard deviation σ_i. As such, the normalized deviates provide a measure of the level of agreement or compatibility between the data and the model (obtained from the fit), relative to the errors σ_i. The deviates z_i should be distributed according to the standard normal distribution if the y_i are Gaussian distributed. Additionally, with known variances σ_i^2, a function $\mu_i(\hat{\theta})$ linear in the parameters $\vec{\theta}$, and with proper functional form (i.e., an appropriate representation of the data), one expects the minimum $\chi^2 = \sum_i z_i^2$ obtained by the $\mathbb{LS}$ method should be distributed according to the χ^2 PDF with $N - m$ degrees of freedom, as shown in §3.13.3. It is consequently appropriate to use the χ^2 value obtained in a fit of data with Gaussian deviates to evaluate the **goodness** of the fit.

Recall from §3.13 that the expectation value of a random variable with a χ^2 PDF is equal to the number of degrees of freedom, N_{DoF}. It is thus convenient (and customary) to quote the χ^2 divided by the number of degrees of freedom, N_{DoF}, as a measure of the goodness of a fit. The number of degrees of freedom N_{DoF} is equal to $N - m$, N being the number of data points or bins, and m the number of fit parameters. If the value χ^2/N_{DoF} is much smaller than one, or near zero, then the fit is much better than expected, on average, given the size of the data set and the number of fit parameters. Very small χ^2/N_{DoF} values are not impossible but have rather low probability of occurring. Fits yielding "very" small χ^2/N_{DoF} values might then signal that the errors (standard deviations), σ_i, used in the fit are overestimated or correlated. Large values of χ^2/N_{DoF}, on the other hand, indicate that the fit is very poor. This is an indication that the model $f(x|\vec{\theta})$ used to fit the data has a very small likelihood of yielding the measured data. The hypothesis that this particular model constitutes an accurate representation of the data (and the phenomenon considered) is thus regarded as having a low probability. Alternatively, it is possible that the errors σ_i are much underestimated and consequently yield a rather large χ^2.

It is customary to report the goodness of a fit in terms of its **significance level** or p-value. The significance level corresponds to the probability that the model hypothesis[4] would lead to a χ^2 value worse (i.e., larger) than that actually achieved:

$$p = \int_{\chi^2}^{\infty} p_\chi(z|N_{\text{DoF}})\, dz, \tag{5.121}$$

where $p_{\chi^2}(z|N_{\text{DoF}})$ is the χ^2-distribution for N_{DoF} degrees of freedom. Integrals of $p_{\chi^2}(z|N_{\text{DoF}})$ are best calculated with numerical routines available in software packages such as Mathematica®, MATLAB®, or ROOT®.

It is important to realize that the choice of minimum p-value used toward the rejection of model hypotheses is rather subjective and may very well depend on the purpose of a particular measurement. See §6.6.5 for a more extensive discussion of this issue. Additionally, it is also important to acknowledge that the errors σ_i may be under- or overestimated, thereby resulting in too large or too small a value of χ^2, respectively. The use of a χ^2 test as a measure of the goodness of a fit may thus be completely unwarranted if the errors have not been properly calibrated.

5.4 Extrapolation

Having obtained estimates of model parameters with the $\mathbb{ML}$, $\mathbb{LML}$, $\mathbb{LS}$, or related methods, it is often necessary to use the estimates to determine values predicted by the model in

[4] The notions of hypothesis and hypothesis testing are discussed at great length in Chapter 6.

regions where no measurements exist. Doing this is fairly easy: one needs only to plug in the parameter estimates, $\hat{\theta}$, obtained from the fit into the model formula, and calculate the function values $y = f(x|\theta)$ at the relevant values of x. Somewhat less easy, however, is the calculation of error estimates δy on values $y = f(x|\theta)$ predicted by or extrapolated from the model. The model parameters are in general not independent. One must then use the covariance matrix U of the model parameters to determine the error δy on an extrapolated value according to

$$\delta y^2 = \left(\frac{\partial y}{\partial \vec{\theta}}\right)^T \mathbf{U} \left(\frac{\partial y}{\partial \vec{\theta}}\right). \tag{5.122}$$

As a practical example, let us consider the implementation of the foregoing formula for a polynomial of order m with coefficients a_i, $i = 0, \ldots, m$. We must first calculate the derivatives $\partial y / \partial a_j$:

$$\frac{\partial y}{\partial \vec{a}_j} = \sum_{k=0}^{m} \frac{\partial a_k}{\partial a_j} x^k = x^j. \tag{5.123}$$

The expression (5.122) may then be written:

$$\delta y^2 = \begin{pmatrix} 1 & x & \cdots & x^m \end{pmatrix} \begin{pmatrix} U_{00} & U_{01} & \cdots & U_{0m} \\ U_{10} & U_{11} & \cdots & U_{1m} \\ \vdots & \vdots & \ddots & \vdots \\ U_{m0} & U_{m1} & \cdots & U_{mm} \end{pmatrix} \begin{pmatrix} 1 \\ x \\ \vdots \\ x^m \end{pmatrix}. \tag{5.124}$$

For a straight-line fit, $y = a_0 + a_1 x$, this expression reduces to

$$\delta y^2 = U_{00} + 2U_{01} x + U_{11} x^2, \tag{5.125}$$

in which $U_{00} = S_{xx}/\Delta$ and $U_{11} = S/\Delta$ are the variances of a_0 and a_1, respectively, while $U_{01} = S_x/\Delta$ is the covariance of a_0 and a_1.

It is important to stress that neglect of the off-diagonal terms, which may be quite large relative to the diagonal terms, can lead to gross misrepresentations of the errors δy on extrapolated values (see Problem 5.12).

5.5 Weighted Averages

It is the hallmark of science that measurements should be reproducible and that advances in technology typically lead to improved measurements of physical quantities. It is often the case that different groups of scientists conduct distinct measurements of a given observable, for instance, the mass of a particle, the value of a fundamental constant such as the speed of light, and so on. Distinct experiments generally have different degrees of reliability, accuracy, precision, and yield different measurement results with distinct errors. Given historical trends, one might expect that more modern experiments yield more accurate and precise results. One might thus be tempted to rely only on the latest or most precise results. But why give up the valuable information provided by other experiments? Why not combine the information gathered to obtain a **world average** that accounts for all data available? This can be accomplished with the **weighted average** ($\mathbb{WA}$) method, which is introduced here as a special case of the $\mathbb{LS}$ method. Weighted averages may additionally be obtained with the $\mathbb{ML}$ method and within the Bayesian inference paradigm discussed in Chapter 7.

Suppose the quantity of interest, with a true value θ, has been measured N times, yielding N independent values (estimates) $\hat{\theta}_i$ and errors σ_i, with $i = 1, 2, \ldots N$. Since a single phenomenon is being considered, it is reasonable to expect all measurements should yield the same value θ. It is thus acceptable to combine them to get a better estimate of θ. This can be readily accomplished with the $\mathbb{LS}$ method by minimization of the χ^2 objective function for a model $f(x|\theta) = \theta$:

$$\chi^2(\theta) = \sum_{i=1}^{N} \frac{\left(\hat{\theta}_i - \theta\right)^2}{\sigma_i^2}. \tag{5.126}$$

We set the derivative of Eq. (5.126) with respect to θ equal to zero to seek an extremum that yields the value θ most compatible with the existing measurements $\hat{\theta}_i$:

$$\frac{d}{d\theta}\chi^2(\theta) = -2\sum_{i=1}^{N} \frac{(\theta_i - \theta)}{\sigma_i^2} = 0. \tag{5.127}$$

Solving for θ yields

$$\hat{\theta}_{\text{WA}} = \sum_{i=1}^{N} \frac{\theta_i}{\sigma_i^2} \Bigg/ \sum_{i=1}^{N} \frac{1}{\sigma_i^2}. \tag{5.128}$$

We use the subscript "WA" to indicate that the preceding estimate is equal to the sum of the estimates θ_i weighted by their respective variances and as such corresponds to a special case of a weighted average procedure defined as

$$\hat{\theta}_{\text{WA}} = \sum_{i=1}^{N} \omega_i \theta_i \Bigg/ \sum_{i=1}^{N} \omega_i, \tag{5.129}$$

with weights $w_i = 1/\sigma_i^2$.

The weights ω_i determine the importance given to estimates θ_i in the average. Measurements with a smaller variance σ_i^2 have a larger weight and thus contribute more to the weighted average $\hat{\theta}_{\text{WA}}$. Note that the factor $\sum_i w_i$ is needed for proper normalization of the weights, unless they are already normalized, in other words, if $\sum_i w_i = 1$.

The second-order derivative of the χ^2-function with respect to θ yields the variance of the estimate

$$\text{Var}\left[\hat{\theta}_{\text{WA}}\right] = \left(\sum_{i=1}^{N} \sigma_i^{-2}\right)^{-1}, \tag{5.130}$$

which amounts to the inverse of the sums of all the weights. The variance $\text{Var}[\hat{\theta}_{WA}]$ is, by construction, smaller than the individual variances σ_i^2. Consequently, there is an obvious advantage in combining the results of several measurements. Three special cases are of interest: first, if all measurements have equal errors, $\sigma_i = \sigma$, then the variance of the estimate simplifies to

$$\text{Var}\left[\hat{\theta}_{\text{WA}}\right] = \left(\sum_{i=1}^{N} \sigma^{-2}\right)^{-1} = \frac{\sigma^2}{N}, \tag{5.131}$$

and the error σ_{WA} on the estimate $\hat{\theta}_{WA}$ equals the measurement error divided by the square root of the number N of measurements:

$$\sigma_{\text{WA}} = \frac{\sigma}{\sqrt{N}} \quad \text{(equal errors)}. \tag{5.132}$$

Second, if one measurement has a much smaller error than the others, it will dominate both the mean and its variance. Third, if a measurement has a much larger error than the others, it will play a negligible role in the evaluation of the mean and its variance.

The foregoing weighted procedure can be generalized to situations in which the measurements θ_i are not independent. This would be the case, for instance, if some or all of the estimates are based in part on the same data. One must then first determine the covariance V_{ij} of the measurements. One can then verify (see Problem 5.13) that the WA is given by Eq. (5.129) with weights replaced by

$$w_j = \frac{\sum_{i=1}^{N} (V^{-1})_{ij}}{\sum_{k,m=1}^{N} (V^{-1})_{km}}. \tag{5.133}$$

Clearly, Eq. (5.133) reduces to Eq. (5.129) if the covariance matrix is diagonal, that is, if the measurements $\hat{\theta}_i$ are uncorrelated.

Averaging of experimental results may also be achieved with products of likelihood functions (as well as sums of log of likelihood functions) of combined datasets [67] and Bayesian inference techniques discussed in Chapter 7.

5.6 Kalman Filtering

Kalman filtering ($\mathbb{KF}$) is a technique that was initially designed and used for radar signal processing. It is quite general, however, and is used nowadays in many applications, including signal processing, signal fitting, pattern recognition, as well as navigation and control.

By design, a Kalman filter operates recursively on one or multiple streams of noisy input data to produce statistically optimal estimates of the underlying state of a physical system. The technique is named after Rudolf E. Kalman[5] (b. 1930), one of the early and primary developers of the theory [122]. It was introduced in high-energy physics by Billoir as a

[5] Hungarian-born American electrical engineer, mathematician, and inventor.

progressive method of track-fitting [41]. The equivalence between progressive methods and Kalman filters was established by Fruhwirth [88].

In nuclear and high-energy physics, Kalman filtering is commonly applied toward track reconstruction in complex detectors, where it usually involves a linear, recursive method of track-fitting shown to be equivalent to a global least-squares minimization procedure (see §5.6.6). It is therefore an optimal linear estimator of track parameters. Provided the track model is truly linear and measurement errors are Gaussian, the Kalman filter is also efficient. It was formally shown that no nonlinear estimator can do better. Extensions and generalizations of the method to nonlinear systems, known as **extended Kalman filters ($\mathbb{EKF}$)**, have also been developed.

Kalman filters have the following attractive features that make them preferable over global least-squares methods under appropriate circumstances:

1. A Kalman filter is recursive and is thus well suited for progressive signal processing, particularly track finding and fitting in large and complex detection systems.
2. A Kalman filter can be extended into a **smoother** and thereby provides for optimal estimates of signals throughout the evolution of a system.
3. A Kalman filter readily enables efficient resolution and removal of outlier points.
4. In contrast to least-squares methods, a Kalman filter does not involve the manipulation or inversion of large matrices.

We motivate and introduce the notion of recursive fitting (filtering) in §5.6.1. The linear Kalman filter algorithm is outlined in §5.6.2. A detailed derivation of the expression of the Kalman gain matrix is presented in §5.6.5, whereas a proof of the equivalence between the Kalman filter method and the least-squares method is sketched in §5.6.6. An example of application for charged particle track reconstruction in complex detectors is presented in §9.2.2. The Kalman filtering techniques presented in this section constitute a small subset of the field of optimal estimation and control theory covered in more specialized works (e.g., see [70, 93, 168, 199]).

5.6.1 Recursive Least-Squares Fitting and Filtering

The least-squares method discussed in §5.2 is ideal for the estimation of model parameters when all data have been acquired and can be fitted all at once. However, there are applications in which a progressive and recursive knowledge of a system's or model's parameters are required. In other words, rather than waiting for the whole dataset to become available, one wishes to obtain an estimate on the basis of existing data and progressively improve the estimate as additional data are collected. Such a task is the domain of recursive least-squares filtering methods.

We saw in §4.5 that the arithmetic mean constitutes an unbiased estimator of the mean of a set of data. One can alternatively obtain the arithmetic mean as the least-squares estimator of a data model involving a zeroth-order polynomial. Indeed, for a dataset

$\vec{x} = (x_1, x_2, \ldots, x_k)$ with measurement errors $\vec{\sigma} = (\sigma_1, \sigma_2, \ldots, \sigma_k)$, the χ^2 of a zeroth-order polynomial is

$$\chi^2 = \sum_{i=1}^{k} \frac{(x_i - a)^2}{\sigma_i^2}. \tag{5.134}$$

Setting the derivative of χ^2 with respect to a equal to zero yields the least-squares estimate:

$$\hat{a} = \frac{\sum_{i=1}^{k} x_i/\sigma_i^2}{\sum_{i=1}^{k} 1/\sigma_i^2}. \tag{5.135}$$

For the sake of simplicity, let us consider a case where all errors σ_i are equal. One writes

$$\hat{a}_k = \frac{1}{k}\sum_{i=1}^{k} x_i, \tag{5.136}$$

where $\hat{a}_k$ denotes the estimate of a obtained with k values x_i. It corresponds to the arithmetic mean of a sample of k values x_i. Adding one value to the sample, one can obviously write

$$\hat{a}_{k+1} = \frac{1}{k+1}\sum_{i=1}^{k+1} x_i. \tag{5.137}$$

It is convenient to formulate the estimate $\hat{a}_{k+1}$ in terms of the estimate $\hat{a}_k$, as follows:

$$\begin{aligned} \hat{a}_{k+1} &= \frac{1}{k+1}\left(k\frac{1}{k}\sum_{i=1}^{k} x_i + x_{k+1}\right) \\ & \frac{1}{k+1}\left(k\hat{a}_k + x_{k+1}\right) \end{aligned} \tag{5.138}$$

where in the second line, we have used the expression (5.136) of the estimator $\hat{a}_k$. Defining $\hat{a}_0 = 0$ and $\hat{a}_1 = x_1$, we shift the indices of Eq. (5.138) by one unit and obtain

$$\begin{aligned} \hat{a}_k &= \frac{1}{k}\left((k-1)\hat{a}_{k-1} + x_k\right) \\ &= \hat{a}_{k-1} + \frac{1}{k}\left(x_k - \hat{a}_{k-1}\right) \end{aligned} \tag{5.139}$$

We find that the *new estimate* $\hat{a}_k$ is equal to the *prior estimate* $\hat{a}_{k-1}$, plus a "correction" proportional to the difference between the new measurement x_k and the prior estimate, known as the kth residue. The weight given to the correction is determined by the factor $1/k$, which we call the *gain* of the filter and denote $K_k^{(1)}$. We thus can write

$$\hat{a}_k = \hat{a}_{k-1} + K_k^{(1)}\left(x_k - \hat{a}_{k-1}\right), \tag{5.140}$$

with the filter gain

$$K_k^{(1)} = \frac{1}{k}. \tag{5.141}$$

Equation (5.140) provides us with a recursive formula to estimate the arithmetic mean of a growing sample of values, x_i. Setting $a_0 = 0$, for $k = 1$, one has $K_1^{(1)} = 1$, and the first

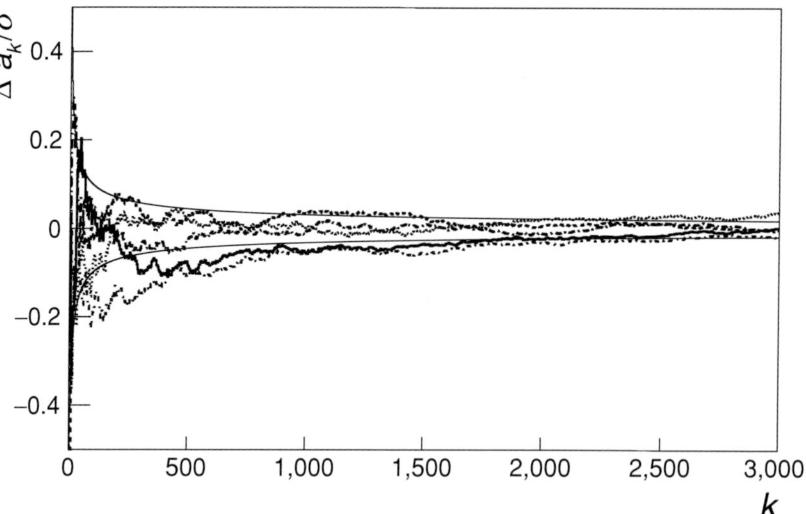

Fig. 5.7 Evolution of the error Δa_k of the estimate $\hat{a}_k$ of five constant signals with Gaussian noise of standard deviation σ. Solid lines show the 68% confidence interval $(1/\sqrt{k})$ of the sample mean for a data sample of size k.

measured value is given maximal weight

$$\hat{a}_1 = K_1^{(1)} x_1 = x_1, \tag{5.142}$$

whereas for increasing values of k, one finds the gain $K_k^{(1)}$ decreases monotonically and vanishes for $k \to \infty$: additional values are progressively given a smaller weight in the calculation of the estimate $\hat{a}$.

Equation (5.140) epitomizes the concept of recursive filtering and fitting. One starts with no information and the first measurement is given maximal weight in the determination of the first estimate $\hat{a}_1$. This and subsequent estimates serve as priors toward the recursive determination of posterior estimates, which are expected to progressively converge toward the true value of the observable. The gain K determines the importance given to new information provided by measurements x_k. It is initially large but tends to decrease as the number of sampled values progressively increases.

Figure 5.7 illustrates how estimates $\hat{a}_k$ of a constant value a progressively converge toward the true value while the gain tends to zero.

We saw in §4.5 that the estimator of the mean is given by Eq. (5.136). This can be verified also for the estimators $\hat{a}_k$ as follows:

$$\mathrm{E}\left[\hat{a}_k\right] = \frac{1}{k}\sum_{i=1}^{k} \mathrm{E}\left[x_i\right] = \frac{1}{k}\sum_{i=1}^{k} a = a. \tag{5.143}$$

In order to examine the variance of estimators $\hat{a}_k$, it is convenient to express the measurements x_k in terms of their expectation value a and a signal noise v_k:

$$x_k = a + v_k, \tag{5.144}$$

where the term v_k represents random numbers with null expectation value, $\mathrm{E}[v_k] = 0$, and variance $\mathrm{E}\left[v_k^2\right] = \sigma^2$. One can then calculate the residue

$$a - \hat{a}_k = a - \hat{a}_{k-1} - \frac{1}{k}\left(a + v_k - \hat{a}_{k-1}\right) \tag{5.145}$$

$$= \left(a - \hat{a}_{k-1}\right)\left(1 - \frac{1}{k}\right) - \frac{1}{k}v_k \tag{5.146}$$

Squaring and taking the expectation value, we obtain the variance of the estimator $\hat{a}_k$:

$$\begin{aligned}\mathrm{Var}\left[\hat{a}_k\right] = \mathrm{E}\left[(a - \hat{a}_k)^2\right] &= \left(1 - \frac{1}{k}\right)^2 \mathrm{Var}\left[\hat{a}_{k-1}\right] \\ &\quad - 2\frac{1}{k}\left(1 - \frac{1}{k}\right)\mathrm{E}\left[\left(a - \hat{a}_{k-1}\right)v_k\right] + \frac{1}{k^2}\mathrm{E}\left[v_k^2\right]\end{aligned} \tag{5.147}$$

The expectation value $\mathrm{E}\left[v_k^2\right]$ is the variance σ^2 of the noise v_k while the expectation value $\mathrm{E}\left[\left(a - \hat{a}_{k-1}\right)v_k\right]$ vanishes because the noise v_k is assumed to be uncorrelated to that of prior measurements. The preceding expression thus provides the variance $\mathrm{E}\left[(a - \hat{a}_k)^2\right]$ in terms of the variance at step $k-1$ and a second term depending on the signal noise. Defining the "covariance matrix," $S_k = \mathrm{E}\left[(a - \hat{a}_k)^2\right]$, we can then rewrite the preceding as

$$S_k = \left(1 - K_k\right)^2 S_{k-1} + K_k^2\sigma^2, \tag{5.148}$$

which gives us an equation for the evolution of the covariance S_k of the estimate $\hat{a}_k$ in terms of prior values S_{k-1} and the variance σ^2 of the measurement noise. Note that for small values of k, the gain is near unity, $K \approx 1$, and the evolution of S_k is dominated by the signal noise, whereas for large k the gain nearly vanishes and the covariance S_k becomes approximately constant. One can verify by simple substitution that the expectation value of the covariances S_k scales as

$$S_k = \frac{\sigma^2}{k^2}. \tag{5.149}$$

The foregoing recursive formalism is readily applicable to polynomial or linear function of all orders. We consider a general extension to all linearizable models in §5.6.2.

5.6.2 The Kalman Filter Algorithm

In the framework of the Kalman filter, a physical system (e.g., a radio signal, a charged particle track traversing a magnetic spectrometer) is represented by a set of n_s parameters, called the Kalman state vector, $\vec{s} = (s_1, s_2, \ldots, s_{n_s})$, which is allowed to vary as a function of some independent variable, t. For live-feed signal processing applications, time is obviously a convenient choice of independent variable. However, other choices are also possible or appropriate depending on the specificities of physical systems under study. For instance, in the case of charged particle reconstruction inside a magnetic spectrometer, the independent variable may be taken as the track length or the position across the spectrometer.

The state of the system is usually regarded as dynamically evolving as a function of the independent parameter t. While primarily deterministic, the dynamic process may also involve a stochastic component, usually called **process noise**. Process noise may arise because of background processes or through the dynamical evolution of the system. For instance, a track propagating through a detector is likely to interact with materials composing the detector. Interactions may lead to energy loss and scattering. The instantaneous properties of the track (e.g., momentum and direction) are thus likely to change stochastically due to such interactions.

Kalman filtering typically involves recursive measurements of n_m dependent parameters, $\vec{m}_k = (m_1, m_2, \ldots, m_{n_m})_k$, determined by the state of the system and the properties of the measurement device. Measurements typically involve fluctuations associated with the granularity and geometry of the devices as well as, ultimately, the intrinsically stochastic nature of the measurement process. Uncertainties associated with the measuring process are commonly known as **measurement noise** (see §12.1 for a more in-depth discussion of process and measurement noises).

Measurements of the dependent parameters are achieved recursively by sampling the system's signal(s). Depending on the applications and systems considered, such sampling might be carried out repeatedly at an arbitrarily large frequency or through finitely many steps. Either way, it is usually the case that both the measurements and the state of the system are expressed as functions of the independent variable t. The frequency of the sampling process being finite by its very nature, it is convenient to discretize all variables of interests in terms of t steps measured with an arbitrary index k. Thus, $\vec{s}_k$ and $\vec{m}_k$ represent the state of the system and measured values at "step" t_k. Some applications involve recursively unlimited measurement of samples and thus have unbound values of step t_k. Spectrometers used in particle physics for measurements of charged particle momenta, however, involve finitely many detection planes and thus feature data processing with finitely many values of (time) steps t_k.

Basic Kalman filters assume that knowledge of the state at step k, noted $\hat{s}_{k|k-1}$ and known as *prior*, can be predicted based on knowledge of the state at t_{k-1} according to a linear model:

$$\vec{s}_{k|k-1} = \mathbf{F}_k \vec{s}_{k-1} + \vec{w}_k, \tag{5.150}$$

where $\mathbf{F}$ is a linear function (an $n_s \times n_s$ matrix) describing the evolution of the state vector between the two times t_{k-1} and t_k. In practical situations, the evolution of the system between two steps may be nonlinear, and one should instead write

$$\vec{s}_{k|k-1} = \phi_k(\vec{s}_{k-1}) + \vec{w}_k, \tag{5.151}$$

where $\phi_k(\vec{s}_{k-1})$ is a nonlinear function of the state vector at step k. The principle of the method remains the same, however, and leads to extended Kalman filters (EKFs) (see §5.6.4).

The vector $\vec{w}_k$ corresponds to process noise and amounts to stochastic variations of the signal (state) associated with background processes accumulated during the interval

$t_k–t_{k-1}$. One assumes the process noise to be unbiased, that is, such that

$$\mathrm{E}\left[\vec{w}_k\right] = 0, \tag{5.152}$$

and characterized by a predictable $n_s \times n_s$ covariance matrix $\mathbf{W}_k$ with elements

$$(\mathbf{W}_k)_{ij} = \mathrm{Cov}\left[(w_k)_i, (w_k)_j\right], \tag{5.153}$$

where $(w_k)_i$ and $(w_k)_j$ are noise values of state parameters $(s_k)_i$ and $(s_k)_j$, $i, j = 1, \ldots, n_s$ at step k.

Uncertainties of the components of the state vector $\vec{s}_k$ are likewise described by an $n_s \times n_s$ covariance matrix denoted $\mathbf{S}_k$[6]:

$$(\mathbf{S}_k)_{ij} = \mathrm{Cov}\left[(s_k)_i, (s_k)_j\right]. \tag{5.154}$$

One represents measurements performed at step k (layer k) with a vector noted $\vec{m}_k$. The dimensionality n_m of $\vec{m}_k$ is smaller or equal to that of the state vector ($n_m \leq n_s$) and is generally limited to just a few parameters, that is, $n_m \ll n_s$. For instance, in a magnetic spectrometer, straw tube chambers would provide a single measurement of position, u or v, yielding $n_m = 1$, while pad chambers or continuous devices such as Time Projection Chambers would yield measurements of two coordinates, y and z, and be represented by a two-element vector, $\vec{z} = (y, z)$. One further assumes that it is possible, given an estimate of the Kalman state, $\vec{s}_k$, to project (predict) a measurement by means of a linear function, $\mathbf{H}_k$, according to

$$\vec{m}_k = \mathbf{H}_k \vec{s}_k + \vec{v}_k. \tag{5.155}$$

The function H_k represents an $n_m \times n_s$ matrix that projects the vector $\vec{s}_k$ onto the measurement coordinates $\vec{m}_k$. It is often possible and convenient to choose some parameters of the state model $\vec{s}_k$ to be values of the measurements $\vec{m}_k$. The matrix H_k thus simplifies trivially. For instance, for $n_m = 2$ and $n_s = 6$, one could write

$$\mathbf{H}_k = \begin{bmatrix} 1 & 0 & 0 & 0 & 0 & 0 \\ 0 & 1 & 0 & 0 & 0 & 0 \end{bmatrix}. \tag{5.156}$$

However, such a simplification is not always possible or desirable. There are also cases where a linear function is not available.

The vector $\vec{v}$ represents the measurement noise associated with the determination of the measurement vector $\vec{m}$. The measurement noise is assumed to be unbiased, in other words, with null expectation value such that

$$\mathrm{E}\left[\vec{v}_k\right] = 0. \tag{5.157}$$

It is characterized by a known error covariance matrix, $\mathbf{V}_k$ defined according to:

$$(\mathbf{V}_k)_{ij} = \mathrm{Cov}\left[(v_k)_i, (v_k)_j\right]. \tag{5.158}$$

[6] In this and following sections, we use lowercase letters for the physical quantities and corresponding capital letters for their respective covariance matrix: $s \to S$; $w \to W$; $v \to V$

It is generally further assumed that the process and measurement noises are strictly independent, and that successive measurement noises are also uncorrelated.

$$\mathrm{Cov}\left[(s_k)_i, (v_k)_j\right] = 0 \tag{5.159}$$

$$\mathrm{Cov}\left[(v_k)_i, (v_{k'})_j\right] = 0 \text{ for } k \neq k' \tag{5.160}$$

The Kalman filter algorithm requires an initial estimate of the system state $\vec{s}_0$ at t_0. This and subsequent estimates of the state vector at t_{k-1}, noted $\vec{s}_{k-1|k-1}$ with $k > 1$, are used to predict the state of the system at the next measurement step t_k.

$$\vec{s}_{k|k-1} = \mathbf{F}_k \vec{s}_{k-1|k-1}. \tag{5.161}$$

One also predicts (projects) the covariance matrix of the state vector as

$$\mathbf{S}_{k|k-1} = \mathbf{F}_k \mathbf{S}_{k-1|k-1} \left(\mathbf{F}_k\right)^T + \mathbf{W}_k. \tag{5.162}$$

The measurement $\vec{m}_k$ is then used to update and improve the knowledge of the Kalman state with

$$\vec{s}_k \equiv \vec{s}_{k|k} = \vec{s}_{k|k-1} + \mathbf{K}_k \left(\vec{m}_k - \mathbf{H}_k \vec{s}_{k|k-1}\right) \tag{5.163}$$

where the quantity $\mathbf{K}_k$ is called the **Kalman gain matrix**. It can be calculated as follows (for a derivation of this result, see §5.6.5):

$$\mathbf{K}_k = \mathbf{S}_{k|k-1} \left(\mathbf{H}_k\right)^T \left(\mathbf{V}_k + \mathbf{H}_k \mathbf{S}_{k|k-1} \left(\mathbf{H}_k\right)^T\right)^{-1}. \tag{5.164}$$

The matrix inversion involved in Eq. (5.164) is typically rather simple because the measurement error covariance matrix has a small dimensionality. In some cases, when $\mathbf{H}_k$ is diagonal, the inversion may even become trivial.

To operate the Kalman filter, one first initializes the covariance matrix $\mathbf{S}_{0|0}$ with large diagonal values and null off-diagonal elements. As the filter progresses from step to step, more information about the system is acquired by added measurements $\vec{m}_k$. The diagonal elements reduce to values representative of the uncertainty on the system parameters. Initially, $\mathbf{S}_{k|k-1}$ dominates the factor $\left(\mathbf{V}_k + \mathbf{H}_k \mathbf{S}_{k|k-1} \left(\mathbf{H}_k\right)^T\right)^{-1}$ so the gain $\mathbf{K}_k$ is near unity. As the number of sampled measurements increases, this factor becomes increasingly dominated by the covariant matrix $\mathbf{V}_k$, and the Kalman $\mathbf{K}_k$ gain becomes progressively smaller. With a large Kalman gain, the addition of a new measurement has a significant impact on the updated system parameters. As the gain reduces, the addition of new measurements has a progressively smaller impact on the updated state of the system.

The filtered (updated) covariance is given by

$$\mathbf{S}_{k|k} = \left(\mathbf{I} - \mathbf{K}_k \mathbf{H}_k\right) \mathbf{S}_{k|k-1}, \tag{5.165}$$

where $\mathbf{I}$ denotes an $n_s \times n_s$ identity matrix. Again, one finds that initially, the Kalman gain being large, the covariance matrix rapidly decreases in magnitude. As more information is added, the gain diminishes and added measurements have diminishing impact on the state covariance.

The recursive operation of the filter is schematically illustrated in Figure 5.9. An example of an application for charged particle track reconstruction in a complex detector is schematically illustrated in Figure 5.8 and discussed in detail in §9.2.3.

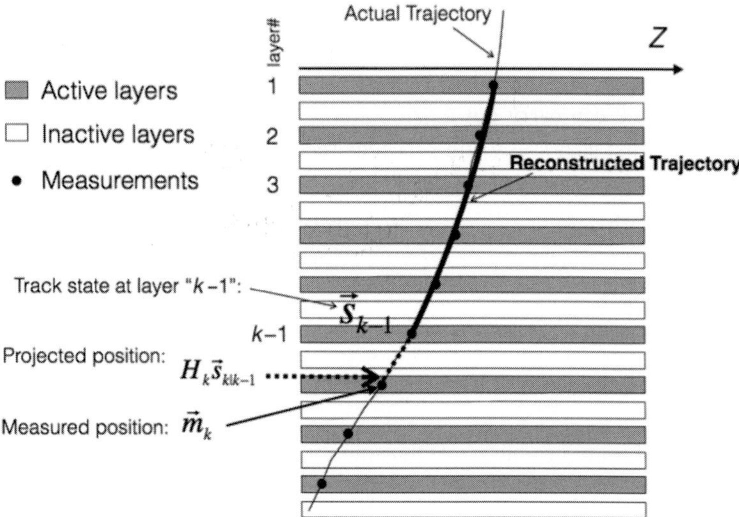

Fig. 5.8 A particle detector may be represented as a succession of material volumes or layers. Measurements $\vec{m}_k$ are carried out in active layers whereas passive volumes contribute no information to the knowledge of the track and may in fact produce degradation of information through stochastic processes such as differential energy loss and multiple Coulomb scattering. The state vector is therefore defined at finitely many layers only. Starting at some base layer i, one proceeds iteratively to predict and measure the state at successive layers. Given the knowledge of a track state $\vec{s}_{k-1}$ at layer $k - 1$, one predicts its state $\vec{s}_k$ at layer k using a linear function. The measurement $\vec{m}_k$ is then used to update and improve the knowledge of the state of the track. The process is repeated iteratively until all (relevant) layers have been traversed.

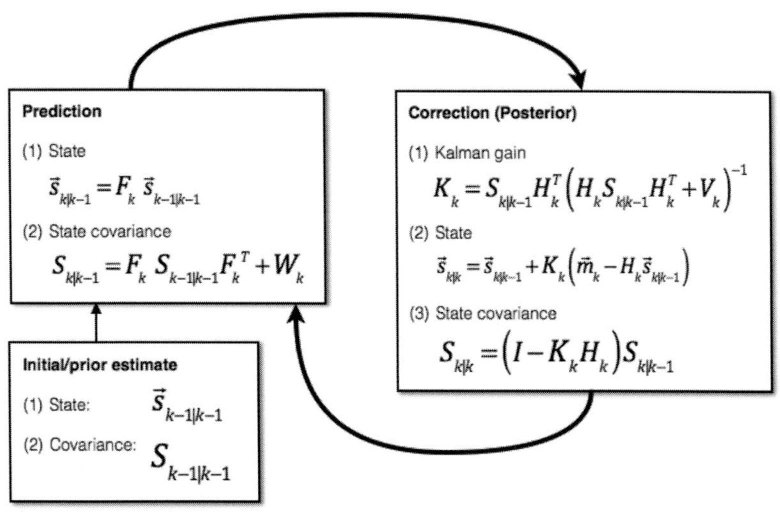

Fig. 5.9 Schematic illustration of the components of the Kalman filter algorithm.

5.6.3 Kalman "Smoother"

In cases where it might be useful to have optimal information on the system at all steps t_k, one can carry out a "smoothing" pass on the data. The smoothing pass begins with the first step $k = n$ and proceeds recursively backward from the last measurement $k = n$. The smoothed state vector at t_k is based on all n measurements steps and is calculated as follows:

$$\vec{s}_{k|n} = \vec{s}_{k|k} + \mathbf{A}_k \left(\vec{s}_{k+1|n} - \vec{s}_{k|n} \right), \tag{5.166}$$

with the smoother gain matrix

$$\mathbf{A}_k = \mathbf{S}_{k|k} \left(\mathbf{F}_k \right)^T + \left(\mathbf{S}_{k+1|k} \right)^{-1}. \tag{5.167}$$

The covariance matrix of the smoothed state vector is

$$\mathbf{S}_{k|n} = \mathbf{S}_{k|k} + \mathbf{A}_k \left(\mathbf{S}_{k+1|n} - \mathbf{S}_{k+1|k} \right) \left(\mathbf{A}_k \right)^T. \tag{5.168}$$

Multiple extensions and variants of the foregoing algorithm are documented in the scientific and engineering literature.

5.6.4 The Extended Kalman Filter

The propagation of the state vector $\vec{s}$ with Eq. (5.150) assumes the evolution of the system may be described with a linear function of the state parameters. This is a rather limiting assumption and, in practice, one often deals with nonlinear state evolution equations such as

$$\vec{s}_k = \vec{f}_k \left(\vec{s}_{k-1}, \vec{w}_{k-1} \right), \tag{5.169}$$

as well as nonlinear state-to-measurement "projection" equations such as

$$\vec{m}_k = \vec{h}_k(\vec{s}_k, \vec{v}_k). \tag{5.170}$$

It may be possible, however, to linearize Eqs. (5.169) and (5.170) and obtain an **extended Kalman filter** ($\mathbb{EKF}$).

Let us define matrices $\mathbf{A}_k$ and $\mathbf{B}_k$ as derivatives of the functions $(\vec{f}_k(\vec{s}_{k-1}, \vec{w}_{k-1}))_i$, $i = 1, \ldots, n_s$, with respect to j components ($j = 1, \ldots, n_s$) of the state vector $\vec{s}_k$ and process noise $\vec{w}_k$, respectively:

$$(\mathbf{A}_k)_{ij} = \frac{\partial (\vec{f}_k)_i}{\partial (\vec{s}_k)_j} \left(\vec{s}_{k-1}, 0 \right), \tag{5.171}$$

$$(\mathbf{B}_k)_{ij} = \frac{\partial (\vec{f}_k)_i}{\partial (\vec{w}_k)_j} \left(\vec{s}_{k-1}, 0 \right). \tag{5.172}$$

Let us additionally define matrices $\mathbf{P}_k$ and $\mathbf{Q}_k$ as derivatives of the functions $(\vec{h}_k(\vec{s}_{k-1}, \vec{v}_{k-1}))_i$, $i = 1, \ldots, n_m$, with respect to j components of the state vector $\vec{s}_k$ and

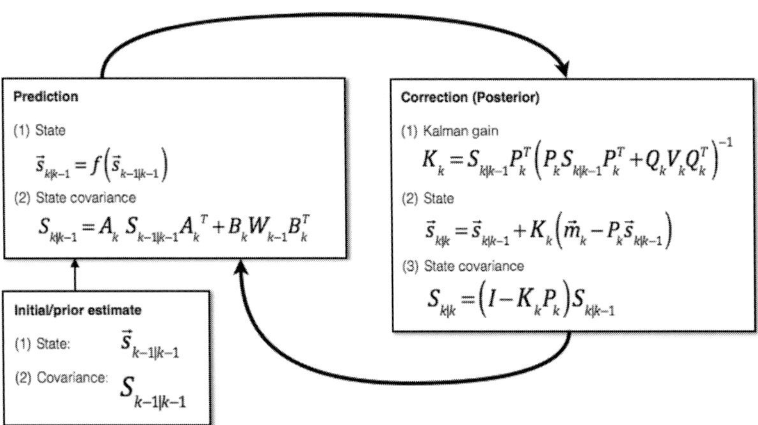

Fig. 5.10 Schematic illustration of the extended Kalman filter algorithm. The algorithm is essentially identical to the discrete algorithm, but the matrices are here replaced by nonlinear functions for the evolution of the state $\vec{s}_k$ and derivatives of these functions for the evolution of the covariance matrix S_k.

measurement noise $\vec{v}_k$, respectively:

$$(\mathbf{P}_k)_{ij} = \frac{\partial(\vec{h}_k)_i}{\partial(\vec{s}_k)_j}(\vec{s}_{k-1}, 0), \tag{5.173}$$

$$(\mathbf{Q}_k)_{ij} = \frac{\partial(\vec{h}_k)_i}{\partial(\vec{v}_k)_j}(\vec{s}_{k-1}, 0). \tag{5.174}$$

The state and state covariance evolution equations may then be written

$$\vec{s}_{k|k-1} = \vec{f}(\vec{s}_{k-1}, 0), \tag{5.175}$$

$$\mathbf{S}_{k|k-1} = \mathbf{A}_k \mathbf{S}_{k-1} \mathbf{A}_k^T + \mathbf{B}_k \mathbf{W}_{k-1} \mathbf{B}_k^T, \tag{5.176}$$

while the measurement update equations are

$$\mathbf{K}_k = \mathbf{S}_{k|k-1} \mathbf{P}_k^T \left(\mathbf{P}_k \mathbf{S}_{k|k-1} \mathbf{P}_k^T + \mathbf{Q}_k \mathbf{V}_k \mathbf{Q}_k^T\right)^{-1}, \tag{5.177}$$

$$\vec{s}_k = \vec{s}_{k|k-1} + \mathbf{K}_k \left(\vec{m} - \vec{h}(s_{k|k-1}, 0)\right), \tag{5.178}$$

$$\mathbf{S}_k = (\mathbf{I} - \mathbf{K_k P_k})\, \mathbf{S}_{k|k-1}. \tag{5.179}$$

As illustrated in Figure 5.10, the extended Kalman filter is quite similar to the regular discrete Kalman filter and proceeds recursively through steps of state update and measurement update based on Eqs. (5.175) and (5.177), respectively.

However, it is important to note, in closing this section, that the linearization of the evolution equations implies that the random variables are no longer Gaussian distributed after undergoing their respective nonlinear transformations.

5.6.5 Derivation of the Kalman Gain Matrix

In this section, we derive the expression (5.164) of the Kalman gain $\mathbf{K}_k$. It may be skipped in a first reading of the material.

Let us assume the state of a system can be represented by an n-elements state vector $\hat{s}_k$ and that its evolution can be described by the linear equation

$$\hat{s}_k = \mathbf{F}_k \hat{s}_{k-1} + \mathbf{K}_k \left(\vec{m}_k - \mathbf{H}_k \mathbf{F}_k \hat{s}_{k-1} \right), \tag{5.180}$$

where the matrix $\mathbf{F}_k$ determines the evolution of the system in the absence of noise, $\hat{s}_{k-1}$ is the prior information on the state of the system, K_k the Kalman gain matrix we wish to determine, $\mathbf{H}_k$ a matrix that projects the state $\hat{s}_k$ onto a prediction of a measurement, and $\vec{m}_k$ represents an actual measurement. A measurement $\vec{m}_k$ of the system may be represented as $\vec{m}_k = \mathbf{H}_k \vec{s}_k + \vec{v}_k$, where $\vec{s}_k$ is the actual value of the system's state at step k and $\vec{v}_k$ a random vector representing the process noise incurred in the evolution of the state from step $k-1$ to step k. We will assume that the state, process noise, and measurement noise are uncorrelated:

$$\text{Cov}\left[\Delta\vec{s}_{k-1}, \vec{w}_k^T\right] = 0, \tag{5.181}$$

$$\text{Cov}\left[\Delta\vec{s}_{k-1}, \vec{v}_k^T\right] = 0, \tag{5.182}$$

$$\text{Cov}\left[\vec{w}_k, \vec{v}_k^T\right] = 0. \tag{5.183}$$

Our goal is to calculate the covariance matrix $\mathbf{S}_k$ of the system state $\vec{s}_k$ and determine the gain matrix $\mathbf{K}$ that simultaneously minimizes all elements of this covariance matrix. In order to obtain the covariance matrix, let us first calculate the residue $\Delta\vec{s}_k$ at each step k of filtering as follows:

$$\begin{aligned} \Delta\vec{s}_k &= \vec{s}_k - \hat{s}_k \\ &= \vec{s}_k - \mathbf{F}_k \hat{s}_{k-1} - \mathbf{K}_k \left(\vec{m}_k - \mathbf{H}_k \mathbf{F}_k \hat{s}_{k-1} \right). \end{aligned} \tag{5.184}$$

We next replace the value of the measurement $\vec{m}_k$ by the sum of the projection of the state $\vec{s}_k$ and measurement noise $\vec{v}_k$:

$$\Delta\vec{s}_k = \vec{s}_k - \mathbf{F}_k \hat{s}_{k-1} - \mathbf{K}_k \left(\mathbf{H}_k \vec{s}_k + \vec{v}_k - \mathbf{H}_k \mathbf{F}_k \hat{s}_{k-1} \right). \tag{5.185}$$

We next also replace $\vec{s}_k$ on the righthand side by its value in terms of the previous state and process noise:

$$\begin{aligned} \Delta\vec{s}_k = {} & \mathbf{F}_k \vec{s}_{k-1} + \vec{w}_k - \mathbf{F}_k \hat{s}_{k-1} \\ & - \mathbf{K}_k \left(\mathbf{H}_k \left(\mathbf{F}_k \vec{s}_{k-1} + \vec{w}_k \right) + \vec{v}_k - \mathbf{H}_k \mathbf{F}_k \hat{s}_{k-1} \right). \end{aligned} \tag{5.186}$$

Noting that the difference $\vec{s}_{k-1} - \hat{s}_{k-1}$ corresponds to the residue at step $k-1$, we obtain after a simple reorganization of Eq. (5.186)

$$\Delta\vec{s}_k = \left(1 - \mathbf{K}_k \mathbf{H}_k\right) \mathbf{F}_k \Delta\vec{s}_{k-1} + \left(1 - \mathbf{K}_k \mathbf{H}_k\right) \vec{w}_k - \mathbf{K}_k \vec{v}_k, \tag{5.187}$$

which expresses the residue $\Delta\vec{s}_k$ in terms of the residue $\Delta\vec{s}_{k-1}$, plus two terms that account for the process and measurement noises.

We define the state vector covariance matrix $\mathbf{S}_k$ as

$$\mathbf{S}_k = \mathrm{E}\left[\Delta\vec{s}_k \Delta\vec{s}_k^T\right], \tag{5.188}$$

where $\Delta\vec{s}_k$ is represented as a column vector while $\Delta\vec{s}_k^T$ corresponds to its transpose, a row vector. The covariance of the process noise and measurement noise are noted $\mathbf{W}_k$ and $\mathbf{V}_k$, respectively:

$$\mathbf{W}_k = \mathrm{E}\left[\Delta\vec{w}_k \Delta\vec{w}_k^T\right], \tag{5.189}$$

$$\mathbf{V}_k = \mathrm{E}\left[\Delta\vec{v}_k \Delta\vec{v}_k^T\right]. \tag{5.190}$$

Substituting the expression (5.187) in the definition (5.188) of the covariance $\mathbf{S}_k$, we obtain after some algebraic manipulations

$$\begin{aligned}\mathbf{S}_k = &((1 - \mathbf{K}_k\mathbf{H}_k)\mathbf{F}_\mathbf{k})\mathrm{E}\left[\Delta s_{k-1}\Delta s_{k-1}^T\right]((1 - \mathbf{K}_k\mathbf{H}_k)\mathbf{F}_\mathbf{k})^T \\ &+ (1 - \mathbf{K}_k\mathbf{H}_k)\mathrm{E}\left[w_k w_k^T\right](1 - \mathbf{K}_k\mathbf{H}_k)^T \\ &+ \mathbf{K}_k\mathrm{E}\left[\vec{v}_k\vec{v}_k^T\right]\mathbf{K}_k^T + \text{cross terms}\end{aligned}$$

where the "cross terms" are proportional to expectation values $\mathrm{E}[\vec{\Delta} s_{k-1}\vec{w}_k]$, $\mathrm{E}[\vec{\Delta} s_{k-1}\vec{v}_k]$, and $\mathrm{E}[\vec{w}_k\vec{v}_k^T]$ and thus null by hypothesis, while the factors $\mathrm{E}[\Delta s_{k-1}\Delta s_{k-1}^T]$, $\mathrm{E}[w_k w_k^T]$, and $\mathrm{E}[\vec{v}_k\vec{v}_k^T]$ correspond to $\mathbf{S}_{k-1}$, $\mathbf{W}_k$, and $\mathbf{V}_k$, respectively. We thus obtain

$$\begin{aligned}\mathbf{S}_k = &(1 - \mathbf{K}_k\mathbf{H}_k)\mathbf{F}_\mathbf{k}\mathbf{S}_{k-1}((1 - \mathbf{K}_k\mathbf{H}_k)\mathbf{F}_\mathbf{k})^T \\ &+ (1 - \mathbf{K}_k\mathbf{H}_k)\mathbf{W}_k(1 - \mathbf{K}_k\mathbf{H}_k)^T + \mathbf{K}_k\mathbf{V}_k\mathbf{K}_k^T.\end{aligned} \tag{5.191}$$

Noting that the transpose of a product of matrices $(\mathbf{AB})^T$ equals the product of their transposes in reverse order, $\mathbf{B}^T\mathbf{A}^T$, we define the matrix $\mathbf{S}_{k|k-1}$ as

$$\mathbf{S}_{k|k-1} = \mathbf{F}_k\mathbf{S}_{k-1}\mathbf{F}_k^T + \mathbf{W}_k. \tag{5.192}$$

The foregoing expression for the covariance matrix $\mathbf{S}_k$ may thus be written

$$\mathbf{S}_k = (1 - \mathbf{K}_k\mathbf{H}_k)\mathbf{S}_{k|k-1}(1 - \mathbf{K}_k\mathbf{H}_k)^T + \mathbf{K}_k\mathbf{V}_k\mathbf{K}_k^T, \tag{5.193}$$

which provides us with a rather complicated formula for the evolution of the state covariance matrix. We will see in the following that this expression greatly simplifies. But first, let us find the value of the Kalman gains $\mathbf{K}_k$ that minimize the covariance $\mathbf{S}_k$. This is readily accomplished by setting derivatives of $\mathbf{S}_k$ with respect to $\mathbf{K}_k$ equal to zero and solving for $\mathbf{K}_k$. Noting that $\mathbf{S}_k$, $\mathbf{S}_{k|k-1}$, and $\mathbf{V}_k$ are symmetric matrices by construction, the derivatives of $\mathbf{S}_k$ readily simplify to

$$0 = \frac{\partial \mathbf{S}_k}{\partial \mathbf{K}_k} = -2(1 - \mathbf{K}_k\mathbf{H}_k)\mathbf{S}_{k|k-1}\mathbf{H}_k^T + 2\mathbf{K}_k\mathbf{V}_k \tag{5.194}$$

which further simplifies to

$$0 = -\mathbf{S}_{k|k-1}\mathbf{H}_k^T + \mathbf{K}_k\left(\mathbf{H}_k\mathbf{S}_{k|k-1}\mathbf{H}_k^T + \mathbf{V}_k\right) \tag{5.195}$$

The optimal Kalman gain matrix is thus

$$\mathbf{K}_k = \mathbf{S}_{k|k-1}\mathbf{H}_k^T\left(\mathbf{H}_k\mathbf{S}_{k|k-1}\mathbf{H}_k^T + \mathbf{V}_k\right)^{-1}. \tag{5.196}$$

Inserting the foregoing expression in Eq. (5.193), we find after some simple manipulations

$$\mathbf{S}_k = (1 - \mathbf{K}_k\mathbf{H}_k)\,\mathbf{S}_{k|k-1}, \tag{5.197}$$

or substituting the definition (5.192) of $\mathbf{S}_{k|k-1}$, we obtain

$$\mathbf{S}_k = (1 - \mathbf{K}_k\mathbf{H}_k)\left(\mathbf{F}_k\mathbf{S}_{k-1}\mathbf{F}_k^T + \mathbf{W}_k\right). \tag{5.198}$$

The covariance $\mathbf{S}_k$ is determined by the linear projection of the prior $\mathbf{S}_{k-1}$ and the process noise determined by the Kalman gain $\mathbf{K}_k$. The gain is initially large and gives more weight to measurements $\vec{m}_k$. It progressively decreases, however, with the addition of data and eventually vanishes for large values of k. The covariance matrix thus tends toward

$$\mathbf{S}_k = \mathbf{F}_k\mathbf{S}_{k-1}\mathbf{F}_k^T + \mathbf{W}_k \tag{5.199}$$

in the limit where the gain vanishes.

5.6.6 Kalman Filter as a Least-Squares Fitter

We demonstrate, in this section, that the Kalman filter technique is equivalent to the least-squares method.

The χ^2 function is defined on the basis of the measurements $\vec{m}_i$, the measurement covariance matrix $\mathbf{V}_i$, and a model $h(t;\vec{s})$ with state parameters $\vec{s}$ to be determined by the fitting procedure:

$$\chi^2 = \sum_{i=1}^{k}\left(\vec{m}_i - \vec{h}(t_i,\vec{s}_i)\right)\mathbf{V}_i^{-1}\left(\vec{m}_i - \vec{h}(t_i,\vec{s}_i)\right)^T \tag{5.200}$$

Rather than trying to evaluate Eq. (5.200) for all values of i, let us consider the contribution of the "last" measurement k and those of the $k-1$ prior measurements. Let χ_k^2 represent the contribution of the last measured point:

$$\chi_k^2 = \left(\vec{m}_k - \vec{h}(t_k,\vec{s}_k)\right)\mathbf{V}_k^{-1}\left(\vec{m}_k - \vec{h}(t_k,\vec{s}_k)\right)^T, \tag{5.201}$$

where $\vec{m}_k$ and $\mathbf{V}_k$ stand for the kth measurement and its covariance matrix, while $\vec{h}(t_k,\vec{s}_k)$ represents the expected measurement value for step t_k and model parameters $\vec{s}_k$. The contribution of all $k-1$ prior measurements is encapsulated in the covariance matrix $\mathbf{S}_{k-1}$ of the state. One can thus write

$$\chi^2_{k-1} = \left(\vec{s}_{k-1} - \hat{s}_{k-1}\right)^T \mathbf{S'}^{-1}_{k-1} \left(\vec{s}_{k-1} - \hat{s}_{k-1}\right). \tag{5.202}$$

The total χ^2 is then simply the sum $\chi^2_{k-1} + \chi^2_k$, which one seeks to minimize in order to determine the optimal state (model) parameters $\vec{s}$. We thus take the derivative of χ^2 relative to $\vec{s}$:

$$\frac{d\chi^2}{d\vec{s}} = 2\mathbf{S'}^{-1}_k \left(\vec{s}_{k-1} - \hat{s}_{k-1}\right) - \nabla_s \vec{h} \mathbf{V}^{-1}_k \left(\vec{m}_k - \vec{h}(t_k, \vec{s}_k)\right). \tag{5.203}$$

Equation (5.203) has a dependency on the unknown parameters $\vec{s}_k$. We thus replace $\vec{s}_k$ by $\hat{s}_k + \Delta\vec{s}_k$ with $\Delta\vec{s}_k$ defined as the residue:

$$\Delta\vec{s}_k = \vec{s}_k - \hat{s}_k \tag{5.204}$$

For small residues, one can expand $\vec{h}(t_k, \vec{s}_k)$ as a Taylor series

$$\vec{h}(t_k, \hat{x}_k + \Delta\vec{s}_k) = \vec{h}(t_k, \hat{s}_k) + \Delta\vec{s}_k \nabla_s \vec{h}(t_k, \hat{s}_k), \tag{5.205}$$

and obtain

$$\begin{aligned} \frac{d\chi^2}{d\vec{s}} &= 2\mathbf{S'}^{-1}_k \Delta\vec{s}_{k-1} \\ &\quad - \nabla_s \vec{h}(\hat{s}) \mathbf{V}^{-1}_k \left(\vec{m}_k - \vec{h}(\hat{s}_k) - \Delta\vec{s}_k \nabla_s \vec{h}(\hat{s})\right), \end{aligned} \tag{5.206}$$

where for brevity we have omitted the dependence on t_k. The first $\nabla_s \vec{h}$ in Eq. (5.206) is a function of the true value $\vec{s}_k$ but it should be legitimate to use a gradient evaluated at $\hat{s}_k$ instead. Defining $\mathbf{H}_k = \nabla_s \vec{h}$, we thus obtain

$$\frac{d\chi^2}{d\vec{s}} = 2\mathbf{S'}^{-1}_k \Delta\vec{s}_{k-1} \tag{5.207}$$

$$+ \mathbf{H}^T_k \mathbf{V}^{-1}_k \mathbf{H}_k \Delta\vec{s}_k - 2\mathbf{H}^T_k \mathbf{V}^{-1}_k \left[\vec{m}_k - \vec{h}(\hat{s}_k)\right]. \tag{5.208}$$

Setting this derivative to zero and solving for $\Delta\vec{s}_k$, one gets

$$\Delta\vec{s}_k = \left[\mathbf{S'}^{-1}_k + \mathbf{H}^T_k \mathbf{V}^{-1}_k \mathbf{H}_k\right]^{-1} \mathbf{H}^T_k \mathbf{V}^{-1}_k \left[m_k - \vec{h}(\hat{s})\right] \tag{5.209}$$

which one readily rewrites

$$\vec{s}_k = \hat{s}_k + \left[\mathbf{S'}^{-1}_k + \mathbf{H}^T_k \mathbf{V}^{-1}_k H\right]^{-1} \mathbf{H}^T_k \mathbf{V}^{-1}_k \left[m_k - \vec{h}(\hat{s})\right] \tag{5.210}$$

to obtain an expression of the form of Eq. (5.180) but with a seemingly different gain matrix

$$\mathbf{K}_k = \left[\mathbf{S'}^{-1}_k + \mathbf{H}^T_k \mathbf{V}^{-1}_k H\right]^{-1} \mathbf{H}^T_k \mathbf{V}^{-1}_k. \tag{5.211}$$

In order to demonstrate this expression is equivalent to Eq. (5.164), we must first show that the inverse of the updated covariance matrix $\mathbf{S}_k$ may be written as

$$\mathbf{S}_k^{-1} = \mathbf{S'}_k^{-1} + \mathbf{H}_k\mathbf{V}_k^{-1}\mathbf{H}_k^T. \tag{5.212}$$

To verify this statement, it suffices to demonstrate that $\mathbf{S}_k\mathbf{S}_k^{-1} = I$ using Eq. (5.197) for $\mathbf{S}_k$ and the preceding expressions for $\mathbf{S}_k^{-1}$ and $\mathbf{K}$.

$$\begin{aligned}\mathbf{S}_k\mathbf{S}_k^{-1} &= (1 - \mathbf{K}_k\mathbf{H}_k)\,\mathbf{S'}_k\left(\mathbf{S'}_k^{-1} + \mathbf{H}_k\mathbf{V}_k^{-1}\mathbf{H}_k^T\right)\\ &= 1 + \mathbf{S'}_k\mathbf{V}_k^{-1}\mathbf{H}_k^T - \mathbf{S'}_k\mathbf{H}_k^T\left(\mathbf{H}_k\mathbf{S'}_k\mathbf{H}_k^T + \mathbf{V}_k\right)^{-1}\mathbf{H}_k\mathbf{S'}_k\mathbf{S'}_k^{-1}\\ &\quad - \mathbf{S'}_k\mathbf{H}_k^T\left(\mathbf{H}_k\mathbf{S'}_k\mathbf{H}_k^T + \mathbf{V}_k\right)^{-1}\mathbf{H}_k\mathbf{S'}_k\mathbf{H}_k\mathbf{V}_k^{-1}\mathbf{H}_k^T,\end{aligned}$$

which can be shown to indeed simplify to unity after a modest amount of matrix algebra (see Problem 9.1). We can then use Eq. (5.212) for $\mathbf{S}_k$ to obtain an alternative expression of the gain matrix. Starting from Eq. (5.211), we write

$$\mathbf{K}_k = \mathbf{S'}_k\mathbf{H}_k^T\left(\mathbf{H}_k\mathbf{S'}_k\mathbf{H}_k^T + \mathbf{V}_k\right)^{-1} \tag{5.213}$$

and insert $\mathbf{S}_k\mathbf{S}_k^{-1}$ and $\mathbf{V}_k^{-1}\mathbf{V}_k$ judiciously in Eq. (5.213):

$$\begin{aligned}\mathbf{K}_k &= \mathbf{S}_k\mathbf{S}_k^{-1}\mathbf{S'}_k\mathbf{H}_k^T\mathbf{V}_k^{-1}\mathbf{V}_k\left(\mathbf{H}_k\mathbf{S'}_k\mathbf{H}_k^T + \mathbf{V}_k\right)^{-1}\\ &= \mathbf{S}_k\mathbf{S}_k^{-1}\mathbf{S'}_k\mathbf{H}_k^T\mathbf{V}_k^{-1}\left(\mathbf{H}_k\mathbf{S'}_k\mathbf{H}_k^T\mathbf{V}_k^{-1} + \mathbf{I}\right)^{-1}.\end{aligned}$$

Inserting the expression (5.212) for $\mathbf{S}_k^{-1}$, one gets

$$\begin{aligned}\mathbf{K}_k &= \mathbf{S}_k\left(\mathbf{S'}_k^{-1} + \mathbf{H}_k\mathbf{V}_k^{-1}H^T\right)\mathbf{S'}_k\mathbf{H}_k^T\mathbf{V}_k^{-1}\mathbf{V}_k\left(\mathbf{H}_k\mathbf{S'}_k\mathbf{H}_k^T + \mathbf{V}_k\right)^{-1}\\ &= \mathbf{S}_k\left(\mathbf{I} + \mathbf{H}_k^T\mathbf{V}_k^{-1}H\mathbf{S'}_k\right)\mathbf{H}_k^T\mathbf{V}_k^{-1}\left(\mathbf{H}_k\mathbf{S'}_k\mathbf{H}_k^T\mathbf{V}_k^{-1} + \mathbf{I}\right)^{-1}\\ &= \mathbf{S}_k\mathbf{H}_k^T\mathbf{V}_k^{-1}\left(\mathbf{I} + \mathbf{H}_k\mathbf{S'}_k\mathbf{H}_k^T\mathbf{V}_k^{-1}\right)\left(\mathbf{H}_k\mathbf{S'}_k\mathbf{H}_k^T\mathbf{V}_k^{-1} + \mathbf{I}\right)^{-1}\\ &= \mathbf{S}_k\mathbf{H}_k^T\mathbf{V}_k^{-1}.\end{aligned}$$

Finally, substituting the expression (5.212) for $\mathbf{S}_k$, we get

$$\mathbf{K}_k = \left[\mathbf{S'}_k^{-1} + \mathbf{H}_k\mathbf{V}_k^{-1}H^T\right]^{-1}\mathbf{H}_k^T\mathbf{V}_k^{-1}, \tag{5.214}$$

which is the expression (5.211) we sought to demonstrate is equivalent to Eq. (5.164) for the Kalman gain. We have thus established that the Kalman filter is formally equivalent to the least-squares method.

Note that while the foregoing expression for the Kalman gain is equivalent to Eq. (5.164), its calculation involves two matrix inversions and is thus more computationally intensive. Use of Eq. (5.164) is thus preferred in general.

Exercises

5.1 Show by direct calculation that the estimator $\hat{\tau}$ given by Eq. (5.14) is an unbiased estimator of the lifetime τ of the exponential PDF, Eq. (5.10). Next, determine whether the estimator $\hat{\tau}$ is biased, unabised, or asymptotically unbiased.

5.2 Show that the expectation value of $\widehat{\sigma^2}$ defined by Eq. (5.22) is $E[\widehat{\sigma^2}] = (n-1)\sigma^2/n$ for a Gaussian distribution, and that $\widehat{\sigma^2}$ is consequently a biased estimator of the variance of this distribution.

5.3 Imagine a researcher is studying the behavior of system and finds it can be represented by a variable x with some PDF $f(x|\vec{\theta})$ in which $\vec{\theta}$ represents a single or multiple unknown parameters. It is often the case that an "experiment" returns a random number, n, of values $\{x_i\}$, with $i = 1, \ldots, n$. Assuming the number n obeys a Poisson distribution of mean ν, show that one can extend the ML method to obtain an extended likelihood function, $L(\nu, \vec{\theta})$, defined as

$$L(\nu, \vec{\theta}) = \frac{\nu^n}{n!} e^{-\nu} \prod_{i=1}^{n} f(x_i|\vec{\theta}) = \frac{e^{-\nu}}{n!} \prod_{i=1}^{n} \nu f(x_i|\vec{\theta}). \tag{5.215}$$

5.4 Maximize the function $L(\nu, \vec{\theta})$ derived in Problem 5.3 to first show that the estimator $\hat{\nu}$ has an expected value of n. Next, consider the application of this extended likelihood function for the determination of the estimator, $\hat{\theta}$, in experiments where n is a random variable.

5.5 Devise an exponential generator and test the convergence of the estimator $\hat{\tau}$ for various combinations of sample sizes and number of samples. Repeat the exercise for a Gaussian PDF.

5.6 Show that the following expression yields an asymptotically unbiased estimator $\hat{\lambda}$ of the decay constant $\lambda = 1/\tau$:

$$\hat{\lambda} = \frac{1}{\hat{\tau}} = n\left(\sum_{i=1}^{n} t_i\right)^{-1}. \tag{5.216}$$

5.7 Show that the estimator $\hat{\mu}$ obtained in Eq. (5.20) is an unbiased estimator of the mean μ of a Gaussian PDF.

5.8 Show that the estimator $\widehat{\sigma^2}$, given by Eq. (5.22), is an asymptotically unbiased estimator of the variance σ^2 of a Gaussian PDF.

5.9 Verify by direct calculation that s^2, given by Eq. (5.22), is an unbiased estimator of the variance σ^2 of a Gaussian PDF.

5.10 Verify that Eqs. (5.27) and (5.30) yield the variances of estimators of the mean and standard deviation of a Gaussian distribution, respectively.

5.11 Derive Eq. (5.78), providing the variance of the coefficients a_0 and a_1 of linear fits obtained with the LS method.

5.12 Imagine a linear fit $y = a_0 + a_1 y$ of 103 measured points $\{x_i, y_i\}$ has produced coefficients $a_0 = 0.55$ and $a_1 = 10.04$ with an error matrix $U_{00} = 0.04$, $U_{11} = 0.11$, and $U_{01} = -0.03$. Extrapolate the value y predicted by the model at $x = 5$ and compare the errors δy on this extrapolation obtained while excluding and including the off-diagonal elements of the error matrix.

5.13 Show that if correlations exist between measurements $\hat{\theta}_i$ and their covariance matrix V_{ij} is known, then the weighted average of these measurements involves the weights given by Eq. (5.133).

5.14 Show that the sum of the weights given by Eq. (5.133) equals unity.

6 Classical Inference III: Confidence Intervals and Statistical Tests

In previous chapters, we presented several methods for the estimation of values of unknown model parameters $\vec{\theta}$ based on measured data. We also discussed how experimental errors on directly measured quantities impact the errors of estimated model parameter values. However, we did not discuss in detail the interpretation or significance of these error estimates. We shall remedy this situation with comprehensive discussions, in this and the next chapter, of the notions of confidence intervals, confidence levels, and how these concepts can be used toward the formulation of scientific hypotheses and hypothesis testing. The discussions presented in this chapter are primarily frequentist in nature, while those of Chapter 7 rely on the Bayesian interpretation of probability.

We begin with a detailed discussion of the notion of error intervals in §6.1 and present practical considerations and examples in §§6.2 and 6.3. We next introduce, in §6.4, the notions of hypothesis, hypothesis testing, and significance level. Key properties of test power, consistency, and test bias are introduced in §6.5. We complete the chapter with a presentation, in §6.6, of statistical tests commonly used in nuclear and particle physics, as well as a discussion, in §6.7, on test optimization and most particularly the Neyman–Pearson test.

6.1 Error Interval Estimation

6.1.1 Basic Goals and Strategy

The physical and measurement processes involved in the observation of physical observables are intrinsically stochastic in nature. Whether a single or several measurements of an observable X are carried out, fluctuations inherent to the measurement process yield observed values that randomly deviate from the actual or true value of an observable. Either explicitly or implicitly, it is generally assumed that these fluctuations may be described by a specific (but possibly partly unknown) probability model $p(x|x_{\text{true}})$. In the frequentist paradigm, this model describes the probability (density) of observing values x in the range $[x, x + dx]$, expected from an arbitrarily large number of observations, and it is the purpose of statistical estimators, introduced in Chapters 4 and 5, to provide estimates of the true value x_{true} based on measured values. Estimates evidently deviate from the true value by some unknown amount[1] but one wishes to use the probability model $p(x|x_{\text{true}})$ to make a

[1] If the actual deviation was known, it could obviously be corrected for and the true value would be revealed. . .

statement about the possible size of such deviations. Clearly, if $p(x|x_{\text{true}})$ is very broad, the dispersion of measured values will be quite large and any particular measured value x may deviate substantially from the actual value x_{true}. On the other hand, if $p(x|x_{\text{true}})$ is narrowly distributed, observed values should be rather close to the true value. One may then use the model $p(x|x_{\text{true}})$ to characterize the error or uncertainty of a measurement. However, the problem arises, typically, that the model $p(x|x_{\text{true}})$ is not perfectly known and that it may feature finite probability density, albeit very small, for arbitrarily large deviates $|x - x_{\text{true}}|$. It is common practice, for instance, to approximate the probability model $p(x|x_{\text{true}})$ by a Gaussian distribution

$$p(x|x_{\text{true}}) \approx \frac{1}{\sqrt{2\pi}\sigma} \exp\left[-\frac{(x - x_{\text{true}})^2}{2\sigma^2}\right], \tag{6.1}$$

which, evidently, features low and high side tails that extended to infinity. For this and similarly behaved models, it is consequently not possible to obtain error bounds that completely enclose the distribution $p(x|x_{\text{true}})$. One must then characterize the uncertainty of an estimate based on a finite interval $[a, b]$ with a specific probability β of containing the true value, given a particular estimate.

To this end, and for simplicity's sake, let us first consider an experiment yielding a single parameter estimate $\hat{\theta}$. We wish to determine the probability, $\beta \equiv P(\hat{\theta}|\theta, \theta_a, \theta_b)$, of finding this estimate within the range $\theta_a \leq \hat{\theta} \leq \theta_b$. In essence, we wish to establish a quantitative method to characterize the errors on the estimate $\hat{\theta}$ based on the probability content β of the interval. Intuitively, it is clear that the larger this probability is, the wider will the range be. Conversely, the wider this range is, the larger should the probability of finding the value $\hat{\theta}$ within it. Evidently, if the probability model of $p(\hat{\theta}|\theta)$ is fully known, there should exist a one-to-one relation between the probability content β and the width of the interval $[\theta_a, \theta_b]$. Unfortunately, there is no unique way of choosing or defining what constitutes a desirable probability value β. Indeed, the choice of a particular value of β is in fact rather arbitrary.

Two approaches are commonly used in the scientific community. For the first, one arbitrarily chooses β to be a convenient and easy-to-remember "round" value, such as 90% or 99%. For the second, one relies on the central limit theorem and assume data tend to follow a Gaussian distribution, with standard deviation σ, and one then typically quotes error intervals in terms of integer multiples of σ.

Recall from §3.9 that for a Gaussian distribution, the probability of obtaining values in the range $\mu \pm n\sigma$ amounts to 68.3%, 95.5%, and 99% for $n = 1$, 2, and 3, respectively. One can then specify error ranges in terms of ± 1, ± 2, or ± 3 standard deviations, with probabilities $\beta = 68.3\%$, 95.5%, and 99%, respectively. However, these probabilities are valid exclusively for Gaussian distributions: random variables obeying a different probability model should in general bear a rather different relation between the width of the range expressed in some multiple of their standard deviation σ and the probability content of the range.

It is worth reiterating that the discussion of errors presented in this chapter is based on the frequentist interpretation of probability. In this context, the use of Bayes' theorem is not strictly possible and one must find a roundabout way to make a statement about θ based on an observed estimate $\hat{\theta}$. We will see in Chapter 7 that posterior probabilities based on Bayes' theorem enable a far more direct reasoning and a much simpler technique to deduce confidence intervals but it is nonetheless convenient and useful to begin with a detailed presentation of frequentist techniques in this and following sections.

6.1.2 Confidence Interval and Confidence Level of Continuous Variables

Let us examine the notion of error quantitatively by considering a set of observations of a *continuous* observable X determined by a PDF $p(x|\theta)$. By definition, $p(x|\theta)\,dx$ is the probability of measuring x in a given interval $[x, x + dx]$ given a parameter θ, and, as such, encapsulates the uncertainty of the measurement. The variable θ may correspond to the actual value of X or some other physical parameter that determines its value. At times fundamentally unknown, $p(x|\theta)$ is often assumed to be Gaussian. In some cases, it can be determined from performance analysis and simulations of the measurement procedure and apparatus. Alternatively, it can also be estimated from a large number of repeated measurements of the same observable X.

Let us assume the function $p(x|\theta)$ represents the density of probability of measurement outcomes of a physical observable X. The probability β of observing (measuring) a value x in the range $[a, b]$ is thus in principle given by

$$\beta \equiv P(a \le x \le b|\theta) = \int_a^b p(x|\theta)\,dx. \tag{6.2}$$

If both the functional form of $p(x|\theta)$ and the parameter θ are known, the integral in Eq. (6.2) completely specified and the probability β easy to compute numerically, if not analytically. In practice, however, the value of θ may not be known a priori and it might in fact be the very purpose of the measurement to determine its value. Without a value for θ, calculation of Eq. (6.2) seems impossible, as well as an assessment of the range $[a, b]$ with probability content β. How, then, can one obtain any meaningful estimate of the error on the parameter θ and the observable X?

It turns out that knowledge of the variable θ is not absolutely required so long as the functional form of $p(x|\theta)$ is known, and, provided it is possible to formulate a function $z(x, \theta)$ of the observation of x and the unknown parameter θ whose PDF is **independent** of θ. Indeed, as we shall describe in the following, if such a function exists, it may become possible to reformulate Eq. (6.2) as a problem of **interval estimation**.

In this context, and given a value β, the problem then consists in finding an (optimal) range $[\theta_a, \theta_b]$ such that

$$P(\theta_a \le \hat{\theta} \le \theta_b) = \beta, \tag{6.3}$$

where $\hat{\theta}$ is an estimate of the true value of the parameter θ. The interval $[\theta_a, \theta_b]$ is then known as a **confidence interval** of $\hat{\theta}$ with a probability β. In other words, the probability of obtaining an estimate within that interval amounts to β. The interval is then said to be a $100 \times \beta\%$ **confidence interval**. For instance, an interval whose content has a probability of $\beta = 0.9$ is said to be a 90% confidence interval.

Formally, a method which produces an interval $[\theta_a, \theta_b]$ satisfying Eq. (6.3) is said to have the property of **coverage**, and, as it happens, an interval that does not possess this property cannot be considered a confidence interval [115]. This notion will become important and much clearer, later in this chapter, when we consider confidence intervals near a physical boundary.

The process of finding a function $z(x, \theta)$ that enables the determination of a confidence interval is best illustrated with examples. We begin our discussion with the determination of the mean of Gaussian deviates.

6.1.3 Confidence Interval for the Mean of Gaussian Deviates

Let the function $p(x|\theta)$, considered in Eq. (6.2), be a Gaussian PDF $p_G(x|\mu, \sigma)$, where the mean, μ, has been substituted for θ and is now the parameter of primary interest. If the mean μ and standard deviation σ are known, the probability β may be evaluated for any arbitrary interval $[a, b]$:

$$\beta \equiv P(a \leq x \leq b|\mu, \sigma) = \frac{1}{\sqrt{2\pi}\sigma} \int_a^b \exp\left[-\frac{(x' - \mu)^2}{2\sigma^2}\right] dx', \tag{6.4}$$

$$= \Phi\left(\frac{b - \mu}{\sigma}\right) - \Phi\left(\frac{a - \mu}{\sigma}\right), \tag{6.5}$$

where $\Phi(z)$ is the **Cumulative Distribution Function** of the normal PDF, given by Eq. (3.131).

The preceding calculation is well defined if μ and σ are known. But what if μ is in fact unknown and the measured value of x intended to obtain an estimate of μ? One would then be interested in calculating the probability β that the value x lies within some interval $[\mu + \Delta_1, \mu + \Delta_2]$. This may be written

$$\beta = P(\mu + \Delta_1 \leq x < \mu + \Delta_2|\sigma) \tag{6.6}$$

$$= \frac{1}{\sqrt{2\pi}\sigma} \int_{\mu+\Delta_1}^{\mu+\Delta_2} \exp\left[-\frac{(x' - \mu)^2}{2\sigma^2}\right] dx'. \tag{6.7}$$

On first inspection, this integral seems impossible because the mean μ is unknown. However, consider a change of variable $z = (x' - \mu)/\sigma$. The preceding integral then becomes

$$\beta = \frac{1}{\sqrt{2\pi}} \int_{\Delta_1/\sigma}^{\Delta_2/\sigma} \exp\left(-\frac{1}{2}z^2\right) dz = \Phi\left(\frac{\Delta_2}{\sigma}\right) - \Phi\left(\frac{\Delta_1}{\sigma}\right), \tag{6.8}$$

which is clearly independent of μ and calculable for specific bounds Δ_1 and Δ_2 provided σ is known.

By construction, the integral (6.8) corresponds to the probability of finding x in the interval $\mu + \Delta_1 \leq x \leq \mu + \Delta_2$. One may, however, rearrange the terms of the two inequalities and get

$$\beta = P(x - \Delta_2 \leq \mu \leq x - \Delta_1|\sigma). \tag{6.9}$$

The value β thus corresponds to the probability of finding the unknown value μ in the range $x - \Delta_2 \leq \mu \leq x - \Delta_1$ given the measured value x, and for a known value of σ. Although, strictly speaking, the integral (6.8) is really a statement about the probability of x given

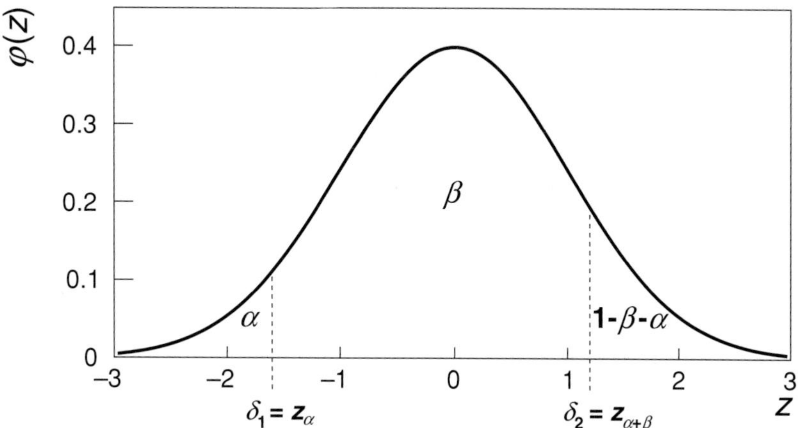

Fig. 6.1 1D confidence interval $[x - \delta_2\sigma, x - \delta_1\sigma]$ for the mean μ of a Gaussian distribution with $\delta_1 = z_\alpha$ and $\delta_2 = z_{\alpha+\beta}$. The probabilities of the three regions delimited by vertical lines are α, β, and $1 - \alpha - \beta$, respectively.

μ, the capacity to reformulate its integrant in terms of $z = (x' - \mu)/\sigma$ enables us to use x to find some bounds on μ with probability β. The interval $x - \Delta_2 \le \mu \le x - \Delta_1$ is thus indeed a **confidence interval** for μ with probability β given by Eq. (6.8).

If neither σ nor μ are known, one may instead calculate $P(\delta_1 \le z \le \delta_2)$ according to

$$\beta \equiv P(\delta_1 \le z \le \delta_2) = \frac{1}{\sqrt{2\pi}} \int_{\delta_1}^{\delta_2} \exp\left(-\frac{1}{2}z^2\right) dz, \tag{6.10}$$

$$= \Phi(\delta_2) - \Phi(\delta_1). \tag{6.11}$$

Converting the inequalities $\delta_1 \le z \le \delta_2$ in terms of μ, one gets

$$P(x - \delta_2\sigma \le \mu \le x - \delta_1\sigma) = \Phi(\delta_2) - \Phi(\delta_1), \tag{6.12}$$

which provides a generic confidence interval $[x - \delta_2\sigma, x - \delta_1\sigma]$ of probability $\beta = \Phi(\delta_2) - \Phi(\delta_1)$ expressed in terms of some multiples (not necessarily integers) δ_1 and δ_2 of the standard deviation of the PDF of x.

There is much latitude in the choice of δ_1 and δ_2 for a given value of β. For instance, choosing $\delta_1 = -1$ and $\delta_2 = 1$, one obtains a symmetric plus/minus one sigma confidence level with $P(x - \sigma \le \mu \le x + \sigma) = 68.3\%$. But arbitrarily many other interval choices also have the same probability. Indeed, recalling the definition of the α–point from Eq. (2.63), one notes that the difference $\Phi(\delta_2) - \Phi(\delta_1)$ in the preceding integral may be written $\Phi(z_{\alpha+\beta}) - \Phi(z_\alpha) = \beta$, where z_α and $z_{\alpha+\beta}$ are the α–point and the $(\alpha + \beta)$–point of the distribution, respectively. The confidence level of probability β can consequently be expressed in any interval $[z_\alpha, z_{\alpha+\beta}]$, as schematically illustrated in Figure 6.1. There are obviously infinitely many such intervals. It is customary, however, to use symmetrical intervals, known as **central intervals**, consisting of integer multiples of the standard deviation or some convenient values of β, as shown in Table 6.1.

Table 6.1 Central confidence intervals commonly used with a standard normal distribution

$\beta = 1 - 2\alpha$	z_α	$z_{\alpha+\beta}$
0.6827	−1.00	1.00
0.9000	−1.65	1.65
0.9500	−1.96	1.96
0.9545	−2.00	2.00
0.9900	−2.58	2.58
0.9973	−3.00	3.00

6.1.4 Central vs. Minimal Error Intervals

For a strongly peaked, symmetric, but not necessarily Gaussian PDF, the notion of central interval is easy to visualize: it suffices to find a symmetric range $\hat{\theta} \pm \Delta$ with a specific probability content β. If the distribution is not symmetric about its peak, however, one must define the interval based on the probability content outside of the interval and require the probability of observing values below and above the central range to be equal. A **central interval** may then be defined by

$$\begin{aligned} \tfrac{1-\beta}{2} &= P(\hat{\theta} < \theta_a|\theta) = \textstyle\int_{-\infty}^{\theta_a} p(x|\theta)\,dx, \\ \tfrac{1-\beta}{2} &= P(\hat{\theta} > \theta_b|\theta) = \textstyle\int_{\theta_b}^{\infty} p(x|\theta)\,dx, \end{aligned} \tag{6.13}$$

which, in general, shall yield bounds θ_a and θ_b with unequal probability density, that is, such that $p(\theta_a) \neq p(\theta_b)$. We show below that such an interval does not correspond to the shortest interval with probability content β. So, rather than a central interval, one may alternatively seek the **shortest interval** consistent with probability content β.

In order to identify the shortest interval with probability content β, consider the intervals $[\theta_a, \theta_b]$ and $[\theta_a', \theta_b']$ displayed in Figure 6.2, and defined such that they both have a probability content β, that is, such that they satisfy

$$\begin{aligned} &\textstyle\int_{\theta_a}^{\theta_b} p(\hat{\theta}|\theta)\,d\hat{\theta} = \beta \\ &\textstyle\int_{\theta_a'}^{\theta_b'} p(\hat{\theta}|\theta)\,d\hat{\theta} = \beta, \end{aligned} \tag{6.14}$$

where the bounds θ_a and θ_b were chosen in order to satisfy the condition

$$p(\theta_a|\theta) = p(\theta_b|\theta), \tag{6.15}$$

illustrated by the horizontal dashed line in Figure 6.2. We proceed to show that these bounds define the most compact (i.e., shortest) interval compatible with Eq. (6.14). To this end, we calculate the integral of the PDF $p(\hat{\theta}|\theta)$ in the range $[\theta_a, \theta_b']$. By construction, we may write

$$\int_{\theta_a}^{\theta_b'} p(\hat{\theta}|\theta)\,d\hat{\theta} = \int_{\theta_a}^{\theta_b} p(\hat{\theta}|\theta)\,d\hat{\theta} + \int_{\theta_b}^{\theta_b'} p(\hat{\theta}|\theta)\,d\hat{\theta} = \beta + \int_{\theta_b}^{\theta_b'} p(\hat{\theta}|\theta)\,d\hat{\theta}, \tag{6.16}$$

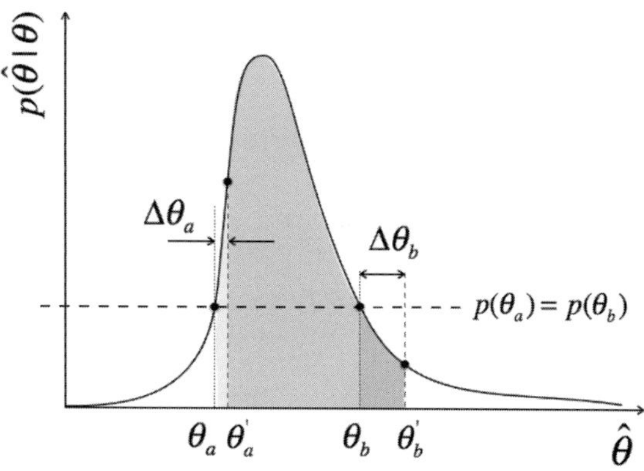

Fig. 6.2 Given a specific probability content β and an asymmetric PDF $p(\hat{\theta}|\theta)$, a minimal interval $[\theta_a, \theta_b]$ is obtained for values θ_a and θ_b that simultaneously satisfy Eqs. (6.14, 6.15).

or alternatively,

$$\int_{\theta_a}^{\theta_b'} p(\hat{\theta}|\theta)\, d\hat{\theta} = \int_{\theta_a}^{\theta_a'} p(\hat{\theta}|\theta)\, d\hat{\theta} + \int_{\theta_a'}^{\theta_b'} p(\hat{\theta}|\theta)\, d\hat{\theta} = \int_{\theta_a}^{\theta_a'} p(\hat{\theta}|\theta)\, d\hat{\theta} + \beta. \tag{6.17}$$

We thus find that

$$\int_{\theta_a}^{\theta_a'} p(\hat{\theta}|\theta)\, d\hat{\theta} = \int_{\theta_b}^{\theta_b'} p(\hat{\theta}|\theta)\, d\hat{\theta}. \tag{6.18}$$

Invoking the mean value theorem, the preceding equality may be written

$$p(\hat{\theta}_{a,0}|\theta)\Delta\theta_a = p(\hat{\theta}_{b,0}|\theta)\Delta\theta_b, \tag{6.19}$$

where we defined $\Delta\theta_a = \theta_a' - \theta_a$ and $\Delta\theta_b = \theta_b' - \theta_b$, while $\hat{\theta}_{a,0}$ and $\hat{\theta}_{b,0}$ are values found in the intervals $[\theta_a, \theta_a']$ and $[\theta_b, \theta_b']$, respectively. Based on the PDF construction shown in Figure 6.2, one knows that the ratio of the densities at $\theta_{a,0}$ and $\theta_{b,0}$ satisfies

$$\frac{p(\theta_{a,0})}{p(\theta_{b,0})} > 1. \tag{6.20}$$

In order to maintain the equality Eq. (6.19), one must then have $\Delta\theta_b > \Delta\theta_a$ which implies that

$$\theta_b' - \theta_a' > \theta_b - \theta_a. \tag{6.21}$$

We thus conclude that the interval $[\theta_a, \theta_b]$ is indeed the most compact interval with a probability content β, and as such provides a unique and consequently meaningful choice of central interval. For PDFs symmetric about their mean μ, such as Gaussian distributions, Eq. (6.15) defines a symmetric interval of width 2Δ such that

$$p(\mu - \Delta) = p(\mu + \Delta), \tag{6.22}$$

which, evidently, amounts to a central and most compact interval.

It is important to acknowledge that the notion of shortest interval is not strictly well defined for continuous variables because it is not invariant under a reparameterization

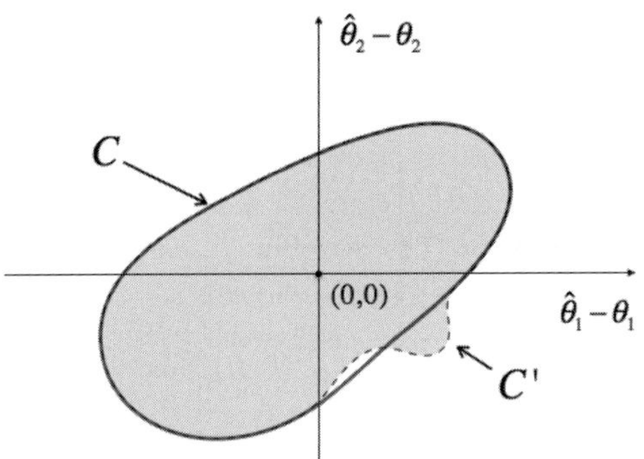

Fig. 6.3 Given a specific probability content β and an asymmetric PDF $p(\hat{\theta}_1, \hat{\theta}_2|\theta_1, \theta_2)$, a minimal confidence region $A_{\min}$ is obtained for values $\hat{\theta}_1$ and $\hat{\theta}_2$ bound within an iso-contour C of probability density $p(\hat{\theta}_1, \hat{\theta}_2|\theta_1, \theta_2) = k$ shown as a solid line. The dashed line illustrates a nonisometric contour C' (i.e., a contour with nonuniform probability density), which, by construction, features an area $A > A_{\min}$.

$\theta \to \xi(\theta)$. Central intervals, defined by Eq. (6.13), are invariant under such transformation, however, and are thus properly defined.

6.1.5 Multiparametric Confidence Intervals: Confidence Regions

The concepts and techniques presented in the previous two subsections are readily extended to multiparametric systems, that is, optimization problems involving $n \geq 2$ unknown parameters. For instance, let us consider an experiment producing n estimators, $\hat{\theta} = (\hat{\theta}_1, \hat{\theta}_2, \ldots, \hat{\theta}_n)$ of model parameters $\vec{\theta} = (\theta_1, \theta_2, \ldots, \theta_n)$. Assuming the probability model of the data is known, one can write

$$\beta \equiv P(\hat{\theta} \in V) = \int_V p(\hat{\theta}|\vec{\theta}) \prod \hat{\theta}_i, \tag{6.23}$$

where V is an n dimensions hyper-volume bounded by an $n - 1$ dimensions hypersurface C defining a **confidence region** for the parameters $\hat{\theta}$ given the true values $\vec{\theta}$. For $n = 2$, V reduces to a two-dimensional bounded surface, and C amounts to a closed contour surrounding V, as illustrated in Figure 6.3. For $n = 3$, C is a closed surface, while for $m > 3$, it amounts to a closed hypersurface.

In two or more dimensions, $n \geq 2$, just as for a single dimension, there is much latitude in selecting boundaries with a specific probability content β. It is legitimate, however, to seek a bounding hypersurface (contour in 2D, surface in 3D, etc.) that circumscribes the smallest "volume" with probability content β. The task of selecting such an enclosing surface is schematically illustrated in Figure 6.3, which for the sake of simplicity is limited to $n = 2$ parameters. The solid curve here represents an iso-contour line corresponding to a locus of constant probability density, that is, where all points $(\hat{\theta}_1, \hat{\theta}_2)$ have the same probability density $p(\hat{\theta}_1, \hat{\theta}_2|\vec{\theta}) = k$, with k chosen such that the bounded area (i.e., the confidence region) has a probability β.

Without repeating in detail the reasoning carried out for a single parameter, it is clear that any contour (with probability content β) deviating from the iso-contour would have a larger area than that of the iso-contour. The condition $p(\hat{\theta}_1, \hat{\theta}_2|\vec{\theta}) = k$ thus defines, for a given β, the bounded region with the least area. The density $p(\hat{\theta}_1, \hat{\theta}_2|\vec{\theta})$ being continuous, the contour is well defined and unique, it thus constitutes a legitimate and sensible choice of bounding region. The reasoning is readily extended in $n \geq 3$ dimensions: bounding surfaces or hypersurfaces may be chosen as iso-surfaces, that is, hypersurfaces corresponding to loci of equal probability density, $p(\hat{\theta}|\vec{\theta}) = k$, where the constant k is determined by the required value of β.

6.1.6 Confidence Region for a Multiparametric Gaussian Distribution

Let us consider a system determined by n parameters $\vec{\theta} = (\theta_1, \theta_2, \ldots, \theta_n)$ for which one has obtained n estimates $\hat{\theta} = (\hat{\theta}_1, \hat{\theta}_2, \ldots, \hat{\theta}_n)$. For illustrative purposes, let us assume the PDF of the estimators may be formulated as a multidimensional Gaussian distribution

$$p(\hat{\theta}|\vec{\theta}) = \frac{1}{(2\pi)^{n/2}|\mathbf{V}|^{1/2}} \exp\left[-\frac{1}{2}(\hat{\theta} - \vec{\theta})^T \mathbf{V}^{-1}(\hat{\theta} - \vec{\theta})\right], \tag{6.24}$$

with covariance matrix $\mathbf{V}$. We are interested in using the estimates $\hat{\theta}$ to establish bounds on the model parameters $\vec{\theta}$. In general, the covariance matrix $\mathbf{V}$ shall be nondiagonal, thereby revealing correlations between the measured estimates $\hat{\theta}_i$. The bounds on each of the model parameters θ_i, $i = 1, \ldots, n$ will then also be correlated. Errors and confidence intervals thus cannot be defined independently for each parameter. Rather than calculating disjoint confidence intervals for all parameters, one must then define and calculate **confidence regions**. Recall from §3.10 that a multiparametric Gaussian distribution can always be transformed in terms of orthogonal coordinates corresponding to the principal axes of the covariance matrix $\mathbf{V}$. The distribution is thus symmetric relative to these principal axes, and based on the discussion on central regions presented in the previous section, one would like to define central confidence regions in terms of iso-contours with probability content β.

Iso-contours with probability content β are readily defined by recalling, from §2.10.4, that the **quadratic form**

$$Q(\hat{\theta}|\vec{\theta}) = (\hat{\theta} - \vec{\theta})^T \mathbf{V}^{-1}(\hat{\theta} - \vec{\theta}) \tag{6.25}$$

has a χ^2 distribution with n degrees of freedom. This implies Q is in fact independent of $\vec{\theta}$. We can then determine the probability β that Q will be found within a domain κ_β^2 based on

$$P\big(Q(\hat{\theta}|\vec{\theta}) \leq \kappa_\beta^2\big) = \beta, \tag{6.26}$$

where the bound κ_β^2 corresponds to the β-point of a χ^2 distribution with n degrees of freedom. By construction, κ_β^2 delimits a hyper-ellipsoidal region in the n-dimensional space of the estimators $\hat{\theta}$ and model parameters $\vec{\theta}$ with probability β:

$$Q(\hat{\theta}|\vec{\theta}) \leq \kappa_\beta^2. \tag{6.27}$$

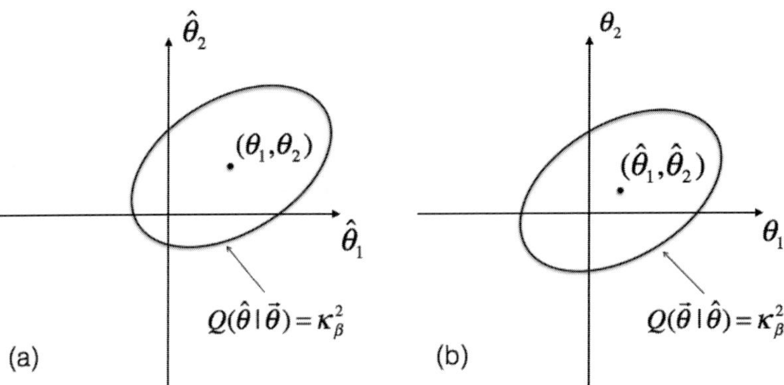

Fig. 6.4 Confidence region of probability β in (a) the space of estimators $\hat{\theta}$ and (b) the model parameter space $\vec{\theta}$ for a two-dimensional Gaussian PDF.

If the model parameter values $\vec{\theta}$ are known a priori, the foregoing condition delimits a hyper-ellipsoidal region in the n-dimensional space of the estimators $\hat{\theta}$, illustrated schematically in Figure 6.4a for a case involving $n = 2$ model parameters. The probability of measuring $\hat{\theta}$ within the ellipsoidal boundary is β, by construction. In the context of an experiment, $\hat{\theta}$ is measured, and then one seeks to invert the above bound to define a bounding region for the model parameters $\vec{\theta}$. By virtue of the symmetry of Q in terms of variables $\hat{\theta}$ and $\vec{\theta}$, this leads to an ellipsoidal domain surrounding the measured values $\hat{\theta}$, as shown schematically in Figure 6.4b.

We proceed to illustrate the quantitative determination of a confidence region in two dimensions, but the technique is readily extended to problems involving more than two parameters. We express the covariance matrix $\mathbf{V}$ as

$$\mathbf{V} = \begin{pmatrix} \sigma_1^2 & \rho\sigma_1\sigma_2 \\ \rho\sigma_1\sigma_2 & \sigma_2^2 \end{pmatrix}, \tag{6.28}$$

where σ_1^2 and σ_2^2 are the variances of $\hat{\theta}_1$ and $\hat{\theta}_2$, respectively, while ρ expresses the Pearson correlation coefficient between $\hat{\theta}_1$ and $\hat{\theta}_2$. Calculation of the inverse of $\mathbf{V}$ yields

$$\mathbf{V}^{-1} = \frac{1}{\sigma_1^2\sigma_2^2(1-\rho^2)} \begin{pmatrix} \sigma_2^2 & -\rho\sigma_1\sigma_2 \\ -\rho\sigma_1\sigma_2 & \sigma_1^2 \end{pmatrix}. \tag{6.29}$$

Defining $\Delta\theta_i = \hat{\theta}_i - \theta_i$, the quadratic form (6.25) may then be expressed as

$$\sigma_2^2\Delta\theta_1^2 - 2\rho\sigma_1\sigma_2\Delta\theta_1\Delta\theta_2 + \sigma_1^2\Delta\theta_2^2 = \kappa_\beta^2\sigma_1^2\sigma_2^2(1-\rho^2), \tag{6.30}$$

which is indeed the equation of an ellipse, as illustrated in Figure 6.5.

Figure 6.5 actually displays three distinct regions. The ellipse is defined by the quadratic form (6.25) and has a probability content of β_1, which means that the probability of measuring a pair of values $(\hat{\theta}_1, \hat{\theta}_2)$ within the ellipse is β_1. The rectangular region is defined by the condition

$$P(|\hat{\theta}_1 - \theta_1| < \kappa_\beta\sigma_1 \text{ and } |\hat{\theta}_2 - \theta_2| < \kappa_\beta\sigma_2) = \beta_2, \tag{6.31}$$

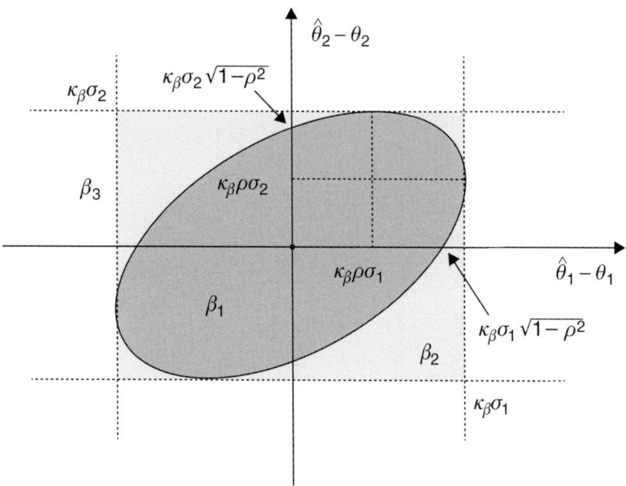

Fig. 6.5 Examples of confidence regions for specific values of σ_1, σ_2, and ρ plotted for arbitrary choices of probability content β_1, β_2, and β_3. See text for definitions of these probabilities.

and has a probability content $\beta_2 > \beta_1$. The third region is delimited by the long horizontal lines and has probability content

$$P(|\hat{\theta}_2 - \theta_2| < \kappa_\beta \sigma_2) = \beta_3, \tag{6.32}$$

which is by construction larger than β_2. One must consequently be careful when quoting error intervals of observables obtained in joint measurements. If the error ellipse can be explicitly reported, it is then possible to quote the region it delimits as a $100 \times \beta_1$ % confidence region. If instead the errors on θ_1 and θ_2 are reported as independent confidence intervals defined by the bounds $\kappa_\beta \sigma_1$ and $\kappa_\beta \sigma_2$ independently, the confidence region then has probability content β_2 and should thus be quoted as such. Finally, if the value of $\hat{\theta}_1$ is ignored and the error on $\hat{\theta}_2$ is reported based on $\kappa_\beta \sigma_2$, than it constitutes a $100 \times \beta_3$ % confidence region.

A practical example of application of these concepts is discussed in Problem 6.1.

6.1.7 Confidence Interval for Discrete Distributions

The technique described in the previous sections applies for the determination of confidence intervals of Gaussian continuous deviates and can be readily extended to any continuous variable distributions using, for instance, the Neyman construction presented in §6.1.8. It must, however, be slightly modified for distributions of discrete random variables.

Let us consider a measurement whose outcome is an integer value n_o determined by a discrete probability distribution $p(n|\theta)$ expressing the probability of measuring a positive definite integer value n given some model parameter θ. One wishes to use the measured value n_o to establish a confidence interval $\theta_\alpha \leq \theta \leq \theta_\beta$ with a specific confidence level CL.

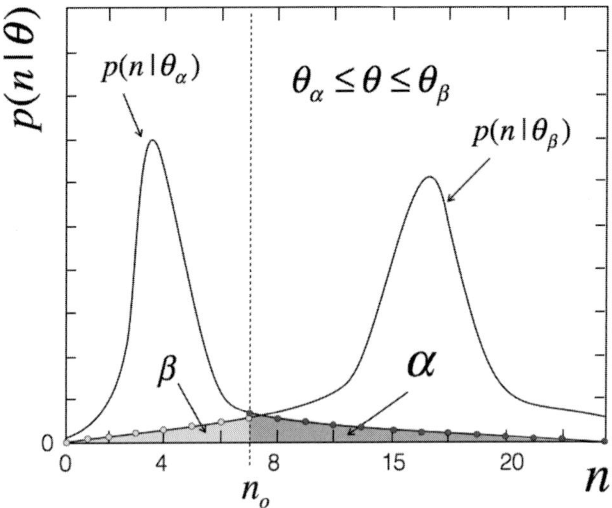

Fig. 6.6 Determination of confidence intervals for discrete PDFs: one seeks parameters θ_α and θ_β such that the probability of observing n with values equal and larger (smaller) to n_o is equal to α and β, respectively.

As for continuous intervals, much flexibility exists in defining such an interval. However, in contrast to continuous variables, one must define the interval on the basis of probabilities of observing values of n smaller and larger than n_o, as illustrated in Figure 6.6. Specifically, one must define an upper bound θ_β by calculating the probability β that n be equal or smaller than the observed value n_o and a lower bound θ_α by calculating the probability α that n be equal or greater than the observed value n_o and such that $\text{CL} = 1 - \alpha - \beta$ according to

$$\alpha = P(n \ge n_o|\theta_\alpha), \tag{6.33}$$

$$\beta = P(n \le n_o|\theta_\beta). \tag{6.34}$$

Assuming n is positive definite, the probability α may be calculated according to

$$\alpha = \sum_{n=n_o}^{\infty} p(n|\theta_\alpha) = 1 - \sum_{n=0}^{n_o-1} p(n|\theta_\alpha), \tag{6.35}$$

while β is given by

$$\beta = \sum_{n=0}^{n_o} p(n|\theta_\beta). \tag{6.36}$$

We present an application of these rules for Poisson distribution in the next section. Intervals for binomial distributions are discussed in §6.1.7.

Confidence Interval for a Poisson Distribution

As a first example of confidence interval for a discrete distribution, let us consider the estimation of confidence intervals for the mean of a Poisson distribution:

$$p_P(n|\lambda) = \frac{\lambda^n}{n!}e^{-\lambda}. \tag{6.37}$$

We saw in §3.3 that the Poisson distribution has equal mean and variance, that is, $\mathrm{E}[n] = \mathrm{Var}[n] = \lambda$. It is also possible to verify that for a given single measurement of n noted n_o, the $\mathbb{ML}$ estimator of λ is equal to n_o. Indeed, an extremum of $p_P(n|\lambda)$ given $n = n_o$ is obtained for

$$0 = \left.\frac{\partial p_P(n|\lambda)}{\partial\lambda}\right|_{n=n_o} = \left.\frac{e^{-\lambda}}{n!}\left(n\lambda^{n-1} - \lambda^n\right)\right|_{n=n_o}, \tag{6.38}$$

which yields the stated result $\hat{\lambda}_o = n_o$. Although the actual value of λ may be any real number ($\lambda \in \mathbf{R}$), the $\mathbb{ML}$ estimate $\hat{\lambda}_o$ obtained from a single measurement is by construction an integer. One is thus in a situation in which one is trying to establish a confidence interval $\lambda_\alpha \le \lambda \le \lambda_\beta$ for a continuous variable based exclusively on an integer value n_o. We thus apply Eqs. (6.35, 6.36) and write

$$\alpha = 1 - \sum_{n=0}^{n_o-1}\frac{\lambda_\alpha^n}{n!}e^{-\lambda_\alpha}, \tag{6.39}$$

$$\beta = \sum_{n=0}^{n_o}\frac{\lambda_\beta^n}{n!}e^{-\lambda_\beta}, \tag{6.40}$$

which, given values α and β, can be solved for both λ_α and λ_β. As noted in ref. [67] §9.4, this can be accomplished using the identity

$$\sum_{k=0}^{n}\frac{\lambda^k}{k!}e^{-\lambda} = \int_{2\lambda}^{\infty} p_{\chi^2}(z|n_d)\,dz = 1 - P\chi^2(2\lambda|n_d), \tag{6.41}$$

where $p_{\chi^2}(z|n_d)$ represents the χ^2 PDF, given by Eq. (3.163), for $n_d = 2(n+1)$ degrees of freedom and $P_{\chi^2}(2\lambda|n_d)$ is its cumulative integral evaluated at $z = 2\lambda$. The values λ_α and λ_β sought for may then be calculated according

$$\lambda_\alpha = \tfrac{1}{2}P_{\chi^2}^{-1}(\alpha|n_d), \tag{6.42}$$

$$\lambda_\beta = \tfrac{1}{2}P_{\chi^2}^{-1}(1-\beta|n_d), \tag{6.43}$$

with $n_d = 2(n_o + 1)$. Given an observed value $\hat{\lambda}_o = n_o$, the probabilities of observing specific ranges of λ can then be characterized according to

$$P(\lambda \ge \lambda_\alpha) \ge 1 - \alpha, \tag{6.44}$$

$$P(\lambda \le \lambda_\beta) \ge 1 - \beta, \tag{6.45}$$

$$P(\lambda_\alpha \le \lambda \le \lambda_\beta) \ge 1 - \alpha - \beta. \tag{6.46}$$

Obviously, a lower limit λ_α cannot be meaningfully defined if $n_o = 0$, and one can only establish an upper limit λ_β. One can then seek the largest λ_β consistent with the observed

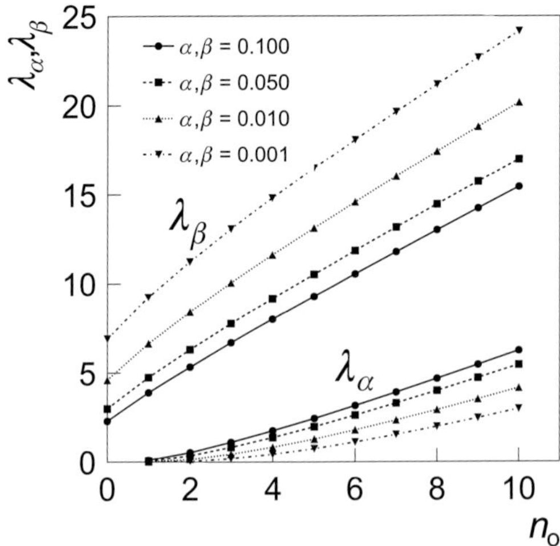

Fig. 6.7 Examples of lower λ_α and upper λ_β Poisson limits for values $\alpha,\ \beta\ = 0.01, 0.05, 0.01, 0.001$ as a function of the observed number of counts.

null value for a specific probability β and write

$$\beta = P(n_o = 0|\lambda_\beta) = \sum_{k=0}^{0} \frac{\lambda_\beta^k}{k!} e^{-\lambda_\beta} = e^{-\lambda_\beta}, \tag{6.47}$$

which yields $\lambda_\beta = -\ln\beta$. For example, requiring the probability of $n_o = 0$ to be 1% one obtains a 99% confidence interval with an upper limit of $\lambda_\beta = -\ln(0.01) \approx 4.605$, thereby setting an upper limit of approximately $\lambda_\beta = 4.6$ given the observed null value. Choosing instead a 99.9% confidence level, one would get an upper limit of $\lambda_\beta = 6.9$. This means that the value of λ could be as large as 6.9 and there would be a probability of 0.001 of observing $n = 0$ counts (of whatever entity is being sought for). Limits λ_α and λ_β are shown in Figure 6.7 for $\alpha, \beta = 0.01, 0.05, 0.01, 0.001$ as a function of the observed number of counts in the range $0 \le n_o \le 10$. Note that significant difficulties may arise in the determination of an upper limit if the measurement of a signal n_o involves subtraction of a background. Since the background may fluctuate, n_o may be negative, and one then deals with an empty confidence interval. Evidently, as for continuous variables in the presence of a physical boundary, one also runs the risk of flip-flopping and incomplete coverage: use of the Neyman construction, discussed in §6.1.8, should thus be considered whenever practical.

Confidence Interval for a Binomial Distribution

For large sample sizes N, the determination of error intervals for the fraction ϵ of the binomial distribution, Eq. (3.1), may be obtained with a normal approximation. Recall that the mean and variance of the binomial distribution are respectively $N\epsilon$ and $N\epsilon(1-\epsilon)$ for a sample size N determined by a binomial fraction ϵ. One can then use an observed number of successes n, given N trials, to determine an estimate $\hat{\epsilon} = n/N$ of the binomial fraction. And for a reasonably large sample size N, an $100 \times \alpha\%$ confidence interval can then be approximated by

$$\hat{\epsilon} \pm z_{1-\alpha/2}\sqrt{\hat{\epsilon}(1-\hat{\epsilon})/N}, \tag{6.48}$$

where $z_{1-\alpha/2}$ is the $1-\alpha/2$ quantile of a standard normal distribution. For instance, for a 95% confidence level, one has $\alpha = 0.05$ and $1 - \alpha/2 = 0.975$, which yields $z = 1.96$.

The normal approximation relies on the central limit theorem, valid for large values of sample size N, and thus yields poor results for small N. The approximation also fails if the estimate $\hat{\epsilon}$ is identically zero or unity since it then yields a null interval. Brown et al. [58] discussed a simple rule of thumb which states that the normal approximation is reasonable when $\epsilon N > 5$ and $N(1-\epsilon) > 5$. Several other approximations are discussed in the literature for such situations: they include the Wilson score interval [196], the Wilson score interval with continuity correction [148], the Jeffreys interval based on the Bayesian credible interval obtained when using a noninformative Jeffreys' prior (see §7.3.2), the Clopper–Pearson interval [65], and various other approximations. Arguably, binomial intervals are best obtained with the unified approach discussed in §§6.1.8 and 6.2.2.

6.1.8 Confidence Interval: The General Case

The Neyman Construction

In previous sections, we described basic techniques for the determination of confidence intervals and confidence regions when measured variables are Gaussian distributed with expectation values equal to the model parameters. While the central limit theorem implies that a vast majority of practical applications involve physics parameters that fluctuate with Gaussian distributions, cases are nonetheless encountered in which parameters fluctuate according to other types of distributions that cannot be approximated by Gaussian PDFs. It is thus desirable to have a general technique, applicable to any PDFs, to determine confidence intervals and confidence regions of model parameters. Such a technique was first developed by Jersey Neyman [149] and is now commonly known as the **Neyman construction**.

In order to introduce the Neyman construction, let us consider a single parameter estimator defined as a function $\hat{\theta} \equiv \hat{\theta}(\vec{x})$ of measured data $\vec{x}$ with a PDF $p(\hat{\theta}|\theta)$, which may be non-Gaussian. For a known value of θ, one may calculate the probability β that a specific estimate $\hat{\theta}$ be found in some interval $[\hat{\theta}_1, \hat{\theta}_2]$ according to

$$\beta = P(\hat{\theta}_1 \le \hat{\theta} \le \hat{\theta}_2|\theta) = P(\hat{\theta}_1(\theta) \le \hat{\theta} \le \hat{\theta}_2(\theta)) \tag{6.49}$$

$$= \int_{\hat{\theta}_1}^{\hat{\theta}_2} p(\hat{\theta}|\theta)\, d\hat{\theta}. \tag{6.50}$$

Obviously, as for the Gaussian case discussed in §6.1.3, there are arbitrarily many values $\hat{\theta}_1(\theta)$ and $\hat{\theta}_2(\theta)$ that satisfy this condition. However, it is convenient and most common to

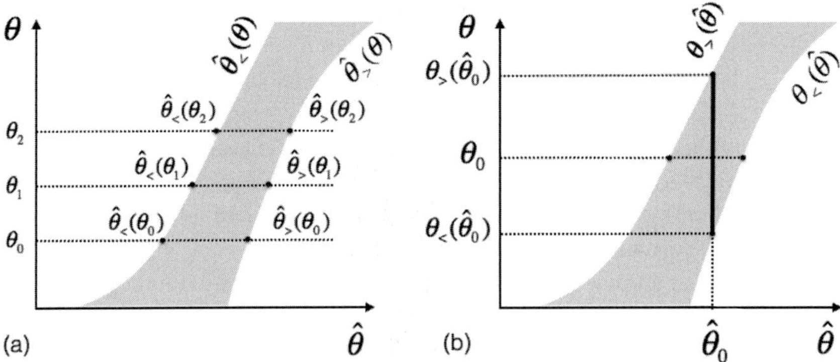

Fig. 6.8 (a) A confidence belt is constructed horizontally by finding the values $\theta_<(\theta)$ and $\theta_>(\theta)$ that satisfy Eq. 6.49. (b) The confidence level is obtained vertically as the range $[\theta_<(\hat{\theta}_0), \theta_>(\hat{\theta}_0)]$ corresponding to a measured value $\hat{\theta}_0$.

define a **central interval** according to

$$\frac{1-\beta}{2} = \int_{-\infty}^{\hat{\theta}_1} p(\hat{\theta}|\theta)\, d\hat{\theta}, \tag{6.51}$$

$$\frac{1-\beta}{2} = \int_{\hat{\theta}_2}^{\infty} p(\hat{\theta}|\theta)\, d\hat{\theta}, \tag{6.52}$$

but one might also use **upper limits**, **lower limits**, or the so-called **unified approach** discussed in §6.1.8.

Let us assume we have values $\hat{\theta}_<(\theta)$ and $\hat{\theta}_>(\theta)$ that satisfy Eq. (6.49). By construction, these are functions of the model parameter value θ and can be plotted as such, as illustrated in Figure 6.8a. The region (shaded area) between the functions $\hat{\theta}_<(\theta)$ and $\hat{\theta}_>(\theta)$ is called a **confidence belt**. It is defined such that for any specific value θ (vertical axis), the interval $[\hat{\theta}_<(\theta), \hat{\theta}_>(\theta)]$ has a probability content β defined by

$$P(\hat{\theta}_<(\theta_0) \leq \hat{\theta} \leq \theta_>(\theta_0)) = \beta. \tag{6.53}$$

The width of the belt along the $\hat{\theta}$ axis is a function of the value θ, which depends on the behavior of the function $p(\hat{\theta}|\theta)$. Our goal is to identify an interval $[\theta_<, \theta_>]$ for θ that has the same probability content β given a measured value of $\hat{\theta}$, as illustrated in Figure 6.8b. If the true but unknown value of the model parameter θ were θ_0, one would expect that a large ensemble of measurements should yield values in the range $\hat{\theta}_<(\theta_0) \leq \hat{\theta} \leq \theta_>(\theta_0)$ with probability β. Assuming that $\hat{\theta}_<(\theta)$ and $\hat{\theta}_>(\theta)$ are both monotonically increasing functions of the parameter θ, as they should if $\hat{\theta}$ is a good estimator, one can then consider their inverse functions and write

$$\theta_< \equiv \theta_<(\hat{\theta}_0) = \hat{\theta}_>^{-1}(\hat{\theta}_0) \tag{6.54}$$

$$\theta_> \equiv \theta_>(\hat{\theta}_0) = \hat{\theta}_<^{-1}(\hat{\theta}_0) \tag{6.55}$$

We can then invert the inequalities in Eq. (6.53):

$$P(\theta_<(\hat{\theta}_0) \leq \theta_0 \leq \theta_>(\hat{\theta}_0)) = \beta, \tag{6.56}$$

and conclude that, given a measured value $\hat{\theta}_0$, the interval $[\theta_<(\hat{\theta}_0), \theta_>(\hat{\theta}_0)]$ is guaranteed, on average, to cover (i.e., include or contain) the true value θ_0 and thus constitutes a confidence interval of the parameter θ_0 with probability β.

It is important to first note that given the estimate $\hat{\theta}_0$ is a random value, the bounds $\theta_<(\hat{\theta}_0)$ and $\theta_>(\hat{\theta}_0)$ are by construction also random values. The bound inversion, Eq. (6.54), indeed yields an estimate of the confidence interval that varies measurement by measurement: just as $\hat{\theta}_0$ fluctuate measurement by measurement, so do the bounds defining the confidence interval. However, the Neyman construction and Eq. (6.56) guarantee that the interval $[\theta_<(\hat{\theta}_0), \theta_>(\hat{\theta}_0)]$ has a probability content β. This implies that if the measurement is repeated a large number of times N, the N confidence intervals obtained from these measurements should cover the actual value of the parameter with probability β (on average or in the large N limit).

One should also note that if the functions $\theta_<(\hat{\theta}_0)$ and $\theta_>(\hat{\theta}_0)$ are not monotonically increasing with $\hat{\theta}_0$, the construction is still usable but one may end up with confidence intervals that are not *simply connected*, that is, which consist of several disconnected ranges.

One-Sided Confidence Intervals

One-sided confidence intervals are of interest when a measured physical quantity lies near a boundary it cannot exceed and lead to lower and upper limit confidence intervals. For instance, the mass of a particle cannot be negative but could be vanishingly small or even null. Defining a central interval in this context might become problematic because part of the confidence interval might extend across the physically forbidden region. Indeed, if the mass is small, or possibly null, it would be preferable to define a confidence level ranging from zero up to some upper limit $m_{\max}$. Upper limits are also of interest when searching for rare phenomena, including, for instance, rare decays of known particles and upper limits on the production cross section of predicted particles whose existence is yet to be established.

Figure 6.9 illustrates the difference between the confidence belt associated with a central interval and a one-sided interval, in this case an upper limit. The intervals are based on a Gaussian distribution with unit standard deviation. The solid lines represent the lower and upper boundaries of the Neyman construction for a 90% confidence belt assuming the model parameter θ cannot be smaller than zero. The dashed line shows the 90% lower limit on the measured value $\hat{\theta}$ given any value of θ. This line translates into a 90% **upper limit** confidence interval on θ. For instance, for a measured value $\hat{\theta} = 1$, one would conclude that θ has a 90% probability of being smaller than $\sim$2.2, whereas the central limit would imply that there is a finite probability for the mean to lie in the negative nonphysical region. Clearly, the use of an upper limit is more appropriate in this case. However, if the measured value is $\hat{\theta} = 4$, the upper limit could only tell us that the mean has an upper bound of $\sim$5.2 with probability of 90% and no statement about a minimal value. It thus appears appropriate to switch methods for the definition of the confidence interval based on the observed value. It is in fact common practice to use an upper-limit interval if the measured value is within 3σ of the origin (or nonphysical region) and a central (symmetric) interval, otherwise. Unfortunately, this switching of methods, sometimes called **flip-flopping**, leads

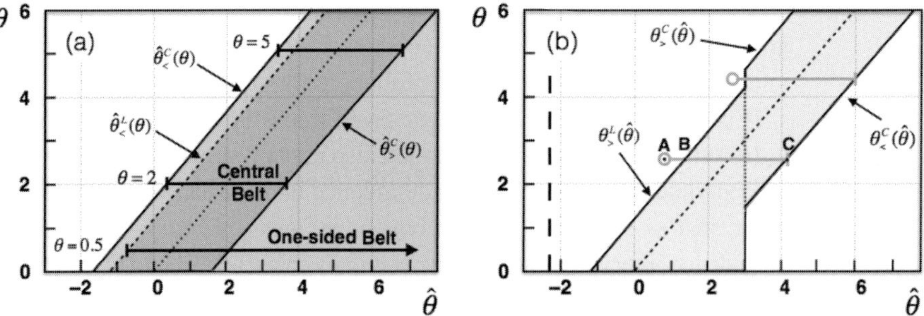

Fig. 6.9 (a) The solid lines $\hat{\theta}^C_<$ and $\hat{\theta}^C_>$ represent the boundaries of a 90% central interval and the dashed lines $\hat{\theta}^L_<$ the lower limit of a 90% one-sided interval, for a standard normal distribution of the parameter θ. (b) The shaded gray region represents the effective confidence belt obtained when flip-flopping between a central limit and an upper limit **after** a measurement $\hat{\theta}$ has been taken. The effective belt lacks coverage for model values in the range $1.35 \leq \mu \leq 4.65$, as illustrated with the light gray segments. The vertical long dashed line represents a measurement $\hat{\theta} = -2.2$ with no allowed physical value and resulting in an empty confidence interval.

to loss of coverage and inconsistent confidence interval definitions, which we describe in the next section.

Flip-flopping, Unphysical Values, and Empty Intervals

The loss of coverage occurring with flip-flopping of confidence belts is illustrated in Figure 6.9(b). The gray area represents the confidence belt obtained when using an **upper-limit method** for a measurement $\hat{\theta}$ within 3σ of the origin or nonphysical region, and switching to a **central interval method** for values larger than 3σ. The flip-flop belt has full coverage for values of $\theta \leq 1.35$ and $\theta > 4.65$ but lacks coverage in the intermediate range. The horizontal gray segments provide examples where the parameter θ lacks coverage. For $\theta = 2.6$, the intervals *AC* and $B\infty$ each contain 90% of the probability but the interval *BC* contains only 85% of the probability. There is consequently a lack of coverage for this and other values of θ in the range $[1.35, 4.65]$. Confidence intervals thus cannot be quoted consistently with the flip-flop method.

Figure 6.9b illustrates an additional problematic issue encountered with confidence belts near the boundary of a physical region. While the model parameter θ is assumed to be nonnegative, a measurement $\hat{\theta}$ of θ may nonetheless yield a negative value. For $\theta = 0$, the 90% central confidence interval amounts to $-1.65 \leq \hat{\theta} \leq 1.65$. If the measurement is repeated a large number of times, one expects that 5% of the measured values will fall below -1.65. Although such values have a smaller probability, they are nonetheless possible, even if the model parameter θ cannot be negative. The problem arises that a vertical line drawn anywhere in this range, say at $\hat{\theta} = -2$, would not intersect the chosen confidence belt in the physical region of θ. Extrapolation into a nonphysical confidence interval for θ obviously does not make sense. One is then dealing with an empty confidence interval for that value, that is, an interval that contains no physically allowed values of θ. So, although the 90% upper limit method has exact coverage, it should nonetheless be avoided since it cannot be used to produce a meaningful confidence interval for θ whenever measurements of $\hat{\theta}$ fall below the 90% confidence interval.

Fortunately, issues of lack of coverage encountered with flip-flopping and empty intervals obtained with one-sided limits may both be remedied with the unified approach discussed in the next section.

Unified Approach for the Definition of Confidence Intervals

The Neyman construction requires that the integral Eq. (6.49) yield the probability β but it does not stipulate which elements should be part of the interval. It is in fact possible to choose which elements of probability should be accepted in the interval. This leads to the notion of an ordering principle.

The purpose of this ordering principle is to produce an optimal confidence belt, with full coverage, no issue associated with flip-flopping, and no empty intervals. Feldman and Cousins found an ordering principle that satisfies these requirements and has now become the basis for the unified approach mentioned earlier [80]. Known as the **likelihood ratio ordering principle**, the ordering is based on the Neyman–Pearson test (discussed in §6.7) and sorts elements of the measurable parameter space on the basis of a likelihood ratio. The construction of intervals for each value of the model parameter θ proceeds by including the elements of probability $P(\hat{\theta}|\theta)$ that maximize the likelihood ratio

$$R(\hat{\theta}) = \frac{P(\hat{\theta}|\theta_0)}{P(\hat{\theta}|\theta_{\text{Best}})}, \tag{6.57}$$

where θ_{Best} corresponds to the value of θ with maximum likelihood $P(\hat{\theta}|\theta)$ for the measured $\hat{\theta}$, while θ_0 is the values one wishes to test for. For a Gaussian distribution, the Feldman and Cousins method yields a confidence belt as illustrated schematically by the dark solid lines in Figure 6.10 (see §6.2.3 for a formal derivation of this result). This unified approach belt is identical to the 90% central interval belt for values of $\theta > 1.65$ but significantly deviates from it below that value. For small values of θ, the most likely θ_{Best} is at zero. The different ordering then produces a belt that extends toward negative values of $\hat{\theta}$. This long negative tail prevents the occurrence of empty intervals that would otherwise occur for $\hat{\theta} < -1.65$. Consistent 90% coverage entails that the high side of the belt must shift to the left relative to the central interval. This leads to the small deviation seen in Figure 6.10 on the high side of the belt. The method then not only eliminates empty intervals but also avoids flip-flopping and provides proper and consistent coverage.

We present calculations of the Felman and Cousins ordering principle applied to Poisson, binomial, and Gaussian processes in §6.2.

6.2 Confidence Intervals with the Unified Approach

6.2.1 Poisson Intervals

Measurements of rare or predicted but as of yet unobserved decay processes constitute a perfect exemplar for the calculation of Poisson intervals: a measurement is designed

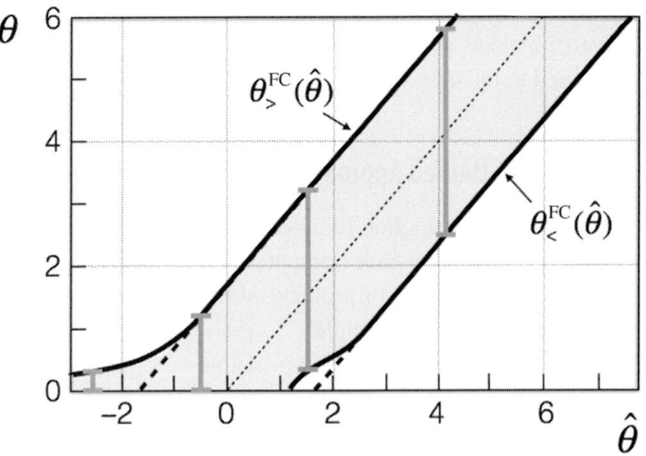

Fig. 6.10 Schematic representation of the 90% confidence belt (light gray) obtained with the Feldman–Cousins unified approach for a Gaussian distribution of unit standard deviation compared to a 90% central interval confidence belt (dashed lines). Dark gray vertical segments represent confidence intervals corresponding to measured values $\hat{\theta} = -2.6, -0.5, 1.5$, and 4.1.

to observe and count the number of decays of a certain particle. Given instrumental and physical process ambiguities, it is possible that the observed decays might originate from other improperly measured physical processes, or decays, as well as random combination of particles mistakenly interpreted as decays. Observed counts of decays thus amount to a combination of a true signal (of unknown strength) and one or several background signals. We here proceed to determine the confidence belt for such Poisson processes with known background using the Neyman construction and the ordering principle of Feldman and Cousins [80].

For a Poisson process with a known background, one can write the probability of observing n events (or counts) as

$$P(n|\mu, b) = \frac{(\mu + b)^n \exp(-(\mu + b))}{n!}, \tag{6.58}$$

where b is the expected (known) background and μ is the unknown mean yield of the Poisson process under consideration. We will assume, for illustrative purposes, that the mean of the background is known to have a specific value.

We build confidence intervals for each value of μ by ordering probabilities of different values of n based on the ratio of likelihoods

$$R(n, \mu) = \frac{P(n|\mu, b)}{P(n|\mu_{\text{best}}, b)}, \tag{6.59}$$

where $P(n|\mu, b)$ is the probability of observing n counts given an actual mean μ and background b, whereas $P(n|\mu_{\text{best}}, b)$ corresponds to the probability of n for the value of $\mu + b$ that yields the largest probability of observing n counts. The construction proceeds as follows. First, one selects ranges of interest for n and μ for a given value of b. Both ranges are divided in finitely many bins with a unit bin width for n (since it is an integer variable)

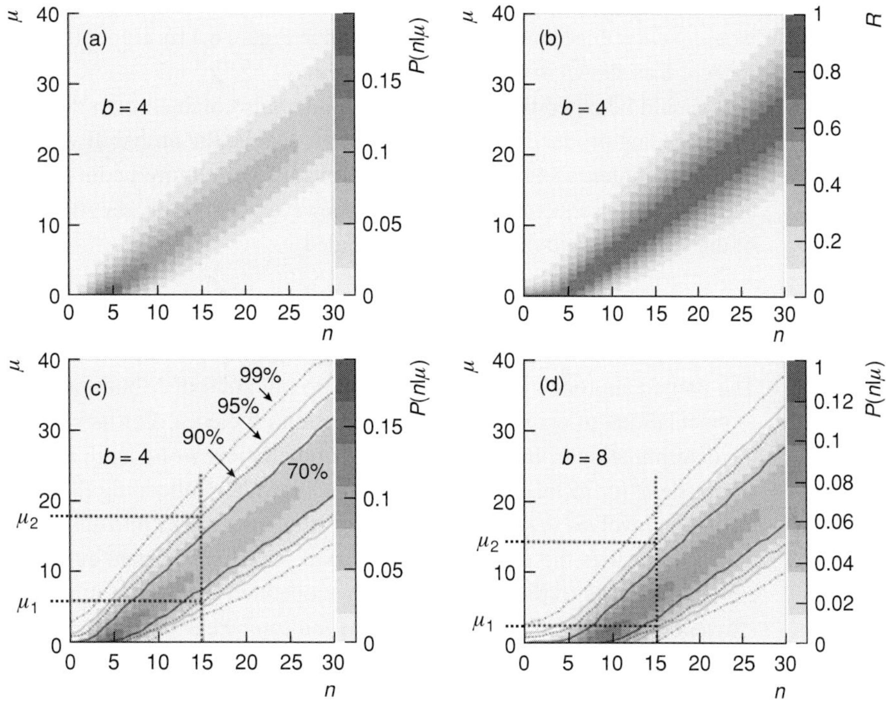

Fig. 6.11 Determination of confidence intervals for Poisson processes with known finite background b. (a) Probability $P(n|\mu, b = 4)$. (b) Ordering likelihood ratio $R(n, \mu|b = 4)$. (c) Confidence intervals calculated with $b = 4$ for CL 70%, 90%, 95%, and 99%. (d) The same confidence intervals calculated with $b = 8$. The 90% confidence intervals for $n = 15$ amount to $[\mu_1, \mu_2] = [5.6, 18.0]$ with $b = 4$ and $[2.2, 14.1]$ for $b = 8$.

and a bin width of order 0.005 for μ. One calculates and stores the probability $P(n|\mu, b)$ in each bin (using the low side of the bin for μ) into an array, or a two-dimensional histogram as illustrated in Figure 6.11a. For each value of n, one scans probabilities vs. μ to find the maximum $P(n|\mu_{\text{best}}, b)$ which can then be used to compute the ratio $R(n, \mu)$ in each bin n, μ. This yields a two-dimensional array, shown as a histogram in Figure 6.11b, which can be used to carry out the ordering of probabilities and the construction of the confidence interval for each value of μ. This is achieved by sorting values of the ratio R in descending order, for each bin in μ. Starting with the most probable bin in n for a given value of μ, one sums probabilities until the desired level of confidence is met or exceeded. The bins in n that were added to reach the desired confidence level constitute the confidence interval for the given value of μ. The sorting procedure is repeated iteratively for each value of μ. The boundaries of the confidence belt can be easily identified by searching the lowest and largest values of μ bins that have been assigned to confidence intervals. Confidence belts for selected confidence levels are shown in Figure 6.11c,d for values $b = 4$ and $b = 8$, respectively.

With an expected background $b = 4$, and no actual signal, observation of $n = 4$ counts would lead to an upper limit of $\mu = 5.1$ at the 90% confidence level ($\mu = 8.4$ at CL 99%).

Observation of fewer counts, consistent with downward fluctuations of the background, would yield reduced upper limits as shown in Figure 6.11c. Upper limits are further reduced for $b = 8$ as shown in panel (d) of the figure.

It should be noted that the confidence intervals obtained with the procedure outlined in the preceding produces slight overcoverage, that is, the probability of values n included in confidence intervals slightly exceeds, typically, the required confidence level. Overcoverage is not a consequence of the method, however, but merely the result of the integer nature of the variable n and thus cannot be avoided.

6.2.2 Binomial Intervals: Detection Efficiency

The unified approach can also be applied to binomial distributions, and most particularly, to calculations of error intervals on detection efficiency. Let us consider, as an example, the determination of the track reconstruction efficiency of a large detector. This can be accomplished, for instance, with a technique known as embedding (§12.4.6). The embedding technique involves the insertion of hits of simulated tracks into actual detector events. One then reconstructs the embedded events as if they were normal events. The ratio of found embedded tracks to the number actually embedded constitutes a measure of the detection efficiency of such tracks. This determination can be carried out as a function of transverse momentum, rapidity, and so on. Other approaches are also available to determine detection efficiencies that depend on the specificities of the measurement of interest. The following discussion also applies to a host of situations in which binomial probabilities determine the outcome of an observation, including the determination of the mortality associated with a disease, the efficacy of a medical treatment or test, or even the success of a marksman in hitting a target.

The efficiency is defined as the ratio of the number of reconstructed tracks, n, by the number of true tracks N:

$$\varepsilon = \frac{n}{N}. \tag{6.60}$$

Since the n reconstructed tracks are a subset of the N embedded tracks, it would be incorrect to use the "usual" error propagation technique. Instead, one can assume N as given and based the evaluation of the error and confidence interval solely on n. One must, however, be cognizant of the fact that the efficiency cannot be larger than unity. So, for instance, if 1000 tracks are embedded and 980 tracks are reconstructed, one might be tempted to assess the error based on $\sqrt{980} \approx 30$. But this would suggest the efficiency might be larger than unity and is thus nonsensical. One might then resort to using a lower limit on the efficiency only for large values of n and a two-sided interval for smaller values. This is technically known as flip-flopping and should be avoided, as per our discussion in §6.1.8. The unified approach provides, thankfully, a better approach to the estimation of the error of the efficiency. As for the calculation of the mean μ in the previous section, one calculates the probability $P(n|N, \varepsilon)$ for a range of multiplicity n, given N and efficiency ε of interest.

$$P(n|N, \varepsilon) = \frac{N!}{(N-n)!n!}\varepsilon^n (1-\varepsilon)^{N-n}. \tag{6.61}$$

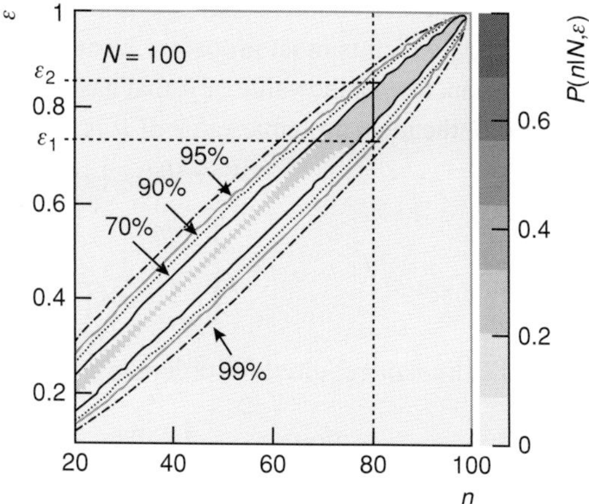

Fig. 6.12 Confidence intervals for the detection efficiency based on n detected objects (e.g., particles) based on $N = 100$ produced objects. Shown is the 90% confidence interval $[\varepsilon_1 = 0.73, \varepsilon_2 = 0.86]$ for $n = 80$ reconstructed objects.

Given a specific value N, one can then proceed to calculate the ratio R defined as

$$R(n, \varepsilon|N) = \frac{P(n|N, \varepsilon)}{P(n|N, \varepsilon_{\max})}, \tag{6.62}$$

where $\varepsilon_{\max}$ corresponds to the most likely efficiency for the given n and N. R values are then used to rank, for given values of ε, the probabilities of n, and establish confidence intervals at the desired confidence level, as per the procedure described in the previous section. Figure 6.11 provides an example of 70%, 90%, 95%, and 99% confidence intervals for a relatively modest number of tracks $N = 100$, plotted as a function of n and the detection efficiency ε.

6.2.3 Gaussian Intervals

As we discussed in §6.1.3, the determination of confidence intervals for Gaussian distributed variables is rather straightforward, given the intrinsic symmetry $\mu \rightleftharpoons x$ of the argument of the exponential defining the Gaussian distribution. The symmetry breaks down, however, in the presence of physical boundaries. For instance, the mass of a particle cannot be negative, but measurements of mass can and often do yield negative values, when the mass being measured is either very small or null. We saw in §6.1.8 that the decision to report a double-sided interval or a one-sided interval (e.g., an upper limit) leads to flip-flopping, inconsistent coverage, or worse empty intervals. The unified approach avoids such problems and is of rather simple application for Gaussian distributions.

Let us consider a measurement involving a physical boundary at $\mu = 0$, that is, such that only positive values are allowed for the physical variable, while measurements can produce both positive and negative values. Let us also assume the standard deviation of the measurement σ is known, so one can rescale measurements by σ to obtain measured values of x with unit standard deviation.

The unified approach requires that one finds, for a given measured value x, the value of the mean μ_{best} which is most probable. Assume x is given. If $x \geq 0$, the most probable value of the mean, that is, the value of μ that has the largest probability is obviously $\mu = x$, while for $x < 0$, the most probable value of μ is zero, since it cannot be negative. One thus writes

$$P(x|\mu_{\text{best}}) = \begin{cases} \frac{1}{\sqrt{2\pi}} & \text{for } x \geqslant 0, \\ \frac{1}{\sqrt{2\pi}} \exp\left(-\frac{x^2}{2}\right) & \text{for } x < 0. \end{cases} \tag{6.63}$$

The ratio of likelihoods required for ordering probabilities is thus simply

$$R(x) = \frac{P(x|\mu)}{P(x|\mu_{\text{best}})} = \begin{cases} \exp\left(-\frac{1}{2}(x-\mu)^2\right) & \text{for } x \geqslant 0, \\ \exp\left(x\mu - \frac{1}{2}\mu^2\right) & \text{for } x < 0. \end{cases} \tag{6.64}$$

The ratio R determines the order in which values of x are added to the acceptance region of a given value μ. Starting with the most probable value of x, one adds values to the left and to the right until

$$\alpha = \int_{x_1}^{x_2} P(x|\mu)\,dx = \frac{1}{2}\left(\text{erf}(x_2/\sqrt{2}) - \text{erf}(x_1/\sqrt{2})\right), \tag{6.65}$$

where α is the desired confidence level. By construction, x_1 and x_2 are given equal ordering priority, so that one has

$$R(x_1) = R(x_2). \tag{6.66}$$

One must thus simultaneously solve Eqs. (6.65) and (6.66). Two cases need to be considered. First, if both $x_1, x_2 \geq 0$, the equality $R(x_1) = R(x_2)$ implies

$$x_1 = 2\mu - x_2, \tag{6.67}$$

whereas if $x_1 < 0, x_2 \geq 0$, one has

$$x_1 = \frac{x_2\,(2\mu - x_2)}{2\mu}. \tag{6.68}$$

Since neither x_2 nor x_1 are known, a priori, one can write $x_2 = \mu + \Delta x$ and seek a numerical solution for Δx iteratively. Starting with a small value Δx, one determines x_2 and x_1 according to (6.67). If x_1 is negative, one recalculates it with (6.68) before calculating the integral (6.65). One iteratively increases Δx until a value is reached that matches the required confidence level α. One plots the values x_1 and x_2 as a function of μ that define the required confidence intervals. This is illustrated in Figure 6.13 for confidence levels of 70%, 90%, 95%, and 95%. Given a measured value x, one finds the desired confidence interval $[\mu_1, \mu_2]$ by drawing a vertical line through the relevant confidence belt. For instance, for $x = 2$ one concludes the observable has a value $\mu = 2$ with a 90% confidence interval [0.58, 3.65], while for a measured value $x = -2$, one would conclude $\mu = 0$ with an upper limit of 0.4 at the 90% confidence level. Note how for smaller values of x still, upper limits are defined for all confidence levels.

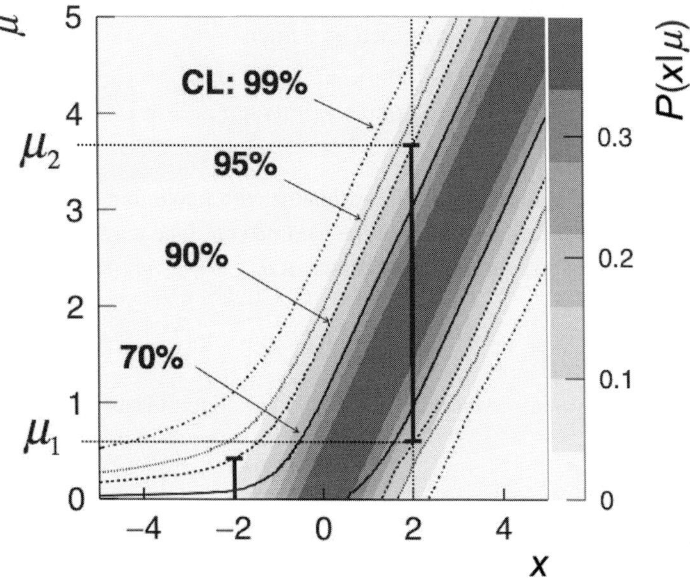

Fig. 6.13 Selected Gaussian confidence belts near a physical boundary, $\mu \geq 0$. Note how for $x \geq 0$, confidence levels are two-sided, whereas for $x < 0$, confidence intervals become one-sided and yield upper limits. Shown is the 90% confidence interval $[\mu_1 = 0.58, \mu_2 = 3.65]$ for $x = 2$, as well as the upper limit $\mu_2 = 0.42$ for $x = -2$.

6.3 Confidence Intervals from $\mathbb{ML}$ and $\mathbb{LS}$ Fits

In §§6.1 and 6.2, we discussed the notions of confidence intervals and confidence region with the implicit assumption that the joint probability density, or likelihood, of measured estimators, $p(\hat{\theta}|\vec{\theta})$, is known ab initio, and could thus be used, for instance, to generate a Neyman construction, thereby enabling the determination of confidence belts from which confidence intervals can be calculated on a case-by-case basis for estimates $\hat{\theta}$. However, the density $p(\hat{\theta}|\vec{\theta})$ may often be a rather complicated multiparameter function of measured data $\vec{x}$ and as such too difficult to model in detail. This is particularly the case in problems of $\mathbb{ML}$ and $\mathbb{LS}$ optimization involving several parameters. In such situations, obtaining an explicit model for $p(\hat{\theta}|\vec{\theta})$ might indeed be technically difficult, tedious, or too fastidious. Fortunately, as we saw in our discussions of the $\mathbb{ML}$ and $\mathbb{LS}$ methods, in §§5.1 and 5.2, the function $p(\hat{\theta}|\vec{\theta})$ may be approximated by a multiparametric Gaussian distribution,

$$p(\hat{\theta}|\vec{\theta}) \approx \frac{1}{(2\pi)^{m/2}\,|\mathbf{V}|^{1/2}} \exp\left[-(\hat{\theta} - \vec{\theta})^T \mathbf{V}^{-1} (\hat{\theta} - \vec{\theta})\right] \tag{6.69}$$

where m is the number of model (fit) parameters. One can show that, by virtue of the central limit theorem, this approximation becomes exact in the large sample limit.

For illustrative purposes, let us consider a fit involving a single parameter θ. The likelihood function $L(\theta)$ may then be written

$$L(\theta) \equiv p(\hat{\theta}|\theta) = L_{\max} \exp\left[-\frac{(\hat{\theta}-\theta)^2}{2\sigma_{\hat{\theta}}^2} \right], \tag{6.70}$$

where $\hat{\theta}$ is the value of the estimator with maximum likelihood $L_{\max}$, and $\sigma_{\hat{\theta}}$ corresponds to the standard deviation of the parameter $\hat{\theta}$. One may readily obtain $n\sigma$ central confidence intervals by computing the values of $\theta_{n\sigma} = \hat{\theta} \pm n\sigma_{\hat{\theta}}$ satisfying

$$\ln L(\theta_{n\sigma}) = \ln(L_{\max}) - \frac{n^2}{2}, \tag{6.71}$$

or equivalently, when using the $\mathbb{LS}$ method with Gaussian errors,

$$\chi^2(\theta_{n\sigma}) = \chi^2_{\min} + n^2. \tag{6.72}$$

For finite datasets, the likelihood function $L(\theta)$ may appreciably deviate from a Gaussian distribution; central confidence intervals $[\theta_<, \theta_>] = [\hat{\theta} - \delta_L, \hat{\theta} + \delta_H]$ may nonetheless be approximated according to

$$\ln L\left(\hat{\theta}^{-\delta_H}_{+\delta_L}\right) = \ln(L_{\max}) - \frac{n^2}{2}, \tag{6.73}$$

where $n = \Phi^{-1}(1-\alpha/2)$ is the quantile of standard normal distribution and the desired confidence level is $CL = 1-\alpha$. Clearly, for $\mathbb{LS}$ fits with Gaussian errors, this translates into

$$\chi^2\left(\hat{\theta}^{-\delta_H}_{+\delta_L}\right) = \chi^2_{\min} + n^2. \tag{6.74}$$

The foregoing confidence interval determination method is readily extended to fits involving a multiparametric model. Iso-contours or iso-surfaces are defined according to the discussion presented in §6.1.6. For a multiparameteric Gaussian model, one requires the quadratic function $Q(\hat{\theta}|\theta)$, given by Eq. (6.25), to satisfy

$$P\left(Q(\hat{\theta}, \vec{\theta}) \le k_\beta^2\right) = \int_0^{\kappa_\beta^2} p_{\chi^2}(z|n)\, dz = \beta, \tag{6.75}$$

where β is the required confidence level, and k_β^2 is the quantile of order β of the χ^2 distribution

$$k_\beta^2 = P_{\chi^2}^{-1}(\beta|n). \tag{6.76}$$

Operationally, for a likelihood multiparametric Gaussian function, the m-dimension confidence region may thus be constructed by finding the locus of parameter values $\vec{\theta}$ that satisfy

$$\ln L(\vec{\theta}) = \ln L_{\max} - \kappa_\beta^2, \tag{6.77}$$

where $L_{\max}$ is the likelihood corresponding to the $\mathbb{ML}$ solution $\hat{\theta}$.

The determination of multiparametric confidence regions may become technically challenging when there are several parameters or if the $\mathbb{ML}$ or $\mathbb{LS}$ solution $\hat{\theta}$ is found near a physical boundary of one or several of the model parameters. A variety of methods and approximation techniques have been developed to deal with such issues. A detailed discussion of these techniques is beyond the scope of this textbook but may be found in the book by F. James [116].

6.4 Hypothesis Testing, Errors, Significance Level, and Power

Progress in the physical sciences, and for that matter in all sciences, is based on a cycle of measurements and hypotheses. A specific model provides a prediction of a certain observable. An experiment is built and carried out to determine the value of the observable. Given the outcome of one or several observations, scientists must then determine whether the predicted and measured values are mutually compatible. If the values are compatible, the model is considered provisionally acceptable and might be used as the lead model until another measurement invalidates its predictions. On the other hand, if the values are not compatible, the model may be abandoned or require substantial revisions before it is considered acceptable. This raises the question of how one can assess whether two values, a measurement and a prediction, or even two distinct measurements, can be considered compatible. Because most observables of interest are continuous variables, one cannot sensibly expect a measured value to exactly equal the predicted value. All experiments are also obviously limited in statistical accuracy and might additionally entail sizable systematic errors. At best, one can hope that the difference between measured and predicted values is small. But how small should the difference be given the measurement errors? Indeed, one needs to quantify how large the difference can be, considering uncertainties in the measurement as well as, possibly, in the formulation of the prediction. At the end of the day, one is interested in assessing whether the model has a large likelihood of being valid given the measured value(s) and their estimated errors. In essence, one wishes to make a decision: Should the model be accepted (provisionally) or rejected? What is required is a statistical test to determine whether the measured and predicted values are compatible. Methods for the formulation of such tests and their properties are within the realm of a branch of statistics called **hypothesis testing**.

Hypothesis testing and decision making are of interest in a wide range of scientific and generic human activities. In car manufacturing, for instance, a manufacturer obviously wishes to reduce the likelihood of car defects and recalls. Parts included in the assembly of cars must then be tested (measured) to determine whether they fall within acceptable limits, or **tolerance limits**, before inclusion in a vehicle. In medicine, a doctor might use a low-cost test to determine the cholesterol level of her patient and decide to prescribe a medication if the measured level is outside a specific range. In particle physics, analyses often require a test to determine the species or type of a specific particle observed in a collision. Is the particle a charged pion, a kaon, or perhaps a proton? The possibilities are as varied as the activities undertaken by humans. It is thus useful to define the notion of hypothesis testing in a broad context based on the methods of statistics.

The notions of hypothesis, hypothesis testing, and test errors are introduced in §6.4.1 while properties of significance, power, consistency, and bias are discussed in §6.5. Methods for the construction of tests, and most particularly the Neyman–Pearson test, are presented in §6.7. Selected examples of applications of statistical tests are presented in the subsections that follow.

6.4.1 Hypothesis Testing

The formulation of a (scientific) test begins with a statement about the expected outcome of a measurement based on a specific model. This statement constitutes the hypothesis to be tested and is commonly called the **null hypothesis**. It is usually represented as H_0.

The notion of null hypothesis is often perceived as counterintuitive by nonscientists, and its interpretation thus requires some elaboration. The point is that it is not possible to scientifically prove that a theory is correct; one can only demonstrate that it is incorrect. This is a principle of **falsifiability**. One can demonstrate that a model is wrong or incomplete under some or all circumstances, but one cannot prove that it is correct because it is not possible to examine all circumstances in which the model might be applied. It is also impossible to test the model with infinite precision. A null hypothesis typically adopts the point of view that an existing model (or theory) is correct but that measurements can be used to falsify the model, that is, to demonstrate the model is in fact incorrect. Consequently, to prove that a new model is required to explain measured data, one must first demonstrate that the data is **incompatible** with the null hypothesis.

As an example, consider a search for a new short-lived particle using the invariant mass technique (§8.5.1). Energy-momentum conservation dictates that the invariant mass of a pair of particles produced by the decay of the purported particle should equal its mass. An invariant mass spectrum obtained from measured particle pairs is thus expected to exhibit a peak at the mass of the decaying particle, as well as some broad-range background. But to prove the existence of the new particle, one must first show that the measured invariant mass spectrum is incompatible with the null hypothesis, which states that there is no new particle, or in other words, that the data can be satisfactorily explained without requiring the existence of a new particle. Discovery of a new particle is achieved when a peak is observed that cannot be explained by a statistical fluctuation. Nuclear and particle physicists typically require the probability a peak might be explainable by statistical fluctuations with a probability smaller than 3×10^{-7}, corresponding to a 5σ deviation from the "background" or null hypothesis.[2]

The null hypothesis H_0 may specify the expected value of an observable or the probability density $f(x)$ of measured values according to the model. It may also represent the status quo (e.g., no need for new physics). If the function $f(x)$ has no free parameters and is unique, the hypothesis H_0 is said to be a **simple hypothesis**. If the functional form of $f(x)$ is known or given but includes parameters to be determined by the measurement, the hypothesis is considered a **composite hypothesis**. We will here focus on the definition and properties of simple hypotheses.

[2] See §6.6.2 for a more in-depth discussion of this notion.

It is often of interest to consider one of several alternative hypotheses $H_1, H_2, \ldots, H_m$. The purpose of the test shall then be to determine whether the null hypothesis H_0 can be rejected in favor of one of these alternative H_i, $i = 1, \ldots, m$.

Testing the null and alternative hypotheses requires making specific predictions about the outcome of one of several measurements of a specific observable x. These predictions are usually formulated in terms of probability densities $f(x|H_i)$, $i = 0, \ldots, m$, the observable x should follow if the hypotheses H_i, $i = 0, \ldots, m$, are considered valid (i.e., true). The PDFs $f(x|H_i)$ would ideally have small or no overlap for a given measurement in order to unambiguously discriminate against the various hypotheses. In practice, these PDFs may have considerable overlap, and one must consequently use boundaries or cuts to establish ranges that might be used to accept or reject particular hypotheses.

A measurement may involve n repeated observations of the same random variable or a single observation of an n-dimensional variable. Testing the agreement between the hypothesis H_0 (and its alternatives, if any) and the measurements $\vec{x} = \{x_i\}$, $i = 1, 2, \ldots, n$ may be investigated using a function of the measured variables called a **test statistic**, denoted $t(x)$. The test statistic is meant to encapsulate all (or selected) measured values and enable a straightforward comparison with model predictions. Given the PDFs $f(x|H_i)$, $i = 0, \ldots, m$, and a specific test statistic $t(\vec{x})$, the m hypotheses determine PDFs, denoted $p(t|H_i)$, which predict the distribution of values of t. Obviously, the test statistic $t(\vec{x})$ could specify a vector of values $\hat{\theta} = \{\theta_1, \theta_2, \ldots, \theta_k\}$, and even the vector $\vec{x}$ itself, or one of its subsets. Test statistics with $k < n$ or with a single value ($k = 1$) are usually preferred, however, because they enable a reduction of the data without, hopefully, a loss of discriminating power, that is, the capacity to keep and reject competing hypotheses.

For the purpose of our discussion, let us assume $t(\vec{x})$ is a scalar function (i.e., that it yields a single number). Let us also limit our considerations, without loss of generality, to a null hypothesis H_0 and one alternative hypothesis H_1 only. In this context, one expects the probability of t to be given by the PDF $p(t|H_0)$ if H_0 is true, and the PDF $p(t|H_1)$ if H_1 is true. The basis for our statistical test is illustrated in Figure 6.14. Let us assume the statistic t spans a domain D. Since the densities $p(t|H_0)$ and $p(t|H_1)$ cover different ranges of t, one selects a value t_{cut} that defines a zone of acceptance D_a for H_0 and a critical region $D_c = D - D_a$. Observations that yield a value t in D_c are considered an indication that the null hypothesis is false, even though there might be a finite probability of observing values of t in that range when H_0 is true. The formulation of a test for H_0 consequently amounts to a choice of a test statistic $t(x)$ and a critical region D_c. In the example presented in Figure 6.14, the region D_c consists of a single range of t, but one can in principle choose zones involving multiple disconnected regions or subsets of D.

The size of the region D_c and the shape of $p(t|H_0)$ determine the probability of finding t in that region if H_0 is true. As such, it determines the **level of significance** α of the test:

$$P(t \in D_c|H_0) = \alpha. \tag{6.78}$$

The value of α corresponds to the probability of rejecting H_0 based on the measurements $\vec{x}$, even if it is in fact true. Obviously, a test can be considered useful insofar as it can

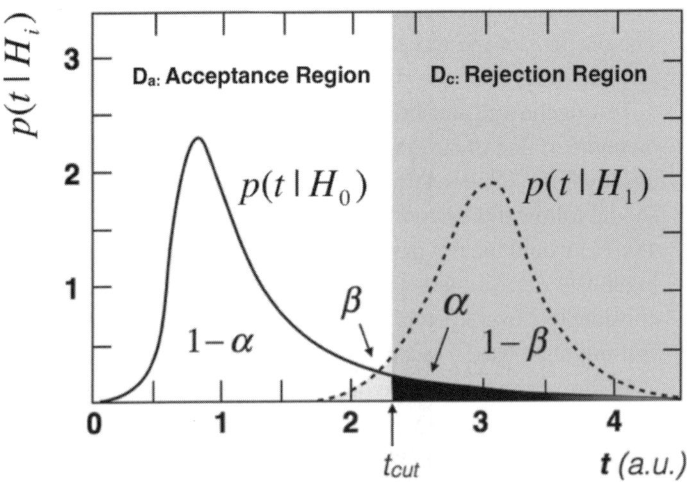

Fig. 6.14 Illustrative examples of test statistics $p(t|H_0)$ and $p(t|H_1)$ determined by PDFs $f(x|H_0)$ and $f(x|H_1)$, respectively. The variable θ_{cut} delineates the zone of acceptance D_a for H_0 and the critical region $D_c = D - D_a$ under which H_0 is rejected with a significance level α. The probability $1 - \beta$ determines the power of the test and the probability that the alternative hypothesis H_1 might be accepted.

discriminate against one or several alternative hypotheses, H_i, $i = 1, 2, \ldots$. The usefulness of the test is commonly expressed in terms of its **power**, defined as the probability $1 - \beta$ of t falling in the critical region D_c if H_1 is true. One thus has

$$P(t \in D_a|H_1) = \beta. \tag{6.79}$$

Given a null hypothesis H_0, alternative hypothesis H_1, and specific choices for the regions D_a and D_c, four distinct situations may be encountered:

1. t falls within D_a and H_0 is true. The hypothesis H_0 is *correctly* accepted with probability $1 - \alpha$.
2. t falls within D_c and H_0 is true. The hypothesis H_0 is *wrongly* rejected with a probability α. This is commonly known as an error of the **1st kind**, although when a test is used for classification purposes it may be more useful to consider this situation as a **loss** of efficiency.
3. t falls within D_a and H_0 is false. The hypothesis H_0 is *wrongly* accepted with a probability β. This is commonly known as an error of the **2nd kind**.
4. t falls within D_c and H_0 is false. The hypothesis H_0 is correctly rejected with a probability $1 - \beta$, known as the **power** of the test.

These four situations and their respective probabilities are summarized in Table 6.2.

Statistical tests may be used in a variety of contexts and various purposes. We consider specific examples of applications in §6.6. But first, let us examine the key properties of statistical tests.

Table 6.2 Definition of errors of the 1st and 2nd kind. Shown are probabilities of a test statistic t to be found in the critical region D_c and used to correctly or falsely reject the null hypothesis H_0 when either H_0 or the alternative hypothesis H_1 is actually correct.

True Hypothesis	Probability & Error Type	
	$t \in D_c$	$t \notin D_c$
	H_0 Rejected	H_0 Accepted
H_0	α	$1-\alpha$
	Error 1st kind	OK
H_1	$1-\beta$	β
	OK	Error 2nd kind

6.5 Test Properties

The capacity of a test to discriminate between a null hypothesis H_0 and an alternative hypothesis H_1 depends on a variety of factors including the measurement protocol and instrumentation performance, the number of measurements n, the choice of test statistic $t(x)$, and the choice of critical region D_c. The extent to which the PDFs $p(t|H_0)$ and $p(t|H_1)$ overlap is key in discriminating between the two hypotheses. The shape and overlap of these two functions are largely determined by the measurement performance embodied in the PDFs $f(x|H_0)$ and $f(x|H_1)$, but also depend on the number of measurements and the specific choice of test statistic $t(x)$ which altogether determine the functions $p(t|H_0)$ and $p(t|H_1)$. The discriminating power of the test and the likelihood of encountering errors of the 1st and 2nd kind obviously also depend on the specific choices of cut(s) used to select the critical region D_c and its complement D_a. All facets of a measurement and the statistical test used to test a hypothesis thus play a role in one's capacity to validate or reject a null hypothesis.

Overall, it is typically the measurement method and protocol that have the most influence on the discriminating power of a test. If the measurement resolution and efficiency are excellent, one surely can expect the measurement to provide a better test of a theory than a measurement with poor resolution and efficiency. Given a particular measurement technique, however, the choice of statistic $t(x)$ and the critical region D_c shall also influence the discriminating capacity of the test. It is thus of interest to examine the properties of a statistical test from a formal point of view and consider how or whether these properties may be optimized.

The most important property of a statistical test is its **power** $1 - \beta$, but one should also be concerned about the **test consistency** and the possibility of a **test bias**. These three properties are briefly discussed in §§6.5.1–6.5.3. A more extensive discussion of these properties is presented by James [116].

6.5.1 Test Power

Definition

Let us consider the determination of a parameter θ based on some measurements $\vec{x} = (x_1, x_2, \ldots, x_n)$ and a statistic $t(\vec{x})$. Let us assume that the null hypothesis stipulates that the parameter's expectation value is $\theta = \theta_0$, whereas the alternative hypothesis implies the value should be $\theta = \theta_1$. We wish to optimize the choice of the statistic $t(\vec{x})$ in order to maximize the power of the test, and minimize the chance of incurring an error of the first kind.

$$H_0 : \theta = \theta_0 \qquad \text{Null hypothesis}$$
$$H_1 : \theta = \theta_1 \qquad \text{Alternative hypothesis}$$

The test statistic t shall have PDFs

$$\begin{aligned} &p(t|H_0(\theta_0)) \quad \text{if } H_0 \text{ is true,} \\ &p(t|H_1(\theta_1)) \quad \text{if } H_1 \text{ is true.} \end{aligned} \tag{6.80}$$

The power of a statistical test is formally defined as the probability the test statistic t might fall in the critical region D_c defined by a cut value t_{cut}, known as a **decision boundary**, as illustrated in Figure 6.14.

$$p = 1 - \beta = \int_{t_{\text{cut}}}^{\infty} p(t|H_1(\theta_1))\, dt. \tag{6.81}$$

The cut value t_{cut} may in principle be selected to achieve a specific significance level α according to

$$\alpha = \int_{t_{\text{cut}}}^{\infty} p(t|H_0(\theta_0))\, dt. \tag{6.82}$$

But for a test to be useful, it should also have as large a power $p = 1 - \beta$ as possible in order to minimize errors of second kind (i.e., the probability of wrongly accepting H_0) and to maximize the probability of identifying H_1 as the correct hypothesis. The choice of t_{cut} is thus often predicated by the discriminating power one wishes to achieve and the losses one can tolerate as determined by the significance level.

A Simple Example with Gaussian Deviates

The notion of power is illustrated with a concrete example in Figure 6.15a, where one considers a measurement of an observable X based on a procedure involving Gaussians fluctuations characterized by a standard deviation σ. One assumes there exist two hypotheses or models predicting the value of X: the null hypothesis H_0 stipulating it has a value $\mu_0 = 0$ and an alternative hypothesis stating it has a value $\mu_1 = \mu_0 + \Delta$, with $\Delta = 2\sigma$. One assumes that it is legitimate to accept or reject the null hypothesis with a 1%

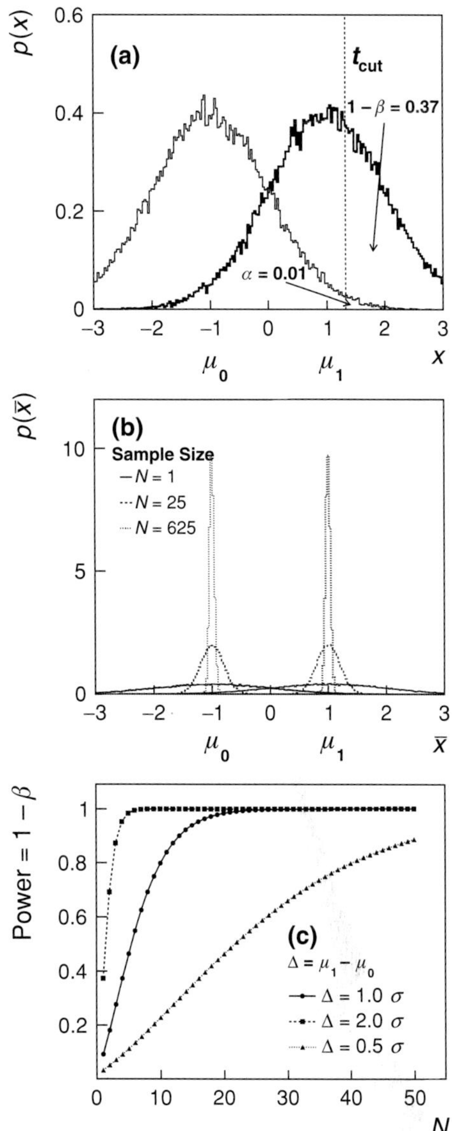

Fig. 6.15 Illustration of the notion of power based on a test involving two hypotheses $x = \mu_0$ and $x = \mu_1$ for a measurement with Gaussian deviates of standard deviation σ. In (a), the test is performed based on a single measurement of x while in (b) it is carried out based on the arithmetic mean of N sampled values. (c) Power of the test as a function of the sample size for three values of relative separation Δ/σ between the two hypotheses.

significance level based on a single measurement of X. The critical region is defined by a cut x_{cut} selected to yield the desired significance level, that is, such that

$$\alpha = 0.01 = \int_{x_{\text{cut}}}^{\infty} \frac{1}{\sqrt{2\pi}\sigma} \exp\left[-\frac{(x-\mu_0)^2}{2\sigma^2}\right] dx = 1 - \Phi(x_{\text{cut}}). \tag{6.83}$$

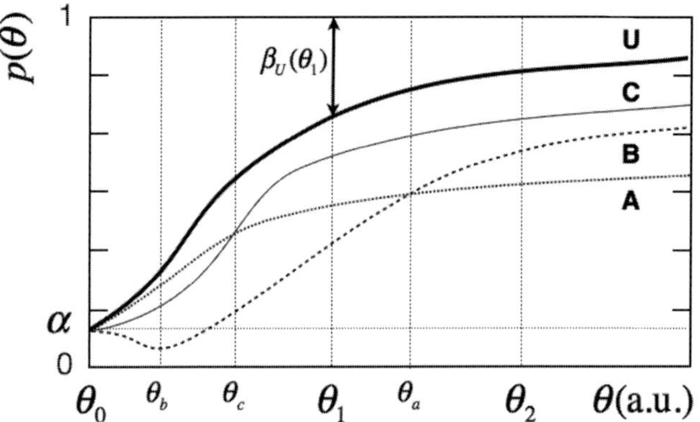

Fig. 6.16 Illustration of the concepts of power function, most powerful test, uniformly most powerful test, and test bias.

Inverting the CDF Φ, one gets $x_{\text{cut}} = \Phi^{-1}(0.99) = 2.326$ shown as a vertical dash line in Figure 6.15a. The power $1-\beta$ of the test is calculated on the basis of the probability distribution of X assuming $x = \mu_1$ is the correct hypothesis. One finds

$$1 - \beta = \int_{x_{\text{cut}}}^{\infty} \frac{1}{\sqrt{2\pi}\sigma} \exp\left[-\frac{(x-\mu_1)^2}{2\sigma^2}\right] dx = 0.37. \tag{6.84}$$

A power of 0.37 is relatively weak and implies that there is a large probability, 63%, that H_0 might be accepted even it is wrong (error of the second kind). Evidently, the power would be larger if the separation $\Delta = \mu_1 - \mu_0$ was larger or if the standard deviation σ of the measurements was smaller, or both.

Power as a Function of the Alternative Hypothesis

The power achievable, for a given significance level α, depends on the value of θ_1 predicted by the alternative hypothesis H_1:

$$p(\theta_1) = 1 - \beta(\theta_1) = \int_{t_{\text{cut}}}^{\infty} p(t'|H_1(\theta_1))\, dt'. \tag{6.85}$$

It is thus useful to regard the power as function of θ:

$$p(\theta) = 1 - \beta(\theta) = \int_{\theta_{\text{cut}}}^{\infty} p(t'|\theta)\, dt'. \tag{6.86}$$

By construction, the power function should yield α for $\theta = \theta_0$:

$$p(\theta_0) = 1 - \beta(\theta_0) = \alpha. \tag{6.87}$$

One can thus characterize a test in terms of its **power function**, which consists of a plot of the test's power as a function of θ. Examples of power functions are displayed in Figure 6.16 for several hypothetical tests. In general, the power of a test increases monotonically (tests A, C, U) when the separation between the null hypothesis, θ_0, and the alternative θ_1 increases, but there can be pathological cases (e.g., test B) where the power decreases in some range of θ (see the discussion of test bias in the next section).

The power function provides a useful tool to compare different tests. If the alternative hypothesis is simple (i.e., θ_1 is specified a priori), the best test of H_0 against H_1 at a given significance level α shall be the test with the maximum power at $\theta = \theta_1$. As an illustrative example of this notion, Figure 6.16 compares the power functions of three hypothetical tests noted **A**, **B**, and **C**. Test **C** exhibits a larger power than **A** and **B** at all values $\theta > \theta_c$, and particularly at $\theta = \theta_1$. It is thus considered the best test to discriminate against that particular hypothesis. However, test **A** would be best for testing against hypotheses with $\theta < \theta_c$, because it has the largest power in that region.

A test which is at least as powerful as any other possible test at a specific value of t is called **most powerful test** at that particular value of t. A test which is most powerful for all values of θ under consideration is known as **Uniformly Most Powerful** (UMP). Figure 6.16 includes an example of a UMP test, noted **U**. A UMP test is obviously the best choice to discriminate against an alternative hypothesis, but the use of a non-UMP test might be preferred for reasons of simplicity and/or robustness (e.g., a test with low sensitivity or dependence on unimportant changes in H_0).

6.5.2 Test Consistency

Definition

Experimentally, one wishes for a test that provides an increasingly better discrimination between a null hypothesis H_0 and an alternative hypothesis H_1 as the number of observations/measurements increases. For instance, the relative error $\delta\theta/\theta$ on the mean of an observable θ typically decreases inversely as the square root of the number N of sampled values. One would thus expect that the PDFs $p(t|H_0)$ and $p(t|H_1)$ become progressively distinct as N increases, thereby leading to an improved power at a fixed significance level.

A test is considered consistent if its power tends to unity as the number of observations N goes to infinity, even though this cannot be achieved in practice. This can be written formally as

$$\lim_{N\to\infty} p(t \in D_c|H_1) = 1, \tag{6.88}$$

where D_c is the critical region with probability α under H_0.

Consistent tests feature a power function that tend to a step function for $N \to \infty$, as illustrated schematically in Figure 6.17.

A Simple Example with Gaussian Deviates (cont'd)

In order to illustrate the notion of consistency, we revisit the test example presented in §6.5.1 and replace a single measurement of the observable X by the mean $\bar{x}$ of a sample consisting of N measurements of X. The error on $\bar{x}$ is $\sigma_N = \sigma/\sqrt{N}$. The relative distance Δ/σ_N between the two hypotheses then scales in proportion to $\sqrt{N}$. The PDFs of $\bar{x}$ for both

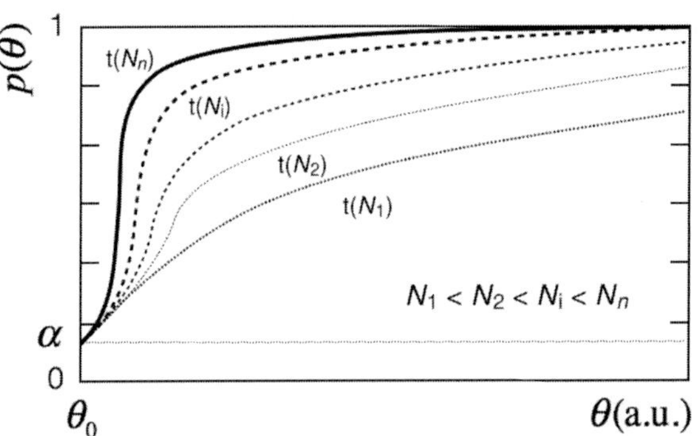

Fig. 6.17 A test is considered consistent if its power tends to unity as the number of observations N increases. Its power function thus approaches a step function as $N \to \infty$.

hypotheses thus narrow proportionally to $\sqrt{N}$ and for large N, distinguishing between the two hypotheses becomes quite straightforward, as illustrated in Figure 6.15b. The power of the test, based on a significance level $\alpha = 0.01$, is easily calculated. One finds

$$1 - \beta = \int_{\mu_0 + 2.326\sigma/\sqrt{N}}^{\infty} \frac{1}{\sqrt{2\pi}\sigma} \exp\left[-\frac{(x-\mu_1)^2}{2\sigma^2/N}\right] dx, \tag{6.89}$$

$$= 1 - \Phi\left(-\sqrt{N}\Delta/\sigma + 2.326\right), \tag{6.90}$$

which is plotted in Figure 6.15c for three values of $\Delta = \mu_1 - \mu_0$. One finds, as anticipated, that the test power increases with $\sqrt{N}\Delta/\sigma$. Distinguishing between hypotheses H_0 and H_1 is thus possible even if Δ/σ is small, provided sufficiently many repeated measurements of the observable X can be carried out. Indeed, the probability $\Phi(-\sqrt{N}\Delta/\sigma + 2.326)$ vanishes in the large N limit, and one concludes the test can be considered consistent. In fact, it should nominally be possible to carry the test for an arbitrary large power $1 - \beta$, for any significance level α, provided measurements of X be carried out for sufficiently large N. In practice, however, limitations also arise from systematic errors inherent to the measurement of X. Unlike the statistical error on the sample mean, systematic errors cannot be reduced or suppressed, usually, by increasing the size N of the data sample. They thus impose a minimum limit on the difference $\Delta = \mu_1 - \mu_0$ which can be resolved in practice. Consequently, although a statistical test may be consistent, its power shall remain smaller than unity, in practice, because of systematic errors involved in the measurement procedure.

6.5.3 Test Bias

A test is generally considered biased if the power function $p(\theta)$ does not rise monotonically with θ. An example of a biased test is provided by test **B** in Figure 6.16. The power of this test decreases below the confidence level α in the range $\theta_0 < \theta < \theta_c$ and reaches a minimum near $\theta = \theta_b$. This implies that H_0 is more likely to be accepted when $\theta = \theta_b$ than for $\theta = \theta_0$, that is, the hypothesis H_0 is more likely to be accepted when it is wrong than when it is actually right, a clearly undesirable situation.

Although undesirable, a biased test might in principle be usable if, for instance, it provides a better power than other tests in a specific region of interest of the test statistic t. This is illustrated in Figure 6.16 by the fact that test **B** has better power than test **A** in the region $\theta = \theta_2$. Test **B** might then be usable in spite of its bias if it is particularly important to discriminate against the hypothesis $\theta = \theta_2$.

6.6 Commonly Used Tests

A simple test can be formulated on the basis of the probability (density) distribution that governs the fluctuations of a variable x of interest in a specific situation. Indeed, given a PDF $f(x|\vec{\theta})$ determined by parameters $\vec{\theta}$, assumed known a priori, one may define a null hypothesis H_0 as stating that observed values of x are consistent with the PDF $f(x|\vec{\theta})$ and parameters $\vec{\theta}$. The desired significance α of the test determines the bounds $x_{\min}$ and $x_{\max}$ beyond which H_0 is rejected. For a two-sided interval, one has

$$\alpha = 1 - \int_{x_{\min}}^{x_{\max}} f(x|\vec{\theta})\, dx. \tag{6.91}$$

The integral is replaced by a sum for discrete probability distributions.

We consider several commonly used tests in the remainder of this section.

6.6.1 Binomial Test

A binomial test is of interest whenever one has reasons to suspect changing conditions in a fixed probability sampling experiment or process. One can, for instance, rely on binomial sampling to monitor the sanity and good performance of the operation of a particle detector. An analysis stream might be set up to identify a specific type of collision or process (e.g., a very high charged particle multiplicity), hereafter called diagnostic events, which happen with a probability p. Once the experimental conditions are fixed, one would expect this particular process to "fire" a certain fraction of the time p and yield, on average, $\langle n\rangle = pN$ diagnostic events in a set of N collected events. If the number n of observed diagnostic events is too small or too large, one might then conclude that the experimental conditions have changed, that is, that the response of the detector has shifted and now yields incorrect results. Periodic measurements of event samples of size N could then be used to determine whether the experimental response is improperly shifting.

Defining a central acceptance interval as $[pN - \Gamma, pN + \Gamma]$, one writes

$$\alpha = 1 - \sum_{n=pN-\Gamma}^{pN-\Gamma} \frac{N!}{n!(N-n)!} p^N (1-p)^{N-n}, \tag{6.92}$$

which (approximately) determines the desired significance level α. For large values n and N, one may use a Gaussian approximation with mean $\mu = pN$ and standard deviation

$\sigma = \sqrt{p(1-p)N}$:

$$\alpha = 1 - \frac{1}{\sqrt{2\pi}\sigma}\int_{\mu-\Gamma}^{\mu-\Gamma} dx \exp\left[-\frac{(x-\mu)^2}{2\sigma^2}\right], \tag{6.93}$$

$$= 1 - \Phi(-\Gamma/\sigma) + \Phi(\Gamma/\sigma). \tag{6.94}$$

Suppose, for instance, that the probability of the diagnostic events is $p = 0.10$. A sample of $N = 1000$ events should thus, on average, yield $\langle n \rangle = 100$ such events. Fluctuations well below or far above this mean could be used to identify a shift in the detector performance. Since one wishes to be on the safe side and identify potential problems early, one might set the significance of the test to 10%. The standard deviation of a binomial distribution is $\sigma = \sqrt{p(1-p)N}$. For $N = 1000$, this yields $\sigma = 9.5$. Considering n and N are both quite large, one can approximate the binomial distribution by a Gaussian distribution and conclude that for a significance level $\alpha = 10\%$, degradation of the performance could be identified if samples featured deviations by more than $1.65 \times 9.5 = 15.7$ from the mean $\langle n \rangle = 100$. One could of course reduce the number of inquiries by reducing the significance level of the test to 5%, but it would be probably be unwise to use a smaller value given that changes in performance might be slow or small and not necessarily yield large shifts in the value of the probability p.

6.6.2 Poisson Test

Whether by design or serendipitous happenstance, the discovery of new entities such as elementary particles, stars, or galaxies, often occurs with the observation of a peak in a probability distribution, spectrum, or histogram. However, such a peak is commonly observed atop a more or less flat background produced by known processes. What peak strength or amplitude shall constitute grounds for a claim of discovery then seemingly stands as the key question. Yet, proper scientific procedure requires that one first formulate a null hypothesis H_0 stipulating that the peak might be simply the result of background fluctuations. The null hypothesis shall indeed be that the observed spectrum is consistent with known processes or backgrounds, while the alternative hypothesis would be that the observed peak corresponds to a new process or entity (e.g., a new particle). The task at hand, the test, shall be to establish whether the observed peak is consistent with the null hypothesis, that is, whether it might be caused by an upward fluctuation of background processes.

As a practical example, consider a search for a new particle based on the invariant mass technique (described in §§8.2.1 and 8.5.1). The technique assumes that the purported particle decays in some specific channel involving two (or three) known particles. Having measured the momenta of the particles, and accessorily identified their species, one can determine the mass of the parent particle based on the invariant mass of the pair:

$$m_p = \sqrt{m_1^2 + m_2^2 + 2|\vec{p}_1||\vec{p}_2|\left(\sqrt{1+\frac{m_1^2}{|\vec{p}_1|^2}}\sqrt{1+\frac{m_2^2}{|\vec{p}_2|^2}} - \cos\theta_{12}\right)},$$

where θ_{12} represents the angle between the momenta $\vec{p}_1$ and $\vec{p}_2$ of the detected particles. Several (known) processes and combinatorial background could yield particle pairs with a mass in the range of interest. Suppose that based on some prior knowledge, one expects the background in a specific mass range will consist of $n_b = 52$ pairs, but that $n_{\text{tot}} = 90$ pairs are actually measured. Can one conclude that the observed yield is incompatible with the expected average, and that there is evidence for a new particle?

Clearly, one must decide, based on some general principles, what constitutes grounds for inconsistency with the null hypothesis and thus a claim of discovery. While a significance level of 10% or 5% is often considered sufficient to identify anomalous behaviors, it does not seem justified to single out evidence for a new particle. Particle and nuclear physicists alike typically require the level of significance to be of order of 3×10^{-7} or less to reject the null hypothesis and announce the discovery of a new particle. In our example, based on Poisson fluctuations one expects the standard deviation of the background yield to be $\sigma = \sqrt{52} = 7.2$. Excess above the expected background would thus be considered significant if it amounts to (approximately) five times the standard deviation, that is, $n = 52 + 5 \times 7.2 = 88$. Observation of 90 counts in the mass window of interest might then be considered grounds for a claim of discovery, provided of course that one can unambiguously demonstrate that the expected background level is correct. Observation of a smaller number of counts (e.g, $n_{\text{tot}} = 73$), though nearly 3σ in excess of the expected number of counts, would not be considered grounds for a claim of discovery. Deviations of this magnitude, though rare, have in fact been observed experimentally with relatively small datasets and later shown, with increased statistics, to indeed be only due to fluctuations.

Obvious complications arise in practice because the expected background may not be known with absolute certainty, and thus, it has an estimated error of its own. The observed counts should then exceed fluctuations from the expected background while accounting for the fact that it might be underestimated. If the uncertainty on the background is $\sigma_{\text{b}} = 3$, one might require the observed number of counts to exceed

$$n = n_b + 3 \times \sigma_{\text{B}} + 5 \times \sqrt{n_b + 3 \times \sigma_{\text{b}}} = 100 \tag{6.95}$$

in order to conclude a new particle has been observed in our example, and then the observed 90 counts would not match this challenge. Note that we here used a factor of 3 toward the estimation of an upper limit of the expected background. Even more conservative factors (i.e., larger) can be used in practice.

6.6.3 Gaussian Test: Comparison of Sample Means with Known Standard Deviations

A common situation encountered in statistical analysis involves the comparison of the sample means $\hat{\mu}_1$ and $\hat{\mu}_2$ obtained from two distinct sets of measurements of the same quantity. One might, for instance, be interested in finding out whether values produced by different experiments or techniques are compatible with one another. Alternatively, one might be interested in assessing whether a change or improvement applied to a process has an impact on the outcome of that process. In general, two samples taken from the same parent distribution will have a different arithmetic mean. One is thus interested in distinguishing samples that are actually different from those that originate from the same parent distribution. The null hypothesis shall be that the two samples and their means are consistent with one another.

Let us first consider the comparison of samples with means $\hat{\mu}_i$, $i = 1, 2$, and known standard deviation σ_i. The two values $\hat{\mu}_i$ are known to be randomly distributed about their true value(s) according to Gaussian distributions with standard deviations σ_1 and σ_2, each determined by the actual standard deviation of the samples divided by the number of measurements in each sample. If the two means $\hat{\mu}_i$, $i = 1, 2$ are representative of the same parent distribution, their difference $z = \hat{\mu}_1 - \hat{\mu}_2$ should be compatible with zero. Actually, the difference z should be distributed according to a Gaussian distribution with a null mean and a variance equal to $V[z] = \sigma_1^2 + \sigma_2^2$. We formulate a null hypothesis stipulating that z should have fluctuations determined by $\sigma_z = \sqrt{V[z]}$ with a null expectation value. One can then test whether the means $\hat{\mu}_i$, $i = 1, 2$ are compatible with one another by assessing whether their difference z exceeds some threshold z_0 considered to identify anomalous differences. The threshold z_0 determines the significance level α of the test (and conversely a choice of α determines z_0) according to

$$\alpha = 1 - [\Phi(z_0/\sigma_z) - \Phi(-z_0/\sigma_z)]. \tag{6.96}$$

Choosing, for instance, the confidence level α to correspond to a difference in excess of 3σ, one would reject the null hypothesis with a probability of 1% if the observed difference $|z|$ exceeds 3σ. Changes in the outcome of a measurement procedure could thus be readily identified if the means differ by more than 3σ. Of course, one can require the test to be less or more stringent based on the circumstances and needs of the situation. If the selected value of α is large, the difference z required to flag an event or difference as anomalous is small. Conversely, if α is chosen to be very small, the acceptance interval will be very large, and very few normal events (or differences) will be considered as anomalous, and if differences in excess of $|z_0|$ are observed, they will be more likely due to genuine differences in the two samples.

6.6.4 Student's *t*-Test: Comparison of Sample Means with Unknown Standard Deviations

While it might be reasonable to assume that two samples have Gaussian distributions, their standard deviations may not be known a priori. A comparison of the means of the two samples thus requires that one estimates their standard deviations according to

$$\hat{\sigma}_i = \sqrt{\frac{1}{n_i - 1}\sum_{k=1}^{n_i}\left(x_k^{(i)} - \hat{\mu}_i\right)^2}, \tag{6.97}$$

where $\hat{\mu}_i$ represents the arithmetic means of the two samples, which consists of n_i, $i = 1, 2$, observed values $x_k^{(i)}$.

Because estimates $\hat{\sigma}_i$ rather than actual values σ of the standard values are available, one cannot make use of the standard normal distribution to determine a confidence interval but must instead use Student's t-distribution:

$$p_t(t|n) = \frac{1}{\sqrt{n\pi}} \frac{\Gamma(n+1/2)}{\Gamma(n/2)} \frac{1}{(1+t^2/n)^{(n+1)/2}}, \tag{6.98}$$

for n degrees of freedom. A derivation of this expression is provided in §6.8. The variable t is nominally defined as

$$t = \frac{(x-\mu)}{\sigma}\frac{\sigma}{\hat{\sigma}} = \frac{z}{\sqrt{\chi^2/n}}. \tag{6.99}$$

In the context of the comparison of two sample means, one has $x = \hat{\mu}_1 - \hat{\mu}_2$, and $\mu = 0$ if the null hypothesis is correct. The variable z is the ratio of x by its standard deviation

$$z = \frac{\hat{\mu}_1 - \hat{\mu}_2}{\sqrt{\sigma_1^2/n_1 + \sigma_2^2/n_2}}, \tag{6.100}$$

while the χ^2 can be written

$$\chi^2 = \frac{(n_1-1)\hat{\sigma}_1^2}{\sigma_1^2} + \frac{(n_2-1)\hat{\sigma}_2^2}{\sigma_2^2} \tag{6.101}$$

for $n_1 + n_2 - 2$ degrees of freedom. By virtue of the null hypothesis, it is legitimate to assume that $\sigma_1 = \sigma_2$ and the variable t simplifies to

$$t = \frac{\hat{\mu}_1 - \hat{\mu}_2}{S\sqrt{1/n_1 + 1/n_2}}, \tag{6.102}$$

with

$$S^2 = \frac{(n_1-1)\hat{\sigma}_1^2 + (n_2-1)\hat{\sigma}_2^2}{n_1+n_2-2}, \tag{6.103}$$

which is a properly weighted estimate of the two samples' standard deviation, known as a "pooled estimate."

Use of Student's t-distribution to establish a confidence interval and/or significance level proceeds as for other distributions. The probability that the difference $|x|$ exceeds a specific value x_0 is given by

$$\alpha = \int_{-t_0}^{t_0} p_t(t|n)\, dt, \tag{6.104}$$

where p_t is computed according to Eq. (6.98) for $n = n_1 + n_2 - 2$ degrees of freedom and t_0 given by

$$t_0 = \frac{x_0}{S\sqrt{1/n_1 + 1/n_2}}. \tag{6.105}$$

Integrals of p_t and values of t_0 for specific significance levels α must nominally be computed for each value of n. Such integrals are tabulated in the literature dating back to times when computing resources were scarce [5]. However, tools to evaluate Student's t-distribution and the foregoing integral are now also available in most statistical software packages, including ROOT [59].

6.6.5 The χ^2 Test: Goodness-of-Fit

Modeling and fitting data is the bread and butter of physicists and scientists alike. Yet, the question remains for every fit: does the model used in the fit provide a suitable representation of the data, or is there evidence of incompatibilities between the data and the model? To answer this question, we formulate a null hypothesis that posits that the model does in fact provide a good description of the data, but that fluctuations and errors may introduce apparent deviations between the two. One then needs to consider the probability that such fluctuations arise and establish, for a given significance level, a measure of what might constitute a meaningful deviation between the data and the model (fit). This measure is formulated in the form of a statistics based on the data sample and the results of the fit. If the statistics deviates "too much" from values expected to be representative of a good fit, one should reject the fit and/or the mathematical model used in the description of the data.

The χ^2 test, by far the most commonly used goodness-of-fit test, is based on the statistics χ^2 defined as

$$\chi^2 = \sum_{i=1}^{N} \frac{(y_i - f(x_i|\vec{\theta}))^2}{\sigma_i^2}. \tag{6.106}$$

The sum is taken over N relevant data points (x_i, y_i) where x_i and y_i are considered independent and dependent variables, respectively. The function $f(x_i|\vec{\theta})$ is a mathematical model meant to describe the relation between the observables x and y and is determined by m parameters θ_k. The values σ_i^2 represent the variance of the measurements y_i, that is, they represent the variance that (infinitely many) repeated measurements of the values y_i would yield if such repeated measurements could be performed.

If the measurements errors of the N data points y_i are correlated, the χ^2 is written

$$\chi^2 = (\vec{y} - \vec{f})^T \mathbf{V}^{-1} (\vec{y} - \vec{f}), \tag{6.107}$$

where $\mathbf{V}^{-1}$ is the inverse of the covariance matrix of the measurements y_i.

The χ^2 test can be used to compare data with a fixed-parameters model or with the outcome of a fit. Fitting techniques to find model parameters that best match the data (i.e., minimize the χ^2) are discussed in §§5.2 and 7.6.

The null hypothesis is that the function $f(x_i|\vec{\theta})$ does describe the data. If that is truly the case, one would expect that each of the differences $y_i - f(x_i|\vec{\theta})$ would be, on average, of the order of σ_i owing to statistical fluctuations. One would then expect the χ^2 to be roughly equal to N on average. If χ^2 is much larger than N, it provides an indication that the fit is bad or that the model function $f(x_i|\vec{\theta})$ does not provide an appropriate description of the data.

We showed in §3.13.3 that if the values y_i fluctuates according to Gaussian distributions with standard deviations σ_i, the χ^2 function is expected to fluctuate according to a χ^2-distribution for n degrees of freedom:

$$p_{\chi^2}(\chi^2|n) = \frac{1}{2^{n/2}\Gamma(n/2)} \chi^{n-2} e^{-\chi^2/2}. \tag{6.108}$$

The number of degrees of freedom n here corresponds to the total number of points if the model has no free parameter, or to the number of fitted points N minus the number of free parameters m used in the fit. The probability $\alpha(\chi_0^2, N)$ of observing a χ^2 in excess of a certain value χ_0^2 with n degrees of freedom is then

$$\alpha(\chi_0^2, N) = \int_{\chi_0^2}^{\infty} p_{\chi^2}(\chi^2|n)\, d\chi^2. \tag{6.109}$$

At this point, two paths are possible. One may decide on a desired level of significance and determine the minimal values χ_0^2 beyond which all observed χ^2 would be flagged as inconsistent with the null hypothesis. Unfortunately, this requires the values χ_0^2 be calculated for each possible values of the number of degrees of freedom n, a rather tedious process. A second path is thus often used in practice, which consists in calculating the integral in Eq. (6.109) with a lower boundary χ_0^2 replaced by the χ^2 obtained in specific fits. If the integral yields a value smaller than a preselected significance level (e.g., $\alpha = 0.01$), one may then reject the null hypothesis and consider the function to be an invalid description of the measured data.

However, it should be stressed that the probability $\alpha(\chi_0^2, N)$ is strictly accurate only if the fluctuations of the variables y_i are actually Gaussian and that the errors σ_i are properly evaluated. If the errors are inflated (i.e., overestimated), the measured χ^2 will be too small and the function might not be appropriately rejected. Conversely, if the errors are underestimated, the measured χ^2 will be too large, and the function might then be improperly rejected. It is thus rather important to calibrate measurement errors if a χ^2 test is to be used to accept/reject fits performed on data sets. A very low χ^2 might alternatively result from "overfitting," that is, the use of a function with too many parameters given the number of data points. For instance, the fit of a distribution involving three data points with a polynomial of order 3 or more would obviously yield a perfect fit and a null χ^2, but it is clear that three points alone do not provide sufficient information to uniquely determine the coefficients of such a polynomial. A small or null χ^2 thus does not always provide a reliable indicator of the goodness of a fit. In this context, indeed, the χ^2 is essentially meaningless because the fit function simply has too many parameters and the data can seemingly be fitted to any degree of accuracy. While it is readily obvious that fitting three points with a polynomial of order 2 (or higher order) shall yield a perfect fit, there are situations in which it might not be readily obvious whether a fit function (i.e., a model) might have too many parameters. In such cases, the χ^2 shall be very low and the uncertainty on some the model parameters very large, but there is no simple recipe to evaluate whether the data quality warrants the given model complexity. Simply put, the frequentist paradigm does not provide a statistic to readily evaluate whether a model might feature too many or too few parameters. In contrast, we will see in Chapter 7 that the Bayesian paradigm provides a natural Occam's razor mechanism to identify unnecessary model parameters.

Figure 6.18 illustrates how the discriminating power of a fit (§6.4.1) and associated χ^2 increases in inverse proportion to the size of measurements errors. For illustrative purposes, data (x_i, f_i) were generated with Gaussian deviates to obey a parabolic dependence $f(x) = b_0 + b_1x + b_2x^2$ with relative measurement errors, σ/f, ranging between 0.05% and 50%, as shown in panels (a) through (e). We then proceeded as if the true functional form of

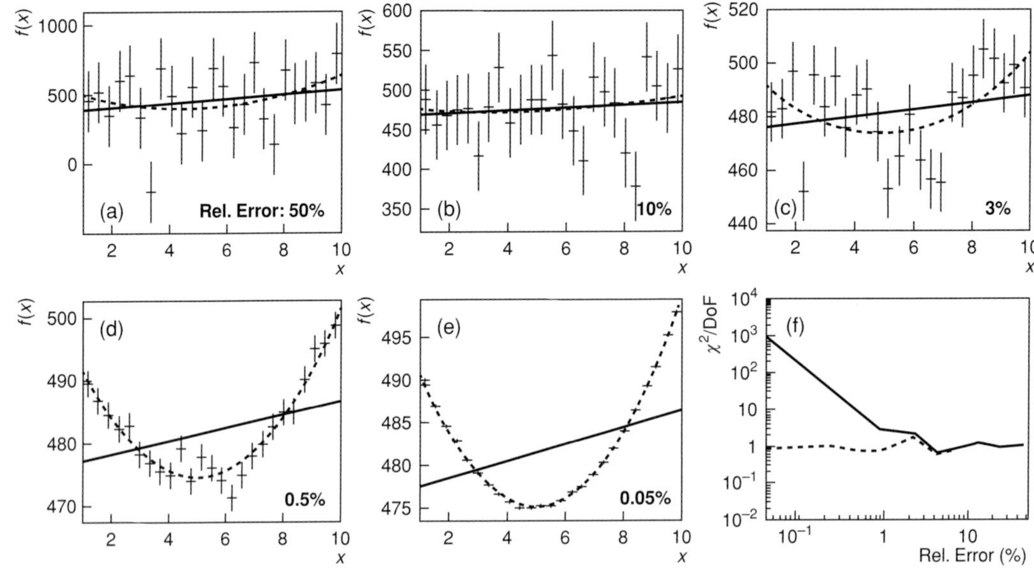

Fig. 6.18 Illustration of the discriminating power of the χ^2-test with decreasing errors. All simulated data were generated with the same quadratic function but such that measurement errors ranged from 50% in panel (a) to 0.05% in panel (e). Panel (f) displays the χ^2/DoF of the linear and quadratic fits to the data. The quadratic fit consistently yields $\chi^2/\text{DoF} \approx 1$ but the χ^2/DoF of the linear fits rises to very large values for relative errors below 1%, and the null hypothesis stipulating that a linear dependency is consistent with the data is readily rejected for such small errors.

the data was unknown and fitted the data with a linear model $f_1(x) = a_0 + a_1 x$ as a null hypothesis, and a quadratic form $f_2(x) = a_0 + a_1 x + a_2 x^2$ as an alternative hypothesis. The χ^2 of both fits were plotted as a function of the relative error in panel (e). One finds that as the relative errors decrease the quality of the linear fit (solid line) progressively worsens and yields large χ^2/DoF, whereas, in contrast, the quadratic fit consistently improves and yields more or less constant values $\chi^2/\text{DoF} \approx 1$. The rejection of the linear model is not warranted for large measurement errors but can be readily accomplished with small errors. The power of the test $1 - \beta$ to reject the linear fit as the alternative hypothesis is plotted in Figure 6.20 as a function of the relative measurement errors. One finds that for relative measurement errors smaller than 0.5%, the χ^2 distributions of the linear and quadratic fits are very well separated, thereby yielding (within the precision of the calculation shown) a power $1 - \beta$ essentially equal to unity. In situations where the variables y_i represents a yield N_i (e.g., number of counts, a number of events, etc.) and the function $f(x_i|\vec{\theta})$ amounts to the expected rate of occurrences $Np_i(x_i)$, with N being the total number of observed events, the test becomes a Pearson's χ^2 test with

$$\chi^2 = \sum_{i=1}^{N} \frac{(N_i - N(x_i|\vec{\theta}))^2}{N(x_i|\vec{\theta})}, \tag{6.110}$$

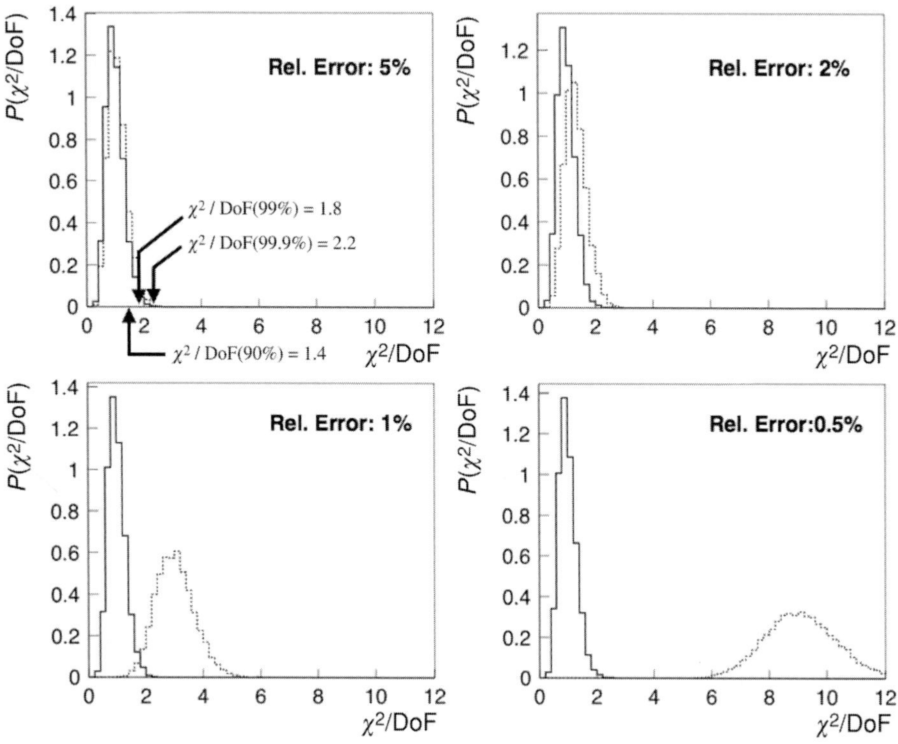

Fig. 6.19 χ^2-Distributions obtained with the linear and quadratic fits shown in Figure 6.18 for 10,000 repeated simulated measurements with relative errors ranging from 0.5% to 5% as shown. The confidence level shown in the top left panel are for 22 degrees of freedom. The rejection power of the linear hypothesis (dotted line) clearly increases when measurement errors are reduced.

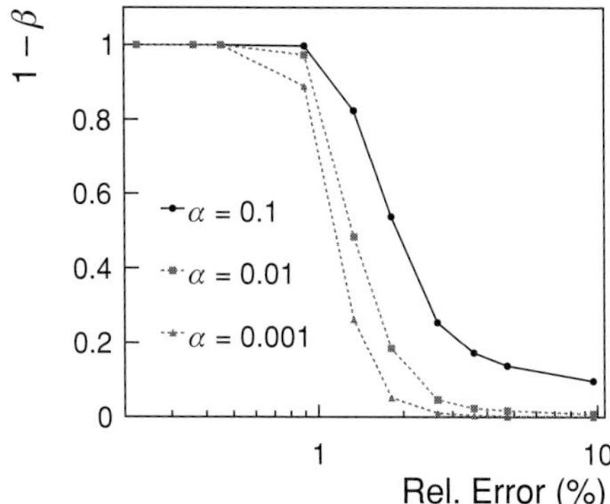

Fig. 6.20 Power of the χ^2 test shown in Figure 6.19 for three significance levels as a function of the relative errors used in the simulations.

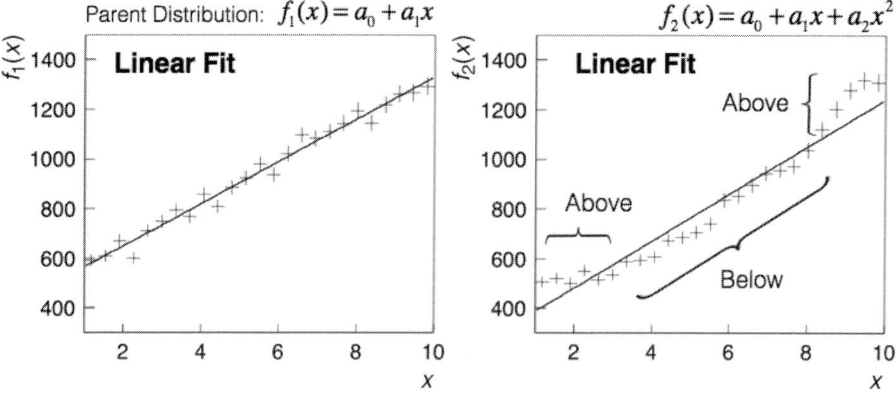

Fig. 6.21 The principle of the **Run Test** is here illustrated with a linear fit of two distinct parent distributions. One expects that statistical fluctuations should yield points randomly distributed below and above the fit if the parameterization used is a reasonable model of the parent distribution, as shown on the left for f_1. This basic expectation is clearly not met with the linear fit of the quadratic function f_2.

where one has replaced the variances σ_i^2 by $N(x_i|\vec{\theta})$. This modified χ^2 functions fluctuates according to a χ^2 distribution provided the values N_i fluctuate according to Gaussian distributions or Poisson fluctuations in the large N limit.

The χ^2-test is often not sufficient to convincingly accept or reject the null hypothesis. For instance, structures in residuals (i.e., difference between the observed and values predicted by a fit) may serve as an indicator of poor fit quality. Indeed, residuals considered as a function of the independent variable x should show a random (Gaussian if the measured values have Gaussian deviates) distribution around zero to be trustworthy. Significant structures or orderings of the residues thus provide an indicator that the null hypothesis might need to be rejected. An additional goodness-of-fit test known as a **run test** may then be used to discriminate situations where the χ^2 is relatively small but the model visually incompatible with the data, as illustrated in Figure 6.21. The run test, discussed for instance by Barlow [26], determines the probability that such seemingly unlikely situation (i.e., large nonrandom residue structures) might occur and thus enables additional power toward the rejection of a fit and the model on which it is based.

6.6.6 Histograms' Compatibility

Just as it is commonly of interest to compare predictions of a model with data, it is often useful to compare two or more histograms against one another. One may, for instance, compare data distributions produced by different experiments and examine their degree of compatibility. Alternatively, one may wish to compare results produced by a single experiment but across the life of the experiment. This is particularly useful in identifying whether the conditions or performance of the experiment evolve over time.

Let us assume we wish to compare two histograms, a control distribution $\vec{H}^{(0)}$ and a sample $\vec{H}^{(1)}$, in order to establish whether the two samples (distributions) have the same parent distribution. Let us assume the two histograms were obtained with the same number of bins m and are both normalized to represent average number of entries per event. If the two histograms have the same parent distribution, their difference calculated bin-by-bin, $\Delta_i = H_i^{(0)} - H_i^{(1)}$, should average to zero with a variance $\sigma_i^2 = (\sigma_i^{(0)})^2 + (\sigma_i^{(1)})^2$ where $(\sigma_i^{(0)})^2$ and $(\sigma_i^{(1)})^2$ are the variances of the respective bins. Let us further assume that the variances are either known or can be estimated with reasonably high precision. The sum $\sum_i (\Delta_i/\sigma_i)^2$ then constitutes a χ^2 variable distributed according to a χ^2-distribution. One can thus use a χ^2-test with m degrees of freedom to evaluate the compatibility of the two histograms to any desired level of significance.

Few remarks are in order. First, note that if the variances $(\sigma_i^{(1)})^2$ of the m bins of sample $\vec{H}^{(1)}$ are not very well known, one can set $\sigma_i^2 = 2(\sigma_i^{(0)})^2$ since the null hypothesis stipulates that the two histograms have same parent distribution. Additionally, note that if the two histograms are normalized to have the same integral rather than being normalized *per event*, the same procedure may be applied but the number of degrees of freedom is then reduced by one unit. If the two distributions have unequal numbers of bins or points, it might be possible to instead use the Kolmogorov test, presented in the next section.

6.6.7 Kolmogorov–Smirnov Test

The **Kolmogorov–Smirnov test** (KS test), also often called the **Kolmogorov test**, is a nonparametric test of the equality of continuous, one-dimensional probability distributions. It can be used to test the compatibility of a sample distribution with a reference probability distribution (one-sample KS test), or to compare two samples (two-sample KS test). Unlike the histogram compatibility test presented in the previous section, the KS test makes no assumption on the number of points (bins) or the variances of the samples.

The test is based on the Kolmogorov–Smirnov statistic, which expresses the distance or difference between the empirical distribution function (EDF) of the sample (defined below) and the cumulative distribution function (CDF) of the reference distribution (or alternatively, the sample and control empirical distributions).

The null hypothesis is that the sample being tested is determined by the reference distribution (one-sample KS test) or that two samples being compared have the same parent distribution (two-sample case). The distributions considered under the null hypothesis are continuous but otherwise unrestricted. Formally, the EDF $F_n(x)$ of a sample is defined as

$$F_n(x) = \frac{1}{n}\sum_{i=1}^{n} I(x_i \le x) \tag{6.111}$$

where n is the size of the sample and $I(x_i < x)$ is called the **indicator function**, defined such that

$$I(x_i < x) = 1 \qquad \text{if } x_i \le x, \tag{6.112}$$

$$0 \qquad \text{otherwise.} \tag{6.113}$$

This is equivalent to sorting the measured values of a sample in ascending order and summing the number of sampled values smaller than x. Alternatively, if the sample is provided

Table 6.3 Critical values d_α of the KS test listed for selected significance levels

α	20%	10%	5%	2.5%	1%	0.5%	0.1%
d_α	1.07	1.22	1.36	1.48	1.63	1.73	1.95

in the form of an m bins histogram $\vec{H} = (H_1, \ldots, H_m)$, where the values of x are implicitly sorted in ascending order, rather than a list of values as earlier, the function $F_n(x)$ becomes

$$F_k = \frac{1}{m}\sum_{i=1}^{k} H_i, \tag{6.114}$$

where $k = 1, \ldots, m$. For a given cumulative distribution function, $F(x)$, the Kolmogorov–Smirnov statistic is defined as

$$D_n = \sup_x |F_n(x) - F(x)|, \tag{6.115}$$

which corresponds to the maximum deviation between the cumulative distributions. One finds that $d_n = \sqrt{n}D_n$ converges to the Kolmogorov distribution [129], and one may reject the null hypothesis at a level of significance α for

$$d_n \equiv \sqrt{n}D_n > d_\alpha. \tag{6.116}$$

Several critical values d_α and their respective levels of significance are listed in Table 6.3.

When the KS test is used to compare to measured samples, the Kolmogorov–Smirnov statistic becomes

$$D_{n,n'} = \sup_x \left|F_{1,n}(x) - F_{2,n'}(x)\right|. \tag{6.117}$$

The null hypothesis is then rejected at level α if

$$D_{n,n'} > d_\alpha \sqrt{\frac{n+n'}{nn'}} \tag{6.118}$$

with values of the coefficient d_α, listed in Table 6.3.

Example of the KS Test

Two samples of 1,000 events each, shown in Figure 6.22b, were generated according to parent distributions

$$f_1(x) = \frac{1}{\sqrt{2\pi}\sigma} e^{-\frac{(x-\mu)^2}{2\sigma^2}} \quad \text{(dashed line)}, \tag{6.119}$$

$$f_2(x) = [1 + 0.2(x - \mu)] \times f_1(x) \quad \text{(dotted line)}, \tag{6.120}$$

with $\mu = 3$ and $\sigma = 0.7$, shown in Figure 6.22a. While it is straightforward to distinguish the two (parent) distributions (or with infinite statistics plotted on a log scale), it is rather difficult to determine whether the two samples shown in Figure 6.22b originate from the

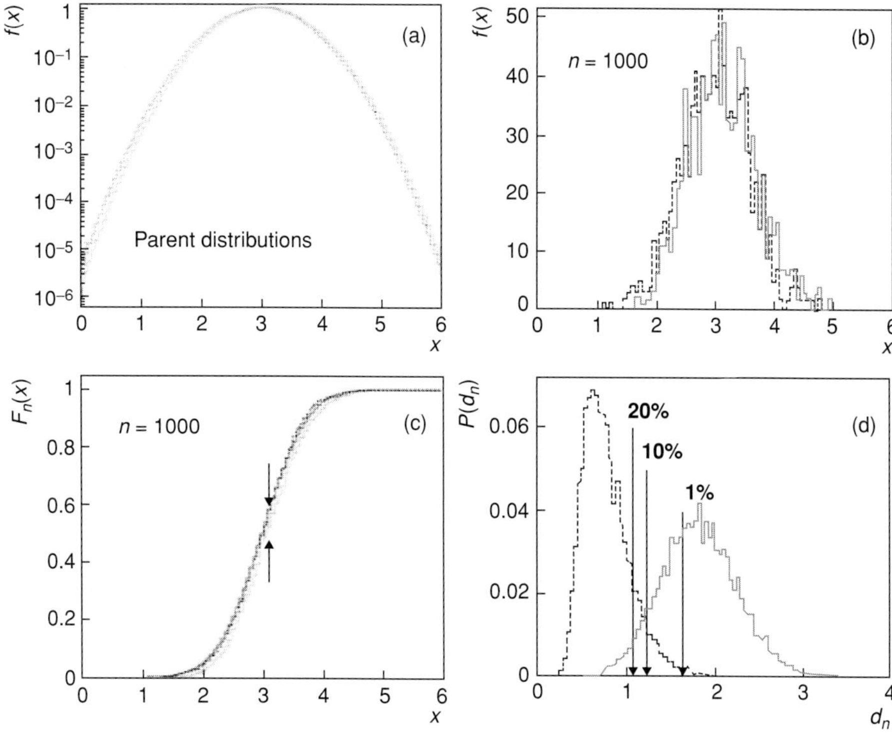

Fig. 6.22 Example of application of the Kolmogorov–Smirnov test. (a) Parent distributions f_1 and f_2 listed in the text. (b) 1,000 events samples generated with distributions f_1 and f_2. (c) Cumulative distributions. (d) Relative probability of the maximum difference d_n for both types of samples.

same or distinct parent distributions. We thus carry out a KS test to establish whether either sample is compatible with the parent f_1. Cumulative distributions of the two samples are compared to the cumulative distribution F_1 of the first parent distribution f_1 in Figure 6.22c. One observes that the cumulative distribution of the first sample essentially overlaps with F_1, whereas the cumulative distribution (dotted line) of the second sample exhibit significant differences with the parent cumulative distribution F_1. The largest difference, of order 0.016, multiplied by $\sqrt{1{,}000}$, is found to be 1.19, which is in excess of the 20% confidence level of 1.07 and thus constitutes grounds for excluding the null hypothesis that the second sample might originate from the same parent distribution as the first sample.

The "experiment" was then repeated 100,000 times: for each experiment, we generated a sample of 1,000 values and created their cumulative distributions. The maximum differences between these and the cumulative distribution F_1 are plotted in Figure 6.22d where arrows indicate the 20%, 10%, and 1%. One finds the distribution of d_n obtained with the first sample (dashed line) covers values typically smaller than unity whereas the second distribution lies mostly above this value. The test thus indeed enables a distinction between parent distributions to be made even when the difference may not be readily obvious to the human eye.

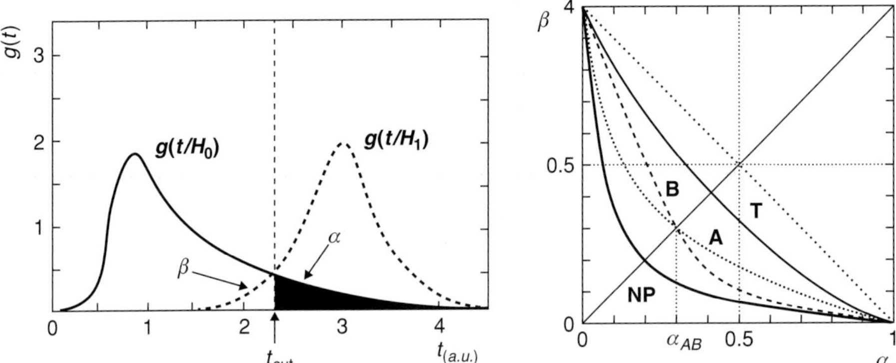

Fig. 6.23 (Left) A symmetry exists between parameters α and β. If t_{cut} is raised β increases while α decreases, and conversely. Right: Tests viewed as functions of both α and β.

6.7 Test Optimization and the Neyman–Pearson Test

6.7.1 Test Optimization

The choice between tests usually relies on a comparison of their power curves for a given value of the significance level α. While the value of α determines the risk of an error of the first kind (false rejection of H_0), the value of $\beta = 1 - p$ controls the risk of an error of the second kind (contamination or incorrect acceptance of H_0). It is common practice to first choose α and thence find which test provides the best power. However, there is clearly a certain form of symmetry between α and β, as shown in the left panel of Figure 6.23, and one obviously wishes to minimize both types of errors. It is thus useful to regard tests as simultaneous functions of both α and β, as illustrated in the right panel of the figure. The function $\beta_\alpha = 1 - \alpha$ corresponds to the dotted diagonal line below which, for a fixed value of α, all other tests have a large power than α, and can thus be considered consistent. By extension, all tests described by curves below this straight line are therefore strictly unbiased (for all values of α). Actual tests tend to follow a curve such as **T** (typical). For specific values of θ_0 and θ_1, decreasing α usually requires raising the value t_{cut}, used to define the critical boundary, and this then simultaneously increases β, thereby reducing the power of the test. Conversely, efforts to reduce β for a specific test will end up producing a larger value of α. The question thus arises as to whether there might exists some tests **A** or **B**, as illustrated in Figure 6.23, that are more acceptable than **T** because they reduce the values of both α and β simultaneously. In 1933, Jerzy Neyman and Egon S. Pearson established that given two hypotheses H_0 and H_1, there exists a most powerful test, now known as the Neyman–Pearson (NP) test, which is at least as good as any other tests for all values of α and β. This test is based on a ratio of likelihood functions and is described below. Because the implementation of the Neyman–Pearson test is often cumbersome or does not exist if hypotheses are composite, then it is commonly acceptable to resort to lower performing tests, such as tests **A** and **B** in Figure 6.23. Obviously, test **A** should be chosen, in this example, if $\alpha < \alpha_{AB}$ is desired given that it enables a higher power in that range of α. Conversely, **B** should be used if a value $\alpha > \alpha_{AB}$ is deemed acceptable. Comparing tests according to their simultaneous performance as a function of α and β permits the selection of the best test for a given job or analysis. In general, deciding on a specific value of α (and consequently β) requires knowing the cost or consequences of a wrong decision. Errors of the first kind might be more acceptable in certain experiments, while others might better tolerate errors of the second kind.

6.7.2 Formulation of the Neyman–Pearson Test

In order to formulate the Neyman–Pearson test, let us consider a measured data sample $\vec{x} = (x_1, \ldots, x_n)$ in its entirety, and let us assume it is determined by a PDF $f_n(\vec{x}|\theta)$. For simplicity's sake, we will also assume that the parameter θ takes a value $\theta = \theta_0$ if the null hypothesis H_0 is correct and a value $\theta = \theta_1$ according to the alternative hypothesis H_1. The goal of the test is to find out whether the data $\vec{x}$ are consistent with the null hypothesis H_0, or if H_0 should be rejected in favor of H_1. We will additionally assume that $f_n(\vec{x}|\theta)$ is a continuous and smooth function of θ between θ_0 and θ_1.

Given a desired significance level α, one expects that there is a portion of the domain of $\vec{x}$ which can be identified as a critical (rejection) region D_c such that

$$\int_{D_c} f_n(\vec{x}|\theta_0)\, d\vec{x} = \alpha, \tag{6.121}$$

$$\int_{D_c} f_n(\vec{x}|\theta_1)\, d\vec{x} = 1 - \beta. \tag{6.122}$$

Our goal is to find the region D_c, given α, which maximizes the power $1 - \beta$. To this end, it is convenient to rewrite Eq. (6.122) as

$$1 - \beta = \int_{D_c} \frac{f_n(\vec{x}|\theta_1)}{f_n(\vec{x}|\theta_0)} f_n(\vec{x}|\theta_0)\, d\vec{x}, \tag{6.123}$$

$$= \int_{D_c} r_n(\vec{x}|\theta_0, \theta_1) f_n(\vec{x}|\theta_0)\, d\vec{x}, \tag{6.124}$$

where we have introduced the ratio $r_n(\vec{x}|\theta_0, \theta_1)$ as

$$r_n(\vec{x}|\theta_0, \theta_1) = \frac{f_n(\vec{x}|\theta_1)}{f_n(\vec{x}|\theta_0)}. \tag{6.125}$$

Inspection of Eq. (6.124) reveals that it corresponds to the expectation value of the function r_n over the domain D_c according to the null hypothesis $H_0(\theta_0)$:

$$1 - \beta = \mathrm{E}_{D_c}\left[r_n(\vec{x}|\theta_0, \theta_1)\right], \tag{6.126}$$

$$= \mathrm{E}_{D_c}\left[\left.\frac{f_n(\vec{x}|\theta_1)}{f_n(\vec{x}|\theta_0)}\right|_{\theta=\theta_0}\right]. \tag{6.127}$$

The power is consequently optimal if and only if the D_c is a fraction of the domain of $\vec{x}$ which contains the maximum value of the ratio r_n. This is easily guaranteed by selecting points such that

$$r_n(\vec{x}|\theta_0,\theta_1) = \frac{f_n(\vec{x}|\theta_1)}{f_n(\vec{x}|\theta_0)} \geq r_\alpha, \tag{6.128}$$

where the constant r_α must be chosen to correspond to the desired level of significance α. The Neyman–Pearson test is thus formulated as follows:

- If $r_n(\vec{x}|\theta_0,\theta_1) > r_\alpha$, reject H_0, and choose H_1.
- If $r_n(\vec{x}|\theta_0,\theta_1) \leq r_\alpha$, choose H_0.

It is important to note that the function $r_n(\vec{x}|\theta_0,\theta_1)$ is a statistic evaluated for specific points $\vec{x}$ and corresponds, in fact, to the ratio of the likelihood of $\vec{x}$ according to the hypotheses H_1 and H_0. Clearly, for the test to be usable it should be possible to calculate the likelihood function $f_n(\vec{x}|\theta)$ for both $\theta = \theta_0$ and $\theta = \theta_1$ at all points of the space of interest. This, in turn, implies that both H_1 and H_0 must be simple hypotheses. If these conditions are verified, the Neyman–Pearson test provides by construction the best power $1-\beta$ against the chosen value of the significance level α and the given data $\vec{x}$. Extensions of the Neyman–Pearson test to composite hypotheses are possible but not guaranteed to yield an optimal test [116].

6.7.3 Comparison of Sample Means with the Neyman–Pearson Test

Let us reexamine the situation, already discussed in §6.6.3, where a change is made to a process that yields entities with a certain property x. The purpose of the test is to determine whether the change has in fact affected the property value x. We once again assume that the mean value of x prior to the change is known to be μ, and that measurement errors are Gaussian deviates with a known standard deviation σ. The test is conducted by measuring a sample of N values x_i and taking their mean $\bar{x}$. The null hypothesis H_0 shall be that the change in process has not altered the mean, that is, the measured mean $\bar{x}$ shall have an expectation value $\mu_0 = \mu$. The alternative hypothesis shall then be that the mean has changed and now has a value μ_1.

Let us proceed to build a Neyman–Pearson test. Given that measurements are assumed to yield Gaussian fluctuations with standard deviation σ, the test may then be formulated based on the ratio r of the likelihood functions of $\vec{x}$ given hypotheses H_1 and H_0:

$$r = \frac{\frac{1}{\sqrt{2\pi}\sigma}\exp\left[-\frac{1}{2\sigma^2}\sum_{i=1}^N (x_i-\mu_1)^2\right]}{\frac{1}{\sqrt{2\pi}\sigma}\exp\left[-\frac{1}{2\sigma^2}\sum_{i=1}^N (x_i-\mu_0)^2\right]}. \tag{6.129}$$

Introducing the sample mean $\bar{x} = \frac{1}{N}\sum_i^N x_i$, Eq. (6.129) simplifies to

$$r = \exp\left[-\frac{N}{2\sigma^2}(\mu_1-\mu_0)(\mu_0+\mu_1-2\bar{x})\right] \tag{6.130}$$

Taking the log on both sides, one writes

$$\frac{2\sigma^2}{N}\ln r = -\left(\mu_1-\mu_0\right)\left(\mu_0+\mu_1-2\bar{x}\right). \tag{6.131}$$

The expectation value of $\bar{x}$ depends on which of the two hypotheses is correct. If H_0 is correct, one gets $\mathrm{E}\left[\bar{x}\right]=\mu_0$, whereas if H_1 is correct, one has $\mathrm{E}\left[\bar{x}\right]=\mu_1$. Either way, the variance of $\bar{x}$ is σ^2/N. It is then useful to define a deviate z according to

$$z=\frac{\bar{x}-\mu_0}{\sigma/\sqrt{N}}, \tag{6.132}$$

which, by construction, is Gaussian distributed with unit standard deviation and expectation values

$$E[z]=\begin{cases}0 & \text{for } H_0\\ \frac{\mu_1-\mu_0}{\sigma/\sqrt{N}} & \text{for } H_1\end{cases} \tag{6.133}$$

Substituting in Eq. (6.131), one gets, after some simple algebra

$$r'\equiv\frac{2\sigma^2\ln r}{N\left(\mu_1-\mu_0\right)^2}=-\left(1-2\frac{z\sigma}{\sqrt{N}(\mu_1-\mu_0)}\right), \tag{6.134}$$

where, for notational convenience, we introduced the statistics r' which clearly has expectations values

$$\mathrm{E}[r']=\mathrm{E}\left[\frac{2\sigma^2\ln r}{N(\mu_1-\mu_0)^2}\right]=\begin{cases}-1 & \text{for } H_0,\\ +1 & \text{for } H_1\end{cases} \tag{6.135}$$

Since z is Gaussian distributed with unit standard deviation, the statistics r' is also Gaussian distributed but its standard deviation decreases proportionally to $\sqrt{N}$. For a given σ, the PDFs $p(r'|H_i)$, $i=0,1$, are thus increasingly separated for increasing N, as illustrated in Figure 6.24. For sufficiently large N, it then becomes possible to distinguishes which of the two hypotheses is the best. The Neyman–Pearson test requires the ratio of likelihood functions for hypotheses H_1 and H_0 to exceed some value r_α, for a given significance level α. Given Eq. (6.134), this translates in a cut on z. The critical region is then defined for values $z>z_\alpha$ (assuming $\mu_1>\mu_0$) where z_α is the α-point of the standard normal distribution. The power of the test is then

$$1-\beta=\int_{z_\alpha}^{\infty}\frac{1}{\sqrt{2\pi}}\exp\left(-\frac{(z-\Delta)^2}{2}\right)dz, \tag{6.136}$$

where

$$\Delta=\frac{\sqrt{N}}{\sigma}\left(\mu_1-\mu_0\right). \tag{6.137}$$

One thus find that the power increases with $\mu_1-\mu_0$ as well as N, as expected from Figure 6.24.

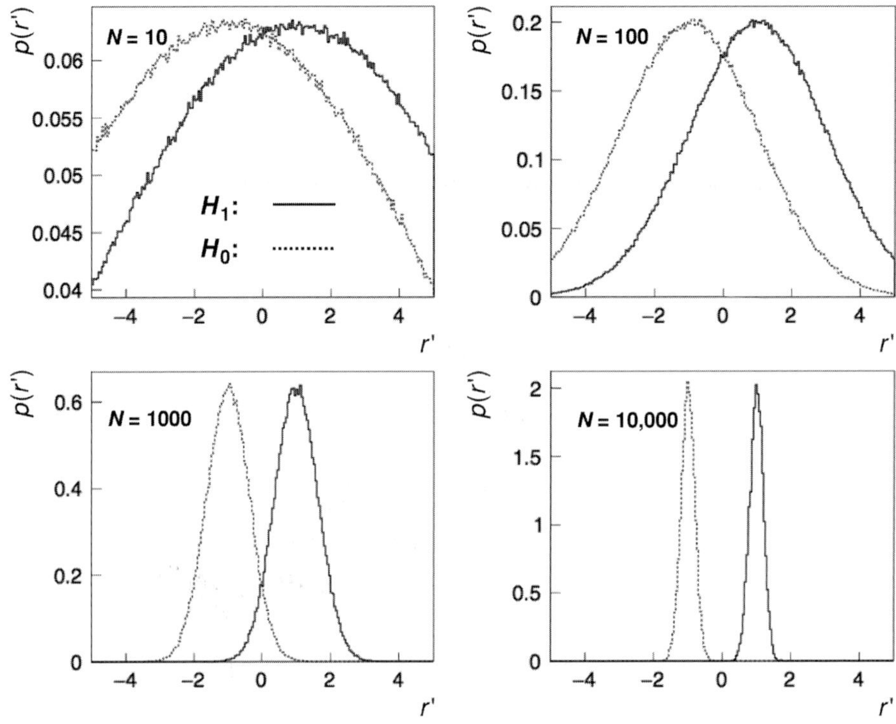

Fig. 6.24 Probability density of r' corresponding to the scaled ratio of likelihood of hypotheses H_1 and H_0, where $r' \equiv 2\sigma^2 \ln r/N(\mu_1 - \mu_0)^2$.

Other tests of the equivalence of two samples may be applied, but one can show that the aforementioned Neyman–Pearson test indeed yields the most powerful test and discrimination power between the null and alternative hypotheses [116].

6.8 Appendix 1: Derivation of Student's *t*-Distribution

Student's t-distribution, Eq. (3.161), expresses the probability (density) of observing a certain value of t defined as

$$t = \frac{x - \mu}{\hat{\sigma}}, \tag{6.138}$$

where x is a continuous random variable Gaussian distributed with a mean μ and a standard deviation σ, while $\hat{\sigma}$ constitutes an estimates of σ obtained from a sample $\{x_i\}$ according to

$$\hat{\sigma}^2 = \frac{1}{N} \sum_{i=1}^{N} (x_i - \mu)^2 \tag{6.139}$$

if the mean μ is known and

$$\hat{\sigma}^2 = \frac{1}{N-1}\sum_{i=1}^{N}(x_i - \hat{\mu})^2 \tag{6.140}$$

if calculated in terms of an estimate $\hat{\mu}$ of the mean.

In order to derive Student's t-distribution, it is convenient to rewrite t as

$$t = \frac{(x-\mu)}{\sigma}\frac{\sigma}{\hat{\sigma}} = \frac{(x-\mu)/\sigma}{\sqrt{(n\hat{\sigma}^2/\sigma^2)/n}} = \frac{z}{\sqrt{\chi^2/n}}, \tag{6.141}$$

where, in the second equality, we introduced $z = (x-\mu)/\sigma$ and $\chi^2 = n\hat{\sigma}^2/\sigma^2$, with n equal to N or $N-1$ as appropriate. By construction, z and χ^2 should be distributed as a standard normal (i.e., $\sigma = 1$) and χ^2-distribution for n degrees of freedom, respectively. The joint probability of observing values z and χ^2 is

$$p(z,\chi^2|n) = \frac{e^{-z^2/2}}{\sqrt{2\pi}} \times \frac{\left(\chi^2\right)^{n/2-1} e^{-\chi^2/2}}{2^{n/2}\Gamma(n/2)}. \tag{6.142}$$

One obtains a distribution for t by integrating all values of z and χ^2 meeting the condition $t = z/\sqrt{\chi^2/n}$:

$$p_t(t|n) = \int d\chi^2 \int dz\, \delta(t - z/\sqrt{\chi^2/n})\frac{e^{-z^2/2}}{\sqrt{2\pi}} \times \frac{\left(\chi^2\right)^{n/2-1} e^{-\chi^2/2}}{2^{n/2}\Gamma(n/2)}, \tag{6.143}$$

$$= \frac{1}{\sqrt{n\pi}}\frac{\Gamma(n+1/2)}{\Gamma(n/2)}\frac{1}{(1+t^2/n)^{(n+1)/2}}. \tag{6.144}$$

Exercises

6.1 Consider an experiment measuring the mass and width of a short-lived resonance based on an invariant mass spectrum. Assume the estimates are normally distributed with a covariance matrix

$$V = \begin{pmatrix} \sigma_M^2 & \rho\sigma_M\sigma_\Gamma \\ \rho\sigma_M\sigma_\Gamma & \sigma_\Gamma^2 \end{pmatrix}.$$

Show that one can establish a symmetric β confidence interval for the mass of the resonance M_0 conditional on its width according to

$$\begin{aligned} \hat{M} - \frac{\rho\sigma_M}{\sigma_\Gamma}\left(\hat{\Gamma} - \Gamma_0\right) - Z_{(1-\beta)/2}\sigma_M\sqrt{1-\rho^2} \le M_0 \\ \le \hat{M} - \frac{\rho\sigma_M}{\sigma_\Gamma}\left(\hat{\Gamma} - \Gamma_0\right) - Z_{(1+\beta)/2}\sigma_M\sqrt{1-\rho^2} \end{aligned} \tag{6.145}$$

where $\hat{M}$ and $\hat{\Gamma}$ are estimates obtained by the experiment for the mass and width of the resonance, respectively; $Z_{(1\pm\beta)/2}$ correspond to the α and β points of the probability of the mass M given by

$$P(Z_\alpha \le Z \le Z_{\alpha+\beta}|\Gamma) = \beta,$$

where

$$Z = \frac{M - \left[M_0 + \frac{\rho\sigma_M}{\sigma_\Gamma}(\Gamma = \Gamma_0)\right]}{\sigma_M\sqrt{1-\rho^2}}.$$

Discuss how this result can be used to constrain the mass M_0 if a prior estimate Γ^* for the width Γ_0 is known from a previous experiment.

6.2 The melting point of two material specimens are determined by actually melting a fraction of each of the specimens. One finds that the first specimen has a melting point of 303 ± 5 K, while the second specimen is determined to have a melting point of 307 ± 4 K. Assuming measurements of temperatures yield Gaussian distributions with the quoted standard deviations, establish if the two samples are likely to be the same substance based on a Gaussian test using a confidence level of 5%.

6.3 A steel company changes the composition and shape of the steel beams it manufactures in order to reduce production costs. However, the company needs to demonstrate that the new beams feature the same quality standards as the older version. If their measurements show that the original beams can withstand a tensile stress of 412 ± 20 MPa[3], and the new beams have a mean strength of 393 ± 18 MPa, would you conclude that the quality of the new beams is necessarily inferior to that of the original beams?

6.4 An experiment searching for proton decays based on a sample of 2,000,000 kg of hydrogen for one year reports the observation of five background events. Calculate the 99% confidence level for the number of proton decays, and determine an upper limit for the half-life of the proton.

6.5 A very high-precision magnetic spectrometer is used to measure the momentum of electrons produced by beta decay of tritium nuclei. Researchers use the end-point of the electrons' momentum spectrum to determine the mass of the neutrino and find a raw mass value of -0.1 eV. Assuming the spectrometer has a Gaussian response with a standard deviation $\sigma = 0.05$ eV, determine a 90% upper limit on the mass of the neutrino using (a) a basic upper limit belt and (b) the unified approach. Also determine the minimum mass the neutrino must have for this experiment to claim a discovery with a 5σ significance level. Compare your estimates to those of the KATRIN experiment [142].

6.6 Consider an experiment for which you must determine the particle detection efficiency as a function of momentum. Assume it is possible to determine the efficient

[3] MPa = Mega-Pascal.

based on the embedded track technique. Use the table below to estimate the detection efficiency and its error using (a) a simple Gaussian approximation and (b) the Feldman–Cousins ordering technique.

$p_\perp$ range	Embedded	Found
0.5–1.0	10,000	9,945
1.0–2.0	9,010	8,456
2.0–3.0	7,890	7,333
3.0–5.0	5,550	5,211
5.0–10.0	4,442	4,012

6.7 An experiment produces 100 data points, which you proceed to fit with linear and quadratic models. You obtain a $\chi^2/\text{DoF} = 2.5$ with the linear fit whereas the quadratic fit yields $\chi^2/\text{DoF} = 2.1$. Can you reject the linear hypothesis? Can you accept the quadratic hypothesis?

7 Bayesian Inference

Concepts and tools commonly used by classical statisticians and scientists in their work, discussed in detail in the previous three chapters, may appear disconnected, and at times, somewhat arbitrary. This uneasy feeling stems largely from the very definition of the frequentist notion of probability as a limiting frequency. By contrast, within the Bayesian paradigm, data are regarded as facts, and it is the models and their parameters, globally regarded as hypotheses, that are given a probability. Indeed, the capacity to assign a degree of plausibility to hypotheses, and the integration of probability as logic enables the elaboration of a systematic and concise inference program that is easy to follow (if not to calculate) and is well adapted to the scientific method.

The basic structure and components of the Bayesian inference process are defined and summarized in §7.2. The following section, §7.3, presents a discussion of technical matters associated with choices of prior probability functions as well as probability models. Basic examples of Bayesian inference in the presence of Gaussian noise, including Bayesian fits with linear models (i.e., models that are linear in their parameters), are discussed in §7.4, while Bayesian fitting in the presence of non-Gaussian noise and nonlinear models is introduced in §7.5. A variety of numerical techniques amenable to the calculation of the mode of posterior probabilities and χ^2 minimization are next discussed in §7.6. Finally, §7.7 presents an introduction to Bayesian model comparison and entity classification with two simple examples of commonly encountered model selection problems.

7.1 Introduction

Thanks to tremendous developments since the mid-twentieth century, frequentist statisticians, and thus by extension all scientists, are now equipped with a rather sophisticated arsenal of tools adapted to a great variety of measurements and inference problems. However, many Bayesian statisticians would argue that the frequentist methods are typically nonintuitive and form a collection of somewhat arbitrary and seemingly disconnected techniques. They would further argue that a more intuitive and better integrated approach to inference is needed and, in fact, possible within the Bayesian paradigm, and most particularly within the context of the interpretation of probability as an extension of logic. While the goals of Bayesian inference are not altogether different from those of classical inference, the Bayesian interpretation of the notion of probability as a level of plausibility (or degree of belief) does in fact enable a more intuitive, systematic, and well-integrated inference process. Indeed, the fact one can associate a prior degree of plausibility to statements or hypotheses about the real world and determine the posterior degree of plausibility of said hypotheses based on data greatly expands the realm of inference and enables a more satisfactory set of procedures, which is arguably better adapted to the scientific method.

The practical difference between frequentist and Bayesian inferences is that a treatment of statistics based on subjective probability enables statisticians to express model parameters and their uncertainties in terms of probabilities. This cannot actually be done in the frequentist approach. For instance, frequentist 95% confidence intervals imply that for repeated measurements, 95% of confidence intervals derived from data would actually cover the parameter being measured. But beyond that point, the notion of probability is not applicable in the frequentist approach. By contrast, Bayesian statistics and inference can compare hypotheses directly and choose (infer) which is best supported by the measured data. Frequentists cannot use a probability in this fashion and must instead consider long-run frequency as extreme or in excess of the observed values, which is intuitively somewhat unclear.

All said, Bayesian statistical inference can be decomposed into three types of considerations and classes of techniques. They include

1. Determination of the posterior probability of a model and its parameters based on measured data
2. Characterization of the posterior to estimate (model) parameters and their errors, as well as compare models
3. Calculation of predictive distributions

By contrast, frequentist inference is focused on the data, which it regards as instances drawn out of a sample space, and its main components include the formulation of a probability model to interpret the data, the calculation of the likelihood to distill key features of the data or model parameter values, and comparison of a null hypothesis with competing hypotheses (hypothesis testing). These three components may obviously be regarded as "special cases" of the broader and more comprehensive Bayesian inference process. Distillation of key features of the data includes, in particular, the estimation of model parameters and associated confidence intervals of interest in either approaches of probability, as is the need to test hypotheses. The Bayesian approach, however, provides a more robust and comprehensive framework for such studies. We will thus endeavor to discuss all aspects of the Bayesian inference process in detail throughout this chapter.

The centerpiece of these considerations is the determination of the posterior probability of one or several hypotheses based on measured data. That involves the identification of a suitable data probability model to describe the measurement process (e.g., fluctuations of measured observables), the choice or evaluation of a prior probability for the physical model and its parameters, the determination of a likelihood function based on the probability model, the measured data and various assumptions about the measurement process, and, finally, a straightforward use of Bayes' theorem. Once the posterior is known, one can then make specific statements about the observables (discrete or continuous hypothesis) or the parameters of the model. Such statements about the posterior, commonly known as **summaries** by Bayesian statisticians, include the determination of the most likely values of the parameters (i.e., the mode of the distribution) as well as the evaluation of credible ranges for the model parameters of interest. Additionally, given the posterior, it is also possible to make predictive inferences and calculate the probability of other observables of interest or predict the outcome of further measurements (predictive inference).

While many scientists still work within the frequentist paradigm and make regular use of "frequentist techniques" to analyze their data, a growing number are embracing the Bayesian paradigm and base their statistical analyses on the subjective interpretation of probability and Bayes' theorem. It was thus tempting to focus this textbook entirely on the most recent developments in Bayesian statistics. We feel, however, that it would be unfair to the young men and women beginning their training as scientists given that a very large fraction of works published, even today, are rooted in the frequentist paradigm or, at the very least, make use of frequentist terminology. It is thus important for young scientists to be equipped with a minimal frequentist toolset so they can understand past works in their field as well as the ongoing debate between Bayesians and frequentists. This said, it is also clear that the framework provided by probability as logic and Bayesian inference enables a robust and well-structured approach to data analysis including both deductive and predictive inferences, which might eventually take precedence over classical inference. There is in fact little doubt that Bayesian inference will steadily grow in popularity within the scientific community over the next decades. It is thus then essential to also include a sound discussion of Bayesian inference.

7.2 Basic Structure of the Bayesian Inference Process

Bayesian inference is defined in the context of Bayes' theorem given by Eq. (2.17) and centrally relies on the Bayesian interpretation of probability. It operates in a hypothesis (or model) space where specific hypotheses, whether discrete or continuous, are associated a probability expressing their plausibility. Measured data, considered as given (facts), are used to update the degree of belief in model hypotheses, and whenever relevant, discriminate between competing hypotheses or models.

Bayesian inference involves the following components:

1. Determination of the posterior probability of a model and its parameters based on measured data:
   a. Formulation of a data probability model based on prior information
   b. Formulation of a prior probability (density) for a working hypothesis of interest
   c. Calculation of the likelihood of the measured data based on the hypothesis
   d. Calculation of the posterior probability of the working hypothesis with Bayes' theorem
2. Distillation of key features of the posterior distribution:
   a. Determination of the mode or expectation value(s) of the posterior relative to the parameter(s) of interest
   b. Calculation of credible ranges for the parameter(s) of interest
3. Comparison of the leading hypothesis with competing hypotheses (hypothesis testing)

4. Calculation of predictive distributions corresponding to the outcome of prospective measurements

Two important and related points are worth stressing. First, as a general scientific rule, the measurement and inference processes should be kept distinct. The gathering of "raw" data must be carried out without interference or biasing from the inference process, that is, the answer obtained by inference should not influence or bias the sampling process. The data acquisition and analysis process must thus formally be considered as distinct and carried out in such a way that one does not affect the other.[1] Second, and for essentially the same reasons, the formulation of a prior probability, the experimental process, and the calculation of a posterior should also be considered as distinct and carried out independently. More specifically, the formulation of the prior should be completed "before" the experiment, or at the very least, without knowledge of the data produced by the experiment. As we will discuss in more detail later in this section, it is the likelihood function, determined by the data, that should influence the posterior. Having the data also influence the prior probability would amount to double counting and is thus scientifically unwarranted.

We first briefly describe each of the aforementioned components of the inference process in the following paragraphs of this section. We then elaborate on selected topics in the following sections of this chapter. Evidently, the main goal of Bayesian statistics is the determination of the posterior probability density or distribution function (PDF) of a hypothesis of interest, which may be viewed as weighted average between the prior PDF and the likelihood function. Once the posterior PDF is known, several basic features of interest can be readily computed, such as the most probable value of a continuous hypothesis, a confidence interval, as well as the plausibility of the hypothesis compared to other hypotheses or models.

7.2.1 Probability Model of the Data

Scientific models are typically formulated to describe properties of systems or entities and their relation with other systems or entities. For instance, in classical mechanics, Newton's third law embodies the notion that the acceleration of an object is strictly proportional to the net force applied on it by external forces. Deterministic as it may be, Newton's law says little about the specificities of measurements that may be carried out to test it. Devices or sensors may be used to determine the net force and the acceleration, or these quantities may alternatively be determined from other measurements such as the compression of a spring (force) and the time evolution of the speed of the object or measurements of its position vs. time, and so on. Each of these measurements may in turn be affected by external conditions, instrumental effects, and so on. Consequently, in addition to the (physical) model of interest, one also needs a measurement model that describes how it is measured, and how it might vary, measurement by measurement. In effect, this requires the formulation

[1] An obvious exception to this rule is the need to monitor data acquisition to ensure all components of a complex apparatus are performing properly.

of a **data probability model** describing the probability of outcomes of the measurement process.

Formulation of a Probability Model

A data probability model embodies the stochastic nature of the measurement process and describe the probability $p(x|x_0)$ of measuring specific values x given the true value x_0 of an observable X. A probability model is formulated on the basis of prior information about the system, which is either known or assumed to be true with a certain degree of belief. Imagine, as a specific example, that you run a nuclear laboratory and that you are given the task of measuring the lifetime of some radioactive compound embedded in a small material sample. The lifetime of a nuclear isotope is evidently not measurable directly. But, if it is reasonable to assume the sample is pure, that is, composed of a single radioactive element, then one expects the activity (rate of decay) of the sample to follow a decreasing exponential with a "slope" determined by the lifetime of the compound. The lifetime may then be determined based on a measurement of the activity of the sample vs. time, or explicit measurements of decay times taken over an extended period of time. A number of complications may obviously arise in the measurement: the compound may not be perfectly pure, and the measurement precision of the activity or decay times shall evidently be limited by the quality of the instrumentation as well as the experimental conditions of the measurements. A proper description of the probability model of the measurement may thus require the inclusion of a smearing function to account for the finite resolution of the measurement and some form of background distribution to reckon for sample impurities which contribute weak but finite activity, and so on.

The formulation of a probability model obviously depends on the specificities of the measurement being considered and is perhaps best described with the detailed examples we discuss in §7.4. However, at the outset, it is useful to briefly discuss the key steps involved in the formulation of such a model. The basic idea is to formulate all information, I, relevant to a system in terms of logical propositions, either known to be true, or whose degree of belief or plausibility can be expressed in terms of a probability distribution or probability density. Reasoning based on these propositions shall then indicate how individual measurement instances will behave in practice. One may then model the probability of an aggregate of n independent measurements in terms of the individual probabilities of each of the measurements.

For illustrative purposes, let us consider a particular model, I, that stipulates that the outcome y of a specific measurement has a PDF $p(y|\theta)$, where θ represents one or several model parameters. The probability of observing values $\vec{y} = \{y_1, y_2, \ldots, y_n\}$ in a series of n *independent* measurements, written $p(\vec{y}|\theta)$, shall then be proportional to the product of the individual probabilities of each of the measurements:

$$p(\vec{y}|\theta) \propto p(y_1|\theta) \times p(y_2|\theta) \times \cdots \times p(y_n|\theta). \tag{7.1}$$

Let us consider two simple examples illustrating the calculation of the probability $p(\vec{y}|\theta)$: the first, presented in §7.2.1, is based on the Bernoulli distribution, while the second, discussed in §7.2.1, makes use of the exponential distribution.

Example 1: Application of the Bernoulli Distribution

Let us first revisit the example discussed in §2.2.2 involving the manufacturing and testing of auto parts by supplier 10P100Bad. Consider that a sample of $n = 500$ manufactured parts are to be examined for defects. Let the outcome y_i of each observation (i.e., each part) be 1 if a defect is found, and 0 otherwise, for $i = 1, \ldots, n$. Let θ represent the probability that a randomly selected part might be defective. Each observation may then be represented with the Bernoulli distribution (see §3.1 for a formal definition of the distribution):

$$p(y_i|\theta) = \theta^{y_i} (1 - \theta)^{1-y_i}, \quad \text{for } i = 1, \ldots, n. \tag{7.2}$$

The sampled data $\vec{y} = \{y_1, y_2, \ldots, y_n\}$ thus has a probability proportional to each of the $p(y_i|\theta)$ of Eq. (7.2), and one obtains the probability model:

$$p(\vec{y}|\theta) \propto \prod_{i=1}^{n} \theta^{y_i} (1 - \theta)^{1-y_i}, \tag{7.3}$$

$$= \theta^{\sum_{i=1}^{n} y_i} (1 - \theta)^{n-\sum_{i=1}^{n} y_i}, \tag{7.4}$$

$$= \theta^{n\bar{y}} (1 - \theta)^{n(1-\bar{y})}, \tag{7.5}$$

where we introduced the arithmetic mean $\bar{y} = \frac{1}{n} \sum_{i=1}^{n} y_i$ of the measured values. The probability model amounts to the likelihood of the observed data given a hypothesis θ, which when combined with a prior probability, as we discuss in more detail later in this chapter, determines the posterior PDF of the (physical) model parameter θ.

Example 2: Application of the Exponential Distribution

Let us formulate in concrete terms the example mentioned in the beginning of this section concerning a measurement of the lifetime of an unknown compound. As discussed in detail in §3.5, one can model the probability $p(t|\tau)$ of a radioactive decay at time t in terms of a decreasing exponential with a parameter τ, corresponding to the lifetime of the nucleus under investigation, as follows:

$$p(t|\tau) = \frac{1}{\tau} \exp\left(-\frac{t}{\tau}\right). \tag{7.6}$$

Observations of n decays at times $\vec{t} = \{t_1, t_2, \ldots, t_n\}$, may then be modeled according to

$$p(\vec{t}|\tau) \propto \frac{1}{\tau^n} \prod_{i=1}^{n} \exp\left(-\frac{t_i}{\tau}\right), \tag{7.7}$$

$$= \frac{1}{\tau^n} \exp\left(-\frac{n\bar{t}}{\tau}\right), \tag{7.8}$$

where $\bar{t} = \frac{1}{n} \sum_{i=1}^{n} t_i$ is the mean of the measured decay times. Here again, the probability model $p(\vec{t}|\tau)$ corresponds to the likelihood of the measured decay times $\vec{t} = \{t_1, t_2, \ldots, t_n\}$ given τ. It must be combined to the prior probability of τ to determine its posterior PDF.

Probability Models in Real Life

In general, observations may involve a number of complications associated with instrumental effects and the presence of background processes. The basic principle of the formulation of a probability model nonetheless remains the same: all relevant information about the system must be explicitly stated and encoded to determine the probability of specific datasets. For instance, in the case of the measurement of radioactive decays, one might have to account for the precision of the time measurements. Assuming, for instance, that the timer used in the measurement exhibits Gaussian fluctuations with a standard deviation σ_t, one may then write the data probability model of each time measurement t as an integral of the product of two PDFs: an exponential to account for the nature of the decay process and a Gaussian to take into account the smearing imparted by the finite resolution of the timer (clock) used in the measurement.

$$p(t|\sigma_t, \tau) = \int_{-\infty}^{\infty} p_{\text{Gaus}}(t|t', \sigma_t) \times p_{\text{Exp}}(t'|\tau)\, dt', \tag{7.9}$$

$$= \int_{-\infty}^{\infty} \frac{1}{\sqrt{2\pi}\sigma_t} \exp\left[-\frac{(t-t')^2}{2\sigma_t^2}\right] \frac{1}{\tau} \exp\left(-\frac{t'}{\tau}\right) dt'. \tag{7.10}$$

The method can be readily extended to joint measurements of two or more observables determined by several model parameters, as we illustrate through several examples later in this chapter.

As we saw in §2.2.4, the introduction of the notion of probability as logic naturally enlarges the scientific discourse to include the probability of models and their parameters. In this context, the choice of a data probability model implicitly involves the formulation of one or several hypotheses about the measurement process, including the choice of a particular PDF to describe fluctuations of the measurement outcome, and parameters of the PDFs. The parameters of the data model PDF may be unspecified a priori and must then be either measured explicitly, or marginalized, as appropriate in the context of a specific measurement. Detailed studies either through actual measurements or computer simulations may then be used to establish the plausibility of the measurement model. Clearly, scientists often assume, based either on their experience or understanding of the measurement process, that the model they use is appropriate, that is, has a large plausibility of properly representing the measurement process. While such an assumption may be viable in most experimental contexts, it may have to be examined in detail if the level of accuracy and precision required of a measurement is very large. As an example, consider that the use of a Gaussian distribution to describe an observable's fluctuations is perfectly adequate as long as one ignores very rare and large fluctuations. By construction, a Gaussian has finite probability density extending to infinity (on either sides of the mean). Since infinities are not possible in practice, the Gaussian model is thus obviously flawed for very large fluctuations and must thus be replaced by a more appropriate PDF whenever extreme deviations (e.g., larger than 5σ) from the mean are considered of interest.

The preceding discussion illustrates that the Bayesian inference framework naturally enables discussions of hypotheses about the measurement process and its modeling. Consideration of the plausibility of a measurement model may thus be naturally and explicitly included in the calculation of physical model parameters. Such inclusion is not readily possible or feasible, strictly speaking, in the context of the frequentist approach.

7.2.2 Formulation of Prior PDFs

Use of Bayes' theorem toward the determination of a posterior probability of a model hypothesis H evidently requires assumptions about the prior probability distribution (or density) of this hypothesis. In many cases, prior distributions may be derived from the posterior distribution of previous experiments. It is indeed the very nature of the scientific process that knowledge begets knowledge and foundational experiments may be used to produce informed guesses on parameters or model hypotheses motivating additional experiments. Prior probabilities may then be assigned directly from the posterior of previous experiments, or through the use of well-established models or frameworks. However, there are also plenty of scientific cases in which no prior experimental data exist or little information is available to construct a prior distribution. One may then establish a prior based on some general guiding principles, if any, or admit total ignorance and encode one's ignorance as a prior that considers all competing hypotheses as equally probable.

Broadly speaking, there are thus two classes of approaches discussed in the literature toward the formulation of prior distributions: they yield prior distributions commonly known as **informative** and **noninformative**. Informative priors, as the name suggests, are meant to convey and encode whatever substantive knowledge may already exist about a particular system, whether derived from other data or by reasonings based on well-established scientific principles. By contrast, noninformative priors are designed to encode one's basic ignorance of a system or a specific set of hypotheses.

Critics of the Bayesian approach argue that there exists too much freedom in the choice of priors, and that as such, the notion of prior introduces a large degree of arbitrariness and subjectivity in the inference process. Although this is true to a degree, many a Bayesian statistician would counteract that a certain level of arbitrariness also exists in frequentist procedures, and more specifically, in the choice of probability models. Supporters of the Bayesian interpretation would also argue that it would be unscientific to neglect or ignore prior information pertaining to a model that can, for instance, narrow the range of plausibility of a given continuous hypothesis. Indeed, neglecting well established facts or (physics) principles would seem rather unsound a scientific approach, much like rejecting data points because they might not support one's preconceived ideas. Bayesians would additionally argue that what really matters is that equivalent states of prior knowledge properly encoded into prior probability should lead to equivalent conclusions. This means that statisticians implementing the same prior knowledge independently, and having access to the same likelihood function, should reach equivalent posteriors and obtain similar conclusions. While such a statement can be made mathematically correct if all statisticians implement prior knowledge with the same functions and parameters, the choice of priors, most particularly uninformative priors, has remained somewhat of a contentious issue for quite some time. This apparent excess of freedom has generated quite a bit of discussions and arguments among statisticians, and much research has gone into the elaboration of techniques that enable sound choices of priors.

Given a lack of prior knowledge, the goal is to represent the ignorance of a parameter before it is measured in an objective and self-consistent manner. It would seem sensible to choose a noninformative prior distribution that encodes this ignorance by assigning equal probabilities to all possible values of a discrete parameter, or a uniform (defined in §3.4) probability density to all values of a continuous parameter.

$$p(\theta|I) = \text{constant}; \qquad \text{Uniform prior.} \tag{7.11}$$

Indeed, one could seemingly argue there is little arbitrariness involved in assigning a constant value to a noninformative prior. Unfortunately, the issue is not so simple, and there are in fact several distinct criteria that may be used to build a noninformative prior based on identical data models and phenomenological contexts. Arguably, these distinct criteria produce distinct priors that may thus lead to different conclusions, numerically, in the assignment of modes, expectation values, error intervals, and hypothesis testing. This simplistic approach may thus lead to troubling inconsistencies.

A central issue is that in many statistical analyses, one has a rather vague prior knowledge of the parameters (continuous hypotheses) of a model. But from a physical perspective, one should expect that the choice of units or scale of a parameter, in particular, should have no bearing on the inference process and the outcome of a statistical analysis. More specifically, consider for instance that if a model formulated in terms of a parameter θ is transformed to depend on a parameter $\rho \equiv \ln\theta$, one should expect the prior distributions of θ and ρ to carry the same information. This is obviously quite problematic, however, because the logarithmic scaling entails that it is, by construction, impossible for both parameters to be characterized by a uniform probability density. Indeed, if θ is chosen to have a uniform prior density, the prior distribution of ρ cannot be uniform, and conversely. The question arises, then, as to which of the two variables, θ or ρ, should be given a uniform prior. One might also wonder how to best choose the parameters that should be uniform. For instance, while using a normal distribution model (§3.9), should the standard deviation or the variance be considered more fundamental and thus assigned a uniform prior? Does the question actually make sense? Indeed, is it meaningful to express lack of knowledge by stating that the width of the normal distribution could be infinitely small ($\sigma = 0$) or infinite? To make matters worse, one should also acknowledge that arbitrarily many forms of scaling might be possible and of interest. Which expression of a physical quantity can thus be deemed most fundamental and given a uniform prior?

Although there does not exist a universal solution to this difficult question, statisticians have developed a number of "rules" to formulate prior probabilities applicable in specific contexts. We explore some of these rules in §7.3. At this point, let us just state that Bayesian statisticians often adopt a strategy that, for instance, involves the construction of nearly (but not perfectly) uniform distributions as noninformative priors. Distributions are chosen such that variations among a set of relatively flat priors should not grossly affect the outcome of the analysis. But if it does, in practice, it is likely an indication that the parameters of interest are in fact rather poorly constrained by the data (likelihood function). Indeed, it should be the data (i.e., measurements of the real world) that constrain model parameters, not a scientist's pre- or misconceptions!

7.2.3 Posterior PDF Calculation

Given a prior distribution, $p(H_i|I)$, for a hypothesis H_i and the likelihood function of measured data, $p(D|H_i, I)$, the posterior probability of the hypothesis, $p(H_i|D, I)$, is readily calculated according to Bayes' theorem:

$$p(H_i|D, I) = \frac{p(H_i|I)p(D|H_i, I)}{p(D|I)}, \tag{7.12}$$

where $p(D|I)$ is the probability of the data given prior information I about the system. While it can in principle be calculated on the basis of the law of total probability,

$$p(D|I) = \int p(H_i|I)p(D|H_i, I)\, dH_i, \tag{7.13}$$

the factor $p(D|I)$ may often be regarded as a normalization constant, which becomes unnecessary when, for instance, computing the ratio of the probability of two hypotheses (§7.7).

We consider practical cases of inference and calculation of posteriors in §§7.4, 7.6, and 7.7.

7.2.4 Posterior PDF Characterization (Summaries)

Bayesian Estimators

The posterior PDF $p(H_i|D, I)$ nominally embodies everything there is to know about a hypothesis H_i based on measured data, D, and prior information, I, about the system. By all accounts, it is the Bayesian estimator of an observable. Although a rather wide spectrum of types of hypotheses may be formulated and put to the test, one is most often concerned with continuous hypotheses $H(\theta)$ about the value of a system or model parameter θ. It is frequently the case that θ might represent a single or specific value (e.g., a constant of nature), but the measurement process and instruments invariably produce smearing effects. The outcome of a measurement of such a parameter is thus typically a posterior probability density with a finite spread across the nominal domain of θ. This finite spread is merely an artifact of the measurement process and not an intrinsic characteristic of the parameter itself. For instance, repeated measurements of the Planck constant or Big G would naturally produce distinct values that gather around the actual values of these constants. The dispersion of values arise from the measurement process, not from fluctuations of these physical parameters over time.[2] One then wishes to use the calculated posterior density $p(H_i|D, I)$ to extract a best estimate of the value of the physical parameter θ. Conceptually, the task of the Bayesian statistician thus becomes rather similar to that of classical statisticians: that of obtaining an optimal estimate of the parameter value and evaluating what might be the error on that value, that is, estimate how far the extracted value might lie from the actual

[2] In the context of commonly accepted and vetted physical theories, there is no reason to believe such variations or fluctuations might occur.

value. Important differences exist, however, in the manner in which Bayesian and classical statisticians might carry out this task. Classical statisticians have no use for a prior or a posterior. Their parameter estimation, as discussed in Chapter 4, is based on a choice of statistics (an estimator actually) in which they plug in measured data values, and while they do need a probability model to calculate the likelihood of the data, the model parameter themselves are not given a probability. Bayesian statisticians, on the other hand, wish to use prior information to constrain the estimation process; they thus carry out their estimates based on posterior probabilities that combine prior information, that is, prior probability distributions of model parameters, as well the likelihood function of the data.

The distinction between frequentist and Bayesian parameter estimation extends well beyond the numerical techniques used to obtain the estimates. In the frequentist interpretation, the probability of a given value represents the frequency with which the value would be obtained if the measurement could be repeated indefinitely. For instance, a 95% frequentist confidence interval (discussed in §6.1) corresponds to a range that should cover the actual value of the parameter 95% of the time (i.e., in 95% of measurement instances), whereas in the Bayesian approach, the true value is actually believed to be within the range with a probability of 95%.

Important differences exist also in the manner in which hypotheses are to be tested. Hypothesis testing in the frequentist paradigm, discussed in §6.4, relies on the rather unintuitive notion of a test statistics exceeding a preset threshold. Indeed, recall that one should reject a hypothesis if its test statistics has a value equal or larger than a specific value determined by the (chosen) significance level of the test. In contrast, in the Bayesian approach, one computes and compares the odds of the hypotheses themselves, that is, their respective probabilities. A particular hypothesis can then be adopted (rejected) if its probability is appreciably larger (weaker) than that of competing hypotheses.

It is important to stress once again that there could be no scientific method without proper assessment of measurement errors. A sound discussion of methods to assess the magnitude of experimental errors is thus paramount. This is not an easy topic, and several aspects of the problem had to be drawn before rigorous methods of error assessment could be considered approachable by beginning students. This is why we used a staged approach and introduced the notion of error as well as error propagation, confidence intervals, and so on, in several steps beginning already in Chapter 2, §2.11, with an appeal to intuition and development of probability distributions in series. The notion of error was formalized in Chapter 4 through the introduction of the variance of estimators and in Chapter 6 with a discussion of classical confidence intervals. In this chapter, we revisit the notion of error in the context of the Bayesian interpretation of probability and introduce the notion of **credible range** based on hypothesis posteriors.

Mode of the Posterior PDF

As already stated in the previous section, a posterior PDF $p(\theta|D, I)$ embodies everything there is to know about the model parameter θ based on the measurement D, and as such constitutes a complete Bayesian estimator of θ. But if θ represents an observable with a single value (or at the very least is believed to have a single and unique value), one shall wish to extract a specific value $\hat{\theta}$ that is most likely to be the true value of θ. Given the posterior probability density $p(\theta|D, I)$ represents the degree of belief that the true value of θ be found in the interval $[\theta, \theta + d\theta]$, it then makes sense to seek and report the mode of this function, that is, the value of θ with the largest probability density, as the best estimate of the parameter or observable. Evidently, if $p(\theta|D, I)$ is a symmetric function, the mode (extremum of the function) shall also correspond to the expectation value of θ. In general, however, the expectation value of θ shall not have the largest probability and thus should not be reported as the value of the observable.

Finding the extremum of a PDF $p(\theta|D, I)$ is in principle straightforward whenever it is obtained from an analytical data probability model and a conjugate prior since, by construction, it is also analytical and a member of the same family of functions as the prior. In general, however, the posterior may be obtained from a nonlinear parametric likelihood function, it may involve several "fit" parameters, or its prior PDF might have been generated by Monte Carlo simulations and thus available in the form of a histogram: finding an extremum of $p(\theta|D, I)$ may then be somewhat arduous and require use of numerical techniques. We discuss various examples of analytical cases in §§7.3.3 and 7.4, while basic principles of extremum searches based on numerical techniques are presented in §7.6.

Credible Range

Almost invariably, a posterior $p(\theta|D, I)$ shall have a finite spread across the domain of the parameter θ. While one might wish for a very narrow distribution or even an infinitely narrow distribution (i.e., a delta function $\delta(\theta - \theta_0)$), the measurement process and instrumental effects invariably lead to a distribution with a finite dispersion. The mode has the highest probability (density) but a range of other values have a finite probability (density), as well, of corresponding to the true value. It is thus meaningful to define a range of values as a **credible range** (CR) within which the value of θ is most likely to fall. Evidently, as for confidence intervals, the breadth, or width, of a credible range shall be defined by its probability content or confidence level α according to

$$\int_{\text{CR}} p(\theta|D, I)\, d\theta = \alpha, \tag{7.14}$$

and there exists an infinite number of ranges CR that satisfy this condition.

Given $p(\theta|D, I)$ represents the degree of belief the observable has a value in the range $[\theta, \theta + d\theta]$, it is most meaningful to include portions of the PDF that have largest values. The credible range must then include the mode and adjoining regions of the domain of θ where $p(\theta|D, I)$ is the largest. For a symmetric distribution, this naturally leads, as for confidence intervals, to the selection of a **central interval** $[\theta_{\min}, \theta_{\max}]$ defined according to

$$\begin{aligned} \int_{-\infty}^{\theta_{\min}} p(\theta|D, I)\, d\theta &= (1-\alpha)/2, \\ \int_{\theta_{\max}}^{\infty} p(\theta|D, I)\, d\theta &= (1-\alpha)/2, \end{aligned} \tag{7.15}$$

and such that $\theta_{\min}$ and $\theta_{\max}$ are at equal distances from the mode of the distribution. While central intervals are also commonly applied for the selection of confidence intervals in the case of asymmetric PDFs, the desire to include elements of the domain of θ with the largest probabilities leads to a different range construction algorithm. Indeed, it appears more sensible to proceed similarly to the Feldman–Cousin algorithm (§6.1.8) and include elements of the domain of θ starting with the mode and with decreasing amplitude thereafter, as illustrated in Figure 7.1a. This then also leads to the shortest interval $[\theta_{\min}, \theta_{\max}]$. One obvious drawback of this approach, however, is that the interval is not invariant under a transformation $\theta \to \xi(\theta)$; that is, an interval in ξ defined with the same algorithm would not map onto the interval for θ.

Use of the posterior $p(\theta|D, I)$, rather than the likelihood distribution $p(D|\theta, I)$, to identify an error interval presents two very important advantages. First, no interval inversion of the type used in the frequentist approach (and discussed in §6.1.8) is required since the "inversion" is already embodied in the posterior. And second, the presence of a physical boundary is handled gracefully. For instance, in a measurement of the a priori unknown mass of a particle (e.g., the neutrino mass), one may use an uninformative and improper prior PDF of the form

$$\begin{aligned} p(m|I) &= 0 \quad \text{for } m < 0, \\ p(m|I) &= 1 \quad \text{for } 0 \le m \le M, \end{aligned} \tag{7.16}$$

where M is a suitably large constant. If the detector response is Gaussian with a precision ξ_0, and given the aforementioned prior, one can show that the posterior (§7.4.1) has the form of a truncated Gaussian distribution

$$\begin{aligned} p(m|D, I) &= 0 && \text{for } m < 0, \\ p(m|D, I) &= N_m \sqrt{\frac{\xi_0}{2\pi}} \exp\left[-\frac{\xi_0}{2}(m - \hat{m})^2\right] && \text{for } 0 \le m \le M, \end{aligned} \tag{7.17}$$

where $\hat{m}$ is the uncorrected mass reported by the measurement, and N_m is a normalization constant equal to

$$N_m = \left(\sqrt{\frac{\xi_0}{2\pi}} \int_0^M \exp\left[-\frac{\xi_0}{2}(m - \hat{m})^2\right]\right)^{-1}. \tag{7.18}$$

Because of instrumental effects, the measured mass $\hat{m}$ may be negative. However, the notion that the mass of a particle cannot be negative is ab initio built in the posterior by virtue of the prior, Eq. (7.16). One can then use the probability density sorting method mentioned previously to establish a credible range for the measured mass: if $\hat{m}$ is large and far exceeds the mass resolution σ, one will naturally obtain a symmetric central interval, as illustrated in Figure 7.1b. However, if $\hat{m}$ is very close to the physical boundary, that is, is smaller or of the order of the mass resolution, then the sorting procedure will automatically yield a credible range with a lower bound $m = 0$ and an upper bound $m_{\max}$ (effectively a one-sided interval) determined by the precision ξ_0 as well as the probability content α, as schematically illustrated in Figure 7.1c. One thus readily avoids technical issues of flip-flopping, lack of coverage, and empty confidence intervals, encountered in the frequentist approach (§6.1.8).

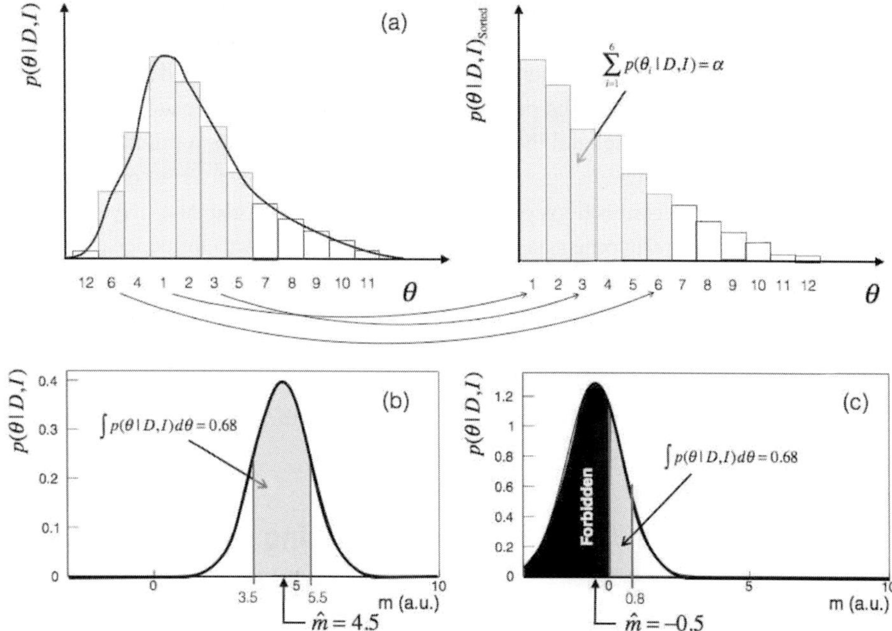

Fig. 7.1 (a) Schematic illustration of the sorting technique used to build credible ranges with the largest probability densities. (b) Central credible region associated with a signal far from physical boundaries. (c) One-sided credible region obtained near a physical boundary (mass measurement).

7.2.5 Predictive Data Distributions

The Bayesian approach readily provides a framework for the computation of predictive distributions. Indeed, suppose a measurement has yielded a dataset with values $\vec{y} = (y_1, \ldots, y_n)$ resulting in the evaluation of posterior probability $p(\theta|\vec{y}, I)$ for the model parameter θ. One may then use this posterior probability as the prior of a new experiment to predict the distribution of values $\vec{y}' = (y'_1, \ldots, y'_n)$ that might be obtained in that measurement. This is known as a **Posterior Predictive Distribution** (PPD) and denoted $p(\vec{y}'|\vec{y}, I)$. It can be regarded as the probability of new data $\vec{y}'$ given old data $\vec{y}$. In this context, the value of the model parameter θ might be considered of minor or no interest, and the sole purpose purpose of the procedure might then be the prediction of future data, or more aptly, the distribution of new data.

The PPD is readily calculated on the basis of the posterior probability of the model parameter(s), $\vec{\theta}$, and the likelihood function according to

$$p(\vec{y}'|\vec{y}, I) = \int p(\vec{y}'|\theta)p(\theta|\vec{y}, I)\, d\theta, \tag{7.19}$$

where the integration is taken over the entire domain of the model parameter(s) $\vec{\theta}$.

Although the expression $p(\vec{y}'|\vec{y}, I)$ might suggest some form of causal connection between the old and the new data, it should be clear that no mechanism for such connection is actually implied. The value of $\vec{y}'$ is not caused by $\vec{y}$, but the knowledge of what it could become is. The PPD of $\vec{y}'$ is **conditioned** by knowledge of the parameter $\vec{\theta}$ deduced from the old data $\vec{y}$. In other words, it becomes possible to effect a prediction of the PDF of new data because the data (old and new) can be described by a model. The old data condition the model and the model so conditioned makes a prediction of the outcome of new measurements.

Causal connections or new data conditioned by old data may occur if the existence of a particular outcome y has an explicit influence (causal) influence on future outcomes y'. Accounting for such connections (conditioning) is possible but requires proper knowledge of the conditioning mechanism. The formalism of Kalman filters, discussed in §5.6, provides a convenient framework for calculations of the evolution of $p(\vec{y}'|\vec{y}, I)$ for cases in which causal connections exist between the old and the new data and can be described based on a model of the evolution of the parameters $\vec{\theta}$ with time or some other independent (ordering) parameter.

7.3 Choosing a Prior

Calculation of the posterior probability $p(H|D, I)$ of a hypothesis H with Bayes' theorem requires a prior probability $p(H|I)$ be formulated before the data are processed (or become available) and the likelihood $p(D|H, I)$ calculated. A difficulty often arises that only a rather limited amount of prior information might be available about a particular hypothesis and the system it describes. In some cases, the prior information available may be sufficient to select an **informative prior**, that is, a definite probability density with specific parameters (e.g., a beta distribution with well informed parameters α and β), but in most cases the amount of information is too limited and, at the outset, the selection of a prior might appear as an impossible challenge. Indeed, what functional dependence should one use? Should one limit or bias the range of a model parameter? And so on. One is then compelled to choose an **uninformative prior**, that is, a prior distribution that actually carries little information about the parameter or observable of interest.

Implementing an arbitrary functional form with ad hoc model parameter boundaries constitutes a rather unsatisfactory course of action toward the selection of an uninformative prior. Fortunately, two foundational methods, guided by simple yet profound principles regarding the information carried by probability distributions, enable rational as well as practical choices of distributions. The first method was devised by Edwin Jaynes (1922–1998)[3] and relies on Shannon's entropy measure of uncertainty, while the second method, credited to Sir Harold Jeffreys (1891–1989)[4], relies on the notion of Fisher information already discussed in §4.7. In actual fact, a number of other methods and principles are discussed and used in the literature for the definition and implementation of prior probabilities.

[3] American physicist famous for his work on statistical mechanics and the foundation of probability and statistical inference.

[4] English mathematician, statistician, geophysicist, and astronomer who played an important role in the revival of the Bayesian view of probability.

Indeed, the topic of prior is still somewhat contentious and much in flux. Regardless of such concerns, we will restrict our discussion to Jaynes' method, in §7.3.1, and Jeffreys' definition of the prior based on the Fisher information of a probability distribution in §7.3.2. Readers should consult more advanced texts for a fully comprehensive discussion of this difficult topic. In particular, see the book by Harney [103] for an extensive discussion of form invariant probability distributions. This said, statistical inference operates in the real word and thus requires much practicality. The choice of priors is indeed often guided by practical needs and considerations. Priors should be relatively easy to define and use; they and the posteriors one derives from them should be integrable, and so on. Families of functions known as **conjugate priors** offer much of this needed practicality. We introduce their definition and provide selected examples of conjugate families in §7.3.3.

7.3.1 Choice of Prior Functional Form and Jaynes' Maximum Entropy Principle

Prior information I about a system might be terse, but in some cases, it might be possible to make a testable statement concerning a specific facet of the system, a particular hypothesis, or model parameter value. For instance, a statement of the type

$$I: \text{ mean value of } \theta \text{ is } \theta_0$$

is readily testable. Indeed, testing I involves repeated measurements of θ to get a better estimate of its value and we will see that under such conditions, Jaynes' method yields an exponential as a prior PDF for θ. However, other types of information, such as θ is smaller than θ_0, might be too vague to identify a specific PDF as a prior.

Jaynes' method is based on the maximization of Shannon's entropy under specific constraints defined by testable statements about the outcomes of a particular measurement. Understanding the method requires a minimal level of familiarity with Shannon's entropy. We thus first present a brief description of the definition of entropy in the context of probability distributions in §§7.3.1–7.3.1 and proceed to describe the basic principles of the method in §7.3.1. Specific applications of the methods are presented in §7.3.1.

Information Entropy

The notion of information entropy stems from the basic observation that stochastic systems are, in general, not equally uncertain. Bizarre as it may seem, this statement can be readily understood by considering the random draw of red and blue balls, otherwise identical in all respect, from a large vat. Imagine the vat contains 1,000 balls, only one of which is blue. It then seems rather likely that a random draw would yield a red ball. Although one is dealing with a stochastic system, the outcome of the draw indeed seems rather certain: a red ball will most likely be picked out.

What if the vat contained 500 red balls and as many blues? The odds would then be 50 : 50, and one would be just as likely to draw a red or a blue ball. The outcome would then be far less predictable. In fact, it would be maximally uncertain! But certainty and uncertainty are relative concepts. Add yellow, white, green, and black balls in equal proportions to the vat. With six colors equally probable, the outcome of a draw is even less certain. The more diverse and numerous the options are, the less certain is the outcome of a draw. This seems rather obvious. But how, then, can one quantify the degree of certainty of a particular draw or, more generally, the outcome of stochastic processes?

In a seminal paper on information theory published in 1948 [172], Claude Shanon demonstrated that the uncertainty of a discrete probability distribution $\{p_1, p_2, \ldots, p_n\}$ is embodied in its **information entropy** $S(p_1, p_2, \ldots, p_n)$, defined as

$$\text{Entropy} \equiv S(p_1, p_2, \ldots, p_n) = -\sum_{k=1}^{n} p_k \ln p_k. \tag{7.20}$$

A convenient way to visualize the meaning of Shannon's entropy is to consider the outcome of a multinomial draw, that is, a stochastic process describable by a multinomial probability distribution (§3.2). As an example of such a process, let us consider N successive rolls of an m-side die. We denote by p_i and by n_i the probability of a given face i and the number of times it is observed in a sequence of n rolls, respectively. All rolls being independent, the probability of a specific sequence of n independent rolls yielding n_1 times face 1, n_2 times face 2, and so on, is proportional to the product $P \equiv p_1^{n_1} p_2^{n_2} \ldots p_m^{n_m}$. Since there are $M \equiv n!/n_1! \ldots n_m!$ ways of generating such a sequence, its probability is thus indeed a multinomial distribution:

$$p(n_1, \ldots, n_m|N; p_1, \ldots, p_m) = MP \tag{7.21}$$

$$= \frac{n!}{n_1! \cdots n_m!} p_1^{n_1} p_2^{n_2} \cdots p_m^{n_m}, \tag{7.22}$$

with

$$\sum_{k=1}^{m} n_k = n, \tag{7.23}$$

$$\sum_{k=1}^{m} p_k = 1. \tag{7.24}$$

The number of sequence permutations, hereafter called **multiplicity**, tells us much about the (un)certainty of a sequence. For instance, let us consider the multiplicity of selected sequences of $n = 8$ rolls of an $m = 4$ face die. The number of sequences yielding eight times face 1 is

$$M_{8000}^{(8)} = \frac{8!}{8!0!0!0!} = 1, \tag{7.25}$$

whereas the number of sequences yielding each face twice is given by

$$M_{2222}^{(8)} = \frac{8!}{2!2!2!2!} = 2520, \tag{7.26}$$

and the number of sequences with $n_1 = 3$, $n_2 = 3$, $n_3 = 1$, and $n_4 = 1$ amounts to

$$M_{3311}^{(8)} = \frac{8!}{3!3!1!1!} = 1120. \tag{7.27}$$

Without prior knowledge of the probabilities p_i, it seems rather certain that sequences producing a single face (e.g., 8000, 0800, etc.) are far less likely than sequences involving an unequal mix of all faces, and less likely still than having all four faces in equal numbers. Actual sequences involving n rolls of course have probabilities also determined by the probabilities p_i of each of the faces. A given selection of n rolls thus provides a means to gather information about these probabilities. Quite obviously, the information shall be of limited value if only a few rolls are realized. Indeed, for a number of rolls smaller or of the order of m, fluctuations should dominate and observed values n_i cannot provide a robust estimate of the probabilities p_i. However, for increasingly large values of the number of rolls n, the numbers n_i shall narrowly cluster about their expectation values $\mathrm{E}[n_i] = \langle n_i \rangle = np_i$ thereby enabling a reasonably robust estimation of the p_i. Parenthetically, this also implies that (prior) statements about the means and variances (or standard deviations) of the n_i are testable and can form the basis for the selection of prior distributions of the parameters p_i.

Let us then consider the multiplicity M in the large sequence size limit. For large values n, calculations of factorials $n!$ yield extremely large numbers that may be impractical to handle even with modern computers. It is then convenient to consider the natural logarithm of these factorials and use the Stirling approximation

$$\ln n! \approx n \ln n - n. \tag{7.28}$$

The log of the multiplicity M may then be written

$$\begin{aligned}
\ln M &= \ln n - \sum_{k=1}^{m} \ln n_k && (7.29)\\
&= n \ln n - n - \sum_{k=1}^{m} (n_k \ln n_k - n_k) && (7.30)\\
&= n \ln n - \sum_{k=1}^{m} n_k \ln n_k, && (7.31)
\end{aligned}$$

where in the second line, we use the Stirling approximation, and in the third, the sum $\sum n_i = n$. In the large n limit here considered, it is legitimate to replace the values n_k by their expectation values, and we get

$$\begin{aligned}
\ln M &= n \ln n - n \sum_{k=1}^{m} p_k \ln(np_k), && (7.32)\\
&= n \ln n - n \ln n \sum_{k=1}^{m} p_k - n \sum_{k=1}^{m} p_k \ln p_k, && (7.33)\\
&= -n \sum_{k=1}^{m} p_k \ln p_k, && (7.34)
\end{aligned}$$

where in the third line, we use the normalization $\sum p_i = 1$. We can then finally write

$$\ln M = nS, \tag{7.35}$$

where S is Shannon's entropy of the distribution

$$S = \sum_{k=1}^{m} p_k \ln p_k. \tag{7.36}$$

The multiplicity M of a sequence may then be written

$$M = \exp(nS). \tag{7.37}$$

Given the multiplicity M of a particular type of sequence tells us something about its (un)certainty, we conclude that the entropy S corresponds to a degree of certainty per draw (so to speak). Indeed S provides a measure of the level of certainty embodied in a specific probability distribution.

The multiplicity has an extremum which we label $M_{\max}$ corresponding to a maximum entropy $S_{\max}$. Defining $\Delta S \equiv S_{\max} - S$, we can then also write

$$M = M_{\max} \exp\left(-n\Delta S\right). \tag{7.38}$$

In practical situations, where n is very large, one finds that the multiplicity can be extremely large for $\Delta S = 0$, but it shall quickly vanish for $\Delta S > 0$. That implies that for a given probability distribution, certain types of outcomes are far more numerous than others, and thus more "uncertain."

Generalized Entropy

Let us once again consider the roll of an m-face die and assume one has estimates $\{q_i\}$ for the probabilities $\{p_i\}$ of each of the faces. Using these prior estimates, one can obtain approximate values of the probability of arbitrary sequences $\{n_i\}$ of n rolls of the die based on the expression of the multinomial distribution where the p_i are replaced by q_i:

$$p(n_1, \ldots, n_m|n; q_1, \ldots, q_m) = \frac{n!}{n_1! \cdots n_m!} q_1^{n_1} q_2^{n_2} \cdots q_m^{n_m}, \tag{7.39}$$

Let us take the logarithm of this expression. Assuming the number of rolls is very large and using the Stirling approximation, one gets

$$\ln p = \ln n! - \sum_{k=1}^{m} \ln n_k! + \sum_{k=1}^{m} n_k \ln q_k \tag{7.40}$$

$$= n \ln n - \sum_{k=1}^{m} n_k \ln n_k + \sum_{k=1}^{m} n_k \ln q_k \tag{7.41}$$

Replacing n_k by their expectation values, one obtains after simplification

$$\frac{1}{n} \ln p = -\sum_{k=1}^{m} p_k \ln p_k + \sum_{k=1}^{m} p_k \ln q_k, \tag{7.42}$$

$$= -\sum_{k=1}^{m} p_k \ln\left(p_k/q_k\right), \tag{7.43}$$

$$= S_{SJ}, \tag{7.44}$$

which defines a generalized entropy S_{SJ} commonly known as Shannon–Jaynes entropy and Kullback entropy[5]. S_{SJ} is related to the Kullback–Leibler[6] divergence, also known as information divergence or information gain, which is a measure of the difference between two probability distributions.

Entropy of Continuous Distributions

Equation (7.43) has the right functional form for an extension of the notion of entropy to continuous distributions

$$S_c = -\int p(x) \ln \left[\frac{p(x)}{m(x)} \right] dx, \tag{7.45}$$

where $p(x)$ is a continuous probability density and $m(x)$ is known as Lebesgue measure. The inclusion of the measure $m(x)$ insures that S_c is invariant under a change of variable $z \equiv z(x)$ because both the probability $p(x)dx$ and the ratio $p(x)/m(x)$ are invariant under such transformation. One may choose the measure $m(x)$ to be a constant k_m across the domain of x. The entropy S_c then becomes

$$S_c = -\int p(x) \ln p(x)\, dx + \ln k_m \int p(x)\, dx, \tag{7.46}$$

$$= -\int p(x) \ln p(x)\, dx + \text{constant}. \tag{7.47}$$

This expression tells us that the entropy of a probability density is defined up to an inconsequential constant value. This constant should indeed have no effect on Jaynes' entropy maximization principle we discuss in the next section.

Maximization of the Entropy

Consider a measurement yielding m distinct outcomes $\{x_i\}$. Let us assume very little information is available about the phenomenon or system considered. Yet, we would like to determine the prior probability distribution $p(x_i|I)$ of observing such outcomes. If very little is known about the system of interest, there is a priori no guidance or reason to choose any particular functional form for $p(x_i|I)$. One must then seek a functional form that yields maximum entropy, that is, a probability distribution with maximum uncertainty about the outcomes $\{x_i\}$. However, if some testable (prior) information is available about the system, one may also use this information as a constraint in the maximization of the entropy. The basic idea of Jaynes' principle of entropy maximization is to carry a variational calculation to find an expression that maximizes the entropy in the presence of finitely many constraints. One thus seeks a maximum of the entropy S by variation of the (unknown) probabilities p_i of the observed values x_i. A variation of these probabilities should yield a

[5] Named after American cryptanalyst and mathematician Solomon Kullback (1907–1994).

[6] Richard A. Leibler (1914–2003), American mathematician and cryptanalyst.

stationary solution for S at the extremum:

$$dS = \sum_{k=1}^{m} \frac{\partial S}{\partial p_k} dp_k \equiv 0. \tag{7.48}$$

In the absence of further information, one might assume that the probabilities p_k are mutually independent, and one would then conclude that all coefficients $\partial S/\partial p_k$ are null. S would then be a constant independent of the probabilities p_k and not much could be said about the functional form of the probability $p(x)$. Suppose, however, that some constraints are imposed on the p_i based on prior information about the system, one should then be able to use the calculus of variation with Lagrange underdetermined multipliers to obtain a functional form for the p_i.

Let us illustrate the idea using a basic constraint. Since the measurement of n values is known to happen, its probability is unity. The sum of the probabilities of the m values x_i must then be unity:

$$\sum_{k=1}^{m} p_k = 1. \tag{7.49}$$

We write the constrained (generalized) entropy as

$$S' = S - \lambda \left(\sum_{k=1}^{m} p_k - 1 \right) \tag{7.50}$$

$$= -\sum_{k=1} p_k \ln p_k - \lambda \left(\sum_{k=1}^{m} p_k - 1 \right), \tag{7.51}$$

where λ is a Lagrange multiplier and we seek a variation of the p_i that yields an extremum of S':

$$dS' = \sum_{k=1}^{m} \frac{\partial S'}{\partial p_k} dp_k \equiv 0. \tag{7.52}$$

Assuming all p_i are a priori independent, the coefficients $\partial S'/\partial p_k$ must all be null and we find

$$0 = \frac{\partial S'}{\partial p_i}, \tag{7.53}$$

$$= -\frac{\partial}{\partial p_i} \left[\sum_{k=1} p_k \ln p_k + \lambda \left(\sum_{k=1}^{m} p_k - 1 \right) \right] = 0, \tag{7.54}$$

$$= \sum_{k=1} \delta_{ik} \ln p_k + \sum_{k=1} \delta_{ik} + \lambda \sum_{k=1} \delta_{ik}, \tag{7.55}$$

$$= \ln p_i + (1 + \lambda). \tag{7.56}$$

We find that the probabilities p_i should be of the form

$$p_i = \exp\left[-(1 + \lambda) \right], \tag{7.57}$$

$$= \exp(-\lambda_0), \tag{7.58}$$

where, for notational convenience, we introduced the constant $\lambda_0 = 1 + \lambda$. In order to determine this modified multiplier, let us insert Eq. (7.58) in the equation of the constraint

$$\sum_{k=1}^{m} \exp(-\lambda_0) = 1. \tag{7.59}$$

This yields

$$\exp(-\lambda_0) = \frac{1}{m}, \tag{7.60}$$

and we conclude that in the absence of prior information, the principle of maximum entropy tells us that all probabilities p_i should be equal

$$p_i = \frac{1}{m}. \tag{7.61}$$

This result reinforces the intuitive notion that in the absence of prior information, all values of a parameter or hypothesis should be considered equally probable. Indeed, with no prior information whatsoever, the most sensible choice for a prior probability of hypothesis, that which is most uncertain, should be taken as a uniform distribution.

Let us next consider what one can learn about the p_k if additional testable information is available about the observable x. Let us write this additional information in the form of s independent constraints

$$\sum_{i=1}^{m} g_j(x_i) p_i = \langle g_j \rangle, \tag{7.62}$$

where the $g_j(x)$ represent s independent functions of x. Adding these constraints with multipliers λ_j to the entropy S', we get

$$S' = -\sum_{i=1} p_i \ln p_i - \lambda \left(\sum_{i=1}^{m} p_i - 1 \right) - \sum_{j=1}^{s} \lambda_j \left(\sum_{i=1}^{m} g_j(x_i) p_i - \langle g_j \rangle \right) \tag{7.63}$$

Seeking once again a stationary solution for $dS' = 0$, we obtain

$$0 = \ln p_i + (1 + \lambda) + \sum_{j=1}^{s} \lambda_j g_j(x_i) \tag{7.64}$$

which yields a generic solution of the form

$$p_i = \exp(-\lambda_0) \exp\left[-\sum_{j=1}^{s} \lambda_j g_j(x_i) \right]. \tag{7.65}$$

The first constraint (i.e., $\sum p_i = 1$) now yields

$$\exp(\lambda_0) = \sum_{i=1}^{m} \exp\left[-\sum_{j=1}^{s} \lambda_j g_j(x_i) \right]. \tag{7.66}$$

The multipliers λ_j can similarly be determined by insertion of the p_i given by Eq. (7.65) into the constraint equations (7.62). Solution of these s equations for the λ_j may in general require numerical algorithms. Analytical solutions are possible in some cases, however, as we discuss next.

Maximization of the Entropy with Simple Constraints

Equation (7.65) provides a generic solution for the probability coefficients p_i in the presence of s constraints of the form given by Eq. (7.62). Let us consider two applications of this equation, each involving two simple constraints.

An Estimate of the Mean as a Constraint

Let us first derive a specific functional form for p_i in the context of a single nontrivial constraint (i.e., one constraint aside from $\sum p_i = 1$) consisting of an estimate of the mean of x. That is, let

$$g_1(x) = x. \tag{7.67}$$

and

$$\sum_{i=1}^{m} g_1(x_i)p_i = \sum_{i=1}^{m} x_i p_i = \langle x \rangle. \tag{7.68}$$

Equation (7.65) then reduces to an exponential distribution:

$$p_i = \exp(-\lambda_0)\exp\left[-\lambda_1 x_i\right]. \tag{7.69}$$

The normalization constraint $\sum p_i = 1$ yields

$$\exp(\lambda_0) = \sum_{i=1}^{m} \exp\left(-\lambda_1 x_i\right). \tag{7.70}$$

and the constraint imposed by the mean gives us

$$\langle x \rangle = \frac{\sum_{i=1}^{m} x_i \exp\left(-\lambda_1 x_i\right)}{\sum_{i=1}^{m} \exp\left(-\lambda_1 x_i\right)}, \tag{7.71}$$

which can be solved numerically for λ_1 and finitely many values x_i. In the continuum limit, sums are replaced with integrals. Restricting the domain of x to $[0, \infty]$, one gets

$$\langle x \rangle = \frac{\int_0^\infty x \exp\left(-\lambda_1 x\right) dx}{\int_0^\infty \exp\left(-\lambda_1 x\right) dx} = \frac{1}{\lambda_1}. \tag{7.72}$$

The maximum entropy principle tells us that in cases for which an estimate $\langle x \rangle$ of the mean of an otherwise unconstrained parameter is known, one should use an exponential prior probability distribution

$$p(x) = \frac{1}{\langle x \rangle} \exp\left(-\frac{x}{\langle x \rangle}\right) \quad \text{for } x \geq 0. \tag{7.73}$$

An Estimate of the Variance as a Constraint

In many studies, the dispersion of a parameter may be more telling than its average. Let us thus consider a constraint based on an estimate of the variance. Let

$$g_1(x) = (x - \langle x \rangle)^2. \tag{7.74}$$

and

$$\sum_{i=1}^{m} g_1(x_i)p_i = \sum_{i=1}^{m} (x_i - \langle x \rangle)^2 p_i = \langle \Delta x^2 \rangle. \tag{7.75}$$

Equation (7.65) then reduces to

$$p_i = \exp(-\lambda_0) \exp\left[-\lambda_1 (x_i - \langle x \rangle)^2\right]. \tag{7.76}$$

which in the continuous limit gives us

$$p(x) = \exp(-\lambda_0) \exp\left[-\lambda_1 \left((x - \langle x \rangle)^2\right)\right]. \tag{7.77}$$

The next step is to insert Eq. (7.77) in the two constraint equations in order to obtain values for the parameters λ_0 and λ_1. In the continuum limit, one has

$$1 = \exp(-\lambda_0) \int_{-\infty}^{\infty} \exp\left[-\lambda_1 (x_i - \langle x \rangle)^2\right] dx, \tag{7.78}$$

$$\langle \Delta x^2 \rangle = \exp(-\lambda_0) \int_{-\infty}^{\infty} ((x - \langle x \rangle)^2 \exp\left[-\lambda_1 ((x - \langle x \rangle)^2\right] dx, \tag{7.79}$$

from which one gets

$$\exp(\lambda_0) = \sqrt{2\pi} \langle \Delta x^2 \rangle^{1/2} \tag{7.80}$$

$$\lambda_1 = \frac{1}{2\langle \Delta x^2 \rangle}. \tag{7.81}$$

Given a constraint determined by an estimate of the variance $\langle \Delta x^2 \rangle$, the principle of maximum entropy yields a Gaussian distribution with a mean $\langle x \rangle$ and standard deviation equal to the square root of this estimate. This means a Gaussian carries the largest uncertain in this context. This is a rather important result. It indicates, that unless additional information is available, the Gaussian distribution is the least certain and thus carries the fewest assumption about the system considered and should thus lead to the most conservative results. If the value of $\langle \Delta x^2 \rangle$ is unknown (i.e., with no available estimates), the Gaussian distribution still provides a safe prior insofar as it can be established that the dispersion of the data is finite and that its variance can be treated as nuisance parameter, that is, it can eventually be marginalized.

In some cases, the range of a model parameter can be limited a priori to a finite domain $x_L \leqslant x < x_H$. The maximum entropy principle can then be applied to the generalized entropy S_{SJ}, Eq. (7.42), with

$$q_i = \begin{cases} \frac{1}{x_H - x_L} & \text{for } x_L \leqslant x < x_H \\ 0 & \text{elsewhere.} \end{cases} \tag{7.82}$$

Because q_i is a constant, the entropy maximization procedure remains essentially the same provided one replaces p_i by p_i/q_i. The integrals of Eq. (7.78) must include a factor $1/(x_H - x_L)$ and the bounds of integration set to x_L and x_H. The integration may then be carried out in terms of error functions, and the resulting PDFs is a truncated Gaussian (see Problem 7.1).

7.3.2 Scalable Priors and Jeffreys' Priors

Issues with Uniform Priors

Let us once again consider the problem of the determination of the detection efficiency of elementary particles in a complex detector. Recall from §4.7.4 that the problem is equivalent to that of an **unfair** coin toss determined by a binomial probability model with a probability ε of yielding "head" (i.e., a success). Let us assume, perhaps dramatically, that no information whatsoever is available about the detection efficiency and that it could have a value anywhere in the range [0, 1]. Given the binomial probability model determining the outcome of N repeated measurements (or coin tosses), it would make sense to use a conjugate prior, that is, a prior probability in the form of a beta distribution[7]. Unfortunately, the assumed lack of knowledge makes such a choice totally arbitrary. How indeed can one justify the use of any particular values for the shape parameters α and β?

Bayes and Laplace proposed that prior (total) ignorance of the value of ε should be represented by a uniform PDF within the range of applicability of the parameter. In the context of our detection efficiency problem, this Bayes' prior takes the form

$$p(\varepsilon) = \begin{cases} 1 & \text{for } 0 \le \varepsilon \le 1 \\ 0 & \text{elsewhere.} \end{cases} \tag{7.83}$$

The posterior probability $p(\varepsilon|D)$ would then be strictly equal to the likelihood $p(D|\varepsilon)$ yielding, in the case at hand, a posterior in the form of a binomial distribution with a mode equal to the maximum likelihood estimate $\varepsilon = n/N$, where n and N are the number of observed and produced particles respectively. Note, parenthetically, that a uniform distribution in a finite range corresponds in fact to a beta distribution with parameters $\alpha = \beta = 1$. If the range of the parameter is unbound, the prior is said to be improper because its integral over the full range of the parameter diverges. It is relatively easy to verify, however, that the posterior remains proper, i.e., with a finite and well-defined normalization.

The notion of using a uniform distribution for a completely unknown parameter is quite appealing and sounds rather straightforward. It is also corresponds, as we saw in §7.3.1, to a functional shape with maximum uncertainty. The obvious problem arises, however, that a different choice of parameterization, $\theta \equiv \theta(\varepsilon)$, shall lead to an arbitrarily shaped prior. For instance, let the total ignorance about ε be represented by Eq. (7.83) and let us choose the new parameterization $\theta = \varepsilon^2$. By definition of the notion of probability, one must have

[7] The notion of conjugate prior is formally introduced in §7.3.3.

$p(\theta)d\theta = p(\varepsilon)d\varepsilon$ and conclude that the PDF of θ is equal to

$$p(\theta) = \frac{1}{2\theta^{1/2}}, \tag{7.84}$$

which is manifestly not a uniform distribution in the range [0, 1]. It then appears that the actual formulation of the parameter matters. Indeed, if one chooses θ to have a uniform prior, this prior shall be clearly incompatible with a uniform prior for ε. The two priors would lead to different and irreconcilable results for the posterior. This seems rather awkward at the outset although in practice, a particular formulation or choice of parameter might be better justified or natural. For instance, in the case of a binomial distribution, it seems natural and reasonable to define a uniform prior in terms of the probability of success ε, and any other parameterization might be considered artificial or unnatural. The choice is considerably less obvious, however, with other commonly used PDFs such as the Gaussian PDF for which one might choose to parameterize a prior in terms of a uniform standard deviation σ or uniform variance σ^2.

Working with a binomial model, J. B. S. Haldane, a geneticist, advocated the use of an improper prior density of the form $p(\varepsilon) \propto \varepsilon^{-1}(1-\varepsilon)^{-1}$, which is a conjugate of the binomial distribution (and more specifically a particular case of the beta distribution with $\alpha = \beta = 1$). Although this distribution yield the right limiting behavior for the mean (mode), it yields an improper and thus problematic posterior if $n = 0$ or $n = N$ are encountered experimentally.

Harold Jeffrey, proposed to instead use

$$p(\varepsilon) \propto \varepsilon^{-1/2}(1-\varepsilon)^{-1/2} \tag{7.85}$$

as a conjugate prior for a binomial probability model. He based this particular form of the beta distribution on a simple invariance principle we discuss in detail in §7.3.2. But first, we consider uninformative location and scaling priors in the next two sections.

Uninformative Location Priors

A model parameter θ may be considered a **location parameter** whenever a probability distribution may be written in the form $p(x-\theta|\theta, I)$. A sensible candidate for an uninformative prior would be a uniform prior $p(\theta|I) \propto 1$. If θ is bounded within a finite interval $\theta_{\min} \leq \theta \leq \theta_{\max}$, the prior may be properly normalized

$$p(\theta|I) = \begin{cases} (\theta_{\max} - \theta_{\min})^{-1} & \text{for } \theta_{\min} \leq \theta \leq \theta_{\max} \\ 0 & \text{elsewhere.} \end{cases} \tag{7.86}$$

However, if the domain of θ is $\mathbb{R}$, the flat prior is said to be **improper** because its integral over the domain diverges. Be it as it may, given this improper normalization appears both in the numerator and the denominator of Bayes' theorem used toward the calculation of a posterior probability for θ, use of such an improper probability distribution remains technically acceptable.

Uninformative Scaling Priors

Consider a physical observable X scalable by an arbitrary factor θ. This factor might correspond to a change of units (e.g., transforming a distance from meters to kilometers) but it is more interesting to consider an actual physical scaling of the observable. Such a scaling should leave the total probability $\int p(x|I)$ of observing X unchanged; one can thus write

$$p_\theta(x|I) = \frac{1}{\theta} p(x/\theta|I). \tag{7.87}$$

The actual scale factor of a phenomenon might be a priori unknown (e.g., a cross-section) and the goal of a measurement might then be to determine this factor. Since the scale factor is totally unknown, the prior on θ should then be invariant under an arbitrary rescaling of θ by a positive constant c, that is,

$$p(\theta) = \frac{1}{c} p(\theta/c). \tag{7.88}$$

This implies that a rescaling of θ should not change the prior, and consequently, the prior provides no information about the physical scale of the process and parameter θ of interest. Equation (7.88) is a functional equation that must admit a single solution for $p(\theta)$ up to an arbitrary scaling factor. The prior must thus be of the form

$$p(\theta) \propto \frac{1}{\theta}. \tag{7.89}$$

Applied without bounds in $\mathbb{R}$, its integral diverges, and it is thus an improper probability distribution. It may also defined with bounds

$$p(\theta) = \begin{cases} \frac{k}{\theta} & \text{for } \theta_{\text{min}} \le \theta \le \theta_{\text{max}} \\ 0 & \text{otherwise.} \end{cases} \tag{7.90}$$

The normalization constant k is obtained by integration

$$1 = \int_{\theta_{\text{min}}}^{\theta_{\text{max}}} p(\theta|H_1, I) d\theta = k \ln\theta|_{\theta_{\text{min}}}^{\theta_{\text{max}}}, \tag{7.91}$$

which yields $k^{-1} = \ln\theta_{\text{max}}/\theta_{\text{min}}$. The bound prior may then be written

$$p(\theta) = \begin{cases} \frac{1}{\ln\theta_{\text{max}}/\theta_{\text{min}}} \frac{1}{\theta} & \text{for } \theta_{\text{min}} \le \theta \le \theta_{\text{max}} \\ 0 & \text{otherwise} \end{cases} \tag{7.92}$$

It is of obvious interest to seek a transformed variable $\xi(\theta)$ endowed with a flat (improper) prior. By definition of the notion of probability, one writes

$$p(\xi) = p(\theta) \left| \frac{d\theta}{d\xi} \right|, \tag{7.93}$$

where $p(\xi)$ is required to be flat and $p(\theta) \propto 1/\theta$. One thus find that

$$\frac{d\xi}{d\theta} = \frac{1}{\theta}, \tag{7.94}$$

which implies that ξ is of the form

$$\xi = \ln \theta. \tag{7.95}$$

We thus conclude that a flat prior formulated in terms of $\log \theta$ provides an arbitrary scalable probability distribution and the logarithm ensures that all orders of magnitude of the scaling factor are treated equally, that is, given equal probability.

As we will see later in this chapter, flat priors in $\log A$, where A is the unknown amplitude of a signal, are quite useful in the study and search of weak signals of unknown amplitude.

Improper Priors

Improper prior are based on density distributions whose integral does not exist or diverges. They are commonly used as noninformative priors. They can usually be viewed as limits of proper priors, with the corresponding posterior being the limit of the posteriors corresponding to those priors. Although they cannot formally be used as probability density, their use is generally deemed acceptable provided the posterior obtained with Bayes' theorem is a proper probability distribution, and when the data are sufficiently informative about the parameter of interest to render the prior essentially irrelevant.

Locally Uniform Priors

Locally uniform priors are based on functions that are essentially constant over the region in which the likelihood is appreciable and do not feature large values outside that range. Because the prior is nearly constant in the range where the likelihood is large, one can write

$$p(\theta|x) = \frac{p(x|\theta)p(\theta)}{\int p(x|\theta)p(\theta)\,d\theta} \approx \frac{p(x|\theta)}{\int p(x|\theta)\,d\theta} = \frac{L(x|\theta)}{\int L(x|\theta)\,d\theta}. \tag{7.96}$$

A locally uniform prior $p(\theta)$ thus effectively provides a convenient uninformative prior for the parameter of interest θ. Locally uniform priors can be defined based on proper density distributions and thus eliminate the need for improper priors.

Conjugate priors (discussed in §7.3.3), such as the beta and Gaussian distributions, may be selected to yield large ranges of parameter space where they are nearly constant and thus provide, effectively, an uninformative prior. Because the posterior of a conjugate prior is in the same family of distributions as the prior, calculations of the posterior and its properties are greatly simplified. Conjugate priors thus provide double convenience: they can be suitably chosen as uninformative priors and provide for simplified calculations of posteriors and their properties.

However, the use of a locally uniform prior may be consider philosophically unsatisfying because it requires a peep at the likelihood distribution to ensure the prior's flat region in fact covers the entire domain where the likelihood is large, thereby violating the principle that a prior should be formulated based on prior knowledge only, that is, without looking at the data for which it is designed to serve as a prior.

Jeffreys' Invariance Principle

Jeffreys argued that equivalent propositions should have the same probability. If two or more parameterizations of a particular proposition are possible, they should be equivalent, that is, they should yield the same end result. Jeffreys additionally argued that the prior probability distribution of a parameter should be determined by the Fisher information of that parameter derived from the data probability model. This makes sense. If a parameter is unknown, its prior probability distribution should be as vague as possible but no more vague than the information that can be extracted from the data. Jeffreys therefore proposed that the prior $p(\varepsilon|I)$ be proportional to some simple power of the Fisher information. But no particular choice of parameterization should have preeminence. More specifically, given some choice of parameter ε, it should be possible to switch to any new model parameterization, $\theta \equiv \theta(\varepsilon)$, provided the densities satisfy

$$p(\theta) = p(\varepsilon)\frac{d\varepsilon}{d\theta}. \tag{7.97}$$

Given the transformation property of the Fisher information under a variable transform derived in §4.7.3, Eq. (4.97),

$$\mathbb{I}(\theta) = \mathbb{I}(\varepsilon)\left(\frac{\partial\varepsilon}{\partial\theta}\right)^2 \tag{7.98}$$

and the preceding requirement, one concludes that a self-consistent prior is obtained with a power one-half, that is,

$$p(\varepsilon) \propto [\mathbb{I}(\varepsilon)]^{1/2}. \tag{7.99}$$

Let us verify that this prior indeed features the desired transformation property. By virtue of the invariance principle, the prior for θ is written

$$p(\theta) \propto [\mathbb{I}(\theta)]^{1/2}. \tag{7.100}$$

Applying the transformation property of the Fisher information, Eq. (7.98), under a change of variable $\varepsilon \to \theta$, one gets

$$p(\theta) \propto [\mathbb{I}(\varepsilon)]^{1/2}\frac{\partial\varepsilon}{\partial\theta}, \tag{7.101}$$

$$= p(\varepsilon)\frac{\partial\varepsilon}{\partial\theta}, \tag{7.102}$$

which indeed satisfies the transformation, Eq. (7.97).

Examples of Jeffreys' Prior

Let us apply Jeffreys' invariance principle with selected probability models to obtain prior probability distributions applicable in various contexts.

Uninformative Scaling Prior Revisited

Let us first verify that the uninformative scaling prior discussed in §7.3.2 is in fact a Jeffreys prior for the amplitude A of (improper) distributions of the form

$$p(x|A) = Af(x), \tag{7.103}$$

where $f(x)$ is a nonnegative, finite, and integrable function of x, independent of the amplitude A. Using the definition, Eq. (4.89), to calculate the Fisher information, we find

$$\begin{aligned}\mathbb{I}(A) &= -\mathrm{E}\left[\frac{\partial^2}{\partial A^2}\ln p\right] \\ &= -\mathrm{E}\left[\frac{\partial}{\partial A}\frac{1}{A}\right] \\ &= \frac{1}{A^2}.\end{aligned} \tag{7.104}$$

Taking the square root of the information,

$$p(A) = \mathbb{I}(A)^{1/2} = \left(\frac{1}{A^2}\right)^{1/2} = \frac{1}{A}. \tag{7.105}$$

we find that Jeffreys prior for A is indeed of the form, Eq. (7.89), obtained for scalable parameters.

Binomial Distribution

Next, consider Jeffreys prior for the success rate, ε, of a binomial distribution. As we saw in §4.109, the Fisher information $\mathbb{I}(\varepsilon)$ of a binomial PDF is given by

$$\mathbb{I}(\varepsilon) = \frac{N}{\varepsilon\,(1-\varepsilon)}. \tag{7.106}$$

We thus verify that Eq. (7.85), reproduced below, indeed corresponds to Jeffreys prior for a binomial PDF, that is, it satisfies Jeffreys invariance principle

$$p(\varepsilon) \propto \varepsilon^{-1/2}(1-\varepsilon)^{-1/2}. \tag{7.107}$$

Gaussian Distribution

Next consider the prior for the parameters of a Gaussian PDF. Given that the off-diagonal elements of the Fisher information matrix of a Gaussian are null, as shown in Eq. (4.116), one can sensibly formulate independent priors for the mean μ and variance σ^2 of the distribution. For the mean, we get

$$p(\mu) \propto \sqrt{\mathbb{I}_{\mu,\mu}(\mu,\sigma^2)}, \tag{7.108}$$

$$= \frac{k}{\sigma}, \tag{7.109}$$

where in the second line, we introduced a normalization constant determined, for proper priors, by the range of applicability $\mu_{\min} \leq \mu \leq \mu_{\max}$ of the mean, that is, $k = \sigma(\mu_{\max} - \mu_{\min})$. In effect, since σ is a constant, albeit of unknown value, we find that the prior for μ is independent of μ, and thus amounts to a flat prior.

For the variance, one gets

$$p(\sigma^2) \propto \sqrt{\mathbb{I}_{\sigma^2,\sigma^2}(\mu, \sigma^2)}, \tag{7.110}$$

$$= \frac{k'}{\sigma^2}. \tag{7.111}$$

One thus concludes that the prior for σ^2 is not flat. However, it is easy to verify (Problem 7.2) that this implies that the prior for $\log \sigma^2$ is a uniform distribution. This is both convenient and meaningful. If a particular observable is really Gaussian distributed, its standard deviation must be finite and bound under some scale. Since the scale may not be known a priori, a uniform prior for $\log \sigma^2$ ensures that smaller values of σ^2 are far more probable that very large values. In essence, it preserves the Gaussian nature of the distribution, which for very large values of σ might otherwise appear flat in a small range of the observable.

Poisson Distribution

Jeffreys prior for the rate parameter λ of a Poisson distribution is similarly calculated.

$$p(\lambda) \propto \sqrt{\mathbb{I}(\lambda)}, \tag{7.112}$$

$$= \sqrt{\mathrm{E}\left[\left(\frac{\partial}{\partial\lambda} \ln p(n|\lambda)\right)^2\right]}, \tag{7.113}$$

$$= \sqrt{\frac{1}{\lambda^2}\mathrm{E}\left[(n-\lambda)^2\right]}. \tag{7.114}$$

Noting that $\mathrm{E}\left[(n-\lambda)^2\right]$ corresponds to the variance of the distribution, $\mathrm{Var}[n] = \lambda$, one concludes that the prior for the rate parameter may be written

$$p(\lambda) \propto \frac{1}{\sqrt{\lambda}}. \tag{7.115}$$

Multinomial Distribution

Finally, let us consider the prior distributions for the rate parameters, $\vec{p} = (p_1, p_2, \ldots, p_m)$, of a multinomial distribution, with the constraint $\sum_{i=1}^{m} p_i = 1$. One can show that Jeffreys prior for the coefficients $\vec{p}$ is the Dirichlet distribution with all of its parameters set to half. One can additionally show that with transformations $p_i = \phi_i^2$, the parameters $\vec{\phi}$ are uniformly distributed on a unit sphere of $m-1$ dimensions.

7.3.3 Conjugate Priors

The choice of a particular functional form for a prior probability density is often guided by practical considerations. One finds, indeed, that certain functional forms are particularly well suited, or convenient, for use with specific probability models. For instance, if a data sample may be described with a Bernoulli probability model, the choice of a beta distribution as prior is quite convenient because, as we demonstrate in the text that follows, the posterior obtained from the product of a beta distribution by the likelihood of a sample of values determined by a Bernoulli distribution is also a beta distribution. The beta distribution (family) is then said to be **conjugate prior** to the Bernoulli distribution (family).

Having a posterior in the same distribution family as the prior is quite convenient because it enables a dynamic and iterative improvement of the knowledge of the parameters of the data model. Indeed, once data are acquired, the product of the likelihood by the prior enables the determination of a posterior of the same family that can then be used as a prior for another sequence of measurements. The process can be iterated indefinitely because the posterior remains in the same family as the prior regardless of the observed values or the number of samples. In that sense, the family of beta distributions is then also said to be be closed under sampling from a Bernoulli distribution.

The notion of conjugate distribution can be extended to virtually any data probability models. Indeed, essentially all data probability models commonly in use in statistical analyses are associated with a family of distributions useable as conjugate priors. In this section, we introduce a basic selection of such conjugate families for illustrative purposes and as a foundation for examples of Bayesian inference discussed in later sections of this chapter. Several additional conjugate prior families, as well as Jeffreys priors, are documented in the very comprehensive compendium produced by D. Fink [82].

Bernoulli Processes

Consider the sum, $y = \sum_{i=1}^{n} x_i$, of n instances x_i of random variables X_i[8] drawn from a common Bernoulli distribution with an unknown success parameter ε defined in the range $0 < \varepsilon < 1$. The likelihood distribution of the sum y is the joint PDF $p(\vec{x}|\varepsilon)$:

$$p(\vec{x}|\varepsilon) = \prod_{i=1}^{n} p(x_i|\varepsilon), \tag{7.116}$$

$$= \prod_{i=1}^{n} \varepsilon^{x_i} (1-\varepsilon)^{1-x_i}, \tag{7.117}$$

$$= \varepsilon^{y} (1-\varepsilon)^{n-y}. \tag{7.118}$$

The functional dependence of the likelihood on ε and $1-\varepsilon$ suggests that the beta distribution (§3.7) might be an appropriate prior for the Bernoulli distribution. The beta distribution may be written

$$p(\varepsilon|\alpha,\beta) = \frac{\Gamma(\alpha+\beta)}{\Gamma(\alpha)\Gamma(\beta)} \varepsilon^{\alpha-1} (1-\varepsilon)^{\beta-1}, \tag{7.119}$$

[8] In a Bernoulli process the variable X equals 1 for successes and 0 for failures.

which indeed features factors in ε and $1-\varepsilon$ similar to those of the likelihood. The posterior is, by construction, proportional to the product of the prior and the likelihood. Keeping exclusively the factors with an explicit dependence on ε, we get

$$p(\varepsilon|x) \propto \varepsilon^{y}(1-\varepsilon)^{n-y}\varepsilon^{\alpha-1}(1-\varepsilon)^{\beta-1}, \quad (7.120)$$

$$\propto \varepsilon^{y+\alpha-1}(1-\varepsilon)^{\beta+n-y-1}, \quad (7.121)$$

which up to a normalization constant may be recognized as a beta distribution of the form Eq. (7.119) with parameters

$$\alpha' = \alpha + y, \quad (7.122)$$

$$\beta' = \beta + n - y, \quad (7.123)$$

and a normalization constant equal to

$$\frac{\Gamma(\alpha+\beta+n)}{\Gamma(\alpha+y)\Gamma(\beta+n-y)}. \quad (7.124)$$

The family of beta distributions thus indeed constitutes a conjugate to the Bernoulli distribution.

The use of a conjugate distribution has an interesting practical advantage. Suppose that the value of ε is unknown but that a beta prior with specific values for α and β is available. A single measurement (trial) would yield either $x = 0$ or $x = 1$. If the value is 1, the posterior of this measurement is a new beta distribution with $\alpha' = \alpha + 1$ and $\beta' = \beta$ whereas if the value is 0, the posterior has $\alpha' = \alpha$ and $\beta' = \beta + 1$. Effectively, α counts the successes while β accounts for the number of failures. As the number of trials N increases, the parameters α and β will tend toward their respective expectation values: $\langle\alpha\rangle = \varepsilon N$ and $\langle\beta\rangle = (1-\varepsilon)N$ but recall from §3.7 that the expectation value of a beta distribution is α/β. That means that for increasing N, the ratio α/β will tend toward $\varepsilon/(1-\varepsilon)$ and, as such, provides an estimator for ε.

$$\varepsilon = \frac{\alpha/\beta}{1+\alpha/\beta}. \quad (7.125)$$

Also recall from §3.7 that the variance of the beta distribution is $\sigma^2 = \alpha/\beta^2$. Substituting the expectation values for α and β, we find

$$\sigma^2 = \frac{1}{N}\frac{\varepsilon}{(1-\varepsilon)^2}, \quad (7.126)$$

which means that the width of the posterior distribution should decrease in proportion to $\sqrt{N}$. As illustrated in Figure 7.2, the posteriors of successive Bernoulli experiments (with a beta prior) provide estimates for ε with an accuracy that improves as $\sqrt{N}$. However, note that the choice of a beta prior is disadvantageous for values of ε very close to zero or unity given the variance of the distribution diverges in these two limits.

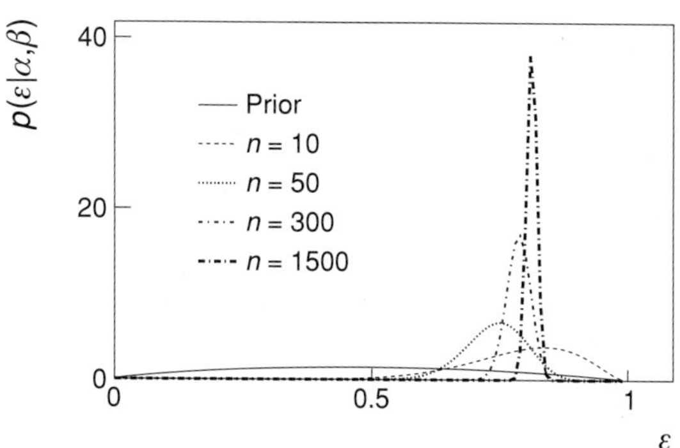

Fig. 7.2 Illustration of the evolution of the beta posterior observed in a succession of Bernoulli measurements. Parameters of the prior (solid line) were arbitrarily set to $\alpha = 1.8$ and $\beta = 1.9$ to obtain a relatively uninformative prior. Beta posteriors are shown for $n = 10, 50, 300$, and 1,500 trials sampling of a Bernoulli distribution with $\varepsilon = 0.8$.

Poisson Processes

Let $x_1, x_2, \ldots, x_n$ represent a random sample from a Poisson distribution with a positive definite rate parameter λ. The likelihood function of the sample is

$$p_n(\vec{x}|\lambda) = \prod_{i=1}^{n} \frac{\lambda^{x_i} e^{-\lambda}}{x_i!}, \quad \text{for } \lambda > 0 \tag{7.127}$$

$$= \frac{\lambda^y e^{-n\lambda}}{\prod_{i=1}^{n} x_i!}, \tag{7.128}$$

where, once again, we defined $y = \sum_{k=0}^{n}$. Choosing a prior PDF for λ in the form of a gamma distribution (§3.6), we write

$$p(\lambda|\alpha, \beta) = \frac{\beta^\alpha}{\Gamma(\alpha)} \lambda^{\alpha-1} e^{-\beta\lambda}. \tag{7.129}$$

The posterior $p(\lambda|\vec{x})$ is thus of the form

$$p(\lambda|\vec{x}) \propto \frac{\beta^\alpha}{\Gamma(\alpha)} \lambda^{\alpha-1} e^{-\beta\lambda} \frac{\lambda^y e^{-n\lambda}}{\prod_{i=1}^{n} x_i!}, \quad \text{for } \lambda > 0$$

$$\propto \lambda^{y+\alpha-1} e^{-(n+\beta)\lambda}, \tag{7.130}$$

which is itself a gamma distribution with parameters

$$\begin{aligned} \alpha' &= \alpha + y, \\ \beta' &= \beta + n. \end{aligned} \tag{7.131}$$

We thus conclude that the family of gamma distributions constitute a conjugate to the family of Poisson distributions.

Normal Processes (Known Variance)

For convenience, we define a **precision** parameter ξ as the multiplicative inverse of the variance of a Gaussian distribution

$$\xi \equiv \frac{1}{\sigma^2}. \tag{7.132}$$

The Gaussian distribution may then be written

$$p(x|\mu,\xi) = \sqrt{\frac{\xi}{2\pi}} \exp\left[-\frac{1}{2}\xi(x-\mu)^2\right], \tag{7.133}$$

and the likelihood of a set of data $\vec{x} = (x_1, x_2, \ldots, x_n)$ is

$$p_n(\vec{x}|\mu,\xi) = \left(\frac{\xi_0}{2\pi}\right)^{n/2} \exp\left[-\frac{1}{2}\xi_0 \sum_{i=1}^{n}(x_i-\mu)^2\right], \tag{7.134}$$

where ξ_0 is the known value of the precision.

Given the Gaussian dependence of the likelihood on μ, let us formulate a conjugate prior for this variable also in terms of a Gaussian distribution

$$p(\mu|\mu_p,\xi_p) = \sqrt{\frac{\xi_p}{2\pi}} \exp\left[-\frac{1}{2}\xi_p\left(\mu-\mu_p\right)^2\right], \tag{7.135}$$

where μ_p is a prior estimate of the mean and ξ_p is the assumed precision of that estimate. The posterior is then of the form

$$p(\mu|\mu_p,\xi_p) \propto \exp\left\{-\frac{1}{2}\left[\xi_0 \sum_{i=1}^{n}(x_i-\mu)^2 + \xi_p\left(\mu-\mu_p\right)^2\right]\right\}. \tag{7.136}$$

We next show that the argument of Eq. (7.136) may be written in the form $-\frac{1}{2}\xi'_p(\mu-\mu'_p)^2$. We first decompose the square $(x_i-\mu)^2$ as follows:

$$(x_i-\mu)^2 = (x_i-\bar{x})^2 + (\mu-\bar{x})^2 + 2\,(x_i-\bar{x})\,(\bar{x}-\mu). \tag{7.137}$$

where $\bar{x}$ represents the arithmetic mean of the sample. A sum of n such terms may be written

$$\sum_{i=1}^{n}(x_i-\mu)^2 = \sum_{i=1}^{n}(x_i-\bar{x})^2 + n\,(\mu-\bar{x})^2 + 2\,(\bar{x}-\ mu)\sum_{i=1}^{n}(x_i-\bar{x}), \tag{7.138}$$

where the last term identically vanishes because the sum of x_i equals $n\bar{x}$. Omitting factors with no dependence on μ, the posterior probability becomes

$$p(\mu|\vec{x}) \propto \exp\left\{-\frac{1}{2}\left[n\xi_0\,(\mu-\bar{x})^2 + \xi_p\left(\mu-\mu_p\right)^2\right]\right\}.$$

In order to further simplify this expression, we introduce

$$Q(\mu) = n\xi_0\,(\mu-\bar{x})^2 + \xi_p\left(\mu-\mu_p\right)^2. \tag{7.139}$$

It is straightforward, although somewhat tedious, to show that $Q(\mu)$ may be transformed in the form

$$Q(\mu) = \left(n\xi_0 + \xi_p\right)\left[\mu - \frac{n\xi_0\bar{x} + \mu_p\xi_p}{n\xi_0 + \xi_p}\right]^2 + K, \tag{7.140}$$

where K represents a constant expression independent of μ that can be relegated to the normalization constant of the posterior (Problem 7.4). We thus finally get

$$p(\mu|\vec{x}) \propto \exp\left(-Q(\mu)/2\right), \tag{7.141}$$

$$\propto \exp\left[-\frac{1}{2}\xi_p'\left(\mu - \mu_p'\right)^2\right] \tag{7.142}$$

where we introduced the updated mean and precision

$$\mu_p' = \frac{n\xi_0}{n\xi_0 + \xi_p}\bar{x} + \frac{\xi_p}{n\xi_0 + \xi_p}\mu_p, \tag{7.143}$$

$$\xi_p' = n\xi_0 + \xi_p. \tag{7.144}$$

Note that μ_p' actually corresponds to the weighted mean of the sample mean, $\bar{x}$, and the prior estimate of the mean μ_p with weights dependent on the actual precision of the measurement ξ_0, the size n of the sample, and the prior precision of the estimate. The precision of the estimate μ_p' is itself greatly improved as it becomes equal to the sum of the precision of n measurements and the prior estimate of the precision. The variance of the posterior mean may then be written

$$\sigma_p'^2 = \frac{\sigma_0^2\sigma_p^2}{n\sigma_n^2 + \sigma_0^2}. \tag{7.145}$$

Normal Processes (Fixed Mean)

Let us next assume that the mean of a process is known (and equal to μ_0) but that its variance (or precision) is not. The likelihood, Eq. (7.134), features an exponent of ξ and an exponential function of ξ. It is thus natural to invoke a gamma distribution as a prior. This yields a posterior of the form

$$p(\xi|\vec{x}) \propto \frac{\beta^\alpha}{\Gamma(\alpha)}\xi^{\alpha-1}e^{-\beta\xi}\left(\frac{\xi}{2\pi}\right)^{n/2}\prod_{i=1}^{n}\exp\left[-\frac{\xi}{2}(x_i - \mu_0)^2\right], \tag{7.146}$$

$$\propto \xi^{\alpha+\frac{n}{2}-1}\exp\left\{-\xi\left[\frac{1}{2}\sum_{i=1}^{n}(x_i - \mu_0)^2 + \beta\right]\right\}, \tag{7.147}$$

which is itself a gamma distribution with parameters

$$\alpha' = \alpha + n/2, \tag{7.148}$$

$$\beta' = \beta + \frac{1}{2}\sum_{i=1}^{n}(x_i - \mu_0)^2. \tag{7.149}$$

We thus conclude that the gamma distribution constitutes a conjugate of the Gaussian distribution for the precision parameter ξ when the mean of the normal process is fixed.

Normal Processes (General Case)

The general case requires special care. Given both the mean μ and the precision ξ are conditioned by the data x, the order in which they are inferred matters. The joint posterior of μ and ξ may be written

$$p(\xi, \mu|\vec{x}) \propto p(\xi) \times p(\mu|\xi, I) \times p(\vec{x}|\xi, \mu, I). \tag{7.150}$$

The prior $p(\xi)$ is a beta distribution (as in the previous section). For $p(\mu|\xi, I)$, we use a Gaussian distribution, as in §7.3.3, with a precision equal to $\xi_p = n_p\xi$ where ξ is the prior precision of the measurement and n_p a multiplicative factor that determines the prior precision on the mean relative to the measurement precision. The joint posterior may then be written

$$\begin{aligned} p(\xi, \mu|\vec{x}) \propto\ & \xi^{\alpha-1} \exp\left[-\beta\xi\right] \\ & \times \xi^{1/2} \exp\left[-\frac{n_p\xi}{2}(\mu - \mu_p)^2\right] \\ & \times \xi^{n/2} \exp\left[-\frac{\xi}{2}\sum_{i=1}^{n}(x_i - \mu)^2\right]. \end{aligned} \tag{7.151}$$

Inserting $\bar{x} - \bar{x}$ in the $x_i - \mu$ terms, we find after some simple algebra

$$\begin{aligned} p(\xi, \mu|\vec{x}) \propto\ & \xi^{\alpha+\frac{n}{2}-1} \exp\left\{-\xi\left[\beta + \frac{1}{2}\sum_{i=1}^{n}(x_i - \bar{x})^2\right]\right\} \\ & \times \xi^{1/2} \exp\left\{-\frac{\xi}{2}\left[n_p(\mu - \mu_p)^2 + n(\bar{x} - \mu)^2\right]\right\}. \end{aligned} \tag{7.152}$$

One may then obtain a posterior for ξ alone by marginalization of μ. Integration of Eq. (7.152) relative to μ yields the posterior for ξ (Problem 7.5):

$$\begin{aligned} p(\xi|\vec{x}) \propto\ & \xi^{\alpha+n/2-1} \exp\left\{-\xi\left[\beta + \frac{1}{2}\sum_{i=1}^{n}(x_i - \bar{x})^2\right.\right. \\ & \left.\left. + \frac{nn_p}{2(n + n_p)}(\bar{x} - \mu_p)^2\right]\right\}, \end{aligned} \tag{7.153}$$

which is a beta distribution, of the form Eq. (7.119), with parameters

$$\alpha' = \alpha + n/2, \tag{7.154}$$

$$\beta' = \beta + \frac{1}{2}\sum_{i=1}^{n}(x_i - \bar{x})^2 + \frac{nn_p}{2(n + n_p)}(\bar{x} - \mu_p)^2. \tag{7.155}$$

The posterior for μ is a Gaussian with a mean $\mu' = \frac{n}{n+n_p}\bar{x} + \frac{n_p}{n+n_p}\mu_p$ and a precision $\xi'_p = (n + n_p)\xi$. Summarizing, given a set of data $\vec{x}$, one can use Eqs. (154) and (155) to get an estimate of ξ and subsequently Eq. (7.139) to get an estimate of the mean.

Multivariate Normal Processes

The choice of a generic conjugate distribution for a multivariate normal distribution is somewhat more involved than the examples presented in previous sections. Several parameterizations are possible and discussed in the literature (e.g., see [82] and references therein).

Brief Epilog on Conjugates

The use of conjugate priors confers several interesting properties to a Bayesian inference analysis. The first obvious advantage derives from the definition of conjugate priors: the posterior remains in the same family as the prior, thus greatly reducing the mathematical complexity of an analysis. Furthermore, since priors of a specific family of functions have a common dependency on finitely many parameters, it is typically possible to obtain closed expressions for the evolution of these parameters with added data, as we have shown in §§7.3.3–7.3.3. This means that one can regard the inference process as an iterative process.

Given a prior and its parameters, the acquisition of new data enables an update of the parameters according to a closed form expression. These updated parameters can then be used for the definition of a prior (of the same family) for another measurement. The procedure may then be iterated arbitrarily as many times as there are new available datasets. The end result is a set of posterior parameters that seamlessly takes into account all measurements pertaining to a specific observable. Additionally note that the explicit dependence of the updated parameters on the measurement precision (e.g., normal process with known precision discussed in §7.3.3) means this iterative analysis process takes into account the actual precision of all measurements in the series. Effectively, the procedure enables full and optimal accounting of all relevant data in a sequence of multiple experiments. This iterative procedure is in fact rather similar to the Kalman filtering technique discussed in §5.6 in which a sequence of measurements are used, one after the other, to improve the knowledge of the state of a system. As for Kalman filtering, one can also treat systems with an evolving state driven by known external parameters. In effect, the notion of Kalman filtering can then be articulated and developed within the Bayesian inference paradigm (see [63] for a recent review), but such a discussion lies far beyond the scope of this textbook.

Conjugate priors may be used even when there is a very little prior information about a system or observable. Indeed, it is typically possible (as illustrated in Figure 7.2) to choose the parameters of a prior to produce a very broad and essentially uninformative prior probability density. In effect, if the parameters are chosen to yield a probability density much broader than the typical precision of measurements, the posterior parameters shall be mainly determined by the measurement and with only a very weak dependency on the prior parameter values.

7.4 Bayesian Inference with Gaussian Noise

Process and measurement noise can often be considered Gaussian or approximately Gaussian; it is thus of great interest to consider applications of Bayesian inference involving

Gaussian data probability model. We begin, in §§7.4.1 and 7.4.2, with examples of Bayesian inference involving the estimation of the mean μ of a signal in the presence of Gaussian noise and the evaluation of the variance such a signal. These two examples set the stage for more elaborate problems involving the determination of model parameters by means of Bayesian fits. Linear model fits are considered in §7.4.3 whereas nonlinear model fits are discussed in §7.6.

7.4.1 Sample Mean Estimation

We first discuss the estimation of the mean μ of a constant signal in the presence of Gaussian noise with known and constant variance in §7.4.1. Given the noise level encountered in measurements may change, we enlarge the discussion to include varying but known noise variance in §7.4.1 and a priori unknown noise variance in §7.4.1.

Sample Mean Estimation with Fixed Variance Noise

Consider n independent measurements $\vec{x} = (x_1, x_2, \ldots, x_n)$ of a constant observable X. As prior information, let us assume measurements of X are known to fluctuate with zero bias but Gaussian noise of fixed and known variance σ_0^2.

Our goal is to determine the posterior probability $p(\mu|X, \sigma_0, I)$. From Bayes' theorem, we get

$$p(\mu|\vec{x}, \sigma_0, I) = \frac{p(\mu|I)p(\vec{x}|\mu, \sigma_0, I)}{p(\vec{x}|\sigma_0, I)}, \tag{7.156}$$

where $p(\vec{x}|\mu, \sigma_0, I)$ is the likelihood of the data for parameters μ and σ_0, $p(\mu|I)$ is the prior of μ, and $p(\vec{x}|\sigma_0, I)$ is the global likelihood of the data.

Let us first determine the likelihood function $p(X|\mu, \sigma_0, I)$. From the prior information, we know that each value x_i, $i = 1, \ldots, n$, represents a constant observable X with Gaussian noise and null bias. We can then model each measurement according to

$$x_i = \mu + e_i, \tag{7.157}$$

where the term e_i represents Gaussian noise with null mean and standard deviation σ_0:

$$p(e_i|\sigma_0, I) = \frac{1}{\sqrt{2\pi}\sigma_0} \exp\left[-\frac{e_i^2}{2\sigma_0^2}\right] \tag{7.158}$$

Substituting $e_i = x_i - \mu$ in Eq. (7.58), one gets a data probability model for the fluctuations of the measurements x_i:

$$p(x_i|\mu, \sigma_0, I) = \frac{1}{\sqrt{2\pi}\sigma_0} \exp\left[-\frac{(x_i - \mu)^2}{2\sigma_0^2}\right] \tag{7.159}$$

For notational convenience, let us define precision factors $\xi_0 = 1/\sigma_0^2$. The likelihood of the data $\vec{x}$ may then be written

$$p(\vec{x}|\mu, \xi_0, I) = \prod_{i=1}^{n} p(x_i|\mu, \xi_0, I) = \prod_{i=1}^{n} \sqrt{\frac{\xi_0}{2\pi}} \exp\left[-\frac{\xi_0}{2}(x_i-\mu)^2\right] \tag{7.160}$$

$$= \left(\frac{\xi_0}{2\pi}\right)^{\frac{n}{2}} \exp\left[-\frac{\xi_0}{2}\sum_{i=1}^{n}(x_i-\mu)^2\right] \tag{7.161}$$

$$= \left(\frac{\xi_0}{2\pi}\right)^{\frac{n}{2}} \exp\left[-\frac{\xi_0}{2}S\right], \tag{7.162}$$

where we defined the sum $S = \sum_{i=1}^{n}(x_i-\mu)^2$. Expanding this sum and introducing the sample mean $\bar{x} = n^{-1}\sum_{i=1}^{n} x_i$, and variance $s^2 = n^{-1}\sum_{i=1}^{n}(x_i-\bar{x})^2 = n^{-1}\sum x_i^2 - \bar{x}^2$, one gets

$$S = n(\mu-\bar{x})^2 + ns^2. \tag{7.163}$$

Defining $\xi_n = n\xi_0$, the likelihood can then be written

$$p(\vec{x}|\mu, \xi_0, I) = \left(\frac{1}{n}\right)^{\frac{n}{2}} \left(\frac{\xi_n}{2\pi}\right)^{\frac{n}{2}-1} \sqrt{\frac{\xi_n}{2\pi}} \exp\left[-\frac{\xi_n}{2}s^2\right] \tag{7.164}$$

$$\times \sqrt{\frac{\xi_n}{2\pi}} \exp\left[-\frac{\xi_n}{2}(\mu-\bar{x})^2\right].$$

Calculation of the posterior for μ requires a prior $p(\mu|I)$. For illustrative purposes, we will consider two cases of prior information on μ. In the first case, we will assume the mean μ is bound in some range $\mu_L \le \mu < \mu_H$ based on some theoretical considerations but without any particular preference within that range. This will require an uninformative prior. In the second case, we will assume previous experiments have reported an estimate μ_p with a standard deviation σ_p. We will then use a Gaussian conjugate prior.

Uninformative Prior (Case 1)

Given the bounds $\mu_L \le \mu < \mu_H$ and lack of preference in that range, we choose a uniform prior distribution

$$p(\mu|I) = \begin{cases} R_\mu^{-1} & \mu_L \le \mu \le \mu_H \\ 0 & \text{elsewhere,} \end{cases} \tag{7.165}$$

where the constant $R_\mu = \mu_H - \mu_L$ is determined by the normalization condition $\int p(\mu|I)\,d\mu = 1$.

Equipped with the prior, Eq. (7.165), and the likelihood, Eq. (7.164), we calculate the global likelihood of the data according to

$$p(X|I) = \int_{\mu_L}^{\mu_H} p(\mu|I)p(X|\mu, \xi_0, I)\,d\mu, \tag{7.166}$$

$$= \frac{I_{LH}}{R_\mu}\left(\frac{1}{n}\right)^{\frac{n}{2}} \left(\frac{\xi_n}{2\pi}\right)^{\frac{n}{2}-1} \sqrt{\frac{\xi_n}{2\pi}} \exp\left[-\frac{\xi_n}{2}s^2\right], \tag{7.167}$$

with

$$I_{LH} = \sqrt{\frac{\xi_n}{2\pi}} \int_{\mu_L}^{\mu_H} \exp\left(-\frac{\xi_n}{2}(\mu - \bar{x})^2\right) d\mu, \tag{7.168}$$

$$= \Phi(z_H) - \Phi(z_L), \tag{7.169}$$

in which the standard normal cumulative distribution $\Phi(z)$ is evaluated at $z_H = \sqrt{\xi_n}(\mu_H - \bar{x})$ and $z_L = \sqrt{\xi_n}(\mu_L - \bar{x})$.

Computation of the posterior of μ with Eq. (7.156) finally yields

$$p(\mu|\vec{x}, \sigma_0, I) = \begin{cases} \frac{1}{I_{LH}} \frac{1}{\sqrt{2\pi}\sigma_n} \exp\left[-\frac{(\mu-\bar{x})^2}{2\sigma_n^2}\right] & \text{for } \mu_L \le \mu < \mu_H, \\ 0 & \text{elsewhere,} \end{cases} \tag{7.170}$$

which corresponds to a truncated Gaussian distribution defined in the range $\mu_L \le \mu \le \mu_H$, with mode $\bar{x}$, standard deviation parameter $\sigma_n = \sigma_0/\sqrt{n}$, and normalization I_{LH}. We thus conclude that the observed sample mean $\bar{x}$ has the highest probability of being the true value of μ and the uncertainty of this estimate is inversely proportional to the square root of the sample size n.

Gaussian Conjugate Prior (Case 2)

Let us next consider a Gaussian prior with mean μ_p and precision ξ_p.

$$p(\mu|I) = \sqrt{\frac{\xi_p}{2\pi}} \exp\left[-\frac{\xi_p}{2}(\mu - \mu_p)^2\right]. \tag{7.171}$$

From our discussion of Gaussian conjugate priors in §7.3.3, we conclude that the posterior of μ is a Gaussian

$$p(\mu|\vec{x}, \xi_0, I) = \sqrt{\frac{\xi_p'}{2\pi}} \exp\left[-\frac{\xi_p'}{2}\left(\mu - \mu_p'\right)^2\right], \tag{7.172}$$

with mean μ_p' and precision ξ_p' given by

$$\mu_p' = \frac{n\xi_0}{n\xi_0 + \xi_p}\bar{x} + \frac{\xi_p}{n\xi_0 + \xi_p}\mu_p, \tag{7.173}$$

$$\xi_p' = n\xi_0 + \xi_p, \tag{7.174}$$

and the updated standard deviation may then be written

$$\sigma_p' = \frac{\sigma_p \sigma_0/\sqrt{n}}{\sqrt{\sigma_0^2/n + \sigma_p^2}}. \tag{7.175}$$

The results obtained in this section are obviously reminiscent of the classical inference properties of estimators for the mean of a sample discussed in §§5.1 and 5.5. Important differences exist, however, between the Bayesian result derived in this section and earlier results. In the frequentist approach, it is not possible to make a statement about the probability of a model parameter, and one obtains a confidence interval based on the measurement. Strictly speaking, the probability content of the interval is not the probability of finding the value of μ within the interval, but the probability of the interval to contain the observable value if and when the measurement is repeated several times. Consider, for instance, a 68% confidence interval. If the experiment is repeated $m = 100$ times, m distinct 68% confidence intervals will be obtained and only 68% of these intervals, on average, are expected to contain the actual value of the observable. By contrast, the Bayesian distribution obtained in this section does state the probability of finding μ in a certain interval, whether the experiment is repeated or not. This important difference stems from the definition of probability used in the Bayesian paradigm. Indeed, the Bayesian interpretation is far simpler and direct. The posterior $p(\mu|X, I)$ gives the probability density of μ and it can then be used to determine the probability of the observable value being any range of interest. Note in particular that if the bounds μ_L and μ_H are very broad and far outside the interval $\bar{x} \pm \sigma/\sqrt{n}$, then the value of I_{LH} quickly converges to unity. An interval $\pm\sigma/\sqrt{n}$ then has a probability of 68% as in the frequentist paradigm. However, if the bounds μ_L and μ_H are near the interval $\bar{x} \pm \sigma/\sqrt{n}$, the integral I_{LH} may deviate significantly from unity, and the probability of μ being found within $[\bar{x} - \sigma/\sqrt{n}, \bar{x} + \sigma/\sqrt{n}]$ may then far exceed 68% and even reach 100%. Note that in this case, the data do not add much to the prior information, and the posterior remains broad within the bounding range $\mu_L \leq \mu \leq \mu_H$.

Sample Mean Estimation with Variable Noise

Let us extend the discussion of the previous section to measurements in which the variance of the noise may vary from measurement to measurement. Let us continue to assume, however, that the variances are known, that is, we assign variances σ_i^2 to each of the n measurements and assume these values to be known based either on previous measurements or some other considerations.

The prior probability $p(\mu|I)$ is indifferent to the new noise conditions but the likelihood function must be slightly modified. For convenience, let us define precision factors $\xi_i = 1/\sigma_i^2$. The likelihood can then be written

$$p(\vec{x}|\mu, \sigma, I) = \prod_{i=1}^{n} \sqrt{\frac{\xi_i}{2\pi}} \exp\left[-\frac{\xi_i}{2}(x_i - \mu)^2\right] \tag{7.176}$$

$$= \left(\prod_{i=1}^{n} \sqrt{\frac{\xi_i}{2\pi}}\right) \exp\left[-\frac{1}{2}S_w\right], \tag{7.177}$$

where n is the number of measurements and the sum S_w is defined as

$$S_w = \sum_{i=1}^{n} \xi_i (x_i - \mu)^2 . \tag{7.178}$$

Introducing the weighted mean $\bar{x}_w$ and the weighted variance $\overline{s_w^2}$ defined respectively as

$$\bar{x}_w = \frac{\sum_{i=1}^{n} \xi_i x_i}{\sum_{i=1}^{n} \xi_i}, \tag{7.179}$$

$$\overline{s_w^2} = \frac{\sum_{i=1}^{n} \xi_i \left(x_i - \bar{x}_w\right)^2}{\sum_{i=1}^{n} \xi_i}, \tag{7.180}$$

as well as the sum of the precision factors

$$\xi_w = \sum_{i=1}^{n} \xi_i, \tag{7.181}$$

it is relatively simple to verify that the sum S_w may be written

$$S_w = \xi_w \left(\mu - \bar{x}_w\right)^2 + \xi_w \overline{s_w^2}, \tag{7.182}$$

from which we obtain the likelihood

$$p(\vec{x}|\mu, \{\xi_i\}, I) = \left(\prod_{i=1}^{n} \sqrt{\frac{\xi_i}{2\pi}}\right) \exp\left(-\frac{\xi_w}{2}\overline{s_w^2}\right) \exp\left[-\frac{\xi_w}{2}\left(\mu - \bar{x}_w\right)^2\right]. \tag{7.183}$$

Defining the average measurement precision $\bar{\xi}$ as

$$\bar{\xi} = \frac{1}{n}\sum_{i=1}^{n} \xi_i, \tag{7.184}$$

the precision parameter ξ_w is written

$$\xi_w = n\bar{\xi}, \tag{7.185}$$

from which we conclude that the precision ξ_w improves in proportion to the number n of measurements. Consequently, the standard deviation of the likelihood,

$$\sigma_w = 1/\sqrt{\xi_w}, \tag{7.186}$$

decreases as the square root of the number of measurements, that is, $\sigma_w \propto 1/\sqrt{n}$.

The posterior PDF $p(\mu|\vec{x}, \{\xi_i\}, I)$ is calculated similarly as in the previous section.

Bounded Uniform Prior (Case 1)

In the case of a bounded uniform prior, one gets

$$p(\mu|X, I) = \begin{cases} \frac{1}{I_w}\sqrt{\frac{\xi_w}{2\pi}} \exp\left[-\frac{\xi_w}{2}\left(\mu - \bar{x}_w\right)^2\right], & \text{for } \mu_L \le \mu < \mu_H, \\ 0 & \text{elsewhere.} \end{cases} \tag{7.187}$$

The posterior PDF of μ is once again a truncated Gaussian distribution with a normalization factor I_w defined by

$$I_w = \sqrt{\frac{\xi_w}{2\pi}} \int_{\mu_L}^{\mu_H} \exp\left[-\frac{\xi_w}{2}(\mu - \bar{x}_w)^2\right] d\mu. \tag{7.188}$$

The expectation value and most probable value of μ is $\bar{x}_w$ and the distribution has a standard deviation parameter σ_w, which, as per Eq. (7.186), varies inversely with the number of points x_i collected in the data sample. Once again, one can directly calculate the probability of the value μ be found in any particular interval with Eq. (7.187).

Gaussian Conjugate Prior (Case 2)

Given the likelihood, Eq. (7.183), and the Gaussian conjugate prior, Eq. (7.171), the posterior PDF for μ is a regular Gaussian with posterior mean μ'_p and precision ξ'_p given by

$$\mu'_p = \frac{\xi_w}{\xi_w + \xi_p}\bar{x}_w + \frac{\xi_p}{\xi_w + \xi_p}\mu_p, \tag{7.189}$$

$$\xi'_p = \xi_w + \xi_p = n\bar{\xi} + \xi_p \tag{7.190}$$

and the updated standard deviation may then be written

$$\sigma'_p = \frac{\sigma_p \bar{\sigma}/\sqrt{n}}{\sqrt{\bar{\sigma}^2/n + \sigma_p^2}}, \tag{7.191}$$

where

$$\bar{\sigma} = 1/\sqrt{\bar{\xi}}. \tag{7.192}$$

For sufficiently large n (and finite $\bar{\xi}$ and ξ_p), the prior parameter ξ_p becomes negligible, and the standard deviation of the posterior scales as

$$\sigma'_p \propto \frac{1}{\sqrt{n}}, \tag{7.193}$$

as expected. For small n, the mode of the distribution is very much influenced by the prior parameter μ_p, but for increasing larger values of n, the second term of Eq. (7.189) becomes progressively smaller and eventually negligible relative to the first term. The mode of the posterior is thus indeed eventually (i.e., for suitably large n) entirely determined by $\bar{x}_w$.

Sample Mean Estimation with Unknown Noise

In the two previous sections, we determined the posterior probability of the mean of a sample assuming a priori knowledge of the variance of the noise. What if the variance of the noise is in fact a priori unknown but can be assumed constant? Or what if the measured x_i also contain a weak but unknown periodic signal? With such limited knowledge, one must conduct a generic analysis that disregards specific unknown details of the observables x_i. The CLT tells us that a sum of several processes should have a Gaussian distribution.

Alternatively, the Max Entropy principle informs us that the most conservative choice for the situation at hand is also a Gaussian distribution. Either way, as long as the noise is finite, we should once again be able use the data model

$$x_i = \mu + e_i \tag{7.194}$$

and assume the e_i are Gaussian distributed with some unknown standard deviation σ_0. Since we now have two unknowns, we must first conduct our analysis to determine a posterior probability of both μ and σ. From Bayes' theorem, we get

$$p(\mu, \sigma|X, I) = \frac{p(\mu, \sigma|I)p(X|\mu, \sigma, I)}{p(X|I)}. \tag{7.195}$$

An estimate of μ can then be obtained by marginalization of this posterior probability against σ.

$$p(\mu|X, I) = \int p(\mu, \sigma|X, I)\, d\sigma. \tag{7.196}$$

Calculation of the posterior proceeds as in previous sections. Given we assumed the standard deviation of the measurement σ_0 is constant, albeit unknown, the likelihood of the data is given by Eq. (7.160). We also need a prior $p(\mu, \sigma|I)$ for μ and σ. Since little is known about the noise, it is safe to factorize the prior into a product

$$p(\mu, \sigma|I) = p(\mu|I)p(\sigma|I). \tag{7.197}$$

Let use the uninformative uniform prior PDF, Eq. (7.165), for μ. Determination of $p(\sigma|I)$ requires additional considerations. The standard deviation σ_0 is positive definite and behaves, for all practical intents, as a scale parameter. It is also very unlikely to be infinite, as this would imply a signal that carries very large energy. It is thus sensible to assume a Jeffreys prior with minimum and maximum bounds σ_L and σ_H, respectively.

$$p(\sigma|I) = \begin{cases} \frac{1}{\sigma_0 \ln(\sigma_H/\sigma_L)} & \text{for } \sigma_L \leq \sigma < \sigma_H \\ 0 & \text{elsewhere.} \end{cases} \tag{7.198}$$

Inserting the expressions for the priors and the likelihood in Eq. (7.196), the posterior $p(\mu|X, I)$ may now be calculated. One gets after simplification

$$p(\mu|X, I) = \frac{\int_{\sigma_L}^{\sigma_H} d\sigma_0 \sigma_0^{-(n+1)} \exp\left[-\frac{S}{2\sigma_0^2}\right]}{\int_{\mu_L}^{\mu_H} d\mu \int_{\sigma_L}^{\sigma_H} d\sigma_0 \sigma_0^{-(n+1)} \exp\left[-\frac{S}{2\sigma_0^2}\right]}. \tag{7.199}$$

In order to evaluate the preceding integrals, it is convenient to define $\zeta = S/2\sigma_0^2$. Insertion in Eq. (7.199) yields after a short calculation,

$$p(\mu|X, I) = \frac{\int_{\zeta_L}^{\zeta_H} S^{-n/2} \zeta^{n/2-1} e^{-\zeta} d\zeta}{\int_{\mu_L}^{\mu_H} \int_{\zeta_L}^{\zeta_H} S^{-n/2} \xi^{n/2-1} e^{-\zeta} d\zeta\, d\mu}, \tag{7.200}$$

which involves integrals that may be evaluated in terms of the incomplete gamma function. Although the integral in ζ depends on μ (through S dependency on μ), one can verify that provided $\sigma_L \ll s$ and $\sigma_H \gg s$, it is essentially constant. Equation (7.200) thus approximately simplifies to

$$p(\mu|X, I) \approx \frac{S^{-n/2}}{\int_{\mu_L}^{\mu_H} S^{-n/2} d\mu}, \tag{7.201}$$

Substituting the value of S given by Eq. (7.163), we get after some additional algebra

$$p(\mu|X, I) \approx \frac{\left[1 + \frac{(\mu-\bar{x})^2}{s^2}\right]^{-\frac{n}{2}}}{\int_{\mu_L}^{\mu_H} \left[1 + \frac{(\mu-\bar{x})^2}{s^2}\right]^{-\frac{n}{2}} d\mu}, \tag{7.202}$$

which, as it happens, features a structure rather similar to Student's t-distribution. To see this, first recall from §3.12.1 that Student's t-distribution is defined according to

$$p_t(t|\nu) = \frac{\Gamma\left(\frac{\nu+1}{2}\right)}{\sqrt{\nu\pi}\,\Gamma\left(\frac{\nu}{2}\right)} \left(1 + \frac{t^2}{\nu}\right)^{-\frac{\nu+1}{2}}, \tag{7.203}$$

where $\nu = n - 1$ is the number of degrees of freedom being sampled. Setting

$$\frac{t^2}{\nu} = \frac{(\mu - \bar{x})^2}{s^2}, \tag{7.204}$$

we find that the numerator of Eq. (7.202) is indeed structurally similar to Student's t-distribution. We can then write

$$p(\mu|X, I) \approx \frac{1}{I_s} \frac{1}{\sqrt{\pi(n-1)}} \frac{\Gamma(\frac{n}{2})}{\Gamma(\frac{n-1}{2})} \left[1 + \frac{(\mu - \bar{x})^2}{s^2}\right]^{-n/2} \tag{7.205}$$

where the extra normalization factor I_s is defined as an integral of Student's t-distribution,

$$I_s = \frac{1}{\sqrt{\pi(n-1)}} \frac{\Gamma(\frac{n}{2})}{\Gamma(\frac{n-1}{2})} \int_{\mu_L}^{\mu_H} \left[1 + \frac{(\mu - \bar{x})^2}{s^2}\right]^{-n/2} d\mu. \tag{7.206}$$

As in the case of the coefficients I_{LH} and I_w defined in previous sections, one finds that I_s is essentially equal to unity for bounds of μ extending to $\pm\infty$, and $p(\mu|X, I)$ can then be strictly identified to a Student t-distribution. For a narrow range $\mu_L \leq \mu \leq \mu_H$, the functional shape remains the same, but the normalization is changed by the presence of the factor I_s. That means that the probability of μ being, say, in the interval $\mu \pm s$ in general exceeds that of the Student t-distribution and can even reach 100% if μ_L and μ_H are both very close to the actual value of μ[9].

7.4.2 Sample Standard Deviation Estimation

Let us now turn to the determination of the variance (or standard deviation) of a sample $\{x_i\}$ of size n. As in the previous section, we can use the data to infer a joint posterior probability $p(\mu, \sigma|\vec{x}, I)$ for μ and σ, but we shall now marginalize this probability against

[9] Similarly to the case of the Gaussian distribution encountered in §7.4.1.

μ to focus our attention on σ. We must then compute

$$p(\sigma|\vec{x}, I) = \frac{p(\sigma|I) \int p(\mu|I)p(\vec{x}|\mu, \sigma, I)\, d\mu}{p(\vec{x}|I)}, \tag{7.207}$$

where the priors and likelihood are the same as in the previous section. Substituting expressions for the priors and likelihoods, and simplifying factors common to the numerator and denominator, one gets

$$p(\sigma|\vec{x}, I) = \frac{\sigma^{-(n+1)} \exp\left(-ns^2/2\sigma^2\right) \int_{\mu_L}^{\mu_H} d\mu \exp\left(-\frac{n(\mu-\bar{x})^2}{2\sigma^2}\right)}{\int_{\sigma_L}^{\sigma_H} d\sigma \sigma^{-(n+1)} \exp\left(-ns^2/2\sigma^2\right) \int_{\mu_L}^{\mu_H} d\mu \exp\left(-\frac{n(\mu-\bar{x})^2}{2\sigma^2}\right)} \tag{7.208}$$

The integrals in μ would yield $\sqrt{2\pi/n\sigma}$ if the bounds μ_L and μ_H extended to $\pm\infty$. For finite values, we may then replace the integrals by $f\sqrt{2\pi/n\sigma}$, where f is a constant in the range $0 \le f \le 1$ determined by μ_L, μ_H, σ, and n. The posterior thus simplifies to

$$p(\sigma|\vec{x}, I) = \begin{cases} K\sigma^{-n} \exp\left(-\frac{ns^2}{2\sigma^2}\right) & \text{for } \sigma_L \le \sigma < \sigma_H \\ 0 & \text{elsewhere,} \end{cases} \tag{7.209}$$

with

$$K^{-1} = \int_{\sigma_L}^{\sigma_H} d\sigma \sigma^{-n} \exp\left(-\frac{ns^2}{2\sigma^2}\right) \tag{7.210}$$

It is illustrated in Figure 7.3 for $s^2 = 1$ and selected values of n. One observes that the function $p(\sigma|\vec{x}, I)$ is peaked at $\sigma = s$ and that it becomes progressively narrower with increasing values of n. Calculating and solving $\partial p/\partial\sigma = 0$ for σ, one verifies the mode of the distribution is indeed

$$\hat{\sigma} = s. \tag{7.211}$$

Moreover, computing the first and second moments of the distribution in terms of the inverse gamma integral, one finds

$$\langle\sigma\rangle = \frac{\sqrt{n}s\Gamma\left[(n-2)/2\right]}{\sqrt{2}\Gamma\left[(n-1)/2\right]} \tag{7.212}$$

$$\langle\sigma^2\rangle = \frac{ns^2}{n-1}. \tag{7.213}$$

One thus concludes that values of the mode $\hat{\sigma}$, the first moment $\langle\sigma\rangle$, and the root mean square $\sqrt{\langle\sigma^2\rangle}$ are distinct. However, one can readily verify that all three values asymptotically converge to s in the large n limit.

7.4.3 Bayesian Fitting with a Linear Model

Problem Definition

The basic Bayesian inference procedure we have used for the estimation of the mean and standard deviation of a dataset can be readily extended for the determination of the

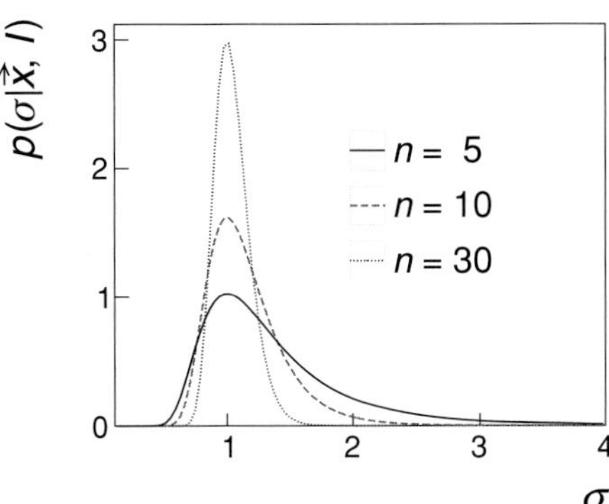

Fig. 7.3 Evolution of the posterior $p(\sigma|\vec{x}, I)$, given by Eq. (7.209), with the sample size n for $s^2 = 1$.

parameters of a model meant to describe the relation between a dependent observable Y and an independent variable X.

Let us consider a set of n measurements $\{x_i, y_i\}$, $i = 1, \ldots, n$. Let us posit that there exists a relation between the measured values x_i and y_i we can model according to

$$y_i = f(x_i|\vec{a}) + e_i, \tag{7.214}$$

where, as in previous sections, the terms e_i represent Gaussian noise with null expectation value and known standard deviation σ_i, while $f(x_i|\vec{a})$ is a (physical) model function expressing the relation between X and Y but with m unknown model parameters $\vec{a} = (a_1, a_2, \ldots, a_m)$.

Calculation of the Likelihood

Bayesian fits with arbitrary models (i.e., models featuring nonlinear dependencies on parameters $\vec{a}$) will be considered in §7.6. In this section, we first consider generalized linear model functions of the form given by Eq. (5.90) encountered in §5.2.5:

$$f(x_i|\vec{a}) = \sum_{k=1}^{m} a_k f_k(x_i), \tag{7.215}$$

where the coefficients $f_k(x_i)$ are linearly independent functions of x that do not depend on the model parameters $\vec{a}$. The noise terms may then be written

$$e_i = y_i - f(x_i|\vec{a}) \tag{7.216}$$

$$= y_i - \sum_{k=1}^{m} a_k f_k(x_i). \tag{7.217}$$

Here again, it is convenient to define the precision ξ_i of each of the n measurements according to

$$\xi_i = \frac{1}{\sigma_i^2}. \tag{7.218}$$

If the noise terms e_i are all mutually independent, the likelihood of the data is the product of each of their probabilities

$$p(\vec{y}|\vec{a}, I) = \prod_{i=1}^{n} \sqrt{\frac{\xi_i}{2\pi}} \exp\left[-\frac{\xi_i}{2}(y_i - f(x_i))^2\right] \tag{7.219}$$

$$= (2\pi)^{-n/2} \left(\prod_{i=1}^{n} \sqrt{\xi_i}\right) \exp\left[-\frac{1}{2}\sum_{i=1}^{n} \xi_i (y_i - f(x_i))^2\right] \tag{7.220}$$

$$= (2\pi)^{-n/2} \left(\prod_{i=1}^{n} \sqrt{\xi_i}\right) \exp\left[-\frac{1}{2}S_w\right] \tag{7.221}$$

where, in the third line, we introduced the sum S_w defined as

$$S_w = \sum_{i=1}^{n} \xi_i (y_i - f(x_i))^2 . \tag{7.222}$$

If the noise terms e_i are correlated with a covariance matrix $\mathbf{V}$, the sum may instead be written

$$S_w = \sum_{i=1}^{n}\sum_{j=1}^{n} \left(y_i - \sum_{k=1}^{m} a_k f_k(x_i)\right) \left(V^{-1}\right)_{ij} \left(y_j - \sum_{k'=1}^{m} a_{k'} f_{k'}(x_i)\right), \tag{7.223}$$

which is identical to the χ^2 function given by Eq. (5.92), for a linear model, discussed in §5.2.5 in the context of classical inference. For notational convenience, we once again introduce coefficients $F_{ik} \equiv f_k(x_i)$ as well as vector notations $\vec{y} = (y_1, \ldots, y_N)$, $\vec{a} = (a_1, \ldots, a_N)$, and matrix notations $\mathbf{F}$ and $\mathbf{V}$. The likelihood $p(\vec{y}|\vec{a}, I)$ then becomes

$$p(\vec{y}|\vec{a}, I) = \frac{1}{(2\pi)^{n/2}} \frac{1}{|\mathbf{V}|^{1/2}} \exp\left[-\frac{1}{2}S_w\right], \tag{7.224}$$

where $|\mathbf{V}|$ represents the determinant of the covariance matrix $\mathbf{V}$, and the quadratic form S_w can now be written

$$S_w = \vec{y}^T\mathbf{V}^{-1}\vec{y} - \vec{a}^T\mathbf{F}^T\mathbf{V}^{-1}\vec{y} - \vec{y}^T\mathbf{V}^{-1}\mathbf{F}\vec{a} + \vec{a}^T\mathbf{F}^T\mathbf{V}^{-1}\mathbf{F}\vec{a}. \tag{7.225}$$

Calculation of the Posterior

With the likelihood $p(\vec{y}|\vec{a}, I)$ and the expression, Eq. (7.225), for S_w in hand, we now need prior probabilities for the parameters $\vec{a}$. Given the Gaussian form of the likelihood, two choices are readily possible and convenient: bounded uniform priors for all a_i and conjugate Gaussian priors.

Bounded Uniform Priors (Case 1)

Let us first consider bounded flat priors for each of the m parameters:

$$p(\vec{a}|I) = \begin{cases} \prod_{i=1}^{m} R_i^{-1} & a_{i,\min} \leqslant a_i \leqslant a_{i,\max} \\ 0 & \text{elsewhere} \end{cases} \tag{7.226}$$

where $R_i = a_{i,\max} - a_{i,\min}$. The posterior probability of parameters $\vec{a}$ may then be written

$$p(\vec{a}|\vec{y}, I) = K \exp\left(-S_w/2\right) \qquad \text{(flat priors)}$$

where K is a constant determined by

$$K = \frac{p(\vec{a}|I)}{p(\vec{y}|I)} \tag{7.227}$$

with

$$p(\vec{y}|I) = \int_{R_i} \prod da_i p(\vec{a}|I)p(\vec{y}|\vec{a}, I) \tag{7.228}$$

$$= \prod_{i=1}^{m} R_i^{-1}(2\pi)^{-n/2}|\mathbf{V}|^{-1/2} \int_{R_i} \exp(-S_w/2). \tag{7.229}$$

The posterior $p(\vec{a}|\vec{y}, I)$, given by Eq. (7.227), constitutes the formal answer to the Bayesian inference optimization problem we set out to solve. It gives the probability density of all values of parameters $\vec{a}$ in parameter space. As such, it contains everything there is to know about the parameters $\vec{a}$ given a particular set of n measured values y_i and their respective measurement noise σ_i.

A full determination of $p(\vec{a}|\vec{y}, I)$ requires the cumbersome calculation of the integral in Eq. (7.229). However, much can be said about the parameters $\vec{a}$ without actually performing this integral. The PDF $p(\vec{a}|\vec{y}, I)$ typically spans a wide range of values in parameter space, but as per the physical model, Eq. (7.214), each of the a_i should have a specific value, that is, $\vec{a}$ should be a specific vector in parameter space. Which vector should one choose? It stands to reason that one should choose the value $\vec{a}$ that maximizes the posterior $p(\vec{a}|\vec{y}, I)$. One is then interested in finding the (global) maximum or mode of this function in the model parameter space defined by the bounds $a_{i,\min} \leqslant a_i \leqslant a_{i,\max}$. This is achieved, as usual, by jointly setting derivatives of $p(\vec{a}|\vec{y}, I)$ to zero and solving for $\vec{a}$.

$$\frac{\partial p(\vec{a}|\vec{y}, I)}{\partial a_i} = 0 \quad \text{for } i = 1, \ldots, m. \tag{7.230}$$

Given our choice of flat prior, $\partial p(\vec{a}|\vec{y}, I)/\partial a_i$ is proportional to an exponential function of $-S_w/2$. There is thus no need to calculate the constant K, and optimization of the exponential amounts to minimization of S_w.

$$\frac{\partial S_w}{\partial a_i} = 0 \quad \text{for } i = 1, \ldots, m. \tag{7.231}$$

The optimization of the posterior $p(\vec{a}|\vec{y}, I)$ by Eq. (7.227) with uniform priors is thus strictly equivalent to the least-squares minimization problem we encountered in §5.2.5 (classical inference). Optimum estimators of the parameters $\vec{a}$ are thus given by

$$\hat{a} = \alpha^{-1}\vec{b}. \tag{7.232}$$

where, as we showed already in §5.2.5, the matrix α and the vector $\vec{b}$ are given by

$$\alpha = \mathbf{F}^T V^{-1}\mathbf{F}, \tag{7.233}$$

$$\vec{b} = \mathbf{F}^T V^{-1}\vec{y}. \tag{7.234}$$

Based on the similitude between the foregoing solution and the frequentist solution discussed in §5.2.5, it may appear as if Bayesian inference brings nothing new to the optimization problem. But that is not quite correct. Bayesian inference in fact provides at least two advantages. The first has to do with the Bayesian interpretation of probability: the posterior $p(\vec{a}|\vec{y}, I)$ expresses the degree of belief of observing the parameters a_i in ranges $[a_i, a_i + da_i]$. One can then directly calculate credible ranges for the parameters a_i by simple integration and marginalization of the posterior $p(\vec{a}|\vec{y}, I)$. These integrals then represent the probability of observing the parameters a_i in such ranges, as we discuss in more detail in §7.4.3. Second, Bayesian inference enables the determination of posteriors when prior information about a system a priori restricts the ranges of the parameters. Bayesian inference then effectively combines old and new information toward the determination of optimum parameters. Such a combination is not formally defined in the context of the frequentist paradigm.

Linear Model Fit with Gaussian Priors (Case 2)

Let us now consider that an experiment might be the successor of one or several prior experiments. Instead of using flat priors for the parameters a_i, one might then use the (combined) outcome of these previous experiments, which we may assume have yielded a Gaussian posterior with optimum values $\vec{a}_0$ and a covariance matrix Q. Let us then define a generic Gaussian prior as

$$p(\vec{a}|I) = \frac{1}{(2\pi)^m |\mathbf{Q}|^{1/2}} \exp\left[-\frac{1}{2}(\vec{a}-\vec{a}_0)^T Q^{-1}(\vec{a}-\vec{a}_0)\right], \tag{7.235}$$

where m is the number of parameters $\vec{a}$ (i.e., the dimension of $\vec{a}$), and $|\mathbf{Q}|$ is the determinant of the covariance matrix $\mathbf{Q}$. The likelihood, Eq. (7.224), remains unchanged but the posterior must be modified to account for the Gaussian priors. One gets

$$p(\vec{a}|\vec{y}, I) = \frac{p(\vec{a}|I)p(\vec{y}|\vec{a}, I)}{p(\vec{y}|I)} \tag{7.236}$$

$$= K \exp\left[-\frac{1}{2}(\vec{a}-\vec{a}_0)^T Q^{-1}(\vec{a}-\vec{a}_0)\right] \exp\left(-S_w/2\right), \tag{7.237}$$

where K is a constant determined by the global likelihood $p(\vec{y}|I)$. The preceding product of exponentials may be reduced to a single exponential by addition of their arguments, which we write

$$p(\vec{a}|Y, I) = K \exp\left(-S'_w/2\right) \tag{7.238}$$

with

$$S'_w = (\vec{a}-\vec{a}_0)^T Q^{-1}(\vec{a}-\vec{a}_0) + S_w. \tag{7.239}$$

Expanding the terms, and rearranging, S'_w may be written

$$S'_w = \vec{y}^T\mathbf{V}^{-1}\vec{y} + \vec{a}_0^T\mathbf{Q}^{-1}\vec{a}_0 - \vec{a}^T\left(\mathbf{Q}^{-1}\vec{a}_0 + \mathbf{F}^T\mathbf{V}^{-1}\vec{y}\right) \tag{7.240}$$

$$- \left(\vec{a}_0^T\mathbf{Q}^{-1} + \vec{y}^T\mathbf{V}^{-1}\mathbf{F}\right)\vec{a} + \vec{a}^T\left(\mathbf{Q}^{-1} + \mathbf{F}^T\mathbf{V}^{-1}\mathbf{F}\right)\vec{a}. \tag{7.241}$$

It is then possible (Problem 7.6) to show that the optimal solution for $\vec{a}$ is of the form of Eq. (7.232) but with modified values of the matrix α and vector $\vec{b}$ as follows:

$$\alpha' = \mathbf{F}^T V^{-1}\mathbf{F} + \mathbf{Q}^{-1} \tag{7.242}$$

$$\vec{b}' = \mathbf{Q}^{-1}\vec{a}_0 + \mathbf{F}^T\mathbf{V}^{-1}\vec{y}. \tag{7.243}$$

Errors and Credible Ranges

Let $\delta a_k = a_k - \hat{a}_k$ represent the difference between an arbitrary value a_k and the optimal estimate obtained by optimization of the posterior $p(\vec{a}|Y, I)$ and let $S_{\min}$ be the value of S_w corresponding to that optimum. In order to make a statement about parameter errors and credible ranges, it is convenient to express the value of S_w for arbitrary parameters $\vec{a}$ as a Taylor expansion:

$$S_w(\delta\vec{a}) = S_{\min} + \sum_{k=1}^{m} \left.\frac{\partial S_w}{\partial a_k}\right|_{\min} \delta a_k + \frac{1}{2}\sum_{k,k'=1}^{m} \left.\frac{\partial^2 S_w}{\partial a_k \partial a_{k'}}\right|_{\min} \delta a_k \delta a_{k'}. \tag{7.244}$$

First note that the second term of this expression is null by construction because the first derivative is evaluated at the minimum. Additionally note that the series does not involve terms of third or higher order because Eq. (7.225) does not contain such terms. Let us define $\Delta S_w = S_w - S_{\min}$. From Eq. (7.244), we get

$$\Delta S_w = \frac{1}{2}\sum_{k,k'=1}^{m} \left.\frac{\partial^2 S_w}{\partial a_k \partial a_{k'}}\right|_{\min} \delta a_k \delta a_{k'}. \tag{7.245}$$

Flat Parameter Priors (Case 1)

For flat parameter model priors, calculation of second derivatives of the solution Eq. (7.225) yields

$$\Delta S_w = \delta\vec{a}^T\mathbf{F}^T\mathbf{V}^{-1}\mathbf{F}\delta\vec{a}. \tag{7.246}$$

We can then write the posterior $p(\vec{a}|Y, I)$ as

$$p(\vec{a}|Y, I) = K' \exp\left(-\frac{1}{2}\Delta S_w\right) \tag{7.247}$$

where K', given by

$$K' = K \exp\left(-\frac{1}{2}S_{\min}\right), \tag{7.248}$$

is just a constant. It then becomes relatively straightforward to determine the covariance matrix $\mathbf{U}$ of the parameters $\vec{a}$ as it may be calculated in terms of the variance $\mathbf{V}$ of the data points $\vec{y}$ as follows:

$$\mathbf{U} = \left[\frac{d\vec{a}}{d\vec{y}}\right]\mathbf{V}\left[\frac{d\vec{a}}{d\vec{y}}\right]^T. \tag{7.249}$$

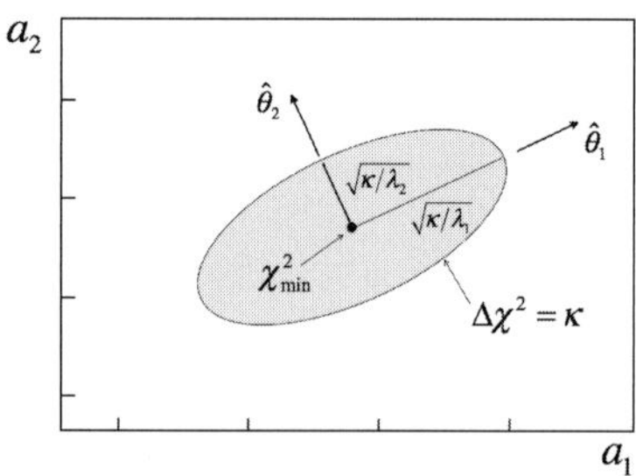

Fig. 7.4 $\Delta\chi^2 = \kappa$ ellipse in a_1 vs. a_2 parameter space.

Using Eqs. (7.233) and (7.234), calculation of the derivative $d\vec{a}/d\vec{y}$ yields

$$\frac{d\vec{a}}{d\vec{y}} = \left(\mathbf{F}^T\mathbf{V}^{-1}\mathbf{F}\right)^{-1}\mathbf{F}^T\mathbf{V}^{-1}. \tag{7.250}$$

Substitution of this expression in Eq. (7.249) yields

$$\mathbf{U} = \left(\mathbf{F}^T\mathbf{V}^{-1}\mathbf{F}\right)^{-1}\mathbf{F}^T\mathbf{V}^{-1}\mathbf{V}\left[\left(\mathbf{F}^T\mathbf{V}^{-1}\mathbf{F}\right)^{-1}\mathbf{F}^T\mathbf{V}^{-1}\right]^T. \tag{7.251}$$

The matrix $\mathbf{V}$ and its inverse are symmetric by construction. Remembering that the transpose of a product of matrices is equal to the product of the transposed in reversed order (e.g., $[\mathbf{ABC}]^T = \mathbf{C}^T\mathbf{B}^T\mathbf{A}^T$), Eq. (7.251) simplifies and we thus conclude that

$$\mathbf{U} = \left(\mathbf{F}^T\mathbf{V}^{-1}\mathbf{F}\right)^{-1}. \tag{7.252}$$

We find that the expression of $\mathbf{U}^{-1} = \mathbf{F}^T\mathbf{V}^{-1}\mathbf{F}$ is identical to α, given by Eq. (7.233). We can then write $p(\vec{a}|\vec{y}, I)$ as a multidimensional Gaussian of the form

$$p(\vec{a}|Y, I) = K' \exp\left(-\frac{1}{2}\delta\vec{a}^T\mathbf{U}^{-1}\delta\vec{a}\right), \tag{7.253}$$

where $\mathbf{U}$ represents the covariance matrix of the model parameters $\vec{a}$ we calculated earlier.

Since $\mathbf{U}$ is a symmetric matrix, the principal axis theorem tells us there exists a transformation $\delta\vec{a} = \mathbf{O}\delta\vec{\theta}$ that transforms $\delta\vec{a}^T\mathbf{U}^{-1}\delta\vec{a}$ into $\delta\vec{\theta}^T\mathbf{\Lambda}\vec{\theta}$ and such that $\mathbf{\Lambda}$ is a diagonal matrix whose diagonal elements are the eigenvalues of $\mathbf{U}$. It is thus always possible to transform the parameters $\vec{a}$ to uncorrelated parameters $\vec{\theta}$. For $m = 2$, this takes the form

$$\Delta S_w = \begin{pmatrix}\delta\theta_1 & \delta\theta_2\end{pmatrix}\begin{pmatrix}\lambda_1 & 0\\ 0 & \lambda_2\end{pmatrix}\begin{pmatrix}\delta\theta_1\\ \delta\theta_2\end{pmatrix} \tag{7.254}$$

$$= \lambda_1\delta\theta_1^2 + \lambda_2\delta\theta_2^2, \tag{7.255}$$

where λ_1 and λ_2 are the eigenvalues of $\mathbf{U}$. Clearly, $\Delta S_w = \kappa$ defines an ellipse

$$1 = \frac{\theta_1^2}{\kappa/\lambda_1} + \frac{\theta_2^2}{\kappa/\lambda_2}, \tag{7.256}$$

as illustrated in Figure 7.4.

In general, fixed values of ΔS_w define credible regions consisting of a simple range in one dimension (i.e., a single parameter), ellipses in two dimensions, ellipsoids in three,

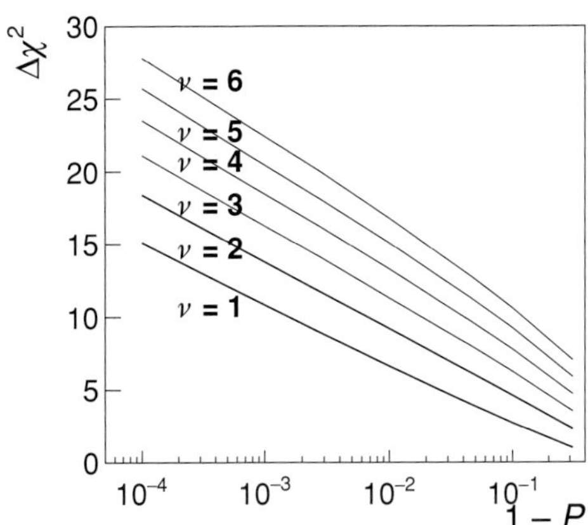

Fig. 7.5 Increment $\Delta\chi^2$ vs. $1 - P$ for selected values of the number of degrees of freedom ν (fit parameters).

and hyper-ellipsoids in more than three dimensions. The probability P associated with such credible regions is obtained by integration of the posterior within the boundary of the ellipse (or ellipsoid in $m = 3$ or hyper-ellipsoid $m \geq 4$ dimensions) defined by $\Delta S_w = \kappa_{\text{crit}}$

$$P = \int_{\Delta S_w < \kappa_{\text{crit}}} p(\vec{a}|Y, I) \prod_k da_i. \tag{7.257}$$

Since S_w is by construction a χ^2 variable, one can show that the probability P is given by

$$P = 1 - \frac{\gamma(m/2, \kappa/2)}{\Gamma(m/2)}, \tag{7.258}$$

where $\gamma(\alpha, \beta)$ is the incomplete gamma function. For instance, a region containing a probability of $P = 68.3\%$ corresponds to a $\Delta\chi^2 = 1$ for $m = 1$ and $\Delta\chi^2 = 2.3$ for $m = 2$, while for $P = 90.0\%$ one has $\Delta\chi^2 = 2.71$ and 4.61 for $m = 1$ and $m = 2$, respectively. Other values of $\Delta\chi^2$ are shown as a function of $1 - P$ for several values of the number of degrees of freedom ν in Figure 7.5.

Recall from our discussion in §6.1.2 that the notion of error is not absolute but predicated by a choice of probability content. In the frequentist approach, with one parameter and Gaussian errors, a probability content of 68.3% determines a $\pm 1\sigma$ interval which may then be interpreted as the error on the parameter. Choosing the probability content to be 95.45% instead, one gets a $\pm 2\sigma$ error interval. In the Bayesian approach discussed in this section, the notion of error is also readily derived from the probability content within selected intervals. However, rather than being called confidence intervals, the selected intervals are typically named **credible intervals** and they are defined by integration of the posterior which yields a given probability content. The interpretation of the credible interval is also far more direct. Because the posterior is the probability density of the parameters a_i, integrals over selected intervals of the parameters (e.g., credible range) of the posterior correspond to the probability of finding the true value of the parameter within the selected interval. It then becomes possible to define credible ranges associated with probability of 68%, 96%, or any other probability. Equation (7.258) then provides a one-to-one relation between the probability content and the increment $\kappa = \Delta\chi^2$ defining its associated credible range. It should be stressed that for $m \geq 2$ parameters, the credible range is not a simple interval but typically a hyper-ellipsoidal locus in parameter space. It is thus not formally possible (or correct) to identify the error on a given parameter by stating a single interval such as $\mu \pm \delta\mu$. One must produce the whole ellipsoidal region, which for $m = 2$ reduces to an ellipse, as illustrated in Figure 7.4. However, it is also possible to obtain the error on a single parameter by marginalizing all the others, as we discuss in §7.4.3.

Gaussian Conjugate Priors (Case 2)

Calculation of the (modified) covariance matrix $\mathbf{U}'$ of the model parameters $\vec{a}'$ obtained with Gaussian priors proceeds similarly to the calculation of $\mathbf{U}$. Using the modified solutions, Eq. (7.242), one finds

$$\mathbf{U}' = \left(\mathbf{F}^T\mathbf{V}^{-1}\mathbf{F} + \mathbf{Q}^{-1}\right)^{-1}\mathbf{F}^T\mathbf{V}^{-1}\mathbf{V}\left[\left(\mathbf{F}^T\mathbf{V}^{-1}\mathbf{F} + \mathbf{Q}^{-1}\right)^{-1}\mathbf{F}^T\mathbf{V}^{-1}\right]^T, \tag{7.259}$$

$$= \left(\mathbf{F}^T\mathbf{V}^{-1}\mathbf{F} + \mathbf{Q}^{-1}\right)^{-1}\mathbf{F}^T\mathbf{V}^{-1}\mathbf{F}\left(\mathbf{F}^T\mathbf{V}^{-1}\mathbf{F} + \mathbf{Q}^{-1}\right)^{-1}. \tag{7.260}$$

Substituting the inverse of the parameter covariance, $\mathbf{U}^{-1} = \mathbf{F}^T\mathbf{V}^{-1}\mathbf{F}$, obtained for flat priors, Eq. (7.252), we get

$$\mathbf{U}' = \left(\mathbf{U}^{-1} + \mathbf{Q}^{-1}\right)^{-1}\mathbf{U}^{-1}\left(\mathbf{U}^{-1} + \mathbf{Q}^{-1}\right)^{-1}. \tag{7.261}$$

Inspection of this expression shows that the fit modifies (shifts) the prior a_0 and reduces the prior variances Q_{ii} of the parameters. These effects are easiest to identify when reducing the scope of the problem to a single parameter. Let $Q \equiv \sigma_0^2$ represent the variance of the parameter observed by the previous experiment (prior knowledge) and let $\mathbf{U} = \left(\mathbf{F}^T\mathbf{V}^{-1}\mathbf{F}\right)^{-1} \equiv \sigma^2$ be the variance of the parameter determined with uniform priors (or from the likelihood function alone). The original solution $\vec{a} = \alpha^{-1}b$ may be written $a = \sigma^2 b$, or $b = a/\sigma^2$. The modified solution is then $a' = (\sigma_0^{-2} + \sigma^{-2})^{-1}(a_0\sigma_0^{-2} + a\sigma^{-2})$ which simplifies to

$$a' = a_0 + \zeta\,(a - a_0), \tag{7.262}$$

where

$$\zeta = \frac{\sigma_0^2}{\sigma^2 + \sigma_0^2}. \tag{7.263}$$

We find, similarly to Kalman filters, that the (new) value a' obtained from the fit is equal to the prior value, a, plus a term proportional to the difference $a - a_0$ times a "gain" factor ζ. Effectively, the parameter value obtained with the inclusion of the prior information is shifted by an amount proportional to the difference between the prior value of the parameter and that obtained on the basis of the likelihood alone. The information carried by the prior carries little weight, however, if the variance of the measurement, σ^2, is much smaller than the variance σ_0^2 of the prior. Indeed, for $\sigma^2 \ll \sigma_0^2$, one gets $\zeta \to 1$ and $a' \approx a$. Similarly, for a single parameter, the variance $\mathbf{U}'$, Eq. (7.261), may be written

$$\sigma'^2 = \left(\sigma^{-2} + \sigma_0^{-2}\right)^{-1} \sigma^{-2} \left(\sigma^{-2} + \sigma_0^{-2}\right)^{-1}, \tag{7.264}$$

which simplifies to

$$\sigma' = \sigma \frac{\sigma_0^2}{\sigma_0^2 + \sigma^2}. \tag{7.265}$$

The posterior variance is determined by a combination of the prior variance σ_0^2 and the variance σ^2 of the measurement. The posterior variance is reduced relative to the prior variance thanks to new knowledge acquired with the measurement. If the procedure is repeated several times, each time with a measurement variance σ^2, the posterior of one measurement can be used as the prior of the next. Initially, the prior variance σ_0^2 is likely to be much larger than the variance of the measurement σ^2 and a single measurement shall then produce a posterior variance σ^2 significantly smaller than the prior's. However, with successive iterations, the posterior variance shall eventually become smaller than measurement variance and successive iterations shall then produce only a small gain in precision. In effect, the precision shall then increase in proportion to the number n of measurements, which means $\sigma' \propto \sigma/\sqrt{n}$.

For fits involving $m \geq 2$ parameters, shifts in parameter values and width reductions are similarly obtained, albeit somewhat less transparently.

Parameter Marginalization and Errors

Marginalization is required whenever one wishes to determine the probability distribution of one parameter (or several) irrespective of the other parameters. It enables the determination of the credible range, and thus the error, of the parameter(s) of interest. For a model involving m parameters, marginalization may be achieved iteratively by integration of Eq. 7.253 one parameter at a time.

Let us exemplify the procedure for the marginalization of a single parameter, which, for the sake of notational simplicity, we choose to be the first one. The posterior of m parameters marginalized against the first parameter is given by

$$p(a_2, \ldots, a_m|\vec{y}, I) = \int_{\Delta a_1} p(a_1, \ldots, a_m|\vec{y}, I)\, da_1, \tag{7.266}$$

where $\vec{y}$ represents the measured data and the integration is carried over the domain Δa_1 of the parameter a_1. We now substitute the posterior, Eq. (7.247), obtained in the previous section and write

$$p(a_2, \ldots, a_m|\vec{y}, I) = K' \int_{\Delta a_1} \exp\left(-\frac{1}{2}\Delta\chi^2\right) da_1, \tag{7.267}$$

$$= K' \int_{\Delta a_1} \exp\left(-\frac{1}{2}\delta\vec{a}^T \mathbf{U}^{-1} \delta\vec{a}\right) d\delta a_1, \tag{7.268}$$

In order to carry out the integral in δa_1, we must factorize the terms in a_1 from other terms. To accomplish this, let us first focus on the argument $\Delta\chi^2$ of the exponential and write

$$\Delta\chi^2 = \delta\vec{a}^T \mathbf{U}^{-1}\delta\vec{a} \tag{7.269}$$

$$= \sum_{k,k'=1}^{m} \delta a_k (U^{-1})_{kk'} a_{k'} \tag{7.270}$$

$$= (U^{-1})_{11}\delta a_1^2 + 2\delta a_1 \sum_{k=2}^{m} (U^{-1})_{1k}\delta a_k + \sum_{k,k'=2}^{m} \delta a_k (U^{-1})_{1k}\delta a_{k'}, \tag{7.271}$$

where in the third line, we have explicitly separated terms in $k, k' = 1$ from others. Completing the square formed by the first two terms of this expression, we get

$$\Delta\chi^2 = (U^{-1})_{11}\delta a_1^2 + 2\delta a_1 \sum_{k=2}^{m}(U^{-1})_{1k}\delta a_k + \frac{1}{(U^{-1})_{11}}\left(\sum_{k=2}^{m}(U^{-1})_{1k}\delta a_k\right)^2 - \frac{1}{(U^{-1})_{11}}\left(\sum_{k=2}^{m}(U^{-1})_{1k}\delta a_k\right)^2 + \sum_{k,k'=2}^{m} \delta a_k (U^{-1})_{1k}\delta a_{k'},$$

$$= (U^{-1})_{11}\left[\delta a_1 + \frac{1}{(U^{-1})_{11}}\sum_{k=2}^{m}(U^{-1})_{1k}\delta a_k\right]^2 + \Delta\chi_r^2, \tag{7.272}$$

where we have introduced a "reduced" $\Delta\chi^2$ defined as

$$\Delta\chi_r^2 = -\frac{1}{(U^{-1})_{11}}\left(\sum_{k=2}^{m}(U^{-1})_{1k}\delta a_k\right)^2 + \sum_{k,k'=2}^{m} \delta a_k (U^{-1})_{kk'} a_{k'}. \tag{7.273}$$

The marginalized posterior we seek is thus

$$p(a_2,\ldots,a_m|Y,I) = K' \exp\left(-\frac{1}{2}\Delta\chi_r^2\right) \tag{7.274}$$

$$\times \int_{\Delta a_1} \exp\left(-\frac{1}{2}(U^{-1})_{11}\left[\delta a_1 + \frac{1}{(U^{-1})_{11}}\sum_{k=2}^{m}(U^{-1})_{1k}\delta a_k\right)^2\right] d\delta a_1.$$

The integrand is a Gaussian in δa_1 with a variance $\sigma^2 = 1/(U^{-1})_{11}$. For an infinite range of integration, the integral yields $\sqrt{2\pi}\sigma = \sqrt{2\pi/(U^{-1})_{11}}$. If the range is finite but far exceeds the region where the integrand is finite, $\sqrt{2\pi/(U^{-1})_{11}}$ constitutes a very good approximation of the true value of the integral. We can then write

$$p(a_2,\ldots,a_m|Y,I) = K'' \exp\left(\frac{1}{2}\Delta\chi_r^2\right), \tag{7.275}$$

with $K'' = K'\sqrt{2\pi/(U^{-1})_{11}}$.

Let us consider a specific case involving $m = 2$ parameters only. The reduced χ^2 is then

$$\Delta\chi_r^2 = -\frac{1}{(U^{-1})_{11}}\left[(U^{-1})_{12}\delta a_2\right]^2 + (U^{-1})_{22}\delta a_2^2 \tag{7.276}$$

$$= \delta a_2^2 \frac{\left[(U^{-1})_{11}(U^{-1})_{22}\right] - (U^{-1})_{12}^2}{(U^{-1})_{11}}. \tag{7.277}$$

Since the inverse of an inverse matrix is the matrix itself, the ratio of the right-hand side may be recognized as the multiplicative inverse of the matrix element U_{22} which one may write σ_2^2. We thus conclude that the marginal probability $p(a_2|Y, I)$ is a Gaussian with a variance $U_{22} = \sigma_2^2$. The matrix U^{-1} is indeed the inverse of the covariance matrix of the parameters a_1 and a_2. After marginalization of a_1, the estimate of a_2 and its associated error may then be written

$$a_2 \pm \sigma_2. \tag{7.278}$$

Repeating the reasoning for marginalization of parameter a_2, we conclude similarly that the estimate of a_1 and associated error may then be written

$$a_1 \pm \sigma_1. \tag{7.279}$$

The error σ_k reflect the credible range obtained for a specific probability content P after marginalization of the other parameter. Indeed, it is important to reiterate that the errors $\pm\sigma_1$ and $\pm\sigma_2$ are applicable only after marginalization, that is, they apply only when a single parameter is considered after the other has been marginalized. If no marginalization is performed, the joint error on the parameters is described by the full ellipse, Eq. (7.256), and the individual ranges $\pm\sigma_1$ and $\pm\sigma_2$ are not strictly applicable as (joint) errors on a_1 and a_2.

7.5 Bayesian Inference with Nonlinear Models and Non-Gaussian Processes

As we saw in the previous section, Bayesian fits of linear models of the form given by Eq. (7.215), with Gaussian measurement noise, reduce to problems of χ^2 minimization that can be expressed in terms of linear matrix equations readily solvable by standard linear algebra techniques. Alas, linear models are not the norm in scientific studies. One indeed often encounters models that cannot be linearized, that is, models whose free parameters $\vec{\theta} = (\theta_1, \ldots, \theta_m)$ cannot be expressed in terms of linear equations of the form Eq. (7.215). Commonly encountered examples of such nonlinear models include Gaussian and Breit–Wigner distributions with polynomial backgrounds. Additionally, one finds that measurements fluctuations, or noise, may also be strictly non-Gaussian. This is the case, for instance, for measurements of rare particle production cross sections in nuclear scattering experiments, which can be described in terms of Poisson and Bernoulli statistics. Poisson statistics is used to describe the stochastic nature of rare particle production processes whereas Bernoulli statistics (binomial distribution) is used to account for instrumental effects such as particle losses (detection efficiency). We must thus seek and develop techniques of Bayesian inference amenable to both nonlinear physical models and non-Gaussian data probability models. The basic goal of Bayesian inference nonetheless remains the same and involves calculation of the posterior probability of the model parameters, $p(\vec{\theta}|D, M, I)$. Unfortunately, the determination of the mode (maximum probability density) of the posterior and credible ranges of the model parameters typically becomes nontrivial problem that must be solved by numerical techniques.

We consider Bayesian inference with nonlinear physical models in §7.5.1, whereas non-Gaussian data probability models are introduced in §7.5.2. General optimization techniques to determine the mode of posterior PDFs are discussed in §7.6.

7.5.1 Bayesian Inference with Nonlinear Models

Consider an experiment reporting n data points (x_i, y_i), $i = 1, \ldots, n$. We shall assume the data may be described with a data model M of the form

$$y_i = f(x_i|\vec{\theta}) + e_i, \tag{7.280}$$

where $f(x_i|\vec{\theta})$ is a physical model expressing a relation between an independent observable X and a dependent observable Y with m free or unspecified model parameters $\vec{\theta} = (\theta_1, \ldots, \theta_m)$. The model $f(x_i|\vec{\theta})$ is considered nonlinear in its parameter $\vec{\theta}$ if it cannot be linearized and expressed in the form of Eq. (7.215). Let us further assume that the noise terms e_i have null expectation values and are Gaussian distributed with a known covariance $\mathbf{V}$. The posterior may then be written

$$\begin{aligned} p(\vec{\theta}|\vec{y}, M, I) = & \frac{p(\vec{\theta}|M, I)}{p(\vec{y}|M, I)} \frac{1}{(2\pi)^{N/2}|\mathbf{V}|^{1/2}} \\ & \times \exp\left[\frac{1}{2}\sum_{i,j}(y_i - f(x_i))\, V_{ij}^{-1}\left(y_j - f(x_j)\right)\right], \end{aligned} \tag{7.281}$$

where $p(\vec{y}|M, I)$ is the global likelihood of the data. This probability density should yield a maximum for the optimal value of the parameters $\vec{\theta}$, denoted $\hat{\theta}$, corresponding to the mode of the distribution. Finding this maximum is an optimization problem, which may be addressed with the numerical techniques presented in §7.6. Given such an optimum $\hat{\theta}$, and based on our treatment of linear models, it is possible to write

$$p(\vec{\theta}|\vec{y}, M, I) = K \exp\left(-\frac{1}{2}\delta\vec{\theta}^{\,T}\mathbf{I}\delta\vec{\theta}\right) + O(3), \tag{7.282}$$

where $\delta\hat{\theta} = \vec{\theta} - \hat{\theta}$, $O(3)$ represents contributions of higher orders admissible for a nonlinear model, while the constant K is determined by the prior $p(\hat{\theta}|M, I)$ and likelihood $\mathbb{L}(\hat{\theta}) \equiv p(\vec{y}|\hat{\theta}, M, I)$ at the mode, as well as the global likelihood $p(\vec{y}|M, I)$:

$$K = \frac{p(\vec{\theta}|M, I)}{p(\vec{y}|M, I)}\mathbb{L}(\hat{\theta}), \tag{7.283}$$

and **I** is known as Fisher information matrix. **I** may be numerically evaluated using Eqs. (7.281, 7.281). One first uses Eq. (7.281), based on the measured data (x_i, y_i), $i = 1, \ldots, n$, to compute $p(\vec{\theta}|\vec{y}, M, I)$. Taking the logarithm of Eq. (7.282), one next gets

$$\ln\left[p(\vec{\theta}|\vec{y}, M, I)\right] = \ln(K) - \frac{1}{2}\left(\delta\vec{\theta}^{\,T}\mathbf{I}\delta\vec{\theta}\right). \tag{7.284}$$

Second derivatives of $\ln[p(\vec{\theta}|\vec{y}, M, I)]$, evaluated at $\hat{\theta}$, then yield

$$\mathbf{I}_{ij} = -\frac{\partial^2}{\partial\theta_i\partial\theta_j}\ln\left[p(\vec{\theta}|\vec{y}, M, I)\right]\bigg|_{\hat{\theta}}. \tag{7.285}$$

In the context of a linear model, as we saw in previous sections, the inverse of **I** would correspond to the covariance matrix of the model parameters. For nonlinear models, Eq. (7.282) is only approximate, and the neglect of higher-order terms implies $\mathbf{I}^{-1}$ is not equal, strictly speaking, to the covariance matrix of the parameters. In practice, however, one finds that for sufficiently large datasets, the Gaussian approximation embodied in Eq. (7.282) may be sufficient and yield an inverse matrix, $\mathbf{I}^{-1}$, which provides a useful approximation of the covariance matrix of the model parameters $\vec{\theta}$.

Equation (7.282) may also be used to evaluate the global hypothesis posterior $p(M|\vec{y}, I)$ by marginalization of all model parameters. This is useful, for instance, in hypotheses testing based on the odds ratio of the global posteriors of competing hypotheses. However, given the Fisher information matrix **I** is in general nondiagonal, integration over all parameters $\vec{\theta}$ may become nontrivial and rather tedious, particularly in the presence of finite bounds on each of the parameters. Integration may be achieved, nonetheless, based on an eigenvalue decomposition of the matrix. Since **I** is by definition real and symmetric, there indeed exists a transformation $\delta\vec{\theta} = \mathbf{O}\delta\vec{\xi}$ such that

$$\delta\vec{\theta}^{T}\mathbf{I}\delta\vec{\theta} = \delta\vec{\xi}^{T}\mathbf{\Lambda}\delta\vec{\xi}, \tag{7.286}$$

where Λ is a diagonal matrix consisting of eigenvalues λ_k of **I**. The integral we seek is thus

$$\begin{aligned}\mathbb{I} &= \int\prod_k d\theta_k \exp\left(-\frac{1}{2}\delta\vec{\theta}^{T}\mathbf{I}\delta\vec{\theta}\right)\\ &= J\int\prod_k d\xi_k \exp\left(-\frac{1}{2}\delta\vec{\xi}^{T}\mathbf{\Lambda}\delta\vec{\xi}\right)\\ &= \prod_k\int_{\xi_{k,\min}}^{\xi_{k,\max}} d\xi_k \exp\left(-\frac{1}{2}\delta\xi_k\lambda_k\delta\xi_k\right),\end{aligned} \tag{7.287}$$

where, in the third line, we used $J = \left|\frac{\partial\theta}{\partial\xi}\right| = \det\mathbf{O} = 1$, and the fact $\mathbf{\Lambda}$ is a diagonal matrix consisting of values λ_k. The boundaries of integration $\xi_{k,\min}$ and $\xi_{k,\max}$ are obtained from those on $\vec{\theta}$ based on the inverse transformation $\delta\vec{\xi}_{\min,\max} = \mathbf{O}^{-1}\delta\vec{\theta}_{\min,\max}$.

7.5.2 Bayesian Inference with Non-Gaussian Processes

There exists a plurality of phenomena (or systems) for which a Gaussian data probability model is not suitable or applicable. Of particular interest in astronomy and high-energy physics are processes involving Poisson, binomial, or multinomial sampling. Unfortunately, the difficulty arises that, for such data probability models, it is typically not possible to reduce the calculation of the posterior to the evaluation of a single quantity such as a χ^2 function, and one must then handle the full posterior probability distribution. Considerable simplifications are possible, however, if the parameters of the probability data model are constant and the prior probability of these parameters may be expressed as conjugate priors. A more general example involving model parameters dependencies on a control variable is considered in §7.5.3.

We limit our discussion to Poisson processes but the techniques introduced in the following can be straightforwardly extended to other types of statistical processes.

Finding the Rate Parameter of a Poisson Process

Poisson sampling describes phenomena with a specific rate, that is, an average number of instances, or events, per unit of time (or dependency on a specific control variable). Let r be the process rate. If observations are carried out over a time period of T, one then expects, on average, to observe $\langle n \rangle \equiv \mu = rT$ instances of the phenomenon. The actual number of instances observed in any given interval, n, should fluctuate (i.e., vary interval by interval) according to a Poisson distribution

$$p(n|\mu, I) = \frac{\mu^n e^{-\mu}}{n!}. \tag{7.288}$$

If the rate r is a priori unknown but can be sensibly expected to be constant over extended time periods, one can use measurements of the number of occurrences, n_k, in m different time intervals of duration T to estimate r. For notational convenience, let us denote these m measurements in terms of a vector $\vec{n} = (n_1, \ldots, n_m)$.

The rate r may be trivially computed on the basis of the mean number of instances, $\hat{\mu}$, and the duration of the time interval T during which events are counted:

$$r = \hat{\mu}/T. \tag{7.289}$$

We thus focus on the determination of $\hat{\mu}$. A frequentist evaluation of $\hat{\mu}$ could of course be readily obtained from the mean of the measured values n_k, but we here consider a Bayesian estimation instead. That means we need to calculate the posterior probability of μ:

$$p(\mu|\vec{n}, I) = \frac{p(\mu|I)p(\vec{n}|\mu, I)}{p(\vec{n}|I)}, \tag{7.290}$$

where $p(\mu|I)$ is the prior probability of μ, $p(\vec{n}|\mu, I)$ is the likelihood of the data given μ, and $p(\vec{n}|I)$ is the global likelihood of the data.

The likelihood $p(\vec{n}|\mu, I)$ is readily evaluated by multiplying the probabilities of each of the observed values n_k, $k = 1, \ldots, m$:

$$\begin{aligned} p(\vec{n}|\mu, I) &= \prod_{k=1}^{m} p(n_k|\mu, I), \\ &= \prod_{k=1}^{m} \frac{\mu^{n_k} e^{-\mu}}{n_k!}, \\ &= \frac{\mu^z e^{-m\mu}}{\prod_{k=1}^{m} n_k!}, \end{aligned} \tag{7.291}$$

where we have introduced the variable z defined as a sum of the measured n_k

$$z = \sum_{k=1}^{m} n_k. \tag{7.292}$$

Recall from §7.3.3 that for a Poisson data probability model, it is convenient to use a conjugate prior for the rate parameter in the form of a gamma distribution (§3.6). We thus write

$$p(\mu|\alpha, \beta) = \frac{\beta^\alpha}{\Gamma(\alpha)} \mu^{\alpha-1} e^{-\beta\mu}, \tag{7.293}$$

where the distribution parameters α and β may be chosen to reflect an uninformative scaling prior (e.g., $\alpha = 1.0$ and $\beta = 1.0$) or may be constrained by prior measurements or knowledge of the phenomenon of interest. And as we showed in §7.3.3, this readily implies that the posterior $p(\mu|\vec{x})$ is then also a gamma distribution. Including proper normalization, one gets

$$\begin{aligned} p(\mu|\vec{n}, I) &= p_\gamma(\mu|z+\alpha, m+\beta) \\ &= \frac{(m+\beta)^{z+\alpha} \mu^{z+\alpha-1} e^{-(m+\beta)\mu}}{\Gamma(z+\alpha)}. \end{aligned} \tag{7.294}$$

with a mode given by

$$\hat{\mu} = \frac{z+\alpha-1}{\beta+m}, \tag{7.295}$$

and a variance equal to

$$\sigma_\mu^2 = \frac{z+\alpha}{(\beta+m)^2}. \tag{7.296}$$

Credible ranges $[\mu_{\min}, \mu_{\max}]$, with probability content C, are determined by simultaneously solving

$$P_\gamma(\mu_{\max}|z+\alpha, m+\beta) - P_\gamma(\mu_{\min}|z+\alpha, m+\beta) = C \tag{7.297}$$

and

$$p_\gamma(\mu_{\min}|z+\alpha, m+\beta) = p_\gamma(\mu_{\max}|z+\alpha, m+\beta). \tag{7.298}$$

$P_\gamma(\mu|\alpha,\beta)$ represents the cumulative density function (CDF) of the gamma distribution with parameters α and β, given by Eq. (3.113), which may be computed according to

$$P_\gamma(\mu|\alpha,\beta) = \frac{\gamma(\alpha,\mu)}{\Gamma(\alpha)}, \tag{7.299}$$

in which $\Gamma(x)$ and $\gamma(\alpha, x)$ are the gamma function and the (lower) incomplete gamma function. Evaluation of these functions is possible within commonly available math packages such as Mathematica®, MATLAB®, or ROOT.

Rate Parameter of a Poisson Process with Fixed Background

In practical situations, the observed process rate r may result from a combination of the signal of interest, with rate r_s, and some background process, with rate r_b, such that $r = r_s + r_b$. In a time interval T, one then expects average numbers of signal and background instances equal to $\mu_s = r_s T$ and $\mu_b = r_b T$, respectively.

If the background rate r_b or number of background instances μ_b are precisely known (and thus fixed), the posterior probability $p(\mu|\vec{n}, I)$ determines the probability distribution of μ_s uniquely:

$$p(\mu_s|\vec{n},\mu_b, I) = p(\mu|\vec{n}, I). \tag{7.300}$$

One can then compute $p(\mu_s|\vec{n},\mu_b, I)$ based on Eq. (7.294) provided one replaces μ by $\mu_b + \mu_s$:

$$\begin{aligned} p(\mu_s|\vec{n},\mu_b, I) &= p_\gamma(\mu_s+\mu_b|z+\alpha, m+\beta) \\ &= \frac{(m+\beta)^{z+\alpha}(\mu_s+\mu_b)^{z+\alpha-1}e^{-(m+\beta)(\mu_s+\mu_b)}}{\Gamma(z+\alpha)} \end{aligned} \tag{7.301}$$

The most probable value of μ_s is then

$$\hat{\mu}_s = \frac{z+\alpha-1}{\beta+m} - \mu_b, \tag{7.302}$$

and the credible range for $\hat{\mu}_s$ is obtained by shifting the range of $\hat{\mu}$ calculated earlier by $-\mu_b$.

Rate Parameter of a Poisson Process with Unknown Background

If the background rate is a priori unknown, it may be possible to turn off the signal in order to estimate the background rate. One should then proceed to carry out m' measurements, with signal turned off, of the number of instances $n_k^{(B)}$ observed during intervals of equal duration T. Evidently, the number of background instances shall fluctuate and the background rate thus cannot be determined with absolute precision. Repeating the aforementioned reasoning, one obtains a posterior probability for the number of background instance μ_b per time interval T:

$$\begin{aligned} p\left(\mu_b|\vec{n}^{(B)}, I\right) &= p_\gamma(\mu_b|z_b+\alpha, m'+\beta) \\ &= \frac{(m'+\beta)^{z_b+\alpha}\,\mu_b^{z_b+\alpha-1}e^{-(m'+\beta)\mu_b}}{\Gamma(z_b+\alpha)}, \end{aligned} \tag{7.303}$$

where $z_b = \sum_{k=1}^{m'} n_k^{(B)}$.

Given the preceding posterior for μ_b, one can then return to the evaluation of the posterior probability of μ_s. For a known background rate, we found that the posterior of μ_s is given by Eq. (7.300). The background rate is not known precisely, however, and one must then account for all possible values of μ_b. To accomplish this, one can think of the samples n_k and $n_k^{(B)}$ as a joint measurement of the background and signal rates, yielding a joint posterior we denote $p(\mu_s, \mu_b|\vec{n}, \vec{n}^{(B)}, I)$. We are not actually interested in the background rate; we thus marginalize this posterior for μ_b and obtain a posterior for μ_s by integration over all values of μ_b. But since the measurements of the samples n_k and $n_k^{(B)}$ are in fact disjoint (or independent), the joint posterior $p(\mu_s, \mu_b|\vec{n}, \vec{n}^{(B)}, I)$ is simply the product of the posteriors for μ_s (at fixed μ_b) and μ_b (with signal turned off). It then suffices to integrate this product over all physically possible values of μ_b. One gets

$$p(\mu_s|\vec{n}, \vec{n}^{(B)}) = \int_0^\infty p(\mu_s|\vec{n}, \mu_b, I)p(\mu_b|\vec{n}^{(B)}, I)\, d\mu_b, \tag{7.304}$$

which can be readily computed with standard numerical techniques.

7.5.3 Bayesian Fitting with Non-Gaussian Processes

Bayesian fitting in the context of non-Gaussian data probability models involving one or several unknown parameters cannot be reduced, in general, to the evaluation and optimization of a χ^2 function but relies on the evaluation of the joint posterior probability of the model parameters. Generally, the evaluation of the posterior proceeds similarly as for cases involving Gaussian noise, and once the mode of the posterior is found, with any of the techniques presented in §7.6, nuisance parameters can be eliminated by marginalization. Overall, the treatment of problems involving non-Gaussian data probability models is thus not very different from the methods already discussed in this chapter. One must acknowledge, however, that the calculation and optimization of non-Gaussian posteriors may be computationally rather intensive and has thus been readily feasible only since the advent of high-performance computers. A number of model-specific methods have been developed to streamline and simplify calculations. Examples of such methods are presented in [97].

7.6 Optimization Techniques for Nonlinear Models

Inference methods seeking the optimal values of a model, that is, the values that best describe measured data, require the optimization of an objective function, often called goal or merit function. Depending on the circumstances, the objective function may be the posterior probability of model parameters, a likelihood function, or a χ^2 function. Either way, one seeks an extremum (also called optimum) of these functions in the n-dimension space defined by the parameters of the model. Evidently, in the case of probabilities, the extremum must be a maximum whereas for χ^2 functions, it must be a minimum.

We saw in §7.4.3 that optimization problems involving linear models in the presence of Gaussian noise can be reduced to linear equations readily solved, at least in principle, by matrix inversion. Data modeling, however, as we discuss in §7.5, often involves nonlinear equations or non-Gaussian probability models not amenable to linear solutions. Finding the extremum of such objective functions must thus, in general, be achieved with numerical techniques.

There is no general solution to the global optimization encountered in classical and Bayesian inference. However, a plurality of methods have been developed over the years to tackle the problem. The techniques most often used include

1. Hill climbing and gradient methods
   a. Extended Newton methods
   b. Levenberg–Marquardt algorithm
   c. Powell's algorithm
   d. Simplex algorithm
2. Stochastic methods
   a. Simulated annealing
   b. Genetic algorithms
3. Hybrid methods consisting of a combination of hill climbing, gradient based, and stochastic methods

We discuss the principle and applicability of these and related techniques throughout this section beginning in §7.6.1 with hill climbing and gradient methods. Stochastic methods are introduced in §7.6.5. Several other methods and variants of the aforementioned methods exist and are documented in the computing literature but a comprehensive discussion of all these methods is beyond the scope of this book. In-depth discussions and implementation of several optimization algorithms are discussed by Press et al. [156], Besset [36], and Gregory [97]. It should also be clear at the outset that many of the methods discussed in this section find applications in a variety of optimization problems and are not restricted to "fitting" problems, but our discussion will focus on optimization of continuous functions in the context of statistical inference and model fitting.

7.6.1 Hill Climbing and Gradient Methods

Consider a function $f(\vec{x})$ defined over a space of n-dimensions, $\mathbb{R}^n$. Our goal is to find a global extremum of the function, that is, the position $\vec{x}$ that globally minimizes or maximizes the function as appropriate. The function $f(\vec{x})$ may be rather generic and is not restricted to objective functions measuring the level of fitness of a model to measured data. Many of the techniques discussed in the text that follows indeed apply to a wide range of functions. In the context of inference problems, however, the vector $\vec{x}$ shall here stand for a collection of model parameters that need to be optimized or fitted to measured data (and not the data itself).

Let us assume the function $f(\vec{x})$ is continuous everywhere in the search space and derivable relative to all components of $\vec{x}$. Extrema of the function may then be sought for by searching values of $\vec{x}$ where the gradient of the function vanishes.

$$\nabla f(\vec{x}) = \frac{df(\vec{x})}{d\vec{x}} = 0 \tag{7.305}$$

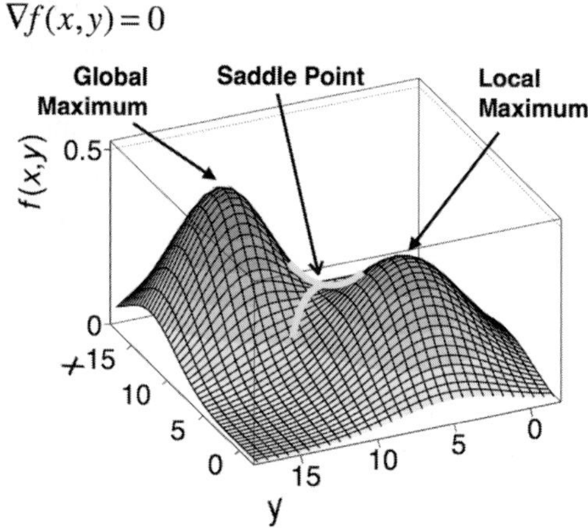

Fig. 7.6 A vanishing gradient, $\nabla f(\vec{x}) = 0$, is a necessary but insufficient condition in the search for a function's global extremum.

Unfortunately, this condition does not guarantee the existence of an extremum. In $n > 1$ dimensions, points or regions where ∇f vanish may indeed include saddle points as well as minima and maxima, as illustrated in Figure 7.6. The condition Eq. (7.305) thus provides a necessary but not sufficient condition for the existence of extrema. Solutions $\vec{x}$ of Eq. (7.305) are, however, easily tested to verify whether they correspond to a minimum, a maximum, or a saddle point region by considering second derivatives of the functions along all of its components. Positive and negative second derivatives identify minima and maxima, respectively, while a mix of negative and positive second derivatives (or all null second derivatives) reveals saddle point regions.[10] A more difficult issue arises from the fact that, in general, the function $f(\vec{x})$ may feature several minima or maxima as well as saddle points. Extrema search algorithms may then end up finding local minima or maxima, not the lowest (global minimum) or highest value (global maximum) of the function. In fitting and inference problems, this entails that the solutions obtained may not be truly optimal, that is, they may not represent the model parameters that best fit the data. We will see that optimization algorithms vary greatly in their capacity to identify a global extremum. Hill climbing and gradient based algorithms are typically very fast and efficient but tend, by their very nature, to settle onto the first extremum they find whether global or not. By contrast, Monte Carlo algorithms are typically designed to provide a better and fuller exploration of the search space, albeit at the cost of requiring many function evaluations and thus greater CPU time.

Newton Optimization Methods

Newton optimization methods implement a generalized version of Newton's zero-finding algorithm to find an extremum of a function $f(\vec{x})$. Starting from an initial position $\vec{x}^{(0)}$,

[10] Note that in one dimension a "saddle point" is observed whenever both the first and second derivative of a function vanish at the same point.

they use derivatives of the function $f(x)$ to quickly home in on an extremum but provide no guarantee whatsoever that the extremum found is a global minimum or maximum.

Newton's method and related techniques are based on a Taylor expansion of the function $f(\vec{x})$ in the vicinity of a point $\vec{x}^{(k=0)}$ believed to be near the sought for optimum of the function:

$$f(\vec{x}) = f(\vec{x}^{(k)}) + \sum_{i=1}^{n} \left.\frac{\partial f(\vec{x})}{\partial x_i}\right|_{\vec{x}^{(k)}} \left(x_i - x_i^{(k)}\right) + O(2), \tag{7.306}$$

where higher-order terms $O(2)$ are neglected. Inserting this expression in Eq. (7.305), we get

$$\sum_{i=1}^{n} \left.\frac{\partial^2 f(\vec{x})}{\partial x_i \partial x_j}\right|_{\vec{x}^{(k)}} \left(x_i - x_i^{(k)}\right) + \left.\frac{\partial f(\vec{x})}{\partial x_j}\right|_{\vec{x}^{(k)}} = 0, \tag{7.307}$$

which can readily be written as an approximate linear equation of the form

$$\alpha^{(k)} \delta\vec{x}^{(k)} = \vec{b}^{(k)}, \tag{7.308}$$

where

$$\delta\vec{x}^{(k)} = \vec{x} - \vec{x}^{(k)}, \tag{7.309}$$

$$\alpha_{ij}^{(k)} = \left.\frac{\partial^2 f(\vec{x})}{\partial x_i \partial x_j}\right|_{\vec{x}^{(k)}}, \tag{7.310}$$

$$b_i^{(k)} = -\left.\frac{\partial f(\vec{x})}{\partial x_j}\right|_{\vec{x}^{(k)}}. \tag{7.311}$$

Multiplying Eq. (7.308) on both sides by the inverse $(\alpha_{ij}^{(k)})^{-1}$, one gets

$$\delta\vec{x}^{(k+1)} = \left(\alpha_{ij}^{(k)}\right)^{-1} \vec{b}^{(k)}. \tag{7.312}$$

Given an initial estimate of the extremum, $\vec{x}^{(k)}$, also called prior estimate, one obtains an updated and more accurate estimate of the extremum's position, $\vec{x}^{(k+1)}$, by addition of $\delta\vec{x}^{(k+1)}$:

$$\vec{x}^{(k+1)} = \vec{x}^{(k)} + \delta\vec{x}^{(k+1)}. \tag{7.313}$$

We labeled the solution with an index $(k+1)$ to indicate that the procedure can be iterated. Indeed, given an initial value $\vec{x}^{(0)}$, the solution provided by Eq. (7.312) is strictly exact only if higher-order terms in Eq. (7.306) are truly negligible. However, one can use the preceding equations iteratively to obtain an approximation of the sought for extremum. Iterations should be terminated when $|\delta\vec{x}^{(k+1)}| < \varepsilon$ or $|f(\vec{x}^{(k+1)}) - f(\vec{x}^{(k)})| < \xi$, where ε and ξ define the required precision of the search.

Newton's algorithm is commonly used in optimization engines and packages such as Minuit [115] for extremum searches, but it is important to be aware of its limitations. The algorithm is efficient and precise but will typically home in on the closest extremum because it does not have built-in capabilities to differentiate between local extrema and a global extremum. Indeed, the algorithm can be rather quickly locked in a local minimum. Other techniques must then be used to first identify a region of global extremum and select a seed value $\vec{x}^{(0)}$ with which to initiate the algorithm. It should also be noted that the matrix α may be ill conditioned and not invertible for seed values $\vec{x}^{(0)}$ "far" from an extremum. Yet another limitation of the technique is that it requires the evaluation of first and second derivatives. If such analytical derivatives are not available or possible, numerical techniques must be used, which may reduce the accuracy and the efficiency of the technique. Numerical calculations of second derivatives, in particular, require several calls to evaluate the function and may then considerably increase the CPU time needed to achieve solutions of required precision. It may then be preferable to make use of techniques that do not require calculations of derivatives such as the hill climbing techniques introduced in §7.6.3.

7.6.2 Levenberg–Marquardt Algorithm

The Levenberg–Marquardt algorithm is a popular and efficient variant of Newton's algorithm used in $\mathbb{LS}$ minimization problems. The objective function is the χ^2 function defined as

$$\chi^2(\vec{\theta}) = \sum_{i=1}^{N} \left[\frac{y_i - f(x_i|\vec{\theta})}{\sigma_i} \right]^2, \tag{7.314}$$

in which N is the number of data points (x_i, y_i), $f(x|\vec{\theta})$ represents the model, and $\vec{\theta}$ are the unknown (free) parameters to be determined by χ^2 minimization. As for Newton's algorithm, one assumes the χ^2-function can be represented by a quadratic form in the vicinity of the optimal values of the parameters $\vec{\theta}_{\text{min}}$:

$$\chi^2(\vec{\theta}) \approx \chi^2_{\text{min}} + \frac{1}{2}\left(\vec{\theta} - \vec{\theta}_{\text{min}}\right)^T \mathbf{D} \left(\vec{\theta} - \vec{\theta}_{\text{min}}\right), \tag{7.315}$$

where χ^2_{min} is the minimum of the function achieved for the optimal value of the parameters $\vec{\theta}_{\text{min}}$; $\mathbf{D}$ is an $m \times m$ symmetric matrix that depends on the data points and the model to be fitted; and $\vec{\theta}$ is an $1 \times m$ column vector one can vary to explore the shape of the χ^2-function.

Let us assume a prior estimate $\vec{\theta}^{(0)}$ of the model parameters is known. If the preceding quadratic form is a reasonable approximation of the actual χ^2-function, then one can use the gradient of the χ^2-function at $\vec{\theta}^{(0)}$ to seek the minimum

$$\left.\frac{\partial \chi^2}{\partial \theta_k}\right|_{\vec{\theta}^{(0)}} = \frac{1}{2} \sum_{ij} \frac{\partial}{\partial \theta_k} \left[(\theta_i - \theta_{\text{min},i}) D_{ij} \left(\theta_j - \theta_{\text{min},j}\right) \right] \Bigg|_{\vec{\theta}^{(0)}}, \tag{7.316}$$

$$= \sum_j D_{kj} \left(\theta_j^{(0)} - \theta_{\text{min},j}\right), \tag{7.317}$$

where we have used $\partial\theta_i/\partial\theta_k = \delta_{ik}$, the symmetric nature of $\mathbf{D}$, and carried sums to eliminate the delta functions. It is convenient to write this expression in matrix form:

$$\nabla \chi^2 \big|_{\vec{\theta}^{(0)}} = \mathbf{D} \cdot \left(\vec{\theta}^{(0)} - \vec{\theta}_{\text{min}}\right). \tag{7.318}$$

One solves for $\vec{\theta}_{\min}$ by multiplying both sides of this expression by the inverse $\mathbf{D}^{-1}$:

$$\vec{\theta}_{\min} = \vec{\theta}^{(0)} - \mathbf{D}^{-1} \nabla\chi^2\big|_{\vec{\theta}^{(0)}} . \tag{7.319}$$

In principle, the values $\vec{\theta}_{\min}$ should correspond to the minimum of the χ^2-function and provide optimal values of the parameters $\vec{\theta}$. In practice, the quadratic form, Eq. (7.315), may be a poor approximation of the actual shape of the χ^2-function. It also relies on numerical calculations of the gradient $\nabla\chi^2\big|_{\vec{\theta}^{(0)}}$ and the matrix $\mathbf{D}$, which may not be perfectly accurate. One may, however, replace the matrix $\mathbf{D}$ by a constant and achieve what is commonly known as the **steepest descent method**. Given a prior estimate $\theta^{(i-1)}$, a posterior (i.e., an improved estimate) can be evaluated as

$$\theta^{(i)} = \theta^{(i-1)} - k \times \nabla\chi^2\big|_{\vec{\theta}^{(i-1)}} , \tag{7.320}$$

in which k is a suitably chosen constant.

Computation of Eq. (7.319) requires knowledge of the gradient $\nabla\chi^2$ and the matrix $\mathbf{D}$. The gradient $\nabla\chi^2$ may be calculated according to

$$\frac{\partial\chi^2}{\partial\theta_k} = -2\sum_{i=1}^{N}\left(\frac{y_i - f(x_i|\vec{\theta})}{\sigma_i^2}\right)\frac{\partial f(x_i|\vec{\theta})}{\partial\theta_k}. \tag{7.321}$$

Calculation of the matrix $\mathbf{D}$, called the **Hessian matrix**[11], involves second-order derivatives of the χ^2-function with respect to θ_k at any value of θ_k:

$$\begin{aligned}\frac{\partial\chi^2}{\partial\theta_k\partial\theta_l} = 2\sum_{i=1}^{N}\frac{1}{\sigma_i^2}&\left[\frac{\partial f(x_i|\vec{\theta})}{\partial\theta_l}\frac{\partial f(x_i|\vec{\theta})}{\partial\theta_k}\right.\\ &\left. - \left(y_i - f(x_i|\vec{\theta})\right)\frac{\partial^2 f(x_i|\vec{\theta})}{\partial\theta_l\partial\theta_k}\right].\end{aligned} \tag{7.322}$$

It is convenient to define coefficients β_k and α_{kl} as follows:

$$\beta_k \equiv -\frac{1}{2}\frac{\partial\chi^2}{\partial\theta_k}, \tag{7.323}$$

$$\alpha_{kl} \equiv \frac{1}{2}\frac{\partial^2\chi^2}{\partial\theta_l\partial\theta_k}. \tag{7.324}$$

The coefficients α_{kl} form a matrix $\alpha = \mathbf{D}/2$. Equation (7.319) can thus be recast as

$$\sum_{l=1}^{m}\alpha_{kl}\delta\theta_l^{(i)} = \beta_k. \tag{7.325}$$

Starting with a prior estimate $\theta^{(i-1)}$, solution of the linear equation (7.325) yields an increment $\delta\theta_l^{(i)}$, which one can use to obtain an improved estimate $\theta^{(i)}$:

$$\vec{\theta}^{(i)} = \vec{\theta}^{(i-1)} + \vec{\delta\theta}^{(i)}. \tag{7.326}$$

[11] The Hessian matrix is named after the nineteenth-century German mathematician Ludwig Otto Hesse (1811–1874), who used the determinant of the matrix as a measure of the local curvature of a function of many variables.

α is commonly called **curvature matrix**, as well as Hessian matrix. Its inverse and Eq. (7.325) provide for the **inverse-Hessian** formula

$$\delta\vec{\theta} = \alpha^{-1}\vec{\beta}. \tag{7.327}$$

As per Eq. (7.322), calculation of α in principle involves both first- and second-order derivatives of the functions $f(x|\vec{\theta})$ with respect to parameters $\vec{\theta}$. Second-order derivatives arise because the gradient, Eq. (7.321), involves first-order derivatives $\partial f/\partial\theta_k$. However, these are often ignored in practice because the term containing second-order derivatives in Eq. (7.322) also includes the coefficients $(y_i - f(x_i|\vec{\theta}))$. These coefficients are expected to be small and oscillate around zero when the fit converges to a minimum χ^2. The sum involving the second-order derivatives thus tends to vanish or at the very least be negligible compared to the first term, which involves a product of first-order derivatives. The calculation of the curvature matrix α is thus reduced to

$$\alpha_{kl} = \sum_{i=1}^{N} \frac{1}{\sigma_i^2}\left[\frac{\partial f(x_i|\vec{\theta})}{\partial\theta_l}\frac{\partial f(x_i|\vec{\theta})}{\partial\theta_k}\right]. \tag{7.328}$$

The inverse-Hessian formula usually converges toward the true minimum of the χ^2-function, but the convergence may be slow because of the relatively small steps produced by the method. The calculation of the Hessian and its inverse can also be computationally intensive. Early computers were not as fast and powerful as those in use today. Finding ways to improve the rate of convergence and reduce the computation involved in numerical fits was thus highly desirable. Clearly, the steepest descent method, Eq. (7.320), has the advantages of simplicity and speed. Calculation of the matrix α is not required, and given a suitably chosen constant provides for large increments $\delta\vec{\theta}^{(i)}$ and a fast descent toward the minimum of the χ^2-function. The use of large increments, however, implies the method can easily overshoot the position of the minimum and end up oscillating around it without ever reaching it. It is thus useful to "turn on" the slower but more precise inverse-Hessian method when nearing the minimum. This idea was first suggested in 1944 by K. Levenberg [136] and further developed by D. W. Marquardt in 1963 [140]. It is thus now commonly known as the **Levenberg–Marquardt algorithm** (LMA) or alternatively as the **Damped Least-Squares** (DLS) method. It forms the basis of many fitting packages commonly available today.

In order to use Eq. (7.320), one must identify a proper scale for the constant k that controls the rate of descent. We first note that the χ^2 is by definition nondimensional; in other words, it is a pure number. The fit parameters θ_k, on the other hand, generally have dimensions and units such as cm, kg, and so on. Since the coefficients β_k are defined as first-order derivatives of χ^2 with respect to to θ_k, they must carry dimensions of $1/\theta_k$. Equation (7.320) thus implies the constants k have units of $1/\theta_k^2$. The diagonal elements of the Hessian matrix α have the same dimension and as such constitute a natural choice to determine the scale of the constant k. The diagonal elements might, however, be too large. Marquadt thus had the insight to include a "fudge" factor λ:

$$\delta\theta_k = \frac{1}{\lambda\alpha_{kk}}\beta_k. \tag{7.329}$$

Since the coefficients α_{kk} are by construction positive definite, the increment $\delta\theta_k$ remains proportional to the gradient β_k and provides the intended behavior. Marquardt also realized he could combine the preceding steepest descent with the inverse-Hessian formula by introducing modified matrix elements α'_{ij} as follows:

$$\alpha'_{ii} \equiv \alpha_{ii}(1+\lambda), \tag{7.330}$$

$$\alpha'_{ij} \equiv \alpha_{ij}, \tag{7.331}$$

and replace α_{ij} by α'_{ij} in Eq. (7.327), thereby yielding

$$\delta\vec{\theta} = \alpha'^{-1}\vec{\beta}. \tag{7.332}$$

When the scale parameter λ is very large, the matrix α' is effectively diagonal and so is its inverse. Equation (7.332) thus yields the steepest descent, Eq. (7.320). On the other hand, if λ vanishes, one recovers the inverse-Hessian formula, Eq. (7.327). The LMA method is iterative. It should converge relatively rapidly to a region of the parameter space $\vec{\theta}$ that yields near-minimum χ^2 values. However, the search should be stopped if (1) it fails to converge or (2) incremental reduction (decreases) of the χ^2 become negligible (e.g., smaller than 0.01 in absolute value), or some small fractional change, like 10^{-3}. One should not stop the procedure when the χ^2 increases, unless, of course, the method fails to converge in a reasonable number of iterations.

The LMA can be implemented as follows:

1. Obtain an initial rough guess $\vec{\theta}^{(i=0)}$ of the fit parameters.
2. Compute $\chi^2(\vec{\theta}^{(i=0)})$.
3. Select a small value for λ, for example, $\lambda = 0.001$.
4. Solve the linear equation (7.332) for $\delta\theta^{(i)}$.
5. Evaluate $\chi^2(\vec{\theta}^{(i)} + \delta\theta^{(i)})$ and calculate the change $\Delta\chi^2 = \chi^2(\vec{\theta}^{(i)} + \delta\theta^{(i)}) - \chi^2(\vec{\theta}^{(i)})$.
6. If $\Delta\chi^2 \geq 0$, increase λ by a factor of 10, and repeat step 3.
7. If $|\Delta\chi^2| < \epsilon$, with ϵ set to be a reasonably small number (e.g., 0.01), stop the search.
8. Otherwise decrease λ by a factor of 10, update the parameters: $\vec{\theta}^{(i+1)} = \vec{\theta}^{(i)} + \delta\theta$, and repeat step 3.

Once an acceptable minimum has been reached, one should set $\lambda = 0$ and compute $C = \alpha^{-1}$ in order to obtain an estimate of the covariance matrix of the standard errors of the final parameters $\vec{\theta}^{(i)}$.

While in general the LMA works rather well, it may be slow to converge when the number of fitted parameters is very large. A variety of modern methods have thus been developed to improve on its basic principles, some of which are presented in other sections of this text. Techniques have also been developed to carry out fits with user-defined constraints on the parameter space. This is particularly useful to avoid nonphysical parameter regions or pathological behaviors. The minimization and fitting package **Minuit** [115], popular in the high-energy physics community, uses a combination of Monte Carlo search methods; the simplex method of Nelder and Mead [147], discussed in §7.6.4; and the variable metric

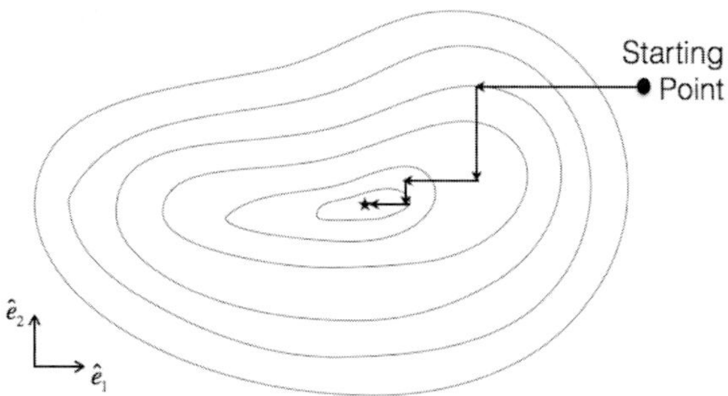

Fig. 7.7 Basic principle of hill climbing methods: Thin solid lines represent iso-contours of the objective function; thick arrows represent paths followed by the hill climbing algorithm along successive directions until a local extremum of the function is reached.

method of Fletcher [84]. It also gives the ability to impose constraints on the parameter space accessible to a search and to control how errors are handled and calculated.

7.6.3 Hill Climbing Methods

Hill climbing methods provide convenient objective function optimization techniques that do not require evaluations of function derivatives. They seek a function's extremum in n-dimensions by sequentially finding extrema along straight lines, as schematically illustrated in Figure 7.7. Indeed, rather than seeking an extremum simultaneously in several dimensions, the hill climbing carries out the search along straight lines of arbitrary direction in n-dimension space. Once a particular extremum is found, the direction is changed, and an improved extremum sought for. The process is repeated iteratively until a true extremum is found within the desired precision. Several distinct techniques may be used for the selection of the direction $\vec{d}^{(k)}$ and the 1D optimization. We will restrict our discussion to a few illustrative techniques in this and the following sections.

The basic idea is to reduce the objective function $f(\vec{x})$ to a single variable function $g(z)$ defined as follows:

$$g(z) = f(\vec{x}^{(k)} + z\vec{d}^{(k)}). \tag{7.333}$$

where the vectors $\vec{x}^{(k)}$ and $\vec{d}^{(k)}$ represent the starting point and direction, respectively, of the kth iteration of the search algorithm. A 1D search algorithm is invoked to find the extremum of $g(z)$ at each iteration. The extremum z is then used to calculate the starting point of the next iteration. The algorithm may thus be summarized as follows:

1. Set $k = 0$, and choose a starting point $\vec{x}^{(k)}$ by any appropriate method.
2. Select a search direction $\vec{d}^{(k)}$ in n-dimensions.
3. Find an extremum $z^{(k)}$ of $g(z)$ using a suitable 1D search technique.
4. Calculate the corresponding point in full space according to $\vec{x}^{(k+1)} = \vec{x}^{(k)} + z^{(k)}\vec{d}^{(k)}$.
5. If convergence is achieved, terminate the search and produce $\vec{x}^{(k+1)}$ as the sought for extremum.
6. Otherwise, set $\vec{x}^{(k+1)}$ as a new starting point, set $k = k + 1$, and proceed to step 2.

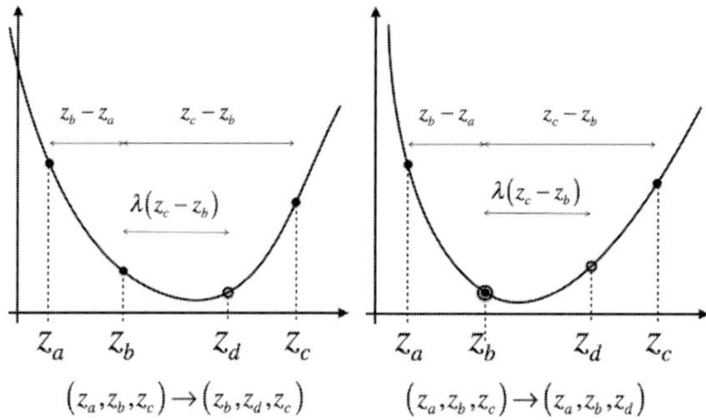

Fig. 7.8 Illustration of the bisection algorithm. The initial triplet (z_a, z_b, z_c) is replaced by (z_b, z_d, z_c) or (z_a, z_b, z_d) depending on whether $g(z_d)$ is better or worse than $g(z_b)$.

Bisection and Bracketing Algorithms

The principle of the bisection algorithm is illustrated in Figure 7.8. One seeks an extremum of a continuous function $g(z)$ assuming its derivatives are unknown or not readily calculable. As a starting point, the algorithm requires a triplet (z_a, z_b, z_c) such that $z_a < z_b < z_c$. Let us assume $g(z_b)$ is "better" than $g(z_a)$ and $g(z_c)$. In this context, better means a higher value if searching for a maximum, or a lower value if searching for a minimum. Given the function is continuous, the fact that $g(z_b)$ is better than $g(z_a)$ and $g(z_c)$ implies the sought for extremum is necessarily within the range $z_a \le z \le z_c$. Let us thus seek a new triplet that narrows this interval. On general grounds, one can assume there is a larger probability to find the extremum in the larger of the two intervals $[z_a, z_b]$ and $[z_b, z_c]$. For illustrative purposes, let us assume $z_b - z_a < z_c - z_b$ and select a point $z_d = z_b + \lambda(z_c - z_b)$ where λ is typically chosen to be the golden ratio $(3 - \sqrt{5})/2$. If $g(z_d)$ is better than $g(z_b)$, the new triplet is identified as (z_b, z_d, z_c); otherwise it is set to (z_a, z_b, z_d). The procedure is then iterated with the new triplet. The search proceeds iteratively and identifies successively narrower intervals. It is terminated when $\Delta z = z_c - z_a \le \varepsilon$, that is, when the width of the triplet is narrower than the required precision ε. By construction, the sought for extremum is somewhere in the interval $[z_a, z_c]$, the solution z_b achieved thus has a precision of order $\pm(z_c - z_a)/2$.

In the preceding discussion, we assumed the starting triplet (z_a, z_b, z_c) featured a point z_b better than both z_a and z_c. If the three points of the triplet are chosen randomly, there is a good chance that z_a or z_c might be better than z_b. There is thus no guarantee that the interval $[z_a, z_c]$ brackets the extremum; that is, the extremum is not necessarily within the interval. One must then seek a new a new triplet (z_a, z_b, z_c), with $g(z_b) < g(z_a), g(z_c)$, that brackets the extremum.

Bracketing of the extremum may be accomplished with a rather simple algorithm as follows. Consider two given points z_b and z_c. Let us assume that $g(z_c)$ is better than $g(z_b)$ as illustrated in Figure 7.9. Let $\Delta z = z_c - z_b$. We then choose a point z_e such that $z_e = z_c + 2\Delta z = 3z_c - 2z_b$ in the direction of the optimum. If $g(z_c)$ is better than $g(z_e)$, then the extremum is bracketed in the interval $[z_b, z_e]$, and one can proceed with the bisection algorithm discussed earlier. Otherwise, iterate the extension, that is, let $z_b^{(k+1)} = z_c^{(k)}$,

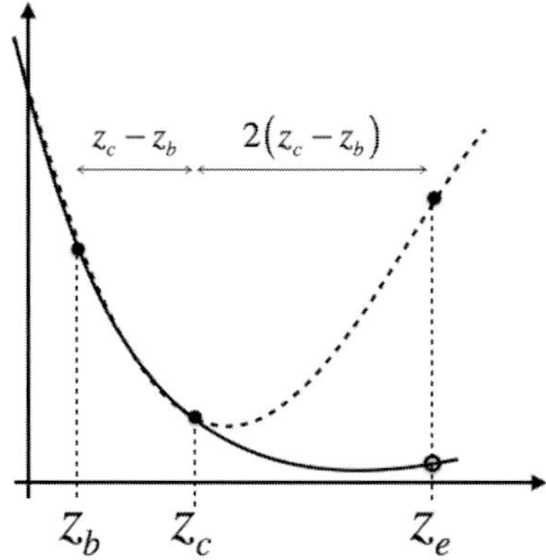

Fig. 7.9 Extension algorithm used to bracket an extremum.

$z_c^{(k+1)} = z_e^{(k)}$, and $z_e^{(k+1)} = 3z_c^{(k+1)} - 2z_b^{(k+1)}$. The width of the interval increases by a factor of 2 at each iteration and is thus guaranteed to eventually contain the extremum, unless an extremum does not exist for finite z values. Implementations of this algorithm should thus check for floating point exceptions.

Insofar as the function $g(z)$ is continuous, the bisection algorithm cannot fail and will always return a valid approximation of the extremum. However, its convergence rate is rather slow, and obtaining a solution of high precision may require a large number of iterations. The required precision ε should thus be set to a practical value reflecting the needs of the application. If extremely high precision is required, one can complete the search with the faster and more precise Newton (gradient) method, which is essentially guaranteed to work well in the immediate vicinity of the extremum.

Powell's Algorithm

Powell's conjugate direction method, commonly known as Powell's algorithm [154], provides a straightforward implementation of the hill climbing method toward the search for a function's extremum. The algorithm relies on the notion that once an optimum has been obtained along a particular direction $\vec{d_k}$, the likelihood for greatest improvement lies in a new direction $\vec{d}_{(k+1)}$ perpendicular to the original direction $\vec{d_k}$, that is, such that $\vec{d}_{(k+1)} \cdot \vec{d_k} = 0$. However, since this condition does not uniquely define $\vec{d}_{(k+1)}$, the algorithm also includes a specific recipe to decide on the next best direction at each step. The algorithm may be summarized as follows:

1. Select the required precision of the search, ε.
2. Let $k = 0$.
3. Given a best point $\vec{x}^{(k)}$ at iteration k, initialize n unit vectors $\hat{e}_i$ to form a complete basis of the search space. The initial set of vectors may be defined as $\hat{e}_1 = (1, 0, \ldots, 0)$, $\hat{e}_2 = (0, 1, \ldots, 0)$, …, $\hat{e}_n = (0, \ldots, 0, n)$.
4. Initiate the search with $k = 1$.
5. Determine the optimum $\vec{x}^{(k)}$ of the function $f(\vec{x})$ along the direction $\hat{e}_k$ starting from $\vec{x}^{(k-1)}$.

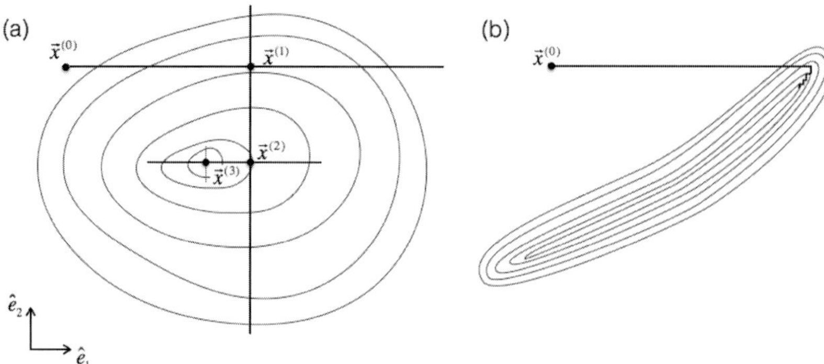

Fig. 7.10 Illustration of Powell's algorithm for (a) a function nearly quadratic in the vicinity of its extremum and (b) a function exhibiting a narrow and elongated extremum structure.

6. Increment k by one unit and if $k \le n$, proceed to step 5.
7. Otherwise, set $\hat{e}_k = \hat{e}_{k-1}$.
8. Set $\hat{e}_n = (\vec{x}^{(n)} - \vec{x}^{(0)})/|\vec{x}^{(n)} - \vec{x}^{(0)}|$.
9. Find a new extremum $\vec{x}^{(n+1)}$ of $f(\vec{x})$ along $\hat{e}_n$.
10. If $|\vec{x}^{(n)} - \vec{x}^{(0)}| < \varepsilon$, terminate the algorithm.
11. Otherwise, set $\vec{x}^{(0)} = \vec{x}^{(n+1)}$, and proceed to step 3.

Powell's algorithm is robust for functions exhibiting a (nearly) quadratic behavior near their extremum, as illustrated in Figure 7.10a, but may be extremely slow to converge (or in some cases not converge at all) if the extremum is located within a very narrow and elongated hill (or valley), as schematically shown in Figure 7.10b.

7.6.4 Simplex Algorithm

The simplex algorithm, invented by Nelder and Mead [147], uses the notion of **simplex** to subdivide a large parameter space and carry out a search for the extremum of a function. Defined in n-dimension space, an n-simplex is a polytope figure formed by joining $n + 1$ vertices (or summits) by straight lines. The $n + 1$ vertices $\vec{x}_k$ of an n-simplex must be affinely independent, that is, the differences $\vec{x}_1 - \vec{x}_0, \vec{x}_2 - \vec{x}_0, \ldots, \vec{x}_n - \vec{x}_0$ must be linearly independent. Formally, a point may be considered a 1-simplex and a line segment a 2-simplex. But the notion of simplex applied to searches for function extremum is of interest mostly for $n \ge 2$. A 3-simplex is a regular triangle in two dimensions, a 4-simplex is a tetrahedron in three dimensions, and so on, as illustrated in Figure 7.11.

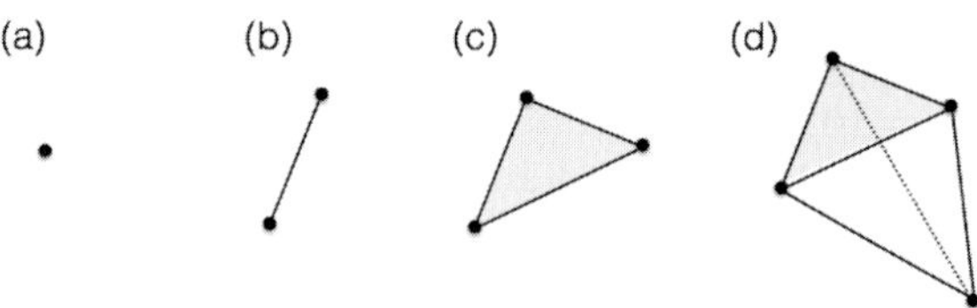

Fig. 7.11 Illustration of basic simplex of order $n = 1$–4: (a) single vertex, (b) line segment, (c) triangle, (d) tetrahedron.

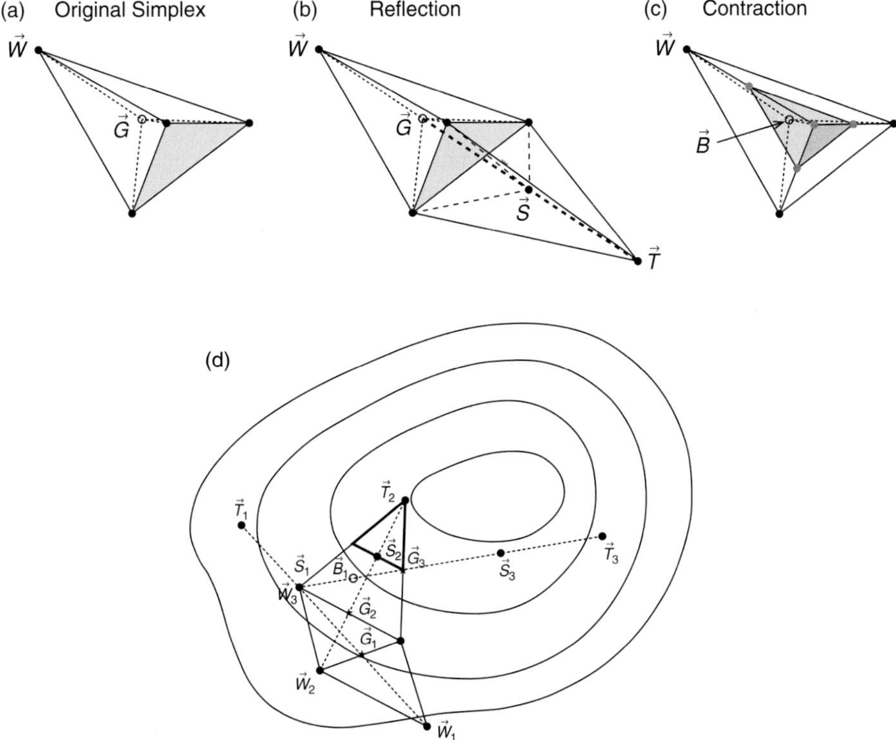

Fig. 7.12 Illustration of the simplex algorithm: (a) original simplex, (b) simplex obtained by reflection of the weakest vertex; (c) simplex contraction relative to the best vertex $\vec{B}$; (d) example of the evolution of a 3-simplex involving two reflections and one contraction.

The bisection technique presented in §7.6.3 has guaranteed success in one dimension but is not applicable in $n \geq 2$ dimensions. It is indeed not possible to robustly bracket the extremum of a function in two or more dimensions. Segmentation of the search space nonetheless remains possible with the notion of simplex. Given the edges and faces of an n-simplex cut across all dimensions of the search space, one may naturally consider how a function of interest behaves across these faces and seek regions where the function grows (search for a maximum) or decreases (search for a minimum).

The algorithm is relatively simple and involves no derivatives. The search for an extremum is based solely on evaluations of the function at the vertices of a simplex while its shape, orientation, and size are varied. The value of the function at a vertex is said to be better if larger (smaller) than those at other vertices in searches for a maximum (minimum). Although bracketing of an extremum is not possible in $n \geq 2$, the shape of the simplex can be varied to stretch or shrink in directions where the function is better. It is thus possible to progressively converge toward an extremum of the function. The simplex algorithm, schematically illustrated in Figure 7.12, may be decomposed as follows:

1. Select a simplex consisting of $n + 1$ vertices $\vec{x}_i$.
   a. If no prior guess of the function's extremum exists, the vertices of the simplex may be selected randomly in the search space of interest. For instance, if the search

consists of an n-dimensions box B (i.e., a hyperrectangle), the vertices may be generated according to

$$\vec{x}_i = r \times (x_{\max,i} - x_{\min,i})\,\hat{e}_i, \tag{7.334}$$

where r is a random number in the range $0 \le r \le 1$; $x_{\max,i}$ and $x_{\min,i}$ represent the upper and lower boundaries of the box, respectively, along dimension i; and $\hat{e}_i$ are unit vectors along each of the dimensions i.

b. Alternatively, given an initial guess $\vec{x}^{(0)}$, one can use $\vec{x}_0 \equiv \vec{x}^{(0)}$ as the first vertex of the simplex and generate n additional vertices according to

$$\vec{x}_i = \vec{x}_0 + k\hat{e}_i, \tag{7.335}$$

where k is constant representative of the scale of the search, and $\hat{e}_i$ are once again unit vectors along each of the dimensions i. If the search space is of considerably different sizes along each of the dimensions, the constant k may be replaced by scale factors for each of the coordinates.

2. Evaluate the objective function at each of the vertices.
3. Tag the worst vertex as $\vec{W}$.
4. If the size of the simplex is smaller than the desired precision ε, terminate the algorithm.
5. Determine the center of gravity, $\vec{G}$, of the simplex according to

$$\vec{G} = \frac{1}{n}\sum_{i \neq W}^{n+1} \vec{x}_i, \tag{7.336}$$

where the worst vertex is excluded.

6. Determine a point $\vec{S}$ located symmetrically to $\vec{W}$ relative to $\vec{G}$,

$$\vec{S} = \vec{G} - (\vec{W} - \vec{G}) = 2\vec{G} - \vec{A}. \tag{7.337}$$

7. Evaluate the function at $\vec{S}$, and skip to step 11 if $f(\vec{S})$ is not better than all other points of the simplex.
8. Otherwise, calculate a new point $\vec{T}$ that is twice as far from $\vec{G}$ than $\vec{S}$

$$\vec{T} = \vec{G} - 2(\vec{S} - \vec{G}) = 3\vec{G} - 2\vec{A}. \tag{7.338}$$

9. Evaluate the function at $\vec{T}$, and if $f(\vec{T})$ is better than $f(\vec{S})$, update the current simplex by replacing $\vec{W}$ by $\vec{T}$ and proceed to step 4.
10. Otherwise, replace $\vec{W}$ by $\vec{S}$ and proceed to step 4.
11. Calculate the point $\vec{B} = (\vec{G} + \vec{W})/2$.
12. If $f(\vec{B})$ is better than all vertices, update the simplex by replacing $\vec{W}$ by $\vec{T}$, and proceed to step 4.
13. Otherwise, contract the simplex by a factor of 2 relative to the best vertex, that is, reduce in half all edges leading to the best vertex, and proceed to step 4.

Step 8 expands the simplex in the direction of the best point and guarantees that the next step will be in a different direction, thereby enabling a complete exploration of the surrounding space. Step 13 shrinks the simplex and thus guarantees an eventual convergence onto an optimum.

The simplex algorithm is reasonably efficient and is guaranteed to converge to an extremum. It is particularly well suited for searches in multidimensional space of large dimensionality, that is, large values of n. Unfortunately, the algorithm may easily get trapped in a local extremum and thus fail to find the sought for global extremum. The algorithm is also found to progress somewhat slowly in the immediate vicinity of an extremum. The simplex algorithm should thus be used with a relatively low precision ε, and once an approximate extremum has been identified, one should switch to more robust and efficient algorithms that work best in the immediate vicinity of an extremum (e.g., Newton algorithm, Levenberg–Marquardt, etc.). The algorithm may also be used in conjunction with the simulated annealing method to reduce the risk of terminating the algorithm at a local extremum.

7.6.5 Random Search Methods

Objective functions involved in modeling of data with nonlinear models often feature a large number of extrema. An extremum search conducted with a gradient or hill climbing type algorithm and initiated totally at random thus runs the risk of homing-in on the "closest" local extremum, thereby missing the true global optimum and the best model parameter set. A technique capable of scanning the entire parameter space that does not get stuck on a local extremum and is capable of finding the best extremum (i.e., the global extremum) of the function is thus needed. Such a global extremum finding method should be particularly useful for modeling problems where the number of parameters is very large or whenever a sensible initialization of the model parameters, either algorithmically or by a user, is not feasible. We discuss two illustrative examples of random model space scanning: simulated annealing and genetic algorithms.

7.6.6 Simulated Annealing Methods

The numerical method of simulated annealing is inspired by the annealing technique used in metallurgy to control the size of crystals and minimize the number and size of defects in materials produced. Studies in metallurgy have shown that the size of crystals and defects are largely determined by the thermodynamic free energy and the rate of cooling used in the production of materials: slow cooling enables uniform temperature decrease and the atoms of the material are then less likely to settle in local configurations that produce "local" minima of the free energy. The numerical method of simulated annealing parallels the slow cooling process by introducing a modified objective function that depends on a temperature parameter T. The technique involves a random exploration of the parameter space in which a migration to better parameters (in the sense of producing a higher likelihood or minimum χ^2) is always accepted, but such that occasional jumps to worse local solutions are also possible. It is those random jumps away from a local minimum that allow, eventually, finding stronger (better) regions of the objective function and, ultimately, the identification of a true global extremum.

Initial developments of the simulated annealing technique are commonly attributed to S. Kirkpatrick et al. [126] and V. Cerny [62], who independently adapted the Metropolis–Hastings algorithm [143] used in Monte Carlo simulations of thermodynamic systems. Several variants of the technique have been developed that find a wide range of applications in physical sciences, life sciences, and manufacturing. A detailed discussion of these variants, their range of applicability, strengths, and weaknesses is beyond the scope of this text and can be found in the computing literature. Our discussion is thus limited to a brief conceptual introduction of the principle and merits of the method.

The basic principle of the simulated annealing technique is to replace the posterior $p(\vec{\theta}|D, I)$ by a modified posterior $p_T(\vec{\theta}|D, I)$ defined as follows:

$$p_T(\vec{\theta}|D, I) = \exp\left(\frac{\ln p(\vec{\theta}|D, I)}{T}\right), \tag{7.339}$$

where T plays the role of a temperature parameter. For $T = 1$, the function $p_T(\vec{\theta}|D, I)$ is obviously strictly equivalent to the posterior $p(\vec{\theta}|D, I)$. However, for increasingly larger values of the temperature, the ratio of the logarithm of $p(\vec{\theta}|D, I)$ to T produces progressively flatter and flatter profiles of probability. Random motions in parameter space thus do not result in large changes in probability. It is then possible to randomly explore the entire parameter space at low cost and run a lower risk of being stuck in a local minimum.

The algorithm achieves an extensive exploration of the space by using random jumps from its "current" position:

$$\vec{\theta}^{(k+1)} = \vec{\theta}^{(k)} + \Delta\vec{\theta}, \tag{7.340}$$

where the size of the jump, $\Delta\vec{\theta}$, is determined randomly for each jump. One then computes the difference of the probabilities at $\vec{\theta}^{(k+1)}$ and $\vec{\theta}^{(k)}$

$$\Delta p = p_T(\vec{\theta}^{(k+1)}|D, I) - p_T(\vec{\theta}^{(k)}|D, I). \tag{7.341}$$

A new value $\vec{\theta}^{(k+1)}$ is considered more advantageous (better), and thus readily accepted as a current estimate of the solution, if $\Delta p > 0$. If it is not better, the jump may nonetheless be randomly accepted with an acceptance rate α determined by the temperature T. The search begins at high temperature and yields a high acceptance rate for bad jumps. This enables a wide (though not systematic) exploration of the parameter space. It is then likely that the algorithm can stumble onto better regions of the parameter space. The temperature is slowly and steadily decreased toward unity as the search proceeds, and with it, the acceptance rate of bad jumps. As the rate of bad jumps decreases, the search becomes more focused on the local region surrounding the current value of the parameter. The hope is then that the random exploration of the space has enabled the identification of the region of the true extremum of $p_T(\vec{\theta}|D, I)$, so that, as the temperature finally converges to unity, the search yields a true global extremum of $p(\vec{\theta}|D, I)$.

The simulated annealing technique involves several parameters and features of its own, including

1. A model for the rate at which T is reduced relative to the number of iterations k completed
2. A model for the bad jumps acceptance rate α
3. A model for the (maximum) size of random jumps $\Delta\vec{\theta}$

Various annealing models are discussed in the computing literature, which are shown to have varying degrees of success and performance. See [156] and references therein for a more in-depth discussion of these models.

An important drawback of the basic annealing algorithm is that random jumps can be quite ineffective at finding a better region of the objective function. Indeed, a large fraction of jumps carried out may be nonadvantageous and the CPU time spent on useless parameter values (or regions) may be quite large. The problem may be particularly acute if the objective function has narrow valleys or peaks. A very large fraction of the jumps may then end up in the flat and nondiscriminating landscape of the function, thus rendering the algorithm rather ineffective. Fortunately, the concept of simulated annealing does not require the jumps to be totally random and one may in fact couple the annealing technique to more structured search algorithms. Press et al. [156] present such an algorithm in which the search parameter $\vec{\theta}^{(k)}$ is replaced by a simplex. The search for an extremum is then largely driven by the structure of the objective function. However, in order to avoid being stuck on local extrema, the annealed simplex of Press et al. accepts disadvantageous changes of the simplex with an acceptance rate that depends on the temperature. It is thus in principle possible to avoid getting stuck in a local extremum of the objective function, while, relatively speaking, limiting the number of useless function calls and steps across the search space. Other variants of the annealing algorithm have also been considered.

7.6.7 Genetic Algorithms

Genetic algorithms (GAs) use a search heuristic that simulates the process of natural selection toward the solution of optimization and search problems. They belong to a larger class of algorithms known as evolutionary algorithms used most particularly in artificial intelligence problems. Many GA variants have been developed and reported in the computing literature. In this section, we focus on the application of GAs in problems of model parameter optimization. As for simulated annealing algorithms, our discussion is meant to be a conceptual introduction: readers interested in implementing and using GAs in practical applications should consult specialized works on this topic [36, 107, 144, 145].

The concept of the GA was introduced in 1975 by John Holland [107] as an alternative optimization technique capable of avoiding getting stuck on local extrema. The technique is called genetic because it mimics the evolutionary process of all known living species. In the context of GAs, parameter values within a search space are considered as chromosomes of individuals. The objective function is then regarded as a measure of

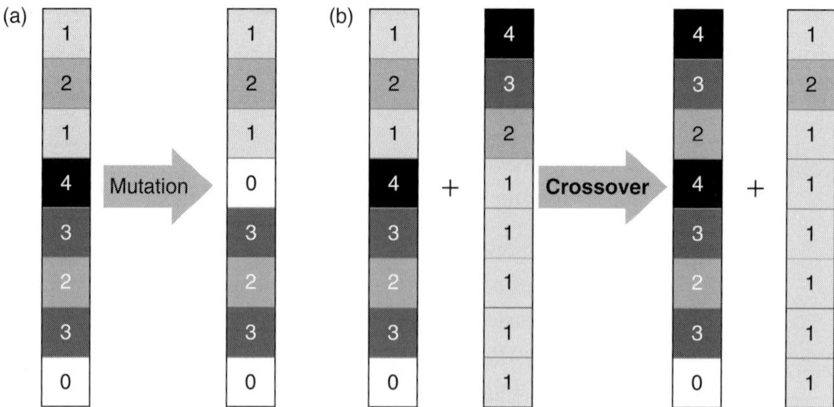

Fig. 7.13 Schematic illustration of the (a) gene mutation and (b) crossover processes involved in genetic algorithms used toward objective function optimization.

the fitness of individuals toward adaptation to their environment. The GA proceeds by successive iterations. At each iteration, gene mutations and cross-overs are produced and examined. The fittest individuals (i.e., in this context, chromosomes with best objective function values) are selected. They "reproduce" and thus make it into the next iteration. A mutation involves a change of one chromosome at reproduction time whereas a crossover occurs when two chromosomes randomly split and recombine with components (genes) of each other, as schematically illustrated in Figure 7.13. The choice of which individuals (i.e., which chromosomes) survive and are selected for reproduction is carried out randomly. This enables a full exploration of the parameter space and local extrema are thus avoided.

In the context of optimization problems, GA is typically implemented by considering individual elements θ_i of a parameter vector $\vec{\theta} = (\theta_1, \ldots, \theta_m)$ as individual genes. The reproduction of an individual involves a perfect copy of all elements of the vector but the selection of which individuals are reproduced and/or genetically modified is a random process. The notion of survival of the fittest is implemented by assigning a selection probability larger for more fit individuals, that is, those with better values of the objective function. Mutations create individuals with genes that span the entire parameter space. Evidently, most mutants are found to have poor objective function value and discarded; that is, they do not make it to the next generation (iteration). However, occasionally, mutations reveal more fit individuals, and those are more likely to make it into the next generation. Similarly, crossovers mix good genes with the intent of producing more fit individuals. They complement mutations toward a full exploration of the parameter space. Their production and selection involves a stochastic process as well. Many crossovers lead to unfit individuals but few produce better adapted individuals that survive into the following generation. Successive iterations eventually yield parameter values that constitute a global extremum of the objective function.

An elegant implementation of the genetic algorithm for parameter estimation in SmallTalk and Java is presented by D. Besset [36].

7.7 Model Comparison and Entity Classification

7.7.1 Problem Definition

Model selection and entity classification problems are encountered in all scientific disciplines. And, although they have rather different practical goals, that of determining which of several models is best and identifying the type or class of a specific observed entity, they are mathematically equivalent. The two types of problems can thus be treated with the same tools and inference framework. We will here restrict our discussion to the Bayesian approach, having already discussed in §6.4 the notion of which test can be used, within the frequentist paradigm, to evaluate the goodness of models and classify entities. One may argue, however, that the Bayesian paradigm provides tools and methods far more intuitive and better adapted to either tasks than the frequentist approach.

As an example of model selection drawn from nuclear physics, consider a measurement of the momenta, $\vec{p}$, of particles produced by some nuclear interaction (e.g., proton on proton at 7 TeV). The goal shall be to determine and use the shape of the momentum spectrum in order to compare and hopefully falsify[12] competing models of the particle production dynamics. Let $D = \{p_1, p_2, \ldots, p_n\}$, where n is a very large integer corresponding to the number of observed particles, represent the set of measured momenta. The inference task shall then be to determine which of $m > 1$ model hypotheses H_k, best represent the data D. For example, one could consider whether the data are best represented by an exponential distribution, a Maxwell–Boltzmann distribution, a Blast–Wave model [170], a power law, and so on. Evidently, the models H_k may have distinct parameters $\vec{\theta}^{(k)}$. The model selection problem shall then consist in finding which of the m models H_k features the largest posterior probability $p(H_k|D, I)$ irrespective of these parameters.

Data classification problems involve a rather similar logic and process. For instance, in nuclear collision studies, one might be interested in identifying the species of measured particles such as pions, kaons, protons, and so on, based on their energy loss in a detector (e.g., a Time Projection Chamber, TPC), or one might wish to use shower shapes observed in a calorimeter system to distinguish photons and electrons from hadrons, and so on. One then formulates $m > 1$ distinct hypotheses H_k corresponding to the several classes, categories, or types the measured entities (e.g., particles or calorimeter showers) might belong to. The probabilities $p(H_k|D, I)$ a given entity might belong to each of the types are then evaluated on the basis of prior probabilities and likelihood functions of each of the models. The hypothesis H_k with the largest posterior probability $p(H_k|D, I)$ may then be selected as the most likely type of the measure entity based on the data D.

7.7.2 Comparing the Odds of Two Hypotheses

Let us consider a specific example of the model selection problem within the Bayesian paradigm. Let us assume that we have carried out a measurement and obtained a dataset

[12] Demonstrate as false or invalid.

D. Our task shall then be to assess which of several models (or model hypotheses) H_k best match or fit the data by comparing the odds, that is, the probabilities, of the models.

The comparison of the odds of the two or more models is readily formulated in terms of ratios of their posterior probabilities. For instance, in order to compare two models H_0 and H_1, with model parameters $\vec{\theta}^{(H_0)}$ and $\vec{\theta}^{(H_1)}$, respectively, we define the odds ratio of model H_1 in favor of model H_0 as

$$O_{10} = \frac{p(H_1|D, I)}{p(H_0|D, I)}, \tag{7.342}$$

where $p(H_0|D)$ and $p(H_1|D)$ represent the posterior probabilities of the models H_0 and H_1, respectively, regardless of the specific parameter values $\vec{\theta}^{(H_j)} = (\theta_1^{(H_j)}, \ldots, \theta_{m_j}^{(H_j)})$ that best fit the data. The posteriors $p(H_j|D, I)$, $j = 1, 2$, are obtained by marginalization of all parameters $\vec{\theta}^{(H_j)}$ of the models according to

$$p(H_j|D, I) = \int_{\Omega_{\vec{\theta}}} d\theta_1^{(H_j)} \cdots d\theta_{m_j}^{(H_j)} p\big(\theta_1^{(H_j)}, \ldots, \theta_{m_j}^{(H_j)}|D, I\big), \tag{7.343}$$

where the integration is carried over all model parameters $\vec{\theta}^{(H_j)}$ and across the entire domain $\Omega_{\vec{\theta}}$ of these parameters. Typically, the hypothesis H_0 shall be considered as the null hypothesis representing a commonly adopted model. A large value of the odds ratio, $O_{10} \gg 1$, shall indicate that the alternative hypothesis H_1 is a far more suitable representation (explanation) of the data than the null hypothesis H_0, while a value of order unity or smaller would indicate H_1 is not particularly favored by the data, and thus should not be adopted in favor of the null hypothesis.

The comparison of the two hypotheses is thus, in principle, rather straightforward as it suffices to take the ratio of their respective probabilities. One may in fact forgo the normalization of the posteriors by the global likelihood of the data, $p(D|I)$, and write

$$O_{10} = \frac{p(H_1|I)p(D|H_1, I)/p(D|I)}{p(H_0|I)p(D|H_0, I)/p(D|I)} = \frac{p(H_1|I)p(D|H_1, I)}{p(H_0|I)p(D|H_0, I)}, \tag{7.344}$$

which involves ratios of priors and likelihoods of the data according to the two hypotheses. Introducing the priors odds ratio

$$O_{10}^{\text{prior}} = \frac{p(H_1|I)}{p(H_0|I)}, \tag{7.345}$$

and the **Bayes factor**

$$B_{10} = \frac{p(D|H_1, I)}{p(D|H_0, I)} \tag{7.346}$$

as the ratio of the likelihoods of the data based on the two hypotheses, one can determine the posterior odds ratio in terms of a product of the priors odds ratio and the Bayes factor

$$O_{10} = O_{10}^{\text{prior}} B_{10}. \tag{7.347}$$

Examples of computation of Bayes factors and odds ratio are discussed later in this section.

7.7.3 Comparing the Odds of Hypotheses with a Different Number of Parameters

A commonly encountered class of problems involves the fitting of data and comparison of models with different numbers of free parameters. For such comparisons, it is often possible (although not essential for the following discussion) to represent measured data with generic linear models of the form

$$f(x|\vec{a}) = \sum_{k=1}^{m} a_k f_k(x), \tag{7.348}$$

where the coefficients a_k are free model parameters and the $f_k(x)$ represent a set of linearly independent functions. Examples of applications of such linear models include fits of polynomials, orthogonal polynomials, and Fourier decompositions. By construction, the inclusion of several functions $f_k(x)$ in a model may enable representation of a wide range of functional dependencies on x, and thus provide the capacity to obtain arbitrarily good fits of the data. Indeed, unless there is prior knowledge dictating how many terms $a_k f_k(x)$ should be included in the sum, one could, in principle, add arbitrarily many such terms and obtain arbitrarily good fits of measured data. However, given finite measurement errors, it may be unclear, a priori, what the true functional shape of the data should be, and what number m of functions $f_k(x)$ should actually be included in the parameterization of the data. One thus wishes for a mathematical procedure capable of enabling an objective decision as to whether the addition of one or several model parameters might be justified.

The problem arises, obviously, that by arbitrarily increasing the number of parameters, one might be able to fit the data exactly. Indeed, a polynomial of order n shall perfectly fit a dataset consisting of $n + 1$ points, but such a perfect fit should not be construed as meaningful or representative of the data. Evidently, the presence of (random) measurement errors can give the illusion of high-frequency components in the measured data. Such high-frequency components should not be included, however, unless they are motivated by a physical understanding of the observed phenomenon. The decision-making mathematical procedure we seek should thus have a built-in mechanism to disfavor complicated hypotheses with too many parameters. The goal of this section is to show that odds ratios, equipped with the notion of Occam factors introduced in the text that follows, do in fact provide a more or less objective procedure to compare the merits of models of the form, Eq. (7.348), that penalizes overly complicated models.

For simplicity's sake, let us begin our discussion with the comparison of a null hypothesis (a model) H_0 determined by a single and fixed parameter, $\theta = \theta_0$, and an alternative hypothesis H_1 functionally identical to H_0, but in which the parameter θ is allowed to vary within some specific domain $\theta_{\min} \le \theta \le \theta_{\max}$. We will see, later in this section, how Bayesian model comparison may be trivially extended to any finite number of parameters and any finite number of models. By virtue of the fact that H_1 has an adjustable parameter θ, one might naively expect that it should readily provide a better fit of the data and thus have a larger probability. However, we shall next calculate the odds ratio O_{10} and show that the posterior probability of H_1 is in fact suppressed, relative to H_0, by the added parameter space.

Let us assume that the models H_0 and H_1 span the entire space of models so one can write

$$p(H_0|D, I) + p(H_1|D, I) = 1. \tag{7.349}$$

Dividing this expression by $p(H_0|D, I)$, and rearranging the terms, one gets

$$p(H_1|D, I) = 1 - p(H_0|D, I) = \frac{1}{1 + (O_{10})^{-1}}, \tag{7.350}$$

which tells us that the probability of the alternative hypothesis is determined entirely by the posterior odds ratio O_{10}. If O_{10} is very large, $p(H_1|D, I) \to 1$, whereas if O_{10} is very small, one gets $p(H_1|D, I) \approx 0$. In the specific context of the comparison of two models of similar functional dependence, but different numbers of free parameters, one may assume there are no a priori reasons to favor one model or the other, and choose $O_{10}^{\text{prior}} = 1$. Equation (7.350) then reduces to

$$p(H_1|D, I) = \frac{1}{1 + (B_{10})^{-1}}. \tag{7.351}$$

Let us thus focus our attention on the Bayes factor B_{10}. This requires marginalization of the posterior probability, $p(\theta|D, H_1, I)$, of the parameter θ. In turn, it also necessitates the choice of a prior for θ and calculation of the likelihood of hypothesis H_1. Let us first consider the prior of θ.

Assuming the range of θ may be sensibly bound to $\theta_{\min} \le \theta \le \theta_{\max}$, and given the prior lack of information about this parameter, it is reasonable to choose a flat prior for θ and write

$$p(\theta|H_1, I) = \begin{cases} \frac{1}{\Delta\theta} & \text{for } \theta_{\min} \leqslant \theta \leqslant \theta_{\max} \\ 0 & \text{elsewhere,} \end{cases} \tag{7.352}$$

where $\Delta\theta = \theta_{\max} - \theta_{\min}$. A calculation of the likelihood $p(D|\theta, H_1, I)$ evidently requires knowledge of the data D and the specificities of the model H_1, but, in general, one may write

$$\int_{\Delta\theta} p(D|\theta, H_1, I)\, d\theta \equiv \delta\theta \times p(D|\hat{\theta}, H_1, I) = \delta\theta \times L_{\max}, \tag{7.353}$$

where $L_{\max} \equiv p(D|\hat{\theta}, H_1, I)$ corresponds to the maximum value of the likelihood function achieved at the mode $\hat{\theta}$, and $\delta\theta$ thence corresponds to a "characteristic width" of the likelihood function, as illustrated in Figure 7.14. For a Gaussian likelihood, the integral in Eq. (7.353) would yield $\delta\theta = \sqrt{2\pi}\sigma_\theta$, where σ_θ is the error on the parameter θ. In the more general case of a non-Gaussian likelihood, one may write $\delta\theta = \alpha\sigma_\theta$, where α is a constant determined by the specific shape of the likelihood function. This is immaterial, however, for the calculation of the global likelihood of the data given H_1

$$\begin{aligned} L(H_1) \equiv p(D|H_1, I) &= \int p(\theta|H_1, I) p(D|\theta, H_1, I)\, d\theta \\ &= \frac{1}{\Delta\theta} \int_{\Delta\theta} p(D|\theta, H_1, I)\, d\theta \\ &= p(D|\hat{\theta}, H_1, I) \frac{\delta\theta}{\Delta\theta}. \end{aligned} \tag{7.354}$$

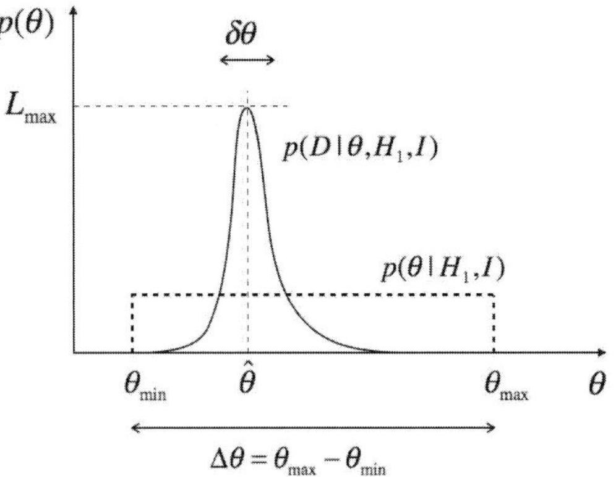

Fig. 7.14 Schematic illustration of the notion of Occam factor, $\Omega_\theta = \delta\theta / \Delta\theta$. The dashed and solid curves represent the prior probability of the model parameter θ and the likelihood distribution of the data D given hypothesis H_1, respectively.

The global likelihood, $L(H_1)$, of model H_1 is thus equal to the maximum likelihood of the model, denoted $L_{\max} \equiv L(\hat{\theta})$, achieved at the mode $\hat{\theta}$ multiplied by a purely geometrical factor $\delta\theta / \Delta\theta$, involving the width $\delta\theta$ of the likelihood distribution as well as the width $\Delta\theta$ of the nominal domain of the parameter θ. Noting that the likelihood of hypothesis H_0 is simply

$$p(D|H_0, I) = p(D|\theta_0, H_0, I) \equiv L(\theta_0) \tag{7.355}$$

We can then complete the calculation of the Bayesian factor B_{10} and write

$$\begin{aligned} B_{10} &= \frac{p(D|\hat{\theta}, H_1, I)}{p(D|\theta_0, H_0, I)} \frac{\delta\theta}{\Delta\theta}, \\ &= \frac{L(\hat{\theta})}{L(\theta_0)} \frac{\delta\theta}{\Delta\theta}. \end{aligned} \tag{7.356}$$

In general, the nominal parameter range $\Delta\theta$ is much wider than the posterior width $\delta\theta$. The ratio $\delta\theta / \Delta\theta \ll 1$ thus penalizes the model H_1 for the presence of its (extra) parameter θ relative to H_0, which has no free parameter. Indeed, the Bayes factor must satisfy

$$\frac{L(\hat{\theta})}{L(\theta_0)} \gg \frac{\Delta\theta}{\delta\theta} \tag{7.357}$$

in order to favor the hypothesis H_1. Since $\delta\theta / \Delta\theta \ll 1$, the maximum likelihood $L_{\max} \equiv L(\hat{\theta})$ must then be much larger than $L(\theta_0)$ to favor H_1. The search domain associated with the parameter θ thus indeed imposes a penalty on H_1 that the model must overcome by featuring a very large maximum likelihood $L(\hat{\theta})$ before it can be deemed a better description of the data. Effectively, the ratio $\delta\theta / \Delta\theta$ plays the role of Occam's razor, as it disfavors the more "complicated model," that is, the model with an extra parameter. The ratio is thus known as the **Occam factor** and commonly denoted Ω_θ. The global likelihood $p(D|H_1, I)$ may then be written

$$p(D|H_1, I) = L(\hat{\theta})\Omega_\theta. \tag{7.358}$$

If the hypothesis H_1 has m extra parameters $\theta_1, \ldots, \theta_m$, one similarly finds

$$p(D|H_1, I) = \int p(D|\theta_1, \ldots, \theta_m, H_1, I) \prod_{k=1}^{m} p(\theta_k|H_1, I)\, d\theta_k \tag{7.359}$$

$$= L(\hat{\theta}_1, \ldots, \hat{\theta}_m) \frac{\delta\theta_1}{\Delta\theta_1} \cdots \frac{\delta\theta_m}{\Delta\theta_m}$$

$$= L(\hat{\theta}_1, \ldots, \hat{\theta}_m) \Omega_{\theta_1} \cdots \Omega_{\theta_m}, \tag{7.360}$$

where $\Omega_1, \ldots, \Omega_m$ represent the Occam factors associated with the parameters $\theta_1, \ldots, \theta_m$. Under ideal circumstances, each Occam factor should be much smaller than 0.1. Consider, for instance, a case involving three (extra) fit parameters, each with an Occam factor of the order of 0.01. The ratio $L(\hat{\theta})/L(\theta_0)$ would then be required to be much in excess of 10^6 to favor the hypothesis H_1. Effectively, the Occam factors collectively disfavor the alternative hypothesis, H_1, unless it is, in fact, correct and well supported by precise data. As we shall discuss with examples in §§7.7.4 and 7.7.4, measurement errors must in general be sufficiently small to render the test conclusive. If the data have large errors, the likelihood $L(\hat{\theta}_1, \ldots, \hat{\theta}_m)$ might be too small to overcome the Occam factors, and model H_1 will remain of questionable value. In the context of a linear model with Gaussian noise, the integral Eq. (7.353) implies $\delta\theta = \sqrt{2\pi}\sigma_\theta$, where σ_θ is the error on the parameter θ. By virtue of Eq. (7.252), one then indeed expects the error σ_θ to scale as $\sigma_y/\sqrt{n}$, where σ_y is the typical measurement error, and n is the number of points in the dataset. The value of the Occam factor $\Omega_\theta = \delta\theta/\Delta\theta$ is thus dependent on the size of the measurement errors as well as the size of the data sample. Availability of better and larger datasets shall thus lead to smaller Occam factors, and thus more stringent demands on the maximum likelihood of the model H_1, and unless the ratio $L(\hat{\theta})/L(\theta_0)$ is sufficiently large, the Bayes factor will consistently favor the simpler model.

In general, there is little interest in calculating the Occam factors explicitly, but because they naturally arise as a result of the marginalization procedure, complex models (with many extra parameters) shall be transparently and automatically disfavored unless strongly supported by the data.

We illustrate the aforementioned principle of model comparison, and the effect of the Occam factors, with two examples: the first, presented in §7.7.4, involves the identification of a (new) signal amidst a noisy background, while the second, in §7.7.5, illustrates the comparison of models that differ by the addition of one or few parameters.

7.7.4 Example 1: Is There a Signal?

Problem Definition

Let us consider a problem often encountered in physics: a search for a predicted signal amidst background noise. For simplicity's sake, we will here neglect much of the experimental considerations involved in such searches and reduce the complexity of the problem to one unknown: the amplitude of the signal (assuming the signal does exist). Let us assume the measurement D is reported in terms of a Fourier spectrum $\vec{n} = (n_1, n_2, \ldots, n_m)$ consisting of m bins (channels) of equal width Δf in the range $f_{\min} = 0$ to $f_{\max} = 100$ (in arbitrary units). Let us further assume, to keep the problem relatively simple, that the signal is expected to have a Gaussian line shape centered at a frequency $f_0 = 50.5$ and of width $\sigma_0 = 1.5$, as illustrated in Figure 7.15a. For illustrative purposes, we will carry out the analysis assuming four different measurement outcomes displayed in Figure 7.15b–e, which feature the same Gaussian noise characterized by a null expectation value, a standard deviation of $\sigma_n = 0.1$, and no bin-to-bin cross correlations. Spectra (b–e) have been generated with a signal of amplitude A equal to 15.0, 5.0, 1.0, and 0.1 (arbitrary units), respectively but we will of course pretend we have no knowledge whatsoever of these amplitudes in our analysis. In fact, we will assume that a prior search of the signal has found the signal is weaker than some upper bound $A_{\max} = 100$ and that the theory stipulates the signal should have a minimum amplitude $A_{\min} = 0.01$ (again in arbitrary units).

Comparing the Odds of Models H_1 and H_0

Our goal is to establish whether the observed spectra (b–e) provide evidence in support of a model H_1 stating the existence of the signal, or a null hypothesis H_0 asserting there is no such new signal. The presence of a signal is clearly obvious in panels b and c but rather difficult to establish visually in panels d and e. Our goal is to show how Bayesian inference can support this visual impression quantitatively when the signal seems obvious but can also provide odds of one model against the other even when the signal is not visually evident. In order to realize this goal, we will compare the posterior probabilities $p(H_1|D, I)$ and $p(H_0|D, I)$ of models H_1 and H_0, respectively, and calculate the posterior odds ratio of the two hypotheses

$$O_{10} \equiv \frac{p(H_1|D, I)}{p(H_0|D, I)}. \tag{7.361}$$

Recall, from §7.7.2, that the odds ratio O_{10} may be expressed in terms of the prior odds ratio O_{10}^{prior} and the Bayes factor B_{10} defined as

$$O_{10} = O_{10}^{\text{prior}} B_{10} \tag{7.362}$$

with

$$O_{10}^{\text{prior}} = \frac{p(H_1|I)}{p(H_0|I)}, \tag{7.363}$$

and

$$B_{10} = \frac{p(D|H_1, I)}{p(D|H_0, I)} = \frac{\mathbb{L}(H_1)}{\mathbb{L}(H_0)}, \tag{7.364}$$

where $\mathbb{L}(H_1)$ and $\mathbb{L}(H_0)$ are the global likelihoods of the models H_1 and H_0, respectively.

Given the limited amount of prior information about the existence of the signal, it appears legitimate to set the prior odds ratio to unity, $O_{10}^{\text{prior}} = 1$. The posterior odds ratio shall thus be entirely determined by the Bayes factor, that is, the ratio of the global likelihood of the two models.

It is important to note that we actually seek posterior probabilities for the models H_1 and H_0 irrespective of their "internal" parameter values. In the context of this problem, only H_1 has an internal parameter, and this parameter is the unknown amplitude A of the purported signal. Since the actual value of this amplitude is irrelevant for the central goal of our study (i.e., establish the existence of the signal), we obtain $p(D|H_1, I)$ by marginalization according to

$$p(D|H_1, I) = \int_{\Delta A} p(A|H_1, I) p(D|H_1, A, I)\, dA \tag{7.365}$$

where $p(D|H_1, A, I)$ is the probability of observing the data D given H_1 is valid and for a specific amplitude A, and $p(A|H_1, I)$ is the prior degree of belief in the strength A of the signal given H_1 is true.

Probability Model and Likelihoods

Let us first identify and discuss each of the components required in our Bayesian analysis of the spectra. The measured data D (one the four spectra in Figure 7.15) are represented as a vector $\vec{n} = (n_1, n_2, \ldots, n_m)$ and values n_i correspond to the number of counts (after background subtraction) in bins $i = 1, \ldots, m$. By construction, the bin content n_i is the sum of the fraction of the signal s_i expected in bin i and the noise e_i in that bin:

$$n_i = s_i + e_i. \tag{7.366}$$

Model H_1 tells us that, for a fixed amplitude A, the fraction of the signal s_i is equal to

$$s_i = A p_i \tag{7.367}$$

where

$$p_i = \int_{f_{i,\min}}^{f_{i,\max}} \frac{1}{\sqrt{2\pi}\,\sigma_0} \exp\left[-\frac{(f - f_0)^2}{2\sigma_0^2}\right] df \tag{7.368}$$

is the relative fraction of the signal in each bin i, not a random number, determined by the model parameters f_0, σ_0, and the bin boundaries $f_{i,\min}$ and $f_{i,\max}$. The signal yield s_i is thus a random number only insofar as A might itself be considered a random number. The noise terms e_i, however, are random Gaussian deviates, which we assume all have the same standard deviation σ_n. The probability density that the noise e_i be found in the range $[e_i, e_i + de_i]$ is

$$p(e_i|\sigma_n, I) = \frac{1}{\sqrt{2\pi}\,\sigma_n} \exp\left(-\frac{e_i^2}{2\sigma_n^2}\right). \tag{7.369}$$

But, given the noise e_i may be written as

$$e_i = n_i - s_i, \tag{7.370}$$

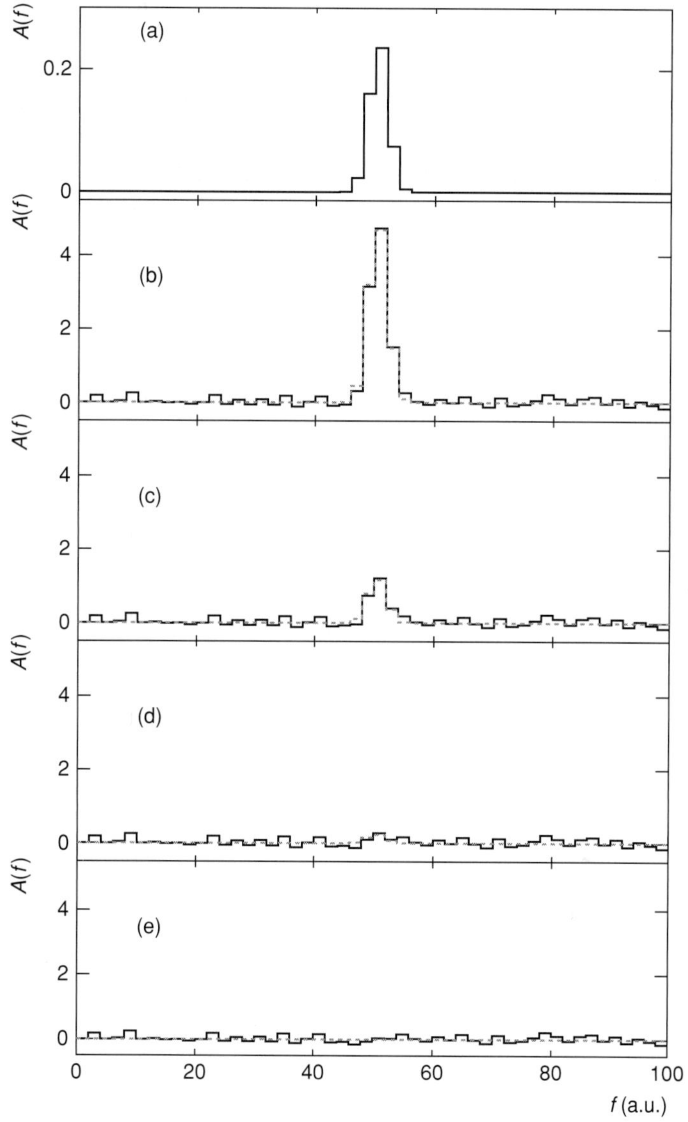

Fig. 7.15 Simulated spectra used in the example of Bayesian inference presented in §7.7.4. (a) Signal line shape and position predicted by model H_1. (b–e) Simulated measurement outcomes with amplitudes $A = 20, 5, 1, 0.005$, and a Gaussian noise with standard deviation $\sigma = 0.1$. Best fits (posterior modes) are shown with dashed lines.

this implies that for a fixed signal amplitude A, the probability that the bin content n_i will be found in the range $[n_i, n_i + dn_i]$ is equal to the probability the noise will be found in $[e_i, e_i + de_i]$, and we thus obtain the data probability model

$$p(n_i|H_1, A, \sigma_n, I) = \frac{1}{\sqrt{2\pi}\sigma_n} \exp\left[-\frac{(n_i - Ap_i)^2}{2\sigma_n^2}\right]. \tag{7.371}$$

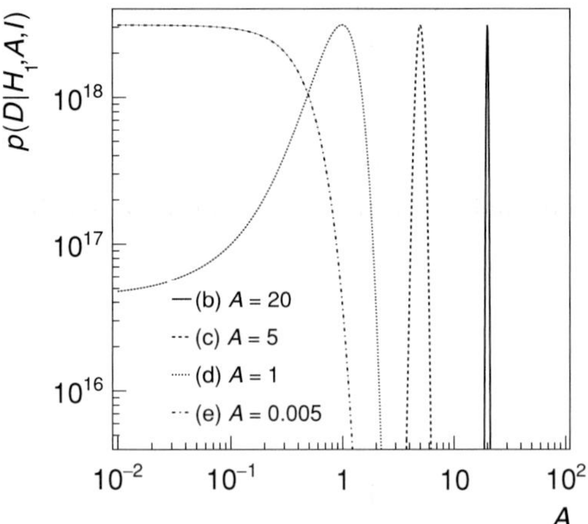

Fig. 7.16 Likelihood distributions $p(D|H_1, A, I)$ vs. A for the measurement outcomes shown in Figure 7.15b–e, with respective true amplitudes 20, 5, 1, and 0.005.

The likelihood of the data D given the model H_1 and a fixed amplitude A may then be written

$$\begin{aligned} p(D|H_1, A, I) &= \prod_{i=1}^{m} \frac{1}{\sqrt{2\pi}\sigma_n} \exp\left[-\frac{(n_i - Ap_i)^2}{2\sigma_n^2}\right], \\ &= (2\pi)^{-m/2}\sigma^{-m} \exp\left[-\frac{\sum_{i=1}^{m}(n_i - Ap_i)^2}{2\sigma_n^2}\right], \\ &= (2\pi)^{-m/2}\sigma^{-m} \exp\left[-\frac{1}{2}\chi^2\right], \end{aligned} \tag{7.372}$$

where we introduced the χ^2 function defined by

$$\chi^2 = \sum_{i=1}^{m} \frac{(n_i - Ap_i)^2}{\sigma_n^2}. \tag{7.373}$$

The likelihood of the model H_0 is readily obtained by setting $A = 0$ in the preceding expressions.

$$p(D|H_0, \sigma_n, I) = (2\pi)^{-m/2}\sigma^{-m} \exp\left[-\frac{1}{2}\chi_0^2\right], \tag{7.374}$$

with

$$\chi_0^2 = \sum_{i=1}^{m} \frac{n_i^2}{\sigma_n^2}. \tag{7.375}$$

The likelihood distributions $p(D|H_1, A, \sigma_n, I)$ corresponding to spectra (b–e) are displayed in Figure 7.16 using a log-log scale to facilitate visualization of the tails of the distributions. Spectra (b) and (c) are characterized by very narrow $p(D|H_1, A, \sigma_n, I)$ distributions, with an approximate Gaussian shape, centered at the expected amplitudes and a maximum likelihood value of 3.1×10^{18}. Distributions associated with spectra (d) and

(e) have similar maximum likelihood values but feature non-Gaussian shapes. Somewhat wider, the likelihood distribution of spectrum (d) features a sizable low-side tail while the likelihood distribution of spectrum (e) is peaked near the origin, thereby indicating it is consistent with a signal of null amplitude (i.e., no signal).

We now wish to determine and compare the posteriors $p(H_0|D,I)$ and $p(H_1|D,I)$. Since H_0 has no amplitude parameter, its posterior may be obtained directly from the likelihood $p(D|H_1,I)$ of the measured spectra. Determination of $p(H_1|D,I)$, however, requires one marginalizes the $p(H_1,A|D,I)$ for A. This implies we must first formulate $p(A|H_1,I)$ expressing the prior probability density of the signal amplitude A given H_1 is assumed valid. Note that to keep everybody honest, the prior should be selected before the measurement is carried out so that one is not tempted to look at the answer (i.e., the data) in order to make that choice. Here, the problem is academic, and we will actually explore what happens under two different choices of prior.

Comparison of Uniform and Scaling Priors for $p(A|H_1,I)$

Obviously, the problem arises that we have only a rather limited amount of information about H_1 and more specifically the signal amplitude A. We do know, based on a prior experiment, that A is smaller than some upper bound $A_{\max}$ and that it should, according to the model H_1, be larger than some minimal value $A_{\min}$. But that's it! How then should we formulate the prior $p(A|H_1,I)$? Given the lack of additional information, it seems natural to express our ignorance about A by choosing a uniform prior

$$p(A|H_1,I) = \begin{cases} \frac{1}{\Delta A} & \text{for } A_{\min} \le A \le A_{\max} \\ 0 & \text{otherwise,} \end{cases} \tag{7.376}$$

with $\Delta A = A_{\max} - A_{\min}$. It is worth noting, however, that this choice implies the top range of values is far more probable than the bottom range. For illustrative purposes, assume $A_{\min} = 0.01$ and $A_{\max} = 10.0$, and let us calculate the ratio of the probabilities of finding A in the ranges $1 \le A \le 10.0$ and $0.01 \le A \le 0.1$:

$$\frac{\int_1^{10} p(A|H_1,I)\,dA}{\int_{0.01}^{0.1} p(A|H_1,I)\,dA} = \frac{\int_1^{10} dA}{\int_{0.01}^{0.1} dA} = \frac{A|_1^{10}}{A|_{0.01}^{0.1}} = 100. \tag{7.377}$$

We find, indeed, that a uniform prior implies the upper range is far more probable than the lower range. But considering the signal sought for has never been observed and thus may qualify as a rare process, it would seem sensible to expect lower ranges of values should be more probable than larger values. A prior that gives higher probability to lower amplitudes therefore sounds far more reasonable. We thus choose[13] to use an uninformative scaling prior of the form, Eq. (7.92), introduced in §7.3.2, and write

$$p(A|H_1,I) = \begin{cases} \frac{1}{\ln(A_{\max}/A_{\min})}\frac{1}{A} & \text{for } A_{\min} \le A \le A_{\max} \\ 0 & \text{otherwise,} \end{cases} \tag{7.378}$$

[13] For comparative purposes, we actually use both uniform and scaling priors in the following.

where the logarithmic factor ensures the proper normalization by integration over the range $A_{\text{min}} \leq A \leq A_{\text{max}}$. Also recall from §7.3.2 that integrals of this prior in the form of logarithms imply the log of the amplitude, $\ln A$, has a uniform distribution. Effectively, ranges such as $0.01 \leq A \leq 0.1$, $0.1 \leq A \leq 1$, and $1 \leq A \leq 10$ have equal (prior) probability, and ranges of higher values are not more probable than ranges of smaller values, as required. Such a prior thus seems far more appropriate, for a search for a weak signal, than the flat prior given by Eq. (7.376).

Posteriors of H_0 and H_1

The hypothesis H_0 has no free parameters. Its posterior is thus obtained simply by application of Bayes' theorem:

$$p(H_0|D, I) = \frac{1}{p(D|I)} p(D|H_0, I), \tag{7.379}$$

with

$$p(D|I) = p(D|H_0, I) + p(D|H_1, I), \tag{7.380}$$

where $p(D|H_0, I)$ and $p(D|H_1, I)$ are the global likelihoods of hypotheses H_0 and H_1, respectively. Note that $p(H_0|I)$ was omitted in the numerator of Eq. (7.379) because we have chosen equal global model priors, that is, $p(H_0|I) = p(H_1|I)$. The global model prior $p(H_1|I)$ is likewise omitted in the following. Calculation of the posterior of H_1 requires marginalization of its free parameter A. With the uniform amplitude prior, one gets

$$p(H_1|D, I) = \frac{1}{p(D|I)} \frac{1}{\Delta A} \int_{A_{\text{min}}}^{A_{\text{max}}} p(D|H_1, A, \sigma_n, I)\, dA \tag{7.381}$$

whereas the scaling prior yields

$$p(H_1|D, I) = \frac{1}{p(D|I)} \frac{1}{\ln A_{\text{max}}/A_{\text{min}}} \int_{A_{\text{min}}}^{A_{\text{max}}} \frac{1}{A} p(D|H_1, A, \sigma_n, I)\, dA. \tag{7.382}$$

Results from the evaluations of the posteriors $p(H_0|D, I)$ and $p(H_1|D, I)$, with Eqs. (7.379, 7.381, 7.382) are shown in Figure 7.17 for the four measurement outcomes introduced in Figure 7.15b–e, while odds ratios calculated from these probabilities are listed in Table 7.1 for both uniform and scaling priors. The posterior probability, $p(H_1|D, I)$, is unequivocally equal to unity for the clear signals shown in Figure 7.16b and c and the choice of prior makes no significant difference in these two cases. However, in the case of the much weaker signals displayed in Figure 7.16d and e, the choice of a scaling prior for the amplitude A appreciably increases the posterior probability $p(H_1|D, I)$ and the odds ratios O_{10}. A prior favoring weaker signals should indeed yield a larger posterior for weak signals. It should be noted, however, that an odds ratio of 8.0 constitutes a tantalizing but somewhat weak indication that there might be a signal in the spectrum (d). Indeed, the use of a scaling prior significantly increases the odds ratio, from 0.6 to 8.0, but the latter value is not sufficiently large to provide incontrovertible evidence for a signal. In the case of spectrum (e), the scaling prior produces a significant increase of the odds ratio in favor

Table 7.1 Odds ratios O_{10} in favor of model H_1 relative to model H_0 calculated with uniform (u) and scaling (s) priors, for data samples shown in Figure 7.16

Dataset	$O_{10}^{(u)}$	$O_{10}^{(s)}$
$A = 20$	∞	∞
$A = 5$	1.6×10^{45}	3.6×10^{45}
$A = 1$	0.60	8.0
$A = 0.005$	0.004	0.38

of model H_1, but most scientists would likely regard a value of 0.38 as far too small to conclude the spectrum contains a signal, even though it actually does in the simulation.

Although not absolutely essential, it is interesting to evaluate the Occam factor of model H_1. Based on Eq. (7.353) and assuming a uniform prior for A, we write

$$\int_{\Delta A} p(D|\theta, H_1, I)\, dA = \delta A \times L_{\max}, \tag{7.383}$$

We carry out numerical evaluations of the integral in Eq. (7.383) for all four spectra ($A = 20, 5, 1, 0.005$) and divide by the observed maximum likelihood values $L_{\max} \approx 3.1 \times 10^{18}$ to find widths $\delta A \approx 0.84$ for spectra $A = 20, 5, 1$ and a width $\delta A \approx 0.41$ for $A = 0.005$. Dividing these by the width of the search domain, $\Delta A = 100$, we find Occam factors $\Omega = \delta A/\Delta A$ of order 0.008 and 0.004, respectively. Using a scaling prior does not appreciably changes these values. Clearly, these small Occam factor values strongly disfavor hypothesis H_1 and given hypothesis H_0 has a likelihood of order 3×10^{18} for null signals (noise only), the posterior of H_1 is given a small probability for spectra (d) and (e).

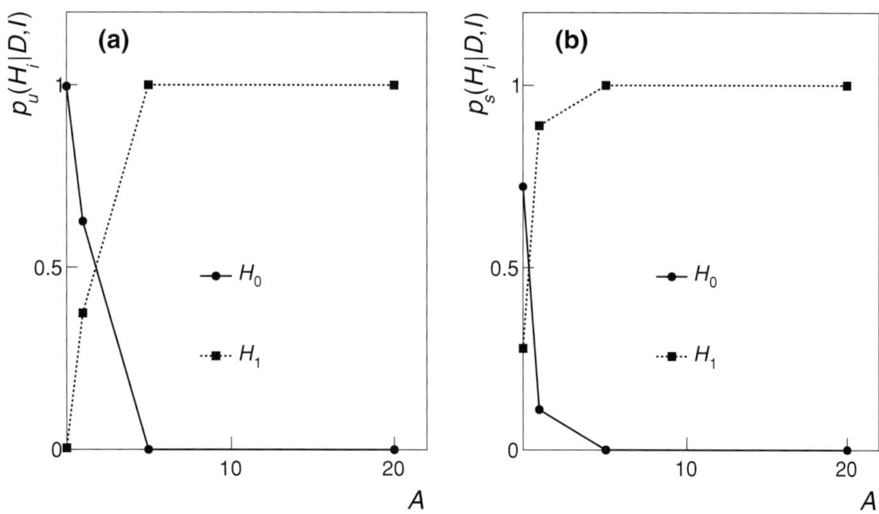

Fig. 7.17 Posteriors $p(H_0|D, I)$ and $p(H_1|D, I)$ for the measurement outcomes shown in Figure 7.15b–e, with respective true amplitudes $A = 20, 5, 1$, and 0.005, based on (a) uniform and (b) scaling priors.

However, the likelihood of H_0 falls dramatically for strong signals and H_1 is then ascribed a high probability in spite of the small value of the Occam factor.

7.7.5 Example 2: Comparison of Linear and Quadratic Models

Problem Definition

An issue often encountered in practical data analyses is whether one or several parameters should be added to a model to obtain a better fit of the data. This is the case, for instance, with the study of Fourier decompositions of angular correlations, or fits of data with polynomials. In either case, one must decide whether the data warrant the addition of one or more additional parameters, that is, higher-order Fourier terms or higher degree terms in a polynomial.

As a practical case, let us revisit the comparison of linear and quadratic models toward the description of a dataset D introduced in §6.6.5 under the frequentist paradigm. For the purpose of this example, we simulated datasets consisting of $n = 25$ measurements $(x_i, y_i \pm \sigma_i)$ of an observable Y measured as a function of a control (independent) observable X generated based on the data model

$$y(x) = \mu_y(x) + R, \tag{7.384}$$

where R represents random Gaussian deviates with standard deviation $\sigma_i = \gamma\sqrt{\mu_y(x)}$, and $\mu_y(x)$ expresses the physical relation between observables Y and X, here taken to be a second-degree polynomial:

$$\mu_y(x) = b_0 + b_1 x + b_2 x^2, \tag{7.385}$$

with fixed coefficients

$$\begin{aligned} b_0 &= 500.0, \\ b_1 &= 10.0, \\ b_2 &= -1.0. \end{aligned}$$

The scaling factor γ was set to arbitrary values $\{2.0, 1.0, 0.5, \ldots\}$ in order to generate datasets of different precision levels. Each of the produced datasets was then analyzed to establish whether they could be properly represented as straight lines or whether a quadratic term should be included. Evidently, in the analysis, we assumed the quadratic dependence of Y on X is not established or known. We thus formulated a null hypothesis, H_0, stipulating that a linear model is sufficient to "explain" the data while the alternative hypothesis, H_1, requires the use of quadratic model, that is, a fit with a second-degree polynomial.

$$\begin{aligned} H_0 &: f_0(x) = a_0 + a_1 x, \\ H_1 &: f_1(x) = a_0 + a_1 x + a_2 x^2. \end{aligned}$$

Determination of the Odds Ratio of Global Posteriors

We saw in the context of the frequentist paradigm (§6.6.5) that the χ^2 of fits based on two distinct hypotheses, a linear and a quadratic model, may be used as a statistical test to challenge and reject the linear hypothesis (null hypothesis) provided its χ^2 exceeds some minimal value determined by the significance level of the test. The null hypothesis shall be retained, however, if the χ^2 falls below the cut. The problem, of course, is that the χ^2 obtained with the quadratic fit might also be relatively small. The test then features a relatively small power and thus cannot be deemed significant.

The necessity to consider both the significance level of a test and its power raises both technical and philosophical issues. Is it possible, in particular, to identify a statistic whose value might combine both the significance level and the power of the test? Can a statistic be found that is capable of rejecting the null hypothesis while showing that the alternative hypothesis has strong merits? Can such a statistic disfavor overly complicated models not properly supported by data? The short answer is that an odds ratio based on the global posteriors of alternative and null hypotheses in fact satisfy (mostly) all these criteria.

Let us indeed reexamine a Bayesian test based on the odds ratio of the global posteriors of hypotheses H_1 and H_0 in the context of the comparison of linear and quadratic models and show that the odds ratio provides an effective tool to meaningfully sustain or reject a null hypothesis challenged by a more complicated model, that is, a model with more free parameters.

The odds ratio is defined and calculated according to

$$O_{10} = \frac{p(H_1|D)}{p(H_0|D)} = O_{10}^{\text{prior}} B_{10} \tag{7.386}$$

with

$$O_{10}^{\text{prior}} = \frac{p(H_1|I)}{p(H_0|I)}, \tag{7.387}$$

and

$$B_{10} = \frac{p(D|H_1, I)}{p(D|H_0, I)} \equiv \frac{\mathbb{L}(H_1)}{\mathbb{L}(H_0)}, \tag{7.388}$$

where $\mathbb{L}(H_1)$ and $\mathbb{L}(H_0)$ are the global likelihoods of the models H_1 and H_0, respectively. These can be calculated by marginalization of the model likelihoods $p(D|H_0, a_0, a_1, I)$ and $p(D|H_1, a_0, a_1, a_2, I)$, respectively.

Given there are no tangible reasons to favor either hypothesis,[14] we set the prior odds ratio O_{10}^{prior} to unity. The posterior odds ratio is thus determined solely by integrals of the likelihood functions $p(D|H_0, a_0, a_1, I)$ and $p(D|H_1, a_0, a_1, a_2, I)$ over the domains of parameters a_0, a_1, and a_2. In order to determine sensible ranges of integration for these parameters, we first plot the likelihood functions $p(D|H_0, a_0, a_1, I)$ and $p(D|H_1, a_0, a_1, a_2, I)$, in Figure 7.18, as a function of their parameters. The likelihood functions are calculated numerically according to

$$p(D|H_0, a_0, a_1, I) \propto \exp\left[-\frac{1}{2}\chi_0^2(a_0, a_1)\right], \tag{7.389}$$

$$p(D|H_1, a_0, a_1, a_2, I) \propto \exp\left[-\frac{1}{2}\chi_1^2(a_0, a_1, a_2)\right], \tag{7.390}$$

[14] Remember that we pretend to ignore the datasets were generated with a quadratic model.

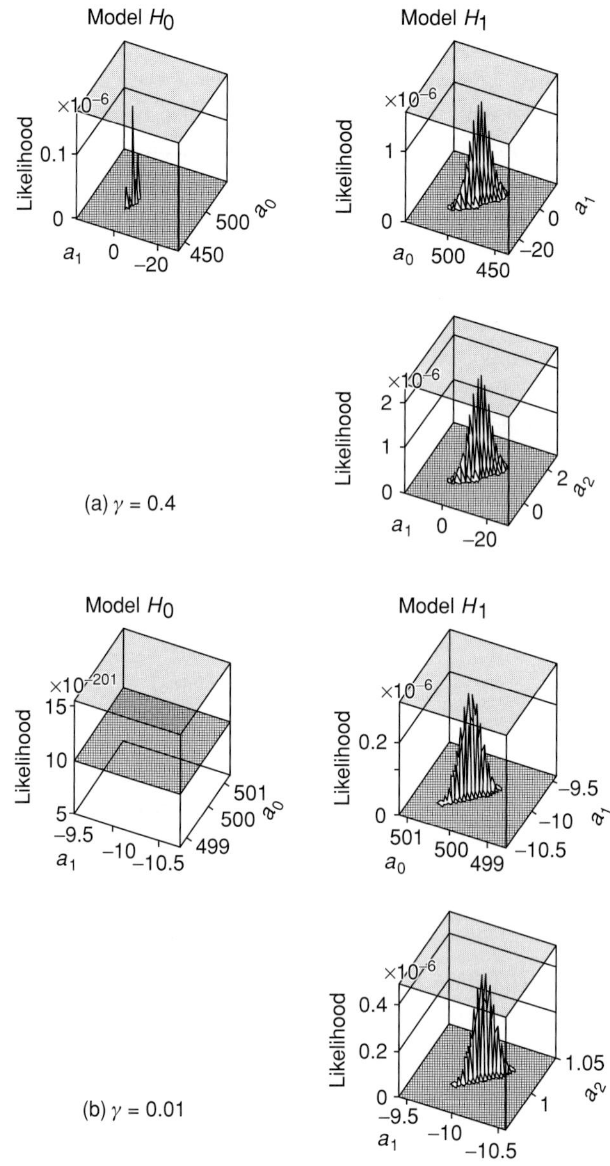

Fig. 7.18 Examples of the likelihoods $p(D|H_0, a_0, a_1, I)$ of a linear model and $p(D|H_0, a_0, a_1, a_2, I)$ of a quadratic model obtained with (a) large noise ($\gamma = 0.4$) and (b) very small noise ($\gamma = 0.01$).

where

$$\chi_0^2(a_0, a_1) = \sum_{k=1}^{n} \frac{\left(y_i - \sum_{n=0}^{1} a_n x^n\right)^2}{\sigma_i^2},$$

$$\chi_1^2(a_0, a_1, a_2) = \sum_{k=1}^{n} \frac{\left(y_i - \sum_{n=0}^{2} a_n x^n\right)^2}{\sigma_i^2}. \tag{7.391}$$

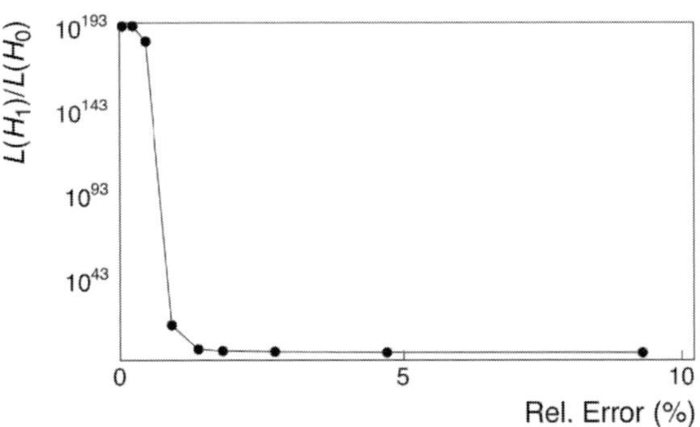

Fig. 7.19 Bayes factor $B_{10} = \mathbb{L}(H_1)/\mathbb{L}(H_0)$ as a function of the relative error size (determined by the error scaling factor γ).

Bounds of integration are chosen to approximately correspond to ± 6 times the widths of the distributions along each axis, and thus such that the functions are negligible (relative to the maximum) beyond the bounds. Marginalization of the likelihood functions $p(D|H_0, a_0, a_1, I)$ and $p(D|H_1, a_0, a_1, a_2, I)$ is obtained by numerical integration within these bounds. The Bayes factors and odds ratios are then computed according to Eqs. (7.388, 7.386). Calculated values of the Bayes factor $B_{10} = \mathbb{L}(H_1)/\mathbb{L}(H_0)$ are presented in Figure 7.19 for simulated datasets produced with γ values ranging from 0.01 to 2. One finds that for large relative errors (i.e., obtained with large values of γ), the odds ratio is smaller than or of order unity and thus favors the null hypothesis (i.e., a linear model of the data). For small errors, obtained with $\gamma \leq 0.3$, the odds ratios increases significantly and unambiguously favors the alternative hypotheses, that is, the notion that the signal Y involves a quadratic dependence on X.

The posteriors $p(H_1|D, I)$ and $p(H_0|D, I)$ are statistics (i.e., functions of the measured data points). As such, they would fluctuate sample by sample, much like a χ^2 function, if it were in fact possible to acquire several distinct data samples. As a ratio of these two statistics, the posterior odds ratio O_{10} is thus also a statistic, and it too shall fluctuate sample to sample. This is illustrated in the three sets of panels of Figure 7.20. The left set of panels displays linear and quadratic model fits of typical simulated spectra obtained with selected values of the error scaling parameter γ, while the central and right sets of panels display χ^2/DoF distributions and histograms of the logarithm of the odds ratio determined for successions of 1,000 Monte Carlo generated samples of $n = 25$ points. From top to bottom, the samples are generated with error scale parameter values of $\gamma = 2.0, 0.4, 0.2$, and 0.01. One finds that for $\gamma = 2.0$ and $\gamma = 0.4$ corresponding to relative errors of the order of 10% and 2%, respectively, the χ^2 distributions of the two hypotheses overlap considerably and the log of the odds ratio of the global posteriors is sharply distributed near zero, implying the odds ratio is of order 1. For smaller relative errors, the χ^2 distributions no longer overlap and the odds ratio grows increasingly large, thereby indicating that the alternative hypothesis H_1 is strongly favored by the data.

It is interesting to note that much like the χ^2 of the fits, the odds ratio fluctuates quite considerably. This stems from the fact that these quantities are in fact not independent of one another. Indeed, Eqs. (7.389,7.390) tell us that for Gaussian deviates, the likelihood of

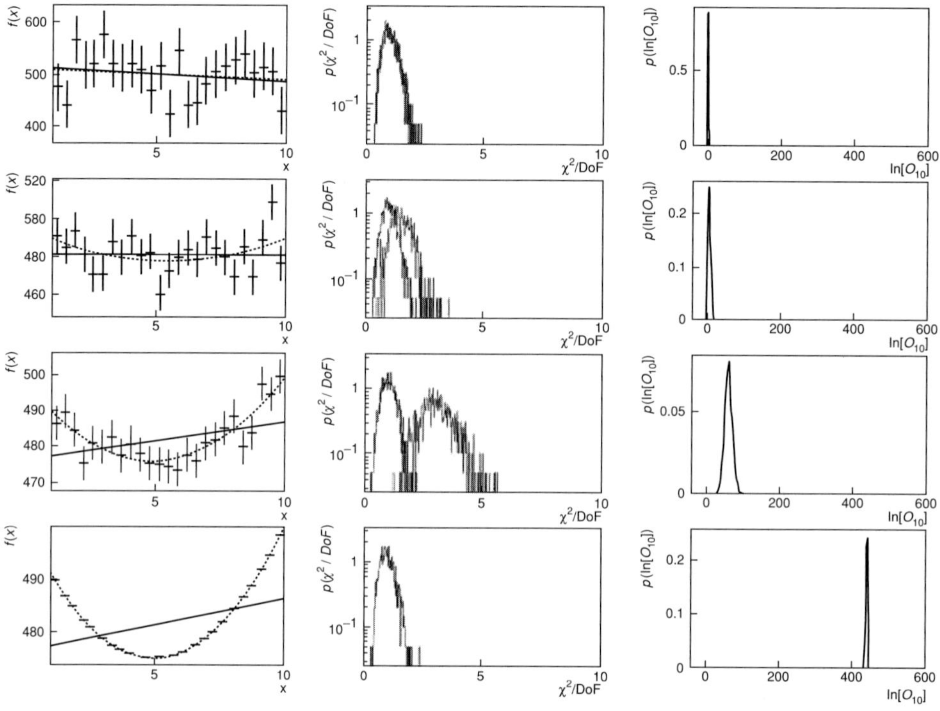

Fig. 7.20 (Left) Representative fits of the linear and quadratic models obtained with, from top to bottom, $\gamma = 2.0, 0.4, 0.2$, and 0.01. (Middle) Histograms of the χ^2/DoF obtained with the two models for 1000 distinct samples of 25 points (as described in the text) generated with same values of γ. (Right) Histograms of the logarithm of the odds ratios obtained with the same samples and γ values.

a dataset, given specific models parameters $a_0, a_1, \ldots$, is strictly related to the χ^2 obtained with those parameters. And, given the models H_0 and H_1 are linear in their coefficients, one can write

$$\begin{aligned}
\chi_1^2(a_0, a_1, a_2) &= \sum_{i=1}^{n} \frac{\left([y_i - a_0 - a_1 x] - a_2 x_i^2\right)^2}{\sigma_i^2} \\
&= \chi_0^2(a_0, a_1) - 2a_2 \sum_{i=1}^{n} \frac{[y_i - a_0 - a_1 x] x_i^2}{\sigma_i^2} \\
&+ a_2^2 \sum_{i=1}^{n} \frac{x_i^4}{\sigma_i^2}
\end{aligned}$$

The Bayes factor may thus be considered a function of the model parameters

$$\begin{aligned}
B_{10} &= \exp\left[-\frac{1}{2}\left(\chi_1^2 - \chi_0^2\right)\right] \\
&= a_2 \sum_{i=1}^{n} \frac{[y_i - a_0 - a_1 x] x_i^2}{\sigma_i^2} - \frac{1}{2} a_2^2 \sum_{i=1}^{n} \frac{x_i^4}{\sigma_i^2},
\end{aligned} \tag{7.392}$$

which, evidently, peak for some specific values of the model parameters a_0, a_1, and a_2. Effectively, there is no new information contained in the odds ratio O_{10} or the Bayes factor B_{10}, given the one-to-one relation between the global likelihoods of hypotheses H_1 and H_0 and their respective χ^2 functions. However, the Bayes factor and odds ratio provide a practical and simple tool to decide whether the null hypothesis should sensibly be rejected.

With a single value, the odds ratio enables the rejection/acceptance of the null hypothesis and effectively provides a minimal statement about the power of the test. If the odds ratio is small (i.e., not much larger than unity), then the null hypothesis cannot be rejected. This can be the case either because the data are very precise and the null hypothesis happens to be correct, or because the data are poor (large errors) and the power of the test is too weak to enable a decision. In this latter case, a single number suffices to make the statement, and one is not required to explicitly calculated the power β of the test, according to Eq. (6.79). If, instead, the value of the odds ratio is very large, one immediately learns that the data are incompatible with the null hypothesis and simultaneously in favor of the alternative hypothesis. Again, there is no need to calculate the power of the test. There is, of course, a possibility that the χ^2 of the null hypothesis might be large due to fluctuations, but the large value of the odds ratio readily indicates that the χ^2 of the alternative hypothesis is in fact very good. It then makes sense to (tentatively) reject the null hypothesis because the large value of the odds ratio guarantees the alternative hypothesis is a good fit to the data.

The calculation of an odds ratio thus has several advantages, both conceptually and in practice. It enables the rejection/acceptance of the null/alternative hypothesis on the basis of a single number without the need for possibly complicated and cumbersome calculations of the power of the test. The decision may then be taken rapidly and efficiently, either by a human or by an algorithm (machine). The value of the odds ratio also readily informs us about the quality of the decision and thus the power of the test. Clearly, an odds ratio of 1000 is far more significant than a ratio of 10 or 100. And likewise, a ratio in excess of 10^6 or 10^9 readily indicates that the alternative hypothesis is very strongly favored by the data.

7.7.6 Minimal Odds Ratio for a Discovery

As we discussed in the context of the frequentist interpretation of probability (see also §6.6.2), scientists commonly require a null hypothesis to be incompatible with the data with a significance level of 3×10^{-7} to claim and announce a discovery. That means the p-value of the null hypothesis must be equal to or less than 3×10^{-7}. For instance, a peak in an invariant mass plot must be deemed incompatible with a background fluctuation because it has a probability equal to or smaller than 3×10^{-7} before it can be considered acceptable to claim a discovery of a new particle. What is then the Bayesian equivalent of this significance level? What indeed should be the minimal odds ratio required to claim a discovery?

Although, ab initio, there is no simple or correct answer to this question, one can establish a minimal criterion, which can be deemed both necessary and sufficient by most scientists, and applicable in most situations. Clearly, the larger the odds ratio is, the more likely is the alternative hypothesis to be correct. One should thus define a minimal value for the odds ratio. What should this minimum value be to provide a significance level of 3×10^{-7}?

In the frequentist approach, a test of the null hypothesis is based on a chosen statistic, called a test statistic. The data are considered incompatible with the null hypothesis if the probability to obtain a statistic in excess of the observed value is smaller than 3×10^{-7}, while the probability of the observed statistic for the alternative hypothesis is large (but often unspecified). The probability p_t to obtain a statistic t value in excess of the observed value t_o is given by

$$p_t = \int_{t_o}^{\infty} p(t|H_i)\,dt, \tag{7.393}$$

where $p(t|H_i)$ represents the PDF of t given the hypothesis H_i is true. The integration to infinity sums over the probability of all suboptimal parameter values. In effect, it corresponds to the marginalization of the model parameters, and thus corresponds to the probability of the null hypothesis given any parameter value.

In the Bayesian paradigm, there is no need for an *extra* statistical test; the global posterior probability of an hypothesis *is* the test. It expresses the probability of a hypothesis to be true by marginalization (as necessary) of all model parameters, that is, given any parameter values. Effectively, there is then a one-to-one correspondence between the probability of a (frequentist) test statistic to be found in excess of a specific value and the posterior probability. It is thus reasonable to require the posterior probability of the null hypothesis to be equal or smaller than 3×10^{-7}. If the alternative and the null hypotheses can be considered an exhaustive set of options, the posterior probability of the alternative hypothesis is equal to $1 - 3 \times 10^{-7}$. The discovery criterion thus corresponds to an odds ratio with a minimum value of 3×10^6:

$$O_{10}^{\text{discovery}} = \frac{1 - 3 \times 10^{-7}}{3 \times 10^{-7}} = 3 \times 10^6. \tag{7.394}$$

Indication of a new phenomenon, which usually requires a 3σ significance level, shall have a correspondingly smaller minimum odds ratio value.

A few remarks are in order. First, one must recognize that an odds ratio is a statistic, as are the respective global posterior probabilities of the hypotheses used in the calculation of the ratio. This means that the odds ratio obtained from several samples extracted from a common parent population (using the same experimental procedure and apparatus) shall have values that appear to fluctuate from sample to sample, as illustrated in Figure 7.20 for the curve fitting example discussed in §7.7.5. It is thus conceivable, owing to the vagaries of the measurement protocol, that the odds ratio of a particular data sample might meet the discovery criterion threshold, while some others do not. Indeed, because of fluctuations (i.e., experimental errors) a given data sample might be less of a match to a specific hypothesis.

Second, it is important to realize that the notion of odds ratio does not require calculation of an additional variable to determine the power of a test. If two hypotheses are truly exhaustive (i.e., with no other options in the model space of a particular phenomenon), the probability of one is the complement of the other (i.e., $p(H_1|D, I) = 1 - p(H_0|D, I)$) and the odds ratio is perfectly well defined and determined. There is no need for a test threshold and thus for the notion of power as defined by Eq. (6.79). In fact, the odds ratio itself provides an assessment of the discriminating power of the test. On the one hand, if the odds ratio is very large, $O_{10} \gg 1$, or very small, $O_{10} \ll 1$, one knows immediately that the test strongly favors a specific hypothesis. The measured data thus provide strong discriminatory power in the selection of the best hypothesis. On the other hand, if the ratio is of order unity, $O_{10} \sim 1$, it is clear that given the experimental errors, neither the null nor the alternative hypotheses are favored by the data. The "test" thus has essentially no discriminatory power toward the selection or adoption of H_0 and H_1.

Third, and last, one may contrast the notion of odds ratio with the Neyman–Pearson test (as a most powerful test for simple hypotheses). Recall from §6.7.2 that a Neyman–Pearson test is based on the ratio of the likelihoods of the alternative and null hypotheses. For instance, for two competing hypotheses about the value of a specific parameter θ, one writes

$$r_n(\vec{x}|\theta_0, \theta_1) = \frac{f_n(\vec{x}|\theta_1)}{f_n(\vec{x}|\theta_0)}, \tag{7.395}$$

where $\vec{x}$ represents a set of n measured points, $f_n(\vec{x}|\theta_0)$ and $f_n(\vec{x}|\theta_1)$ are the likelihoods of these data points given parameter values $\theta = \theta_0$ (hypothesis H_0), and $\theta = \theta_1$ (hypothesis H_1). The power of the test is then calculated as the expectation value of $r_n(\vec{x}|\theta_0, \theta_1)$ given the null hypothesis, H_0, is true, a rather unintuitive notion to say the least. However, note that the notion of using a ratio of likelihoods (for the alternative and null hypotheses) is totally in line with a Bayesian test of the null hypothesis based on the odds ratio O_{10}, defined as a ratio of the global posterior probabilities of the two hypotheses. The advantage of the odds ratio is that it is conceptually far easier to visualize and interpret. The only difference, of course, is that posterior probabilities may also include nontrivial (i.e., nonuniform) prior probabilities of the model parameters, which are subsequently marginalized to obtain the global posteriors. In essence, the odds ratio thus provides a Bayesian extension of the Neyman–Pearson test, and because its calculation involves all facets of the model and all of the data, it is guaranteed to be the most powerful test. Consequently, if an odds ratio of order unity, $O_{10} \sim 1$, is obtained in the comparison of two hypotheses H_1 and H_0, no manipulation of the data or use of other statistics shall improve the power of the test and discrimination of the hypotheses. Rejection of the null hypothesis H_0 in favor of the alternative H_1 thus rests entirely on the quality of the data (i.e., the size of the errors) used to carry out the test.

Exercises

7.1 Show that the use of the measure q_i given by Eq. (7.82), with a constraint on the variance according to Eq. (7.74), leads to a truncated Gaussian prior PDF of the form

$$p(x|\mu, \sigma) = \begin{cases} \frac{1}{\Phi(z_L)-\Phi(z_H)} \frac{1}{\sigma} \exp\left[-\frac{(x-\mu)^2}{2\sigma^2}\right] & \text{for } x_L \le x \le x_H \\ 0 & \text{elsewhere,} \end{cases} \tag{7.396}$$

where $z_L = (x_L - \mu)/\sigma$ and $z_H = (x_H - \mu)/\sigma$, corresponding to the boundaries within which the observable is allowed (or defined).

7.2 Verify that if the variance parameter σ^2 of a Gaussian PDF is given a prior of the form $p(\sigma^2) = k/\sigma^2$, then the log of the variance has a flat prior.

7.3 Show that Jeffreys' prior for a multinomial distribution with rate parameters $\vec{p} = (p_1, p_2, \ldots, p_m)$, with the constraint $\sum_{i=1}^{m} p_i = 1$, is the Dirichlet distribution (§3.8) with all its parameters set to half.

7.4 Demonstrate that the expression for $Q(\mu)$, given by Eq. (7.139), may be transformed to yield

$$Q(\mu) = \left(n\xi_0 + \xi_p\right) \left[\mu - \frac{n\xi_0\bar{x} + \mu_p\xi_p}{n\xi_0 + \xi_p}\right]^2 + K,$$

where K represents a constant expression independent of μ.

7.5 Derive the expression, Eq. (7.153), for the posterior of the precision ξ based on Eq. (7.152).

7.6 Derive the expression, Eqs. (7.240), (7.242), and (7.243), for the Bayesian solution of a linear model fit with Gaussian priors for the model parameters.

PART II

MEASUREMENT TECHNIQUES

8 Basic Measurements

Measurements in particle and nuclear physics span a wide range of observables, conditions, and collision systems. They may involve the determination of cross section, momentum spectra, angular distributions, particle lifetimes, particle mass, and so on. This chapter introduces several of these types of measurements and observables. We begin, in §8.1, with definitions and commonly used notations used by practitioners of the field and follow up in §8.2 with notions of particle decay and cross section. Measurements of elementary observables such as the momentum and the energy of particles produced in scattering experiments are discussed §8.3 whereas techniques for the identification of long- and short-lived particles are described in §§8.4 and 8.5, respectively. Lastly, §8.6 discusses issues involved in searches and discoveries of new particles and elementary processes.

8.1 Basic Concepts and Notations

8.1.1 Particle Scattering and Definition of Cross Section

The scattering of α-particles off a thin foil of gold (illustrated schematically in Figure 8.1) by H. Geiger and E. Marsden in the early 1910s led E. Rutherford to the discovery of the atomic nucleus and a first estimate of its size. Geiger and Marsden observed that while most of the α-particles were little deflected by their passage through this gold foil, a small but significant fraction was backscattered (i.e., scattered by more than 90°). Rutherford understood that only a massive object could deflect the charged and heavy alpha particles. He formulated a model based on the Coulomb interaction by which he could predict the number of scattered α as a function of their scattering angle θ relative to their initial direction, and he was able to accurately reproduce the observations of Geiger and Marsden. He thus concluded the existence of a massive and extremely small atomic nucleus. This discovery gave birth to the fields of nuclear and particle physics.

Rutherford relied on a quantitative description of the number of deflected particles, dN, detected in a small solid angle $d\Omega$, as a function of their scattering angle θ, as illustrated in Figure 8.1. In his model, the scattering angle θ is determined by the impact parameter b of the incoming α-particle onto the target nucleus, thereby defining a ring of area $2\pi b\, db$, as shown in Figure 8.2. Classically, this defines the **differential cross section** of the nucleus $d\sigma$ associated with a specific scattering direction

$$\frac{d\sigma}{d\Omega}(\theta) = -\frac{b}{\sin\theta}\frac{db}{d\theta}, \tag{8.1}$$

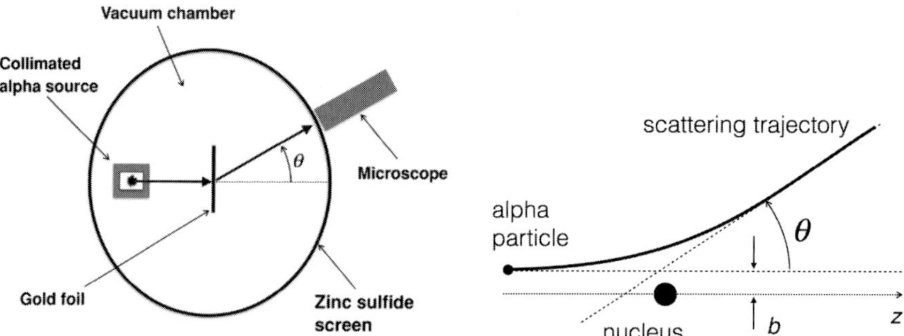

Fig. 8.1 (Left) Schematic of the apparatus used by Rutherford, Geiger, and Marsden to study the scattering of α-particles off a gold foil. (Right) Model formulated by Rutherford to explain the observation of back-scattered particles.

where b is the impact parameter of the collision, θ is the deflection angle of the particle relative to its incident direction, and $d\Omega$ the solid angle in which the particle is measured. For a purely Coulomb (electrostatic) interaction, one has

$$\frac{d\sigma}{d\Omega}(\theta) = \left(\frac{ZZ'e^2}{4E}\right)^2 \frac{1}{\sin^4 \theta/2}, \tag{8.2}$$

where Z and Z' are the atomic number of the colliding nuclei, e the electron charge, and E the kinetic energy of the incoming particle. Integration over all rings $2\pi b\, db$, corresponding to the observation of all scattering angles, leads to the total effective area of the target or **total scattering cross section**.

In a scattering experiment involving a fixed-target and an incoming beam, the number of particles dn deflected in a solid angle $d\Omega$, per unit time, is proportional to the incoming flux N_0 (particles per unit time) of beam particles, the number of target nuclei per unit area,

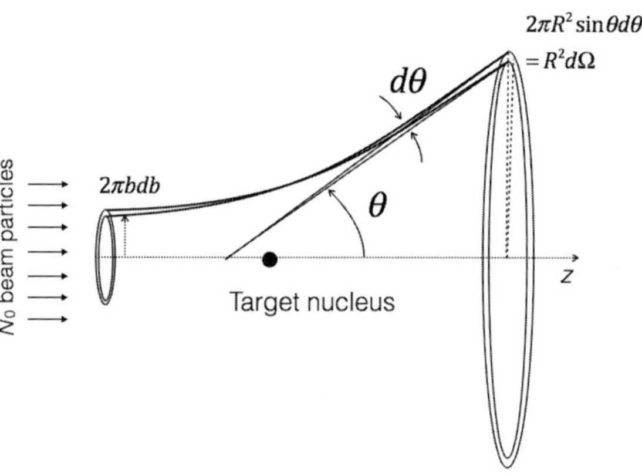

Fig. 8.2 Geometric definition of cross section in the context of Rutherford's model of the scattering of α-particles off gold nuclei.

N/S, and the differential scattering cross section:

$$\frac{dn}{d\Omega} = N_0 \frac{N}{S} \frac{d\sigma}{d\Omega}(\theta, \phi). \tag{8.3}$$

The number of target nuclei per unit area may be written

$$\frac{N}{S} = \frac{\rho t}{A} A_0, \tag{8.4}$$

where ρ is the target material density, t is the physical thickness of the target foil, A is the atomic mass of the material, and A_0 is Avogadro's number. The differential scattering cross section may then be determined by precise measurements of the incoming flux, the target thickness, and the outgoing flux of particles $dn(\theta, \phi)$ in a specific solid angle $d\Omega$ at scattering angle θ relative to the direction of the incoming particle:

$$\frac{d\sigma}{d\Omega}(\theta, \phi) = \frac{1}{N_0} \frac{A}{\rho t A_0} \frac{dn(\theta, \phi)}{d\Omega}. \tag{8.5}$$

The scattering cross section, between a projectile and a target, depends on the nature of the interaction that takes place, as well as the size and internal structure of colliding particles. The Coulomb interaction of two point-like objects leads to a definite and specific angular dependence of the scattering cross section. Deviations from the differential scattering cross section expected from a Coulomb interaction at very large deflection angle and higher collisional energies revealed the existence of an object of finite size within the atom, the nucleus, and eventually enabled study of the nuclear force. Scattering experiments, and more specifically measurements of differential cross sections were thus instrumental in the discovery and study of the atomic nucleus, its structure, the nuclear force, and so on.

The notions of differential and total cross sections are readily extended to all types of elementary collisions, whether elastic or inelastic, and are determined by the number of particles produced at specific angles or momenta relative to the number of incoming projectiles. Given its $1/r^2$ dependence, the total elastic interaction cross section associated with the Coulomb force is infinite for point-like objects. Other types of processes and interactions have different behaviors and accordingly different differential and total cross sections. It is consequently of much interest to measure partial cross sections (both differential and total) associated with specific subprocesses of an interaction between two colliding particles or nuclei. The cross section of a specific process or interaction expresses its effective surface probability and, as such, says much about the nature and strength of the interaction.

It is customary to break down the cross section of elementary processes according to their type and properties, for instance, elastic scattering cross section, diffractive cross section, interaction or inelastic cross section, and so on. It is also common to report cross sections corresponding to specific particle production processes whether exclusive, semi-exclusive, or inclusive. One can, for example, measure the production cross section of pions, kaons, or other particle species, in the context of a specific interaction. By extension, one can also consider how such cross sections depend on the colliding partners (whether protons, neutrons, or nuclei, etc.) and the beam energy. It is also possible to break down cross section measurements in terms of other global collision characteristics such as the (estimated) impact parameter of collisions or the presence/absence of specific particles or other specific features in the final state of interactions.

In closing this section, it is important to remark that the cross section of particular process is a property of that process and should thus be independent of experimental conditions. Evidently, experimental conditions may hinder or alter measurements of this cross section (integral or differential) and it is thus necessary to consider how experimental conditions may affect or alter the outcome of a specific measurement. This very important topic is discussed in detail in Chapter 12.

8.1.2 Relativistic Kinematics

Modern experiments in particle and nuclear physics are commonly carried out at high collisional energies and involve measurements of particles with large momenta. The use of relativistic kinematics is thus usually advised, if not required. In this and the following sections, we use natural units: $c = \hbar = 1$.

In special relativity, and even more so in general relativity, one distinguishes vectors according to their properties of transformation when expressed in two different coordinate systems. Space–time coordinates are denoted with a contravariant four vector with components x^μ:

$$x^\mu = (x^0, x^1, x^2, x^3) \equiv (t, \vec{x}). \tag{8.6}$$

Similarly, the components of the contravariant four momentum of a particle are denoted p^μ:

$$p^\mu = (p^0, p^1, p^2, p^3) \equiv (E, \vec{p}). \tag{8.7}$$

The time-like component p^0 corresponds to the *total* energy of the particle while the space-like components p^1, p^2, p^3 correspond to the momentum components, commonly denoted $\vec{p} = (p_x, p_y, p_z)$.

Many of the problems of interest in nuclear and particle physics involve collisions of beam particles (also called projectiles) onto a fixed target (see Figure 8.6) or collisions of two beams traveling colinearly, but in opposite directions. We thus adopt a right-handed coordinate system with the z-axis, or third component of the momentum vector, corresponding to the beam axis (and direction). In this context, there are three natural and convenient choices of reference frames: the target rest frame, the projectile rest frame, and the center of mass (CM) rest frame. These correspond to reference frames where the target, the projectile, and the CM are at rest, respectively.

Given our choice of z-axis, coordinate transformations between these three frames of reference leave the first and second components of the momentum invariant, while the third component and energy are transformed according to a Lorentz transformation (discussed in the text that follows). It is thus convenient to introduce the notion of **transverse momentum** vector $\vec{p}_T$, defined as

$$\vec{p}_T \equiv (p_x, p_y), \tag{8.8}$$

and its norm (magnitude)

$$p_T = \sqrt{p_x^2 + p_y^2}. \tag{8.9}$$

The four-momentum vector of a particle may thus also be written $p = (E, \vec{p}_T, p_z)$.

We adopt the space–time metric tensor convention $g_{\mu\nu}$, defined as

$$g_{\mu\nu} = g^{\mu\nu} = \begin{pmatrix} 1 & 0 & 0 & 0 \\ 0 & -1 & 0 & 0 \\ 0 & 0 & -1 & 0 \\ 0 & 0 & 0 & -1 \end{pmatrix}. \tag{8.10}$$

The covariant vectors (also called one-form) x_μ and p_μ are related to their contravariant counterparts according to

$$x_\mu = g_{\mu\nu}x^\nu = (t, -x_1, -x_2, -x_3), \tag{8.11}$$

$$p_\mu = g_{\mu\nu}p^\nu = (E, -p_1, -p_2, -p_3), \tag{8.12}$$

where we used the Einstein convention that a repeated index is summed (unless indicated otherwise). Conversely, one also has

$$x^\mu = g^{\mu\nu}x_\nu, \tag{8.13}$$

$$p^\mu = g^{\mu\nu}p_\nu. \tag{8.14}$$

The scalar product of two four-vectors a and b is defined as

$$a \cdot b = a^\mu b_\mu = g_{\mu\nu}a^\mu b^\nu = g^{\mu\nu}a_\mu b_\nu. \tag{8.15}$$

In particular, the modulus of the four momentum vector p is written

$$p^2 = p \cdot p = p_o^2 - p_1^2 - p_2^2 - p_3^2 = E^2 - |\vec{p}|^2 = m^2, \tag{8.16}$$

where we introduced the mass m of the particle (recall that we are using natural units in which $c = 1$).

8.1.3 Lorentz Transformation

A Lorentz transformation along the beam axis leaves the transverse momentum invariant. Consider the momentum, $p = (E, \vec{p}_T, p_z)$, of a particle in a reference frame K (e.g., the target rest frame). The components $p' = (E', \vec{p}_T{}', p_z{}')$ in a frame K' (e.g., projectile rest frame) moving along $+z$ at a velocity β relative to K are written

$$E' = \gamma(E - \beta p_z), \tag{8.17}$$

$$\vec{p}_T{}' = \vec{p}_T, \tag{8.18}$$

$$p_z{}' = \gamma(p_z - \beta E), \tag{8.19}$$

where $\gamma = (1 - \beta^2)^{-1/2}$.

Transformations for motion along x- or y-axes are trivially obtained by substitution of the appropriate momentum components. Transformations for motion along an arbitrary direction,[1] $\hat{\beta}$, at velocity β, requires one explicitly separates the momentum components parallel and perpendicular to $\hat{\beta}$. One writes

$$\vec{p} = \vec{p}_T + \vec{p}_{\parallel}, \tag{8.20}$$

with

$$\vec{p}_{\parallel} = (\vec{p} \cdot \hat{\beta})\hat{\beta}, \tag{8.21}$$

$$\vec{p}_T = \vec{p} - \vec{p}_{\parallel}. \tag{8.22}$$

The $\vec{p}_T$ component is unmodified by the transformation along $\hat{\beta}$, but the parallel component is boosted according to

$$|\vec{p}_{\parallel}{}'| = \gamma \left(|\vec{p}_{\parallel}| + \beta E\right). \tag{8.23}$$

The boosted moment can consequently be written

$$\vec{p}\,' = \vec{p} + \left[(\gamma - 1)\vec{p} \cdot \hat{\beta} + \gamma E\right]\hat{\beta}. \tag{8.24}$$

8.1.4 Rapidity

Equation (8.16) reduces the number of degrees of freedom of a free particle from four to three. Given the invariance of $\vec{p}_T$ relative to Lorentz transformations along the beam axis (z), one can regard Eq. (8.16) as introducing a special connection between the energy and the z-component of the momentum. In this context, it is convenient to introduce the notion of particle rapidity, noted y, and defined according to

$$y = \frac{1}{2} \ln \left(\frac{E + p_z}{E - p_z}. \right) \tag{8.25}$$

The rapidity is by construction a dimensionless quantity, and it can be either positive, negative, or null. In the nonrelativistic limit, $v \approx 0$, it is equal to the velocity in units of the speed of light (see Problem 8.1), thereby providing a justification for the name "rapidity."

While y is not an invariant under Lorentz transformations, one finds its transformation law is particularly simple. Consider, for instance, the rapidity of a particle in frame K' introduced in Eq. (8.25):

$$y' = \frac{1}{2} \ln \left(\frac{E' + p_z{}'}{E' - p_z{}'} \right), \tag{8.26}$$

[1] Breaking with earlier chapters where the caret indicates statistical estimators, the caret is used here to denote unit vectors.

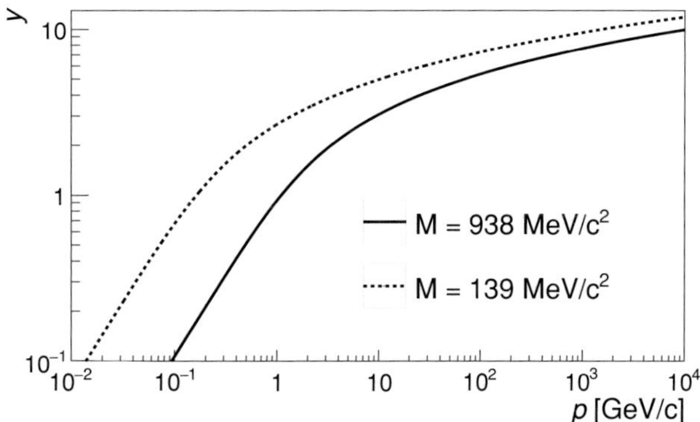

Fig. 8.3 Beam rapidity as a function of the momentum of beam particles plotted for protons (solid line) and pions (dashed line).

and substitute the expressions for E' and p_z' from Eq. (8.17).

$$y' = \frac{1}{2}\ln\left(\frac{\gamma(E-\beta p_z)+\gamma(p_z-\beta E)}{\gamma(E-\beta p_z)-\gamma(p_z-\beta E)}\right), \tag{8.27}$$

$$y' = \frac{1}{2}\ln\left(\frac{1-\beta}{1+\beta}\times\frac{E+p_z}{E-p_z}\right), \tag{8.28}$$

$$y' = y - \frac{1}{2}\ln\left(\frac{1+\beta}{1-\beta}\right), \tag{8.29}$$

$$y' = y - y_\beta, \tag{8.30}$$

where in the last line, we introduced the rapidity of the moving frame K' relative to K, noted y_β. The simplicity of this transformation relation makes it very convenient for transformation of the particle kinematics between the target, projectile, and CM rest frames.

The rapidity of particles propagating along the z-axis (e.g., $p = p_z$, the beam axis) is plotted in Figure 8.3 as a function of their momentum p.

8.1.5 Pseudorapidity

Experimentally, the determination of the rapidity of a particle requires that one identifies its mass. But as we will see later in this chapter, although it is relatively straightforward to measure the momentum of a particle (or its energy), it is not always possible to measure both quantities directly or to readily identify the species of the particle detected. Thus, the masses of detected particles are often unknown, and calculation of their rapidity from, say, measured momenta, is consequently not feasible. Fortunately, there is a convenient substitute, known as **pseudorapidity**, defined according to

$$\eta = -\ln\left(\tan(\theta/2)\right), \tag{8.31}$$

where θ is the polar angle between the particle momentum $\vec{p}$ and the beam axis. It is easy to verify (see Problem 8.3) that η equals y in the limit where the mass of a particle is negligible compared to its energy, that is, for $m << E$.

Evidently, in the limit $m = 0$, which hold for (real) photons, one has (in natural units)

$$E = p \tag{8.32}$$

$$m_T = p_T, \tag{8.33}$$

and consequently

$$y = \eta. \tag{8.34}$$

8.1.6 Useful Kinematic Relations

The energy and momentum of a particle of known mass may be expressed in terms of its velocity, $\vec{\beta} = \vec{v}/c$, according to (in natural units)

$$E = \gamma m, \tag{8.35}$$

$$\vec{p} = \gamma m \vec{\beta}, \tag{8.36}$$

with the Lorentz factor $\gamma = (1 - \beta^2)^{-1/2}$. The velocity, β, can consequently be determined from the ratio

$$\beta = \frac{|\vec{p}|}{E}. \tag{8.37}$$

It is useful to introduce the notion of **transverse mass**:

$$m_T = \sqrt{p_T^2 + m^2}. \tag{8.38}$$

One may then express the energy and the *longitudinal* (or z) component of the momentum of a particle in terms of y and m_T, as follows (see Problem 8.4):

$$E = m_T \cosh y, \tag{8.39}$$

$$p_z = m_T \sinh y. \tag{8.40}$$

Alternatively, one may also obtain the modulus of the momentum $|\vec{p}|$ and the z-component based on the transverse momentum p_T and the pseudorapidity (see Problem 8.7):

$$|\vec{p}| = p_T \cosh \eta, \tag{8.41}$$

$$p_z = p_T \sinh \eta. \tag{8.42}$$

Measurements of momentum components are commonly carried out in terms of *spherical coordinates* (p, θ, ϕ), defined in Figure 8.4. The momentum components (p_x, p_y, p_z) may then be obtained with the following relations:

$$p_x = p_T \cos\phi, \tag{8.43}$$

$$p_y = p_T \sin\phi, \tag{8.44}$$

$$p_z = p \cos\theta, \tag{8.45}$$

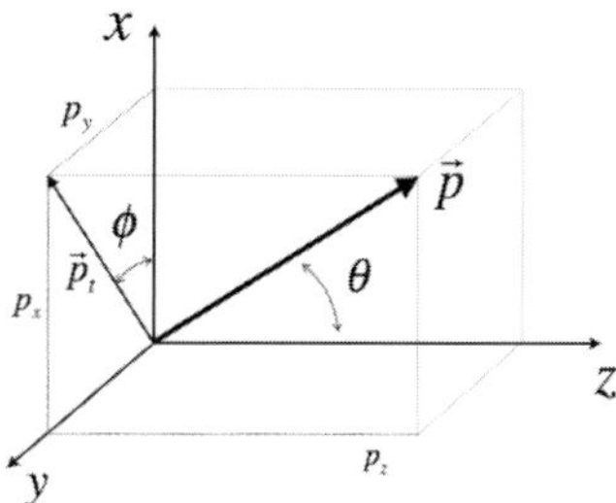

Fig. 8.4 Definition of spherical coordinates relative to the beam direction (z-axis).

where $p_T = p\sin\theta$. The angles are obtained with the relations

$$\phi = \tan^{-1}(p_y/p_x), \tag{8.46}$$

$$\theta = \tan^{-1}(p_T/p_z). \tag{8.47}$$

8.1.7 Mandelstam Variables

The Mandelstam variables, denoted s, t, and u, were introduced by Stanley Mandelstam in 1958. They are defined in the context of nuclear collisions involving two incoming and two outgoing particles, with four momenta denoted p_1, p_2, p_3, and p_4 as illustrated in Figure 8.5:

$$s = (p_1 + p_2)^2 = (p_3 + p_4)^2, \tag{8.48}$$

$$t = (p_1 - p_3)^2 = (p_2 - p_4)^2, \tag{8.49}$$

$$u = (p_1 - p_4)^2 = (p_2 - p_3)^2, \tag{8.50}$$

where the right-hand side equalities are determined by virtue of energy and momentum conservation. By construction as squares of four-momentum sums or differences, the three variables s, t, and u are Lorentz invariants and thus have the same value in all reference frames.

The variable s is particularly useful to quantify the collision energy. For instance, in fixed-target geometry, one has $p_1 = (E_1, \vec{p}_1)$, and $p_2 = (m_2, 0)$. One then gets

$$s = (E_1 + m_2)^2 - p_1^2 = m_1^2 + m_2^2 + 2E_1 m_2, \tag{8.51}$$

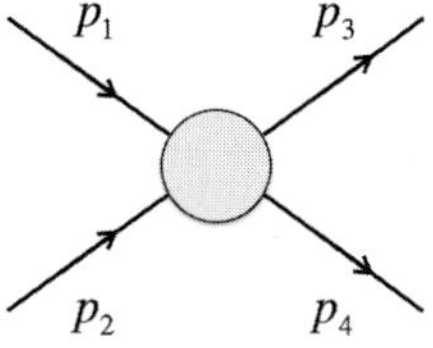

Fig. 8.5 The Mandelstam variables s, t, and u are defined based on the four-momenta p_1, p_2, p_3, and p_4 as shown. The circle represents an energy and momentum conserving interaction between two incoming particles with 4-momenta p_1 and p_2, and producing two particles with momenta p_3, and p_4.

which is proportional to the beam energy E_1. On the other hand, in symmetric collider geometry, that is, with colliding beams of equal momentum but opposite directions, one has $p_1 = (E, \vec{p})$ and $p_2 = (E, -\vec{p})$, thereby yielding

$$s = (E + E)^2 - (\vec{p} - \vec{p}) = 4E^2, \tag{8.52}$$

which is proportional to the square of the beam energy. In symmetric collider geometry, the CM is at rest relative to the laboratory. All the energy of the incoming beams is consequently available in the CM frame of the collision. In contrast, in fixed-target geometry the CM is moving toward the target with a velocity $\beta_{CM} = |\vec{p}_{1,\text{lab}}|/(E_{1,\text{lab}} + m_2)$. Only a small fraction of the beam energy is therefore available in the CM frame of the collision. Considering that the acceleration of high-energy beams is an energy-intensive and costly endeavor, it is thus advantageous to collide particles in a collider geometry to maximize the usefulness of the beam energy delivered toward interactions and particle production.

Mandelstam variables are particularly useful also in the calculation of scattering amplitudes because they are frame independent. The variable t, in particular, determines the momentum transferred to particle 3 by the collision.

Finally, note that the tree variables s, t, and u are not independent, and are, in fact, constrained by the relation

$$s + t + u = m_1^2 + m_2^2 + m_3^2 + m_4^2. \tag{8.53}$$

8.1.8 Basic Classifications of Scattering Experiments

Research conducted in the last half century has involved a wide range of collisional energies, beam particles, and target materials. Experiments can be broadly divided into **fixed-target** and **collider geometry** experiments. Fixed-target experiments involve accelerated beams consisting of specific particle species and transport of these beams to bombard a fixed target, which may consist of a "thin" foil or a gas contained in a special vessel. Collider experiments require the acceleration of two beams brought to collide in a colinear geometry, moving in opposite directions (clockwise and counterclockwise). Existing colliders typically use two concentric acceleration/storage rings. The beams intersect and collide at one or more crossing points along the rings, as illustrated in Figure 8.7. Colliders are said to operate in symmetric collider mode if the two beams being accelerated consist of identical particle species (as well as particle and antiparticle combinations, e.g., e^+e^-, $p\bar{p}$) at the same energy. They are said to be asymmetric if the two beams have different energies. It is also possible to produce asymmetric colliding systems, such as $p + Pb$ or $d + Au$.

Huge advances have been realized since the modest beginnings of early accelerators that produced protons and alpha particles of a few MeV. Past and existing facilities were developed to accelerate a wide variety of beams, including electrons, protons, light and heavy stable nuclei, radioactive nuclei with very short half-lives, as well as elementary

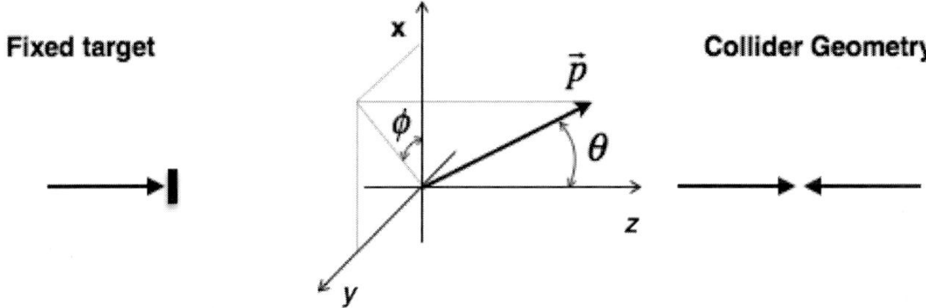

Fig. 8.6 Illustration of the collision geometry of fixed-target and collider mode experiments.

particles such as pions or muons. Figure 8.8 presents a summary of trends in accelerators at facilities worldwide since the 1930s.

Collisions between two elementary particles (or nuclei) can be broadly divided into elastic and inelastic interactions. Elastic collisions, as the name suggests, are nondissipative and involve the deflection of the incoming particles only. At low energy, proton on proton (p + p) inelastic collisions involve the excitation of one or both colliding particles

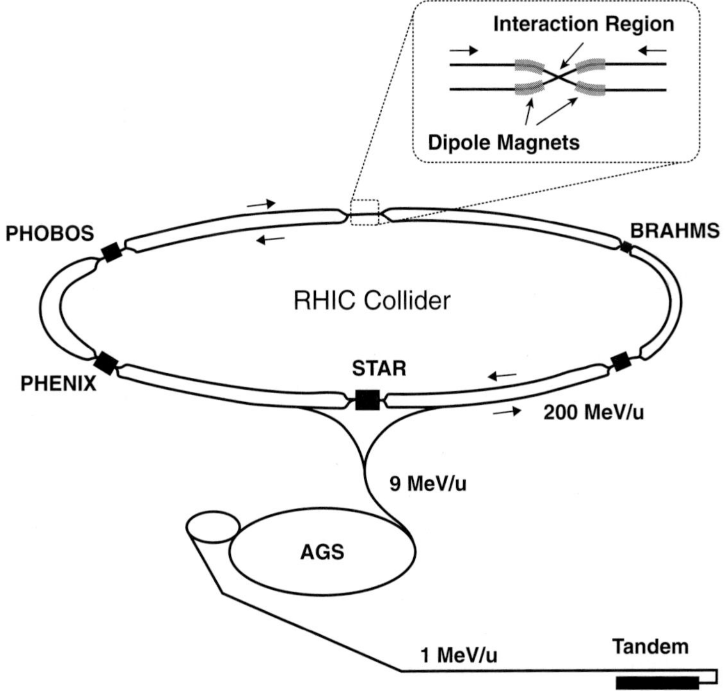

Fig. 8.7 Schematic of the BNL/RHIC collider geometry. Beams traveling counterclockwise (CCW) and clockwise (CW) are tuned to collide at marked intersections points.

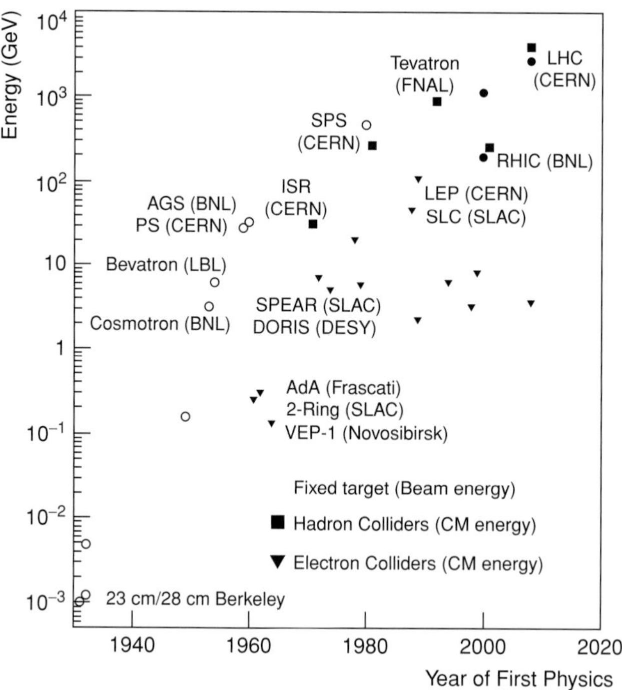

Fig. 8.8 Trend in accelerator energies. Open and solid symbols show fixed-target and collider facilities, respectively, since 1935.

and typically lead to the production two or more particles. At higher energy, p + p collisions produce several tens or even hundreds of particles, while head-on heavy-ion collisions may produce several thousands. Diffractive p + p collisions are inelastic collisions involving a modest energy dissipation and momentum transfer between the colliding partners. One or both of the protons are produced in an excited state from which they decay by emission of one or several hadrons. Such interactions are said to be singly and doubly diffractive depending on whether only one or both protons end up in an excited state and produce particles. Nondiffractive inelastic collisions involve substantial energy dissipation and typically the production of large particle multiplicities (number of particle produced).

It is useful to classify produced particles according to their source. For instance, particles produced by the breakup of the projectile, a process known as beam fragmentation, are observed at rapidities near the projectile rapidity, y_P, and are said to be part of the **projectile fragmentation region**. Likewise, particles found near the target rapidity, y_T (e.g., $y_T \sim 0$ in fixed-target geometry or $y_T \sim -y_P$ in a symmetric collider mode) are commonly associated with the **target fragmentation region**, whereas particles produced in between these two extremes, particularly those around the CM rapidity, y_{CM}, are said to be part of the **central rapidity region**. Clearly, the gap between the projectile and target rapidities grows with increasing beam energy. At Relativistic Heavy-Ion Collider (RHIC) top heavy-ion energy, $\sqrt{s_{NN}} = 200$ GeV, the beams have rapidities of ± 5.3, while at CERN, the

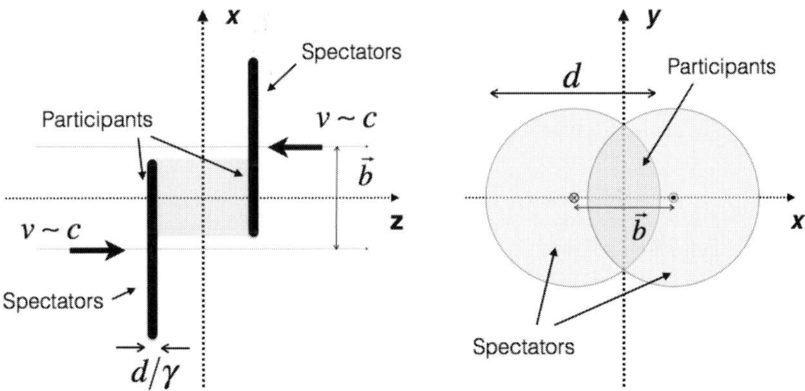

Fig. 8.9 Longitudinal (left) and transverse (right) profiles of two heavy-ion nuclei shortly before a collision at relativistic ($\beta \sim c$) energy. The diameter d of the nuclei is Lorentz contracted by a factor γ in the beam direction. The light gray area between the nuclei (longitudinal profile) is where most of the particle production takes place after the nuclei have passed through each other.

7 TeV proton beams have a rapidity of ± 9.6. Observations indicate that projectile and target fragmentation produce particles clustered within approximately one unit of rapidity of the projectile and target rapidity, respectively. The central rapidity region thus spreads considerably with increasing beam energy. This means that at large beam energies, it is rather challenging to design and build detector systems capable of measuring all three regions, and experiments are consequently designed to focus on only particular aspects of the particle production. For instance, detectors such as STAR, PHENIX at RHIC, DO, and CDF at the Tevatron, and CMS, ATLAS, and ALICE at the LHC are best equipped to measure the central rapidity region, whereas TOTEM and LHCf at the LHC were designed to study the production of "forward" particles.

In the context of heavy-ion collisions, one can further extend this source classification based on the impact parameter of the collision between two nuclei. Figure 8.9 schematically illustrates the longitudinal and transverse profiles of two large nuclei colliding with an impact parameter b. In the longitudinal profile, the nuclei are represented in the CM reference frame (assuming a symmetric collision) just before the collision. They appear as thin slices because their longitudinal size is contracted by the Lorentz factor $\gamma = (1 - \beta^2)^{-1/2}$, with β being their velocity in the CM frame. In the transverse profile, the nuclei are represented as they pass through one another. At collider energies, the nuclei have speeds approaching the speed of light. The penetration time (measured in the CM frame) is thus of the order of $d/\gamma c$, with d being the diameter of the nuclei. At RHIC top energy, this amounts to $\sim 4 \times 10^{-25} s$, or 0.12 fm/c. On the other hand, transverse propagation of a signal from one side (for instance, the top of the transverse profile in Fig. 8.9) to the other side (the bottom) of a nucleus requires a minimum of $\sim$ 12 fm/c. In the transverse plane, one can consequently divide the colliding nuclei into two regions, as illustrated in Figure 8.9. The darker shade region, consisting of the two nuclei overlapping area, is

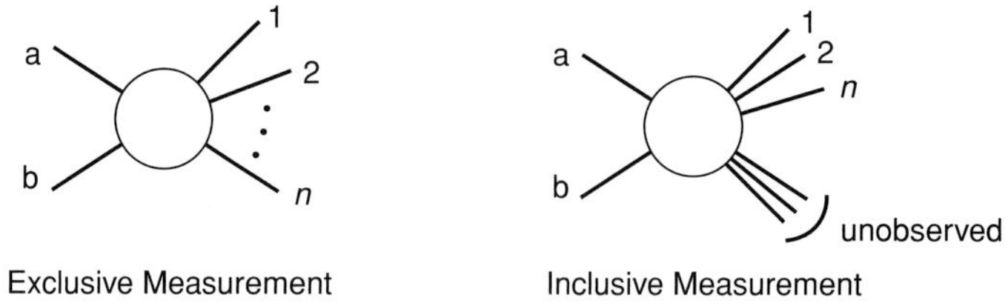

Fig. 8.10 Exclusive measurements involve all particles produced whereas semiexclusive and inclusive measurements specify only a small fraction of the final state of collisions (or decays).

directly and instantly affected by the collision. Nucleons in this region are thus called **participants**. Their interactions may produce particles at all rapidities, that is, in both the central rapidity region and the beam/target fragmentation regions. Nucleons in the lighter shaded areas of the two nuclei are referred to as spectators. Signals that nucleons of the participant region have undergone interactions with another nucleus do not reach them until well after the two nuclei have passed through one another. These nucleons are therefore not directly involved in the main part of the interaction and can thus indeed be considered spectators. The two colliding nuclei are completely shattered, but the spectator nucleons continue their path forward/backward at the beam/target rapidity. They lead to particle production in the beam/target fragmentation regions. Particle produced by the participants, typically deemed of more interest, dominate the central rapidity region.

Experiments can be further classified according to the type of measurements they carry (Figure 8.10). Measurements are said to be exclusive if they involve the totality of produced particles. Such measurements were possible in the early days of particle physics with, for instance, bubble chamber detectors. However, modern experiments conducted at high-beam energy typically do not cover all regions of particle production and thus cannot measure all produced particles (charged or neutral). They are consequently called semi-exclusive or inclusive. In an inclusive measurement, only one or a few of the produced particles are detected and their properties (e.g., momentum, species type, and so forth) measured. The final state of a collision thus cannot be completely specified or identified.

8.2 Particle Decays and Cross Sections

8.2.1 Particle Decays

The decay of an elementary particle of mass M and four-momentum $p = (E, \vec{p})$ is a stochastic phenomena determined by an exponential distribution (§3.5)

$$P(t) = \frac{1}{\gamma\tau} \exp\left(-\frac{t}{\gamma\tau}\right), \tag{8.54}$$

where τ is the **proper lifetime** of the particle determined by the **decay rate** Γ,

$$\tau = \hbar/\Gamma, \tag{8.55}$$

and $\gamma = E/M$. The function $P(t)$ gives the probability the particle lives for a time t or greater before decaying. For a particles with n decay channels, the rate Γ is the sum of the partial decay rates Γ_i corresponding to each of the decay channels:

$$\Gamma = \sum_{i=1}^{n} \Gamma_i. \tag{8.56}$$

The partial decay rate of a particle of mass M into n particles, in its rest frame, is given by

$$d\Gamma = \frac{(2\pi)^4}{2M} |\mathfrak{M}|^2 \, d\Phi_n(P; p_1, ..., p_n), \tag{8.57}$$

where $|\mathfrak{M}|$ represents the amplitude of the decay process. The factor $d\Phi_n$ is an element of n-body phase space equal to

$$d\Phi_n(P; p_1, ..., p_n) = \delta^4\left(P - \sum_{i=1}^{n} p_i\right) \prod_{i=1}^{n} \frac{d^3 p_i}{(2\pi)^3 2E_i}, \tag{8.58}$$

where the delta function $\delta^4(P - \sum_{i=1}^{n} p_i)$ encapsulates energy conservation between the parent and decay particles. It is useful to note that the n-body phase space factor $d\Phi_n$ can be calculated recursively

$$\begin{aligned} d\Phi_n(P; p_1, ..., p_n) &= d\Phi_j(P; p_1, ..., p_j) \\ &\quad \times d\Phi_{n-j+1}(P; q, p_{j+1}, ..., p_n)(2\pi)^3 dq^2, \end{aligned} \tag{8.59}$$

where

$$q^2 = \left(\sum_{i=1}^{j} E_i\right)^2 - \left(\sum_{i=1}^{j} \vec{p}_i\right)^2. \tag{8.60}$$

Two-Body Decays

In the rest frame of a particle of mass M, a two-body decay yields $\vec{p}_1 = -\vec{p}_2$. Energy conservation implies $M = E_1 + E_2$. One thus finds

$$E_1 = \frac{M^2 - m_2^2 + m_1^2}{2M}, \tag{8.61}$$

$$= \frac{\left[(M^2 - (m_1 + m_2)^2)(M^2 - (m_1 - m_2)^2)\right]^{1/2}}{2M}, \tag{8.62}$$

from which we conclude

$$d\Gamma = \frac{1}{32\pi^2} |\mathfrak{M}|^2 \frac{|p_1|}{M^2} d\Omega, \tag{8.63}$$

where $d\Omega = d\phi_1 d(\cos\theta_1)$ is the solid angle within which particle 1 is emitted. Experimentally, M^2 can be obtained from the invariant mass $(p_1 + p_2)^2$ of the pair and thus provides a basis for short-lived particle reconstruction and identification (see §8.5).

Three-Body Decays

In the rest frame of a three-body decaying particle of mass M, momentum conservation constrains the momenta $\vec{p}_1$, $\vec{p}_2$, and $\vec{p}_3$ of the produced particles to lie in the same (decay) plane. Energy conservation further constrains the energy of the three particles, and one has

$$m_{12}^2 = (P - p_3)^2 = M^2 + m_3^2 - 2ME_3, \tag{8.64}$$

where we defined $m_{ij}^2 = p_{ij}^2$ with $p_{ij} = p_i + p_j$. Additionally, one has

$$m_{12}^2 + m_{13}^2 + m_{23}^2 = M^2 + m_1^2 + m_2^2 + m_3^2. \tag{8.65}$$

One can then write the decay rate as

$$d\Gamma = \frac{1}{16M(2\pi)^5} |\mathfrak{M}|^2 \, dE_1 dE_2 d\alpha d(cos\beta) d\gamma, \tag{8.66}$$

where α, β, and γ represent three Euler angles defining the orientation of the decay plane. The decay rate may also be written (see Problem 8.10)

$$d\Gamma = \frac{1}{16M(2\pi)^5} |\mathfrak{M}|^2 \, |p_1^*||p_3| dm_{12} d\Omega_1^* d\Omega_3, \tag{8.67}$$

where $|p_1^*|$ and $d\Omega_1^*$ represent the momentum and solid angle of emission of particle 1 in the rest frame of particles 1 and 2, while $d\Omega_3$ is the angle of the third particle in the rest frame of the decaying particle. The momenta $|p_1^*|$ and $|p_3|$ are given by

$$|p_1^*| = \frac{\left[(m_{12}^2 - (m_1 + m_2)^2)(m_{12}^2 - (m_1 - m_2)^2)\right]^{1/2}}{2m_{12}}, \tag{8.68}$$

$$|p_3| = \frac{\left[(M^2 - (m_{12} + m_3)^2)(M^2 - (m_{12} - m_3)^2)\right]^{1/2}}{2M}, \tag{8.69}$$

For angular momentum $J = 0$ decaying particles, or averaging over all spin states of a $J \neq 0$ particle, the rate simplifies to

$$d\Gamma = \frac{1}{8M(2\pi)^3} \overline{|\mathfrak{M}|^2} dE_1 dE_2, \tag{8.70}$$

$$= \frac{1}{8M(2\pi)^3} \overline{|\mathfrak{M}|^2} dm_{12}^2 dE_{23}^2, \tag{8.71}$$

which provides a basis for Dalitz plots (see §8.5.2).

8.2.2 Cross Section Measurements

Invariant Cross Section

Because the longitudinal momentum component, p_z, of produced particles is reference-frame dependent, the expression of the differential cross section is also reference-frame dependent. It is thus useful to seek an expression for the cross section in terms of a relativistic invariant. Given $E^2 = m^2 + p_T^2 + p_z^2$, and for fixed p_T, one gets $EdE = p_z dp_z$. A longitudinal boost of the differential dp_z thus yields

$$dp_z{}^* = \gamma\,(dp_z - \beta dE) = \gamma\left(dp_z - \beta\frac{p_z}{E}dp_z\right),$$
$$= \gamma\frac{dp_z}{E}(E - \beta p_z) = \frac{dp_z}{E}E^*, \tag{8.72}$$

from which we find that $dp_z{}^*/E^* = dp_z/E$. Since p_T is invariant under a longitudinal boost, we thus conclude that d^3p/E is a Lorentz invariant. One can thus express cross sections in a frame-independent way using the invariant cross section

$$E\frac{d^3\sigma}{dp^3} = E\frac{d^3\sigma}{d\phi p_T dp_T dp_z}, \tag{8.73}$$
$$= \frac{d^2\sigma}{\pi dp_T^2 dy}, \tag{8.74}$$

where, in the second line, we used the fact that $dp_z/E = dy$ and averaged over ϕ.

As stated earlier, it is not always possible to identify the species or type of detected particles. It is thus convenient to also define cross section in terms of the pseudorapidity variable η. At a fixed or given p_T, one writes

$$\frac{dN}{d\eta} = \frac{dN}{dy}\frac{dy}{d\eta}. \tag{8.75}$$

The derivative $dy/d\eta$ is calculated at fixed p_T as follows:

$$\frac{dy}{d\eta} = \frac{dy/dp_z}{d\eta/dp_z} = \frac{p}{E} = \sqrt{1 - \frac{m^2}{m_T^2\cosh^2 y}}. \tag{8.76}$$

The differential cross section may thus be written

$$\frac{dN}{d\eta dp_T} = \sqrt{1 - \frac{m^2}{m_T^2\cosh^2 y}}\frac{dN}{dydp_T},$$
$$= \sqrt{1 - (1 + p_T^2/m^2)^{-1}\cosh^{-2}y}\frac{dN}{dydp_T}. \tag{8.77}$$

The Jacobian $J = \sqrt{1 - (1 + p_T^2/m^2)^{-1}\cosh^{-2}y}$, illustrated in Figure 8.11 as a function of y for selected values of the ratio p_T/m, tends to unity for $p_T \gg m$. It may thus be neglected when the mass of particles are very small compared with the energy or momentum scale of the process.

n-Body Cross Section

The **exclusive** differential cross section of a collision producing n particles (illustrated in Figure 8.12) may be written as

$$d\sigma = \frac{(2\pi)^4\,|\mathfrak{M}|^2}{4\sqrt{(p_1 \cdot p_2)^2 - m_1^2 m_2^2}}d\Phi_n(p_1 + p_2; p_3, \dots, p_n + 2), \tag{8.78}$$

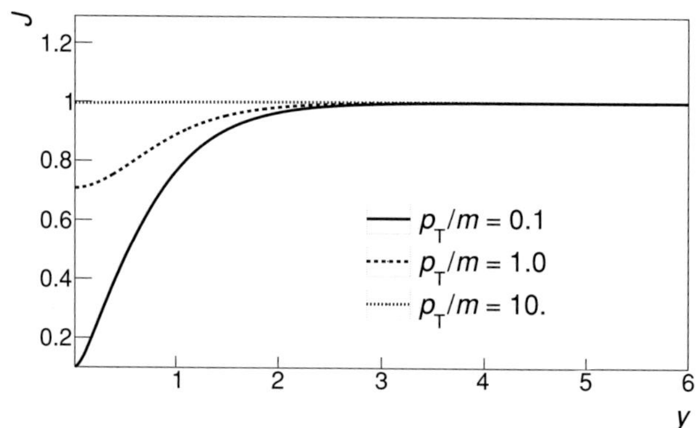

Fig. 8.11 Jacobian $J = \partial y/\partial \eta$ for selected values of the ratio p_T/m.

with

$$d\Phi_n = \delta^{(4)}\left(\vec{P} - \sum_{j=1}^{n} \vec{p}_j\right) \prod_{i=1}^{n} \delta\left(p_i^2 - m_i^2\right) d^4 p_i, \tag{8.79}$$

where $|\mathfrak{M}|$ represents the amplitude of the scattering process and Φ_n is a phase space factor defined by Eq. (8.58). In the rest frame of m_2 (the lab frame), one has $p_1 = (E_1, \vec{p}_1)$ and $p_2 = (m_2, 0)$, which yields

$$\sqrt{(p_1 \cdot p_2)^2 - m_1^2 m_2^2} = m_2 p_1. \tag{8.80}$$

In the center-of-mass frame, $p_1 = (E_1, \vec{p}_{1,\rm cm})$ and $p_2 = (E_2, -\vec{p}_{1,\rm cm})$, Eq. (8.80) becomes

$$\sqrt{(p_1 \cdot p_2)^2 - m_1^2 m_2^2} = p_{1,\rm cm}\sqrt{s}. \tag{8.81}$$

Two-Body Reactions

Cross sections for two-body processes are conveniently expressed in terms of the Mandelstam variable t defined in Eq. (8.48):

$$\frac{d\sigma}{dt} = \frac{1}{64\pi s}\frac{1}{|p_{1,\rm cm}|^2}|\mathfrak{M}|^2 \tag{8.82}$$

In the CM frame, one has

$$t = (E_{1,\rm cm} - E_{3,\rm cm})^2 - p_{1,\rm cm}^2 - p_{3,\rm cm}^2 + 2p_{1,\rm cm}p_{3,\rm cm}\cos\theta_{\rm cm}, \tag{8.83}$$

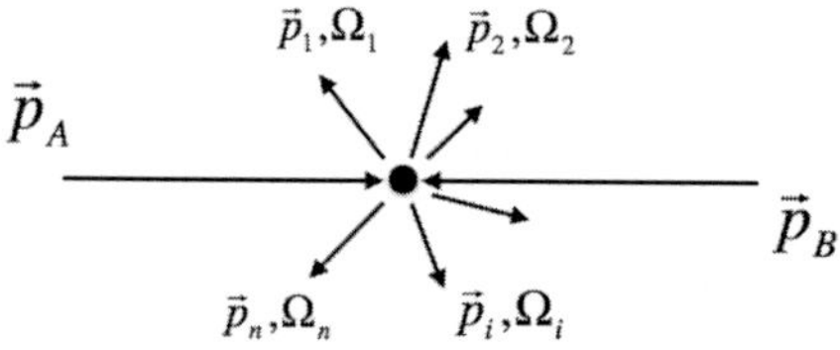

Fig. 8.12 Differential cross section of an n-body process.

where $\theta_{\rm cm}$ is the angle of emission, in the CM frame, relative to the direction of particle 1. Since $\cos u = 1 - 2\sin^2 u/2$, one gets

$$t = (E_{1,\rm cm} - E_{3,\rm cm})^2 - (p_{1,\rm cm} - p_{3,\rm cm})^2 - p_{1,\rm cm}p_{3,\rm cm}\sin^2\theta_{\rm cm}/2. \tag{8.84}$$

For $2 \to 2$ scatterings, limiting values for t_0 and t_1, corresponding to $\theta_{\rm cm} = 0$ and $\theta_{\rm cm} = \pi$ respectively, are given by (see Problem 8.8):

$$t_0(t_1) = \left[\frac{m_1^2 - m_3^2 - m_2^2 + m_4^2}{2\sqrt{s}}\right]^2 - (p_{1,\rm cm} \mp p_{3,\rm cm})^2. \tag{8.85}$$

Luminosity

In a fixed-target experiment, the rate of interactions of a specific type, hereafter noted W_i, is proportional to the flux of incoming beam particles, F, the total number of target particles, N_T, and the cross section, σ_i, of the process, so that

$$W_i = \sigma_i N_T F, \tag{8.86}$$

where the rate W_i has units of inverse time, $[T^{-1}]$, the number of targets has units of inverse area $[L^{-2}]$, and the flux has units of inverse time $[T^{-1}]$. In contrast, in a collider experiment, there is no "solid target," one must then replace the product of the number of target nuclei (per unit area) and the flux of incoming particles by a beam parameter known as **instantaneous luminosity**, L. The measured rate is thus

$$W_i = \sigma_i L, \tag{8.87}$$

where L has units of inverse area, inverse time $[T^{-1}L^{-2}]$.

The time-integrated luminosity multiplied by the cross section yields the number of observable interactions, N_i, of a given type over the life of an experiment:

$$N_i = \sigma_i \int L\,dt. \tag{8.88}$$

The luminosity L delivered by a collider depends on the number of beam particles, the frequency of beam crossings, as well as beam focusing according to

$$L = \frac{nfB_1B_2}{A} \tag{8.89}$$

where n is the number of beam bunches, f the revolution frequency, B_1 and B_2 are the number of beam particles contained per bunch, and A is the cross-sectional area of the overlapping crossing beams. While the number of bunches n, and the frequency f are typically fixed by an **accelerator's design**, the parameters B_1, B_2, and A are functions of the **accelerator tuning** and may thus vary over time. Since tuning an accelerator is an intricate and complex task, accelerator facilities typically begin their existence with small delivered luminosities. However, as beam physicists master the operation of their machine, they generally raise the delivered luminosity to eventually achieve **design luminosity** within weeks or months of operation.

The design luminosity of a machine is limited by various conditions, including the maximum number of bunches, maximum number of ions per bunch, and focusing. The LHC was originally designed to achieve a luminosity of 10^{34} cm^{-2}s^{-1}. The total interaction cross section of proton–proton collisions is of the order of 60 mb (6×10^{-26} cm^2) at LHC energy. This makes for an interaction rate of about 6×10^8 s^{-1} or 600 MHz.

Given a process cross section σ_i, the number of observable interactions N_i is intrinsically limited by the integrated luminosity $\int L\,dt$ (as well as the efficiency of the measurement). Accelerator facilities thus typically report *delivered* integrated luminosities in terms of the inverse of cross section units. For instance, an integrated luminosity of one inverse picobarn corresponds to the minimal luminosity required to observe (on average) one event with a cross section of one picobarn, while an integrated luminosity of one inverse femtobarn, which is one thousand times larger, corresponds to the minimal integrated luminosity needed to observe a one femtobarn process. Laboratories also customarily report plots of the delivered integrated luminosity as a function of (operation) time as an indicator of the performance of their accelerator facilities.

8.2.3 Measurements of Differential Cross Sections

Differential measurements of cross section may be carried out for single or multiple particles concurrently. Whether one wishes to obtain differential cross sections as a function of transverse momentum, p_T, rapidity, y, or other kinematic variables, it is necessary to scan a dataset and fill one or several histograms corresponding to the kinematic variables and particles of interest. One uses one-dimensional histograms (§4.6) for differential quantities with dependency on a single variable and multidimensional histograms (§4.6.3) for cross sections involving several kinematic variables, several particles, or both.

Consider, as an example, the determination of the transverse momentum differential cross section for the production of charged particles in A + A collisions. Let us assume one has a dataset consisting of a known number of events, N_{ev}, satisfying some event conditions and data **quality criteria** (commonly called **quality cuts**). The dataset must then be scanned for all events satisfying the event condition and quality criteria of interest. Each event is itself scanned for particles, with appropriate particle selection and quality selection criteria. Each particle satisfying the selection criteria produce entries in a p_T histogram, $H_{p_T}(i)$, in bin i determined by the momentum of the particle and the partition of the histogram (i.e., its minimum and maximum value boundaries and the number of bins). A raw differential spectrum S_{p_T} is obtained by scaling the histogram by the number of events included in the dataset scan. That is, for each bin i, one calculates

$$S_{p_T}(i) = \frac{1}{N_{\text{ev}}} H_{p_T}(i), \tag{8.90}$$

where N_{ev} is the actual number of events satisfying the event selection and quality selection criteria. One obtains an estimator of the particle density $\hat{\rho}_1(p_T) = dN/dp_T = \sigma^{-1} d\sigma/dp_T$ as function of p_T by further dividing $S_{p_T}(i)$ by the bin width, $\Delta p_{T,i}$, of each bin i:

$$\hat{\rho}_1(p_T) \equiv dN/dp_T = \frac{S_{p_T}(i)}{\Delta p_{T,i}}, \tag{8.91}$$

$$= \frac{1}{N_{ev}\Delta p_{T,i}} H_{p_T}(i). \tag{8.92}$$

Technically, the quantity $\hat{\rho}_1(p_T)$ obtained with this formula is "still" a histogram. To obtain an actual function, one must account for bin width averaging, as discussed in §12.3.5. Additionally, various instrumental conditions typically limit the efficiency of the measurement and introduce p_T smearing effects, as well as backgrounds. Techniques to correct for such effects are presented in §12.3.1. Neglecting such effects, or assuming they have been properly accounted for, one obtains the differential cross section by dividing the density $\rho_1(p_T)$ by the luminosity $\mathcal{L}$ integrated by the experiment and the event selection efficiency ε:

$$d\sigma/dp_T = \frac{1}{\mathcal{L}\varepsilon}\rho_1(p_T). \tag{8.93}$$

Techniques and issues related to the evaluation and correction of instrumental inefficiencies are discussed in §12.3.1.

The aforementioned procedure can be readily extended for differential cross section measurements involving two or more kinematical variables. For instance, one obtains the density $\rho_1(\eta, p_T)$ by filling a two-dimensional histograms with the particles pseudorapidity η and transverse momentum p_T:

$$\hat{\rho}_1(\eta, p_T) \equiv \frac{d^2N}{d\eta dp_T} = \frac{1}{\sigma}\frac{d\sigma}{d\eta dp_T} = \frac{S_{\eta,p_T}(i)}{\Delta\eta_i\Delta p_{T,i}}, \tag{8.94}$$

$$= \frac{1}{N_{ev}\Delta\eta_i\Delta p_{T,i}}H_{\eta,p_T}(i,j), \tag{8.95}$$

where the indices i and j span bins in pseudorapidity and transverse momentum, respectively. The double differential density $d^2\sigma/d\eta dp_T$ is obtained, as for single differential cross section, by dividing by the integrated luminosity $\mathcal{L}$ and event efficiency. The procedure is similar for two- or multiparticle cross sections, except that one must then scan each event for pairs, triplets, or n-tuplets of particles, as appropriate. This involves two, three, or more, nested loops over all particles of each event. The number of pairs and triplets scale as $n(n-1)$ and $n(n-1)(n-2)$, respectively. The process can thus become particularly CPU-intensive and fastidious for very large multiplicity events such as those produced in high-energy nucleus nucleus collisions at RHIC or LHC. However, note that for indistinguishable particles, it is unnecessary to examine all pairs (triplets or n-tuplets). For pairs, one can use a loop over all particles of the event for the first particle, that is, $i = 1, \ldots, n$, while the loop for the second particle is restricted to $j = i + 1, \ldots, n$. Subsequently, one must then appropriately symmetrize the histograms used to evaluate two-particle densities.

8.3 Measurements of Elementary Observables

Studies of interest in nuclear and particle physics include measurements of total and differential cross sections, correlation functions, particles mass and lifetimes, and much more. At their very core, all these measurement involve elementary measurements of particle momentum, energy, production angles, and so on. We discuss basic measurement techniques

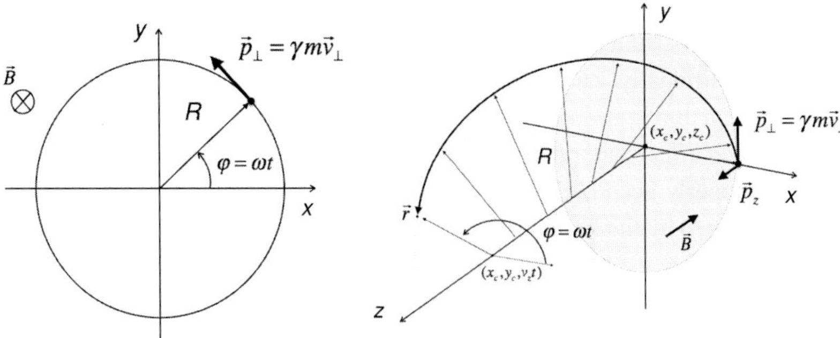

Fig. 8.13 (Left) Principle of measurement of particle momenta with a magnetic spectrometer. (Right) illustration of helicoidal trajectories described by Eqs. (8.98–8.100).

in this section and describe more advanced techniques in remaining section of this Chapter as well as in Chapters 10 and 11. Several data correction techniques are presented in Chapter 12.

8.3.1 Momentum Measurements

The momentum of a particle can be measured by observing its trajectory under the influence of a force of known magnitude and direction. For charged particles, this can be readily accomplished with static electric or magnetic fields. Here, we focus our discussion on magnetic spectrometers, which are far more common than electric field based momentum analyzers.

A charged particle, q, traveling in a magnetic field $\vec{B}$ at speed $\vec{v}$ is deflected by the Lorentz force, which acts as a centripetal force. The particle thus follows (in the absence of other forces) a helicoidal path of radius R such that

$$|\vec{F}_{\text{Lorentz}}| = kq|\vec{v} \times \vec{B}| = kqv_T B = \frac{mv_T^2}{R}, \tag{8.96}$$

for v_T perpendicular to $\vec{B}$. We then conclude that the particle momentum transverse to the field (as illustrated in Figure 8.13) is given by

$$p_T = kqBR, \tag{8.97}$$

where $k = 0.3$ when the charge q is expressed in unit charge (e.g., ± 1), the momentum p_T in GeV, the magnetic field B in Tesla, and the trajectory radius R in meters. For a uniform magnetic field oriented along the z-direction, the momentum component p_z is a constant, and the trajectory may be described parametrically as a helicoidal trajectory

$$x(t) = x_c + R\cos(\omega t + \phi_0), \tag{8.98}$$

$$y(t) = y_c + R\sin(\omega t + \phi_0), \tag{8.99}$$

$$z(t) = z_c + v_z t, \tag{8.100}$$

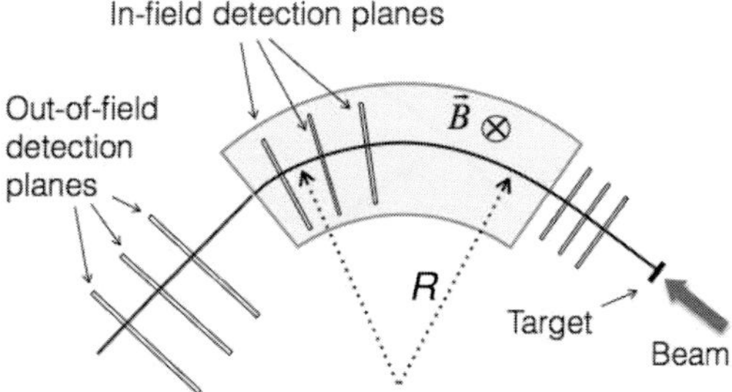

Fig. 8.14 Particle momentum determination with a dipole magnetic spectrometer.

where the coordinates (x_c, y_c) represent the center of the circular path and z_c is the position of the particle along the z-axis at $t = 0$. The particle's angular velocity is

$$\omega = \frac{v_T}{R} = \frac{p_T}{mR} \tag{8.101}$$

and the pitch λ of the helicoidal trajectory depends on the ratio of the momentum components parallel and transverse to the magnetic field

$$\tan\lambda = \frac{p_z}{p_T}. \tag{8.102}$$

Reconstruction of the momentum of a particle can be achieved by detecting the trajectory of the particle $\vec{r}(t)$ and evaluating its radius R and pitch angle λ. Equation (8.97) provides the magnitude p_T while Eq. (8.102) enables the determination of p_z. The magnitude of the momentum is thus $p = \sqrt{p_T^2 + p_z^2}$.

The determination of the trajectory $\vec{r}(t)$ is typically accomplished using position-sensitive sensors, discussed in §8.3.3, and requires sophisticated pattern recognition techniques to identify the charged particle trajectories measured by these sensors. Examples of these techniques are discussed in Chapter 9.

Charged Particles Detection and Tracking

Several magnetic field and detector geometries are commonly used to carry out magnetic analysis of particle momenta, including dipole, solenoidal, and toroidal field configurations. A plurality of detector technologies have been developed to measure charged particle trajectories within these field configurations and determine momenta and identify particle species. Detector technologies much in use include scintillator slats, straw tube chambers, wire chambers, pad chambers, time projection chambers, as well as solid-state detectors such as pixel detectors, strip detectors, or silicon drift detectors. Detectors may be placed either within or outside the field region, as schematically illustrated in Figure 8.14. A full discussion of tracker topologies and associated technologies is well beyond the scope of this textbook. We thus limit our discussion to few illustrative examples of the methods used in charged particle detection with a solenoidal field configuration (Figure 8.15).

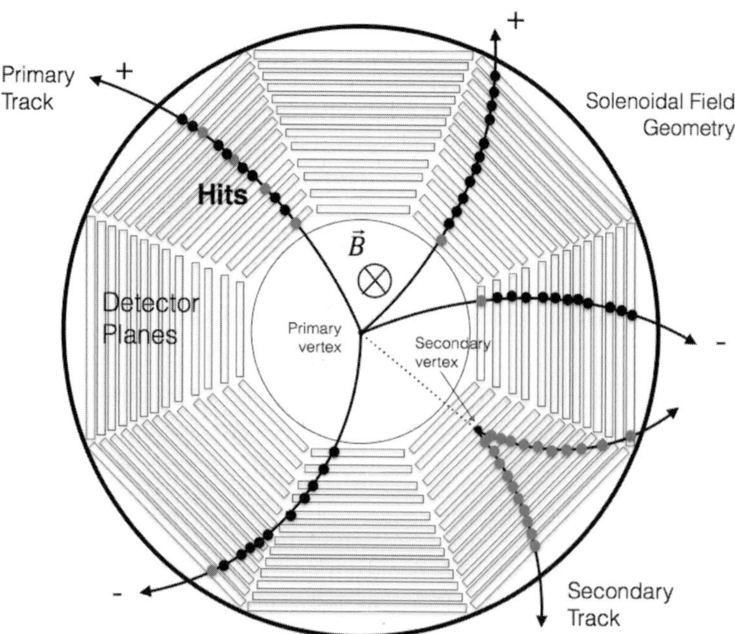

Fig. 8.15 Charged particle detection in a multilayer charged particle detector within a solenoidal field configuration. Open rectangles represent detector layers and solid dots represent energy deposition "hits" within these layers. Solid lines represent primary charged particle tracks, whereas the dotted line represents a neutral short-lived particle that decayed into two charged particles within the detector volume.

High-momentum charged particles passing through matter lose energy via interactions with (mostly) electrons of the medium. Electronic excitations resulting from these interactions can be exploited toward ionization of the medium or light emission (fluorescence). Tracking detectors are thus designed to either detect the presence and position of ionization or light emission within a specific detection plane or volume. A combination of several detection planes, as illustrated in Figure 8.15, then yield **hits** that provide an "image" or snapshot of the particle trajectory called **track**. Tracking detectors measure the position and energy deposited at these hits. Pattern recognition software can then be used to find series of hits that belong to a given track, and the curvature and trajectory of each track are used to determine the momentum of the particles as per Eq. (8.97). Extrapolation of the trajectories to their point of origin then yields the momentum vector of the particles at their point of production, called **production vertex** or just **vertex**.

Tracks may be fitted for momentum vector determination with or without the inclusion of their production vertex position. Tracks reconstructed and fitted without the inclusion of the primary vertex are often known as **global tracks**, whereas those fitted with the primary vertex are loosely called primary tracks, although they may, technically, originate from unresolved secondary decays taking place near the primary vertex. Such unresolved secondary tracks constitute a systematic background in the determination of particle production cross sections and correlation functions.

A wide variety of pattern recognition techniques are commonly used to reconstruct hits, tracks, and production vertices. We here limit our discussion to some of the most basic techniques in Chapter 9 and refer the interested reader to specialized works [18, 41, 42, 43, 60, 86, 88, 89, 94, 95, 102, 127, 139, 178].

Tracks are usually classified based on their vertex of origin and mechanism of production. Tracks originating from the main collision vertex, known as a **primary vertex**, are called **primary tracks**, while those produced by decays of long-lived particles (e.g., K_s^0, Λ^0) are called **secondary tracks** and are associated with a **secondary vertex**, as illustrated in Figure 8.15. Tracks produced within the detection apparatus by particle interactions with detector materials may be referred to as **tertiary** or **background tracks**.[2] Tracks obtained by wrong or accidental associations of hits are generally known as **faked tracks** or **ghost tracks**. It is important to realize that secondary tracks are a byproduct of the scattering process under study and cannot be avoided. Measurements of the production cross section of the short-lived particle decays that produce secondary tracks are actually of interest to fully characterize the scattering processes under study. Tertiary and ghost tracks, however, are detector artifacts and thus ought to be minimized by careful design of the tracking system and the software used in the reconstruction of the data. This calls for detection systems that reduce the amount of materials that can cause tertiary tracks. It also requires the number of tracking planes be optimized to reduce ambiguities in the extrapolation and extension of tracks. Also note that minimization of materials reduces **Multiple Coulomb Scattering** (MCS), which otherwise limits the achievable momentum resolution (see discussion that follows).

Momentum Resolution

The magnetic fields used in magnetic spectrometers are typically quite stable and measured with very high precision. The momentum resolution achieved in measurements is thus almost entirely determined by tracking resolution and more specifically by the resolution δC of the track curvature C defined as the multiplicative inverse of the radius R. Fluctuations of the curvature are driven by process noise consisting predominantly of MCS and measurement noise associated with the spatial or hit resolution (RES) of measuring devices (discussed in §8.3.3)[3]:

$$\delta C^2 = \delta C^2_{\text{res}} + \delta C^2_{\text{MCS}}. \tag{8.103}$$

Let us first examine the term δC_{res} dependent on position resolution. At a minimum, three position measurements are required to estimate the radius of a circular trajectory, as illustrated in Figure 8.16a. The center of the circle is readily obtained by finding the intersection of segment bisectors S_1 and S_2. Let us define midpoints $\vec{r}_A = \vec{r}_1 + \Delta\vec{r}_{12}/2$ and $\vec{r}_B = \vec{r}_2 + \Delta\vec{r}_{23}/2$, with $\Delta\vec{r}_{12} = \vec{r}_2 - \vec{r}_1$ and $\Delta\vec{r}_{23} = \vec{r}_3 - \vec{r}_2$. The bisectors may be written $\Delta\vec{r}_{AC} = \vec{r}_A - \vec{r}_C$ and $\Delta\vec{r}_{BC} = \vec{r}_B - \vec{r}_C$, with $\vec{r}_C$ defining the position of the center of the

[2] Note that background tracks are often referred to as secondary tracks also.

[3] In a wider context, C_{res} is known as measurement noise while C_{MCS} corresponds to process noise. These concepts are discussed in §12.1.

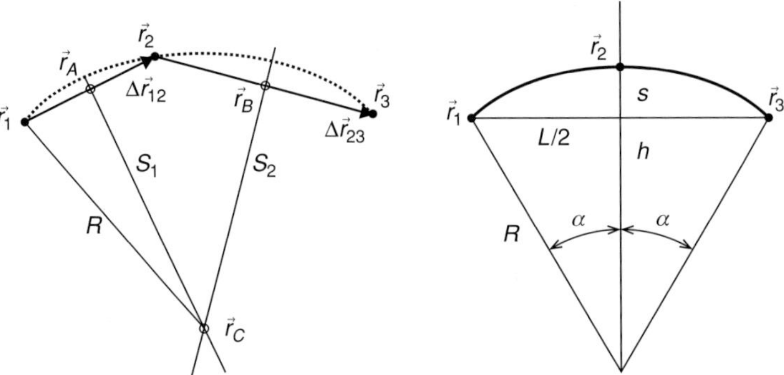

Fig. 8.16 (a) Determination of a track center and *f* radius. (b) Definition of the track sagitta *s*.

circle. By definition of the bisectors, one has

$$\Delta\vec{r}_{12} \cdot \Delta\vec{r}_{AC} = 0, \tag{8.104}$$

$$\Delta\vec{r}_{23} \cdot \Delta\vec{r}_{BC} = 0, \tag{8.105}$$

which reduces to

$$\Delta\vec{r}_{12} \cdot \vec{r}_C = \frac{1}{2}\left(r_1^2 - r_2^2\right), \tag{8.106}$$

$$\Delta\vec{r}_{23} \cdot \vec{r}_C = \frac{1}{2}\left(r_2^2 - r_3^2\right). \tag{8.107}$$

These two equations can be solved (see Problem 8.11) for the components $\vec{r}_C = (x_c, y_c)$. The radius R of the circle is then

$$R = \sqrt{(\vec{r}_1 - \vec{r}_C) \cdot (\vec{r}_1 - \vec{r}_C)}. \tag{8.108}$$

In order to estimate the curvature resolution based on the position resolution, consider the diagram shown in Figure 8.16b, which defines the **track sagitta** $s = x_2 - (x_1 + x_3)/2$ in terms of three position measurements. For a track length L much smaller than the radius, $L << R$, one has

$$s = R(1 - \cos\alpha) \approx R\frac{\alpha^2}{2}. \tag{8.109}$$

Introducing the momentum of the track transverse to the field, and noting $L \approx R\alpha$, one obtains

$$s \approx \frac{R^2\alpha^2}{2R} = \frac{qBL^2}{8p_T}. \tag{8.110}$$

The transverse momentum can then be expressed in the terms of the length of the track L and its sagitta s:

$$p_T = \frac{qBL^2}{8s}. \tag{8.111}$$

Assuming all three measurements have the same position resolution σ_x, and with $L >> s$, the momentum resolution δp is thus entirely determined by the resolution on the sagitta.

For a measurement involving three hits, each with a position resolution σ_x, one gets $\sigma_s = \sqrt{3/2}\sigma_x$. The relative momentum resolution is

$$\frac{\delta p_T}{p_T} = \frac{\sigma_s}{s} = \sqrt{96}\sigma_x \frac{p_T}{qBL^2}. \tag{8.112}$$

The momentum resolution can of course be improved by carrying out N uniformly distributed measurements of the track trajectory. Assuming all measurements have the same resolution σ_x, one finds [150]

$$\frac{\delta p_T}{p_T} = \frac{\sigma_s}{s} = \sqrt{\frac{720}{N+4}}\sigma_x \frac{p_T}{qBL^2}. \tag{8.113}$$

One thus concludes that the relative momentum uncertainty $\delta p/p$ is proportional to the position resolution σ_x and grows linearly with the particle momentum, inversely as the strength of the field B and inversely as the square of the measured track length. There is thus an interest in carrying out momentum measurements with large magnetic fields and large spectrometer enabling large path lengths L determined with very high precision σ_x.

The momentum resolution is also affected by MCS, which amounts to a process noise. Its impact on the relative momentum uncertainty can be estimated as

$$\frac{\delta p_T}{p_T} = \frac{13.6\sqrt{X/X_0}}{qBL}, \tag{8.114}$$

where X/X_0 is the material thickness expressed in units of radiation lengths X_0 [35].

In practice, the usable field strength is limited by the smallest momentum one wishes to detect. Indeed, if the field is very high, low p_T tracks end up having very small radii and may not be measurable. The resolution σ_x is determined by the technologies employed. Drift chambers (wire) have resolutions in the range 50–200 μm, whereas Si pixel and Si strip detectors can achieve resolutions down to a few microns.

The momentum resolution of modern complex detectors is commonly estimated based on Monte Carlo simulations of track propagation and reconstruction involving a detailed description of all detector components and materials, as well as simulators of the sensor responses and resolution. It may also be studied using cosmic rays and the reconstruction of decaying particles with a very narrow (or very well known) decay width (§8.5).

Track Quality Characterization

Not all reconstructed tracks have the same quality. Indeed, various instrumental effects may degrade and smear the quality of tracks randomly. Fortunately, several parameters may be used to assess the quality of reconstructed tracks and reject tracks of lesser quality. Quality criteria common to most reconstruction techniques and experiments include the track χ^2, the number of hits on a track, and the track's distance of closest approach to the primary collision vertex (or secondary vertex in the case of secondary tracks). It is in general possible to carry out analyses with quality criteria that are tailored to the needs of a specific physics analysis.

The track χ^2 is defined as

$$\chi^2 = \sum_{i=1}^{n} \left(\vec{r}_i - \vec{f}_i(x_i|\hat{\theta})\right)^T \mathbf{V}_i^{-1} \left(\vec{r}_i - \vec{f}_i(x_i|\hat{\theta})\right), \tag{8.115}$$

where $\vec{r}_i = (y_i, z_i)$, $i = 1, \ldots, n$, are measured hit positions, $\mathbf{V}_i = \mathrm{Cov}[y_i, z_i]$, and $\vec{f}_i(x_i)$ is the position predicted by the model representing the track in terms of fit parameters $\hat{\theta}$. Fit parameters include the momentum (or track curvature) and angles of emission of the track. If the measurements y_i and z_i can be considered Gaussian deviates and yield a well-calibrated covariance matrix $\mathbf{V}$, the χ^2 values obtained for genuine tracks can be characterized by a χ^2 distribution (see §3.13). It is then possible to use the track χ^2 as a reliable and predictable quality criterion to identify and eliminate bad tracks. Indeed, one can select a maximum $\chi^2_{\max}$ value and reject all tracks that exceed this criterion with a well-defined selection criterion efficiency given by

$$\varepsilon = \int_0^{\chi^2_{\max}} f_{\chi^2}(z; n)\, dz, \tag{8.116}$$

where $f_{\chi^2}(z; n)$ is the χ^2 distribution for n degrees of freedom given by Eq. (3.163). A track χ^2 thus nominally constitutes a robust, reliable, and well-defined quality selection criterion. In practice, however, the determination of hit errors is often met with considerable challenges, and a proper and precise assessment of the size and correlations of the errors may not be possible. Although a χ^2 may still be used to reject tracks, the efficiency (or losses) associated with a particular maximum χ^2 value may no longer be obtained from Eq. (8.116) but must instead be determined by means of Monte Carlo or embedding techniques (see §12.4.6).

The number of found hits associated with a track, n_F, also provides powerful control of the quality of tracks, particularly when compared to the total number of "possible" hits, $n_{\max}$, that could be on the track given its specific geometry, that is, the number of active sensors (detection planes) physically traversed by the track. Given a particular sensor technology, one expects there is a finite probability $\varepsilon_{\mathrm{Hit}}$ for finding hits produced by a particular track. For illustrative purposes, let us assume this efficiency is identical for all tracking planes and independent of kinematical track parameters. Let us further assume that the probability of finding a track hit on a particular plane is independent of the probability of finding one on other planes. The probability of finding n_F hits on a track with a possible maximum of $n_{\max}$ may then be modeled with a binomial distribution (§3.1):

$$P_B(n_F|n_{\max}, \varepsilon_{\mathrm{Hit}}) = \frac{n_{\max}!}{n_F!(n_{\max} - n_F)!} (\varepsilon_{\mathrm{Hit}})^{n_F} (1 - \varepsilon_{\mathrm{Hit}})^{n_{\max} - n_F}. \tag{8.117}$$

This is illustrated in Figure 8.17 for selected values of efficiencies $\varepsilon_{\mathrm{Hit}}$ and a maximum value $n_{\max} = 150$.

Because the number of hits actually detected on a track affects the momentum resolution, it is desirable to impose a minimum number of hits, $N_{\min}$, requirement on reconstructed tracks. One must, however, realize that for fixed hit efficiency and number of detector layers, the number of hits detected will then influence the track reconstruction efficiency,

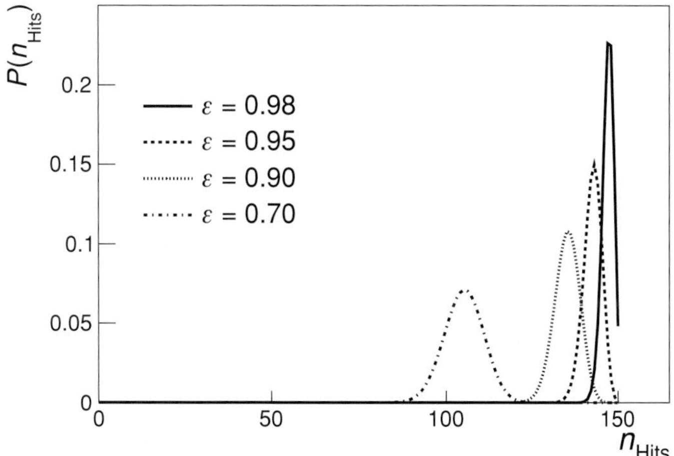

Fig. 8.17 Probability distribution of the number of hits, n_{Hits}, associated with reconstructed tracks modeled according to a binomial distribution with selected hit-finding efficiencies ε, and for a maximum number of hits $n_{\max} = 150$.

which amounts at best[4] to

$$\varepsilon_{\text{Track}}(\varepsilon_{\text{Hit}}) = \sum_{n_F=n_{\min}}^{n_{\max}} P_B(n_F|n_{\max}, \varepsilon_{\text{Hit}}), \tag{8.118}$$

and is shown in Figure 8.18 as a function of the hit-finding efficiency ε_{Hit} for selected values of $N_{\min}$. Since short tracks are expected to yield inferior momentum resolution and wider distance of closest approach (DCA) distributions, and are also more likely to consists of faked tracks, it is legitimate to require tracks to have a specific minimum number of hits. Equation (8.118) nominally determines the efficiency of the selection criterion (requirement). However, several additional factors may influence the likelihood of adding a found hit to a track. The hit-finding efficiency used in the preceding formula should thus obviously include all these effects.

A selection criterion on the number of hits is particularly useful to eliminate **split tracks**, that is, charged particle trajectories that for one reason or another are reconstructed as two or more track segments rather than a single track. Inclusion of split tracks in measurements leads to overestimation of the particle production cross section as well as spurious features in correlation functions. Fortunately, one can greatly reduce the double counting associated with split tracks by requiring that tracks accepted in an analysis feature a ratio $n_F/n_{\max}$ larger than 50%. Indeed, requiring a minimum length of 50% ensures the longest split track segment is utilized while shorter segment(s) are eliminated. One can optionally use a higher ratio requirement (e.g., 60%) if there is a concern that spurious hits may randomly be added to split tracks.

[4] This expression accounts only for hit-finding efficiency; other effects may reduce the track-finding efficiency also.

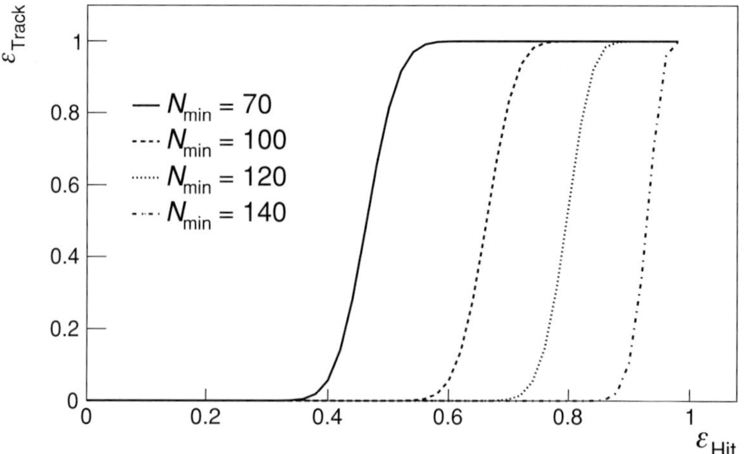

Fig. 8.18 Track-finding efficiency as a function of the hit-finding efficiency for selected values of the minimum of hits requirement N_{min}, assuming a binomial hit-finding distribution and $n_{max} = 150$.

In practical analyses, it is often the case that the hit-finding efficiency may vary from one detection plane/sensor to the next. It may also depend on the track parameters, and more specifically on the energy deposition of the track determined by the momentum of the track and the angle at which it crosses a specific detection plane. It is also the case that various detector components may malfunction intermittently or become completely nonoperational. An exact calculation of the probability of the track-finding efficiency as determined by hit-finding efficiencies may thus become impractical. Experimenters commonly resort to estimations of track-finding efficiency based on Monte Carlo simulation of the detector performance or embedding techniques (§12.4.6).

The DCA to the collision primary vertex, defined in Figure 8.19, is yet another track parameter routinely used to assess the quality of tracks. It may be expressed as a single quantity or broken into two components, one along the beam axis, and one in the plane transverse to the beam axis. Track DCAs are particularly useful to distinguish primary, secondary, and background tracks. Indeed, secondary and tertiary tracks are produced either by decays or interactions throughout the detector. They are thus unlikely to project

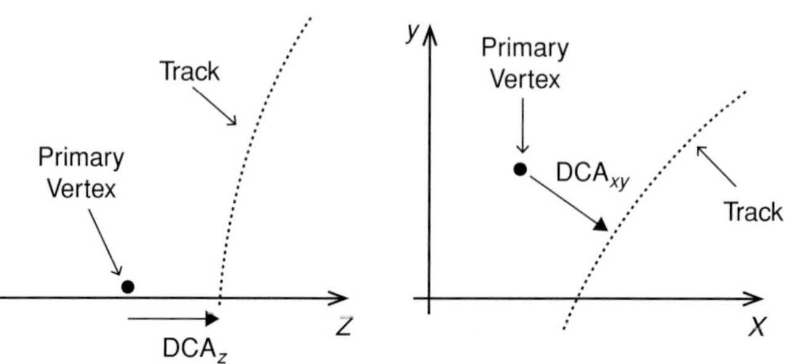

Fig. 8.19 Definition of the distance of closest approach (DCA) to the primary vertex.

back to the primary vertex. A measurement of DCA thus nominally provides an indicator of the origin of the track: secondary or tertiary tracks are expected to have large DCA, while primary tracks should have null DCA.

In practice, finite resolution of hit reconstruction and MCS lead to finite resolution in momentum and angle determination. Primary tracks projected to the main vertex thus have a finite DCA also. The width of the primary track DCA distribution depends on the vectoring resolution and is typically a function of the track momentum and rapidity. It is of course determined largely by the position resolution of the tracking detectors used to identify the tracks. The discrimination of secondary tracks is complicated by the fact that short-lived particles decay stochastically on a time scale determined by their mean life τ. For instance, kaon-short (K_s^0) have a mean life of $(8.954 \pm 0.004) \times 10^{-11}$ s and travel a mean distance $c\tau = 2.69$ cm at the speed of light before decaying. The probability of decaying on a shorter distance is, however, finite. In practice, this means that the secondary particles produced by a decay may look as if they were produced very near the primary vertex. Since primary tracks have a finite width DCA distribution, it becomes strictly impossible, for small DCA values, to distinguish whether a particle is a true primary or actually a secondary. One must then carry out a statistical analysis to determine the likelihood of secondary particles being reconstructed and identified as primary particles (and reciprocally).

Short-lived particles have typically a much smaller production cross section, and secondary tracks thus constitute a small background to primary tracks. It is in fact often sufficient to eliminate secondary tracks on the basis of DCA selection criteria in the longitudinal and transverse directions. The use of specific maximum DCA selection criteria obviously implies losses of primary particles (efficiency) and contamination by secondary particles. One expects the DCA resolution to be determined by MCS effects as well as by the detector granularity and hit precision. The DCA width is thus typically largest at the lowest track momenta and decreases with increasing momentum. It eventually rises, however, when the track curvature is small relative to the position resolution of the detector. Understanding primary track losses and contamination by secondaries is thus typically a nontrivial task best accomplished on the basis of Monte Carlo simulations.

It is also worth pointing out that the track χ^2, the number of hits on a track, and the DCA are typically highly correlated because all three parameters depend on the momentum resolution. The efficiency associated with selection criteria on the χ^2, the number of hits n_F and the DCA thus cannot be represented by a product of three independent efficiencies. Monte Carlo and embedding techniques are thus best for estimating losses and contamination effects.

8.3.2 Calorimeters and Energy Measurements

Calorimeters are devices designed to measure the total energy of particles produced in elementary collisions. They involve complete absorption of particle energy by means of either hadronic or electromagnetic processes. As such, calorimeters tend to be rather bulky and massive. As for tracking devices, several types of calorimeters have been developed and tailored toward various kinds of applications. One generally distinguishes between electromagnetic and hadronic calorimeters.

Electromagnetic and Hadronic Calorimeters

Electromagnetic calorimeters are designed and optimized to measure the energy of leptons (i.e., photons and electrons) by means of electromagnetic interactions. For instance, high-energy electrons penetrating through matter lose energy by scattering and by emission of radiation in the form of photons called *bremsstrahlung radiation*. High-energy photons, in turn, interact through matter via photoelectric effect, Compton scattering, or pair production. These processes lead to the production of electron and photons that also interact with the medium bulk and thereby produce a large flurry of particles known as an **electromagnetic shower**.

Hadronic calorimeters are optimized for the detection of high-energy hadrons but are also generally sensitive to electromagnetic energy. They rely on hadronic interactions of hadrons (e.g., protons, neutrons, pions, etc.) within the bulk material. These interactions produce large numbers of charged and neutral pions. Neutral pions decay and initiate electromagnetic showers, while charged pions have further hadronic interactions that develop into a mixed shower of hadrons and leptons.

Calorimeters are commonly characterized in terms of the e/h ratio, which measures their relative response to leptons and hadrons. A calorimeter is said to be perfectly **compensated** if it features a ratio $e/h = 1$, which means it is equally sensitive (with same gain) to both hadrons and leptons.

Calorimeters are built either with a single and homogeneous material or a combination of absorbing and sensing (shower sampling) materials. **Homogeneous calorimeters** rely on the same material for both the initiation of showers and the production of signals. They usually consist of inorganic crystals (e.g., lead-glass) transparent to light and are read out either by photomultiplier tubes or photodiodes. As such they provide a poor response to hadron and are thus used as electromagnetic calorimeters exclusively.

Sampling calorimeters are typically constructed with alternating layers of absorbing and sampling materials. Absorber materials commonly used include high-density, large atomic number (Z) materials such as lead, tungsten, or uranium. The sampling material may be solid or liquid and chosen to either transmit light (e.g., plastic scintillators) or electric currents (e.g., liquid argon, liquid krypton).

Calorimeters may also be segmented into cells and modules both longitudinally and transversely. Not too common, longitudinal segmentation enables measurements of the longitudinal shape and depth of showers. It may then be used to discriminate hadronic from electromagnetic showers, or high-energy electrons from photons. Transverse segmentation, on the other hand, is used almost universally because it enables measurements of the position (direction) of produced particles based on energy sharing between calorimeter cells.

Electromagnetic showers have a shape and depth that fluctuates within comparatively narrow limits and feature an overall scale determined by the radiation length of the materials composing the calorimeter. Hadronic interactions have smaller cross sections than electromagnetic processes. Hadronic showers thus exhibit larger shower shape fluctuations both transversely and longitudinally. They typically start at a depth determined by the interaction length of the material but their size do not have a simple dependence on the interaction length and are partly determined by the radiation length of the material.

Although calorimeters do not typically contain or stop muons, they are nonetheless sensitive to energy deposition by these particles.

Bolometric Calorimeters

Bolometric calorimeters measure incoming energy fluxes based on the change in temperature ΔT they produce. Assuming the incoming flux is entirely converted into heat, the temperature rise amounts to the ratio of incoming energy by the heat capacity of the heat-sensing material C:

$$\Delta T = \frac{E}{C}. \tag{8.119}$$

Since the heat capacity scales as the cube of the temperature T^3, best sensitivities are achieved by cooling calorimeter materials at cryogenic temperatures.

Bolometric devices are used in measurements of cosmic microwaves and in searches of dark matter.

Basic Calorimeter Signal Characterization

Hadronic and electromagnetic shower developments are stochastic processes. This implies that the signals produced by calorimeters exhibit relatively large fluctuations. The relative energy resolution of a calorimeter, $\delta E/E$, is thus an intricate function of the material used in its construction, its geometry, as well as the readout process. It is commonly described in terms of the empirical formula

$$\frac{\delta E}{E} = a \oplus \frac{b}{\sqrt{E}} \oplus \frac{c}{E}, \tag{8.120}$$

where the constants a, b, and c are determined by physical characteristics of the calorimeter, including the type of materials used and geometry, the technique used to generate and read out a signal, the amplification of the signal, and so on. The term a arises from nonuniformities in the construction and response of different modules, nonlinearities, while the term inversely proportional to $\sqrt{E}$ originates from the stochastic nature of the shower process. Roughly speaking, this stems from the fact that the shower contains a number of particle n proportional to the energy of the particle that initiated it. Each particle deposits a small amount of energy with some variance. The variance of the energy of the whole shower thus decreases with $\sqrt{n}$ and is inversely proportional to $\sqrt{E}$. The third component results mainly from instrumental effects that are energy independent (e.g., electronic noise). Its relative contribution thus decreases inversely with the energy.

Electromagnetic and hadronic showers have very different responses and resolutions. The resolution of electromagnetic showers is intrinsically limited by the variance of the path length of charged particles within the shower. For homogeneous calorimeters, it has been shown to be

$$\frac{\delta E}{E} \propto \frac{0.005}{\sqrt{E\,[\text{GeV}]}}. \tag{8.121}$$

Fluctuations are quite a bit larger in sampling calorimeters, and one typically gets

$$\frac{\delta E}{E} \propto 0.04\sqrt{\frac{1000E_0}{E\ [\text{GeV}]}}, \tag{8.122}$$

where E_0 represents the energy loss of a single charged particle in one sampling layer. Energy resolutions below the percent level are routinely achieved with homogeneous devices, but percent resolution appears to be a lower limit for E&M sampling calorimeters [81].

The resolution of hadronic shower calorimeters is larger still owing to fluctuations in the fractional energy loss of shower particles. Hadronic showers involve several processes, some of which produce muons, electrons, positrons, and slow neutrons. Because these particles have very different rates of energy deposition in the material, there are considerable fluctuations in the development and energy deposition of hadronic showers with resolution of the order of

$$\frac{\delta E}{E} \propto \frac{0.25 - 0.45}{\sqrt{E\ [\text{GeV}]}}, \tag{8.123}$$

where the lower limit is achieved with perfect compensation for nuclear effects.

Hadronic showers tend to be spatially large, both transversely and longitudinally. The aforementioned performance is in fact possible only when the showers are fully contained within the active calorimeter volume.

Calorimeters are an essential component of high-energy physics detectors and have thus been the subject of considerable developments and performance studies [81, 194].

8.3.3 Measurements of Hit Positions

Measurements of particle hit position are an essential component of measurements of track momenta in magnetic spectrometers and particle detection with calorimeters. Hit position detectors are extremely varied in design, construction, and operation. At a basic level, the determination of positions relies on particle energy deposition in energy-sensing detector modules and components.

Charged Particle Energy Loss in Materials

Charged particles passing through matter interact with individual electrons and nuclei as well as the medium as a whole under appropriate conditions, leading to scattering, energy loss, Cherenkov radiation, and coherent bremsstrahlung. Scattering and energy loss dominate at intermediate energies: light projectiles (e.g., electrons) colliding with heavier target particles (e.g., nuclei) randomly deflect but lose relatively little energy unless they undergo inelastic collisions, whereas heavy projectiles colliding with lighter targets lose energy without being appreciably deflected. Repeated stochastic deflections of particles traversing a medium are predominantly caused by Multiple Coulomb Scattering (MCS). MCS degrades a spectrometer's capacity to determine the direction of a particle and thus introduces uncertainties in the determination of both the magnitude and direction of its momentum vector. Energy losses and fluctuations in energy loss degrade and smear the

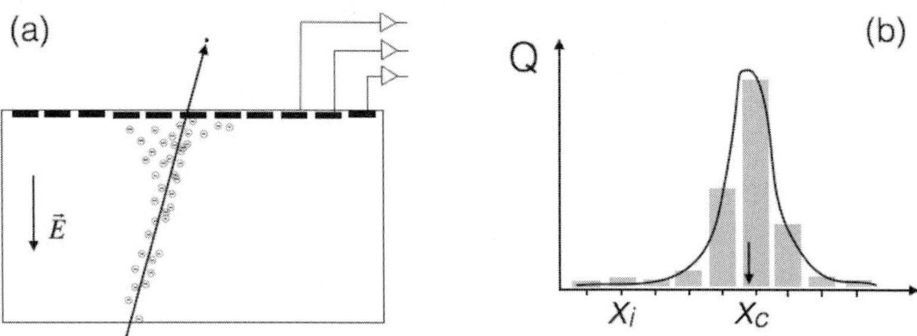

Fig. 8.20 Illustration of hit position determination based on charge sharing between contiguous energy deposition sensing components or cells.

momentum of particles and thus contribute to an additional loss of information and resolution. However, energy loss is a necessity in order to detect and analyze the trajectory of particles through detector layers, which enables tracking and momentum reconstruction. Additionally, because the energy loss of a particle depends on both its charge and momentum, measurements of energy deposition along a particle's trajectory make its identification possible (see §8.4).

Broadly speaking, position sensitive detectors can be grouped into two categories: those that sense the energy deposition but do not report its amplitude and those that do. This said, while hit sensing may be accomplished with either gas or solid state devices, basically all technologies rely on charge sharing between modules, illustrated in Figure 8.20, and discussed in the next section, to optimize the determination of hit position resolution.

A wide variety of hit-sensing technologies have been developed and are used in modern particle and nuclear physics experiments.

Characterization of the Performance of Hit-Sensing Detectors

Hit-sensing or position-sensitive devices that do not report the amount of energy deposited rely on a signal discriminator designed to "fire" and produce a logical pulse indicating the presence of a particle when the energy deposition in a cell (or module) exceeds a preset threshold. Detectors of this type do not require analog-to-digital converters (ADCs) and are thus considerably less expensive than more sophisticated devices that sense both the position and the amount of energy deposited. Their position resolution σ_x is, however, intrinsically limited by the width Δx of the sensing modules. Indeed, the variance of a uniform distribution (§3.4) across a finite width Δx determines the position resolution:

$$\sigma_x = (\text{Var}[x])^{1/2} = \frac{\Delta x}{\sqrt{12}}. \tag{8.124}$$

The position resolution achievable with such devices is thus strictly limited by the granularity (width) of the modules.

Better performance can be achieved with devices of similar segmentation that also explicitly measure the energy deposition or charge collected. Improvement in the position resolution relies on energy or charge sharing between contiguous detector cells, illustrated schematically in Figure 8.20.

In gas and solid state detectors, energy deposition leads to the creation of electron–hole pairs. The number of electrons (or holes) produced may be sizable, ranging from several hundreds to tens of thousands. The charge diffuses and ends up spreading across several anode segments with an approximate Gaussian profile. Denoting the cell center positions as x_i and the energy or charge deposited in a given cell as w_i, an estimate of the hit position may be obtained from the distribution centroid

$$\bar{x} = \frac{\sum_i w_i x_i}{\sum_i w_i}, \tag{8.125}$$

where the sums are taken on contiguous cells forming a **cluster** of charge (or energy) deposition.

Considering a particle entering the segmented readout device at position x_0, let us assume the energy deposition and charge diffusion processes lead to a Gaussian distribution centered at the position x and with a width σ_x. Let Δx represent the width of the segments and x_i the center position. The expectation value of the charge N_i measured in segment i is thus given by

$$p_i = N \int_{x_i-\Delta x/2}^{x_i+\Delta x/2} e^{\frac{(x-x_0)^2}{2\sigma^2}}\, dx. \tag{8.126}$$

For a fixed number of electrons N, the diffusion process leads to charges $\vec{n} = (n_1, n_2, \ldots, n_m)$ on the m segments, forming a cluster with $\sum_i n_i = N$. The probability of observing $\vec{n}$ charges is given by a multinomial distribution $P(\vec{n}|N, \vec{p})$. The weights are $w_i = n_i$. The expectation value of the position $\bar{x}$ is thus

$$E[\bar{x}] = \frac{1}{N}\sum_i x_i E[n_i] = \frac{1}{N}\sum_i x_i p_i = x_0, \tag{8.127}$$

and its variance given by (see Problem 8.12)

$$\delta\bar{x}^2 \equiv \text{Var}[\bar{x}] = \frac{1}{N}\sum_i x_i^2 p_i - x_0^2 = \frac{1}{N}\sigma_x^2, \tag{8.128}$$

where $\sigma_x^2 = \sum_i x_i^2 p_i - x_0^2$ is the variance of the charge distribution. We thus conclude that the resolution of the charge cluster centroid is

$$\delta\bar{x} = \frac{\sigma_x}{\sqrt{N}}. \tag{8.129}$$

The segment width Δx should be chosen to be of order of the width σ_x. The centroid resolution thus improves in proportion to the square root of the number of electrons N. This implies that the resolution can be far better than $\Delta x/\sqrt{12}$ for large values of N. In practice, however, the presence of electronic noise limits the achievable precision, and the position resolution may be expressed as

$$\delta\bar{x} = a \oplus \frac{\sigma_x}{\sqrt{N}} = a \oplus \frac{b}{\sqrt{\langle E\rangle}}, \tag{8.130}$$

where $\langle E \rangle$ represents the typical energy deposition in a segment and a and b are two constants determined by the level of noise and the average ionization energy of the medium.

Time Projection Chambers (TPCs) achieve position resolutions of the order of 1 mm, while gas wire chambers routinely achieve position resolutions of the order of 200–1000 μm. Pixel and strip silicon detectors achieve even better resolution still with precision down to 5 μm.

8.4 Particle Identification

Although particle production models are constrained a great deal by measurements of inclusive (unidentified) particle production cross sections and correlation functions, studies of flavor dependencies of the particle production requires measurements of the cross section and correlation functions of precisely identified particle species (e.g., $\pi^{\pm}$, $K^{\pm}$, p, $\bar{p}$, etc.). While positively charged particles are readily distinguishable from negatively charged particles on the basis of the sign of their track curvature in a magnetic spectrometer, discriminating among particle species is a far more difficult task. Fortunately, several techniques and associated detector technologies have been developed over the years to achieve reliable Particle Identification (PID) on a particle-by-particle basis and enable specific measurements of particle-identified spectra and correlation functions.

In this section, we focus our discussion on PID techniques based on differential energy loss (§8.4.1) and time-of-flight (§8.4.2), whereas particle detection by invariant mass reconstruction of two- and three-body decays is presented in §8.5.1. Introductions to PID techniques based on Cherenkov radiation and transition radiation may be found in specialized texts [33, 21]. Discrimination of electrons against hadrons based on E&M calorimeter measurements is also possible [165].

8.4.1 PID by Specific Energy Loss

The energy loss of particles traversing thin media may be estimated with the Bethe–Bloch equation [35] for the specific energy loss dE/dx:

$$-\left\langle \frac{dE}{dx} \right\rangle = Kz^2 \frac{Z}{A} \frac{1}{\beta^2} \left[\frac{1}{2} \ln \left(\frac{2m_e c^2 \gamma^2 \beta^2 T_{\max}}{I^2} \right) - \beta^2 - \frac{\delta}{2} \right], \tag{8.131}$$

where $K/A = 4\pi N_A r_e^2 m_e c^2 / A = 0.307075\ \text{MeVg}^{-1}\text{cm}^2$ for $A = 1\ \text{g mol}^{-1}$; Z and A are the (average) atomic number and atomic mass of the absorber material; $N_A = 6.0221415 \times 10^{23}\ \text{mol}^{-1}$ is the Avogadro number; $m_e c^2 = 0.510998$ MeV is the mass of the electron ($\times c^2$); z is the charge (in units of the proton charge) of the incoming particle; β is the ratio of the speed of the incoming particle to the speed of light; $\gamma = (1 - \beta^2)^{-1/2}$; I is the mean excitation of the absorber material (expressed in eV); and $\delta(\beta\gamma)$ is a density effect correction, which is typically small but becomes significant at very low and very

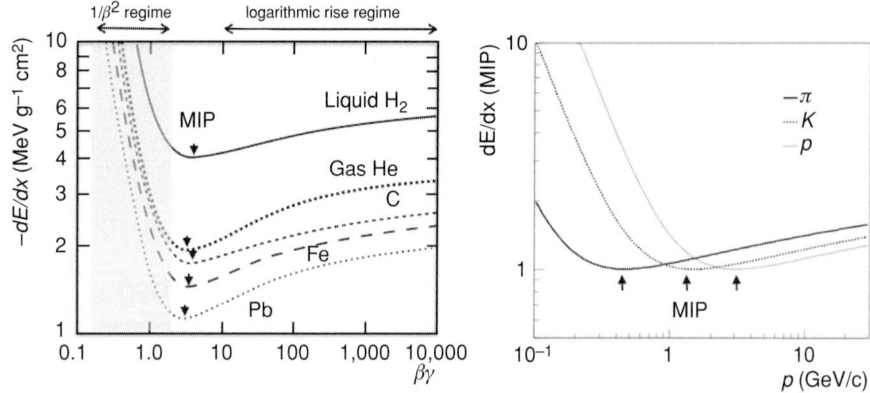

Fig. 8.21 (a) Specific energy loss of charged particles as a function of $\beta\gamma$ in selected materials. (b) Specific energy loss as function of particle momenta for pions, kaons, and protons. Arrows indicate minimum ionizing particle (MIP) values of $\beta\gamma$ and momentum. (Adapted with permission from J. Beringer et al. Review of particle physics. *Physical Review D*, 86:324–325, 2012.)

high momenta. The maximum energy $T_{\max}$ that can be transferred to an electron is given by [35]

$$T_{\max} = \frac{2m_e c^2 \beta^2 \gamma^2}{1 + 2\gamma m_e/M + (m_e/M)^2}, \tag{8.132}$$

where M is the mass of the incident particle. The mean excitation energy, I, is of the order of 10 ± 1 eV $\times Z$ for elements heavier than oxygen but it is recommended to use the well established values available in the scientific literature [35]. The specific energy loss predicted by the Bethe–Bloch equation is illustrated in Figure 8.21a as a function of $\beta\gamma$ for selected materials. It was verified experimentally that the formula provides estimates of the average energy losses that are accurate to within a few percents, for heavy particles with velocities in the range $0.1 \lesssim \beta\gamma \lesssim 1,000$ [35]. For values $\beta\gamma \lesssim 1$, the average energy loss is dominated by the first term of the Bethe–Bloch equation and thus features an approximate $1/\beta^2$ dependence on the particle's velocity. In this range, known at the $1/\beta^2$ energy loss regime, particles of different masses but equal speed have different momenta. It is thus possible to distinguish them on the basis of a plot of their energy loss vs. momentum, as illustrated in Figure 8.21b. The average energy loss reaches a minimum near $\beta\gamma \approx 3$, commonly referred to as the **minimum ionizing particle** (MIP). Distinguishing particles of same z but different masses in this range is, however, not possible based on their energy loss because they all feature essentially the same specific energy loss. For higher momenta, $\beta\gamma > 5$, the energy loss exhibits a slow rise with logarithmic dependence on $\beta\gamma$. Particle identification based on joint measurements of energy loss and momentum is thus possible in this **logarithmic rise regime**. The energy loss eventually saturates at very high momenta and the technique has therefore only a finite range of applicability.

Basic Principle

Identification by energy loss requires joint measurements of the momentum of a particle and energy loss through one or more energy loss sensitive devices. These two functions are often accomplished by distinct and dedicated detectors but can also be combined into a single device, such as a time projection chamber. The principle of the method is simple: measure the momentum and energy deposition of a particle in one or several detector layers, compute the average energy loss, and compare it to values expected for different species at the measured momentum. The species of the particle is that which most closely matches the measured average energy loss. This strategy works best in the low-momentum regime where the $1/\beta^2$ factor dominates the dependence of the energy loss and creates a large energy loss differential between particles of different masses. It also works in the logarithmic range regime. However, the technique is essentially ineffective for the MIP, that is, particles of intermediate momentum for which the $1/\beta^2$ saturates and the logarithmic term is not sufficiently large to enable a distinction of species.

Fluctuations and Straggling Functions

The identification of particle species based on their energy loss is complicated by two sources of noise commonly known as **process noise** and **measurement noise**. Process noise is associated with the physical process used to measure the physical observable at hand, whereas measurement noise refers to the noise present in the actual measurement, signal amplification, and readout. While the measurement noise can be minimized by careful design of the readout electronics, the particle energy loss and signal generation processes involve large event-by-event fluctuations that cannot be suppressed on an individual basis because they are largely determined by the physics of the collisions that engender energy losses.

Fluctuations in the energy deposition can be approximately modeled with the Landau distribution. Defining ξ as

$$\xi = 0.1535 \frac{z^2 Z}{A\beta^2} \rho(x), \tag{8.133}$$

the **Landau distribution** may be written

$$f_L(x, \xi) = \frac{\phi_L(\lambda)}{\xi}, \tag{8.134}$$

where

$$\phi_L(\lambda) = \frac{1}{2\pi j} \int_{r-j\infty}^{r+j\infty} \exp\left(u \ln(u) + \lambda u\right) du. \tag{8.135}$$

The variable r is an arbitrary constant and

$$\lambda = \frac{1}{\xi} \left(\Delta - \langle \Delta \rangle\right) - \beta^2 - \ln\left(\frac{\xi}{E_m}\right) - 1 + C_E, \tag{8.136}$$

with C_E the Euler constant. The function ϕ_L must be evaluated numerically, but is actually implemented in the ROOT package [59].

The Landau distributions shown in Figure 8.22a, calculated for three arbitrary values of mean and width, typify the large fluctuations in energy deposition that particles are subjected to as they pass through relatively thin detector layers. The large high side tail of the distribution, in particular, implies that the energy loss of low-mass particles (e.g., pions) may be mistakenly interpreted as being produced by heavier particles (e.g., kaons or protons). Since charged pion production typically dominates over kaon and proton production, the presence of a high side tail on the pion energy-loss distribution renders the identification of particles quite challenging, most particularly at the top of the $1/\beta^2$ range where energy-loss distributions produced by different pions, kaons, and protons considerably overlap.

The Landau distribution, derived from the Rutherford cross section, is suitable to illustrate the basic principle of particle identification by energy loss, but practical applications must account for a number of physical and instrumental effects. The energy lost ΔE by a particle in a detector volume is not necessarily entirely deposited within the sensing volume of the detector component. One must consider the conversion of the energy deposited D into ionization of the medium (whether gas or solid state). One must also consider the transport of electrons to the anode and the signal amplification of the proportional counter used to read out the signal. And finally, effects may arise in the digitization of the analog signal. Fluctuations in the energy loss and deposition have, however, been demonstrated to dominate the integrated charge Q measured in thin detector layers [39]. The measured distribution of energy Q are commonly referred to as **straggling functions** in the instrumentation literature. Bichsel has shown that finite corrections must be made to both the Bethe-Bloch formula and the Landau distribution to properly account for instrumental effects, particularly fluctuations in the energy deposition [39]. He published detailed parameterizations of the energy deposition and its fluctuations now commonly known as **Bichsel functions**. However, given the availability of computing functions for the calculation of Landau distribution (e.g., in ROOT), we illustrate the PID method based on the Bethe–Bloch and Landau distribution rather than the more complicated Bischel functions in the following sections.

Truncated Mean Technique

Particle identification may be accomplished based on a maximum-likelihood fit of the measured energy sample Q_i through the layers $i = 1, \ldots, m$ of energy-loss sensing devices. However, for the sake of simplicity, we here focus on a somewhat easier technique involving a truncated mean of the samples Q_i.

The standard deviation of the energy deposition in thin medium layers is typically rather large. However, by virtue of the central limit theorem (§2.10.4), the mean of a sample of m Q_i measurements should have a standard deviation of the order of $\sigma/\sqrt{m}$, where σ is the width of the parent distribution of energy loss of measured particles. Although the second moment of the Landau distribution is undefined (because the corresponding integral

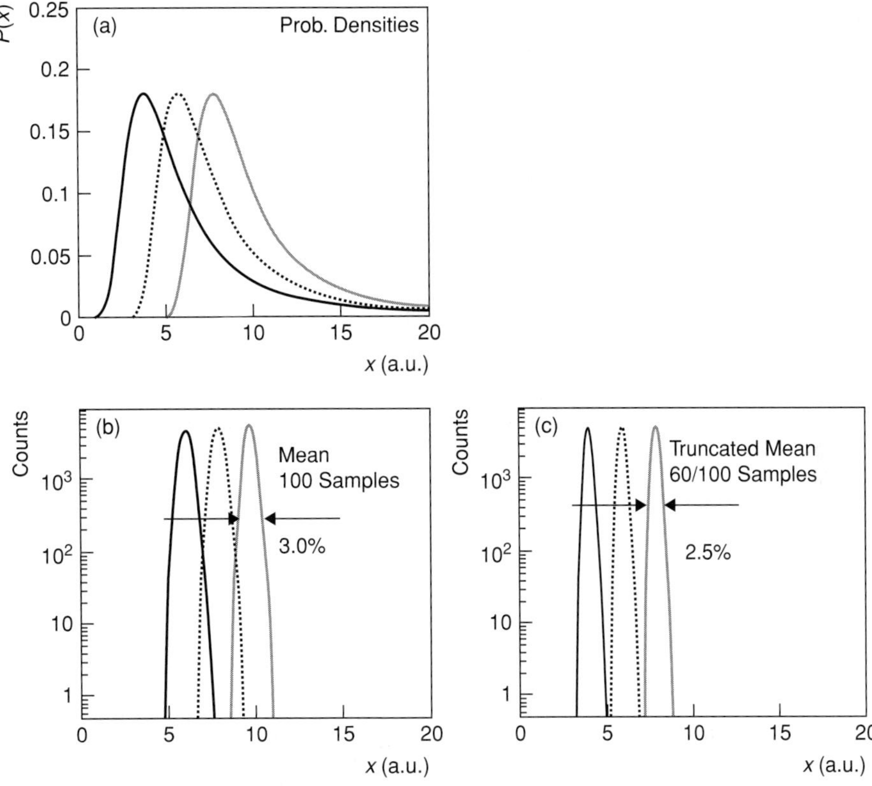

Fig. 8.22 Schematic illustration of the dE/dx truncated mean technique. (a) Probability densities of three particle species (with arbitrary means). (b) Distribution of the arithmetic means of 100,000 samples, each consisting of 100 values produced randomly with the distributions shown in (a). (c) Truncated mean distribution obtained with the 60 lowest values of each sample. Note that while the truncation introduces a bias on the value of the mean (i.e., a downward shift), it enables a sizable reduction of the dispersion of the mean which permits better separation of particle species shown in (c) relative to (b).

diverges), it is nonetheless possible to benefit from the $1/m$ reduction if a truncated mean of the sample values is used rather than the mean of the full sample. The principle of the truncated mean technique is simply to reject a suitably well chosen fraction f of the sampled values with the largest measured energy-loss values. Algorithmically, this is easily achieved by first sorting the n values of a sample of energy-loss measurements. One can then identify and eliminate the $f \times n$ largest values and calculate the arithmetic mean of the remaining values.

The rejection of the $f \times n$ largest values effectively truncates the high side tail of the Landau distribution, thereby producing a truncated average that features a nearly Gaussian distribution, as illustrated in Figure 8.22. The reduced fluctuations then yield dE/dx vs. p distributions that may provide good discrimination of species up to but excluding the MIP regime. The optimal fraction f is that which yields a truncated mean with the smallest standard deviation.

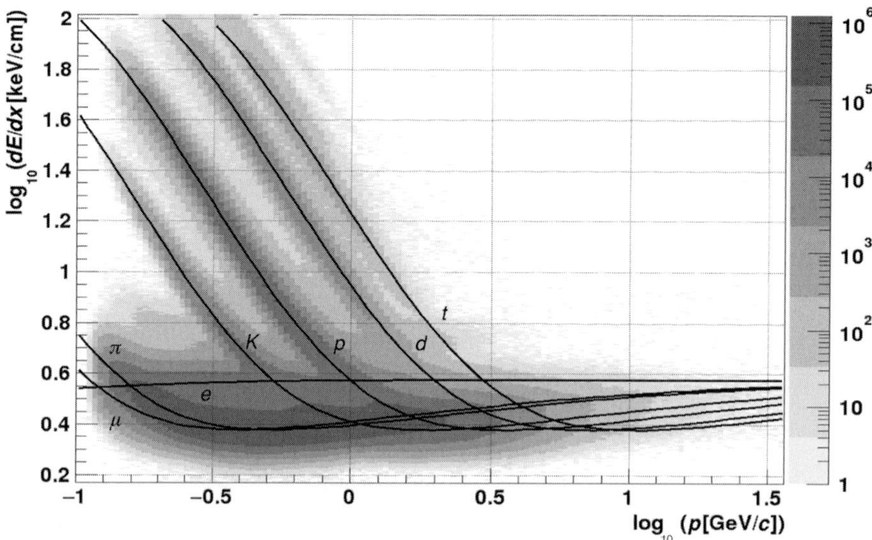

Fig. 8.23 dE/dx vs. p plot measured with the STAR Time Projection Chamber. One can readily distinguish bands corresponding to electrons, muons, charged pions and kaons, protons and antiprotons, deuterons and tritons. (Courtesy of Y. Fisyak and Z. Xu from the STAR Collaboration [83].)

The truncated mean technique is used in conjunction with several modern detectors. It is particularly useful for energy-loss measurements in gas detectors, such as time projection chambers, because fluctuations in energy deposition in such detectors can be rather large. As an example, Figure 8.23 displays an energy loss vs. p plot obtained with the STAR TPC, which exhibits excellent particle identification at low momenta [83].

Some Technical Considerations

It should be noted that the energy loss is a function of $\beta\gamma$: the measured truncated mean $\langle Q\rangle$ should thus be plotted as function of the momentum, not the transverse momentum. It is also important to account for the track geometry. The crossing angle of the track relative to the energy deposition sensing detector layer must be accounted for to properly calculate the energy loss per unit length within the sensing volume. For low-momentum tracks, it may also be necessary to account for the finite track curvature which may effectively increase the path length of a charged particle ΔL through the sensing volume, as illustrated in Figure 8.24.

8.4.2 Identification by Time-of-Flight

Basic Principle

The momentum of a particle is determined by its mass and velocity:

$$p = \gamma m\beta c, \tag{8.137}$$

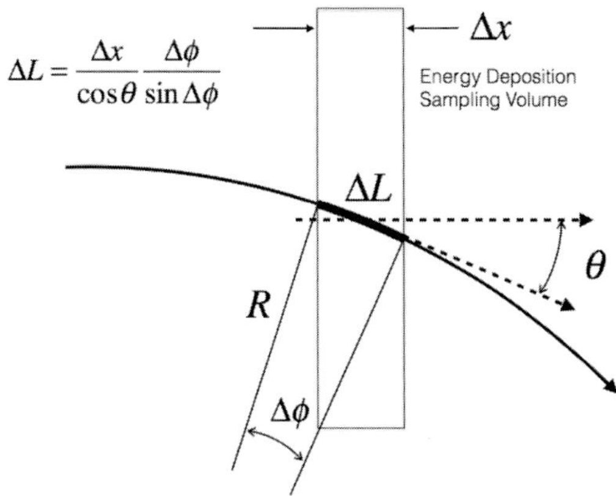

Fig. 8.24 For purposes of particle identification based on energy loss, the energy deposition measured in a detector volume must be scaled to account for the actual path length ΔL through the sensing volume of thickness Δx. The angle θ denotes the angle between the track direction and a normal to the detector plane. It may as such involve a component outside the bending plane. R and $\Delta\varphi$ represent the radius of the track and the angle subtended by the width of the module being traversed as measured from the center of the circular trajectory followed by the charged particle.

with $\beta = v/c$ and $\gamma = (1 - \beta^2)^{-1/2}$. One can then use simultaneous or joint measurements of the momentum and speed of a particle to obtain its mass:

$$m = \frac{p}{\beta\gamma c}. \tag{8.138}$$

Such simultaneous measurements could be used to determine the mass of "new" particles (e.g., search for ultra-heavy nuclei), but in modern particle physics experiments they mostly serve to identify particle species, that is, to determine whether a given particle is a pion, a kaon, a proton, and so on. We here focus on such measurements designed to identify the species of observed charged particles in a magnetic spectrometer.

The momentum (or transverse momentum) is readily obtained from the radius of curvature of the charged particle in the bend plane. One must also obtain a measurement of the particle's speed. The most straightforward technique to determine the speed of a particle relies on a time-of-flight (TOF) measurement along a well-determined trajectory or flight path d. The trajectory is obtained from tracking, while a TOF measurement requires at least two timing detectors: one providing a start time t_{start} and the second a stop t_{stop}. The TOF of the particle, Δt, is thus nominally given by the difference of the times

$$\Delta t = t_{\text{stop}} - t_{\text{start}}, \tag{8.139}$$

and the speed is estimated according to

$$\beta = \frac{d}{c\Delta t}. \tag{8.140}$$

In principle, one can then use Eq. (8.138) to determine the mass of particles. In practice, various instrumental effects limit the precision to which Δt may be determined and β may then exceed unity. It is then preferable to calculate the square of the mass of the particle according to

$$m^2 = p^2 \left(\frac{1}{\beta^2} - 1 \right). \tag{8.141}$$

Obviously, the technique assumes that negligible energy/momentum losses occur along the particle's flight path and thus that the particle's momentum and speed remain constant throughout the measurement. In practice, corrections may be required to account for such effects.

TOF Measurements

In fixed-target experiments based on a magnetic dipole spectrometer, TOF measurements can be readily accomplished with two or more **TOF hodoscopes** as schematically illustrated in Figure 8.25a. In collider experiments, the need to reduce the thickness of materials along the path of measured particles and cost considerations typically dictate a design based on a single barrel-shaped array of TOF counters providing stop times and some form of interaction counter providing a **common start** for all TOF measurements, as schematically shown in Figure 8.25b. The flight path d relevant for the speed measurement is determined based on the geometry of the particles' trajectory. If two or more TOF hodoscopes are used, the relevant distance could be the flight path between the start and stop counters. In a collider geometry, one would typically use the trajectory from the collision primary vertex and the TOF counter. This requires that the start time, which is derived from interaction counters, be adjusted for the vertex position.

TOF measurements are based on the time of passage of a charged particle through timing detectors. In high-energy experiments, particles of interest have large momenta and thus speeds amounting to significant fractions of or even very close to the speed of light. The speed of light amounts to (approximately) 3×10^8 m/s. That corresponds to 30 cm per nanosecond. Given that typical detectors have sizes ranging from a few meters to several tens of meters, TOFs are rather short, ranging from several to tens of nanoseconds. Fast and highly accurate timing sensors are thus required. Diverse technologies may be used to conduct fast timing measurements, including fast scintillator detectors read out with **photomultipliers** (PMTs) or fast **avalanche photodiodes** (APDs), as well as multigap resistive plate chambers (MRPCs). The choice of technology is driven by cost and practicality. We here focus on the concept rather than the technical details of such measurements.

The principle of scintillator and MRPC counters is illustrated in Figures 8.26 and 8.27. Charged particles passing through a thin slab of scintillator material produce molecular excitations, which result in rapid photon emission. For an appropriately chosen material thickness, the number of emitted photons can be rather large and thus usable to establish a fast timing pulse. Well-designed scintillation counters feature geometry enabling fast light collection with high photon yield, minimal timing dispersion, coupled with fast and high-precision light sensors. The leading edge of the signal may then be used to produce a logic signal corresponding to the time of passage of the particle through the detector, as illustrated in Figures 8.26 and 8.27. Similarly, charged particles traversing an MRPC produce

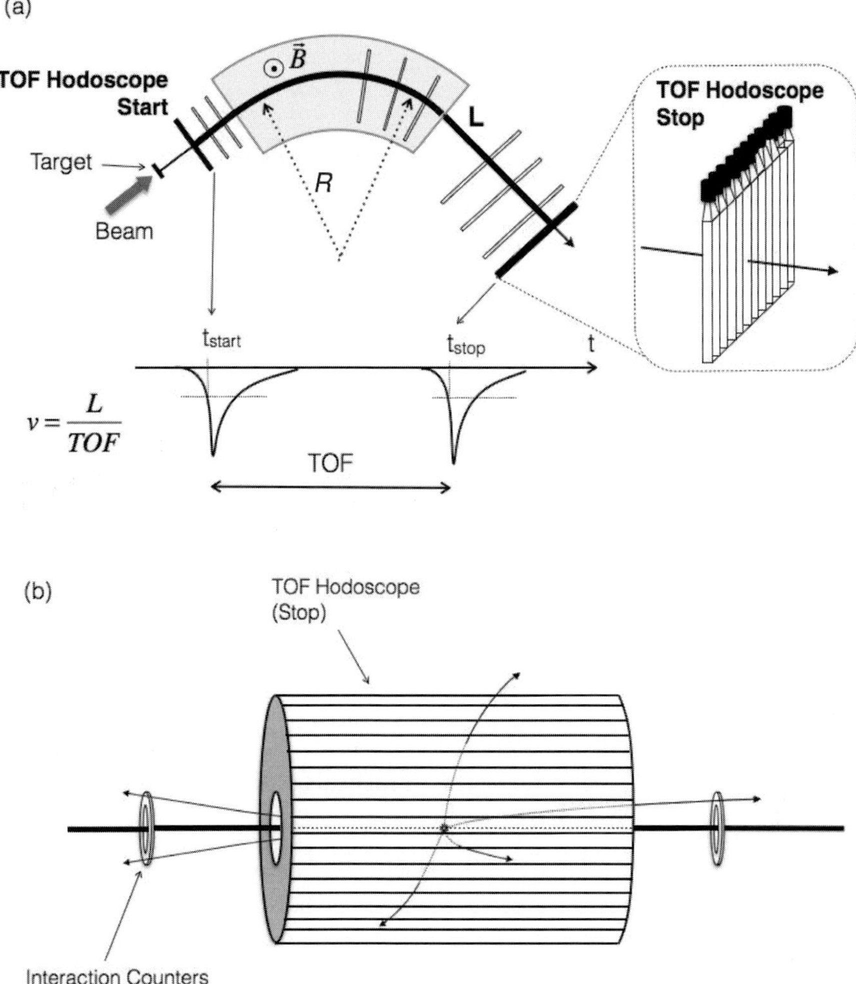

Fig. 8.25 Particle TOF hodoscopes in (a) fixed-target and (b) collider geometries may consist of sets of contiguous TOF scintillator modules read out by one or two light sensors (e.g., PMT, or APD). Alternatively, they may be based on multigap resistive plate chambers (MRPC). In collider geometry, interaction counters are commonly used to provide a TOF start, while barrel hodoscope counters provide a TOF stop, as schematically shown.

ionization within the gaps separating glass plates. The high voltage across gaps produces fast charge avalanches read out as image charge on segmented anodes. Low noise amplification of these signals produces fast rise time, which can also be utilized to produce logic signals fed into a time-to-digital converter (TDC) to obtain an actual timing measurement. A detailed discussion of TDC designs and technologies is beyond the scope of this text but may be found in several publications [99, 195].

TOF hodoscopes used in large particle and nuclear physics experiments typically involve from hundreds to thousands of distinct counters read out by fast electronics. They are

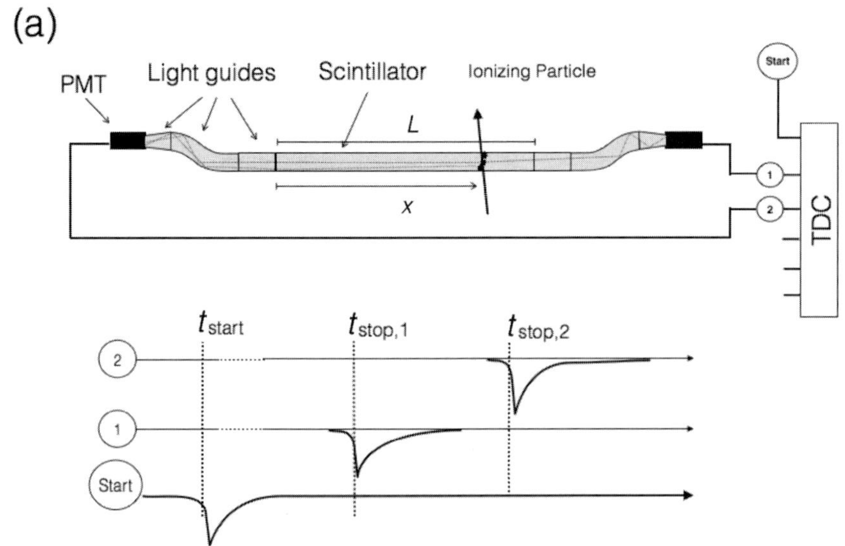

Fig. 8.26 Schematic illustration of the design of dual readout TOF scintillator counter modules.

usually equipped with TDCs featuring a digitization resolution of 25 ps and achieve TOF resolutions in the 50–100 ps range.

Practical Considerations

Signal collection, amplification, and transmission through cables and electronic components add an arbitrary but nominally fixed delay to the logic signals used to start and stop a TDC. In fact, delay cables often need to be added to electronics to produce start and stop logic signal inputs that fall within the counting range of TDCs. Changes in environmental conditions may, however, induce variations in signal propagation speed and must thus be accounted for in practice. The digitized times produced by TDCs then need to be calibrated to compensate for arbitrary and slowly varying offsets. This procedure, commonly known as t_0 calibration, is usually carried out based on a well-established reference, consisting either of identified particles at a precisely known momentum, or very high momentum particles that can effectively be assumed to travel at the speed of light. This calibration process must usually be carried out for each TOF counter individually and thus requires a fair amount of events. Also note that the calibration must be achieved with charged particle tracks acquired under normal running conditions (i.e., with the nominal magnetic field and trigger conditions).

Once the offsets are properly calibrated, the raw TOF is then obtained according to

$$\Delta t = t_{\text{stop}} - t_{\text{start}} + t_0. \tag{8.142}$$

Depending on the detector geometry and the type of readout, a number of additional corrections and calibrations may be required.

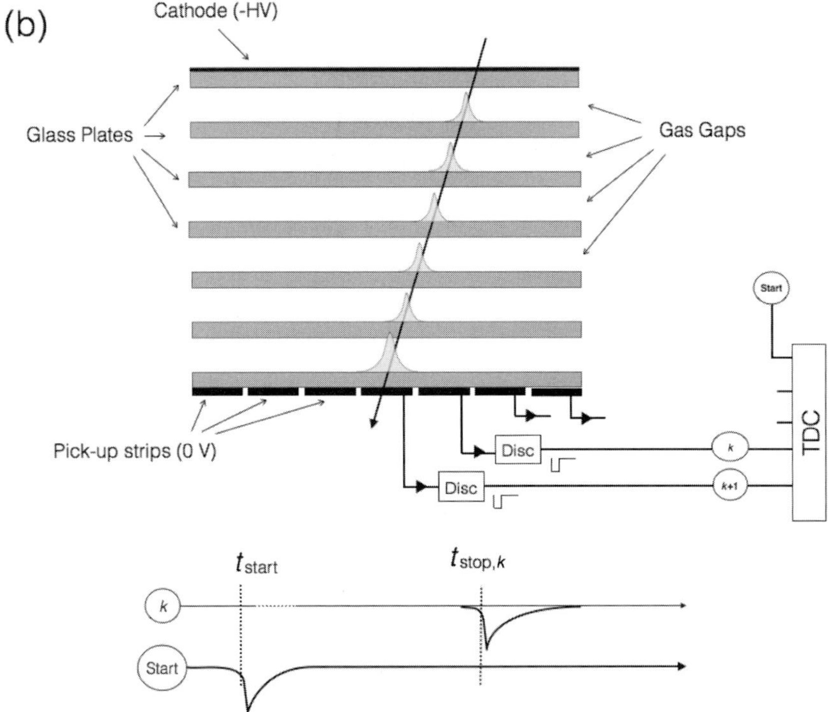

Fig. 8.27 Schematic illustration of the design of multigap resistive plate chambers (MRPCs).

In modern detectors, the generation of logic signals used by TDCs for TOF determination is achieved with **leading edge** discriminators whose principle of operation is illustrated in Figure 8.28. Leading edge discriminators use a voltage comparator with a fixed voltage threshold. The logic signals they produce thus exhibit a dependence on the amplitude of the incoming analog signal, known as "slewing": high-amplitude signals fire the discriminator early, whereas small-amplitude signals result in a later signal. Fluctuations in the energy deposition, charge (MRPC), or light (scintillators) production and collection result into timing fluctuations known as "jitter." While the slewing associated with pulse height dependence of the timing signal can be compensated by applying a correction based on the amplitude or leading edge slope of the signal, signal jitter is an intrinsic component of the signal and thus cannot be corrected for. Properly calibrated MRPC signals enable TOF resolution in the 50–100 ps range whereas scintillator counters achieve TOF resolution in the 70–100 ps range.

Particle Identification with TOF

Particle identification based on TOF and momentum measurements may be achieved by direct calculation of m^2 according to Eq. (8.141) or with selection criteria in the $1/\beta$ vs. p plane.

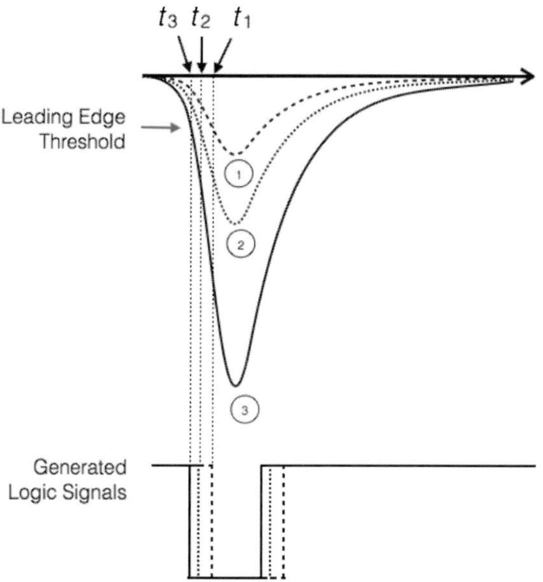

Fig. 8.28 Principle of leading edge discriminators used in the generation of timing signals in TOF measurements. A comparator issues a fast logic pulse when the input signal crosses a preset threshold. The use of a fixed threshold produces a delay, known as slewing, which depends on the signal amplitude. This undesirable effect is easily calibrated based on the signal amplitude and removed, in software, during the data analysis.

At a basic level, if the time-zero calibrations and corrections for hit position, light attenuation in scintillator counters, and amplitude dependence (slewing) are properly accounted for, Eq. (8.141) can be used to estimate m^2 and identify particle species. Evidently, the main technical challenge of such measurements is that high-momentum particles have high speeds whose accurate determination is largely determined by the TOF resolution and, to a lesser extent, by the momentum and path length resolution.

The mass resolution σ_m is simultaneously determined by the momentum resolution, σ_p, and the speed resolution, σ_β, which in turn depends on the TOF and path length resolution:

$$\sigma_m^2 \equiv \delta m^2 = \left(\frac{\partial m}{\partial p}\right)^2 \sigma_p^2 + \left(\frac{\partial m}{\partial \beta}\right)^2 \sigma_\beta^2 + \left(\frac{\partial m}{\partial p}\frac{\partial m}{\partial \beta}\right) \mathrm{Cov}[p, \beta], \tag{8.143}$$

where σ_p, σ_β, and $\mathrm{Cov}[p, \beta]$ are the momentum resolution, the speed resolution, and the covariance between the momentum and the speed, respectively. Using Eq. (8.138), one finds

$$\frac{\sigma_m^2}{m^2} = \left(\frac{\sigma_p}{p}\right)^2 + \gamma^4 \left(\frac{\sigma_\beta}{\beta}\right)^2 + \gamma^2 \left(\frac{\mathrm{Cov}[p, \beta]}{p\beta}\right). \tag{8.144}$$

Note that $\mathrm{Cov}[p, \beta]$ may be nonzero owing to the fact that the path length d, used in the calculation of β, depends on the momentum p. Such correlations are small, however, because the relative path length resolution is typically below 1%. The speed resolution is

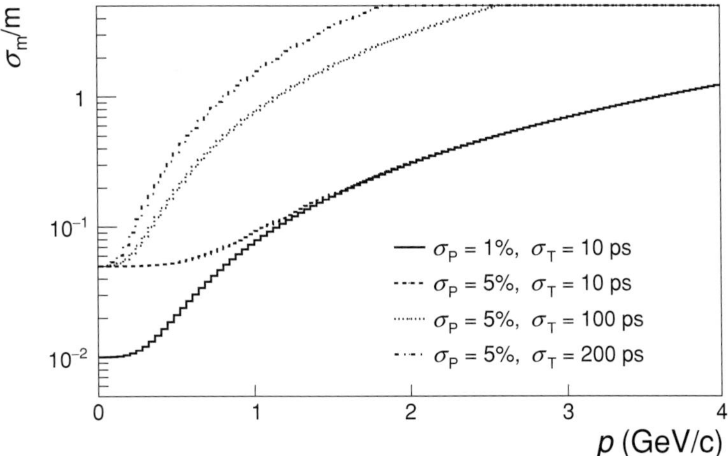

Fig. 8.29 Pion mass resolution obtained by TOF measurements as a function of momentum for selected resolution figures.

determined by the path length resolution and the TOF resolution

$$\frac{\sigma_\beta^2}{\beta^2} = \frac{\sigma_d^2}{d^2} + \frac{\sigma_T^2}{\Delta t^2}. \tag{8.145}$$

Overall, one finds, as illustrated in Figure 8.29, that the mass resolution is dominated by the momentum resolution for small values of β and $\gamma \approx 1$ and overwhelmingly driven by the (TOF) resolution for $\beta \approx 1$ and $\gamma \gg 1$.

Figure 8.30 displays the PDF of the mass of charged pions, kaons, and protons in the momentum range from 0.2 to 4 GeV/c for a flight path of 2 meters and selected momentum

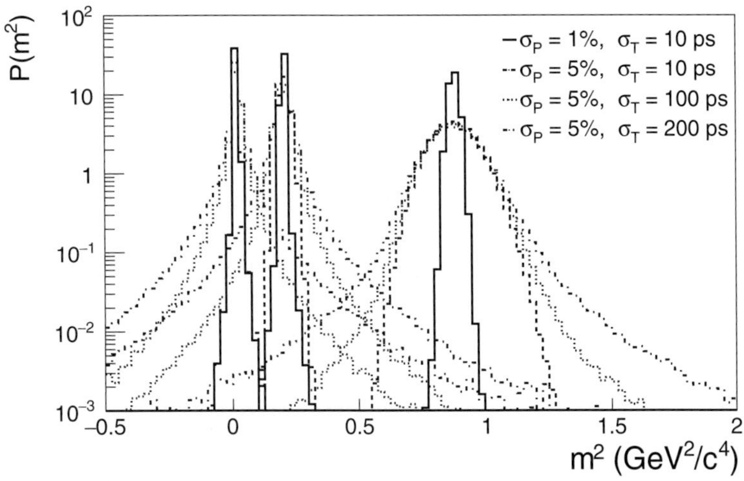

Fig. 8.30 Simulation of mass spectra obtained by joint measurements of momentum and TOF based on resolution figures shown for a 2-meter flight path.

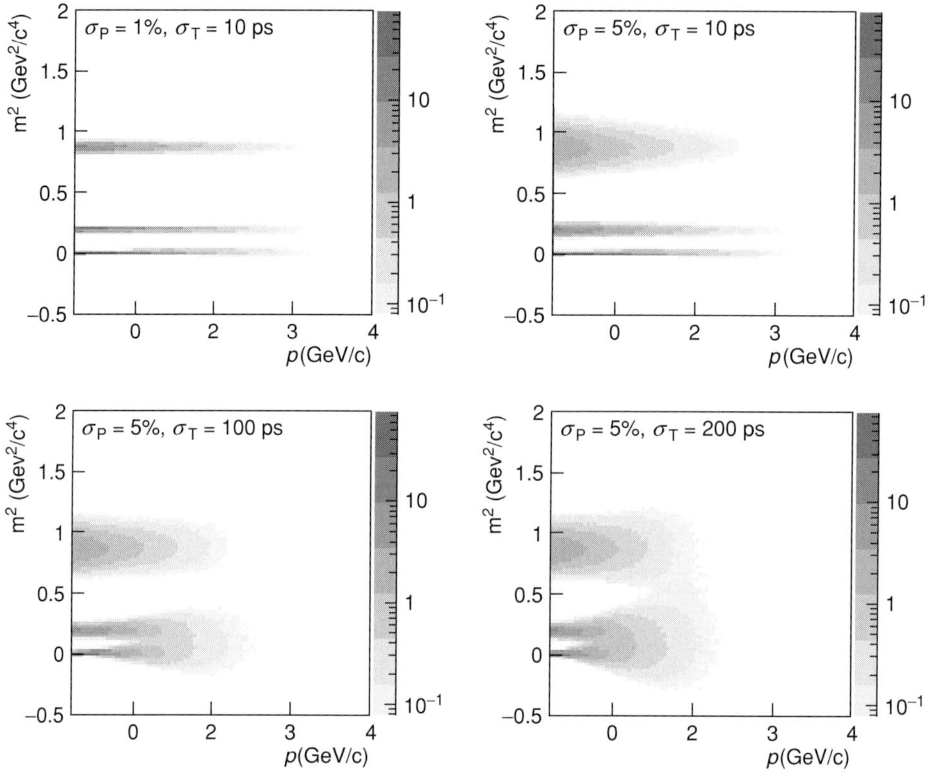

Fig. 8.31 Simulation of mass spectra vs. p obtained by joint measurements of momentum and TOF based on resolution figures shown.

and TOF resolution figures. The PDFs were generated by Monte Carlo simulations of the mass reconstruction assuming exponential spectra in momentum, with an average momentum of 0.5 GeV/c. The non-Gaussian profiles of the mass distribution stem from the momentum dependence of the mass resolution, as illustrated in Figure 8.31, which shows PDFs of the mass distribution as a function of the particle's momenta. One observes that at low momenta the mass resolution is indeed largely dominated by the momentum resolution, whereas at high momenta the saturation of speeds toward c implies that the finite TOF resolution leads to a significant broadening of the mass distributions.

Various biases in the determination of the particle's TOF may worsen the mass resolution embodied by Eqs. (8.143) and (8.145). It is then useful to plot the inverse of the measured speed, $1/\beta$, as a function of the momentum. In such a plot, particles of different mass populate different bands, as illustrated in Figure 8.32. It is thus possible to identify particle species on the basis of selection criteria in the $1/\beta$ vs. p plane. Losses of particles associated with these selection criteria or based on the reconstructed mass, Eq. (8.138), may be determined on the basis of realistic simulations of the detector response (see §14.2), as well as analyses of matched reconstructed tracks in the TOF counters.

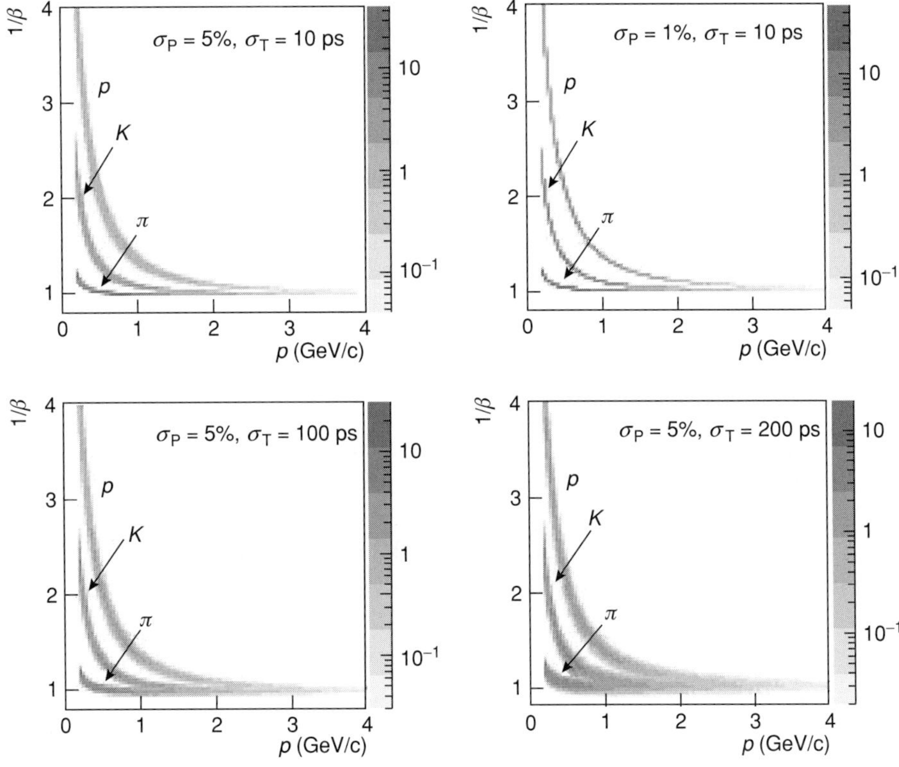

Fig. 8.32 PID based on plots of $1/\beta$ vs. p for selected momentum and TOF resolution figures.

8.4.3 Statistical PID and Semiexclusive Cross Sections

In studies of semiexclusive or species dependent cross sections, it seems natural to first utilize the information provided by energy-loss or TOF detectors to identify the species of particles, and subsequently proceed to increment relevant histograms as a function of the transverse momentum, rapidity, or azimuthal angle. One would then, in principle, obtain differential particle production cross sections that are specific to each particle type of interest (e.g., pions, kaons, protons, and so on). The problem arises, however, that unambiguous identification of particle species, particle-by-particle, is not always possible. As we saw in previous sections, unambiguous PID is difficult or impossible via dE/dx for particle momenta in the minimum ionization range or via TOF for particles with velocities $v \to c$. Measurements of semiexclusive (species dependent) cross sections as a function of transverse momentum (as well as rapidity and azimuthal angle) are nonetheless possible provided the dE/dx (or TOF) profiles of distinct species are sufficiently different and well known. Assuming, indeed, that the energy loss ΔE profile of each species k is known and can be described with a PDF $f_k(\Delta E|p)$, which is specific to each particle species being measured at a particular momentum p, one can then describe the inclusive energy loss profile according to

$$f(\Delta E|p) = \sum_{k=1}^{m} \mu_k(p) f_k(\Delta E|p), \tag{8.146}$$

where the parameters μ_k correspond to the relative contributions (or yield) of species $k = 1, \ldots, m$ of interest at the momentum p. So, rather than attempting identification on a particle-by-particle basis, it becomes possible to obtain cross sections by measuring a histogram of the energy loss as a function of the particle momenta. One then use the extended maximum likelihood method (§5.1.7) to carry out a fit to determine the coefficients μ_k, for each species, as a function of momentum (and other kinematic variables of interest). The coefficients $\mu_k(p)$ then effectively provide a measurement of the production yield of each of the species as a function of p or other kinematical variables of interest.

The technique is most straightforwardly applied to measurements of single particle cross section but can be extended to measurements of two- or multiparticle cross sections also.

8.5 Detection of Short-Lived Particles

While many particles produced in nuclear collisions are sufficiently long-lived to be measurable, for instance with a magnetic spectrometer, many particle species decay too rapidly to be observed directly. Their existence as collision products can nonetheless be identified on the basis of their decay particles. Two basic strategies are available to achieve their identification, each of which are discussed at a conceptual level in some details in the subsections that follow. The first strategy, known as the **invariant mass reconstruction** technique, relies on energy-momentum conservation and utilizes the measured momenta of decay products to calculate the mass of the decaying or **parent particle**. The second strategy relies on the **decay topology** and the presence of a **displaced vertex**. It is usable when the decaying particle is sufficiently long-lived to travel a measurable distance from its production location (primary vertex) before it decays (secondary vertex). This method is known as the **displaced vertex** technique and alternatively as the **decay topology** technique.

8.5.1 Invariant Mass Reconstruction Technique

Consider the two-body decay process illustrated in Figure 8.33. The parent particle has an unknown four-momentum p_p, and the decay products have 4-momenta p_1 and p_2. Energy-momentum conservation dictates that the sum of the 4-momenta of the decay products equals the four-momentum of the parent particle:

$$p_p = p_1 + p_2 = (E_1 + E_2, \vec{p}_1 + \vec{p}_2). \tag{8.147}$$

The square of the 4-momentum p_p yields the square of the mass, m_p, of the parent particle. A measurement of the momenta of the decay products and calculation of the square

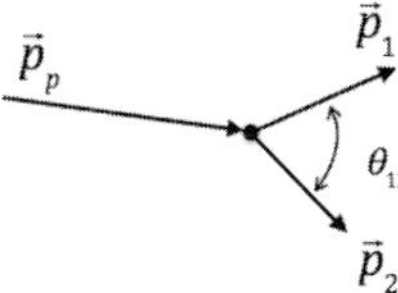

Fig. 8.33 Kinematics of two-body decays and invariant mass reconstruction.

of the sum of their 4-momenta thus enables identification of the parent particle species. If the PIDs of the decay products are known, one can indeed express these particles' energy in terms of the measured momenta and known masses:

$$E_i = \sqrt{|\vec{p}_i|^2 + m_i^2}. \tag{8.148}$$

However, if the PIDs are not known, one can formulate a specific hypothesis as to what their masses are and calculate energy estimates according to Eq. (8.148). The mass of the parent particle is thus given by the square of p_p:

$$\begin{aligned} m_p^2 &= E_1^2 + 2E_1E_2 + E_2^2 - |\vec{p}_1|^2 - |\vec{p}_2|^2 - 2\vec{p}_1 \cdot \vec{p}_2, \\ &= m_1^2 + m_2^2 + 2E_1E_2 - 2\vec{p}_1 \cdot \vec{p}_2, \end{aligned} \tag{8.149}$$

where in the second line we have used $m_i^2 = E_i^2 - |\vec{p}_i|^2$. Expressing the energies of the decay products in terms of their momenta, we get

$$m_p^2 = m_1^2 + m_2^2 + 2|\vec{p}_1||\vec{p}_2| \left(\sqrt{1 + \frac{m_1^2}{|\vec{p}_1|^2}} \sqrt{1 + \frac{m_2^2}{|\vec{p}_2|^2}} - \cos\theta_{12} \right),$$

where θ_{12} represents the angle between the momenta $\vec{p}_1$ and $\vec{p}_2$. For $|p_i| \gg m_i$, the preceding simplifies to

$$m_p^2 = m_1^2 + m_2^2 + 2|\vec{p}_1||\vec{p}_2| (1 - \cos\theta_{12}), \tag{8.150}$$

and the mass resolution is then typically limited by the momentum resolution, as illustrated in Figure 8.34, and to a lesser extent by the precision of the angle θ_{12}.

Combinatorial Background

In principle, the technique described in the previous section enables the identification of decaying particles of specific mass. In practice, nuclear collisions, particularly those of relativistic heavy ions, produce a large number of particles, and there is no way (other than the technique described in §8.5.3) to identify which particle pair was actually produced by a decaying particle.

Consider, for instance, an event consisting of n particles in which two (and only two) particles were produced by the decay of a short-lived particle. To find out which two particles might originate from the decay, one must examine all $n(n-1)/2$ particle pairs that can be formed by combining n particles, and calculate their invariant mass according to

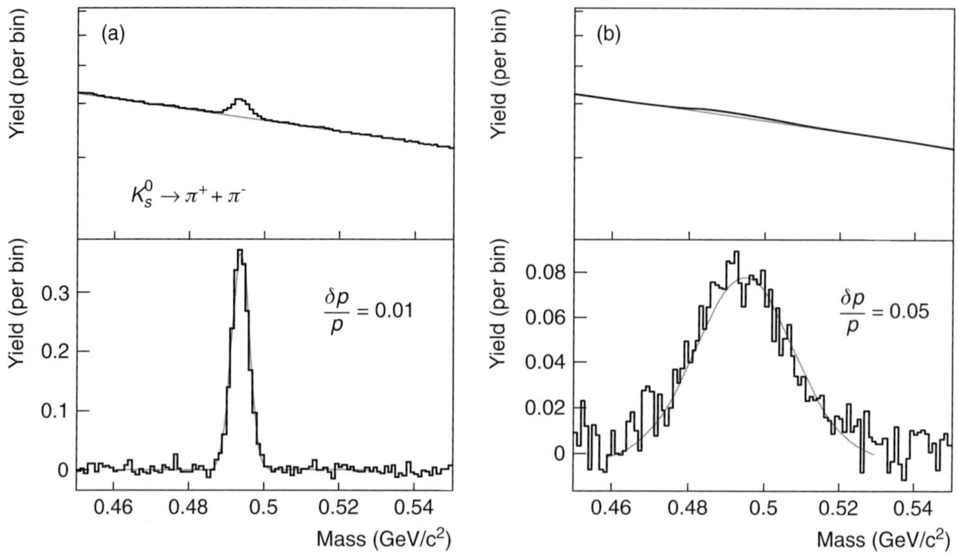

Fig. 8.34 (Top) Simulations of 100,000 "real events" (solid line) and "mixed events" (dotted lines) invariant mass distributions of $K_s^0 \to \pi^+ + \pi^-$ decays with finite momentum resolution, $\delta p/p = 0.01\%$ (left) and $\delta p/p = 0.05\%$ (right). Simulated events feature an average of two measurable kaons decays and 50 background pions per event. The mixed-event distribution is normalized to have equal yield per event as the real distribution at $m = 0.6$ GeV/c^2. (Bottom) net kaon mass spectrum determined from the subtraction of actual and mixed events distributions, where the solid line shows a Gaussian+constant background fit to the kaon mass spectrum.

Eq. (8.150). However, particles not produced by a decay have no specific relation: they can be separated by an arbitrary angle and have arbitrary momenta. Their invariant mass is not constrained and can then cover a very wide range of values. The $n(n-1)/2-1$ unrelated pairs thus constitute a **combinatorial background** to the one true correlated pair (by decay) that covers a wide range of masses, as illustrated in Figure 8.34. The shape and spread of the background are largely determined by the steepness of the single particle spectrum, the kinematical (p_T) range, and the angular range of the acceptance. They may also be influenced by resolution (smearing) effects associated with the reconstruction of the particles constituting the event. All in all, distinguishing one particular pair out of a group of $n(n-1)/2-1$ unrelated pairs consequently amounts to an impossible challenge if attempted for a single event. Fortunately, one is generally not interested in detecting a particular decaying particle but rather in measuring the production yield of such particles as a function of transverse momentum, production angle, and so on. It is then possible, at least in principle, to obtain a large sample of decays by examining a large number of distinct events. There remains the task of determining and subtracting the combinatorial background in order to determine the production cross section of the short-lived particles. We describe techniques to achieve this result in the next paragraph. We additionally describe a technique based on the decay topology, usable for decays with a displaced vertex, in §8.5.3.

Background Subtraction and Signal Determination

The determination of the decaying particle yield may be achieved by a variety of methods, including the bin summing method, the fit method, or the combinatorial background removal method, each discussed in the following.

Bin Summing Method

If the background can be properly subtracted (with either of the methods described in the text that follows), it may be sufficient to sum the yields observed in invariant mass bins surrounding the expected mass peak:

$$Y = \sum_{\text{peak}} y_i, \tag{8.151}$$

where y_i represent the yield in bins surrounding the mass peak after background subtraction. The statistical error, δY, on the integrated yield Y can then be estimated as the sum, in quadrature, of the errors in each bin included in the sum:

$$\delta Y = \sqrt{\sum_{\text{peak}} \delta y_i^2}. \tag{8.152}$$

Proper selection of the number of mass bins included in the sum is important and must reflect the actual (or expected) peak width, including smearing effects. On the one hand, if the number of bins included in the sum far exceeds the actual width of the signal peak, the yield may end up being biased by improper background subtraction and the yield uncertainty is likely to be overestimated. On the other hand, if the number of bins is too small, only a fraction of the signal can be integrated. This results in an efficiency loss that can be estimated, and thus corrected for, if the signal shape and its width can be properly estimated.

Fitting Method

The fitting method relies on a line fit of the signal and (combinatorial) background. In cases where the signal line shape is dominated by smearing effects, it is often possible to model the signal peak with a Gaussian distribution and the background with a polynomial, as illustrated in Figure 8.34:

$$f(m) = \sum_{n=0}^{n_{\max}} a_n m^n + \frac{Y}{\sqrt{2\pi}\sigma} \exp\left(-\frac{(m-m_o)^2}{2\sigma^2}\right), \tag{8.153}$$

where the coefficients a_n must be adjusted to fit the combinatorial background, while Y represents the yield of decaying particles, m_o is the centroid of the mass peak, and σ its width determined by the resolution of the mass measurement. If smearing effects are negligible, the peak line shape can generally be modeled with a Breit–Wigner distribution (Eq. 3.179), whereas if smearing effects are finite but not dominant, one must fold the Breit–Wigner line shape with the detector smearing function (typically a Gaussian):

$$f(m) = \sum_{n=0}^{n_{\max}} a_n m^n + \frac{Y}{\sqrt{2}\pi^{3/2}\sigma} \int e^{-\frac{(m-m')^2}{2\sigma^2}} \frac{\Gamma/2}{\Gamma^2/4 + (m' - m_o)^2} dm'. \tag{8.154}$$

The line shape may optionally have to be adapted to various kinematical conditions.

Given that invariant mass spectra are usually obtained with finite bin width histograms, fits should be carried out using integrals of the preceding functions across the width of the mass bins. An option to accomplish fits with such integrals is conveniently provided in the ROOT fitting package [59]. There are cases where multiple peaks may be present, and one must then model them accordingly using multiple peak functions.

Combinatorial Background Removal:

When the number of combinatoric pairs, $n(n-1)$, is very large, a direct fit to the spectral line shape may become impractical. This is the case, in particular, when the background shape is complicated and cannot be readily modeled with a simple function (as earlier). It is then necessary to model the background based on the data itself. This can be accomplished, for instance, by producing mass spectra obtained from mixed events (§12.4.5). Particles from mixed events are uncorrelated by construction and should thus not produce a mass peak. However, they should provide a reasonable model of the combinatorial background. Since the shape of the combinatorial background may depend on the shape of the particle spectrum, kinematical selection criteria, and so on, it is important to carry out subtraction based on mixed events that have similar characteristics (e.g., multiplicity, kinematical, and quality selection criteria) as the actual events.

Additionally, since the multiplicity of real and mixed events may differ, so will the number of pairs that can be formed in real and mixed events. It is thus necessary to normalize the amplitude of the mixed events "combinatorial" background to match the amplitude of the real background. This can be typically accomplished by arbitrarily scaling the mixed-event mass spectrum such that its yield at a mass m_s several standard deviations away from the peak of interest equals the yield of the real pair spectrum at the same mass m_s. If the production of mixed events is not possible, another technique involves randomly rotating tracks in azimuth or reshuffling the particles in pseudorapidity (i.e., randomly setting the rapidities but keeping the same p_T values), or both.

It is important to note that the mixed- and real-event combinatorial background shapes may differ substantially. The background subtraction shall consequently be imperfect and constitutes a source of systematic error. One must then typically utilize a parameterization of the background to fit and remove the remaining background, as discussed earlier. Differences may occur owing to a variety of physical and instrumental effects, including correlations between particles which are present in the final state of collisions because of (broad) resonance decays, cluster or string decays, jets, or correlated backgrounds. The latter may be generated in the apparatus by correlated electron–positron pairs produced by photon conversion, and possibly also event-to-event variations in the acceptance or efficiency of the detection system. Such effects can in principle be studied with Monte Carlo simulations.

Statistical and Systematic Errors on Yields

The statistical uncertainty on yields obtained with the bin summing method is given by Eq. (8.152). Estimates of the error on the yield obtained with the fit method can in principle be obtained directly from the fit if the fit includes both the signal and the background. If the fit does not include the background, one must add the uncertainty on the yield ($\delta Y = \sqrt{Y}$) and the uncertainty on the background. The background uncertainty may involve a global uncertainty and bin-to-bin uncertainties.

Systematic errors involved in the determination of short-lived particle yields are diverse and may include both physical backgrounds and instrumental effects. Physical backgrounds include broad underlying resonances and invariant mass structures caused by particle correlations in the final state; instrumental effects include uncertainties in the evaluation of the particle reconstruction efficiencies and acceptance, invariant mass spectrum structures caused by resolution effects or discretization of the kinematical parameters used in the invariant mass reconstruction (particularly angles), and instrumental sources of background such as electron–positron production and generation of tertiary particles within the detection apparatus. In practice, many of these sources of uncertainties become linked or correlated within the context of the background subtraction, and it may not be appropriate to quote them as distinct and independent sources added in quadrature.

Systematic uncertainties associated with the background subtraction may be estimated by generating several distinct but equally plausible backgrounds and signal line shapes. The mean and variance of the resulting yields may then be used to estimate the signal strength and systematic uncertainties associated in its determination. For instance, if fits of the background with, say, three different models (e.g., first-, second-, or third-order polynomials) and three different ranges of mass yield similarly acceptable χ^2 values, it is legitimate to carry out a measurement of the yield above background with either of these models and ranges. One can then use the mean of the yields obtained in each of these cases as the estimate of the production yield and obtain a systematic error based on their standard deviation. If the yields obtained with the different backgrounds and fit models are Gaussian distributed and can be considered uncorrelated, then their standard deviation can be considered as one σ systematic error with a confidence level of 68%.

8.5.2 Three-Body Decays and the Dalitz Plot

A three-body decay involves the disintegration of a particle of mass M into three lighter particles of mass m_i, $i = 1, 2, 3$, as illustrated in Figure 8.35a. Unlike two-body decays where the momenta of the decay products are completely specified by energy-momentum conservation, three-body decays involve additional degrees of freedom. Different values of momenta are indeed possible depending on the decay configuration as shown in Figure 8.35b. Dalitz [74, 75] realized that the angular momentum of the parent particle largely

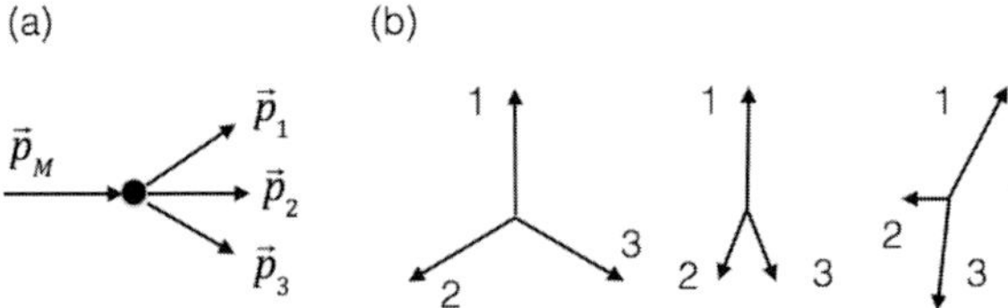

Fig. 8.35 (a) Kinematics and (b) topologies of three-body decays.

determines its decay configuration. Observations of the relative frequency of different configurations of decay products thus provides insight as to the spin of the decaying particle.

In a two-body decay, the four momenta of the decay products nominally amount to eight free parameters. However, energy-momentum conservation reduces this number to four, and the relations between energy and momentum, $E^2 = p^2 + m^2$, further reduces this number to only one free parameter, which can be taken as the invariant mass of the decay particles. In a three-body decay, there are three four momenta and thus twelve parameters. Momentum conservation reduces this number to eight, and the relation between energy-momentum further reduces the number to four parameters. In the rest frame of the parent particle, the three decay products are within a plane of arbitrary orientation. Given that it takes two angles to determine the orientation of this plane, the number of free parameters is further reduced to two units: a three-body decay has thus two degrees of freedom. Many options exist in the selection of two variables for the description of a three-body decay. Dalitz [74] originally used variables x and y, defined as

$$x = \sqrt{3}\frac{T_1 - T_2}{Q}, \tag{8.155}$$

$$y = \frac{2T_3 - T_1 - T_2}{Q}, \tag{8.156}$$

where T_i, $i = 1, 2, 3$ represent the kinetic energy of the three decay products and Q is the energy released by the decay:

$$Q = M - m_1 - m_2 - m_3. \tag{8.157}$$

However, it is now more common to analyze three-body decays in terms of the square of invariant masses m_{12} and m_{13}, defined as

$$m_{ij}^2 = p_{ij}^2 = \left(E_i + E_j\right)^2 - \left(\vec{p}_i + \vec{p}_j\right)^2. \tag{8.158}$$

One may then represent decays in a **Dalitz plot**, according to their position in the plane m_{23} and m_{12}, as schematically illustrated in Figure 8.36. In the rest frame of the parent particle, the invariant mass of a given pair, say m_{12}, has a minimum value $m_1 + m_2$ if particles 1 and 2 are produced at rest relative to each other, and it takes a maximum value $M - m_3$ if particle 3 is produced at rest relative to the parent. These and similar conditions for m_{23} define the boundaries shown as dashed lines in Figure 8.36. In between these extremes, a specific value of m_{12} limits the energy accessible to the pair 23 and thus defines minimal and maximal values shown by arrows in the figure.

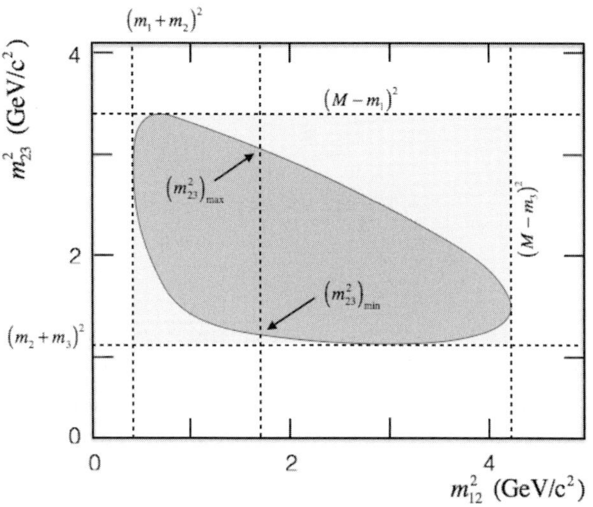

Fig. 8.36 Schematic illustration of a Dalitz plot used in the study of three-body decays. The dark gray area indicates kinematically allowed values while the dashed lines delimitating the light gray area define basic kinematic limits achieved when two daughter particles are emitted at rest relative to each other, or one produced at rest relative to the parent particle.

A Dalitz plot constitutes a **ternary plot**, that is, a plot involving three variables constrained by a condition. Choosing, for instance, m_{12}, m_{13}, and m_{23}, one readily finds (see Problem 8.9)

$$m_{12}^2 + m_{13}^2 + m_{23}^2 = M^2 + m_1^2 + m_2^2 + m_3^2. \tag{8.159}$$

One additionally finds

$$m_{12}^2 = (P - p_3)^2 = M^2 + m_3^2 - 2ME_3, \tag{8.160}$$

where E_3 is the energy of the third particle in the rest frame of the decaying particle.

In the absence of angular correlations (e.g., parent particle with $J = 0$), the distribution of the decay products in m_{12} vs. m_{13} is expected to be flat. However, symmetries may impose various restrictions on the distribution of decay products. Additionally, resonant decays in which the parent particle first decays into two particles, with one of the decay products readily decaying into two additional decay products, often dominate three-body decay processes and are manifested by the presence of peaks around the mass of the resonant decay, as shown in Figure 8.37. As such, the Dalitz plot provides a useful tool for investigating the dynamics of three-body decays. The technique can be adapted to four-body decays as well [171].

8.5.3 V^0 Topological Reconstruction Technique

Several "short-lived" elementary particles (e.g., K_s^0, Λ^0, Σ, Ds, and their antiparticles) have lifetimes sufficiently long to be observable based on their decay topology and displaced

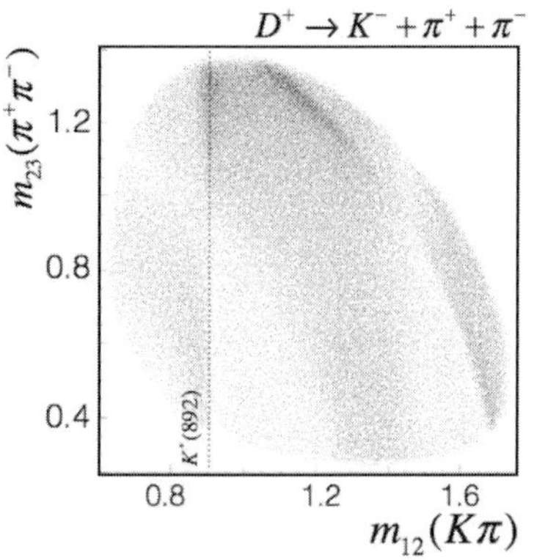

Fig. 8.37 Example of Dalitz plot for $D^{+-} \to K^- + \pi^+ + \pi^+$ decays measured by the CLEO Collaboration in $e^+ + e^-$ collisions at $\sqrt{s} = 3.770$ GeV. The peaked structure of the plot along the m_{12} axis stems from the fact that D-mesons may decay into K^*, which subsequently decays into $K + \pi$. (Courtesy of G. Bonvicini from the CLEO Collaboration [51].)

vertex. Kaons (K_s^0), lambdas (Λ^0), cascades (Ξ^0), and omegas (Ω^-) were in fact discovered in bubble chamber experiments based on their decay topology. For instance, Figure 8.38 displays the first observed Ω decay, by Barnes et al. [27], with the BNL 80-inch hydrogen bubble chamber on the basis on its unique sequence of charged-track topologies. In general, decays of neutral particles produce V-shaped topologies of charged particle tracks and are, for this reason, commonly called V^0 **particles** or V^0 **decays**. The phrase "V^0 particle" is

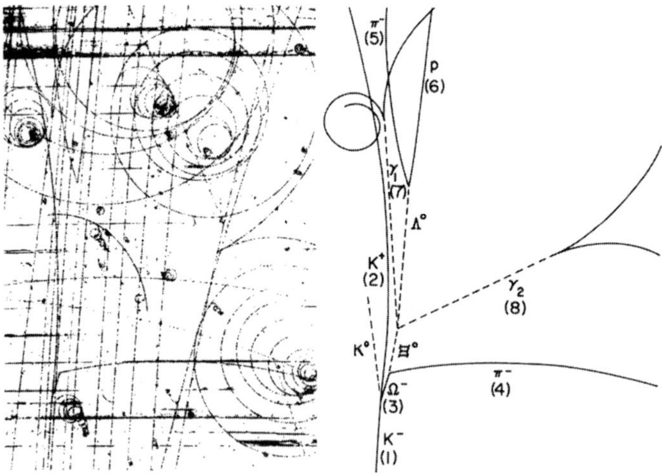

Fig. 8.38 (Left) Photograph of a bubble chamber event showing the first reported $^-$ decay [27]. (Right) Analyzed trajectories: solid lines represent observed charged particles while dashed lines indicate reconstructed trajectories of decayed neutral particles. The $^-$ decays into a π^- and a/^{0}, which subsequently decays into a ||^0 and two photons (γ_1 and γ_2). The ||^0 itself decays into a proton (p) and a π^-. (Reproduced with permission from V. E. Barnes, et al. Observation of a hyperon with strangeness minus three. *Physical Review Letters*, 12:204–206, 1964.)

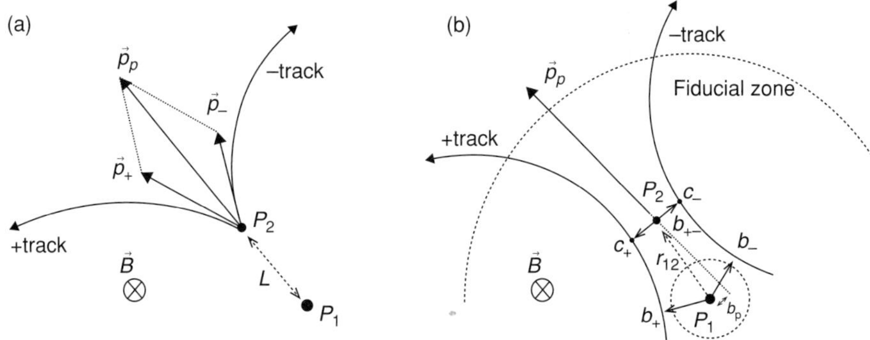

Fig. 8.39 (a) Nominal and (b) actual V^0 decay topologies in a uniform magnetic field with definitions of selection criterion variables typically used in V^0 analyses.

also often used as a generic term that includes all particles decaying with a measurable displaced vertex, whether neutral or not.

Principle of the V^0 Reconstruction Technique

The nominal spatial configuration of a V^0 four decay in a uniform magnetic field is illustrated in Figure 8.39 (a). A neutral particle of four-momentum p_P produced at the primary vertex P_1 propagates a certain distance L before it decays at the **secondary vertex** P_2 into a pair of charged particles with four-momenta p_+ and p_-. By virtue of energy-momentum conservation, one has $p_P = p_+ + p_-$, and Eq. (8.150) provides for an estimate of the mass $m_P = (p_P \cdot p_P)^{1/2}$ of the neutral particle. For sufficiently large values of L, such decay topologies are readily visualized in bubble chamber detectors or photo emulsions. They can also be reconstructed and identified with multilayered tracking detectors, as schematically illustrated in Figure 8.39b. However, finite position resolution and scattering effects smear the reconstruction of charged particle trajectories, and one consequently ends up with tracks that do not feature perfect decay topologies; that is, they do not intersect at the decay point P_2. In a very low track multiplicity environment, even imperfect track topologies such as the one schematically illustrated in Figure 8.39b may be readily identifiable as V^0. In practice, uncertainties arise due to finite decay lengths L, combinatorial effects associated with large track multiplicities, and limited track resolution. The identification of V^0 must then rely on extensive sets of topological and kinematical selection criteria designed to distinguish true displaced vertex decays from **combinatorial decays** (also called fake decays).

We discuss the definition and use of topological selection criterion in the next paragraph, and consider technical details of their calculations in the following paragraphs.

Topological and Kinematical Selection Criteria

V^0 reconstruction analyses typically involve all or a subset of selection criteria based on variables introduced in Figure 8.39b.

DCA b_{+-} of the Daughter Tracks

Tracks produced by a decay must originate from the secondary vertex P_2. The trajectories of these **daughter tracks** must then intersect or nearly intersect near P_2. The distance of closest approach (DCA) between the two tracks, b_{+-}, as defined in Figure 8.39b characterizes how closely any two tracks intersect. Ideally, this distance should be null for true daughter tracks. In practice, given smearing effects associated with track reconstruction resolution, b_{+-} is finite and characterized by a PDF $P_{\text{DCA}}(b_{+-})$ of width σ_{+-} for actual daughter tracks; in contrast, combinatorial pairs of tracks feature a much wider, nearly uniform distribution of values. It is thus legitimate to use a selection criterion $b_{+-} \le b_{\text{max}}$ to eliminate combinatorial track pairs. The nominal efficiency of the selection criterion is then

$$\varepsilon_{+-} = \int_0^{b_{\text{max}}} P_{\text{DCA}}(b_{+-})\, db_{+-}. \tag{8.161}$$

The level of contamination leaking through this selection criterion depends on the track reconstruction resolution, the proximity of the decay point P_2 to the primary vertex P_1, as well as the overall track multiplicity of the event. Other selection criteria are thus necessary.

Distance r_{12} between the Secondary and Primary Vertices

One estimates the position of the secondary vertex P_2 as the bisection of a segment joining points C_+ and C_- corresponding to the points of closest approach between two tracks. The variable r_{12} characterizes the distance between the estimated position of the secondary vertex and that of the primary vertex. Nominally, for true daughter pairs the distance r_{12} should exhibit an exponential distribution with a decay constant $\gamma c\tau$ determined by the lifetime τ and the speed distribution of the parent particle. For sufficiently large values of r_{12}, the number of combinatorial pairs is null or negligible, but for small values, primary particles may contribute a sizable background. It is thus useful to require that r_{12} exceeds a minimum value $r_{12,\text{min}}$ determined largely by the DCA resolution of primary tracks. In practice, one may also use an upper limit $r_{12,\text{max}}$. The values $r_{12,\text{min}}$ and $r_{12,\text{max}}$ thus define the fiducial volume of the measurement. Given that particles decay stochastically with an exponential distribution, these two criteria produce a loss of decaying particles with an efficiency of the order of

$$\varepsilon_r = \frac{1}{\lambda} \int_{r_{12,\text{min}}}^{r_{12,\text{max}}} \exp\left(-\frac{r_{12}}{\lambda}\right) dr_{12}, \tag{8.162}$$

where $\lambda = \bar{\gamma} c\tau$ corresponds to the effective decay length determined by the (average) speed and lifetime of the decaying particles.

DCA of the Parent Relative to the Main Vertex

The sum of the momentum vectors of the two daughter tracks nominally adds to the momentum of the parent particle, $\vec{P}$. The momentum vector $\vec{P}$ and the point P_2 define the trajectory of the parent particle relative to the primary vertex. If the parent is neutral, this trajectory is simply a straight line which should nominally intersect the primary vertex. However, finite resolution effects in the reconstruction of $\vec{P}$ as well as the positions of P_1 and P_2 imply that the straight line has a finite DCA, b_P, relative to the primary vertex. Real daughter pairs are expected to yield a small b_P values with a finite width PDF $P(b_P)$, whereas combinatorial pairs may have arbitrarily large b_P values. Applying a maximum value selection criterion $b_P \le b_{P,\max}$ is thus in principle useful to further suppress combinatorial pairs. This selection criterion has an efficiency that is determined by the shape and width of the PDF $P(b_P)$. However, be warned that the impact of this selection criterion may be highly correlated to the effects of other selection criteria, so it is typically incorrect to estimate the overall efficiency of the selection criteria as the product of the efficiencies of each of the selection criteria taken separately.

DCA of the Daughter Tracks Relative to the Main Vertex

In very high track multiplicity environments, particularly in the study of relativistic heavy-ion collisions, it is usually effective to also use minimal value selection criteria on the DCA of daughter particles d_+ and d_- relative to the position of the main vertex to suppress the number of combinatorial pairs. Note that while symmetric selection criteria may be reasonable, for instance, for K_s^0 that decay into a pair $\pi^+\pi^-$, asymmetric selection criteria may be better suited for Λ^0 that produce a heavy and a light particle because the heavy particle typically carries a larger fraction of the momentum of the parent (in the lab reference frame) and thus tends to point directly at the main vertex.

Other Selection Criteria

Various backgrounds and instrumental effects may hinder measurements of V^0. Analyses are thus often limited to specific transverse momentum ranges of the daughter particles and the parent. Various track quality selection criteria may also be applied to the daughter particles to minimize contamination by primary particles and fake tracks.

Topological Variable Calculations

Calculation of the DCA between a charged particle track and the primary vertex requires the track be projected near the origin of the track $x = 0$, corresponding to the region of the primary vertex. One can then estimate the DCA based on a nonlinear distance minimization method or use a linear approximation of the track in the vicinity of the primary vertex. Both techniques are briefly outlined in §9.4.

The calculation of the distance between two tracks produces points $\vec{C}_1$, $\vec{C}_2$, and $\vec{P}_2$. The points $\vec{P}_2$ is used in combination with the primary vertex position $\vec{P}_1$ to estimate the path length r_{12} of the parent particle before its decay

$$r_{12} = \left[(\vec{P}_2 - \vec{P}_1) \cdot (\vec{P}_2 - \vec{P}_1)\right]^{1/2}, \tag{8.163}$$

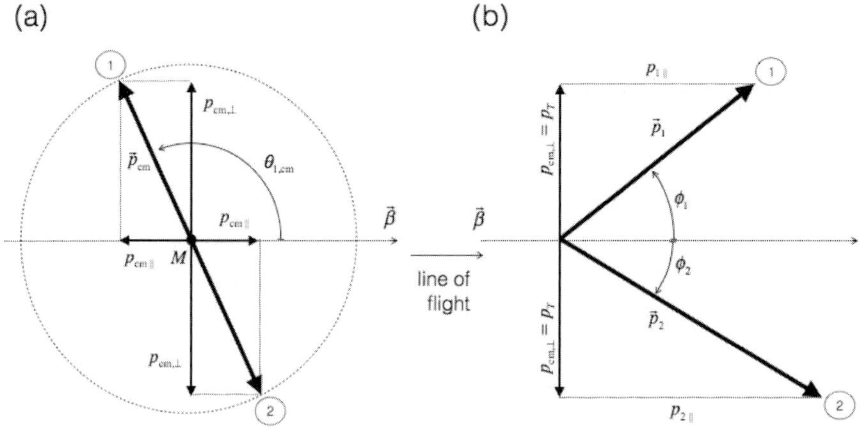

Fig. 8.40 Two-body decay kinematics in (a) the rest frame of parent particle and (b) the laboratory used in the definition of Armenteros–Podolansky variables. The parent particle has a mass M and the two daughter particles have masses m_1 and m_2. Note how the momentum components transverse to the line of flight are invariant under a Lorentz boost between the center of mass and laboratory frames.

whereas the points C_1 and C_2 are used to project the tracks and obtain the momentum vectors $\vec{p}_1$ and $\vec{p}_2$ at the decay vertex $\vec{P}_2$. The momentum of the parent particle is thus simply the sum of these two vectors:

$$\vec{P} = \vec{p}_1 + \vec{p}_2. \tag{8.164}$$

It is optionally possible to carry out a constrained refit of the tracks assuming a specific particle decay model (e.g., $K_s^0 \to \pi^+\pi^-$). Techniques to carry out such fits are documented in the literature (§9.3.2) [25, 45, 87, 178] . The fit yields an updated vertex position $\vec{P}_2$ and moments $\vec{p}_1$ and $\vec{p}_2$.

The vector $\vec{P}$ determines an estimate of the parent direction before the decay. Using the position $\vec{P}_2$ and this vector, one can then next determine the impact parameter (or DCA) of the parent at the primary vertex (provided it can be assumed the parent in fact originates from the main vertex of the interaction) using the DCA determination techniques outlined in §9.4.

The Armenteros–Podolansky Plot

In the rest frame of a two-body decay, the locus of points spanned by the momentum vectors of the two daughter particles form a sphere of radius $p_{\rm cm}$ (Figure 8.40):

$$p_{\rm cm} = \frac{1}{2\sqrt{s}} \left\{ \left[s - (m_1 - m_2)^2 \right] \left[s - (m_1 + m_2)^2 \right] \right\}^{1/2}. \tag{8.165}$$

This sphere becomes an ellipsoid under a boost in the lab frame and must satisfy

$$E_1E_2 - p_1p_2 \cos\phi = \sqrt{m_1^2m_2^2 + M^2p_{\rm cm}^2}, \tag{8.166}$$

where E_i and p_i represent the energy and momenta of the two daughter particles in the lab frame and ϕ is the relative angle between their momentum vectors. Unfortunately, the ellipsoidal nature of the decay sphere is not readily apparent from this expression. However, Armenteros and Podolansky figured out that the ellipsoid can be visualized in terms of two variables ε and α calculated based solely on the momentum vectors of the daughter particles. Since the masses of the parent and the two daughter particles determine the magnitude of p^*, the size and position of the ellipse constitute a unique signature of decays that can be used to analyze data and distinguish different decay types. The technique is now known as the Armenteros–Podolansky plot (or often only the Armenteros plot) [153]. The CM momentum $p_{\rm cm}$ of the daughter particles may be decomposed in terms of longitudinal and transverse components:

$$p_{\rm cm}^2 = p_{{\rm cm},\parallel}^2 + p_{\rm cm,T}^2. \tag{8.167}$$

The transverse components are invariant under a boost in the direction of motion of the parent particle. Let $p_{\rm T} = p_{\rm cm,T} = p_{\rm Lab,T}$. By definition of the polar angles ϕ_1 and ϕ_2 shown in Figure 8.40, one has

$$p_{\rm T} = p_1 \sin\phi_1 = p_2 \sin\phi_2, \tag{8.168}$$

while conservation of momentum implies the momentum of the parent particle, P, satisfies

$$P = p_1 \cos\phi_1 + p_2 \cos\phi_2. \tag{8.169}$$

Defining $\phi = \phi_1 + \phi_2$, and given $\sin(\phi_1 + \phi_2) = \sin\phi_1 \cos\phi_2 + \cos\phi_1 \sin\phi_2$, one can use the preceding two expressions to obtain

$$p_{\rm T} = \frac{p_1p_2 \sin\phi}{P}. \tag{8.170}$$

It is then convenient to define the variable ϵ as

$$\epsilon = \frac{2p_{\rm T}}{P} = \frac{2p_1p_2 \sin\phi}{P^2} = \frac{2\sin\phi_1 \sin\phi_2}{\sin\phi}. \tag{8.171}$$

However, $p_{\rm T}$ does not uniquely characterize $p_{\rm cm}$, and one thus needs a way to estimate $p_{{\rm cm},\parallel}$. This can be achieved based on the difference $p_{1,\parallel} - p_{2,\parallel}$. It is then convenient to define a variable α according to

$$\alpha = \frac{p_{1,\parallel} - p_{2,\parallel}}{p_{1,\parallel} + p_{2,\parallel}} = \frac{p_{1,\parallel} - p_{2,\parallel}}{P}. \tag{8.172}$$

Given $\sin(\phi_2 - \phi_1) = \sin\phi_2 \cos\phi_2 - \cos\phi_1 \sin\phi_2$, one finds

$$\alpha = \frac{\sin(\phi_2 - \phi_1)}{\sin\phi}, \tag{8.173}$$

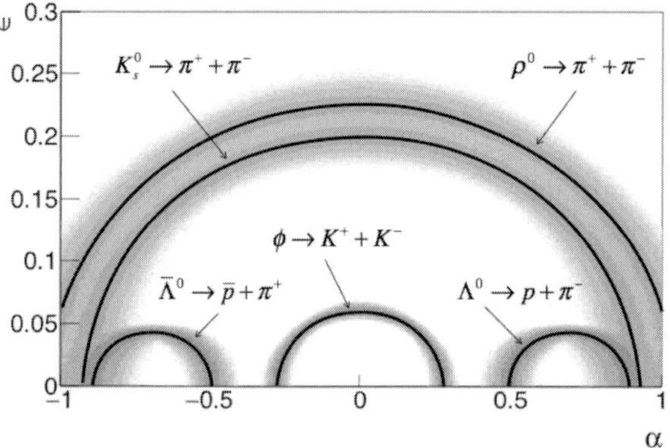

Fig. 8.41 Simulation of an Armenteros–Podolansky plot for selected decays with perfect momentum resolution (solid line) and 3% momentum resolution (shaded scatterplot).

which is uniquely determined by the vectors $\vec{p}_1$, $\vec{p}_2$, and the relative angle ϕ. One can show that the average value of α, noted $\bar{\alpha}$, is equal to (see Problem 8.13 or ref. [153])

$$\bar{\alpha} = \frac{m_1^2 - m_2^2}{M^2} \tag{8.174}$$

and

$$p_T^2 + \frac{1}{4\left(M^{-2} + P^{-2}\right)}\left(p_{1,\parallel} - p_{2,\parallel} - \frac{m_2^2 - m_2^2}{M^2}\right)^2 = p_{\rm cm}^2. \tag{8.175}$$

This expression may then be rearranged to yield

$$\frac{\epsilon^2}{4\left(p_{\rm cm}/P\right)^2} + \frac{(\alpha - \bar{\alpha})^2}{4\left[(p_{\rm cm}/M)^2 + (p_{\rm cm}/P)^2\right]} = 1, \tag{8.176}$$

which is the expression of an ellipse in the α vs. ϵ plane centered at $\alpha = \bar{\alpha}$, $\epsilon = 0$, and with semi-major axes $2\left[(p_{\rm cm}/M)^2 + (p_{\rm cm}/P)^2\right]^{1/2}$ and $2p_{\rm cm}/P$, respectively.

The CM momentum $p_{\rm cm}$ and average $\bar{\alpha}$ are determined solely by the masses M, m_1, and m_2. Given that α and ϵ are both determined relative to P, the ellipse is completely specified by and constitutes a unique signature of the decay $M \to m_1 + m_2$.

Figure 8.41 presents examples of ϵ vs. α distributions for selected decays and clearly illustrates the fact that the Armenteros plot is a useful analysis tool for identifying different decaying particles. Indeed, one finds that different decays cover different loci in the ϵ vs. α space and are as such easily recognizable. It is thus rather tempting to use band selection criteria in ϵ vs. α space to select specific decays. One must realize, however, that such selection criteria are strictly equivalent to invariant mass selection criteria. More importantly, such selection criteria cannot distinguish between actual decays and backgrounds caused by instrumental effects. One must then exercise great care in the use of such selection criteria or ideally refrain from their use.

8.6 Searches, Discovery, and Production Limits

Searches for new particles, rare decays, or new particle production processes are an essential component of the work of nuclear and particle physicists. Models designed to explain the structure of matter and fundamental forces, most particularly in searches for physics beyond the standard model of particle physics, abound and predict a vast array of new particles, decay processes, or particle production processes. Not all models can be right, however, and it is in part the role of experimental physicists to test the validity of (credible) model predictions by seeking experimental evidence for the proposed new particles or processes.

8.6.1 Searches for New Particles and Processes

It is fair to say that there are usually more incorrect models than correct ones. So, while many search experiments do identify the particles or processes they seek, many do not. But lack of discovery should not be construed as the failure of an experiment. It should be the models that fail, not the experiments. Search experiments are thus typically designed with great care and attention to detail in order to achieve as high a sensitivity as possible to the sought after particles or processes. The sensitivity of a search is determined by the number of "events" it can examine and its capacity to recognize the sought after phenomenon against background processes. Searches for new particles or low-probability decay modes are typically based on scattering experiments with high-rate capabilities. The idea is to observe as many collisions as possible that are susceptible of producing the predicted new particles or rare decay. This typically requires long data acquisition periods, high beam intensity and target thickness (or luminosity), high granularity and fast detectors capable of handling the large particle production rates and detector occupancy. It usually also requires specialized triggers to select potentially interesting processes, as well as robust and ultrafast data acquisition systems to handle the large throughput of data. Search experiments also usually involve detection redundancy in order to identify and reject background processes.

In the remainder of this section, we first consider the statistical conditions that may justify a claim of discovery and publication in scientific journals. We next discuss how to report the negative findings of experimental searches for new particles or processes.

8.6.2 Discovery: 5σ

Searches for new particles or rare decay processes are largely based on invariant mass reconstruction methods (involving either particles or jets) and require that a sufficiently strong signal (e.g., a peak) be observed relative to background processes. In order to quantify the possibility that an observed peak might be caused by an uninteresting statistical fluctuation, a statistical test must be performed. One formulates the **null hypothesis** that the observed peak or signal might correspond to a background fluctuation. The purpose of the test is thus to assess the probability of this null hypothesis. This requires the choice of a statistic (a test), and one must estimate the **significance level** (p-value) of the observed signal. This corresponds to the probability that given the background rate b from "known" sources, the observed number of events (e.g., in the mass range of interest) would fluctuate up to n_0 or a larger value. A small value of p shows that the data are not compatible with the expected background and suggests that the null hypothesis has an extremely low probability. The observed signal then provides evidence for a new particle (or process).

It is customary in particle physics to convert the value of p into a number of standard deviations σ of a Gaussian distribution, beyond which the one-sided tail has a probability p. Particle physicists typically consider that a new signal or particle can be reported as **discovered** if p is smaller than 3×10^{-7} corresponding to a value of, or in excess of, 5σ (one-sided confidence interval). Note that the correspondence to 5σ is valid only if the signal/background are Gaussian distributed, which is not always the case in practice. The 5σ value is nonetheless useful, though somewhat arbitrary, because it is quite easy to remember. Indeed, no fundamental principle of science or statistics dictates that a 5σ interval (or p-value of 3×10^{-7}) should define a discovery any more than any other "large" $n\sigma$ (small p) value. However, a probability of 3×10^{-7} is construed by most scientists as sufficiently small a value to warrant a claim of discovery, and the "round" number 5 indeed makes it easy to remember. It has thus become the de facto standard requirement in particle and nuclear physics for the announcement of new particle (or signal) discoveries.

This said, many a statistician might question the need for as stringent a requirement as 5σ. Why indeed is 3σ or 4σ not deemed sufficient? As pointed out by Lyons [138], perhaps the most compelling answer to this question is that a number of exciting and interesting "signals" reported at the 3σ or even 4σ level have been later found to disappear once more data were collected. It must also be noted that typical particle/nuclear physics analyses involve many histograms, each with several bins. The chance of one particular bin showing a 5σ effect is thus not that small, let alone a 4σ or 3σ effect. It is also important to note another mitigating factor, the fact that physicists tend to subconsciously make prior assumptions that favor the notion they have discovered something new rather then being in the presence of some undetected procedural or systematic effect(s). While it may not be necessarily equitable or fair to expect both large-purpose experiments (such as the LHC experiments) and smaller, more dedicated experiments to use the same standard, who is to decide which experiment is to apply what standard? Most physicists and journal editors thus prefer abiding by the strict and somewhat arbitrary 5σ rule. However, given that data analyses are often challenged by the use of too small datasets, it is also common to announce and report new signals of weaker significance (e.g., 4.5σ) using a carefully chosen wording (e.g., indication, suggestive evidence, and so on) that properly reflects the status of an observation. However, purported signals observed at less than the 3σ level are typically frowned upon.

The calculation of p-values is often complicated by the existence of nuisance parameters such as uncertainties in the estimated background. Various techniques to handle the calculation of the p-value are discussed in the review article of Lyons [138].

8.6.3 Production Upper Limit

If an experiment is properly designed and operated, a lack of discovery is indeed not a failure, and limits on the production cross section of new particles or branching ratios of nonobserved decays are worth publishing in scientific journals given that they inform models and constrain their predictions. In fact, models can often be rejected altogether by the lack of evidence for the particles or decays they predict.

Negative search results are generally reported in terms of **production limits**. In the search for new particles, such production limits are usually expressed in terms of upper limits on the production cross section of the sought after particle(s), whereas in rare decays, they are most often formulated in terms of an upper limit on the **branching fraction** of the decay process.

Production limits are calculated on the basis of the number of observed events, N_{obs}, as a function of the experimental acceptance and efficiency, ε, for the sought after particle (or process). Determination of the acceptance requires a model of the expected differential cross section ($d\sigma_p/dydp_T$). It is convenient to express the cross section (or branching ratio) of the sought after particle relative to the cross section of a known process, or the number of collisions susceptible of producing the particle or decay. For instance, if the search is conducted on the basis of N–N collisions with a known cross section σ_{NN}, the expected number of observed particles, N_{ex}, may be written

$$N_{\text{ex}} = \mathcal{L} f_{\text{trig}} f_p \sigma_{NN} \varepsilon_p, \tag{8.177}$$

where $\mathcal{L}$ is the integrated luminosity delivered to the experiment; f_{trig} is the fraction of the N–N cross section sampled by the experiment due to trigger selection and efficiency, as well as analysis selection criteria; f_p is the fraction of the cross section (upper limit) that produces the particle of interest; and ε_p is the detection efficiency (of the particle or decay process):

$$\varepsilon_p = \frac{1}{\sigma_p} \int_\Omega \varepsilon_p(y, p_T) \frac{d\sigma_p}{dydp_T}, \tag{8.178}$$

where $\varepsilon_p(y, p_T)$ is the particle detection efficiency expressed as a function of rapidity and transverse momentum. So, if the number of observed particles N_{obs} is actually finite, the particle production cross section $\sigma_p = f_p \sigma_{NN}$ can be estimated on the basis of

$$\sigma_p = \frac{N_{\text{obs}}}{\mathcal{L} f_{\text{trig}} \varepsilon_p}. \tag{8.179}$$

For rare decay searches, one replaces $\mathcal{L} f_{\text{trig}}$ by the number of decays produced and observable by the experiment.

If the search produces no candidate particles (or rare decays), one might be tempted to state that the production cross section of the particle vanishes, or that the particle does not exist. But that would be incorrect. Indeed, even if the search uncovers no candidate, it is still conceivable that the particle *does* exist but that its production cross section is too small to be observable in the experiment. Since the production cross section is at best small, it

Table 8.1 Poisson upper limits N_{up} for n observed events

n	CL = 90%	CL = 95%	n	CL = 90%	CL = 95%
0	2.30	3.00	6	10.53	11.84
1	3.89	4.74	7	11.77	13.15
2	5.32	6.30	8	13.00	14.44
3	6.68	7.75	9	14.21	15.71
4	7.99	9.15	10	15.41	16.96
5	9.27	10.51	11	16.60	18.21

Reproduced from R. M. Barnett, et al. Review of particle physics. *Physical Review D*, 54:1–708, 1996.

is legitimate to assume the number of observed particles n can be described by a Poisson distribution:

$$f_P(n|N_{\rm ex}) = \frac{N_{\rm ex}^n e^{-N_{\rm ex}}}{n!}, \tag{8.180}$$

where $N_{\rm ex}$ represents the number of counts that one should expect to find, on average, if the experiment was repeated several times. Since no candidates were found, one seeks the largest value $N_{\rm up}$ compatible with $n = 0$ at a given confidence level (CL). For instance, choosing CL= 10%, one gets

$$f_P(n = 0|N_{\rm ex}) = \frac{(N_{\rm up})^0 e^{-N_{\rm up}}}{0!} = e^{-N_{\rm up}} = 0.1, \tag{8.181}$$

which implies an upper limit $N_{\rm up} = -\ln(0.1) = 2.3$. This value can then be inserted in Eq. (8.179), in lieu of $N_{\rm obs}$, to determine an **upper limit** σ_p for the particle production cross section (or rare decay branching fraction).

If the number of observed candidates n remaining after all selection criteria is not zero but relatively small, it is likely that they correspond to background particles (events) that could not be eliminated by the selection criteria. One must then determine an upper limit $N_{\rm up}$ compatible with the observed background n. Choosing CL, one then has

$$f_P(0 \le k \le n|N_{\rm ex}) = \sum_{k=0}^{n} \frac{(N_{\rm up})^k e^{-N_{\rm up}}}{k!} = {\rm CL}, \tag{8.182}$$

which must be solved numerically for $N_{\rm up}$. Table 8.1 lists upper limits $N_{\rm up}$ for small values of n corresponding to the most used confidence levels of 10% and 5%.

Exercises

8.1 Show that in the nonrelativistic limit, $v \approx 0$, the rapidity of a particle, y, is equal to its velocity, $\beta = v/c$.

8.2 Show that the rapidity of particle, defined by Eq. (8.25) transforms according to Eq. (8.27). Hint: Compound two Lorentz transformations and determine the rapidity of the particles.

8.3 Show that for $E \gg m$, the rapidity of a particle, given by Eq. (8.26), may be approximated by the pseudorapidity defined as $\eta = -\ln(\tan(\theta/2))$ (Eq. 8.31).

8.4 Demonstrate Eqs. (8.39) and (8.40) using the definitions of the cosh and sinh functions and the definition (8.26) of the rapidity of a particle.

$$\cosh x = \frac{e^x + e^{-x}}{2}$$

$$\sinh x = \frac{e^x - e^{-x}}{2}$$

8.5 Demonstrate Eqs. (8.41) and (8.42) using the definitions of the cosh and sinh functions (given in the previous problem) and the pseudorapidity of a particle expressed as

$$\eta = \frac{1}{2}\ln\left(\frac{|\vec{p}| + p_z}{|\vec{p}| - p_z}\right).$$

8.6 Consider the TOF slat geometry illustrated in Figure 8.26 and obtain expressions for the velocity of the particle (assuming a flight path L) and its position x along the slat as defined in the figure.

8.7 Verify the expressions (8.41) and (8.42).

8.8 Derive the expression (8.85) for the limits on t for two-body processes.

8.9 Verify the two expressions below by substitution of the values of m_{ij} defined by Eq. (8.158):

$$m_{12}^2 + m_{13}^2 + m_{23}^2 = M^2 + m_1^2 + m_2^2 + m_3^2. \tag{8.159}$$

Additionally demonstrate Eq. (8.160):

$$m_{12}^2 = (P - p_3)^2 = M^2 + m_3^2 - 2ME_3, \tag{8.160}$$

where E_3 is the energy of the third particle in the rest frame of the decaying particle.

8.10 Derive the expression (8.67) for the three-body decay rate.

8.11 Show that the radius of a circle is determined by three points and can be expressed as Eq. (8.108).

8.12 Show that the position resolution achievable with charge sharing between sensors is given by Eq. (8.128).

8.13 Verify the expression (8.174) for the average value of the Armenteros variable α and use this result to demonstrate Eqs. (8.175) and (8.176).

9 Event Reconstruction

Modern nuclear and particle physics detectors are typically very complex and produce large datasets that must be analyzed **offline** after an experiment is completed. The analysis of the experimental data is typically articulated in two main stages known as **event reconstruction** and **physics analysis**. Basic elements of physics analyses have already been discussed in Chapter 8 and advanced techniques, mostly pertaining to measurements of fluctuations and correlations, will be presented in Chapters 10 and 11. In this chapter, we focus on key aspects of event reconstruction beginning with a brief overview of basic reconstruction tasks in §9.1, and followed in §9.2, with a discussion of track reconstruction techniques, including a detailed example of track reconstruction with a Kalman filter, as well as presentations in §9.3 of various vertex finding and fitting techniques.

9.1 Event Reconstruction Overview

We begin this section with a discussion of the types and structures of data commonly acquired and handled by large experiments in §9.1.1. We next present a general overview of typical data analysis workflows in §9.1.2 and conclude with a brief discussion of basic tasks of detector and signal calibration in §9.1.3.

9.1.1 Types and Structure of Experimental Data

Data produced by large modern experiments typically involve a mix of **slow control data** characterizing the operating conditions (e.g., power supply settings, magnetic field, currents, time, etc.) and performance of the detector, as well as **physics data** produced by the many sensors and detector components of the experiment, as schematically illustrated in Figure 9.1. Physics data include sensor readout in the form of digitized information produced by **analog-to-digital converters (ADCs)** and a variety of **geographical data** indicating the components that produce and report signals. Sensor data may be read out as single values, time sequences, or arrays of various types. They may be **sparsified**, that is, zero suppressed, or read out in their entirety. Speed is usually of the essence and data are assembled by one or several processors (known as **event builders**) into data buffers of various complexities and then dispatched for storage and possibly **online** analysis. Event storage was traditionally accomplished with high-speed magnetic tape drives but is increasingly based on high-performance and high-volume magnetic hard-disk storage systems, as well as optical disks.

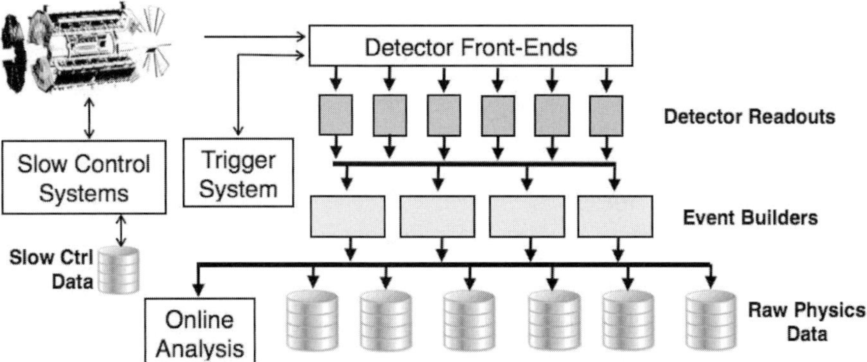

Fig. 9.1 Basic components of modern detector data acquisition systems include detector front-end electronics (analog pipeline, ADC, digital pipelines, sparsification processors), optical readout components, event builder systems, high-speed/capacity storage systems, as well as slow control systems, trigger systems, and an online analysis farm. In this and the following figures, detector diagrams are included for illustrative purposes only. This and Figures 7.2, 7.3, and 7.4 are meant to illustrate the basic concepts rather than actual implementations of the detectors shown.

9.1.2 Data Analysis Workflow

Data produced by large experiments are typically rather complex and usually require multiple stages of analysis, as schematically illustrated in Figure 9.2. There are essentially three basic stages involved in the data analysis of large experiments: (1) event reconstruction,

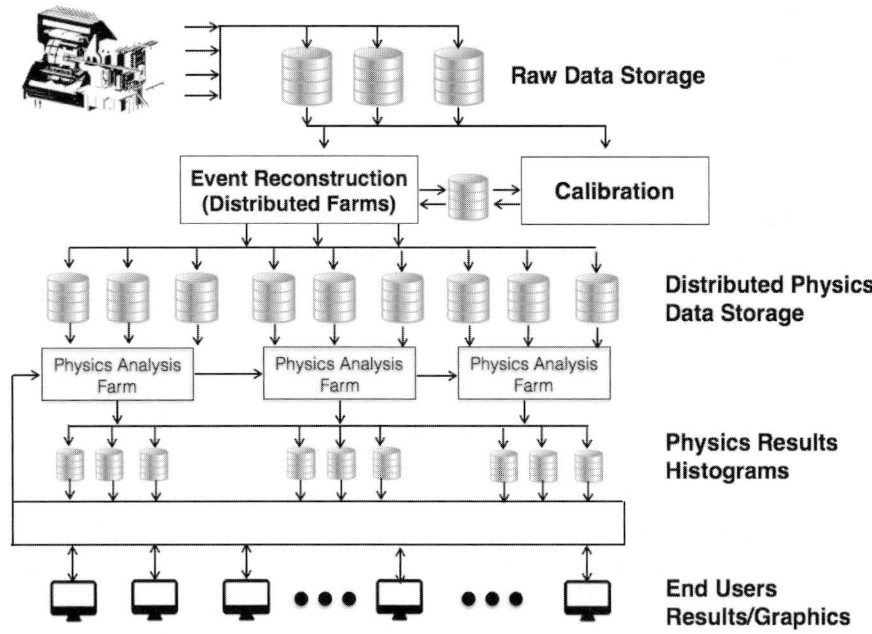

Fig. 9.2 Basic stages of data reconstruction and analysis of large modern experiments.

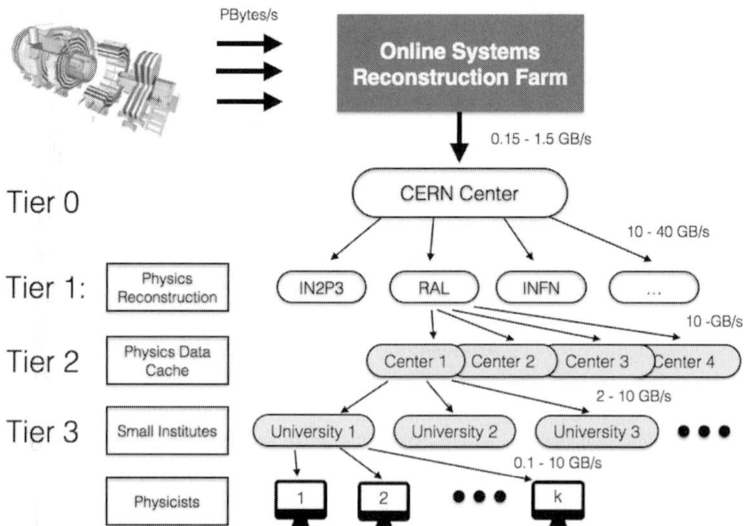

Fig. 9.3 Large-scale modern experiments use multitier grid-based computing architecture to reconstruct and analyze data. Shown here is a schematic layout of grid organization into multiple tiers and stages. Tier 0 and 1 proceed to data calibration and data reconstruction and produce reconstructed events that are then reproduced, archived, and migrated to several Tier 2 centers, where the data are then available for retrieval and analysis by physicists working at their home institution. Different variants of the design shown are used by experiments. Modern large experiments also make increasing use of commercial cloud computing, storage, and data migration.

(2) calibration, and (3) physics analyses. These three stages may be computed at a single facility, or carried out on a multitier computing grid infrastructure, as schematically illustrated in Figure 9.3. Grid computing is now an essential component of large experiments and involves the development and deployment of grid distributed analysis software; Grid management and monitoring; as well as end user tasks that include job submission, monitoring, and result retrieval. A discussion of these concepts and tools lies far beyond the scope of this text. Note, however, that most large experiments have ample documentation describing their specific grid toolset and analysis protocols.

Event reconstruction is usually a time-consuming and resource-intensive task carried out by a dedicated group of scientists (involving physicists and software engineers) on behalf of their collaboration. The main components of the event reconstruction process are schematically illustrated in Figure 9.4, and the object model they are based upon is shown in Figure 9.5. Event reconstruction typically involves the unpacking of the raw data produced by the detector; application of sensor and detector calibrations to obtain physical signals; and various stages of pattern recognition to reconstruct hits, tracks, collision, and decay vertices, as well as other constructs such as V^0 or jets. It may also involve the calculation of various observables deemed representative of the collisions of interest and that are used by most physicists analyzing the experiment's data. The hits, tracks, vertices, and other data entities produced are then assembled into **reconstructed events**, filtered, and stored in easily retrievable data formats (Figure 9.5).

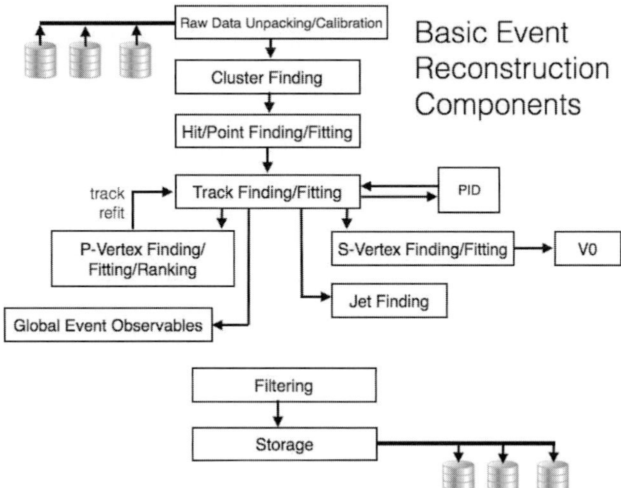

Fig. 9.4 Typical event reconstruction workflow used in modern particle physics experiments.

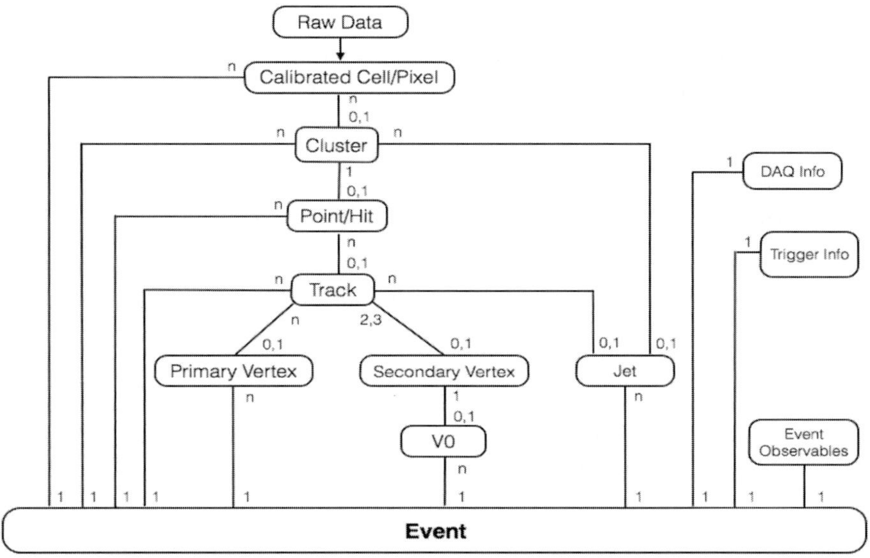

Fig. 9.5 Physics object model and structures commonly used in event reconstruction and physics analyses. Lines and numbers indicate the relation among components of the data model. An *n* indicates several instances of the given type that may be associated with the connected object instance, while 0,1 indicate that at most one instance (one or none) may be involved in the relation. For instance, several tracks, but not necessarily all of them, may be associated with a given vertex. Note that some event reconstruction schemes allow many-to-many, or *n* to *n*, object relations: for example, a point might be allowed to belong to several tracks concurrently, or a track might be associated with several vertices.

Although the responsibility for event reconstruction is typically reserved for a few experts in each collaboration, it is important to be familiar with the different components and stages of event reconstruction at a conceptual level in order to understand the "features" and limitations of data produced by one's detector.

The tasks of determining the basic kinematic parameters of charged particles at their point of production as well as the location of these production points are commonly called **track reconstruction** and **vertex reconstruction**, respectively. Both these reconstruction tasks may be carried **online**, especially in the high-level trigger, where for the sake of speed, applications frequently use simplified reconstruction methods. However, our discussion will concentrate on methods usually employed in offline analyses, where precision and quality are the major goals, while speed is rarely a decisive factor. Track and vertex reconstruction components are arguably the most important, and we consequently dedicate two sections of this chapter to these tasks. It should be understood, at the outset, that our discussion of these topics is meant to be illustrative of the methods and computations involved only, and as such cannot cover the vast array of methods and techniques that have been developed in the subfield of nuclear and particle physics. Readers particularly interested in these reconstructions should consult the recent literature on these topics, and most particularly recent advances by the CMS Collaboration on the particle flow algorithm [34].

The calibration stage typically involves a plurality of tasks that operate either on raw unpacked data or physical variables derived from them. Virtually all signals produced by a detector must be calibrated in some fashion. Calibration tasks indeed range from gain calibration of charge, light, or energy sensors, calibration of timing signals, as well as geometry and detector alignment. Some of the basic calibration tasks involved in modern experiments are presented in §9.1.3. Here again, we emphasize that it is impossible to cover all types of calibration procedures used in experimental measurements. We thus limit our discussion in §9.1.3 to a few illustrative examples only.

The physics analysis stages involves a wide variety of computing tasks that are often segregated in different groups based either on the data employed or the analysis techniques used to extract physics results. Groups often produce sub-datasets consisting of a fraction of events or distilled events that store only a subset of observables. This enables faster and more efficient analysis of the data of interest. This said, at this stage speed and computing efficiency, although important, are usually not a decisive factor, and analyses may be repeated numerous times with various tweaks of control parameter values to extract physical observables of interest and study systematic errors (discussed in detail in §12.5).

9.1.3 Signal Calibration

Particle detectors measure a wide variety of physical observables, ranging from energy deposition in calorimeters, gas cells, or pads, or solid-state devices, currents, and voltages, timing signals including those of time-of-flight detectors and clocks, various types of registers and hit arrays, and so on. All detector analog signals are digitized and recorded for future analyses. A vast number of techniques and technologies have been developed for signal amplification, filtering, analog storage (pipeline), digitization, sparsification, event building and packing, and data storage. A discussion of such topics is beyond the scope of this text but may be found in various other publications. We here restrict our discussion to the main types of digital signal processing required to extract physically meaningful information out of digitized data.

Analog-to-Digital Conversion

An analog-to-digital converter, or ADC, is a device that converts a physical signal, such as a voltage or an integrated charge, to a digital number that represents the quantity's size or amplitude. Conversion of signals into digital numbers involves quantization and digital representation of the signals with finitely many bits. The number of bits used in analog-to-digital conversion is determined by the needs and specificities of the physical signals being handled. For instance, a single bit, 0 or 1, is sufficient to indicate that a signal has crossed a minimal threshold, but several bits are needed to provide meaningful resolution in the digitization of energy deposition or timing signals. The conversion from a continuous-time or continuous-amplitude analog signal thus results in a sequence of discrete-time or discrete-amplitude digital signals that feature a finite granularity. Quantization thus involves loss of information and the introduction of quantization errors. Such errors may, however, be rendered negligible by judicious choices of dynamic range and number of bits used in the conversion.

ADCs can fulfill a wide variety of purposes and thus have a wide range of design and performance characteristics:

1. Bandwidth: The bandwidth of an ADC corresponds to the range of signal frequencies it can measure.
2. Number of bits: The number of bits used in the analog-to-digital conversion.
3. Signal-to-noise ratio: How accurately an ADC can measure a signal relative to the noise and information loss it introduces.
4. Sampling rate: The speed at which an ADC can carry out conversions.
5. Aliasing: How an ADC handles the errors introduced by digitization.
6. Dynamic range: Ratio between the largest and smallest possible values an ADC can convert.
7. Linearity: The degree to which there is a linear relation between the analog and digitized signals.
8. Accuracy: The degree to which quantization levels match the true analog signal.
9. Effective number of bits (ENOB).

The bandwidth of an ADC is determined primarily by its sampling rate, and to a lesser extent by how it handles errors such as aliasing, whereas the dynamic range is largely determined by its resolution, that is, the number of "levels" it can convert a signal to, its linearity, and its accuracy. The dynamic range is often summarized in terms of an effective number of bits (ENOB), the number of bits produced that are on average not noise. An ideal ADC has an ENOB equal to its resolution.

The design of ADCs is usually constrained by the desire to match the bandwidth of the signal and the required signal-to-noise ratio as well as engineering (e.g., size, power

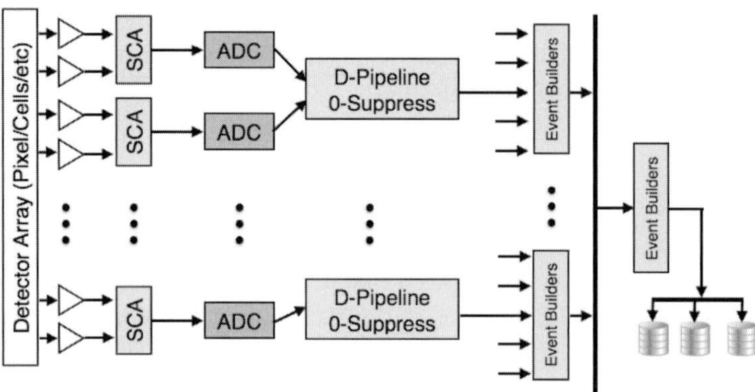

Fig. 9.6 Schematic illustration of the stages involved in detector signal read out and digitization.

required, etc.) and cost considerations. An ADC operating at a sampling rate greater than twice the bandwidth of the signal can in principle achieve perfect signal reconstruction, but the presence of quantization error limits the dynamic range of even an ideal ADC. This said, if the dynamic range of an ADC exceeds that of its input signal, the effective quantization error may be negligible, thereby resulting in an essentially perfect digital representation of the input signal.

The input of an ADC may consist of a single signal to be digitized once a trigger is received, or involve a continuous signal stream to be sampled and processed uninterruptedly. Signals may also be received and processed from multiple interleaved streams or from switched capacitor arrays used to hold analog signals from one or several signal streams until a trigger is received (see Figure 9.6). Consequently, the digitized output may consist of a single number or a stream of numbers to be stored directly into some computer memory or held momentarily onto a digital pipeline device. Digital pipeline devices are used, in particular, toward the sparsification of data, that is, the elimination of null or vanishing signals as well as triggering purposes, thereby considerably reducing the volume of data produced collision by collision.

ADCs can be used to convert voltage amplitude, integrated charge (charge-to-digital conversion, often called QDC), as well as timing signals consisting of start and stop logical pulses (time-to-digital conversion, or TDC). Many ADC designs and implementations have been developed over the years that match the wide range of scientific, engineering, and commercial applications in use today [90, 134, 186].

Basic Digital Signal Analysis

Pedestal Determination and Removal

No matter what technology is used to build them, ADCs are usually designed to produce a non-null number, called a **pedestal**, even in the presence of a vanishing signal (see Figure 9.7). This enables the identification of slightly negative signals as well as determination of the noise characteristics of the detector signals being converted. Effectively, the production of a pedestal is achieved by leaking a finite amount of charge or current at the input. In the

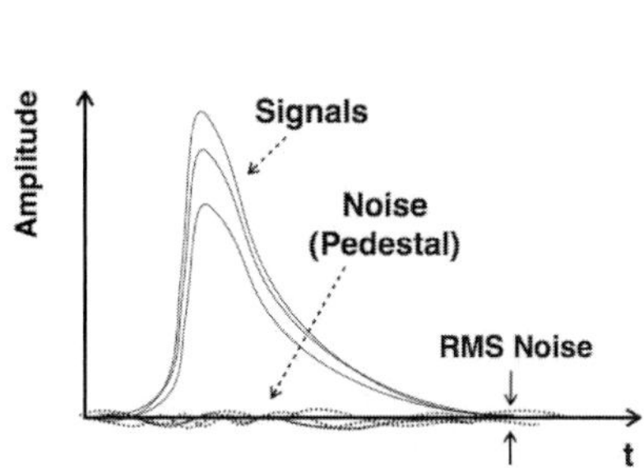

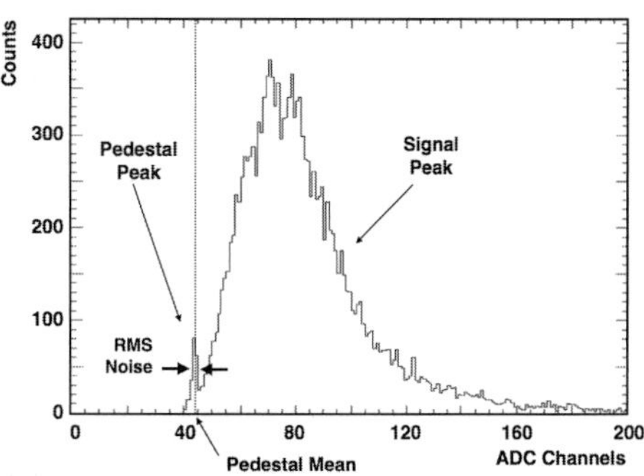

Fig. 9.7 Illustration of the notion of ADC pedestal: A small but finite current is fed into an ADC's input in order to produce a digital output with constant nonzero offset known as a pedestal. The RMS width of the pedestal peak is determined primarily by the noise characteristics of the readout chain.

presence of a truly vanishing and noise-free signal, this should ideally produce a pedestal value consisting of a single number. In practice, ADC pedestals may have a finite width, which is usually characterized in terms of a root mean square number of bins (levels), as schematically illustrated in Figure 9.7. Obviously, the incoming signal has a small quiescent current and noise of its own, which usually broadens the width of the pedestal and may also change its position.

The determination of the centroid and width of ADC pedestals is thus the most basic and elementary operation to be conducted on measured signals. The evaluation of pedestals requires data be accumulated and recorded in the absence of beams (collisions) and with veto against spurious events such as cosmic ray detection. Given a sample consisting of n signals s_i produced by a single ADC, it is usually sufficient to obtain the expectation value, s_0, and standard deviation, s_{std} of these signals to fully characterize a given ADC channel:

$$s_0 = \frac{1}{n}\sum_{i=1}^{n} s_i, \tag{9.1}$$

$$s_{std} = \sqrt{\frac{1}{(n-1)}\sum_{i=1}^{n}(s_i - s_o)^2}. \tag{9.2}$$

The average pedestal value s_o is used in the determination of calibrated signals discussed in the text that follows, whereas s_{std} serves as a basic characterization of the resolution that can be achieved with the given signal.

The performance of ADCs is known to "drift" over time. While their gains are typically quite stable, pedestals are usually observed to shift by few ADC channels over periods ranging from several hours to few days. It is thus necessary to recalibrate ADC pedestals on a regular basis, over a period commensurate with the rate at which they are observed to drift in practice. It is also important to note that AC couplings may lead to 60 Hz (or other frequencies such as the frequency of power supplies) leakage resulting in correspondingly low frequency variations of the pedestal values. Such variations can in principle be eliminated or suppressed by proper grounding of the electronics. However, one can also use an empty ADC slot to monitor the AC couplings in real time and effect a correction based on the phase of the signal measured by the empty ADC slot. More sophisticated techniques are also possible when couplings involve a large number of ADC signal streams (see, e.g., [152]).

Large modern experiments typically make use of dedicated electronic components (e.g., ASICs) mounted directly onboard with the amplification + ADC electronics to accomplish semi-automated pedestal determination and storage. They often also employ a variety of sparsification techniques to eliminate null data. Physicists carrying out analyses of data produced by large modern experiments thus typically do not handle the tasks of pedestal determination and monitoring.

Gain Calibration

The voltage signals produced by detectors are typically very weak and thus require amplification, which may take place within the detection device itself (e.g., gas amplification) or "external" components involving typically two steps (preamplifier and an amplifier). One way or another, the signals s_i produced by ADCs typically have a somewhat arbitrary amplitude which must be calibrated before they can be of any use in physics analyses. Procedures for gain calibration are as varied as the devices used in modern experiments. It is thus not possible in the context of this textbook to offer a full and comprehensive discussion of all techniques used toward gain calibration. However, it can be pointed out that calibration can in general be achieved by measuring a precalibrated signal, that is, a signal of known strength or value. For calorimeters or other types of energy-sensing devices, this can be done by exposing detector modules to beam particles of specific and known energy, cosmic rays, or through the use of radioactive sources of precisely known activity. Various calibration techniques are also based on prior knowledge of the mass of short-lived particles. It is indeed possible to use the known mass of narrow resonances (e.g., ϕ-meson) to obtain energy or momentum calibrations. Techniques to accomplish such calibrations are abundantly documented in the scientific literature (see, e.g., [24, 76, 79, 190]).

Calibrations can be determined either in absolute terms (and units) or relative to convenient references easily obtained or replicated in an experimental environment using, for instance, calibrate laser signals or physics data.

Given a raw ADC signal, s_i, the calibrated signal, E_i (e.g., an energy signal) produced by a specific data channel (i.e., corresponding to a specific detector module or component) is obtained according to the expression

$$E_i = g_i \times (s_i - s_o) \qquad \text{Linear response} \tag{9.3}$$

where s_o is the average pedestal value of the given module i.

This simple linear relation assumes the detection, amplification, and digitization processes are all linear. If either of these components fails to be perfectly linear, a detailed study of the response of the "faulty" component(s) is required, and one must apply a gain correction that is channel- and signal-strength specific:

$$E_i = g_i(s_i) \times (s_i - s_0) \qquad \text{Nonlinear response.} \tag{9.4}$$

It should be noted that large (voltage) signals fed into an ADC might exceed its dynamic range. This may be indicated either by a special bit being set to one (1) or as a specific saturation value. The preceding calibration equation fails for inputs exceeding the saturation threshold of an ADC, and therefore, one must be cognizant of the mechanism used by a specific type of ADC to indicate its signal saturation and handle such signals accordingly. Additionally, note that some ADCs are designed to have dual gain functions: a voltage comparator is first used to evaluate the amplitude of the incoming signal and decide which of two gain settings should be used in the digitization of the signal. This produces ADCs with a large effective number of digitization bits, but usually at much lower cost. Obviously, in the calibration and analysis, one must check which of the low- or high-gain modes was used to digitize a specific signal so the appropriate gain g_i may be used to produce the calibrated signal E_i.

Timing Signal Calibration

A **time-to-digital converter (TDC)**, is an ADC designed to convert a time interval into a digital binary output. The time interval to be converted is usually specified by two events (or signals) in the form of logical pulses. TDC devices, much like regular ADCs, usually incorporate several channels and typically operate either on the basis of a common start or common stop (Figure 9.8). In common start mode, a signal produced by a dedicated trigger is used to start the TDC clock. Each channel input into a TDC module then supplies a stop time corresponding to the time elapsed between the trigger and the detection of a signal (above a preset threshold) in a certain detector module. Alternatively, in common stop mode, detector modules each produce their own start signal and the time-to-digital conversion stops when a common stop signal is received by the TDC. In common stop mode, an arming signal may have to be supplied to ready the TPC for conversion before individual starts are received. Whether in common start or common stop mode, the lack of an input signal (stop or start) is indicated by a time-out or saturation bit.

Basic TDC implementations use a high-frequency counter (oscillator) to increment clock cycles. Counting is initiated and stopped when receiving start and stop signals, respectively. An analog method using a voltage ramp and comparator is used for very short time intervals for which a high-frequency counter cannot be used. Either way, one must realize that the measured time interval may not exactly correspond to the time interval between events of interest (e.g., a particle traversing start and stop counters) because the cabling used to carry both the start and stop signals to the TDC has, in general, arbitrary lengths, which add delays to the instantiation of both the start and stop signals. These arbitrary delays must consequently be calibrated out.

Calibrations in common start or common stop modes are essentially carried out in the same way. In essence, one needs to identify either an external time reference that provides an absolute time calibration (e.g., using calibrated laser pulses) or a data-based indicator (e.g., particles traveling nearly at the speed of light on a known flight path) which enables the determination of what is effectively a time pedestal, t_0, but is commonly called "time zero offset" or simply "time zero." The start-to-stop time, which might correspond to a particle's time-of-flight (TOF) between two specific detector components, may then be

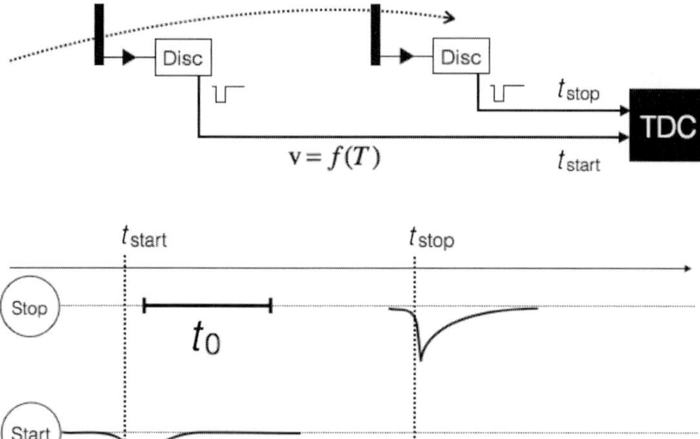

Fig. 9.8 Illustration of timing measurements and the notion of time zero offset.

written

$$t = t_{\text{stop}} - t_{\text{start}} + t_0. \tag{9.5}$$

Few remarks are in order. First, note that the precision (resolution) δt achieved on t depends on the resolution of both the start and top signals:

$$\delta t = \sqrt{\delta t_{\text{start}}^2 + \delta t_{\text{stop}}^2}. \tag{9.6}$$

However, the accuracy of the result, that is, whether there is a bias, also depends on the value of t_0. Indeed, if t_0 is not properly calibrated, t will, on average, be either over- or underestimated. All quantities derived from t, such as particle speeds and masses, will thus also be biased.

TOF resolutions required and achieved in modern high-energy experiments range between 50 and 100 ps. Such performances are typically achieved with TOF detectors based on MRPCs or ultrafast scintillator materials (see §8.4.2). The resolution achieved depends on the design and quality of the MRPC or scintillation materials: the stability of amplification, discriminators, and cabling components used in the generation and propagation of timing signals; as well as the resolution and stability of the TDC used for time-to-digital conversion (see Figure 9.8). Much like ADC pedestals, the time calibration t_0 is often observed to slowly drift over time owing to changes in sensor or amplifier gain or due to slight temperature variations. Timing delays produced by cabling or electronic components may indeed change over periods of several hours as a result of temperature variations, for instance. They thus produce slowly varying timing signal drifts that must be compensated for by recalibrating the value of the t_0 of the TOF components on a time scale commensurate with the time scales over which such variations take place.

Position Calibration

The determination of hit positions, that is, the locations where particles produce energy deposition in detector components used in the reconstruction of trajectories, often rely on weighted sums such as

$$x_{\text{hit}} = \frac{\sum_i E_i^\alpha x_i}{\sum_i E_i^\alpha}, \tag{9.7}$$

where the sums are taken over the same range of module or components, with x_i and E_i representing the nominal positions and energy read out by these components, respectively. The position x_{hit} is commonly referred to as **centroid** of the energy deposition. An exponent value $\alpha = 1$ is typically appropriate for segmented readout detectors with a small dynamic range (e.g., position sensitive detectors such as silicon strip detectors) but non-linear weights ($\alpha \neq 1$) are typically more appropriate for segmented calorimeter read-outs. The precision and accuracy achieved in the reconstruction of hit positions depend on proper measurements of both the energies E_i and the position x_i of the components as well as optimum selection of the exponent α (either derived from detailed simulations or data driven analysis of the detector response). These measurements consequently depend on the accurate calibration of signal pedestals and gains, discussed earlier in this section, proper determination of the position of the components (geometrical calibration), as well as precise modeling (studies) of the energy response of the detector modules.

While the design and construction of detector components should in principle dictate their positions relative to one another and relative to the beam axis, target, or interaction region, deviations from the nominal design may occur in practice because of a variety of circumstances including, improper detector placements, shifts in positions due to temperature variations, or intense magnetic fields. It is thus important to properly "calibrate" the position of the detector components before attempting precise data analyses.

Precise determination of detector component positions involves a variety of techniques, including in situ component surveys, laser-based measurements, as well as a host of **detector alignment** techniques based on physics data. Detector alignment techniques based on data rely on the fact that charged particles, in particular, travel on average on well-defined trajectories that can be modeled with precision. Analysis of systematic deviations of layer positions relative to measured particle trajectories thus enables fine adjustments of the component's positions and proper detector alignment. A discussion of techniques used in detector alignment is beyond the scope of this textbook but may be found in various publications [47, 111, 179].

9.2 Track Reconstruction Techniques

Charged particle tracks produced by particle detectors are the basis of much of the analyses carried out in modern nuclear and particle physics. Given that trajectories of charged particles may be significantly altered by the media they traverse largely as the result of multiple coulomb scattering, it is important to use detector technologies that are as thin as possible. Quite a few such technologies have been developed over the years, and very high momentum resolution detectors are now routinely used in the study of particles produced by elementary nuclear collisions. However, there remains the task of identifying these trajectories (hereafter called **tracks**) based on recorded data in order to obtain the kinematic parameters of the particles that produced them. This is the domain of the subfield known as **track reconstruction**, traditionally divided into two stages called **track finding** and **track-fitting**. Track finding involves processes of pattern recognition and classifications, while track-fitting aims at extracting the kinematic parameters of tracks at their point of origin with as much precision as permitted by recorded data and the hardware technologies on which they are based.

A large number of mathematical techniques and algorithms have been developed for both the finding and fitting stages of track reconstruction. There are even techniques that gracefully integrate the two tasks in a single process. We discuss track finding and pattern recognition commonly used with large detectors in §9.2.1 and elaborate on high-efficiency and precision track-fitting techniques in §9.2.2. The underlying theory of how charged particles interact with matter and the basic detection principles used in modern experiments are described in several publications [20, 50, 135] and are not covered here.

9.2.1 Track-Finding Techniques

Track finding is a pattern recognition process that consists of associating recorded hits produced by particles traversing a detection system into groups or classes believed to originate from a common particle. Associated hits form a **track candidate**, which must then be fitted and tested in order to determine whether it constitutes a genuine track or a fake track resulting from incorrect associations of actual or fake hits. Track finders are usually designed to be rather conservative and thus tend to produce and keep a large number of track candidates: eliminating track candidates after the fitting stage is relatively straightforward, while a track candidate discarded at an early stage is not recoverable at the fitting stage. This said, many finders use an iterative procedure involving several track-finding passes either with a single or several distinct algorithms.

Unambiguous track reconstruction and identification is readily achievable when the distance between tracking planes is considerably smaller than the typical distance between hits on a given detection plane. Tracking in a high hit density environment where this condition is not verified may be nonetheless possible if the origin and overall topology of tracks is known, as schematically shown in Figure 9.9.

Track-finding techniques are broadly divided into two classes commonly referred to as **global** and **local** methods. Global methods process all hits "simultaneously," whereas local methods proceed through lists of hits sequentially and build tracks iteratively. Conformal mapping, Hough transforms, and Legendre transforms, presented in the following paragraphs, are examples of global methods, whereas track road and track-following methods, discussed later in this section, are local methods. Additionally, some finding techniques require a track model and prior knowledge of the propagation of particles within the detection system. Others use combinatorial or hit proximity criteria. The use of simplified track

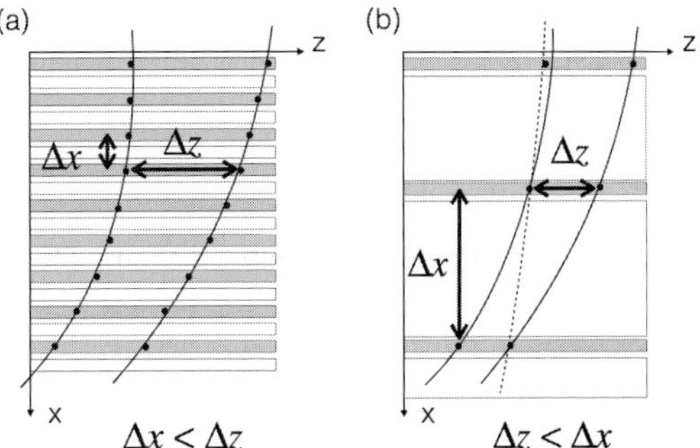

Fig. 9.9 Problematic of pattern recognition and track reconstruction. (a) Track identification is readily achievable when the distance between tracking planes is considerably smaller than the typical distance between hits on a given detection plane. (b) Ambiguities arise in the reverse situation but may be in some cases avoided if the origin or other external constraints exist on the shape and momentum of the track.

models and fast-tracking methods are important in online applications, particularly those where track finding is an integral part of the event-triggering process. We will not consider such applications and refer the reader to specialized publications on this topic. We here focus on techniques used in offline analysis.

A track may be fitted after all its hits have been identified (global track fit) or iteratively as successive hits are added to the track. The Kalman filter and adaptive techniques introduced in §9.2.3 constitute the basis for the best and most robust track reconstruction techniques available to date. Given their mathematical complexity and the fact they are computationally rather intensive, simpler techniques are often used in practice to effect a rapid reconstruction of events. Global track reconstruction techniques such as conformal mapping, Hough transforms, and track templates may be considerably faster and easier to implement. However, they typically require vast amounts of computer memory and fare poorly for secondary tracks.

It is important to note that proper reconstruction of tracks and their kinematic parameters require the tracking device be accurately calibrated. This means that their physical positions and the procedures used to extract hit positions must be accurately determined. An important component of the calibration of hit-sensing devices is a procedure known as **detector alignment**. A discussion of detector alignment techniques is beyond the scope of this textbook but may be found in the relatively recent review of Strandlie and Frühwirth [178].

Adaptive fit methods incorporate competition between several (track) hypotheses, such that the outcome of the competition depends on the current observations. They also feature the possibility of making soft decisions and of combining hypotheses according to their posterior weights. Some adaptive methods have a strong Bayesian character, reflected for example in the presence of posterior probabilities or weights attached to the different hypotheses. In general, weight values change after the inclusion of additional information from the data. Methods can also exhibit different degrees of adaptivity, depending on the level of prior assumptions inherent to the actual approach. Good adaptive methods achieve maximum flexibility and robustness with as few assumptions about the data as possible.

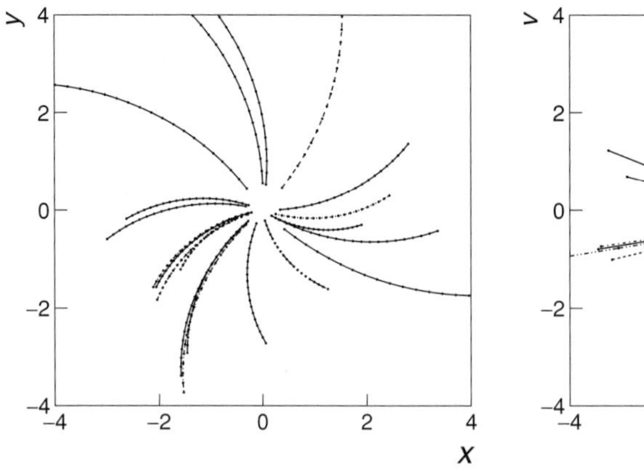

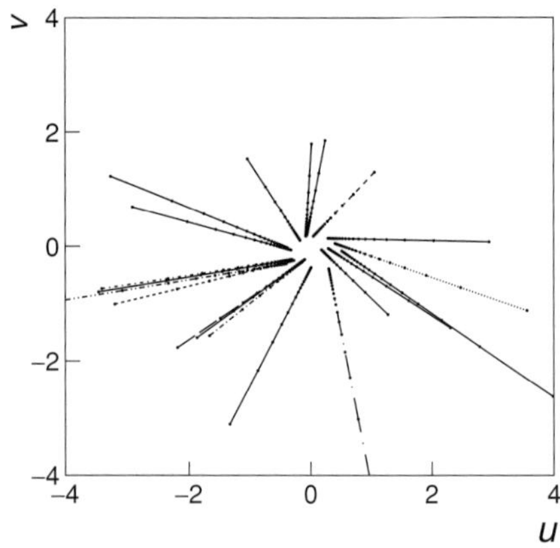

Fig. 9.10 Track finding by conformal mapping $(x, y) \to (u, v)$.

Conformal Mapping

The track-finding and fitting conformal mapping method [102] is based on the notion that circles going through the origin of a two-dimensional x–y coordinate system map onto straight lines in a u–v coordinate system defined by the conformal transformation:

$$u = \frac{x}{x^2 + y^2} \tag{9.8}$$

$$v = \frac{y}{x^2 + y^2} \tag{9.9}$$

where the circles are defined by

$$(x - a)^2 + (y - b)^2 = r^2 = a^2 + b^2 \tag{9.10}$$

One can verify (see Problem 9.2) that straight lines in the u–v planes may be written as

$$v = \frac{1}{2b} - \frac{a}{b}u \tag{9.11}$$

As illustrated in Figure 9.10, for large-momentum tracks (i.e., large values of the radius r) the conformal mapping yields straight lines passing relatively near the origin of the u–v coordinate system, which can be easily identified. Fits to straight lines in u–v space are fast and yield parameters a and b which, when combined according to Eq. (9.10), provide the radius of a trajectory and its center.

Hough and Legendre Transforms

Conformal mapping, presented in the previous section, is useful when the experimental interest is focused on large momentum particles but becomes less practical for low-momenta tracks which yield straight lines not passing through, or near, the origin of the u–v plane. The Hough transform [109] provides a useful remedy to this problem.

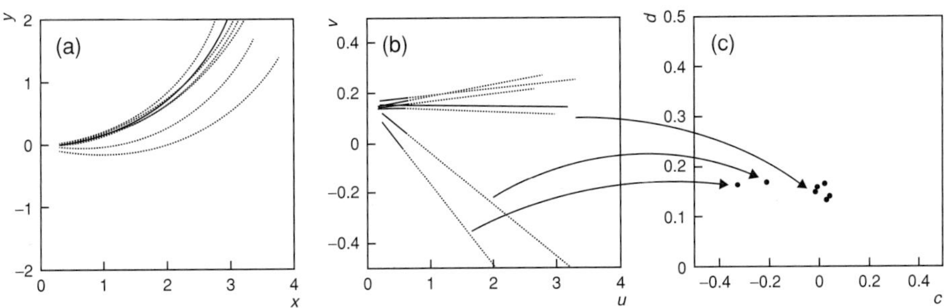

Fig. 9.11 Example of the Hough transform used in the reconstruction of tracks. (a) Hits from charged particles in the bending plane (x–y). (b) Conformal mapping of the hits onto u–v space. (c) Hough transform onto c–d space. Hits belonging to given tracks yield peaks, here represented as solid dots.

The Hough transform is based on a simple mapping of the equation of a straight line in the u–v plane, written $v = cu + d$ to another straight line in the c–d plane, $d = -uc + v$. This means that points along a line in the c–d plane correspond to all possible lines passing through the point (u, v) in the u–v plane. Points located on a straight line in the u–v plane thus tend to create peaks in the c–d plane, which cross at the point that specify the actual parameters of the line in the u–v plane. One can then discretize the c–d plane and increment a histogram for each hit value (x, y). Peaks in this space therefore identify curves in the x–y space. As illustrated in Figure 9.11, one gets a two-dimensional histogram that can be quickly scanned for peaks, which correspond to circular tracks in the x–y plane.

The Hough transform can be adapted to track finding with drift tubes using Legendre transforms [18]. It has also been extended to a variety of pattern recognition problems.

Road Finders

Road finders, also called track template finders, are an example of a global finder technique based on specific track models. The idea is to map each measured point or hit onto all of the track roads (or track templates) to which they could belong. Once all hits of an event have been processed, one then counts how many hits each track template gets. Track candidates are then those templates that have a required minimum number of entries, as illustrated in Figure 9.12. The track candidates must next be fitted to obtain precise kinematic parameters and test their quality.

The road finder technique is quite useful and efficient for finding primary tracks with a specific vertex point. However, it becomes computationally onerous for searching secondary tracks because the number of roads to be considered is likely to be extremely large. In very large track multiplicity environments with a high yield of secondaries, track-following techniques are then more likely to be effective.

Track-Following Technique

The **track-following** technique is a rather generic track-finding method that can be applied with or without track models. Its use in the absence of a track model is possible in a low

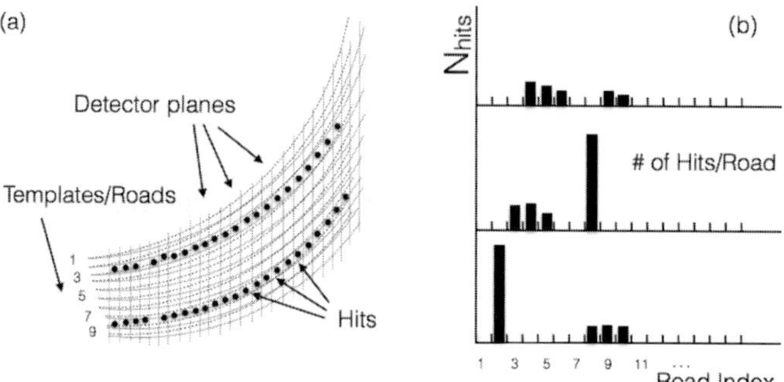

Fig. 9.12 Schematic illustration of the road finder technique. All hits in all detector planes are scanned. For each hit, one increments all bins corresponding to the road they cross. Tracks are those roads with the largest number of hits.

hit density environment and requires the distance between tracking planes be much smaller that the typical distance between hits on a given plane. However, the use of track models enables track reconstruction in much higher density environments.

Track following starts with a **track seed** and proceeds iteratively to the extension of the track, as illustrated in Figure 9.13. Seeds can be formed in the inner region of the tracking detector (e.g., close to the interaction region) if high precision hit measurements are available and the hit density is sufficiently modest. The finder is then described as an inside-out finder. Alternatively, the seed may be first constructed in the outer region of the detector, where hit and track densities are usually much lower. Track extension proceeds in three steps: extrapolation to the next layer, search for matching hits, and choice of best

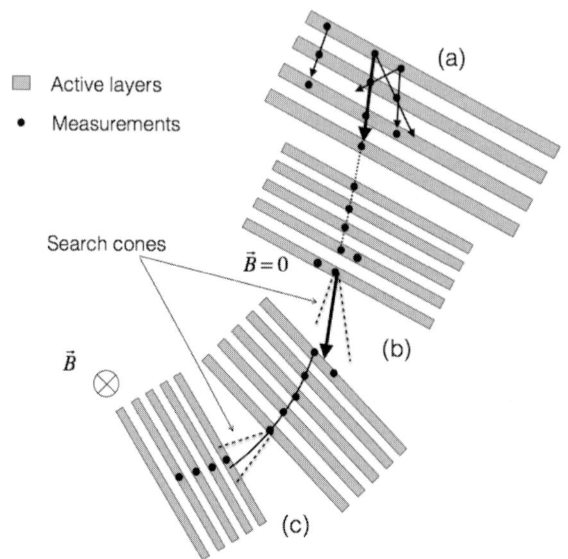

Fig. 9.13 Schematic illustration of the track-following technique. (a) Formation of track seeds in the outermost layers of the detection system. (b) Linear extrapolation in field free regions. (c) Circular extrapolations in finite field regions.

matching hit if multiple matching hits are found. If no hit is found on a particular layer, the extrapolation is carried to the next layer. The extension procedure is iterated until the number of layers with a missing hit exceeds a preset threshold or until the boundary of the detection system is reached.

9.2.2 Track-Fitting Techniques

The track-fitting stage is subject to challenging constraints. It should be as computationally fast as possible, particularly in the study of relativistic heavy-ion collisions that produce thousands of tracks,[1] but it should not sacrifice the quality of reconstruction. It should be robust against mistakes (e.g., inappropriate hit association and missed hits) made during the track-finding procedure; it should be numerically stable and as accurate as possible based on the precision of the hits supplied by the tracking detectors. It should also be possible to explicitly test whether a track candidate is a valid track. The detection and addition of hits to a track are stochastic processes. Track finding and fitting are thus bound to have limited efficiency and accuracy and may be hampered by the inclusion of hits that do not actually belong to the track.

The testing of tracks can be accomplished, for instance, on the basis of the χ^2 obtained during the fit. The χ^2 is determined by the sum of the squared standardized differences between the measured positions of the track candidate and the estimated positions of the track at the points of intersection of the detector devices. If the χ^2 is very large (given the number of degrees of freedom), the set of associated hits is not statistically compatible with the hypothesis of having been produced by a single particle. The incompatibility may arise from a number of circumstances: for example, extraneous hits might have been wrongly added to the track candidate or a totally illegitimate candidate was built from a random set of hits produced by several distinct particles. Sometimes, the incompatibility results from detector noise or background (e.g., ghost tracks). An outlier hit removal stage and track refit are typically included in track-fitting algorithms. Other track selection criteria are often applied (see §8.3.1). The covariance matrix obtained from the fit additionally provides an estimate of the uncertainty of the track kinematic parameters.

A few elementary considerations are in order, including the choice of a track parameterization, track propagation model, error propagation, effect of detector materials, and track seed finding. These are discussed in following paragraphs of this section.

Track Parameterization

While the goal of track reconstruction consists in the determination of kinematic parameters $\vec{p} = (\vec{p}_T, \phi, p_z)$ of a particle at its point of production (i.e., a primary or secondary vertex), the reconstruction process and association of a track to its proper production vertex (see §9.3.1 for a discussion of vertex finding and track association to vertices) require accurate determination of its actual trajectory through the detection apparatus, which involves finding all hits belonging to a track.

[1] Or in the context of LHC run 3 expected to deliver unprecedented data volumes.

The trajectory of a charged particle in a static magnetic field $\vec{B}(\vec{r})$ is determined by the Lorentz force:

$$\frac{d\vec{p}}{dt} = \frac{d(m\gamma\vec{v})}{dt} = kq\vec{v}(t) \times \vec{B}(\vec{r}), \tag{9.12}$$

where $\vec{v}$ is the velocity of the particle, q its charge in units of elementary charge, and k a constant whose value depends on the units used to express the magnetic field and the particle momentum $\vec{p}$. The constant k equals 0.3 if the field and momenta are expressed in Tesla and GeV/c, respectively. Given the magnitude of the velocity v is a constant, the Lorentz factor $\gamma = (1 - v^2/c^2)^{-1/2}$ is also a constant, and Eq. (9.12) may be rewritten

$$\frac{d^2\vec{r}}{d^2l} = \frac{kq}{p}\frac{d\vec{r}}{dl} \times \vec{B}(\vec{r}(l)), \tag{9.13}$$

where l is the path length of the track and $dl/dt = v$. Choosing $\vec{B}$ to be parallel to the z-axis, a solution of this equation for uniform $\vec{B}$ corresponds to a helicoidal trajectory with an axis parallel to z:

$$\begin{aligned} x &= x_0 + R\left[\cos\left(\Phi_0 + \frac{hl\sin\Theta}{R}\right) - \cos\Phi_0\right], \\ y &= y_0 + R\left[\sin\left(\Phi_0 + \frac{hl\sin\Theta}{R}\right) - \sin\Phi_0\right], \\ z &= z_0 + l\cos\Theta, \end{aligned} \tag{9.14}$$

where $\vec{r}_0 = (x_0, y_0, z_0)$ is the starting point of the helix at $l = 0$, and $\Theta = \arcsin(\partial z/\partial l)$ determines the **pitch** of the helix. The radius of the helix is $R = p\sin\Theta/|kqB|$ while $h = -\text{sign}(qB_z)$ determines the direction of rotation of the helix in the transverse plane. The angle Φ_0 corresponds to the azimuth angle of the starting point in cylindrical coordinates with respect to the axis of the helix, as illustrated in Figure 9.14. The helix trajectory is thus completely specified by only five parameters: x_0, y_0, z_0, R, and Φ_0. This choice is not unique. Many other combinations of five parameters may also be used to uniquely define a helix. For instance, one could have alternatively chosen the center of the helix in the transverse plane ($x_c = x_0 - R\cos\Phi_0, y_c = y_0 - R\sin\Phi_0$), the radius R, and the helix pitch λ. Other choices are also possible.

There indeed is much freedom in the choice of the five track parameters used to express the state of particle $\vec{s}$ at a given detection surface. In practice, the choice of a specific set of parameters is usually determined by convenience and the geometry of the tracking detector. The product of the track radius and azimuthal angle of hits, $R\Phi$, is a convenient parameter choice for cylindrical detector layers, but Cartesian coordinates are more frequently used because most tracking devices involve a multilayered planar detector geometry. The preceding parameterization, Eq. (9.14), while convenient for tracking in a uniform magnetic field is somewhat cumbersome for experiments where the field varies through the apparatus. A more detailed discussion of track parameterizations may be found in the book by Frühwirth et al. [89].

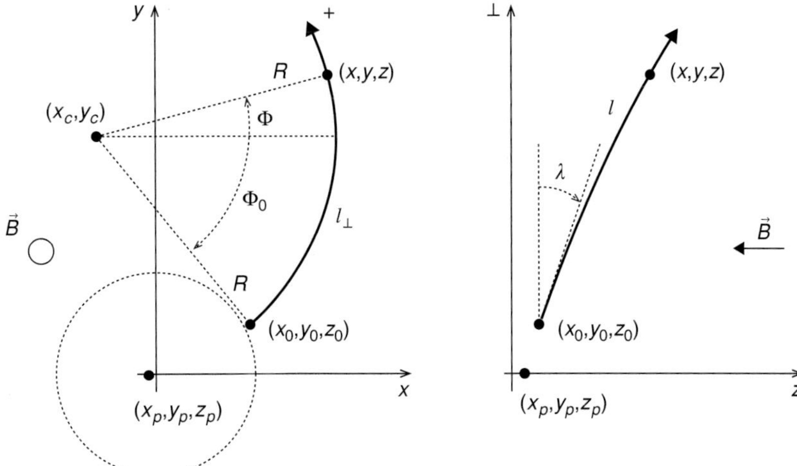

Fig. 9.14 Schematic illustration of a basic helicoidal track parameterization: R and (x_0, y_0, z_0) are the radius and first point measured on the helix; (x, y, z) represents an arbitrary point along the helix. (x_p, y_p, z_p) and (x_c, y_c) are the primary vertex and the geometrical center of the helix in the transverse plane $\perp$, respectively.

Track Propagation Model

As illustrated in Figure 9.15, a **track propagation model** or simply **track model** $\vec{f}_{j|i}(\vec{s_i})$ is required to describe how the track parameters, $\vec{s}$, or **state vector**, can be propagated from one detector surface i to a different surface j.

$$\vec{s_j} = \vec{f}_{j|i}(\vec{s_i}) \tag{9.15}$$

An analytical track model can usually be formulated for simple detection surfaces (e.g., planar and cylindrical surfaces) operated in a vanishing (straight lines) or homogeneous magnetic field (helicoidal trajectories). However, propagation of trajectories in inhomogeneous magnetic fields and nonelementary surfaces requires the use of numerical methods such as Runge–Kutta integration techniques.

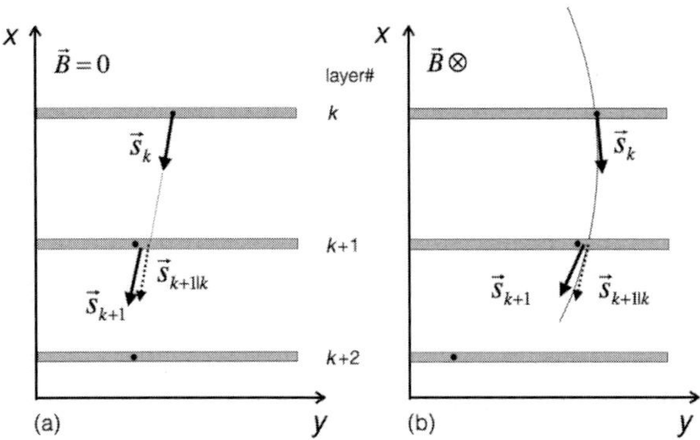

Fig. 9.15 Schematic illustration of a track propagation models for (a) $\vec{B} = 0$ and (b) finite + uniform field.

Error Propagation

Propagation of the track parameter covariance matrix, $\mathbf{S}_i$, along with its state $\vec{s}_i$, is required in Kalman fit and adaptive fit methods. This can be usually accomplished using linear error propagation techniques (see §2.11). For instance, given the error covariance matrix at the detection surface i, one may extrapolate the state of the track to surface j with Eq. (9.15), and the extrapolated covariance matrix can then be written

$$\mathbf{S}_{j|i} = \mathbf{F}_{j|i}\mathbf{S}_i(\mathbf{F}_{j|i})^T \tag{9.16}$$

where the transformation matrix $\mathbf{F}_{j|i}$ is defined according to

$$\mathbf{F}_{j|i} = \frac{\partial \vec{s}_j}{\partial \vec{s}_i} \tag{9.17}$$

The functions $\mathbf{F}_{j|i}$ are analytical for analytical track models, but numerical techniques or semianalytical techniques must be used when propagation through inhomogeneous fields is involved [60].

Effect of Detector Materials

Charged particles propagating through a detector undergo interactions with materials present along their path. Such interactions are typically dominated by ionization energy loss and multiple Coulomb scattering. Energy loss by emission of Bremsstrahlung radiation also plays an important role for light particles such as electrons. In well-designed detectors, fluctuations in energy loss are relatively small and can be treated as deterministic corrections to the state vector. Fluctuations associated with Bremsstrahlung radiation are typically large and thus affect both the state vector and its covariance matrix. Multiple coulomb scattering is an elastic and stochastic process. With thin or low density scatterers, it only disturbs the direction of motion of charged particles, but with thick scatters one must also account for a change in position in the plane of scattering materials. However, since the expectation value of the scattering process is null, only the covariance matrix of the state vector needs updating. An example of treatment of such effects is presented in §9.2.3.

Measurement Model

The measurement model, denoted $\vec{h}(\vec{s})$, expresses the functional dependence of a measurement vector, $\vec{m}_k$, in a specific detector layer k on the state vector $\vec{s}_k$ of a particle in that same layer:

$$\vec{m}_k = \vec{h}(\vec{s}_k) \tag{9.18}$$

Much flexibility is available in the definition of measurement vector $\vec{m}$. While position measurements (y_k, z_k) in the detection plane k are most commonly used, other choices are possible that may involve measurements of direction or even the momentum of the particle.

It is convenient to define a projection matrix $\mathbf{P}_k$ as

$$\mathbf{P}_k = \frac{\partial \vec{m}_k}{\partial \vec{s}_k}. \tag{9.19}$$

This projection matrix is typically needed for state and covariance matrix propagation which account for measurement errors. The follow-your-nose algorithm uses the track parameters at layer k to carry out a projection to layer $k-1$ (or other inner layers) to obtain an estimate of the position $(x_{k-1}, y_{k-1}, z_{k-1})$ of the "next" hit. A search is thus conducted for matching hits in the vicinity of this point with a maximum search radius r_{k-1}. The search radius may be kept constant throughout the extension (i.e., for all layers) or may be tailored to progressively shrink as the length of the track increases. This is most naturally accomplished in the context of a Kalman filter algorithm, which provides layer-by-layer, estimates of the track parameter covariance matrix. If two or more matching hits are found within the search radius, one may elect to use the closest, or attempt track extensions with all matching candidates and choose the one that eventually leads to the longest and highest quality track. The extension of a track stops once no more hits can be found or when the inner-most layer is reached. The track can be optionally extended outward toward other detectors, if any. It is eventually saved/stored if it meets required minimal quality criteria. The track reconstruction then switches to seed finding for a new track. Once all hits have been used or no more tracks are found, one can optionally initiate a second pass with looser search criteria.

Once track reconstruction is completed, one can then proceed to combine all tracks and identify the position of the primary vertex. When the collision vertex is identified, an additional pass is typically performed on all tracks to attempt a projection to this vertex and carry out a track refit that includes the vertex. Since the primary vertex is the origin of many tracks, particularly in heavy-ion collisions, it is possible to achieve very high precision in the determination of its position. A track refit including this very precise position thus enables an improved determination of the momentum and direction of primary tracks. It is also possible to carry out a refit that excludes outlier hits based on some preset maximum fit residue Δr.

A fit of the track and projection at the primary vertex yields its (transverse) momentum and direction at the point of point of production. In collider geometries, the direction is usually specified in terms of an azimuthal angle and a pitch angle or pseudorapidity. The use of polar angles measured to the beam axis tends to be reserved for fixed-target experiments or forward-going particles in collider geometries.

Track Seed Finding

The principle of seed finding algorithms is shown in Figure 9.16, where we use the variable x to express the position of tracking planes, and the variables y (and z) to identify the position of hits in a given plane.

Track reconstruction is often initiated in a region of low hit density, typically the detection planes farthest from the production vertex, and proceeds "inward" toward regions of higher hit and track density. The algorithm proceeds iteratively: a seed is identified, and

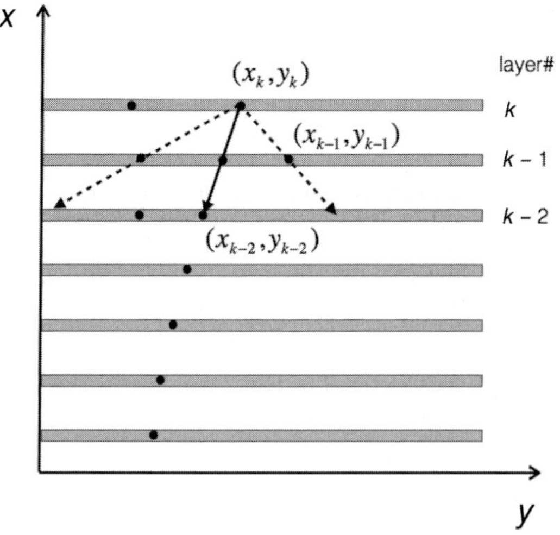

Fig. 9.16 Illustration of the linear seed-finding technique.

it is progressively extended using the method described in the text that follows. Once the innermost track layer is reached or preset number of tracking planes are traversed without successful addition of hits, the track may be fitted and stored if it meets minimal quality criteria or otherwise simply abandoned and deleted. Most tracking algorithms limit the use of hits to one track per hit, that is, they do not allow hit sharing between tracks. Hits associated with a track are thus flagged as "taken" and removed from the pool of hits to be associated with tracks. However, track reconstruction with hit sharing is also possible.

Track seeds are formed by associating hits from two to five (rarely more) contiguous outer tracking layers, as illustrated in Figure 9.13. Beginning, say, on the outermost layer, $k = n$ (where n is the number of tracking planes), one selects a hit and search on layer $k - 1$ for matching hits. In order to limit random associations, hits belonging to neighboring layers are usually required to be within a preset maximal distance along the y- and z-axes. One may also begin the search with nearly "vertical" tracks by setting the y–z ranges to be very narrow, and recursively increase the search radius up to its maximum preset value. Once a second hit is found, one proceeds to calculate the parameters of straight lines in the x–y and x–z planes formed by these two hits according to

$$\begin{array}{ll} a_{1,y} = \dfrac{y_k - y_{k-1}}{x_k - x_{k-1}} & a_{0,y} = y_k - a_{1,y} x_k, \\ a_{1,z} = \dfrac{z_k - z_{k-1}}{x_k - x_{k-1}} & a_{0,z} = z_k - a_{1,z} x_k \end{array} . \tag{9.20}$$

The track parameters are used to extrapolate the tracklet to layer $k - 2$ according to

$$y_{k-2} = a_{0,y} + a_{1,y} x_{k-2}, \tag{9.21}$$

$$z_{k-2} = a_{0,z} + a_{1,z} x_{k-2}. \tag{9.22}$$

Then, one searches for hits in the vicinity of the point (y_{k-2}, z_{k-2}) within a preset maximum search radius. If the search is not successful, one may optionally attempt to find matching hits on layer $k-3$ or $k-4$. If a matching hit found, it is added to the tracklet. The search can be optionally extended to additional inner layers. If the tracking planes used to form the seed are located in a field-free region, one proceeds with a simple linear fit to obtain straight line parameters in the x–y and x–z planes. If the planes are within a finite magnetic field, one carries a fit using a helicoidal track model. Either way, one gets an initial estimate of the track parameters. The tracklet may then be passed to the main tracking routine for an extended search through the rest of the detector.

9.2.3 Kalman Filter Track Finding and Fitting

Most track-fitting techniques are based on linear least-squares methods. The linear, global least-square method is optimal [178] if the track model itself is linear, that is, if the propagation function defined by Eq. (9.15) is a linear function of the state vector $\vec{s}$. In cases where the functions $\vec{f}(\vec{s})$ are nonlinear, the linear least-squares method remains the optimal linear estimator but may lack robustness [166]. Least-squares track-fitting methods are discussed in a generic context in ref. [178]. Here, we focus our discussion on an example of the Kalman filter method largely developed for track and vertex finding by Billoir [42] and Frühwirth [86].[2] The principle and properties of Kalman filters (KFs) were already discussed in §5.6, where we showed, in particular, that KFs are strictly equivalent to least-squares methods. KF methods also present the advantage, particularly in the context of track finding and fitting, that only small number of matrix inversions are required to carry out track fits, while global fit methods may require repeated inversion of large matrices [178].

As an example of application of the Kalman filter, we consider the reconstruction of charged particle tracks within a uniform solenoid magnetic field spectrometer, as schematically illustrated in Figure 8.15. The detection system consists of n_s identical azimuthal sectors each divided into n_p measurement planes. Collisions take place near the origin of the coordinate system and produce particles streaming out at high momentum. Charged particles are deflected by the magnetic Lorentz force and follow helicoidal trajectories one describes according to Eq. (9.14).

The coordinates x_c, y_c represent the center of the circle formed by projecting the helicoidal trajectory in the x–y plane traverse to the beam–beam collision axis, which defines the z-axis. R, φ, and λ are the radius, the phase angle, and the pitch of the helix, as defined in Figure 9.14. We arbitrarily choose the x-axis to be local to each sector and along a radius normal to the measurement planes. The y-axis is chosen along the measurement planes and perpendicular to the beam direction. In this context, the position of the detection planes along the x-axis constitutes a natural choice for the independent variable t. Helicoidal trajectories (9.14) involve nine variables and three independent equations. Choosing x as the independent variable leaves $(9-3-1)=5$ independent track state parameters. The choice of these five parameters is somewhat arbitrary. Given that position measurements in each detection plane are performed in the y–z plane, it is natural to use

[2] The example presented is based on the software developed by the author for the STAR experiment at the Relativistic Heavy-Ion Collider.

these two coordinates as part of the Kalman state. The measurement vector is set to be

$$\vec{m}_k = \begin{bmatrix} y_k \\ z_k \end{bmatrix}, \tag{9.23}$$

where y_k and z_k are the coordinates of track hits on measurement plane $k = 1, \ldots, n_p$. We similarly choose those two coordinates to be the first two elements of the Kalman state vector. This leaves us with only three more parameters to define or identify. Many choices are still possible. For illustrative purposes, we choose variables η, C, and ξ. C is the curvature defined as the multiplicative inverse of the radius R of the track,

$$C \equiv \frac{1}{R}. \tag{9.24}$$

The variable ξ is defined as the tangent of the pitch angle of the track which yields

$$\xi = \tan\lambda = \frac{p_z}{p_T}, \tag{9.25}$$

where p_T and p_z are the transverse and longitudinal momentum of the charged particle that produced the track, respectively. Defining the azimuthal angle φ of the track as

$$\tan\varphi = \frac{p_y}{p_x}, \tag{9.26}$$

one calculates η according to

$$\eta = Cx - \sin\varphi. \tag{9.27}$$

The track Kalman momentum state on plane k, denoted $\vec{s}_k$, is thus

$$\vec{s}_k = \begin{bmatrix} y_k \\ z_k \\ \eta_k \\ C_k \\ \xi_k \end{bmatrix}. \tag{9.28}$$

With the preceding choices for $\vec{m}_k$ and $\vec{s}_k$, the measurement matrix, $\mathbf{H}_k$, is reduced to a trivial form:

$$\mathbf{H}_k = \begin{bmatrix} 1 & 0 & 0 & 0 & 0 \\ 0 & 1 & 0 & 0 & 0 \end{bmatrix}. \tag{9.29}$$

Inserting this specific form for $\mathbf{H}_k$ in Eq. (5.164), we get

$$K_k = \mathbf{S}_{k|k-1} (H_k)^T \tag{9.30}$$

$$\times \left(\mathbf{V}_k + \begin{bmatrix} 1 & 0 & 0 & 0 & 0 \\ 0 & 1 & 0 & 0 & 0 \end{bmatrix} \right)^{-1} \begin{bmatrix} C_{11} & C_{12} & C_{13} & C_{14} & C_{15} \\ C_{21} & C_{22} & C_{23} & C_{24} & C_{25} \\ C_{31} & C_{32} & C_{33} & C_{34} & C_{35} \\ C_{41} & C_{42} & C_{43} & C_{44} & C_{45} \\ C_{51} & C_{52} & C_{53} & C_{54} & C_{55} \end{bmatrix}_{k|k-1} \begin{bmatrix} 1 & 0 \\ 0 & 1 \\ 0 & 0 \\ 0 & 0 \\ 0 & 0 \end{bmatrix}.$$

The propagation (projection) of the state from plane k to $k+1$ is accomplished by incrementing the independent variable x from x_k to $x_{k+1} = x_k + \Delta x_{k,k+1}$, where $\Delta x_{k,k+1}$ is the distance between the two detector planes. The projected state vector is written

$$\vec{s}_{k+1} = \mathbf{F}\left(\vec{s}_k\right). \tag{9.31}$$

Since the state projection involves a displacement of the position of the track in a uniform longitudinal magnetic field, the parameters η, C, and ξ remain unchanged:

$$\eta_{k+1} = \eta_k, \tag{9.32}$$

$$C_{k+1} = C_k, \tag{9.33}$$

$$\xi_{k+1} = \xi_k. \tag{9.34}$$

Only the positions y and z need updating. Using the geometry illustrated in Figure 9.14, one finds after some simple algebra

$$y_{k+1} = f_1(\vec{s}_k) = y_k + \frac{\sin\varphi_k + \sin\varphi_{k+1}}{\cos\varphi_k + \cos\varphi_{k+1}}, \tag{9.35}$$

$$\begin{aligned} z_{k+1} &= f_2(\vec{s}_k) = z_k + \xi_k\left(\varphi_{k+1} - \varphi_k\right)/C_k, \\ &= m_k + \frac{\xi_k}{C_k}\sin^{-1}\left(\frac{(\sin\varphi_{k+1} - \sin\varphi_k)(\sin\varphi_{k+1} + \sin\varphi_k)}{\sin(\varphi_k + \varphi_{k+1})}\right), \end{aligned} \tag{9.36}$$

where the sine functions can be expressed in terms of η according to Eq. (9.27).

The propagation of the state error matrix is accomplished with

$$\mathbf{S}_{k|k-1} = \mathbf{F}_k\mathbf{S}_{k-1|k-1}(\mathbf{F}_k)^T + \mathbf{W}_k, \tag{9.37}$$

where $\mathbf{S}_{k-1|k-1} \equiv \mathbf{S}_{k-1}$ is the state error matrix obtained after the $(k-1)$th step of Kalman filtering. The propagation of $\vec{s}$ determined by Eq. (9.31) is nonlinear and makes the propagation of the error matrix rather cumbersome. Fortunately, since the propagation of errors does not have to be as precise as the propagation of the state itself, one can then use a linearization of Eq. (9.31) to obtain the projection matrix $\mathbf{F}_k$ required in Eq. 9.37. Consider the error at step k is of order $\Delta\vec{s}_k$. The error at step $k+1$ may then be written

$$\Delta\vec{s}_{k+1} = \mathbf{F}\left(\vec{s}_k + \Delta\vec{s}_k\right) - \mathbf{F}\left(\vec{s}_k\right). \tag{9.38}$$

Using a truncated Taylor expansion of the first term on the right-hand side, one gets

$$\Delta\vec{s}_{k+1,j} = \sum_{j=1}^{5}\Delta\vec{s}_{k,j}\frac{\partial f_j(\vec{s}_k)}{\partial s_j}. \tag{9.39}$$

The covariance matrix $\mathbf{S}_{k+1}$ is defined as the expectation value of $\langle\Delta\vec{s}_{k+1,i}\Delta\vec{s}_{k+1,j}\rangle$ over an ensemble of measurements at step $k+1$. This can be written

$$\langle\Delta\vec{s}_{k+1,i}\Delta\vec{s}_{k+1,j}\rangle = \sum_{m,n=1}^{5}\frac{\partial f_i(\vec{s}_k)}{\partial s_m}\frac{\partial f_j(\vec{s}_k)}{\partial s_n}\langle\Delta\vec{s}_{k,i}\Delta\vec{s}_{k,j}\rangle. \tag{9.40}$$

We thus obtain the matrix $\mathbf{F}$ as

$$(\mathbf{F}_k)_{i,m} = \frac{\partial f_i(\vec{s}_k)}{\partial s_m}. \tag{9.41}$$

In order to calculate the preceding derivatives, we first note

$$\frac{\partial \sin\varphi}{\partial \eta} = \frac{\partial (Cx-\eta)}{\partial \eta} = -1, \tag{9.42}$$

$$\frac{\partial \cos\varphi}{\partial \eta} = \tan\varphi, \tag{9.43}$$

$$\frac{\partial \sin\varphi}{\partial C} = x, \tag{9.44}$$

$$\frac{\partial \cos\varphi}{\partial C} = -x\tan\varphi. \tag{9.45}$$

We next define $u = \sin\varphi_{k+1}\cos\varphi_k - \cos\varphi_{k+1}\sin\varphi_k$ and compute the following derivatives:

$$\frac{\partial z_{k+1}}{\partial u} = \frac{\xi}{C\sqrt{1-u^2}}, \tag{9.46}$$

$$\frac{\partial z_{k+1}}{\partial \sin\varphi_{k+1}} = \frac{\xi}{C\sqrt{1-u^2}}\cos\varphi_k, \tag{9.47}$$

$$\frac{\partial z_{k+1}}{\partial \sin\varphi_k} = -\frac{\xi}{C\sqrt{1-u^2}}\cos\varphi_{k+1}, \tag{9.48}$$

$$\frac{\partial z_{k+1}}{\partial \cos\varphi_k} = \frac{\xi}{C\sqrt{1-u^2}}\sin\varphi_{k+1}, \tag{9.49}$$

$$\frac{\partial z_{k+1}}{\partial \cos\varphi_{k+1}} = -\frac{\xi}{C\sqrt{1-u^2}}\sin\varphi_k. \tag{9.50}$$

Derivatives of z_{k+1} with η and C can then be written

$$\frac{\partial z_{k+1}}{\partial \eta} = \frac{\xi}{C\sqrt{1-u^2}}(\xi\sin\varphi_{k+1} - \xi\sin\varphi_k + \xi\cos\varphi_{k+1} - \xi\cos\varphi_k), \tag{9.51}$$

$$\frac{\partial z_{k+1}}{\partial C} = \frac{\xi}{C\sqrt{1-u^2}} \times (-x_1\xi\sin\varphi_{k+1} + x_2\xi\sin\varphi_k - x_1\xi\cos\varphi_{k+1} + x_2\xi\cos\varphi_k). \tag{9.52}$$

Based on these, we can next calculate the elements of the matrix $\mathbf{F}$. All diagonal elements F_{ii} are equal to unity. Other non-null elements are as follows:

$$F_{13} = \Delta x_{k+1,k}\frac{-2(\cos\varphi_k + \cos\varphi_{k+1}) - (\sin\varphi_k + \sin\varphi_{k+1})(\tan\varphi_k + \tan\varphi_{k+1})}{(\cos\varphi_k + \cos\varphi_{k+1})^2},$$

$$F_{14} = \Delta x_{k+1,k}\frac{(x_k + x_{k+1})(\cos\varphi_k + \cos\varphi_{k+1}) + (\sin\varphi_k + \sin\varphi_{k+1})(\tan\varphi_k + \tan\varphi_{k+1})}{(\cos\varphi_k + \cos\varphi_{k+1})^2},$$

$$F_{23} = \frac{\xi}{C\sqrt{1-u^2}}(\xi\sin\varphi_{k+1} - \xi\sin\varphi_k + \cos\varphi_{k+1} - \cos\varphi_k),$$

$$F_{24} = \frac{\xi}{C\sqrt{1-u^2}}(-x_k\xi\sin\varphi_{k+1} + x_{k+1}\xi\sin\varphi_k - x_k\cos\varphi_{k+1} + x_{k+1}\cos\varphi_k),$$

$$F_{25} = \Delta x_{k+1,k}\frac{\sin\varphi_k + \sin\varphi_{k+1}}{\sin(\varphi_k + \varphi_{k+1})}.$$

An energy loss correction may also be included in the propagation of the tracks through the different detection planes. Propagating tracks inward, that is, toward the production vertex, one must add the average loss of tracks traversing the detection plane. The energy of the track is calculated according to

$$E_{k+1} = E_k + \Delta E_{k,k+1} = E_k + \Delta x_{k,k+1} \left| dE/dx \right|, \tag{9.53}$$

where dE/dx is calculated with the Bethe–Bloch formula, Eq. (8.131). The added energy implies an increase in momentum, which translates into a change of curvature:

$$C_{k+1}(E_{k+1}) = C_k(E_k) + \Delta E_{k,k+1} \left.\frac{dC}{dE}\right|_{E_k}. \tag{9.54}$$

The radius of curvature of the track, $R = 1/C$, is proportional to its momentum. One gets

$$\frac{dC}{dE} = -C\frac{\sqrt{p^2c^2 + m^2c^4}}{p^2}, \tag{9.55}$$

which yields the following correction for the curvature

$$C_{k+1}(E_{k+1}) \approx C_k(E_k)\left[1 - \frac{\sqrt{p^2c^2 + m^2c^4}}{p^2}\Delta E\right]. \tag{9.56}$$

Given the explicit dependence on the (unknown) mass of the particle, one must carry the correction with an arbitrary choice of particle identification, for instance, the mass of the pion, which is abundantly produced in high-energy collisions. However, the choice of mass used in Eq. (9.56) has typically a relatively small impact on the reconstruction of the charged particle tracks with very high momenta.

Multiple coulomb scattering through various material layers of the detector contributes a process noise to the propagation of the state covariance matrix (§9.5). In the local reference frame of a track, the scatterings cause tracks to be deflected vertically and horizontally by angles we label as γ_1 and γ_2, as illustrated in Figure 9.17.

For a thin scatterer, the noise process covariance matrix can then be estimated with the expression

$$\mathbf{W} = \langle\gamma^2\rangle \left(\frac{\partial(\eta, C, \tan\lambda)}{\partial(\gamma_1, \gamma_2)}\right) \begin{bmatrix} 1 & 0 \\ 0 & 1 \end{bmatrix} \left(\frac{\partial(\eta, C, \tan\lambda)}{\partial(\gamma_1, \gamma_2)}\right)^T, \tag{9.57}$$

which after considerable algebra reduces to

$$\mathbf{W}_k = \langle\gamma^2\rangle \begin{bmatrix} x^2C^2\xi^2 + (1+\xi^2)\cos^2\varphi & x^2C^2\xi^2 & xC\xi(1+\xi^2) \\ xC^2\xi^2 & C^2\xi^2 & C\xi(1+\xi^2) \\ xC\xi(1+\xi^2) & C\xi(1+\xi^2) & (1+\xi^2) \end{bmatrix}, \tag{9.58}$$

with

$$\langle\gamma^2\rangle = \left(\frac{14.1}{p\beta}\right)^2 \frac{X}{X_o}, \tag{9.59}$$

where X and X_o stand for the thickness and the radiation length of the scatterer, respectively.

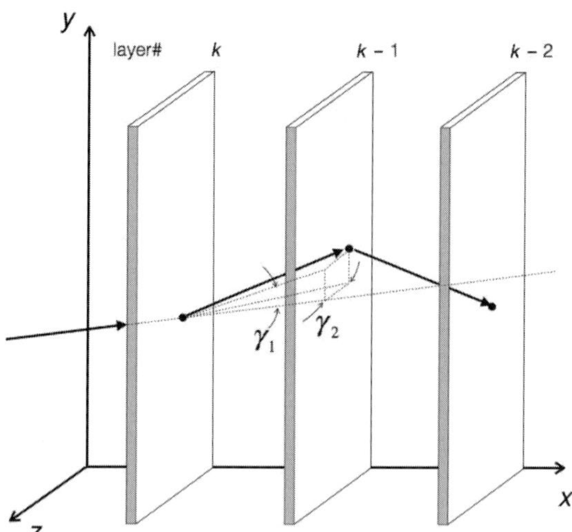

Fig. 9.17 Track geometry and multiple coulomb scattering. The scattering angles γ_1 and γ_2 have been exaggerated for the sake of clarity.

9.2.4 Removal of Hit Outliers and Competition for Hits

Track candidates produced by a track-finder can involve one or more spurious hits that belong to another track or are the result of detector noise. Given that such hits may steer or distort a track, it is appropriate to consider a third track reconstruction stage, known as **outlier removal**, where one proceeds to eliminate outlier hits, that is, hits that appear to lie too far from the track obtained by a fit.

Rejection of hit outliers is in principle readily accomplished using a χ^2 cut. Indeed, if the hit errors are known and well calibrated, the χ^2 of a fit including an outlier point is expected to be large. One can, for instance, compare the χ^2 obtained with and without a hit under consideration. If the χ^2 that includes the hit exceeds a preset value (relative to the one that excludes it), the point is simply removed from the track. One may equivalently consider the normalized fit residue of the hit based on a fit that excludes it. However, the presence of several outliers may bias or skew a track so much that they render either of these procedures completely inadequate.

An alternative approach is to increase the robustness of the fit, thereby reducing the influence of potential outliers. This can be accomplished using adaptive estimation techniques, discussed in the next section, which automatically reduce the weight of outlying observations. Yet another technique involves the use of Tukey's bi-square function (see. e.g., [95, 101]).

Another common issue arising in track finding and fitting is that two tracks may have one or more hits in common. This is particularly the case in track pairs produced by the Hanbury-Brown and Twiss (HBT) effect: two particles of same charge sign may be emitted with nearly the same momenta and angles and thus propagate near one another through the detector. Finite hit resolution may then cause two tracks to share one or several hits.

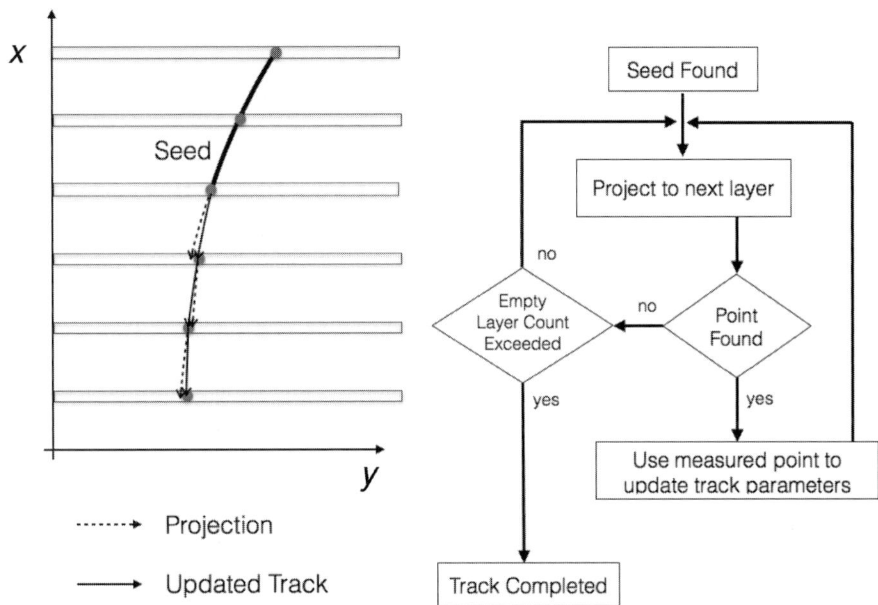

Fig. 9.18 Example of progressive track recognition with a Kalman filter algorithm. A seed formed in the outer layers of the detector is extended inwards through a recursive process involving projection of the track to the next inner layer and use of a matching point to update the track parameters.

Such tracks may be considered incompatible or at the very least biased, since shared hits are likely to be merged hits with suboptimal accuracy because the actual two hits they correspond to cannot be resolved. A technique based on graphs in which every track corresponds to a node is presented in Das [77]. Such technique may result in the production of several solutions of the same size (i.e., number of tracks). It may then be desirable to choose tracks also based on their quality. This is accomplished most generally by assigning each track with a quality index (based on its χ^2, length, distance of closest approach to the primary vertex, etc.) The best node is the one that maximizes the sum of quality indices. A technique to find the best node based on Hopfield networks is discussed in [178] and references therein.

9.2.5 Hybrid Track Finding and Fitting Techniques

The Kalman filter method can conveniently be used to achieve track finding and track-fitting concurrently [40, 43] with the algorithm illustrated in Figure 9.18. The process starts with a track seed obtained from hits located in contiguous layers. The seed is extended iteratively to adjacent layers based on the track parameters determined by the Kalman fit. A track candidate is abandoned if too few compatible hits are found and added to the track.

The χ^2 of the residual $\vec{r}_{k|k-1}$ of the measurement $\vec{m}_k$ with respect to the predicted state can be used as an indicator of the compatibility of a given hit to a track:

$$\chi^2_{k,+} = \vec{r}^{\,T}_{k|k-1}\mathbf{V}^{-1}_{k|k-1}\vec{r}_{k|k-1}, \tag{9.60}$$

where

$$\vec{r}_{k|k-1} = \vec{m}_k - \vec{h}_k(\vec{s}_{k|k-1}), \tag{9.61}$$

and the covariance matrix of the residuals is

$$\mathbf{V}_{k|k-1} = \mathbf{V}_k + \mathbf{H}_k \mathbf{S}_{k|k-1} \mathbf{H}_k^T. \tag{9.62}$$

In cases where several hits on a given detector layer might be compatible with a track projection, one should choose the one that yields the lowest χ^2, given by Eq. (9.60), add it to the track candidate, update the Kalman state, and use it to extend (predict) the track further.

A generalized Kalman filter technique, known as **combinatorial Kalman filter**, may be used to deal with high track density environments or a high density of noisy measurements [139]. Global/local hybrid track finding techniques based on cellular automata have also been developed to cope with such situations [94, 127]. A generic Kalman finder/fitter named GenFit is reported in [108].

9.3 Primary Vertex Reconstruction Techniques

A vertex consists of a point where particles are produced either as a result of a collision between beam and target particles, by the decay of a particle, or by an interaction of a particle with the material of the detector. The vertex corresponding to a beam + target leads to the production of **primary particles** and is known as a **primary vertex**, whereas vertices corresponding to decays are known as **secondary vertices**. While there are many commonalities between the identification of primary and secondary vertices, we focus in this section on the reconstruction of primary vertices.

The primary vertex of a collision corresponds to the point (location) where the collision between the beam and target particles took place. In fixed-target experiments, the position of primary vertices is largely constrained by the transverse cross section of the beam and the physical thickness of the target. By contrast, in collider experiments the position of the vertex may vary considerably from collision to collision owing to the longitudinal extension of the colliding beam bunches. The volume spanned by collisions is commonly known as the **collision diamond**. The longitudinal and transverse sizes of the collision diamond are determined by the beam crossing angle and the bunches' size, two parameters essentially fixed by the design of a collider. While the transverse size of the diamond is typically rather small (e.g., few hundred microns to few millimeters), the longitudinal extension of the diamond may be rather long. For instance, at RHIC the Au – Au diamond initially had an root mean square of approximately 30 cm. This implies one cannot rely on the collision longitudinal diamond geometry, and one must determine the primary vertex position collision by collision (event by event). Knowledge of the position of the primary vertex of a collision is required first and foremost in order to be able to distinguish primary, secondary, and tertiary particles. It is also useful to establish cuts that suppress pile-up events (i.e., corresponding to two or multiple overlapping collisions), useful particularly in high-luminosity environments such as the LHC proton–proton running. The primary vertex is typically also used in the refit of primary tracks, given that, as we shall see in the text that follows, the precision of the position of the primary vertex usually far exceeds that of hits on tracks and thus enables a very precise constraint in the determination of the curvature of tracks. Precise knowledge of the primary vertex also enables, once secondary vertices are found, the measurement of decay length of short-lived particles.

Similarly to track reconstruction, the task of vertex reconstruction nominally involves two stages known as **vertex finding** and **vertex fitting**. Vertex finding typically begins with all valid tracks produced by the track reconstruction, supplied as a list of track parameter vectors. We discuss the basic algorithms and methods used for vertex finding in §9.3.1, which by and large all yield the same outcome consisting of a list of vertex candidates with their associated tracks. These candidates are then fed to a vertex fitter, which attempts to use all associated tracks to obtain an optimal vertex position. Selected vertex-fitting techniques are discussed in §9.3.2.

9.3.1 Vertex Finding

Vertex finding is a task involving the association and classification of reconstructed tracks of an event into one or several vertex candidates. Three types of vertices are nominally found. A **primary vertex** corresponds to the point of interaction between two beam particles (in a collider geometry) or a beam and a target (in a fixed-target geometry), while particles produced by the decay of an unstable particle (e.g., $\Lambda^0 \to p + \pi^-$) are said to originate from a **secondary vertex**. A secondary vertex (rarely called tertiary vertex) may also correspond to the point of production of background particles via the interaction of primary or secondary particles with the material of the detector.

Primary vertices are usually straightforward to find, particularly in heavy-ion collisions where they are associated with tens to thousands of particles. Exceptions arise in electron-positron colliders, for instance, for collisions that produce two short-lived neutral particles (e.g., $\Upsilon(4S) \to B^0\bar{B}^0$). Secondary vertices are usually limited to two or three particles and are thus more challenging to identify, particularly if they are produced in close proximity to a primary vertex in central heavy-ion collisions.

A large number of techniques have been devised and implemented in experiments to determine the position of vertices and their associated tracks. We present few illustrative examples of such methods and refer the reader to a recent comprehensive review by Strandlie for additional methods and details [178].

Track Clustering Methods

Clustering methods use the distance or proximity between objects (e.g., tracks) or measures of similarity to classify objects into clusters or groups. A cluster consists of a group of objects located within a small distance from one another (or with many shared features), but at a larger distance (or dissimilar features) relative to members of other clusters. There are a number of techniques to form clusters, but given the relative nature of the notion of proximity, it is useful to think in terms of hierarchical clustering or a tree of clusters. Each object is a leaf on the tree. Objects that are closest together form a small branch, and objects of two small branches that are close to one another are associated into a bigger branch, and so on. Clustering methods that begin with the leaves are said to be agglomerative or associative, whereas those that start at the root, that is, with all the objects, are known as divisive methods.

In high-luminosity experiments (e.g., in p–p collisions at the LHC), recorded events may involve a plurality of primary vertices. In some instances, it may be possible to record and analyze all vertices and their associated tracks, but in general, selection criteria are applied to choose which vertex is actually of interest. In proton–proton collisions, this could for instance be the interaction vertex with the largest number of high transverse momentum tracks (determined based on a sum of the p_T of all the tracks associated with a vertex), while in heavy-ion experiments, one is typically interested in the collision with the largest multiplicity of associated tracks.

However, note that in TPC-based experiments such as STAR or ALICE, the long drift and readout time of the TPC may lead to the pile-up of many collisions, and the collision that triggers the recording of an event is not necessarily the one featuring the largest multiplicity. Additionally, note that at very high luminosity, there is a finite probability that two distinct collisions may have primary vertices located so close to one another as to be virtually indistinguishable.

Various hierarchical and nonhierarchical clustering methods have been evaluated and documented in the literature (see [187] and references therein). In hierarchical agglomerative clustering, each track starts out as a single cluster. Clusters are then merged iteratively on the basis of a distance measure. It is important to realize that if two tracks a and b are close, and tracks b and c are also close, this does not imply that tracks a and c are necessarily close to one another. The distance between two clusters of tracks must therefore be defined as the maximum of the individual pairwise distances, known as complete linkage in the clustering literature. In contrast, divisive clustering starts out with a single cluster containing all tracks. This and subsequent clusters are iteratively divided based on some maximum distance criterion identifying the presence of outliers. Outliers are removed from a cluster and become the seed of one or several new clusters. The procedure is iterated until all produced clusters are smaller than a certain preset size and have no outlier tracks.

The **histogramming method** (also known as the **mapping method**) is an agglomerative technique based on hits from (pixel) detector layers located in close proximity to the beam axis (or nominal interaction region). It was implemented by the CMS experiment [71] and various other prior experiments but is otherwise little discussed in the literature, perhaps because of its intrinsic simplicity of design and implementation.

The technique consists in histogramming the point of intersection of tracklets, m_{ip}, (track consisting of three hits) with the nominal beam axis. Empty bins are discarded and nonempty bins are scanned. A cluster is defined as a contiguous set of bins separated by a distance smaller than a preset upper limit $\Delta m_{\max}$. Clusters are ranked according to their number of associated tracklets, and only clusters with a number of tracklets exceeding a

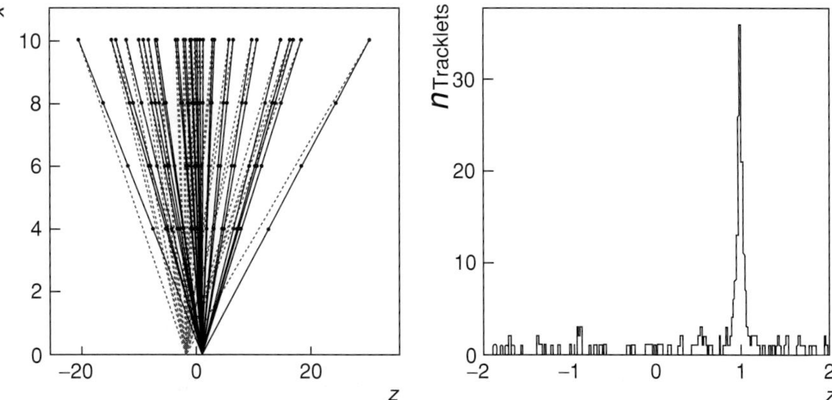

Fig. 9.19 Illustration of the principle of the vertex histogramming method. Fifty tracks were generated by Monte Carlo to originate from a vertex position at $z = 1$ cm and to produce hits on four detector layers as shown on the left. To find this vertex, one scanned the z-axis in small steps, as explained in the text, to find the position that yields the largest number of tracklets, as shown on the right, where one recovers the actual position of the vertex as a peak in the number of tracklets reconstructed at $z = 1$.

preset threshold are considered. Their positions and estimated errors are calculated based on a weighted average.

A variant of the technique, shown in Figure 9.19, consists in scanning the z-vertex axis in small steps. For each step, one determines the pitch angle, $\lambda = \tan^{-1}(z/x)$, of the hits and fills a histogram with these angles. High p_T tracks produce hits with similar pitch angle in the detector layers near the beam axis. One then counts the number of bins with more than three hits and fills a histogram with the number of tracklets with three or more hits. Primary vertices are those positions with a number of tracklets exceeding a preset threshold. This variant presents the advantage of being extremely fast since the number of operations scales as the number of hits while the first histogramming technique discussed earlier scales roughly as the number of hits to the 3rd (or 4th) power for 3 (4) hit tracklets.

Other approaches are also possible. For instance, a divisive clusterizer based on the intersection of tracklets, m_{in}, with the z-axis has been implemented by the CMS experiment [71]. Tracks are sorted by increasing value of m_{in} and the ordered list is scanned. Clusters are initiated and terminated when the gap between two consecutive track intersections exceeds a nominal threshold. For each initial cluster, an iterative procedure is applied to remove incompatible tracks. The discarded tracks are recovered to form a new cluster, and the same procedure is applied iteratively until there are fewer than two remaining tracks.

Linear Finders

Linear finders use charged particle track extrapolations near the beam axis (collider geometry) and local linear approximations. Rather than seeking a position P that minimizes the (average) distance of closest approach to all primary charged tracks, it is simpler and much

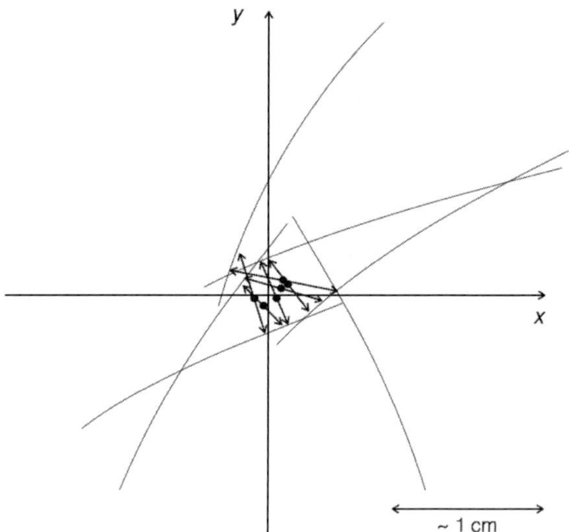

Fig. 9.20 Determination of the primary vertex position based on track pair DCAs. The vertex position is obtained as the three-dimensional average of all track pair DCA midpoint positions. Only a handful of tracks and pair midpoints are shown for the sake of clarity.

faster to calculate the mean position of the DCA midpoint of all primary track pairs, as illustrated in Figure 9.20.

The DCA midpoint, $\vec{r}_{ij}$ of a pair of tracks (i and j), corresponds to the bisection point of a segment $\Delta\vec{r}$ joining points at closest approach. A technique to calculate the position of the DCA midpoint is presented in §9.4. The position of the primary vertex $\vec{S}_v$ is estimated as an average of the DCA midpoint position of all pairs of charged tracks:

$$\hat{S}_v = \frac{2}{n(n-1)} \sum_{i,j=1}^{n} \vec{r}_{ij}, \tag{9.63}$$

where n is the number of tracks considered in the estimation of the vertex position. Given an event typically contains a combination of primary, secondary, and tertiary tracks, and possibly several vertices, one must ensure that the position determined with all tracks is not arbitrarily pulled toward specific secondary or tertiary particles. This can be accomplished in part by effecting a maximum DCA cut on the pairs included in the sum. One can also use an iterative procedure to eliminate outlier tracks from the sample used to evaluate the primary vertex position. Once an estimate of the primary vertex position is obtained, one calculates the DCA of all tracks of the sample (using the track-to-point DCA technique described in §9.4) to eliminate tracks that feature a DCA to the primary vertex in excess of a selected cut value. The vertex position can then be recalculated using the estimator, Eq. (9.63), using only tracks that satisfy the DCA cut. The procedure can be iterated until no further tracks are removed or too few tracks are left in the sample. This method is found to produce usable and reliable primary vertices when the number of tracks exceeds three.

Topological Vertex Finders

A general topological vertex finder method, known as ZVTOP, was introduced by Jackson [112]. It is related to the Radon transform, which is a continuous version of the Hough transform used for track finding (§9.2.1). The method and its recent developments are discussed in [178].

9.3.2 Vertex Fitting

Vertex fitting methods commonly used in particle and nuclear physics experiments include least-squares fitters, notably the Billoir vertex fitter [45], Kalman filter based fitters, as well as Kalman filters extended into adaptive vertex fitters. The Billoir vertex fitter uses all tracks associated to the vertex at once and estimates the vertex position by means of a global least-squares fit. By contrast, Kalman vertex fitters use a seed and proceed iteratively to add tracks to the vertex. Adaptive vertex fitters are an extension of the Kalman filter method and assign tracks incrementally to a vertex according to their compatibility with the vertex. A detailed description of Kalman filter vertex fitters and adaptive vertex fitters is well beyond the scope of this textbook, but the reader is referred to the review by [178] for details on these methods. For illustrative purposes, we here describe the Billoir vertex fitter following the original development of the method by its authors [45].

The Billoir vertex fitting method is based on a local parameterization of tracks in the vicinity of the vertex point. For convenience, one defines two different coordinate systems: a global coordinate system defined as the interaction point, and a local coordinate system in the neighborhood of the vertex point. After the vertex fit, this origin moves to the vertex point. The goal is to estimate the vertex position in the global reference frame and use it to recalculate the parameters of tracks in the local (vertex) frame. We demonstrate the applicability of the method for a uniform magnetic field of intensity $|\vec{B}|$ and the helix model stated in Eq. (9.14) as a starting point. The basic idea of the method is to project the helix trajectories near the local origin of the coordinate system. One may then use a first-order approximation to describe the local curvature of the tracks and employ all tracks to obtain an estimate of the vertex position by minimizing a χ^2 function. In so doing, one will obtain an "optimal" vertex position and updated track states and momentum vectors $\vec{p}_i$. The parameters relevant for the determination of the DCA to the z-axis are illustrated in Figure 9.21, in which d_0 is the transverse impact parameter of the projected helix. It corresponds to the DCA, or perigee, of the helix to the z-axis. m_0 is defined as the longitudinal impact parameter of the helix and corresponds to the z position of the particle at perigee. ϕ_0 is the azimuth angle of the momentum vector at the perigee. It is by convention measured in the range $[-\pi, \pi]$; θ is the polar angle, measured in the range $[0, \pi]$; and q/p the ratio of the charge to the momentum of the particle. The distance d_0 is signed according to the convention illustrated in Figure 9.21. It is defined as positive if $\phi - \phi_0 > 0$ and negative otherwise. In the vicinity of a vertex, the trajectory $\vec{r}$ of the particle may be locally expressed as

$$\vec{r}(l) \approx \vec{r}(l_0) + (l - l_0) \left.\frac{d\vec{r}}{dl}\right|_{l_0} + \frac{(l - l_0)^2}{2} \left.\frac{d^2\vec{r}}{dl^2}\right|_{l_0} \tag{9.64}$$

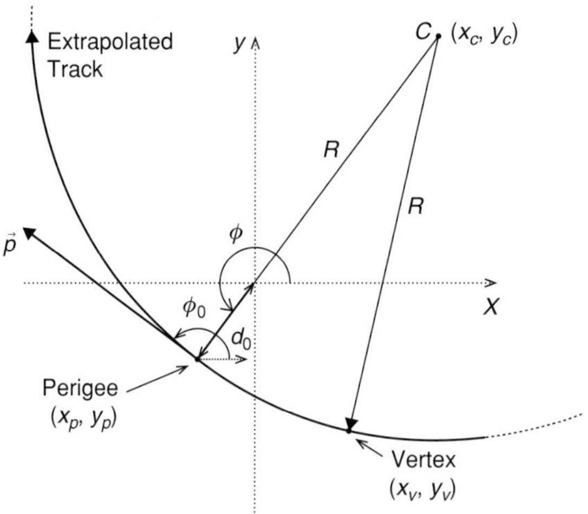

Fig. 9.21 Track perigee parameters used in the determination of production vertices with the Billoir vertex fitting method [45].

The parameters d_0, m_0, ϕ, θ, and q/p of the point of closest approach to the local z-axis are shown in Appendix §9.4.3. One finds

$$d_0 = R_0 + \frac{L^2}{2hR}, \tag{9.65}$$

$$z_0 = z_P - L \cot \theta_P, \tag{9.66}$$

$$\phi = \phi_P - \frac{L}{hR}, \tag{9.67}$$

$$\theta = \theta_P, \tag{9.68}$$

$$\frac{q}{p} = \frac{q}{p_P}. \tag{9.69}$$

Given prior estimates of the vertex position $\vec{V}_0$, measured track states $\vec{s}_{i,0}$ and momenta $\vec{p}_{i,0}$, we wish to find new momenta $\vec{p}_i = \vec{p}_{i,0} + \delta\vec{p}_i$ that correspond to the "real" vertex position $\vec{V} = \vec{V}_0 + \delta\vec{V}$. We then seek to minimize a χ^2 function defined as

$$\chi^2 = \sum_{i=1}^{n} \Delta\vec{s}_i^{\,T} \mathbf{W}_i \Delta\vec{s}_i, \tag{9.70}$$

where n is the number of tracks nominally associated with the vertex and $\Delta\vec{s}_i = \vec{s}_{i,0} - \vec{F}(\vec{V}, \vec{s}_i)$. The function $\vec{F}$ determines the state a track should have given a vertex position $\vec{V}$ and momentum vector $\vec{p}$. By minimizing the preceding χ^2 function, one finds the changes $\delta\vec{V}$ and $\delta\vec{s}_i$ that are most compatible with the measured states $\vec{s}_{i,0}$. In principle, the function $\vec{F}$ has a nonlinear dependency on both $\delta\vec{V}$ and $\delta\vec{s}_i$. However, if the initial estimate $\vec{V}_0$ is reasonably close to the real value $\vec{V}$, it is sufficient to use a linear approximation of the function $\vec{F}$, and one writes

$$\vec{F}(\vec{V}, \vec{s}_i) \simeq \vec{F}(\vec{V}_0, \vec{s}_{i,0}) + D_i\delta\vec{V} + E_i\delta\vec{s}_i, \tag{9.71}$$

where the coefficients D_i and E_i are derivatives of $F(\vec{V}, \vec{s}_i)$ with respect to $\vec{V}$ and $\vec{s}_i$ evaluated at $\vec{V}_0$ and $\vec{s}_{i,0}$. Substituting the preceding expression in Eq. (9.70), one gets the quadratic function

$$\chi^2 \simeq \sum_{i=1}^{n} \left(\delta\vec{s}_i - D_i\delta\vec{V} - E_i\delta\vec{p}_i\right)^T \mathbf{W}_i\left(\delta\vec{s}_i - D_i\delta\vec{V} - E_i\delta\vec{p}_i\right), \tag{9.72}$$

where $\delta\vec{s}_i = \vec{s}_{i,0} - \vec{F}(\vec{V}_0, \vec{s}_{i,0})$. Minimization of the χ^2 must be accomplished simultaneously relative to $\vec{V}$ and all $\vec{p}_i$. Setting derivatives with respect to these quantities equal to zero, one gets

$$0 = \sum_{i=1}^{n} \left(D_i^T W_i D_i \delta\vec{V} + D_i^T W_i E_i \delta\vec{p}_i - D_i^T W_i \delta\vec{s}_i\right), \tag{9.73}$$

$$0 = E_i^T W_i D_i \delta\vec{V} + E_i^T W_i E_i \delta\vec{p}_i - E_i^T W_i \delta\vec{s}_i. \tag{9.74}$$

For convenience, in order to solve for $\delta\vec{V}$ and $\delta\vec{p}_i$, let us define the following coefficients and vectors

$$A = \sum_i D_i^T W_i D_i, \tag{9.75}$$

$$B_i = D_i^T W_i E_i, \tag{9.76}$$

$$C_i = E_i^T W_i E_i, \tag{9.77}$$

$$\vec{T} = \sum_i D_i^T W_i \delta\vec{s}_i, \tag{9.78}$$

$$\vec{U}_i = E_i^T W_i \delta\vec{s}_i, \tag{9.79}$$

where the sums are taken over particles $i = 1, \ldots, n$. Equations (9.73) and (9.74) may then be written

$$0 = A\delta\vec{V} + \sum_i B_i \delta\vec{p}_i - \vec{T}, \tag{9.80}$$

$$0 = B_i^T \delta\vec{V} + C_i \delta\vec{p}_i - \vec{U}_i. \tag{9.81}$$

Solving for $\delta\vec{V}$ and $\delta\vec{p}_i$, one obtains the updated vertex and momenta according to

$$\begin{aligned} \vec{V} &= \vec{V}_0 + \delta\vec{V} \\ &= \vec{V}_0 + \left(A - \sum_i B_i C_i^{-1} B_i^T\right)^{-1} \left[\vec{T} - \sum_i B_i C_i^{-1} \vec{U}_i\right], \end{aligned} \tag{9.82}$$

$$\begin{aligned} \vec{p}_i &= \vec{p}_{i,0} + \delta\vec{p}_i \\ &= \vec{p}_{i,0} + C_i^{-1}\left(\vec{U}_i - B_i^T \delta\vec{V}\right). \end{aligned} \tag{9.83}$$

Covariance matrices $\mathrm{Cov}[\vec{V}, \vec{V}]$, $\mathrm{Cov}[\vec{p}_i, \vec{p}_i]$, and $\mathrm{Cov}[\vec{V}, \vec{p}_i]$ are provided in [45].

Equations (9.82,9.83) together constitute a $(3n + 3) \times (3n + 3)$ matrix equation that may be rather tedious solve, particularly if the number of tracks n is very large. However, Billoir and Qian developed a perigee parameterization that alleviates the inversion of such a large matrix and uses only small matrices. Discussions of this and more advanced techniques based on Kalman and adaptive Kalman filter methods are beyond the scope of this work but may be found in the recent literature [44, 45, 71, 88, 112, 178, 187].

9.3.3 Performance Characterization

A primary vertex finder can be characterized on the basis of its efficiency, the associated track multiplicity, and the position resolution of the vertices it produces.

The vertex-finding efficiency, ϵ_V, is defined as the ratio of the number of vertices actually found, n_{found}, divided by the actual number of vertices, n_{produced}.

$$\epsilon_V = \frac{n_{\text{found}}}{n_{\text{produced}}} \tag{9.84}$$

The vertex-finding efficiency typically depends on a wide variety of experimental factors, including the experimental trigger, the track reconstruction efficiency and resolution, as well as the specific technique used to reconstruct vertices. While the reconstruction efficiency may be roughly estimated based on the data, it is generally obtained via careful Monte Carlo modeling (simulations) of the experimental apparatus and its performance.

The number of tracks associated with a particular vertex, or track multiplicity, provides a basic selection criterion of the quality of reconstructed vertices. It can be used to rank multiple primary vertices found in events recorded in very high luminosity environments such as the proton–proton beams at the LHC. It can additionally provide a selection criterion for reaction types of interest. A maximum multiplicity can be applied, for instance, to identify quasi diffractive collisions in proton–proton collisions or grazing heavy-ion collisions involving a single or multiple photon exchanges.

The vertex position resolution improves as the number of pairs included in a measurement of the vertex position increases. Let δr_{ij} represent the resolution of the midpoint positions $\vec{r}_{ij}$. Assuming the track sample consists of n primary tracks exclusively, and neglecting correlations in measurements of the positions, one estimates the resolution on $\hat{S}_v$ as

$$\delta S_v = \sqrt{\frac{2}{n(n-1)} \sum_{i,j=1}^{n} \delta r_{ij}^2}. \tag{9.85}$$

9.4 Appendix 1: DCA Calculations

9.4.1 DCA Straight Line to Fixed Point

The DCA between a straight line l_1 and a fixed point P_2 can be obtained using the geometrical construction illustrated in Figure 9.22a.

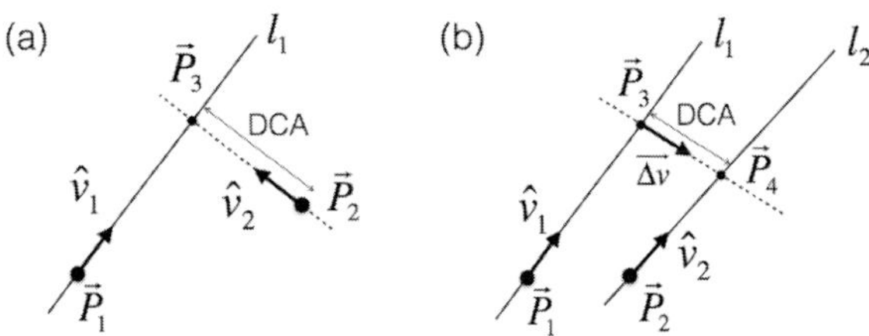

Fig. 9.22 Calculation of the DCAs between (a) a line l_1 and a fixed point S_1 and between (b) two lines l_1 and l_2.

The line l_1 is determined by a point $\vec{P}_1$ and a direction $\hat{v}_1$ (with $\hat{v}_1 \cdot \hat{v}_1 = 1$) and can thus be parameterized as

$$\vec{l}_1 = \vec{P}_1 + t\hat{v}_1, \tag{9.86}$$

where $t \in \mathbb{R}$. The DCA is achieved for a point $\vec{P}_3$ which belongs to l_1 and lies along a line l_2 determined by

$$\vec{l}_2 = \vec{P}_2 + s\hat{v}_2, \tag{9.87}$$

with $s \in \mathbb{R}$, $\hat{v}_1 \cdot \hat{v}_2 = 0$ and $\hat{v}_2 \cdot \hat{v}_2 = 1$. One thus seeks the values t and s such that $\vec{l}_1 = \vec{l}_2$:

$$\vec{P}_1 + t\hat{v}_1 = \vec{P}_2 + s\hat{v}_2. \tag{9.88}$$

Multiplying both sides of this equation by $\hat{v}_1$, one gets

$$\vec{P}_1 \cdot \hat{v}_1 + t = \vec{P}_2 \cdot \hat{v}_1, \tag{9.89}$$

where we used the fact that $\hat{v}_1$ is a unit vector by construction perpendicular to $\hat{v}_2$. Solving for t, we get

$$t = \left(\vec{P}_2 - \vec{P}_1\right) \cdot \hat{v}_1, \tag{9.90}$$

which uniquely determines the position of the point $\vec{P}_3$ of l_1 closest to P_2.

$$\vec{P}_3 = \vec{P}_1 + \left[\left(\vec{P}_2 - \vec{P}_1\right) \cdot \hat{v}_1\right]\hat{v}_1. \tag{9.91}$$

The *DCA* is then

$$DCA = \left[\left(\vec{P}_3 - \vec{P}_2\right) \cdot \left(\vec{P}_3 - \vec{P}_2\right)\right]^{1/2} \tag{9.92}$$

9.4.2 DCA between Two Straight Lines

We calculate the DCA between two straight line l_1 and l_2 using the geometrical construction illustrated in Figure 9.22b. The two lines are determined by the points P_1 and P_2 as well as the unit vectors $\hat{v}_1$ and $\hat{v}_2$. The points P_3 and P_4 corresponding to the DCA between the two lines may be written

$$\vec{P}_3 = \vec{P}_1 + t_1\hat{v}_1 \tag{9.93}$$

$$\vec{P}_4 = \vec{P}_2 + t_2\hat{v}_2, \tag{9.94}$$

where $t_1, t_2 \in \mathbb{R}$. The segment joining these two points is expressed as

$$\vec{P}_3 - \vec{P}_4 = d\Delta\hat{v}, \tag{9.95}$$

where $d \in \mathbb{R}$ is the DCA between the two lines and $\Delta\hat{v}$ is a unit vector. Substituting the expressions (9.93) for $\vec{P}_3$ and (9.94) for $\vec{P}_4$, we get

$$\vec{P}_1 + t_1\hat{v}_1 - \vec{P}_2 - t_2\hat{v}_2 = d\Delta\hat{v}, \tag{9.96}$$

which we in turn multiply by $\hat{v}_1$ and $\hat{v}_2$:

$$\vec{P}_1 \cdot \hat{v}_1 + t_1 - \vec{P}_2 \cdot \hat{v}_1 - t_2\hat{v}_1 \cdot \hat{v}_2 = 0, \tag{9.97}$$

$$\vec{P}_1 \cdot \hat{v}_2 + t_1\hat{v}_1 \cdot \hat{v}_2 - \vec{P}_2 \cdot \hat{v}_2 - t_2 = 0, \tag{9.98}$$

where we have use the fact that $\Delta\hat{v}$ is by construction perpendicular to both $\hat{v}_1$ and $\hat{v}_2$. A solution for t_1 and t_2 may be written

$$\begin{bmatrix} t_1 \\ t_2 \end{bmatrix} = \begin{bmatrix} 1 & -\hat{v}_1 \cdot \hat{v}_2 \\ \hat{v}_1 \cdot \hat{v}_2 & -1 \end{bmatrix}^{-1} \begin{bmatrix} (\vec{P}_1 - \vec{P}_2) \cdot \hat{v}_1 \\ (\vec{P}_1 - \vec{P}_2) \cdot \hat{v}_2 \end{bmatrix}, \tag{9.99}$$

which determines the position of $\vec{P}_3$ and $\vec{P}_4$. The DCA is then

$$DCA = \left[(\vec{P}_3 - \vec{P}_4) \cdot (\vec{P}_3 - \vec{P}_4)\right]^{1/2}. \tag{9.100}$$

9.4.3 DCA between a Helix and the Local z-Axis

Based on Figure 9.21, the position of the perigee may be written

$$x \simeq -d_0 \sin\phi_P + (l \sin\theta_P)\cos\phi_P + \frac{l^2 \sin^2\theta_P}{2hR} \sin\phi_P, \tag{9.101}$$

$$y \simeq -d_0 \cos\phi_P + (l \sin\theta_P)\sin\phi_P + \frac{l^2 \sin^2\theta_P}{2hR} \cos\phi_P, \tag{9.102}$$

$$z \simeq z_P + l\cos\theta_P. \tag{9.103}$$

It is convenient to define $L = l\sin\theta_P$, which can also be written as

$$L = x\cos\phi + y\sin\phi. \tag{9.104}$$

Writing the radius R of the helix as

$$R_0 = y\cos\phi_P - x\sin\phi_P, \tag{9.105}$$

one obtains

$$R_0 = d_0 - \frac{L^2}{2hR}. \tag{9.106}$$

The parameters of the perigee may then be written

$$d_0 = R_0 + \frac{L^2}{2hR}, \tag{9.107}$$

$$z_0 = z_P - L\cot\theta_P, \tag{9.108}$$

$$\phi = \phi_P - \frac{L}{hR}, \tag{9.109}$$

$$\theta = \theta_P, \tag{9.110}$$

$$\frac{q}{p} = \frac{q}{p_P}, \tag{9.111}$$

where d_0 and m_0 correspond to the DCA of the track relative to the estimated vertex.

9.5 Appendix 2: Multiple Coulomb Scattering

Charged particles passing through a detector medium undergo multiple coulomb scattering and are thus very likely to change direction. For sufficiently thick media (but not too thick), the scattering angle θ has a probability density that can be well approximated with a Gaussian distribution with an RMS angle $\theta^{\rm rms}_{\rm plane}$ given [35] by

$$\theta^{\rm rms}_{\rm plane} = \theta_0 = \frac{13.6\text{MeV}}{\beta cp} z\sqrt{\frac{x}{X_0}}\left[1 + 0.038\ln\left(\frac{x}{X_0}\right)\right], \tag{9.112}$$

where x is the thickness of the material and X_0 is the radiation length of the material. The radiation lengths of selected materials often used in the construction of detectors are listed in [35].

A well-designed detector should be thick enough to produce a signal that can be detected with high efficiency but thin enough to limit multiple coulomb scattering effects. Optimization of the detector is contingent on many factors and depends, in particular, on the signal-to-noise ratio.

Exercises

9.1 Show that the expression (5.213) simplifies to **I**.

9.2 Show that the conformal mapping defined by Eqs. (9.8) and (9.9) yields a straight line (9.11) for circular trajectories given by Eq. (9.10).

10 Correlation Functions

Correlation observables constitute an essential component of the toolset used by researchers in the study of the dynamics of elementary and nuclear collisions, and for measurements of the properties of the matter produced in these collisions. As such, correlation functions span a wide variety of forms and provide sensitivity to a broad range of phenomena and nuclear matter properties. Although they seemingly take many different forms, correlation functions all share a core definition and have common properties.

We introduce and motivate the notion of correlation function as an extension of the concept of covariance between two random variables in §10.1. The concept of correlation function is, however, not limited to two variables (or fields) and can be readily developed for an arbitrary number of random variables in the form of cumulant functions, discussed in §10.2. Cumulants play a very important role, in particular, in studies of collective properties, such as flow, of the medium produced in high-energy collisions. They are also essential for proper studies of multiparticle correlations.

Correlation functions can be broadly divided into differential and integral correlation functions. Differential correlations can be studied as a function of selected kinematical variables of two or more particles. They may be averaged over all interactions measured for a given type of collision or studied as a function of global event observables, such as the total transverse energy or the charged particle multiplicity measured in a specific kinematic range. The basic definition of inclusive correlation functions, that is, correlation averaged over all events, is introduced in §10.2. Semi-inclusive correlations, measured as a function of global observables, are discussed in §10.3. (Specific examples involving correlation measurements as a function of the relative emission angle and rapidity are discussed in §11.1.1.) The concept of differential correlation functions is extended, in §10.8, to include correlation functions weighted by particle properties. An example of such weighted correlation functions involving transverse momentum deviates is presented in §11.1.5.

Integral correlation functions are introduced in §10.4 in the form of factorial cumulants. They are useful, in particular, for the study of fluctuations of particle multiplicity, net charge, transverse momentum, or the relative yield of production of different particle species, discussed in Chapter 11. Measurements of correlation functions also play a central role in studies of collective flow, particularly in heavy-ion collisions. Flow constitutes a subfield of study of its own that encompasses a wide variety of techniques, introduced here in §11.4.

Correlation functions are subject, like any other observables, to instrumental effects that must be properly accounted for to extract meaningful physics results. Techniques to

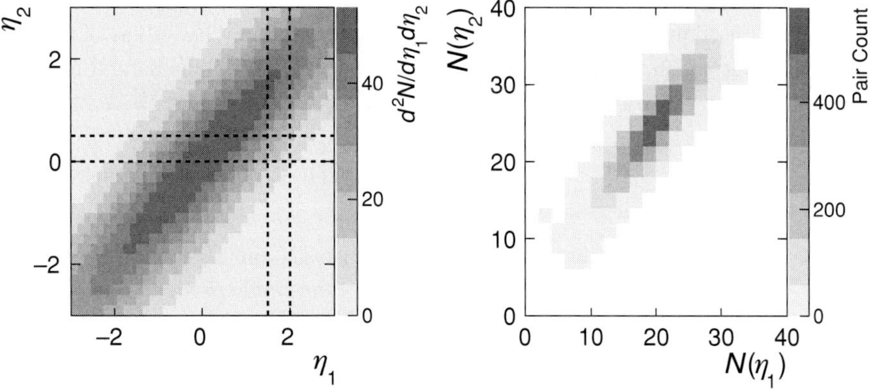

Fig. 10.1 Illustration of the notion of covariance between the particle production yields measured at two points (η_1 and η_2) in momentum space with a simple Gaussian correlation model. The left-hand plot shows the pair yield vs. the pseudorapidities of the particles, average over a large number of events. The right-hand plot displays the number of particles in the wide rapidity bins shown on the left with dashed lines. It reveals that the particle yields in bins η_1 and η_2 are tightly correlated.

account for the finite acceptance of measurements, particle detection efficiency, and various other instrumental effects are discussed in §12.4.

10.1 Extension of the Notion of Covariance

We introduce the notion of **correlation function** on the basis of the covariance of the number of particles detected event-by-event at two distinct points $\vec{p}_1$ and $\vec{p}_2$ in momentum space, as illustrated in Figure 10.1.

Let N_i represent the number of particles produced in the volumes Ω_i, $i = 1, 2$, defined by the ranges $p_{T,i}^{\min} \le p_{T,i} < p_{T,i}^{\max}$, $\eta_i^{\min} \le \eta_i < \eta_i^{\max}$, and $\phi_i^{\min} \le \phi_i < \phi_i^{\max}$ centered at $\vec{p}_i$. Given the stochastic nature of particle production, the yields N_i are expected to fluctuate event-by-event. For a given type of particle (and a specific projectile, target, collision energy, and possibly several other collision system parameters), the yields will have average values $\langle N_i \rangle$ that depend on the cross section of the process considered:

$$\langle N_i \rangle = \int_{\Omega_i} \frac{d^3 N_i}{dp_T d\phi d\eta} \, dp_T \, d\phi \, d\eta. \tag{10.1}$$

Fluctuations about each of these means $\langle N_i \rangle$ are characterized by the variances

$$\text{Var}[N_i] = \langle N_i^2 \rangle - \langle N_i \rangle^2. \tag{10.2}$$

However, it is usually more informative to study the covariance of these two yields:

$$\text{Cov}[N_1, N_2] = \langle N_1 N_2 \rangle - \langle N_1 \rangle \langle N_2 \rangle \tag{10.3}$$

The covariance Cov$[N_1, N_2]$ is quite obviously a function of the size of the bins Δ_1 and Δ_2 used to measure the yields N_1 and N_2, respectively, as well as of the coordinates $\vec{p}_1$ and $\vec{p}_2$ at which the particle emission is considered. It is thus natural to introduce the notion of **correlation function**

$$C(\vec{p}_1, \vec{p}_2) = \frac{1}{\Omega_1\Omega_2}\left\{\langle N(\vec{p}_1)N(\vec{p}_2)\rangle - \langle N(\vec{p}_1)\rangle\langle N\vec{p}_2)\rangle\right\}, \tag{10.4}$$

defined in the limit in which the bin sizes Ω_1 and Ω_2 vanish. For finite bin sizes, the ratios $\langle N(\vec{p}_i)\rangle/\Delta_i$ provide estimates $\hat{\rho}_1(\vec{p}_i)$ of the single particle density $\rho_1(\vec{p}_i)$, defined as

$$\rho_1(\vec{p}_i) = \frac{d^3N_i}{p_T dp_T d\phi d\eta}(\vec{p}_i). \tag{10.5}$$

The term $\langle N(\vec{p}_1)N\vec{p}_2)\rangle$ represents the average number of particle pairs detected jointly, that is, in the same event, at momenta $\vec{p}_1$ and $\vec{p}_2$, in bins of size Δ_1 and Δ_2, respectively. The ratio $\langle N(\vec{p}_1)N\vec{p}_2)\rangle/(\Delta_1\Delta_2)$ thus provides an estimate $\hat{\rho}_2(\vec{p}_1, \vec{p}_2)$ of the two-particle density, $\rho_2(\vec{p}_1, \vec{p}_2)$, defined as

$$\rho_2(\vec{p}_1, \vec{p}_2) = \frac{d^6N_{\text{pairs}}}{dp_{T,1}d\phi_1 d\eta_1 dp_{T,2}d\phi_2 d\eta_2}(\vec{p}_1, \vec{p}_2). \tag{10.6}$$

In the limit $\Delta_1, \Delta_2 \to 0$, one thus gets

$$C(\vec{p}_1, \vec{p}_2) = \rho_2(\vec{p}_1, \vec{p}_2) - \rho_1(\vec{p}_1)\rho_1(\vec{p}_2). \tag{10.7}$$

In its most general form, the two-particle correlation function $C(\vec{p}_1, \vec{p}_2)$ is defined as a function of six coordinates, that is, three momentum coordinates for each particle. However, a measurement of correlation function can obviously be reduced to a smaller number of coordinates of interest by integrating (marginalization) over variables that are not of interest. For instance, it is common to study the correlation functions of produced particles as a function of the relative angle $\Delta\phi = \phi_1 - \phi_2$, or the difference in pseudorapidity $\Delta\eta = \eta_1 - \eta_2$, or both, for specific types of particles (e.g., all charge hadrons, positive particles only, or only pions, etc.), and within a specific range of transverse momentum, and for events (i.e., collisions) satisfying specific conditions. We will discuss examples of such correlation functions in §§11.1 and 11.2. However, it is appropriate to first define correlation functions formally without specific recourse to the notion of covariance, and we do so in §10.2.

It is important to realize that while particle yields are by definition nonnegative (i.e., positive or null), the function $C(\vec{p}_1, \vec{p}_2)$ may be positive, null, or even negative. As for covariances, a positive value indicates that a rise of the particle yield at $\vec{p}_1$ is, on average, accompanied by a rise of the yield at $\vec{p}_2$, whereas a negative value corresponds to anti-correlation, so that the rise of the yield at one momentum is accompanied by a decline at the other momentum point. A null value, of course, implies that the two yields, at the given momenta $\vec{p}_1$ and $\vec{p}_2$, are seemingly independent.

10.2 Correlation Function Cumulants

10.2.1 Multiparticle Densities and Factorial Moments

Let us consider the distribution of particles produced by a beam a and a target b at a specific collisional energy $\sqrt{s}$ in a specific subvolume Ω of the total phase space Ω_{tot}. For convenience, and without loss of generality, we will here denote all particle kinematic variables (e.g., p_x, p_T, η, ϕ, etc.) of a given particle with a single variable y, which we shall use to identify its position in Ω. For simplicity's sake, we will first treat all particles as if they were of the same type or species, and the kinematic variables of different particles will be denoted y_1, y_2, and so on. We will further assume that the probability of finding m particles at specific values $y_1, y_2, \ldots, y_m$, can be described with an **exclusive** continuous probability density function $P_m(y_1, y_2, \ldots, y_m)$. Since the particles are assumed to be of the same species, the PDF $P_m(y_1, y_2, \ldots, y_m)$ will be taken to be fully symmetric in the coordinates y_1, $y_2, \ldots, y_m$. This means that the value of the probability should remain invariant under exchange of any two variables, for instance:

$$P_m(y_1, y_3, y_2, \ldots, y_m) = P_m(y_1, y_2, y_3, \ldots, y_m). \tag{10.8}$$

The next step is to connect the joint-probabilities $P_m(y_1, y_3, y_2, \ldots, y_m)$ to particle densities or invariant cross sections. By construction, inclusive number densities ρ_n yield a sequence of inclusive differential functions:

$$\frac{1}{\sigma_{\text{inel}}} d\sigma = \rho_1(y)\, dy, \tag{10.9}$$

$$\frac{1}{\sigma_{\text{inel}}} d^2\sigma = \rho_2(y_1, y_2)\, dy_1\, dy_2,$$

and so on.

Let us first consider the lowest order $n = 1$. Clearly, the single particle density can be factorized in terms of the average number of particles emitted, $\langle N\rangle$, and the probability of finding a given particle at a specific point y (recall that y here stands for (y, ϕ, p_T). One can then write

$$\rho_1(y) = \langle N\rangle P_1(y). \tag{10.10}$$

The pair density can be likewise factorized into the average number of pairs and the probability of finding particles at y_1 and y_2 (see §10.3 for alternative derivation of the preceding and following relations):

$$\rho_2(y_1, y_2) = \langle N(N-1)\rangle P_2(y_1, y_2). \tag{10.11}$$

Similarly, for higher-order densities, one can write

$$\rho_n(y_1, \ldots, y_n) = \langle N(N-1)\cdots(N-n+1)\rangle P_n(y_1, \ldots, y_n). \tag{10.12}$$

Integration of densities over the momentum volume Ω thus yields the following important relations:

$$\int_\Omega \rho_1(y)\, dy = \langle N \rangle \tag{10.13}$$

$$\int_\Omega \int_\Omega \rho_2(y_1, y_2)\, dy_1\, dy_2 = \langle N(N-1) \rangle \tag{10.14}$$

$$\cdots \tag{10.15}$$

$$\int \cdots \int_\Omega \rho_n(y_1, \ldots, y_n)\, dy_1 \cdots dy_n = \langle N(N-1)\cdots(N-n+1) \rangle \tag{10.16}$$

which provide the average number of particles, the average number of pairs of particles, and the average number of n-tuplets of particles produced in the volume Ω, respectively. The averages $\langle N(N-1)\cdots(N-n+1)\rangle$ are commonly known as **factorial moments** of order n.

10.2.2 Definition of Cumulants

In general, inclusive n-particle densities $\rho_n(y_1, \ldots, y_n)$ are the result of a superposition of several subprocesses. Indeed, although the n particles might be produced by a single and specific subprocess, it is also quite possible that they originate from two or more distinct subprocesses. It is in fact possible that the n particles originate from n distinct and uncorrelated subprocesses. The measured n-tuplets of particles may then feature a broad variety of correlation sources associated with a plurality of dynamic processes. It is thus a common goal of multiparticle production measurements to identify and study these correlated emission as distinct (sub)processes. This is best accomplished by invoking correlation functions known as (factorial) cumulant functions, expressed either in terms of integral correlators or as differential functions of one or more particle coordinates. Cumulants of order m, hereafter noted C_m, are defined as m-particle densities representing the emission (production) of m correlated particles originating from a common production process. The emission of n particles, with $n > m$, can thus be regarded as a superposition (sum) of several processes that together concur to produce a total of n particles. There are obviously several ways to cluster n particles. An n-particle density can then be expressed as a sum of several terms yielding n particles, but each with its own **cluster decomposition** into products of m-cumulants, as illustrated schematically in Figure 10.2.

Particle densities of the two lowest orders can be expressed in terms of correlation functions cumulants as follows:

$$\rho_1(1) = C_1(1) \tag{10.17}$$

$$\rho_2(1,2) = C_1(1)C_1(2) + C_2(1,2). \tag{10.18}$$

For orders 3 and 4, one gets the much lengthier expressions:

$$\begin{aligned}\rho_3(1,2,3) = {} & C_1(1)C_1(2)C_1(3) + C_1(1)C_2(2,3) \\ & + C_1(2)C_2(1,3) + C_1(3)C_2(1,2) + C_3(1,2,3)\end{aligned} \tag{10.19}$$

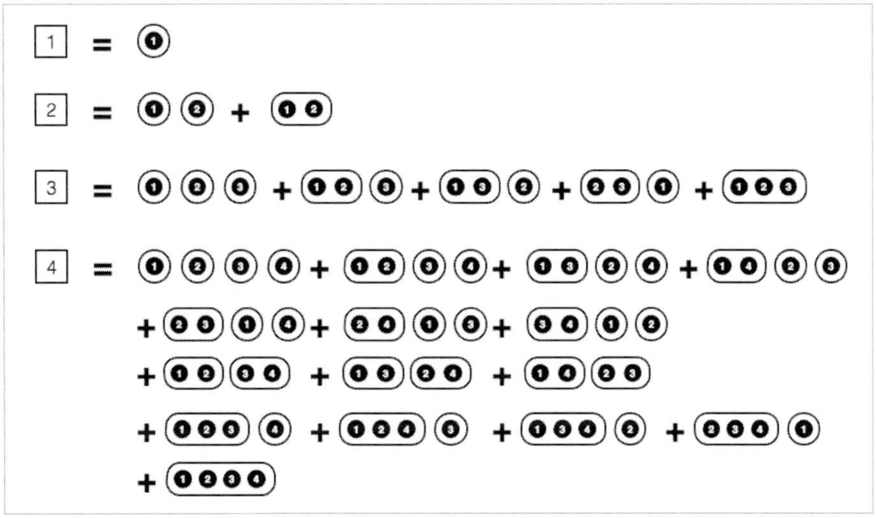

Fig. 10.2 Diagrammatic expression of n-particle densities in terms of cumulants. n-particle densities are represented as squares, while cumulants are denoted by circles and ovals.

$$\begin{aligned}\rho_4(1,2,3,4) = {} & C_1(1)C_1(2)C_1(3)C_1(4) \\ & + C_2(1,2)C_1(3)C_1(4) + C_2(1,3)C_1(2)C_1(4) \\ & + C_2(1,4)C_1(2)C_1(3) + C_2(2,3)C_1(1)C_1(4) \\ & + C_2(2,4)C_1(1)C_1(3) + C_2(3,4)C_1(1)C_1(2) \\ & + C_2(1,2)C_2(3,4) + C_2(1,3)C_2(2,4) \\ & + C_2(1,4)C_2(2,3) + C_3(1,2,3)C_1(4) \\ & + C_3(1,2,4)C_1(3) + C_3(1,3,4)C_1(2) \\ & + C_3(2,3,4)C_1(1) + C_4(1,2,3,4),\end{aligned} \tag{10.20}$$

where we used a shorthand notation indicating the index of the particles rather than kinematical variables y_i. Higher-order densities may be obtained based on the following expression:

$$\begin{aligned}\rho_m(1,\ldots,m) = {} & C_m(1,\ldots,m) + \sum_{\text{perm}} C_1(1)C_{m-1}(2,\ldots,m) \\ & + \sum_{\text{perm}} C_1(1)C_1(2)C_{m-2}(3,\ldots,m) \\ & + \sum_{\text{perm}} C_2(1,2)C_{m-2}(3,\ldots,m) + \cdots + \prod_{i=1}^{m} C_1(i)\end{aligned} \tag{10.21}$$

where "perm" indicates permutations of all particle indexes yielding distinct terms. Theoretically, cumulants naturally arise as a byproduct of calculations of the cross section of specific processes yielding specific particle multiplicities. They can be calculated, in particular, on the basis of cumulant generating functions, such as those introduced in §2.13 and discussed in more details in §10.4 (see also [56] and references therein). Experimentally, the extraction of cumulants is not as direct. They are not readily available experimentally as **raw** measurements but must be derived from (efficiency corrected) measurements of n-particle densities. One finds cumulants of order n must be recursively calculated from lower order cumulants and densities. For instance, from Eqs. (10.17) and (10.18), one gets

$$C_2(1,2) = \rho_2(1,2) - \rho_1(1)\rho_1(2), \tag{10.22}$$

where, once again, we used a shorthand notation indicating the index of the particles rather than kinematical variables y_i. Inserting the preceding expression for C_2 into Eq. (10.19), and accounting for particle indices, one gets

$$C_3(1,2,3) = \rho_3(1,2,3) - \sum_{(3)} \rho_1(1)\rho_2(2,3) + 2\rho_1(1)\rho_1(2)\rho_1(3). \tag{10.23}$$

The symbol (3) is used to indicate that the sum is carried over three permutations of the indices.[1] The fourth-order cumulant is obtained by substitution of second- and third-order cumulants into Eq. (10.20). One gets

$$\begin{aligned} C_4(1,2,3,4) = \rho_4(1,2,3,4) &- \sum_{(4)} \rho_1(1)\rho_3(2,3,4) \\ &- \sum_{(3)} \rho_2(1,2)\rho_2(3,4) + 2\sum_{(6)} \rho_1(1)\rho_1(2)\rho_2(3,4) \\ &- 6\rho_1(1)\rho_1(2)\rho_1(3)\rho_1(4), \end{aligned} \tag{10.24}$$

where the symbols (q) indicate the number of indices permutations over which the sums are taken. The preceding expressions for cumulants C_2, C_3, and C_4 are illustrated diagrammatically in Figure 10.3. Expressions for higher-order cumulants can be obtained recursively or from generic relations such as those provided in ref. [123].

In closing this section, it is important to reiterate that cumulants may be negative as well as positive, or null. Indeed, while the n-densities $\rho_n(1,\ldots,n)$ are by definition either positive or null, their combinations into cumulants may yield negative values as well as positive or null values. As for covariances, positive values indicate the yields at coordinates $y_1,\ldots,y_m$ are correlated, that is, collectively grow or decrease together, while negative values imply that upward fluctuations of the yields at some coordinate y_i are on average accompanied by downward fluctuations at other points $y_{j\neq i}$.

10.2.3 Cumulants Scaling with Source Multiplicity

The cumulants $C_n(y_1,\ldots,y_n)$ feature a simple scaling property for collision systems consisting of a superposition of m_s independent (but otherwise identical) subsystems. To demonstrate this property, consider, for instance, a collision of two large nuclei (A–A collisions) at a specific energy, and let us assume that it can be reduced, to first-order approximation, to a superposition of m_s proton-proton (p–p) interactions, which each produce clusters consisting of n correlated particles. Let us further assume that the production of

[1] Permutations are $\rho_1(1)\rho_2(2,3)$, $\rho_1(2)\rho_2(1,3)$, and $\rho_1(3)\rho_2(1,2)$.

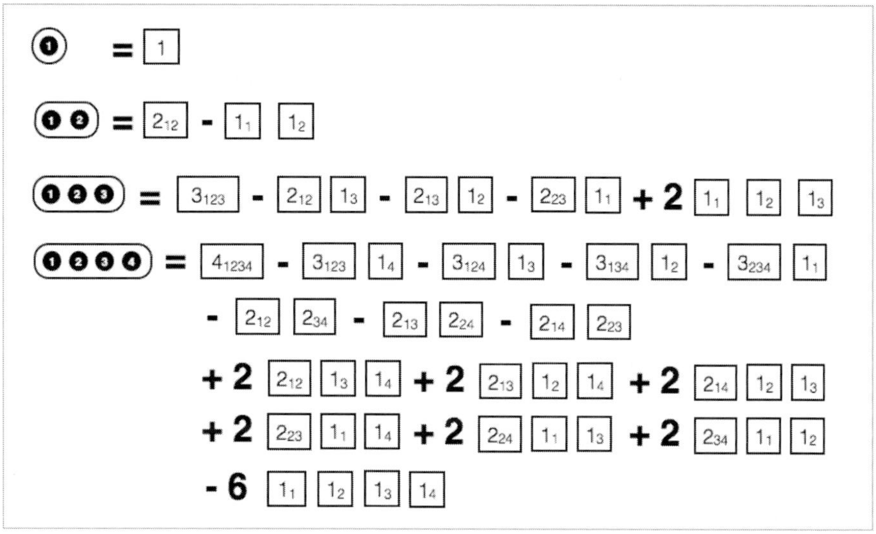

$$(1) = \boxed{1}$$

$$(1\,2) = \boxed{2_{12}} - \boxed{1_1}\,\boxed{1_2}$$

$$(1\,2\,3) = \boxed{3_{123}} - \boxed{2_{12}}\,\boxed{1_3} - \boxed{2_{13}}\,\boxed{1_2} - \boxed{2_{23}}\,\boxed{1_1} + 2\,\boxed{1_1}\,\boxed{1_2}\,\boxed{1_3}$$

$$\begin{aligned}(1\,2\,3\,4) = {} & \boxed{4_{1234}} - \boxed{3_{123}}\,\boxed{1_4} - \boxed{3_{124}}\,\boxed{1_3} - \boxed{3_{134}}\,\boxed{1_2} - \boxed{3_{234}}\,\boxed{1_1} \\ & - \boxed{2_{12}}\,\boxed{2_{34}} - \boxed{2_{13}}\,\boxed{2_{24}} - \boxed{2_{14}}\,\boxed{2_{23}} \\ & + 2\,\boxed{2_{12}}\,\boxed{1_3}\,\boxed{1_4} + 2\,\boxed{2_{13}}\,\boxed{1_2}\,\boxed{1_4} + 2\,\boxed{2_{14}}\,\boxed{1_2}\,\boxed{1_3} \\ & + 2\,\boxed{2_{23}}\,\boxed{1_1}\,\boxed{1_4} + 2\,\boxed{2_{24}}\,\boxed{1_1}\,\boxed{1_3} + 2\,\boxed{2_{34}}\,\boxed{1_1}\,\boxed{1_2} \\ & - 6\,\boxed{1_1}\,\boxed{1_2}\,\boxed{1_3}\,\boxed{1_4}\end{aligned}$$

Fig. 10.3 Diagrammatic expression of cumulants in terms of particle densities. Cumulants are denoted by circles and ovals, and densities by squares. Sub-labels indicate particle indices.

such clusters in p–p may be described by cumulants $C_n^{\rm pp}$. At a given impact parameter, A–A collisions should involve an average of $\langle m_s\rangle$ p–p interactions. This number m_s is obviously anticipated to fluctuate collision by collision, but for a given value of m_s, one expects that the number of clusters of correlated particles of size n should be, on average, m_s times larger than in p–p collisions. The n-cumulant for A–A collisions, at fixed m_s, may thus be written

$$C_n^{AA}(y_1,\ldots,y_n|m_s) = m_s C_n^{\rm pp}(y_1,\ldots,y_n) \tag{10.25}$$

Given that m_s fluctuates event by event, averaging over all A–A collisions consequently yields

$$C_n^{AA}(y_1,\ldots,y_n) = \langle m_s\rangle C_n^{\rm pp}(y_1,\ldots,y_n) \tag{10.26}$$

for A–A collisions consisting of a superposition of independent and unmodified p–p collisions, and such that produced particles do not interact with one another.

The total multiplicity of particles produced in A–A collisions consisting of m_s independent and unmodified p–p collisions also features the same simple scaling with m_s. Indeed, if the m_s p–p collisions are independent and unmodified, the average multiplicity obtained in A–A for a given (fixed) value of m_s should simply be the product of m_s by the average particle multiplicity produced in p–p, which we can thus write

$$\rho_1^{AA}(y) = m_s\rho_1^{\rm pp}(y) \tag{10.27}$$

and

$$\langle n\rangle_{AA} = m_s\langle n\rangle_{\rm pp}. \tag{10.28}$$

This simple property arises because $\rho_1(y) = C_1(y)$ and the scaling of the first cumulant $C_1^{AA}(y_1) = m_s C_1^{\rm pp}(y_1)$ is trivial. The treatment of pairs, triplets, and higher n-tuplets is, however, more complicated because higher-order densities, ρ_n, involve sums of several terms with varied products of cumulants.

Let us first consider pairs of particles. In an A–A collision consisting of m_s independent p–p interactions, one can form m_s times the pairs from individual p–p collisions. But one can also mix particles from different p–p interactions. Since there are $m_s(m_s - 1)$ ways of doing that, one can write

$$\rho_2^{AA}(y_1, y_2) = m_s \rho_2^{\rm pp}(y_1, y_2) + m_s(m_s - 1)\rho_1^{\rm pp}(y_1)\rho_1^{\rm pp}(y_2). \tag{10.29}$$

Note that one can obtain the same result using the cumulant decomposition shown in Eq. (10.22), as follows:

$$\begin{aligned} \rho_2^{AA}(y_1, y_2) &= C_1^{AA}(y_1)C_1^{AA}(y_2) + C_2^{AA}(y_1, y_2) \\ &= m_s^2 C_1^{\rm pp}(y_1)C_1^{\rm pp}(y_2) + m_s C_2^{\rm pp}(y_1, y_2) \\ &= m_s^2 \rho_1^{\rm pp}(y_1)\rho_1^{\rm pp}(y_2) + m_s\left[\rho_2^{\rm pp}(y_1, y_2) - \rho_1^{\rm pp}(y_1)\rho_1^{\rm pp}(y_2)\right] \\ &= m_s(m_s - 1)\rho_1^{\rm pp}(y_1)\rho_1^{\rm pp}(y_2) + m_s \rho_2^{\rm pp}(y_1, y_2). \end{aligned} \tag{10.30}$$

At fixed value of m_s, integration over y_1 and y_2 yields

$$\langle n(n-1)\rangle_{AA} = m_s\langle n(n-1)\rangle_{\rm pp} + m_s(m_s - 1)\langle n\rangle_{\rm pp}^2. \tag{10.31}$$

For large m_s, the scaling of the number of pairs produced in A–A is thus dominated by the term in $m_s(m_s - 1)$, which involves uncorrelated, combinatorial pairs from particles produced by different p–p collisions.

Based on the assumption that A–A collisions consist of m_s independent p–p interaction, it might be tempting to seek a measure of particle correlation in A–A by writing

$$Z_2^{AA}(y_1, y_2) = \rho_2^{AA}(y_1, y_2) - k\rho_1^{\rm pp}(y_1)\rho_1^{\rm pp}(y_2), \tag{10.32}$$

where the constant k is adjusted so the maximum of the product $\rho_1^{\rm pp}(y_1)\rho_1^{\rm pp}(y_2)$ matches the minimum of $\rho_2^{AA}(y_1, y_2)$. A particular application of this approach is known, in the recent literature, as zero yield at minimum (ZYAM) approximation [16]. However, it should be clear from the preceding discussion that although $Z_2(y_1, y_2)$, and in particular the ZYAM approximation, do provide a technique to assess whether $\rho_2^{AA}(y_1, y_2)$ differs from what is expected from a trivial scaling, it does NOT constitute a logically consistent measure of two particle correlations in A–A collisions. Only $C_2^{AA}(y_1, y_2)$ defined by Eq. (10.22) does, although there are several different ways of expressing and measuring C_2, which we discuss in the next section.

Let us next consider the scaling of triplets. We use the decomposition (10.19) and our shorthand notation to write

$$\begin{aligned} \rho_3^{AA}(1,2,3) &= C_1^{AA}(1)C_1^{AA}(2)C_1^{AA}(3) \\ &\quad + C_1^{AA}(1)C_2^{AA}(1,2) + C_1^{AA}(2)C_2^{AA}(1,3) \\ &\quad + C_1^{AA}(3)C_2^{AA}(2,3) + C_3^{AA}(1,2,3). \end{aligned} \tag{10.33}$$

We next apply the scaling property (10.25) and the expressions (10.22) and (10.23) to get

$$\begin{aligned}\rho_3^{AA}(1,2,3) &= m_s^3 C_1^{\rm pp}(1) C_1^{\rm pp}(2) C_1^{\rm pp}(3) \\ &\quad + m_s^2 \sum_{\rm perms} C_1^{\rm pp}(1) C_2^{\rm pp}(2,3) + m_s C_3^{\rm pp}(1,2,3) \\ &= \left(m_s^3 - m_s^2 + 2m_s\right) \rho_1^{\rm pp}(1)\rho_1^{\rm pp}(2)\rho_1^{\rm pp}(3) \\ &\quad + \left(m_s^2 - m_s\right) \sum_{\rm perms} \rho_1^{\rm pp}(1)\rho_2^{\rm pp}(2,3) + m_s \rho_3^{\rm pp},\end{aligned} \tag{10.34}$$

where the sums are carried over distinct permutations of the three particle coordinates. Integration over the coordinates y_1, y_2, and y_3 yields

$$\begin{aligned}\langle n(n-1)(n-2)\rangle_{AA} &= \left(m_s^3 - m_s^2 + 2m_s\right)\langle n\rangle_{\rm pp} \\ &\quad + 3\left(m_s^2 - m_s\right)\langle n(n-1)\rangle_{\rm pp}\langle n\rangle_{\rm pp} \\ &\quad + m_s \langle n(n-1)(n-2)\rangle_{\rm pp}\end{aligned} \tag{10.35}$$

The average number of triplets in A–A collisions is thus determined mostly by combinatorics and essentially scales as $m_s^3\langle n\rangle_{\rm pp}^3$, while the role of correlated triplets (from individual p–p collisions) is suppressed by m_s^2. By extension, we conclude that the average number of n-tuplets in A–A collisions scales as the nth power of m_s times the average multiplicity measured in p–p interactions, and the influence of truly correlated n-tuplets is suppressed by a factor m_s^{n-1}.

10.2.4 Normalized Cumulants and Normalized Factorial Moments

It is convenient to divide the densities ρ_n and cumulants C_n by products of one-particle densities. This leads to the definition of **normalized inclusive densities** and **normalized cumulants** as follows:

$$r_n(y_1, \ldots, y_n) = \frac{\rho_n(y_1, \ldots, y_n)}{\rho_1(y_1)\cdots\rho_1(y_n)} \tag{10.36}$$

$$R_n(y_1, \ldots, y_n) = \frac{C_n(y_1, \ldots, y_n)}{\rho_1(y_1)\cdots\rho_1(y_n)} \tag{10.37}$$

The use of ratios of factorial moments $\langle N(N-1)\cdots(N-n+1)\rangle$ to the nth power of the average particle multiplicity $\langle N\rangle$ is also common:

$$f_n = \frac{\langle N(N-1)\cdots(N-n+1)\rangle}{\langle N\rangle^n} \tag{10.38}$$

Unfortunately, there is no universally agreed upon notation for either of these quantities. Indeed, while the quantities are unique and well defined, the labels or notations used to describe them typically vary from one author to the next. The functions r_n and R_n are often referred to as **reduced densities** and **reduced cumulants**, respectively. Likewise, the **normalized factorial moments** are often called **reduced factorial moments**. In this text, we will use **normalized** as the qualifier.

It is interesting to consider the scaling behavior of the normalized cumulants $R_n(y_1, \ldots, y_n)$ for systems consisting of a superposition of m identical subprocesses. Based on the scaling property (10.25), one gets

$$R_n^{(m)}(y_1, \ldots, y_n) = \frac{C_n^{(m)}(y_1, \ldots, y_n)}{\rho_1^{(m)}(y_1) \cdots \rho_1^{(m)}(y_n)} = \frac{1}{m^{n-1}} R_n^{(1)}(y_1, \ldots, y_n). \tag{10.39}$$

The normalized n-cumulant of a composite system consisting of m (identical) subsystems scales inversely as the power $m^{(n-1)}$ times the n-cumulant of the subsystems. The normalized cumulants $R_n^{(m)}$ are consequently said to be **diluted** by a power m^{n-1} relative to the subsystems' normalized cumulants $R_n^{(1)}$. This dilution stems from the fact that the cumulant of order n scales as the number of distinct sources or subsystems while the denominator is proportional to m^n. For instance, the strength of a two-particle cumulant measured in A–A collisions consisting of m sources is m times the strength of the two-particle cumulant of each of these sources while the total number of pairs scales as m^2. The ratio of C_2 by $\rho_1 \otimes \rho_1$ thus scales as $1/m$. This implies that, in general, measurements of correlation functions in large colliding systems require a level of statistical precision commensurate with this dilution effect. Obviously, the impact of combinatorial terms increases as powers of m. Measurements of higher-order cumulants thus require very large datasets.

It is also interesting to consider the scaling behavior of the normalized factorial moments f_n defined by Eq. (10.38). The scaling of the number of n-tuplets involves a combination of several terms. We have seen, however, from Eqs. (10.31) and (10.35) that the scaling is dominated by combinatorial effects that scale as m^n. The normalized factorial moments f_n are thus expected to converge toward unity for increasing values of the number of subsystems m contributing to an A–A collision.

We will show in §12.4 that the normalized densities r_n, the normalized cumulants R_n, and the normalized factorial moments all share the property of being robust under particle losses associated with particle detection efficiencies. We thus recommend their use, experimentally, against other observables.

It is useful to also note that a simple relationship exists between the normalized densities r_n and the normalized R_n. Indeed, based on the definitions of the normalized densities and normalized cumulants, one finds

$$r_2(1,2) = 1 + R_2(1,2) \tag{10.40}$$

$$r_3(1,2,3) = 1 + \sum_{(3)} R_2(1,2) + R_3(1,2,3) \tag{10.41}$$

$$\begin{aligned} r_4(1,2,3,4) = 1 &+ \sum_{(6)} R_2(1,2) + \sum_{(3)} R_2(1,2)R_2(3,4) \\ &+ \sum_{(3)} R_3(1,2,3) + R_4(1,2,3,4) \end{aligned} \tag{10.42}$$

$$\begin{aligned} r_5(1,2,3,4,5) = 1 &+ \sum_{(10)} R_2(1,2) + \sum_{(15)} R_2(1,2)R_2(3,4) \\ &+ \sum_{(10)} R_3(1,2,3)R_2(4,5) + \sum_{(10)} R_3(1,2,3) \\ &+ \sum_{(5)} R_4(1,2,3,4) + R_5(1,2,3,4,5), \end{aligned} \tag{10.43}$$

with similar expressions for higher-order densities. Sums are to be carried on the indicated number, (m), of permutations of the particle indices.

10.2.5 Particle Probability Densities

The integration (10.13) of particle densities $\rho_n(y_1, \ldots, y_n)$ over the momentum volume Ω provides a natural and convenient normalization to define particle probability densities:

$$P_n(y_1, \ldots, y_n) = \frac{\rho_n(y_1, \ldots, y_n)}{\langle N(N-1)\cdots(N-n+1)\rangle}, \tag{10.44}$$

which expresses the probability (per unit volume) of finding n particles jointly, or simultaneously, at coordinates $y_1, \ldots, y_n$. It is also interesting to consider the normalization of these probability densities by products of single particle probability densities $P_1(y)$, as follows:

$$q_n(y_1, \ldots, y_n) = \frac{P_n(y_1, \ldots, y_n)}{P_1(y_1)\cdots P_1(y_n)}. \tag{10.45}$$

The functions $q_n(y_1, \ldots, y_n)$ equal unity if the production of particles 1 to n is statistically independent, that is, if the probability densities $P_n(y_1, \ldots, y_n)$ factorize into products $P_1(y_1)\cdots P_1(y_n)$. Substituting the expression (10.44) for n-particles and single particle densities into the expression of normalized densities, one gets

$$r_n(y_1, \ldots, y_n) = \frac{\langle N(N-1)\cdots(N-n+1)\rangle}{\langle N\rangle^n} q_n(y_1, \ldots, y_n), \tag{10.46}$$

which tells us that correlated particle production occurs in part because of genuine correlations, that is, for $q_n(y_1, \ldots, y_n) \neq 1$ and in part because of simple multiplicity fluctuations, that is, for $\langle N(N-1)\cdots(N-n+1)\rangle / \langle N\rangle^n \neq 1$. Evidently, $q_n \neq 1$ is required, no matter what, to yield nonvanishing normalized cumulants R_n. As an example, consider that the normalized two-particle cumulant may be written

$$R_2(y_1, y_2) = \frac{\langle N(N-1)\rangle}{\langle N\rangle^2} q_2(y_1, y_2) - 1. \tag{10.47}$$

The strength of two-particle correlations is thus determined both by the function $q_2(y_1, y_2)$ and the amplitude of multiplicity fluctuations, which yield $\langle N(N-1)\rangle / \langle N\rangle^2 \neq 1$.

10.3 Semi-inclusive Correlation Functions

Elementary hadron collision processes are typically rather complex and lead to stochastic fluctuations of the produced particle multiplicity, m. The probability P_m of encountering an event with a specific value of multiplicity m depends on the cross section of the process that produces the m particles relative to processes that lead to other values of multiplicity:

$$P_m = \sigma_m / \sum_{m'} \sigma_{m'}. \tag{10.48}$$

Given the laws of momentum and energy conservation, charge conservation, and other quantum numbers, it is reasonable to expect that the n-particle densities associated with collisions that produce different multiplicities, m, might be quite different. It is thus of interest to consider n-particle densities for collisions that produce multiplicities m explicitly for "all" values of m. This leads to the notion of **semi-inclusive** n-densities and n-cumulants measured for specific values of m. For simplicity's sake, we restrict our discussion to single- and two-particle densities only, but the notions introduced here can be extended straightforwardly to all orders n.

As in prior sections, we use the symbol y_i to represent all momentum coordinates of a particle (i.e., transverse momentum, rapidity, and azimuthal angle) in aggregate. But, again for simplicity's sake and without loss of generality, one may also regard the variables y_i as representing rapidity variables exclusively.

Single- and two-particle density distribution, expressed as a function of y_i, may be written

$$\begin{aligned} \rho_1^{(m)}(y) &= \frac{1}{\sigma_m}\frac{d\sigma_m}{dy} \\ \rho_2^{(m)}(y_1, y_2) &= \frac{1}{\sigma_m}\frac{d^2\sigma_m}{dy_1 dy_2}. \end{aligned} \tag{10.49}$$

The inclusive single- and two-particle densities are averages of the semi-inclusive distributions determined by the relative probabilities of these processes:

$$\begin{aligned} \rho_1(y) &= \sum_m P_m \rho_1^{(m)}(y) \\ \rho_2(y_1, y_2) &= \sum_m P_m \rho_2^{(m)}(y_1, y_2). \end{aligned} \tag{10.50}$$

The two-cumulants of the semi-inclusive processes of multiplicity m are by definition

$$C_2^{(m)}(y_1, y_2) = \rho_2^{(m)}(y_1, y_2) - \rho_1^{(m)}(y_1)\rho_1^{(m)}(y_2). \tag{10.51}$$

It is convenient to define density deviates $\Delta\rho^{(m)}(y_i) = \rho_1^{(m)}(y_i) - \rho_1(y_i)$, and introduce two new functions, noted $C_S(y_1, y_2)$ and $C_L(y_1, y_2)$, defined as

$$C_S(y_1, y_2) = \sum_m P_m C_2^{(m)}(y_1, y_2) \tag{10.52}$$

$$C_L(y_1, y_2) = \sum_m P_m \Delta\rho^{(m)}(y_1)\Delta\rho^{(m)}(y_2). \tag{10.53}$$

It is then straightforward to verify (see Problem 10.2) that the inclusive two-cumulant $C_2(y_1, y_2)$ may be written

$$C_2(y_1, y_2) = C_S(y_1, y_2) + C_L(y_1, y_2). \tag{10.54}$$

The function $C_S(y_1, y_2)$ corresponds to the average of the semi-inclusive two-cumulants and as such is quite sensitive to the actual collision dynamics, while the term $C_L(y_1, y_2)$ arises from the mixing of different event topologies (multiplicity) in the two-cumulant calculation and has consequently little to do with actual particle correlations. The label S

and L have a historical origin and are often misleadingly referred to as "short-" and "long-" range correlation terms.

Given the expression (10.52), a normalized form of C_S may be defined as

$$R_S(y_1, y_2) = \frac{C_S(y_1, y_2)}{\sum_m P_m \rho_1^{(m)}(y_1)\rho_1^{(m)}(y_2)} = \frac{\sum_m P_m \rho_2^{(m)}(y_1, y_2)}{\sum_m P_m \rho_1^{(m)}(y_1)\rho_1^{(m)}(y_2)} - 1. \tag{10.55}$$

Analogous expressions may be derived for three- or more particle correlations [128].

10.4 Factorial and Cumulant Moment-Generating Functions

We saw in §2.10 that it is useful to introduce the notions of moment-generating function and characteristic function to facilitate the calculation of the moments of probability density functions. Indeed, recall that the moment-generating function

$$M_x(t) = E\left[e^{tx}\right] = \int_{-\infty}^{\infty} e^{tx} f(x)\, dx, \qquad t \in \mathrm{R} \tag{10.56}$$

yields moments of the function $f(x)$ at all orders:

$$\mu'_k = \left.\frac{d^k}{dt^k} E\left[e^{tx}\right]\right|_{t=0} = \left.\frac{d^k}{dt^k} M_x(t)\right|_{t=0}. \tag{10.57}$$

It is also convenient to introduce generating functions for factorial moments and for cumulant moments. We derive expressions for these moments in the following paragraphs. Examples of application are discussed later in this section.

The probability P_m of a semi-inclusive process yielding m particles is determined by its cross-section σ_m relative to the total inelastic cross section

$$P_m \equiv \frac{\sigma_m}{\sum_m \sigma_m} = \frac{\sigma_m}{\sigma_{\text{inel}}}. \tag{10.58}$$

Inclusive densities of order n may consequently be written

$$\rho_n(y_1, \ldots, y_n) = \sum_m P_m \rho_n^{(m)}(y_1, \ldots, y_n) \tag{10.59}$$

where the functions $\rho_n^{(m)}(y_1, \ldots, y_n)$ are n-particle densities for processes that produce exactly m particles ($m \geq n$). Integration of the single density $\rho_1^{(m)}(y)$ over the fiducial range y is by construction equal to the multiplicity m:

$$\int_\Omega \rho_1^{(m)}(y)\, dy = m. \tag{10.60}$$

Integration of the inclusive density over the range Ω consequently yields

$$\int_\Omega \rho_1(y)\, dy = \sum_m P_m \int_\Omega \rho_1^{(m)}(y)\, dy = \sum_m P_m m = \langle m \rangle. \tag{10.61}$$

Likewise, integration of two-particle densities $\rho_2^{(m)}(y_1, y_2)$ yields $m(m-1)$, which is the number of particle pairs one can form with m particles. And for higher-order densities, one gets

$$\int_\Omega \rho_n^{(m)}(y_1, \ldots, y_n)\, dy_1 \cdots dy_n = m(m-1)\cdots(m-n+1). \tag{10.62}$$

Integration of inclusive functions consequently yield the averages

$$\tilde{F}_n \equiv \int_\Omega \rho_n(y_1, \ldots, y_n)\, dy_1 \cdots dy_n \tag{10.63}$$

$$= \sum_m P_m \int_\Omega \rho_n^{(m)}(y_1, \ldots, y_n)\, dy_1 \cdots dy_n \tag{10.64}$$

$$= \sum_m P_m m(m-1)\cdots(m-n+1) \tag{10.65}$$

$$= \langle m(m-1)\cdots(m-n+1)\rangle \tag{10.66}$$

$$\equiv \langle m^{[n]}\rangle, \tag{10.67}$$

where we have introduced the alternative notation $\langle m^{[n]}\rangle$ for (nonnormalized) factorial moments. The aforementioned development suggests the introduction of a moment-generating function as follows:

$$G(z) = 1 + \sum_{n=1}^{\infty} \frac{z^n}{n!} \tilde{F}_n \tag{10.68}$$

$$= 1 + \sum_{n=1}^{\infty} \frac{z^n}{n!} \int_\Omega \rho_n(y_1, \ldots, y_n)\, dy_1 \cdots dy_n \tag{10.69}$$

Given a generating function $G(z)$, the moments $\tilde{F}_n$ may then be obtained from (see Problem 10.4)

$$\tilde{F}_n = \left.\frac{d^n G(z)}{dz^n}\right|_{z=0} \tag{10.70}$$

We next use Eq. (10.65) to derive expressions for the probabilities P_n. Assuming there is a value $n = N$ beyond which all probabilities P_n vanishes, one can write

$$P_N = \frac{\tilde{F}_N}{N!} \tag{10.71}$$

since terms in $P_{n<N}$ cannot contribute to $\tilde{F}_N$ and by hypothesis, there are no terms with $n > N$. The $(N-1)$th factorial moment may be written

$$\tilde{F}_{N-1} = P_{N-1}(N-1)! + P_N \frac{N!}{1!} \tag{10.72}$$

from which we find

$$P_{N-1} = \frac{1}{(N-1)!} \left\{ \tilde{F}_{N-1} - \frac{\tilde{F}_N}{1!} \right\} \tag{10.73}$$

Proceeding recursively, one finds

$$P_n = \frac{1}{n!}\sum_{k=0}^{N-n}(-1)^k \frac{\tilde{F}_{k+n}}{k!} \quad \text{for } n = 0, 1, \ldots, N. \tag{10.74}$$

Introducing factorial cumulants

$$f_n = \int_\Omega dy_1 \cdots \int_\Omega dy_n C_n(y_1, \ldots, y_n) \tag{10.75}$$

where $C_n(y_1, \ldots, y_n)$ are cumulants of order n, one can show that that factorial moments $\tilde{F}_n$ of order $n \le 5$ may be written

$$\begin{aligned} \tilde{F}_1 &= f_1 \\ \tilde{F}_2 &= f_2 + f_1^2 \\ \tilde{F}_3 &= f_3 + 3f_2 f_1 + f_1^3 \\ \tilde{F}_4 &= f_4 + 4f_3 f_1 + 3f_2^2 + 6f_2 f_1^2 + f_1^4 \\ \tilde{F}_5 &= f_5 + 5f_4 f_1 + 10 f_3 f_2 + 10 f_3 f_1^2 + 15 f_2^2 f_1 + 10 f_2 f_1^3 + f_1^5 \end{aligned} \tag{10.76}$$

and for higher moments,

$$\tilde{F}_n = n! \sum_{\{l_i\}_n} \prod_{j=1}^{n} \left(\frac{f_j}{j!}\right)^{l_j} \frac{1}{l_j!} \tag{10.77}$$

where the summation is done for permutations satisfying $\sum_{l=1}^{q} il_i = q$.

Factorial cumulant generating functions can be introduced similarly to the generating functions of basic cumulants discussed in §2.13.1. One writes

$$\ln G(z) = \langle n \rangle\, z + \sum_{k=2}^{\infty} \frac{z^k}{k!} f_k \tag{10.78}$$

and consequently,

$$f_n = \left.\frac{d^n \ln G(z)}{dz^n}\right|_{z=0} \tag{10.79}$$

The preceding equations are useful if the generating function $G(z)$ can be calculated directly. For instance, for a Poisson distribution [128],

$$P_n = e^{-\langle n \rangle} \frac{\langle n \rangle^n}{n!}, \tag{10.80}$$

the generating function is

$$\begin{aligned} G(z) &= \sum_{n=0}^{\infty} P_n (1+z)^n = e^{-\langle n \rangle} \sum_{n=0}^{\infty} \frac{\langle n \rangle^n}{n!} (1+z)^n \qquad (10.81) \\ &= \exp\left(\langle n \rangle\, z\right). \qquad (10.82) \end{aligned}$$

Equation (10.70) then yields

$$\tilde{F}_m = \langle n \rangle^m. \tag{10.83}$$

One can also verify from (10.79) that $f_1 = \langle n \rangle$ and that for $m > 1$, one gets $f_m \equiv 0$, as expected, since Poisson statistics implies the production of uncorrelated particles and cumulants of order $m \geq 2$ must vanish.

10.5 Multivariate Factorial Moments

The factorial moments $\tilde{F}_n$ and cumulant moments f_n characterize multiplicity fluctuations in a single phase-space region or cell Ω. Short of considering fully differential correlation functions such as those presented in §10.2, one can also consider multivariate factorial moments based on the multiplicities measured in two or more phase space regions. For nonoverlapping regions Ω_m and $\Omega_{m'}$, one may thus define twofold factorial moments, or correlators, as

$$\tilde{F}_{pq} = \left\langle n_m^{[p]} n_{m'}^{[q]} \right\rangle, \tag{10.84}$$

where n_m and $n_{m'}$ are the number of particles in region m and m', respectively. Normalized correlators are defined according to

$$\tilde{f}_{pq} = \frac{\left\langle n_m^{[p]} n_{m'}^{[q]} \right\rangle}{\tilde{F}_p \tilde{F}_q}. \tag{10.85}$$

Extensions to higher-order correlators are also possible.

Multifold correlators of this type find applications in high-energy physics for the study of self-similar particle production behavior (intermittency), in radio- and radar-physics as well as in quantum optics. They are discussed in more detail in ref. [128].

10.6 Correlation Functions of Nonidentical Particles

We have so far considered correlation cumulants, factorial moments, and normalized cumulants for the study of correlation of identical or indistinguishable particles. However, these functions may also be used for the study of correlations of nonidentical particles. In the context of data analyses, nonidentical particles could be particles of different species, different charges, or those belonging to distinct kinematical ranges. For instance, one may be interested in studying correlations between pions and kaons, or correlations between positively and negatively charged particles, or even correlations between low p_T and high p_T particles.

Multiparticle densities are once again denoted $\rho_n(y_1, \ldots, y_n)$, but it is now understood that the coordinates $y_1, \ldots, y_n$ apply to distinguishable particles, either because of their species type or because of their kinematical variable attribute or range.[2] As such, the densities $\rho_n(y_1, \ldots, y_n)$ are no longer symmetric under interchange or permutation of the coordinates of the $y_1, \ldots, y_n$. One must in particular recognize the functions $\rho_1(y_1)$ and $\rho_1(y_2)$ as distinct single particle densities. Their integration over fiducial volumes Ω_1 and Ω_2 yields the average number of particles of type 1 and type 2, respectively,

$$\int_{\Omega_1} \rho_1(y_1)\, dy_1 = \langle N_1 \rangle \tag{10.86}$$

$$\int_{\Omega_2} \rho_1(y_2)\, dy_2 = \langle N_2 \rangle.$$

Similarly, integration of the two-particle density $\rho_2(y_1, y_2)$ yields the average number of pairs

$$\int_{\Omega_1} dy_1 \int_{\Omega_2} dy_2 \rho_2(y_1, y_2) = \langle N_1 N_2 \rangle \tag{10.87}$$

produced within the acceptance $\Omega_1 \otimes \Omega_2$. Likewise, integration of higher-order densities $\rho_n(y_1, \ldots, y_n)$ yield the average number of n-tuplets. For triplets, one has

$$\int_{\Omega_1} dy_1 \int_{\Omega_2} dy_2 \int_{\Omega_3} dy_3 \rho_3(y_1, y_2, y_3) = \langle N_1 N_2 N_3 \rangle, \tag{10.88}$$

where N_1, N_2, and N_3 represent the number of particles of type 1, 2, and 3 measured event by event. Note, however, that it may often be of interest to mix distinguishable and indistinguishable particles. For instance, one might wish to study the correlation function of particle triplets involving one high p_T particle, emitted above a specific minimum threshold, and two low p_T particles. The integration of $\rho_3(y_1, y_2, y_3)$ would then yield

$$\int_{\Omega_1} dy_1 \int_{\Omega_2} dy_2 \int_{\Omega_2} dy_3 \rho_3(y_1, y_2, y_3) = \langle N_1 N_2 (N_2 - 1) \rangle, \tag{10.89}$$

where N_1 and $N_2(N_2 - 1)$ represent the number of high p_T particles and the number of low p_T particle pairs measured event by event, respectively. Similar considerations apply to higher-order densities.

Correlation cumulants for distinguishable particles of all orders may be defined similarly to those of indistinguishable particles. Second-order cumulants are written

$$C_2(y_1, y_2) = \rho_2(y_1, y_2) - \rho_1(y_1)\rho_1(y_2), \tag{10.90}$$

where it is now understood that the variables y_1 and y_2 apply to particles of different species, different types (e.g., positive and negative charges), or different kinematic ranges (e.g., forward and backward production, or high and low transverse momentum).

[2] Once again in this section, and unless otherwise noted, y_i is meant to represent all kinematical variables of particles, such as p_T, η, and ϕ.

Normalized densities and cumulants are defined as for distinguishable particles as follows:

$$r_n(y_1, \ldots, y_n) = \frac{\rho_n(y_1, \ldots, y_n)}{\rho_1(y_1) \cdots \rho_1(y_n)} \tag{10.91}$$
$$R_n(y_1, \ldots, y_n) = \frac{C_n(y_1, \ldots, y_n)}{\rho_1(y_1) \cdots \rho_1(y_n)}.$$

Given their identical structure, the functions $C_n(y_1, \ldots, y_n)$ and $R_n(y_1, \ldots, y_n)$ obey the same scaling relations with the number of sources as for distinguishable particles:

$$C_n^{(m)}(y_1, \ldots, y_n) = mC_n^{(1)}(y_1, \ldots, y_n) \tag{10.92}$$
$$R_n^{(m)}(y_1, \ldots, y_n) = \frac{1}{m^{n-1}} R_n^{(1)}(y_1, \ldots, y_n),$$

where m stands for the number of identical sources producing correlated n-tuplets.

All in all, correlation functions of indistinguishable particles have very similar properties and behaviors as those constructed for distinguishable particles. However, there is also a common practice of defining so-called "per trigger" correlation functions. These are usually obtained by counting the number of particles in a given low p_T range associated with a "trigger" particle found in a higher p_T range as a function of some relative kinematic variable such as $\Delta\phi$, which is the difference between the azimuthal angle of the particles, or $\Delta\eta$, corresponding to the difference between their pseudorapidities. A measurement of per trigger correlation thus amounts to a number of trigger–associate pairs, N_{pairs}, normalized by the number of trigger particles, N_T. This can be done in two ways. The event-by-event (ebye) method consists in calculating the number of pairs per trigger on an event-by-event basis:

$$C_{T,\text{ebye}} = \left\langle \frac{N_{\text{pairs}}}{N_T} \right\rangle. \tag{10.93}$$

By contrast, the *inclusive* method involves a ratio of averages.

$$C_{T,\text{inc}} = \frac{\langle N_{\text{pairs}} \rangle}{\langle N_T \rangle}. \tag{10.94}$$

While the event-by-event method seems more intuitive, it hides and complicates the fact that not all particle pairs are correlated. The inclusive method, on the other hand, readily translates into simple integrals of the single- and two-particle cross sections. For instance,

$$C_{T,\text{inc}}(y_T, y_A) = \frac{\int_{\Omega_T} dx_T \int_{\Omega_A} dx_A \rho_2(x_T, y_T, x_A, y_A)}{\int_{\Omega_T} dx_T dy_T \rho_1(x_T, y_T)} \tag{10.95}$$

is a per trigger measure of the average number of pairs of trigger (T) and associate (A) particles per trigger, expressed as a function of the coordinates y_T and y_A. The number of pairs is integrated over kinematic variables x_T and x_A, and the number of triggers is obtained by integration over both y_T and x_T. In a practical case, y could stand for the pseudorapidity and x for both p_T and azimuthal angle ϕ of the particles.

The inclusive method has the additional advantage of being trivially related to simple integrals of single- and two-particle density functions defined earlier in this section. For instance, the per trigger function (10.95) may be written

$$C_{T,\mathrm{inc}}(y_T, y_A) = \frac{\rho_2(y_T, y_A)}{\langle N_T \rangle} \tag{10.96}$$

$$C_{T,\mathrm{inc}}(y_T, y_A) = \frac{\rho_1(y_T)\rho_1(y_A) r_n(y_T, y_A)}{\langle N_T \rangle}.$$

It is to be noted, however, that neither $C_{T,\mathrm{inc}}$ nor $C_{T,\mathrm{ebye}}$ are actual correlation functions since they are not based on cumulants, and as such do not account for nor remove combinatorial contributions to the pair production cross section. In other words, both functions could be nonvanishing even in the absence of actual correlations. Additionally, note that the common practice of subtracting the minimal value of $C_{T,\mathrm{inc}}(y_T, y_A)$ from the function to seek an estimate of the "true" correlation is *not* logically consistent, since it may actually over- or undersubtract the combinatorial contributions. A better and logically consistent estimate of the number of correlated associates is obtained by using cumulants. One can, for instance, define the per trigger cumulant

$$K_{T,\mathrm{inc}}(y_T, y_A) = \frac{C_2(y_T, y_A)}{\langle N_T \rangle} \tag{10.97}$$

where the cumulant $C_2(y_T, y_A)$ is obtained according to Eq. (10.90). Note, however, that since $C_2(y_T, y_A)$ is not necessarily positive definite, $K_{T,inc}(y_T, y_A)$ may also end up being negative for selected ranges of the variables y_T and y_A. That simply reflects the fact that the notion of per trigger correlated associates is not a very well defined quantity in general, although it may apply in the context of two-component particle production models.

The notion of per trigger correlation function may be generalized to higher-order densities or cumulants:

$$K_n(y_1, y_2, \ldots, y_n) = \frac{C_n(y_1, y_2, \ldots, y_n)}{\langle N_1 \rangle} \tag{10.98}$$

where, for instance, particle 1 is assumed to be the **trigger** and particles 2 to n are the **associates**.

10.7 Charge-Dependent and Charge-Independent Correlation Functions

Conservation laws play a central role in shaping and constraining particle production. Charge conservation, in particular, dictates that the production (or destruction) of a negative particle must necessarily be accompanied by the production (or destruction) of a positive particle. It applies both locally, where particles are created in space, as well as globally (i.e., for an entire collision). While the dynamics of collisions (momentum transfer, pressure, and so on) largely drive the transverse momenta, rapidity, and angle at which particles are produced, they cannot erase the common origin of positive and negative particle pairs. It is thus interesting to define **charge-independent** (CI) and **charge-dependent** (CD) correlation functions cumulants. In the study of two-particle correlations, one may separate the pairs according to their (electric) charge, and measure two-particle densities and cumulants for $(+,+)$, $(-,-)$, and $(+,-)$ separately. The CI and CD cumulants are thus defined as follows:

$$\begin{aligned} C_{\text{CI}}(y_1,y_2) &= \frac{1}{4}\{C_{+-}(y_1,y_2)+C_{-+}(y_1,y_2) \\ &\quad + C_{++}(y_1,y_2)+C_{--}(y_1,y_2)\} \\ C_{\text{CD}}(y_1,y_2) &= \frac{1}{4}\{C_{+-}(y_1,y_2)+C_{-+}(y_1,y_2) \\ &\quad - C_{++}(y_1,y_2)-C_{--}(y_1,y_2)\}. \end{aligned} \tag{10.99}$$

One may likewise define CI and CD normalized cumulants as follows:

$$\begin{aligned} R_{\text{CI}}(y_1,y_2) &= \frac{1}{4}\{R_{+-}(y_1,y_2)+R_{-+}(y_1,y_2) \\ &\quad + R_{++}(y_1,y_2)+R_{--}(y_1,y_2)\} \\ R_{\text{CD}}(y_1,y_2) &= \frac{1}{4}\{R_{+-}(y_1,y_2)+R_{-+}(y_1,y_2) \\ &\quad - R_{++}(y_1,y_2)-R_{--}(y_1,y_2)\} \end{aligned} \tag{10.100}$$

In very high-energy collisions, particle production at mid-rapidities ($y \approx 0$) yields approximately equal numbers of positive and negative particles, $\rho_+(y) \approx \rho_-(y)$. Defining $\rho_1(y) = \rho_+(y) + \rho_-(y)$, one expects

$$\begin{aligned} R_{\text{CI}}(y_1,y_2) &\approx \frac{1}{\rho_1(y_1)\rho_1(y_2)}\{C_{+-}(y_1,y_2)+C_{-+}(y_1,y_2) \\ &\quad + C_{++}(y_1,y_2)+C_{--}(y_1,y_2)\} \\ &= \frac{C_2(y_1,y_2)}{\rho_1(y_1)\rho_1(y_2)} = R_2(y_1,y_2). \end{aligned} \tag{10.101}$$

where C_2 is obtained by summing over all pairs, i.e., $(+,+)$, $(+,-)$, $(-,+)$, and $(-,-)$. This relationship is, however, not strictly valid at "low" energy or in regions of momentum space where the yields of positive and negative particles differ appreciably.

The functions C_{CI} and C_{CD} have properties and scaling behaviors similar to those of the more basic functions C_2. In particular, their amplitude is expected to be proportional to the number of sources m in a system consisting of independent sources. Consequently, the functions R_{CI} and R_{CD} are expected to scale as $1/m$, similarly to the generic normalized cumulant R_2 functions.

It is also of common practice to study **balance functions** to determine how the production of positively charged particles is balanced by the production of negatively charged particles. Balance functions are typically defined and integrated in a fixed kinematic range (e.g., the pseudorapidity η). Steffen et al. [30], for instance defined a balance function as

$$B(\Delta\eta) = \frac{1}{2}\left\{\frac{\langle N_{+-}(\Delta\eta)\rangle - \langle N_{++}(\Delta\eta)\rangle}{\langle N_+\rangle} + \frac{\langle N_{-+}(\Delta\eta)\rangle - \langle N_{--}(\Delta\eta)\rangle}{\langle N_-\rangle}\right\}, \tag{10.102}$$

where $\langle N_{\alpha\beta}(\Delta\eta)\rangle$ is the number of particle pairs of type (α, β) measured at a given pseudorapidity interval $\Delta\eta$, while $\langle N_+\rangle$ and $\langle N_-\rangle$ are the average number of positive and negative particles produced in that same interval, respectively. The balance function $B(\Delta\eta)$ can be expressed in terms of the normalized cumulant $R_{\rm CD}(y_1, y_2)$ (see Problem 10.1). A more detailed discussion of the balance function is presented in §11.1.3.

While the correlation functions C_{CD} and R_{CD} and the balance function B were first introduced for the study of charged particles, their use can be narrowed to specific particle species, that is, π^+ relative to π^-, or K^+ vs. K^-. They can also be extended to any conserved quantities, for instance, the net strangeness or the net baryon number. For baryons, one could, for instance, write

$$R_{\rm Net-B}(y_1, y_2) = \frac{1}{4}\left\{R_{B\overline{B}}(y_1, y_2) + R_{\overline{B}B}(y_1, y_2) - R_{BB}(y_1, y_2) - R_{\overline{B}\overline{B}}(y_1, y_2)\right\} \qquad (10.103)$$

$R_{\rm Net-B}(y_1, y_2)$ quantifies the correlation between produced baryons (e.g., protons) and anti-baryons (e.g., anti-protons). As such, it provides an invaluable tool to study both the production and transport of baryons in nuclear collisions.

10.8 Generalized (Weighted) Correlation Functions

Just as it is possible and meaningful to study the expectation value of a function $g(x)$ of a continuous random variable x determined by a PDF $f(x)$

$$\mathrm{E}[g(x)] = \int g(x)f(x)\,dx, \qquad (10.104)$$

one can also study generalized correlation functions that involve functions $g(x_1, \ldots, x_n)$ of the particle coordinates $x_1, x_2, \ldots, x_n$. Effectively, such generalized correlation functions are a measure of the average of the function $g(x_1, \ldots, x_n)$. Since it is not necessary to integrate all particle coordinates, we label coordinates to be integrated over as $x_1, \ldots, x_n$, and those not to be integrated over as $y_1, \ldots, y_n$. By strict definition of an average, one can consequently define weighted densities functions as follows:

$$\langle g(y_1, \ldots, y_n)\rangle = \frac{\int_\Omega g(x_1, \ldots, x_n)\rho_n(x_1, y_1, \ldots, x_n, y_n)\prod_i^n dx_i}{\int_\Omega \rho_n(x_1, y_1, \ldots, x_n, y_n)\prod_i^n dx_i}. \qquad (10.105)$$

If the integration is taken over all particle coordinates, that is, for both the x_i and y_i coordinates, the preceding function become an integral correlator defined over the fiducial volume Ω:

$$\langle g\rangle_n = \frac{\int_\Omega g(x_1, \ldots, x_n)\rho_n(x_1, y_1, \ldots, x_n, y_n)\prod_i^n dx_i dy_i}{\langle N(N-1)\cdots(N-n+1)\rangle}. \qquad (10.106)$$

As a practical example of a generalized correlation function, consider a two-particle average of the product $\Delta x_1 \Delta x_2$, in other words, with

$$g(x_1, x_2) = \Delta x_1 \Delta x_2, \qquad (10.107)$$

where $\Delta x_i = x_i - \langle x\rangle(y_1, y_2)$, and $\langle x_i\rangle(y_1, y_2)$ stands for the average value of x evaluated as

$$\langle x_i\rangle(y_1, y_2) = \frac{\int_\Omega x_i \rho_2(x_1, y_1, x_2, y_2)\, dx_1\, dx_2}{\int_\Omega \rho_2(x_1, y_1, x_2, y_2)\, dx_1\, dx_2}. \tag{10.108}$$

Additionally, defining $\langle x_1 x_2\rangle(y_1, y_2)$ as

$$\langle x_1 x_2\rangle(y_1, y_2) = \frac{\int_\Omega x_1 x_2 \rho_2(x_1, y_1, x_2, y_2)\, dx_1\, dx_2}{\int_\Omega \rho_2(x_1, y_1, x_2, y_2)\, dx_1\, dx_2}, \tag{10.109}$$

one finds that Eq. (10.105) may be written, for $n = 2$, and $g(x_1, x_2) = \Delta x_1 \Delta x_2$,

$$\langle \Delta x_1 \Delta x_2\rangle(y_1, y_2) = \langle x_1 x_2\rangle(y_1, y_2) - \langle x_1\rangle(y_1, y_2)\langle x_2\rangle(y_1, y_2), \tag{10.110}$$

which defines the covariance of variables x_1 and x_2 evaluated as a function of the particle coordinates y_1 and y_2. Integration over y_1 and y_2 yields the system wide covariance $\langle \Delta x_1 \Delta x_2\rangle$. Practical implementations of differential and integral correlation function, with $x \equiv p_T$, are discussed in §§11.1.5 and 11.3.5.

10.9 Autocorrelations and Time-Based Correlation Functions

Autocorrelation and time-based correlation functions are commonly used for the study of signals that vary with time and to identify repetitive behaviors, time ordering, and/or dependencies.

An autocorrelation is usually defined as the cross-correlation between a signal and itself at prior time. Consider a random process X that may be measured repeatedly in several "runs" indexed by an integer i. Further assume the process is measured as a function of time and yields a signal $x_i(t)$, where the time t may be expressed either in discrete steps or as a continuous variable. Let $\mu(t)$ and $\sigma(t)$ represent the expectation value and the standard deviation of the signal X at a given time t (or in a given time step), respectively. The autocorrelation of the signal between two times t_1 and t_2 is defined as

$$R_x(t_1, t_2) = \frac{E\left[(x(t_2) - \mu(t_2))\,(x(t_1) - \mu(t_1))\right]}{\sigma(t_1)\sigma(t_2)}. \tag{10.111}$$

Obviously, this definition is meaningful provided the variances $\sigma(t_1)$ and $\sigma(t_2)$ are finite. Variants of this definition includes applications where the means are not subtracted, and no division by the standard deviations is carried out. If the signal is stable, the strength of the correlation becomes exclusively a function of the time difference .

A time-based correlation function between two signals $X(t)$ and $Y(t)$ is similarly defined as

$$R_{xy}(t_1, t_2) = \frac{E\left[(x(t_2) - \mu_x(t_2))\,(y(t_1) - \mu_y(t_1))\right]}{\sigma_x \sigma_y}, \tag{10.112}$$

where $\mu_x(t)$ and $\mu_y(t)$ are the expectation values of the signals x and y, respectively.

Exercises

10.1 Express the balance function $B(\Delta\eta)$ given by Eq. (10.102) in terms of the normalized cumulant $R_{\rm CD}(y_1, y_2)$ given by Eq. 10.101), and single particle densities $\rho_1(y_i)$, or their integrals.

10.2 Derive the expression of the inclusive two-cumulant, Eq. (10.54), in terms of the functions C_S and C_L defined by Eqs. (10.52) and (10.53).

10.3 Derive an expression for the inclusive three-cumulant.

10.4 Verify that the expression (10.68) yields the factorial moments $\tilde{F}_n$ by explicitly calculating the lowest four-order moments.

11 The Multiple Facets of Correlation Functions

The notion of correlation function, defined in Chapter 10, can be extended and adapted toward the study of a wide range of observables designed to investigate specific aspects of nuclear collision dynamics or properties of the matter produced in elementary particle and nuclear collisions. Figure 11.1 displays an overview of the many correlation functions and fluctuation measures commonly used by particle and nuclear physics in their study of nuclear collisions.

In this chapter, we examine several types of correlation functions in detail, beginning with a discussion of two-particle differential correlation functions in §11.1 and a cursory look at three-particle correlations in §11.2. Several integral correlators commonly used in high-energy nuclear for measurements of event-by-event fluctuations are presented in §11.3. The chapter ends with a comprehensive discussion of flow measurement techniques.

11.1 Two-Particle Correlation Functions

Two-particle correlation functions play an important role in the study of collision dynamics in nucleus-nucleus collisions as well as in elementary particle interactions (e.g., pp, $\bar{p}p$, e^+e^-, ep, and so on). Azimuthal correlations, in particular, have been instrumental in the discovery of jet quenching by the opaque and dense medium (QGP) formed in central A–A collisions studied at the Relativistic Heavy-Ion Collider (BNL) and the Large Hadron Collider (CERN) [11]. Such correlations may be studied between indistinguishable particles in identical kinematic ranges, or with particles of different types, species, or belonging to distinct kinematic ranges.

We begin with a description of two-particle azimuthal correlations in §11.1.1 and extend the discussion to joint azimuthal and rapidity correlations in §11.1.2.

11.1.1 Two-Particle Azimuthal Correlation Functions

Momentum and energy conservation impart a specific relation between particles produced by elementary processes. For instance, a ρ^0-meson decaying at rest in the laboratory is expected to produce a pair of π^+ and π^- that fly back-to-back in this frame of reference. This means the two pions should be emitted with an angle of 180° relative to one another, as illustrated in Figure 11.2a. However, if the ρ-meson is in slow motion, the pions will not be seen exactly back-to-back but with a slightly smaller angle than 180°. And if the ρ^0 travels at high velocity in the lab frame, the pions will appear to be focused in nearly the same

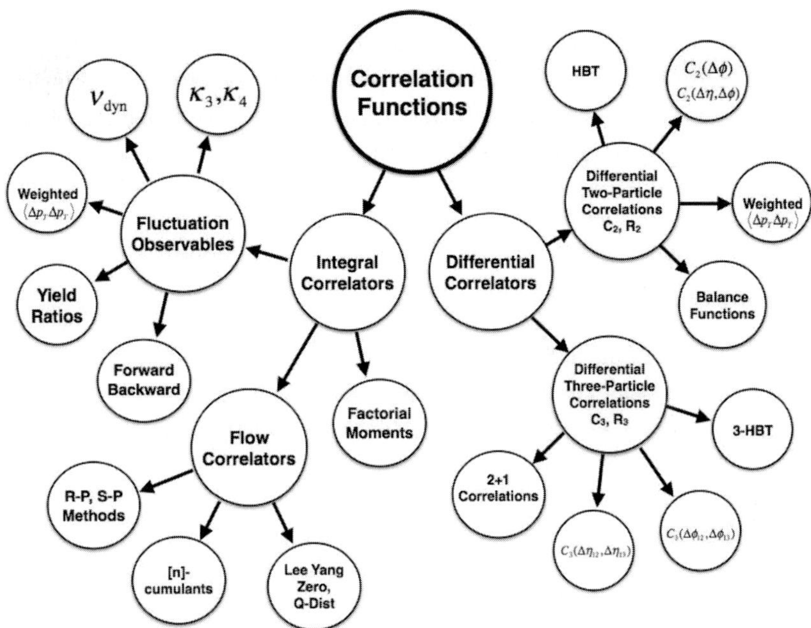

Fig. 11.1 The multiple facets of correlation functions. Differential and integral correlations have a common basis in terms of cumulants, and enable the definition of several types of correlation observables, many of which are shown in this figure and discussed in this chapter.

direction and will thus be observed with a very small relative angle. Two-particle decays of other resonances produce qualitatively similar results. Three-particle decays at rest might produce Mercedes topologies, termed this because of their similarity to the Mercedes-Benz logo, as shown in Figure 11.2d, while parton fragmentation yields collimated particle production in the forms of jets consisting of several particles emitted in relatively narrow cone,

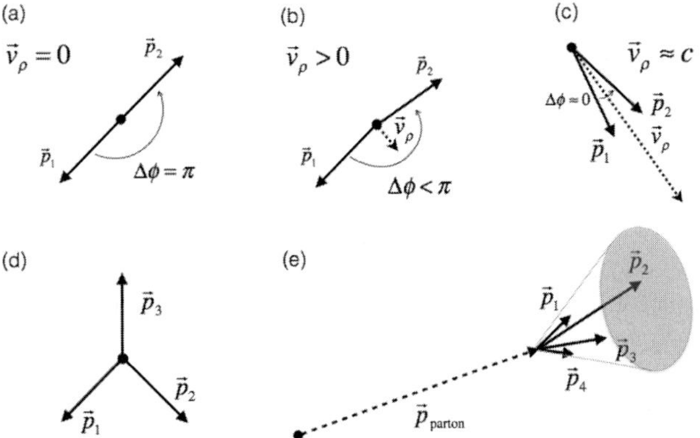

Fig. 11.2 Momentum vectors of π-mesons produced by ρ-meson decays (a) at rest, (b) at small velocity, and (c) at very large velocity (near speed of light). (d) 3-prong resonance decay at rest. (e) Jet: emission of several hadrons in a narrow cone surrounding the direction of a fragmenting parton.

as illustrated in Figure 11.2e. Of course, a typical elementary collision might involve a random selection or/and superposition of several of these and other processes. Two-particle distributions in relative angle may thus become arbitrarily complex. Still, understanding the specific mechanisms that lead to particle production may be possible if one can identify specific correlation features. This said, since the number of particles produced in elementary processes (and even more so in nucleus–nucleus collisions) can be rather large, and given that not all particle pairs may be correlated (i.e., result from a common process), it is useful to consider the use of two-particle cumulants to identify the strength of two particle correlations. Recall from §10.2 that while two-particle densities $\rho_2(y_1, y_2)$ are sensitive to the number of particle pairs produced, they do not readily provide an unambiguous indication of the degree of correlation of these pairs. Different reaction mechanisms may lead to different correlation topologies (e.g., in terms of relative azimuthal angle), and thus it is indeed necessary to use cumulants to properly gauge the degree of correlation between measured pairs.

Observable Definition: Two-Particle Cumulant

We begin our discussion for pairs of "identical" particles measured in the same kinematic range. In §10.2, we defined generic two-particle cumulant C_2 in terms of kinematic variables y_1 and y_2 as

$$C_2(y_1, y_2) = \rho_2(y_1, y_2) - \rho_1(y_1)\rho_1(y_2), \tag{11.1}$$

where $\rho_1(y_i)$ and $\rho_2(y_1, y_2)$ represent single and pair densities expressed as function of kinematical variables y_1 and y_2, respectively. We also introduced normalized cumulants R_2,

$$R_2(y_1, y_2) = \frac{\rho_2(y_1, y_2)}{\rho_1(y_1)\rho_1(y_2)} - 1, \tag{11.2}$$

which effectively carry the same information about the correlations between produced particles. We here restrict the two variables y_1 and y_2 to specifically represent the azimuthal production angles ϕ_1 and ϕ_2 of the two particles of interest, and write

$$C_2(\phi_1, \phi_2) = \rho_2(\phi_1, \phi_2) - \rho_1(\phi_1)\rho_1(\phi_2) \tag{11.3}$$

$$R_2(\phi_1, \phi_2) = \frac{\rho_2(\phi_1, \phi_2)}{\rho_1(\phi_1)\rho_1(\phi_2)} - 1, \tag{11.4}$$

where the densities $\rho_1(\phi_i)$ and $\rho_2(\phi_1, \phi_2)$ are measured/calculated for specific ranges $p_{T,\min} \le p_T < p_{T,\max}$ and $\eta_{T,\min} \le \eta_T < \eta_{T,\max}$. From a theoretical standpoint, in the absence of polarization or other discriminating direction or axis, one expects the single particle yield should have no dependence on ϕ,

$$\rho_1(\phi_1) = \rho_1(\phi_2) \equiv \bar{\rho}_1, \tag{11.5}$$

while the strength of the correlation function C_2 should depend only on the relative angle $\Delta\phi = \phi_1 - \phi_2$. It is thus of interest to recast the correlation function in terms of $\Delta\phi$ exclusively. Toward that end, consider the change of variable

$$\phi_1, \phi_2 \;\rightarrow\; \Delta\phi = \phi_1 - \phi_2, \quad \bar{\phi} = (\phi_1 + \phi_2)/2 \tag{11.6}$$

with the Jacobian $J = \left|\partial\left(\Delta\phi, \bar{\phi}\right)/\partial\left(\phi_1, \phi_2\right)\right| = 1$. This yields a correlation function C_2 that can in principle depend on both $\Delta\phi$ and $\bar{\phi}$. But given the collision system is not polarized and the reaction plane of the colliding particles (nuclei) is not explicitly determined, particle production must be invariant under rotation in $\bar{\phi}$. It is thus legitimate to average out (marginalize) this coordinate. One consequently gets a correlation function C_2 that depends exclusively on the relative angle $\Delta\phi$ of the particles:

$$\begin{aligned} C_2(\Delta\phi) &= \rho_2(\Delta\phi) - \frac{1}{2\pi}\int_0^{2\pi} \bar{\rho}_1^2\, d\bar{\phi}, \\ &= \rho_2(\Delta\phi) - \bar{\rho}_1^2, \end{aligned} \tag{11.7}$$

where, in the second line, we used the fact that ρ_1^2 is a constant and can be taken out of the integral. Given R_2 is a ratio, one can consider two distinct approaches to average out the $\bar{\phi}$ coordinate. The first involves a ratio of averages while the second consists of the average of a ratio. In general, one can in fact use two approaches for the determination of C_2 and R_2. Indeed, one can first proceed to estimate these correlation functions in terms of both ϕ_1 and ϕ_2, with a subsequent average over $\bar{\phi}$ to obtain functions of $\Delta\phi$ only. Alternatively, one can seek estimates of either function directly in terms of $\Delta\phi$. The latter approach is more common and is therefore referred as **Method 1** ($M1$) in the following:

Method 1 ($M1$)

$$C_2^{(M1)}(\Delta\phi) = \rho_2(\Delta\phi) - \rho_1 \otimes \rho_1(\Delta\phi) \tag{11.8}$$

$$R_2^{(M1)}(\Delta\phi) = \frac{\rho_2(\Delta\phi)}{\rho_1 \otimes \rho_1(\Delta\phi)} - 1, \tag{11.9}$$

where the term $\rho_1 \otimes \rho_1(\Delta\phi)$ may be evaluated either through an event-mixing technique or by averaging the product $\rho_1(\phi_1)\rho_1(\phi_2)$ over $\bar{\phi}$.

Method 2 ($M2$) first requires the determination of both the single and pair densities in terms of ϕ_1 and ϕ_2 explicitly. The two functions are subsequently obtained as averages over $\bar{\phi}$:

Method 2 ($M2$)

$$\begin{aligned} C_2^{(M2)}(\Delta\phi) = &\iiint_0^{2\pi} \left\{\rho_2(\phi_1, \phi_2) - \rho_1(\phi_1)\rho_1(\phi_2)\right\} \\ &\times \delta(\Delta\phi - \phi_1 + \phi_2)\delta(\bar{\phi} - (\phi_1 + \phi_2)/2)\, d\phi_1 d\phi_2 d\bar{\phi} \end{aligned} \tag{11.10}$$

$$\begin{aligned} R_2^{(M2)}(\Delta\phi) = &\iiint_0^{2\pi} R_2(\phi_1, \phi_2) \\ &\times \delta(\Delta\phi - \phi_1 + \phi_2)\delta(\bar{\phi} - (\phi_1 + \phi_2)/2)\, d\phi_1 d\phi_2 d\bar{\phi}. \end{aligned} \tag{11.11}$$

where

$$\begin{aligned} R_2(\phi_1, \phi_2) &= \frac{C_2(\phi_1, \phi_2)}{\rho_1(\phi_1)\rho_1(\phi_2)} \\ &= \frac{\rho_2(\phi_1, \phi_2)}{\rho_1(\phi_1)\rho_1(\phi_2)} - 1 \end{aligned} \tag{11.12}$$

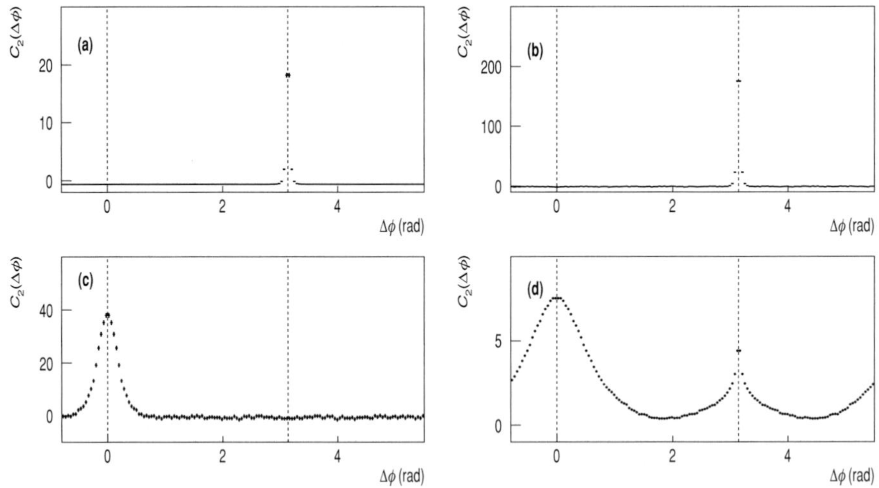

Fig. 11.3 Correlation function C_2 of ρ^0-meson decaying (a, b) near rest, (c) at very large velocity in the laboratory reference frame, and (d) over a "realistic" range of velocities spanning values from zero up to nearly the speed of light. In (a), the number of ρ^0 was fixed to one per event, while in (b–d), it is set to fluctuate randomly from five to ten ρ^0 per event. Distributions have been "shifted" in $\Delta\phi$ for purely esthetic reasons. Dashed lines are drawn at $\Delta\phi = 0$ and $\Delta\phi = 0$ to guide the eye.

For angles ϕ_1 and ϕ_2 in the range $[0, 2\pi]$, one obtains a difference $\Delta\phi = \phi_1 - \phi_2$ in the range $[-2\pi, 2\pi]$. However, by virtue of the system's symmetry, this range may be reduced to $[0, \pi]$ and other components of the $[-2\pi, 2\pi]$ range are redundant and usually shifted onto $[0, \pi]$. It should thus be understood that the preceding integrals over ϕ_1 and ϕ_2 are evaluated onto $0 \leq \Delta\phi \leq \pi$, and often symmetrized about π and plotted in the range $[0, 2\pi]$, or shifted and plotted in $[-\pi, 3\pi/2]$ for ease of visualization. From a theoretical standpoint, it is easy to show that the two methods yield identical results for $\Delta\phi$ correlations, since the single particle yield is invariant under azimuthal rotation, that is, $\rho_1(\phi_1) = \rho_1(\phi_2) \equiv \bar{\rho}_1$. We will see in §12.4, where we discuss methods to account for instrumental effects, that both $M1$ and $M2$ yield robust[1] measurements of $R_2(\Delta\phi)$. Measurements of C_2, however, require efficiency corrections in either method. Additionally, while $M2$ may be subject to aliasing because of the finite size of the bins used to obtain estimates of the densities ρ_1 and ρ_2, we will also show in §12.4 that $M1$ is not strictly robust when measurements are extended to involve dependencies on other coordinates such as the rapidity of the particles, or their differences.

Example 1: Resonance Decays

A great variety of dynamical processes such as jet production, resonance decays, and collective flow can shape azimuthal correlation functions. Figure 11.3 presents examples of correlation functions, $C_2(\Delta\phi)$, obtained with Monte Carlo simulations of collisions

[1] Unaffected by instrumental effects, most particularly detection efficiencies.

producing charged pion pairs ($\pi^+ + \pi^-$) by decays of ρ^0-mesons. Figure 11.3a displays the correlation function of events involving a single ρ^0 decaying near rest in the lab frame yield. One observes a prominent away-side peak centered at $\Delta\phi = \pi$, which corresponds to pion pairs being emitted essentially back-to-back. Emission from ρ^0 at rest would produce a narrow peak at $\Delta\phi = \pi$ exactly, but moving ρ^0s produce pairs that are not strictly back-to-back and consequently lead to a finite width peak centered at $\Delta\phi = \pi$. Also note that because the number of ρ^0 is fixed, so is the number of pions because no efficiency losses or acceptance cuts were accounted for in the generation of this plot. A fixed number of particles effectively produces a multinomial behavior. The total multiplicity is fixed, and the pions can fall in a wide variety of bins. The covariance of the yields for $\Delta\phi$ values outside the peak thus tend to be negative. This effect disappears if the ρ^0 multiplicity fluctuates, as illustrated in Figure 11.3b, which was computed with a uniformly distributed random number of ρ^0 in the range 5–10. However, as illustrated in Figure 11.3c, the correlation peak shifts to the origin (i.e., $\Delta\phi = 0$) if the ρ^0-decays take place at very large momentum in the lab frame. In practice, resonances such as the ρ^0 are produced over a large range of momenta, one then observes a more complicated correlation function, as shown in Figure 11.3d, which combines peaks at both the origin and at $\Delta\phi = \pi$.

Example 2: Correlations from Anisotropic Flow

Two-particle correlations are also very much influenced by collective effects. For instance, collisions of heavy nuclei at finite impact parameter, illustrated in Figure 11.4, may lead to the production of a very dense but inhomogeneous and anisotropic medium or system. The expansion of this medium, through pressure or momentum transfer gradients, is understood to lead to collective motion and anisotropic emission of low to medium p_T particles in the plane transverse to the beam axis. High-energy partons produced within this medium are also believed to be subject to differential energy loss according to the pathlength they traverse to exit the medium, thereby leading to complementary high p_T anisotropic emission patterns. Both effects are commonly known as **anisotropic flow**. While general techniques to measure flow are discussed in §11.4, we demonstrate, in the remainder of this section, that collective particle motion may readily be identified with simple two-particle cumulants $C_2(\Delta\phi)$ and/or normalized cumulants $R_2(\Delta\phi)$.

On general grounds, let us assume that nucleus–nucleus collisions produce, on an event-by-event basis, systems that are inhomogeneous and anisotropic, as illustrated in Figure 11.4a. One can model the system (energy density) spatial anisotropy in terms of a simple Fourier decomposition relative to the origin O:

$$\rho(\phi, r) = f(r)\left(1 + \sum_{n=1}^{\infty} \epsilon_n \cos(n\phi)\right). \tag{11.13}$$

The dynamics of the collisions leads to system expansion and particle emission that reflect the magnitude of the spatial anisotropy coefficients ϵ_n. One can then model the collective motion of particles produced by the system as a Fourier expansion in momentum space,

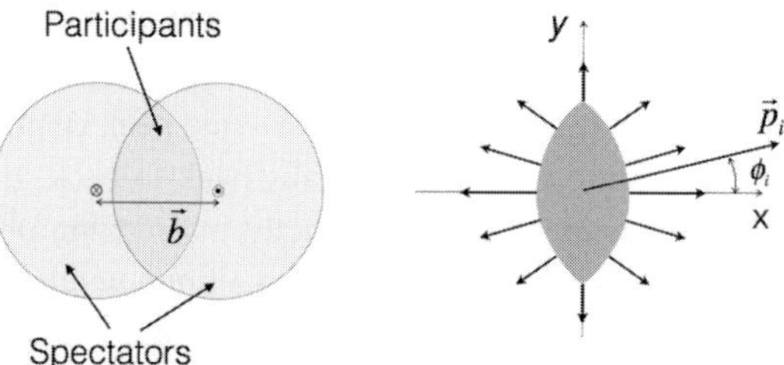

Fig. 11.4 Schematic illustration of the transverse profile of the participant matter produced in high-energy heavy-ion collisions. The participant region features pressure gradients which propel particle outward anisotropically in the transverse plane.

relative to the collision impact parameter vector, $\vec{b}$:

$$\rho(\phi_i|\psi) = \bar{\rho}\left\{1 + 2\sum_{n=1}^{\infty} v_n \cos(n(\phi_i - \psi))\right\}, \tag{11.14}$$

where $\bar{\rho}$ is the average particle density, ϕ is the angle of emission of the particles, and ψ is the orientation angle of the reaction plane in the laboratory frame of reference. Since the impact parameter is not readily observed, one must average over all possible orientations of the reaction plane to get the observed single particle density:

$$\rho_1(\phi_i) = \int_0^{2\pi} d\psi\, \rho_1(\phi_i|\psi)P(\psi) = \bar{\rho}. \tag{11.15}$$

The orientation of the reaction plane, ψ, is assumed to vary collision by collision uniformly in the range $[0, 2\pi]$ and is given a probability $P(\psi) = (2\pi)^{-1}$. The integration in (11.15) thus indeed yields $\bar{\rho}$, the average particle density. The density of particle pairs has a less trivial behavior, however. At fixed ψ, we model the two-particle density as

$$\begin{aligned}\rho_2(\phi_1, \phi_2|\psi) = \bar{\rho}^2 &\left\{1 + 2\sum_{n=1}^{\infty} v_n \cos(n(\phi_1 - \psi))\right\} \\ &\times \left\{1 + 2\sum_{n=1}^{\infty} v_n \cos(n(\phi_2 - \psi))\right\}.\end{aligned} \tag{11.16}$$

Given ψ is not explicitly measured, the two-particle density must be averaged over all equally probable values of this angle. One finds

$$\begin{aligned}\rho_2(\phi_1, \phi_2) &= (2\pi)^{-1}\int_0^{2\pi} \rho_2(\phi_1, \phi_2|\psi)\, d\psi \\ &= \bar{\rho}^2\left\{1 + 2\sum_{n=1}^{\infty} (v_n)^2 \cos(n(\phi_1 - \phi_2))\right\}\end{aligned} \tag{11.17}$$

and concludes that the spatial anisotropy of the collisions may lead to observable anisotropies in momentum space. The harmonic coefficients v_n are commonly known as **flow coefficients**. Their magnitudes have been measured in a variety of collision systems

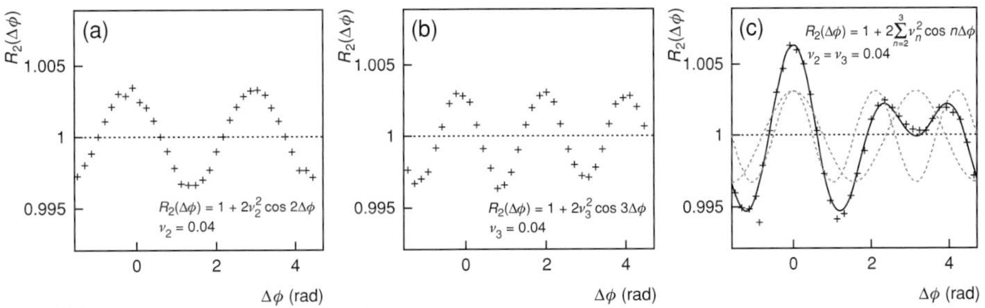

Fig. 11.5 Simulations of $R_2(\Delta\phi)$ correlation functions determined by anisotropic flow. (a) Simulated elliptic flow with $v_2 = 0.04$. (b) Simulated triangular flow with $v_3 = 0.04$, (c) combination of elliptic and triangular flow with $v_2 = v_3 = 0.04$, which produces an away-side dip at $\Delta\phi = \pi$ similar to those observed in central Heavy-Ion collisions by several RHIC and LHC experiments [1]. The dashed and solid lines display the Fourier components used in the simulation and their sum, respectively.

at several beam energies. In high-energy nucleus-nucleus the v_2 coefficients are typically the largest. The v_3 are also found to be significant, even in symmetric collision systems (e.g., Au–Au, or Pb–Pb), most likely the result of fluctuations in the initial spatial configurations of the colliding nuclei. Higher-order coefficients, $v_{n\geq 4}$, are typically much smaller than either v_2 or v_3 coefficients [182].

Figure 11.5 displays simulated $R_2(\Delta\phi)$ correlation functions obtained for selected values of flow coefficients v_2 and v_3. Note how the particular combination of coefficients $v_2 = v_3 = 0.04$, shown in panel (c), produces an away-side "dip" featured in many observed distributions measured in central A–A collisions at RHIC and LHC energies [1].

Semi-inclusive and Species Dependent Correlation Functions

As we saw earlier, resonance production and flow lead to varied correlation shapes in azimuth. To further complicate the interpretation of correlation functions, consider that jet fragmentation and other particle production processes such as "string fragmentation" have their own signatures also in two-particle azimuthal correlations. The fragmentation of a single jet in particular produces a strong and broad peak centered at $\Delta\phi = 0$, while the production of di-jets features both a near-side peak ($\Delta\phi = 0$) and an away-side peak ($\Delta\phi = \pi$). The strength, shape, and width of these peaks are known to depend on the jet transverse momentum and particle multiplicity (number of particles composing the jet). All in all, the structure of correlation functions measured in p–p collisions are typically the result of a random superposition of several production processes that depend on the collision energy, as well as the number and species of particles produced. Correlation functions produced in heavy-ion collisions are further "enriched" by collective processes, such as the anisotropic flow discussed earlier, and radial (or outward) flow effects discussed in [181]. Inclusive measurements of two-particle azimuthal correlation functions are consequently insufficient, typically, to fully pinpoint the collision dynamics. Researchers have thus sought to study more exclusive correlation functions (e.g., azimuthal correlation as a function of particle multiplicity), for specific particle species or by adding additional kinematic constraints. One may also gain further insight into the nature of the particle production processes by studying correlations as a function of the particle momenta and rapidity (pseudorapidity).

Correlation functions of distinguishable particle types or species are readily defined on the basis of Eq. (11.3), and both $M1$ and $M2$ may be used to carry out measurements of these functions. The correlation measurements discussed earlier in this section can thus be straightforwardly extended to correlation studies of particles of different charge or species (e.g., π^+ vs. π^-, or π vs. K, and so on), or for particles in different momentum (or rapidity) ranges, discussed in the next subsection.

Triggered Correlation Functions

Correlation functions involving one high- and one low-p_T particle are particularly useful to study and characterize the production of jets in both p–p and A–A collisions without actually resorting to a full-fledged jet-finding analysis. Correlation studies of this type are commonly known as triggered correlation functions, even though no actual trigger is actually involved in such studies. Triggered correlation functions have been instrumental in unravelling the existence of jet quenching in Au–Au collisions at RHIC [11].

The study of triggered correlation functions typically requires the trigger particle to be at much higher p_T than the second particle, known as the associate. The production of the trigger particle may consequently be **rare**. It is thus legitimate to count the number of associates found in a given event, relative to the trigger particle. Labeling the trigger particle as T and the associates as A, one may define a triggered correlation as

$$C_{T,A}(\Delta\phi) = \frac{\rho_2(\Delta\phi)}{\langle N_T\rangle}, \tag{11.18}$$

where

$$\begin{aligned} \rho_2(\Delta\phi) &= \int_{\Omega_T}\int_{\Omega_A} \rho_2(\eta_T, \phi_T, p_{\perp,T}, \eta_A, \phi_A, p_{\perp,A}) \\ &\quad \times \delta(\Delta\phi - \phi_T + \phi_A)\, d\eta_T d\phi_T dp_{\perp,T} d\eta_A d\phi_A dp_{\perp,A} \end{aligned} \tag{11.19}$$

$$\langle N_T\rangle = \int_{\Omega} \rho_T(\eta_T, \phi_T, p_{\perp,T})\, d\eta_T, d\phi_T, dp_{\perp,T}, \tag{11.20}$$

The integration is taken over the volumes Ω_i, $i = T, A$, that specify the kinematic range in which the trigger and associates are measured. The numerator $\rho_2(\Delta\eta)$ corresponds to the average number of trigger–associate pairs detected as a function of the relative azimuthal angle $\Delta\phi$, while the denominator $\langle N_T\rangle$ corresponds to the average number of trigger particles altogether detected, in a given dataset, or under some specific event selection criteria. Effectively, $C_{T,A}(\Delta\phi)$ provides the average number of associates, vs. $\Delta\phi$, per trigger particle. A measurement of $C_{T,A}(\Delta\phi)$ may be implemented by taking the ratio of the average number of trigger–associate pairs per event by the average number of trigger particles per event:

$$C_{T,A}(\Delta\phi) = \frac{\sum_{\alpha=1}^{N'_{ev}} N_{T,A}(\Delta\eta)}{\sum_{\alpha=1}^{N'_{ev}} N_{T,\alpha}} \tag{11.21}$$

where the sums are computed only for events containing a trigger particle. Triggered correlations are typically of greatest interest for jet-like trigger particles. Those have a low production cross section and only a small fraction of the events are thus expected to contain a trigger particle. If there is at most one trigger particle per event, one may write

$$C_{T,A}(\Delta\phi) = \frac{1}{N'_{ev}} \sum_{\alpha=1}^{N'_{ev}} N_{T,A}(\Delta\eta) \tag{11.22}$$

where the sum is taken over events containing one trigger particle. For kinematical conditions such that there is typically more than one trigger particle per event, there is a common practice that consists in defining $C_{T,A}(\Delta\phi)$ as

$$C_{T,A}^{\text{ebye}}(\Delta\phi) = \frac{1}{N'_{ev}} \sum_{\alpha=1}^{N'_{ev}} \frac{N_{T,A}(\Delta\eta)}{N_T} \tag{11.23}$$

where N_T stands for the number of trigger particle measured in each event (the sum being carried out only on events with at least one trigger particle). In general, this formulation is expected to yield values in qualitative agreement to those obtained with Eq. (11.21). Quantitative difference of several percent may, however, occur. And while this latter formulation of the correlation function may seem more natural, it does not lend itself to a simple definition in terms of single- and two-particle densities such as Eq. (11.21). The inclusive formulation, though seemingly less intuitive, is thus usually preferred.

It is also important to realize that $C_{T,A}(\Delta\phi)$ is not a correlation function in the strict sense of the word, since it does not involve subtraction of purely combinatorial contributions. This is not a serious problem, however, as long as the trigger particle rate is small and the kinematic range considered for associates actually involves mostly true associates. This condition is easily satisfied in p–p collisions but is typically violated in high-energy A–A collisions, unless trigger particles are selected in a very high p_T ranges.

Correlation Function Scaling

By construction as a two-particle cumulant, the correlation function $C_2(\Delta\phi)$ is expected to scale in proportion to the number of "sources," while $R_2(\Delta\phi)$ should be inversely proportional to this number. For a collision system consisting of m independent identical sources, one should consequently get

$$C_2^{(m)}(\Delta\phi) = mC_2^{(1)}(\Delta\phi) \tag{11.24}$$

$$R_2^{(m)}(\Delta\phi) = \frac{1}{m} R_2^{(1)}(\Delta\phi). \tag{11.25}$$

The triggered correlation $C_{T,A}$ has yet a different scaling property. If the number of trigger and associate particles are truly rare, both the numerator and the denominator of $C_{T,A}$ are expected to scale in proportion to the number of particle sources, and their ratio should consequently be invariant under variations of this number. In practice, particularly in heavy-ion collisions, both the number of trigger and associates can become rather large in central collisions, and this simple scaling is therefore violated. However, many researchers attempt to recover this property by removing combinatorial and collective correlation backgrounds using a technique known as zero yield at minimum (ZYAM) [16]. This technique assumes a two-component (two sources) particle production model, which may be approximately valid in the study of jets, provided the coupling (correlation) between the jet and the underlying background is small. This hypothesis may, however, be difficult to justify in analyses seeking to study the influence of the medium (bulk of produced particles) on the produced jets. It is thus preferable and logically consistent to define triggered correlation functions on the basis of $C_2(\Delta\phi)$ rather than $\rho_2(\Delta\phi)$ (see Problem 11.11).

11.1.2 Two-Particle $\Delta\eta$ Correlation Functions

Correlation functions measured in terms of the pseudorapidity difference of produced particles provide another powerful tool to explore the collision dynamics of both elementary particle and nuclear collisions. They have been used extensively, in particular, to study hadron production in p–p collisions and jet production in heavy-ion collisions.

As for the azimuthal correlations discussed in the previous section, we proceed to define two-particle correlation functions vs. $\Delta\eta$ in terms of two-particle cumulants and normalized cumulants, as defined by Eq. (10.22),

$$C_2(\eta_1, \eta_2) = \rho_2(\eta_1, \eta_2) - \rho_1(\eta_1)\rho_1(\eta_2) \tag{11.26}$$

$$R_2(\eta_1, \eta_2) = \frac{\rho_2(\eta_1, \eta_2)}{\rho_1(\eta_1)\rho_1(\eta_2)} - 1. \tag{11.27}$$

The η_1 vs. η_2 dependence of these functions cannot, however, be readily reduced to a single variable $\Delta\eta$, such as in the case of azimuthal correlations. Indeed, it is quite possible that the correlation dynamics may be an arbitrary function of both η_1 and η_2. In fact, measurements conducted in p–p collisions at the ISR and in p–$\bar{\text{p}}$ collisions at the Tevatron have shown that correlation functions may have a rather intricate dependence on both η_1 and η_2, particularly when the particles have rapidities approaching the beam rapidity [85, 193]. However, in collider experiments with a focus on central rapidities ($\eta \approx 0$), one may expect the correlations to depend primarily on the rapidity difference of the two particles considered, it is thus reasonable to recast the C_2 and R_2 correlation functions in terms of variables $\Delta\eta = \eta_1 - \eta_2$ and $\bar{\eta} = (\eta_1 + \eta_2)/2$, average out the dependence on $\bar{\eta}$, and obtain correlation functions in terms of $\Delta\eta$ exclusively. One must, however, properly account for the finite acceptance of the detector.

As an example, let us assume, as illustrated in Figure 11.6, that both particle 1 and 2 are measured in the same range $-\eta_0 \leq \eta < \eta_0$, and let us calculate the two-particle cumulant $C_2(\Delta\eta)$ assuming the cumulant $C_2(\eta_1, \eta_2)$ is known. Averaging out over $\bar{\eta}$ may

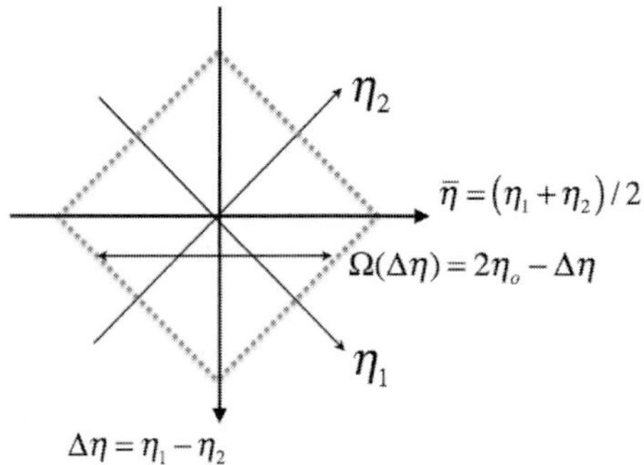

Fig. 11.6 Illustration of the rapidity acceptance of two-particle correlation measurements.

be accomplished as follows:

$$C_2(\Delta\eta) = \frac{1}{\Omega(\Delta\eta)} \iiint_{\Omega} C_2(\eta_1, \eta_2) \times \delta(\Delta\eta - \eta_1 + \eta_2)\delta(\bar{\eta} - (\eta_1 + \eta_2)/2)\, d\eta_1 d\eta_2 d\bar{\eta}, \tag{11.28}$$

where the factor $\Omega(\Delta\eta)$ accounts for the width of the $\bar{\eta}$ acceptance at a given value of $\Delta\eta$. For the square and symmetric acceptance illustrated in Figure 11.6, one gets

$$\Omega(\Delta\eta) = 2\eta_o - \Delta\eta. \tag{11.29}$$

This factor accounts for the fact that, with a square acceptance $|\eta_i| < \eta_0$, it is far less likely to observe pairs with a rapidity difference $|\Delta\eta| \approx 2\eta_0$ than with $\Delta\eta \approx 0$. Failure to account for this simple geometric effect results in an apparently strong triangular correlation shape. The division by $\Omega(\Delta\eta)$ is thus commonly known as the **triangle acceptance correction**. This is an unfortunate misnomer, however, because the division by $\Omega(\Delta\eta)$ is not strictly speaking a correction of the data but a factor born out of the acceptance averaging over $\bar{\eta}$.

As for measurements of $C_2(\Delta\phi)$, two basic techniques, hereafter referred to as Method 1 ($M1$) and Method 2 ($M2$), may be used to obtain an estimator of $C_2(\Delta\eta)$.

$M1$ involves the determination of the two particle density $\rho_2(\Delta\eta)$ directly using a 1D pair histogram, H_2, binned and filled at $\Delta\eta = \eta_1 - \eta_2$ for all relevant pairs:

$$\hat{\rho}_2(\Delta\eta) = \frac{1}{\Omega(\Delta\eta)} \frac{k}{(N_{ev} - 1)\Delta_{\Delta\eta}} H_2(\Delta\eta). \tag{11.30}$$

The histogram H_2 must be filled over all N_{ev} events satisfying relevant event cuts and all particle pairs satisfying track quality cuts and kinematic selection criteria of interest. The factor $\Delta_{\Delta\eta}$ represents the bin width used to define and fill the histogram H_2. The factor k is unity for nonidentical particle pairs (e.g., kaons vs. pions, or particles in different p_T ranges), and equal to 2 for identical particles analyzed with two nested loops $i < j$ over all particles (i.e., with $1 \leq i \leq n_{\text{parts}}$ and $i + 1 < j \leq n_{\text{parts}}$). The factor $\Omega(\Delta\eta)$ must be determined numerically and accounts for the finite acceptance in $\bar{\eta}$.

The second term of the correlation function may be obtained by numerical integration of the product $\hat{\rho}_1(\eta_1)\hat{\rho}_1(\eta_2)$, where $\hat{\rho}_1(\eta_i)$ is an estimator of the single particle density as a function of η:

$$\hat{\rho}_1(\eta) = \frac{1}{(N_{ev})\Delta_\eta} H_1(\eta). \tag{11.31}$$

The factor Δ_η represents the bin width used to fill the histogram H_1. Two distinct histograms $H_1^{(1)}(\eta)$ and $H_1^{(2)}(\eta)$ (and corresponding single particle densities) must obviously be used if the analysis is carried out on distinguishable particles. Numerical techniques to carry out the numerical integration of $\hat{\rho}_1(\eta_1)\hat{\rho}_1(\eta_2)$ to obtain a function $\hat{\rho}_1 \otimes \hat{\rho}_1(\Delta\eta)$ are discussed in §11.5. An estimator of $\hat{C}_2(\Delta\phi)$ of the correlation function may thus be written

$$\hat{C}_2(\Delta\phi) = \hat{\rho}_2(\Delta\eta) - \hat{\rho}_1 \otimes \hat{\rho}_1(\Delta\eta). \tag{11.32}$$

One may alternatively estimate the second term $\rho_1 \otimes \rho_1(\Delta\eta)$ of the correlation function on the basis of mixed events rather than a product of single particle densities. This is accomplished by filling 1D histogram $H_2^{(\text{Mixed})}$ from pairs of particles obtained from mixed events. One then gets a mixed event density $\hat{\rho}_2^{(\text{Mixed})}(\Delta\eta)$

$$\hat{\rho}_2^{(\text{Mixed})}(\Delta\eta) = \frac{1}{\Omega(\Delta\eta)} \frac{k}{(N_{ev}-1)\Delta_{\Delta\eta}} H_2^{(\text{Mixed})}(\Delta\eta), \tag{11.33}$$

which should provide a legitimate estimate of the density product, $\rho_1 \otimes \rho_1(\Delta\eta)$, since particles produced in different collisions (events) should be physically uncorrelated. An estimator of $\hat{C}_2(\Delta\phi)$ of the correlation function may thus also be obtained with

$$\hat{C}_2(\Delta\eta) = \hat{\rho}_2(\Delta\eta) - \hat{\rho}_2^{(\text{Mixed})}(\Delta\eta). \tag{11.34}$$

Another approach, hereafter known as Method 2 ($M2$), consists of obtaining $C_2(\Delta\eta)$ as an integral of $C_2(\eta_1, \eta_2)$. That is, one first obtains an estimate $\hat{C}_2(\eta_1, \eta_2)$ based on histograms $H_2(\eta_1, \eta_2)$ and $H_1(\eta_i)$ accumulated over all relevant pairs of particles (tracks) and single particles, respectively, and estimate C_2 as

$$\hat{C}_2(\eta_1, \eta_2) = \hat{\rho}_2(\eta_1, \eta_2) - \hat{\rho}_1(\eta_1)\hat{\rho}_1(\eta_2) \tag{11.35}$$

with

$$\hat{\rho}_1(\eta_i) = \frac{1}{N_{ev}\Delta_\eta} H_1(\eta_i), \tag{11.36}$$

$$\hat{\rho}_2(\eta_1, \eta_2) = \frac{1}{(N_{ev}-1)\Delta_\eta^2} H_2(\eta_1, \eta_2), \tag{11.37}$$

where Δ_η is the bin size used to define and fill histograms H_1 and H_2. The estimator $\hat{C}_2(\Delta\eta)$ is subsequently obtained by numerical integration of $\hat{C}_2(\eta_1, \eta_2)$

$$\hat{C}_2(\Delta\eta) = \frac{1}{\Omega(\Delta\eta)} \sum_{\eta_1, \eta_2, \bar{\eta}} \hat{C}_2(\eta_1, \eta_2) \times \delta(\Delta\eta - \eta_1 + \eta_2)\delta(\bar{\eta} - (\eta_1 - \eta_2)/2). \tag{11.38}$$

Alternatively, the term $\hat{\rho}_1(\eta_1)\hat{\rho}_1(\eta_2)$ may be determined with pairs from mixed events:

$$\hat{\rho}_1(\eta_1)\hat{\rho}_1(\eta_2) = \frac{1}{(N_{ev}-1)\Delta_\eta^2} H_2^{(\text{Mixed})}(\eta_1, \eta_2), \tag{11.39}$$

Note that the use of finite width bins leads to a smearing of the correlation across the width of the $\Delta\eta$ bin. Indeed, a specific bin in $\Delta\eta$ obtained from two specific bins in η_1 and η_2 corresponds to an actual range $\eta_1 - \eta_2$ which is larger than the width of the bin and effectively smears the correlation signal across bins.[2] This effect, often referred to as **aliasing**, can be reduced by oversampling the $\rho_1(\eta_i)$ and $\rho_2(\eta_1, \eta_2)$ distributions, in other words, by using bins in η that are two or more times smaller than the width $\Delta\eta$.

As for C_2, the normalized cumulant $R_2(\Delta\eta)$ may in principle be determined with either $M1$ or $M2$. $M1$ estimates $R_2(\Delta\eta)$ as the ratio

$$\hat{R}_2^{(M1)}(\Delta\eta) = \frac{\hat{C}_2(\Delta\eta)}{\hat{\rho}_1 \otimes \hat{\rho}_1(\Delta\eta)} \tag{11.40}$$

while $M2$ requires an average of $\hat{R}_2(\eta_1, \eta_2)$ over $\bar{\eta}$:

$$\begin{aligned}\hat{R}_2^{(M2)}(\Delta\eta) = \frac{1}{\Omega(\Delta\eta)} &\sum_{\eta_1,\eta_2,\bar{\eta}} \hat{R}_2(\eta_1, \eta_2) \\ &\times \delta(\Delta\eta - \eta_1 + \eta_2)\delta(\bar{\eta} - (\eta_1 + \eta_2)/2).\end{aligned} \tag{11.41}$$

Unfortunately, the two methods may yield substantially different results. Indeed, unlike the case of azimuthal correlations, for which the single densities $\rho_1(\phi)$ are invariant under rotation in ϕ, the single densities $\rho_1(\eta_i)$ are arbitrary functions of η_i, that is, they may exhibit substantial variations throughout the acceptance of the measurement. This is a problem because the integral of the ratio of functions is in general not equal to the ratio of the integrals of these functions:

$$\int_\Omega \frac{\rho_2(\Delta\eta, \bar{\eta})}{\rho_1 \otimes \rho_1(\Delta\eta, \bar{\eta})} d\bar{\eta} \neq \frac{\int_\Omega \rho_2(\Delta\eta, \bar{\eta})\, d\bar{\eta}}{\int_\Omega \rho_1 \otimes \rho_1(\Delta\eta, \bar{\eta})\, d\bar{\eta}} \tag{11.42}$$

The issue may be exacerbated by detection inefficiencies that strongly depend on the coordinates η_1 and η_2. This is discussed in detail in §12.4.3, where we argue that although less intuitive and direct from an experimental standpoint, $R_2^{(M2)}(\Delta\eta)$ constitutes a more robust and meaningful measure of the normalized cumulant. In practice, at collider experiments, both the single particle ρ_1 and pair densities ρ_2 appear to have only rather modest dependence on rapidity in fiducial ranges measured. Constants can, of course, be factorized out of the preceding integrals and the two methods consequently yield very similar results in symmetric A–A collisions. Rapidity dependencies may, however, be larger in asymmetric collisions such as p–A collisions.

By construction as cumulants, the functions C_2 and R_2 exhibit the scaling properties embodied in Eqs. (10.25) and (10.39) discussed for generic cumulants in §10.2.4 and for azimuthal correlation functions in §11.1.1.

Triggered correlation functions $C_{TA}(\Delta\eta)$ may be defined similarly to the functions $C_{TA}(\Delta\phi)$ introduced §11.1.1. Caveats discussed for $C_{TA}(\Delta\phi)$ apply to $C_{TA}(\Delta\eta)$ also

[2] For instance, combining yields from bins $0.9 < \eta_1 \leq 1.1$ and $0.9 < \eta_2 \leq 1.1$ corresponds to a range of $-0.2 \leq \eta_1 - \eta_2 \leq 0.2$ being projected onto a $\Delta\eta$ bin of width 0.2.

and are further complicated by issues associated with Method 1 discussed earlier. Precision measurements should be conducted exclusively with cumulants calculated based on Method 2.

The $R_2(\Delta\eta)$ correlation function introduced in this section and the $R_2(\Delta\phi)$ function introduced earlier can be trivially combined to obtain correlation functions with joint dependencies on $\Delta\eta$ and $\Delta\phi$. Measurements with Method 2 yield

$$R_2^{(M2)}(\Delta\eta, \Delta\phi) = \frac{1}{\Omega(\Delta\eta)} \int_\Omega R_2(\eta_1, \phi_1, \eta_2, \phi_2) \times \delta(\Delta\eta - \eta_1 + \eta_2)\delta(\bar{\eta} - (\eta_1 + \eta_2)/2)\, d\eta_1 d\eta_2 d\bar{\eta}, \times \delta(\Delta\phi - \phi_1 + \phi_2)\delta(\bar{\phi} - (\phi_1 + \phi_2)/2)\, d\phi_1 d\phi_2 d\bar{\phi}, \tag{11.43}$$

where

$$R_2(\eta_1, \phi_1, \eta_2, \phi_2) = \frac{\hat{\rho}_2(\eta_1, \phi_1, \eta_2, \phi_2)}{\hat{\rho}_1(\eta_1, \phi_1)\hat{\rho}_1(\eta_2, \phi_2)} - 1, \tag{11.44}$$

and the densities are obtained from multidimensional histograms filled by processing all relevant events and particles:

$$\hat{\rho}_1(\eta_i, \phi_i) = \frac{1}{N_{ev}\Delta_\eta\Delta_\phi} H_1(\eta_i, \phi_i), \tag{11.45}$$

$$\hat{\rho}_2(\eta_1, \phi_1, \eta_2, \phi_2) = \frac{k}{(N_{ev} - 1)\Delta_\eta^2\Delta_\phi^2} H_2(\eta_1, \phi_1, \eta_2, \phi_2), \tag{11.46}$$

where, as in the preceding, k is unity for distinguishable particles and equal to 2 for indistinguishable particles analyzed with two-particle nested loops with an $i < j$ condition.

11.1.3 Balance Functions

The production of particles in elementary particles is constrained by energy/momentum conservation and several other conservation laws, such as charge conservation, baryon number conservation, and so on. A charge balance function provides a tool to emphasize effects associated with charge conservation on the charge particle production. For instance, in the case of pion production, one can assume that energy–momentum considerations affect the production of negative and positive pions in essentially the same way. However, one expects on general grounds that correlations between unlike-sign pions should be stronger than those between like-sign pions because charge conservation dictates that particles must be created in unlike-sign pairs. The charge balance function is designed to isolate the correlation strength associated with charge conservation specifically by measuring the number of unlike-sign pairs relative to the number of like-sign pairs in a given momentum volume. In heavy-ion collision studies, the charge balance function was also proposed by Bass et al. [31, 155] as a tool to identify the presence of a qualitative change in charged particle production versus collision centrality. The production of a long lived phase of quark gluon plasma, in particular, is expected to lead to delayed hadronization manifested by a narrowing of the balance function measured as a function of the rapidity (or pseudorapidity) difference between produced particles [155].

The balance function introduced in ref. [31] is written

$$\begin{aligned} B(\Delta\eta) = & \frac{\langle N_- N_+\rangle(\Delta\eta)}{\langle N_-\rangle} + \frac{\langle N_+ N_-\rangle(\Delta\eta)}{\langle N_+\rangle} \\ & - \frac{\langle N_+(N_+ - 1)\rangle(\Delta\eta)}{\langle N_+\rangle} - \frac{\langle N_-(N_- - 1)\rangle(\Delta\eta)}{\langle N_-\rangle}, \end{aligned} \tag{11.47}$$

where $\langle N_- N_+\rangle(\Delta\eta)$, $\langle N_+(N_+ - 1)\rangle(\Delta\eta)$, and $\langle N_-(N_- - 1)\rangle(\Delta\eta)$ represent the number of $(+-)$, $(++)$, and $(--)$ pairs observed in a fixed range of pseudorapidity $\Delta\eta$, while $\langle N_+\rangle$ and $\langle N_-\rangle$ are the average numbers of positively and negatively charged particles detected in the measurement acceptance, respectively. Introducing normalized two-particle densities $r_{\alpha,\beta}(\Delta\eta)$, defined as

$$\begin{aligned} r_{\alpha,\beta}(\Delta\eta) &= \frac{\langle N_\alpha N_\beta\rangle(\Delta\eta)}{\langle N_\alpha\rangle\langle N_\beta\rangle} \quad \text{for } \alpha \neq \beta \\ &= \frac{\langle N_\alpha(N_\alpha - 1)\rangle(\Delta\eta)}{\langle N_\alpha\rangle^2} \quad \text{for } \alpha = \beta, \end{aligned} \tag{11.48}$$

where the indices α and β represent charges $+$ and $-$, one may write the balance function as

$$\begin{aligned} B(\Delta\eta) = & \langle N_+\rangle r_{-+}(\Delta\eta) + \langle N_-\rangle r_{+-}(\Delta\eta) \\ & - \langle N_+\rangle r_{++}(\Delta\eta) - \langle N_-\rangle r_{--}(\Delta\eta). \end{aligned} \tag{11.49}$$

At high collisional energy, one expects $\langle N_+\rangle \approx \langle N_-\rangle$ and $r_{++}(\Delta\eta) \approx r_{--}(\Delta\eta)$ at central rapidities. Additionally, for a symmetric collision system (e.g., Pb + Pb), one also expects $r_{-+}(\Delta\eta) = r_{+-}(\Delta\eta)$. The balance function may then be expressed in terms of the charge-dependent correlation function, R_{CD}, defined in §10.7 (see also Problem 11.13).

$$B(\Delta\eta) = \langle N_{\text{ch}}\rangle R_{\text{CD}}(\Delta\eta) \tag{11.50}$$

where $N_{ch} = N_+ + N_-$. Use of this expression for a measurement of the balance function, rather than Eq. (11.47), has two obvious advantages: (1) the observable R_{CD} is by construction robust against particle losses associated with detection efficiencies, and (2) the charge particle multiplicity has a simple dependency on detection efficiency that is usually simple to correct.

Pratt et al. have shown that the balance function is sensitive to radial flow effects that are independent and distinct from delayed hadronization effects. Radial flow effects may be modeled using a variety of techniques. The balance function thus constitutes an excellent observable to investigate the relative effects associated with charge transport (flow) and delayed hadronization [64]. The balance function (and related correlation functions) may be adapted to studies of net baryon number or net strangeness transport by substituting the number of baryons (strange particles) and anti-baryons (anti-strange) for the number of positively and negatively charged particles, respectively.

11.1.4 Forward–Backward Correlations

The production of particles in elementary collisions at very high energy spans a very large range of rapidity. In order to understand the mechanisms that yield particles over such large range, researchers have sought to measure the correlation level between particles emitted forward and backward with a large rapidity gap. This type of correlation measurement is commonly known as **forward–backward correlation**. We first describe the technique in some detail, and next discuss its merits and limitations.

The forward–backward correlation technique essentially consists of a linear regression between the number of particles produced at **forward rapidities**, within a narrow range $\delta\eta$ centered at $+\eta$, and the number of particles emitted at **backward rapidities**, $-\eta$, within an equivalent narrow range $\delta\eta$. Given fixed values of η and $\delta\eta$, one determines and plots the average number of particles produced forward, $\langle n_F \rangle$, for a given value (binned) of the number produced backward, n_B. One then fits the measured $(n_B, \langle n_F \rangle)$ data points using a simple linear parameterization (i.e., a linear regression):

$$\langle n_F(n_B) \rangle = a + b \times n_B. \tag{11.51}$$

The measurement is repeated, and the coefficients a and b determined, for several values of the **rapidity gap** 2η. The coefficient b, which describes the strength of the linear regression (i.e., the forward–backward correlation) may then be plotted as a function of the rapidity gap 2η. While conceptually simple, the aforementioned measurement recipe is tedious of execution, and thus an alternative, more straightforward method is commonly used. This method relies on the hypothesis of a linear relationship between the particle yield emitted forward and backward. One writes

$$n_F - \langle n_F \rangle = b \times (n_B - \langle n_B \rangle) + r, \tag{11.52}$$

where b is the correlation strength to be determined, while r represents a random variable uncorrelated to the value of n_B, and with a vanishing expectation value, $\langle r \rangle = 0$. One may then calculate the expectation value of the product $n_F n_B$ as follows:

$$\langle n_F n_B \rangle = b \left(\langle n_B^2 \rangle - \langle n_B \rangle^2 \right) + \langle n_F \rangle \langle n_B \rangle + \langle r n_B \rangle. \tag{11.53}$$

Given r is assumed to be uncorrelated to n_B, the last term on the right vanishes, since $\langle r n_B \rangle = \langle r \rangle \langle n_B \rangle$ and $\langle r \rangle = 0$. The slope coefficient b may consequently be written

$$b = \frac{\langle n_F n_B \rangle - \langle n_F \rangle \langle n_B \rangle}{\langle n_B^2 \rangle - \langle n_B \rangle^2}, \tag{11.54}$$

which provides a simple and rapid method to determine the coefficient b in terms of the covariance, $\text{Cov}[n_F, n_B] = \langle n_F n_B \rangle - \langle n_F \rangle \langle n_B \rangle$, and the variance, $\text{Var}[n_B] = \langle n_B^2 \rangle - \langle n_B \rangle^2$.

Although the implementation of Eq. (11.54) is seemingly simple, accounting for detector effects, most particularly particle losses, is not. We will see in §12.4.2 that for measurements with limited efficiency, ϵ, the variance of the particle multiplicity, n, detected in a given kinematical range scales as $\epsilon^2 \text{Var}[n] + \epsilon n$. The ratio b expressed by Eq. (11.54) consequently does not constitute a robust observable, and its correction to account for detection efficiencies, which may differ at forward and backward rapidity, is therefore not trivial. However, since the information being sought lies within the strength of the covariance $\langle n_F n_B\rangle - \langle n_F\rangle\langle n_B\rangle$, a measurement of forward-backward correlation may then be obtained with a normalized two-particle cumulant:

$$C_{BF} = \frac{\langle n_F n_B\rangle - \langle n_F\rangle\langle n_B\rangle}{\langle n_F\rangle\langle n_B\rangle} \tag{11.55}$$

which is, by construction, robust under particle losses associated with detection efficiencies.

11.1.5 Differential Transverse Momentum Correlations

Measurements of two- and multiple-particle correlations constitute powerful tools to study heavy-ion collision dynamics and enable the identification in nucleus–nucleus collisions of new correlation features not found in proton–proton interactions. These tools can be further enhanced by inserting weights dependent on the kinematical properties of the particles into the correlation function definition. We consider, as an example, the definition of two-particle differential transverse momentum correlation functions. As for integral transverse momentum fluctuation correlations (see §11.3.5), there is a certain latitude in the definition of such correlation functions. Indeed, as for integral correlations, one can use both **inclusive** and **event-wise** definitions [173]. One can also consider dynamical fluctuations as the difference between measured fluctuations and statistical fluctuations, that is, fluctuations expected for a purely Poisson system. Extension of the fluctuation variable Φ_{p_T} [92, 146] is also possible. In this section, we use the notations introduced in prior sections and focus our discussion only on an **inclusive** definition. Extensions to event-wise and other differential fluctuation measures are also possible [173].

The $\langle \Delta p_T \Delta p_T\rangle$ correlation function is defined as a **pair-averaged** product of deviates $\Delta p_T \Delta p_T$,

$$\begin{aligned}
\langle \Delta p_T \Delta p_T\rangle(\Delta\eta, \delta\phi) = \frac{1}{\Omega}\int_\Omega & \frac{\rho_2^{(\Delta p_T \Delta p_T)}(\eta_1, \phi_1, \eta_2, \phi_2)}{\rho_2(\eta_1, \phi_1, \eta_2, \phi_2)} \\
& \times \delta(\Delta\eta - \eta_1 + \eta_2)\delta(\bar{\eta} - (\eta_1 + \eta_2)/2) \\
& \times \delta(\Delta\phi - \phi_1 + \phi_2)\delta(\bar{\phi} - (\phi_1 + \phi_2)/2) \\
& \times d\phi_1 d\eta_1 d\phi_2 d\eta_2 d\bar{\eta} d\bar{\phi}
\end{aligned} \tag{11.56}$$

where

$$\begin{aligned}
\rho_2^{(\Delta p_T \Delta p_T)}(\eta_1, \phi_1, \eta_2, \phi_2) &= \int_\Omega \rho_2(\vec{p}_1, \vec{p}_2) \\
& \quad \times \Delta p_{T,1} \Delta p_{T,2} dp_{T,1} dp_{T,2}, \\
\rho_2(\eta_1, \phi_1, \eta_2, \phi_2) &= \int_\Omega \rho_2(\vec{p}_1, \vec{p}_2) dp_{T,1} dp_{T,2},
\end{aligned} \tag{11.57}$$

The function $\rho_2(\vec{p}_1, \vec{p}_2)$ is the pair density expressed with respect to the particles' rapidity, azimuth, and transverse momentum; $\Delta p_{T,i} = p_{T,i} - \langle p_{T,i}\rangle$ and $\langle p_{T,i}\rangle$ represents the inclusive average of the particle momenta in the fiducial acceptance of the measurement.

Given its explicit dependence on the product of deviates $\Delta p_{T,1}\Delta p_{T,2}$, the $\langle\Delta p_T\Delta p_T\rangle$ correlation function provides a qualitatively different measure of particle correlations (relative to R_2) that is sensitive to the particle momenta. Indeed, consider that a given pair of particles may involve two particles below the average momentum $\langle p_{T,i}\rangle$, two above, or one above and one below. The product $\Delta p_{T,1}\Delta p_{T,2}$ may consequently be either positive or negative. A correlated particle pair involving two particles below (or above) the mean momentum will have a positive contribution to the correlation average, while correlated pairs consisting of one particle above and one particle below the momentum average should have a negative contribution to the correlation average. The $\langle\Delta p_T\Delta p_T\rangle$ correlator thus provides a different way to probe particle production processes. For instance, processes such as Bose–Einstein condensation yield correlated pairs in close momentum proximity and should have a strong positive contribution to the $\langle\Delta p_T\Delta p_T\rangle$ correlation function. On the other hand, particle decays that lead to the production of one slow particle and one high-momentum particle should have a negative contribution to $\langle\Delta p_T\Delta p_T\rangle$. The study of this correlation function thus provides an additional tool to probe and understand the particle production dynamics in elementary and nuclear collisions.

A measurement of $\langle\Delta p_T\Delta p_T\rangle$ can be straightforwardly implemented using Method 2 discussed in §11.1.2. The average momentum should be obtained as an inclusive average over all particles of interest:

$$\langle p_{T,i}\rangle = \frac{1}{N_{\text{part}}}\sum_{\text{parts}} p_{T,i} \tag{11.58}$$

Estimates of the density ρ_2 and weighted density $\rho_2^{(\Delta p_T\Delta p_T)}$ may be obtained using four dimension histograms:

$$\hat{\rho}_2(\eta_1,\phi_1,\eta_2,\phi_2) = \frac{1}{N_{ev}}\frac{1}{\Delta_\eta^2\Delta_\phi^2}H_2(\eta_1,\phi_1,\eta_2,\phi_2) \tag{11.59}$$

$$\hat{\rho}_2^{(\Delta p_T\Delta p_T)}(\eta_1,\phi_1,\eta_2,\phi_2) = \frac{1}{N_{ev}}\frac{1}{\Delta_\eta^2\Delta_\phi^2}H_2^{(\Delta p\Delta p)}(\eta_1,\phi_1,\eta_2,\phi_2), \tag{11.60}$$

where $H_2(\eta_1,\phi_1,\eta_2,\phi_2)$ is incremented by 1 at $\eta_1,\phi_1,\eta_2,\phi_2$ for each pair measured while $H_2^{(\Delta p\Delta p)}(\eta_1,\phi_1,\eta_2,\phi_2)$ is incremented by the product $\Delta p_{T,1}\Delta p_{T,2}$.

11.2 Three-Particle Differential Correlation Functions

Two-particle correlations enable a rather extensive study of particle correlations and have been instrumental in the study of jet production, flow, and several other phenomena encountered in both elementary and nuclear collisions at high-energy. At times, however, the interpretation of two-particle correlations may be somewhat ambiguous. For instance, in 2004, the PHENIX and STAR experiments reported two-particle correlation functions, with asymmetric p_T ranges (one high- and one low-p_T particle), that indicated the

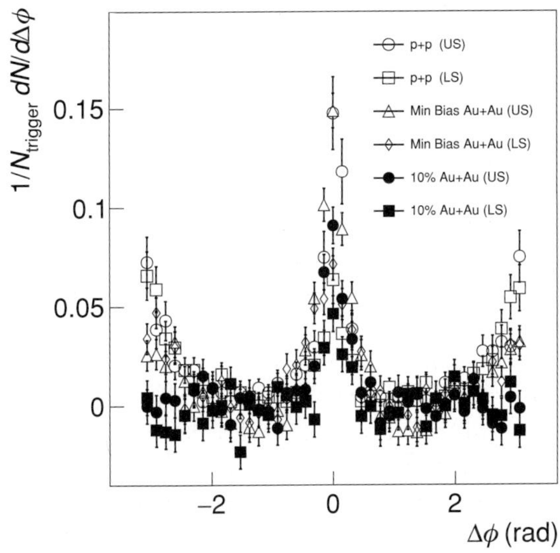

Fig. 11.7 Back-to-back suppression in like-sign (LS) and unlike-sign (UL) two-particle correlations measured in central (0–10%) Au + Au collisions relative to correlations measured in minimum-bias (Min Bias) Au+Au and in p–p collisions. Data from STAR collaboration. (Adapted from C. Adler, et al. Disappearance of back-to-back high *pT* hadron correlations in central Au+Au collisions at $\sqrt{s_{NN}}$ = 200-GeV. *Physical Review Letters*, 90:082302, 2003.)

presence of a depletion in back-to-back emission at top RHIC energies, as illustrated in Figure 11.7 [11].

Theoretical analyses of these correlation functions suggested that several distinct production mechanisms might explain the suppressed and flattened away-side emission structure revealed by the two experiments. Unfortunately, the two-particle correlation measurements could not readily distinguish whether the broadening of the away-side was due to medium induced deflection of the leading partons with no actual broadening of the jet, or due to the interactions and dispersion of jet fragments by the medium, or due to a new phenomenon known as Mach cone emission, or perhaps another mechanism as yet unknown. Measurements of three-particle correlations were then conceived to attempt a resolution of this ambiguity.

If the jet structure is unchanged except for initial scattering of the leading jet parton, then the width of the away-side particles remains unchanged between p–p and Au + Au collisions. If, on the other hand, the jet fragments are dispersed (scattered) by the medium, then the width of the away-side should indeed increase, as schematically illustrated in Figure 11.8b. Three-particle azimuthal correlations could also be useful for identifying Mach cone emission or Cerenkov radiation. In the case of the predicted Mach cone emission, the propagation of the away-side parton in the dense medium formed by the A–A collisions was expected to lead to the production of a wake at an angle determined by the ratio of parton speed and sound velocity in the medium. There were predictions of a sound velocity of the order of $0.33c$, whereas the parton speed is near c. This might then have led to particle emission at 60–70° from the away-side direction, as illustrated in Figure 11.8d, e.

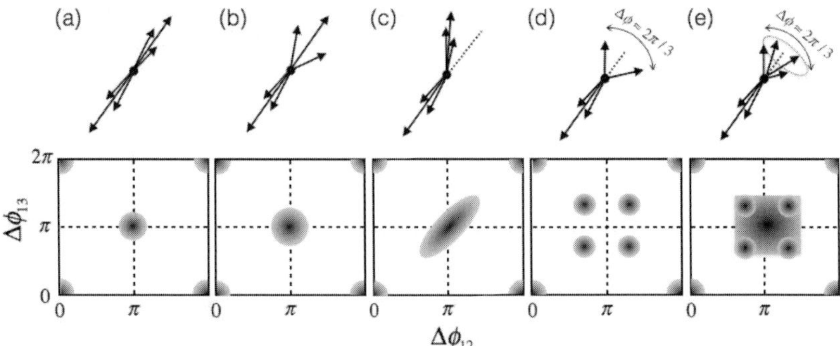

Fig. 11.8 Searching for Mach cone emission with three-particle correlations. Schematic illustration of three-particle azimuthal correlation patterns expected for (a) back-to-back jets; (b) medium induced broadening of the away-side jet; (c) deflected jets; (d) transverse plane emission of Mach cone; and (e) 3-D Mach cone.

Mach cone emission was then anticipated to yield four-side structures at 60–70° from the away-side direction, as schematically illustrated in Figure 11.8. In contrast, parton deflection (scattering) should have instead led to an elongation along the diagonal of the away-side jet peak (Fig. 11.8c). Measurements of three-particles were subsequently carried out by various groups to seek evidence for either of the correlated emission structures described earlier. Measurements by the STAR collaboration, in particular, involved a medium to high-p_T particle, hereafter labeled as particle 1, and two lower p_T particles, labeled 2 and 3. Measurements were done in terms of the relative angles $\Delta\phi_{12} = \phi_1 - \phi_2$ and $\Delta\phi_{13} = \phi_1 - \phi_3$ between the particles.

Recall from §10.2 that measured n-particle densities involve combinatorial contributions from uncorrelated particle emission as well as correlated emission. It is thus strongly advisable to carry out an analysis based on normalized three-particle cumulants, $R_3(\Delta\phi_{12}, \Delta\phi_{13})$ [158], which eliminate, by construction, combinatorial and uncorrelated triplets of the form $\rho_1\rho_1\rho_1$ and $\rho_2\rho_1$. Normalized cumulants R_3 also naturally lend themselves to particle-yield corrections required to account for instrumental effects, discussed in §12.4. In the context of the $\Delta\phi_{12}$ vs. $\Delta\phi_{13}$ three-particle correlation described earlier, the normalized cumulant R_3 may be written:

$$R_3(\Delta\phi_{12}, \Delta\phi_{13}) = \frac{C_3(\Delta\phi_{12}, \Delta\phi_{13})}{\rho_1 \otimes \rho_1 \otimes \rho_1(\Delta\phi_{12}, \Delta\phi_{13})}. \tag{11.61}$$

As for measurements of second-order normalized cumulants, R_2, one has the option of measuring R_3 as function of $\Delta\phi_{12}$ vs. $\Delta\phi_{13}$ directly, using for instance a mixed-event technique, or via averages of R_3 measured as functions of the three angles ϕ_1, ϕ_2, and ϕ_3 explicitly. In this latter case, the cumulant C_3 may be written

$$\begin{aligned} C_3(\phi_1, \phi_2, \phi_3) = {} & \rho_3(\phi_1, \phi_2, \phi_3) - \rho_2(\phi_1, \phi_2)\rho_1(\phi_3) \\ & - \rho_2(\phi_1, \phi_3)\rho_1(\phi_2) - \rho_2(\phi_2, \phi_3)\rho_1(\phi_1) \\ & + 2\rho_1(\phi_1)\rho_1(\phi_2)\rho_1(\phi_3). \end{aligned} \tag{11.62}$$

In practice, the determination of $\rho_3(\phi_1, \phi_2, \phi_3)$ requires three dimensional histograms, and thus may be rendered difficult by the size of the dataset, or by computational and storage issues. One may therefore opt to determine the three-particle density directly in terms of the relative angles $\Delta\phi_{12}$ and $\Delta\phi_{13}$. The determination of C_3 then involves the computation of three terms $\rho_2 \otimes \rho_1(\Delta\phi_{12}, \Delta\phi_{13})$ and $\rho_1 \otimes \rho_1 \otimes \rho_1(\Delta\phi_{12}, \Delta\phi_{13})$. Numerical techniques to estimate these terms on the basis of histograms with finitely many bins are discussed in §11.5. Direct estimations of $\rho_3(\Delta\phi_{12}, \Delta\phi_{13})$ and terms of the form $\rho_2(\Delta\phi_{ij})$ with 2D and 1D histograms, respectively, proceed with the same techniques as discussed in prior sections.

Three-particle correlation studies in pseudorapidities can be done, similarly, by substituting rapidities η_i to azimuthal angle ϕ_i and differences $\Delta\eta_{ij}$ to $\Delta\phi_{ij}$ in Eq. (11.62). Note, however, that for correlation studies in rapidity (or pseudorapidity), no periodic boundary applies, and one must consequently consider the integration and averaging of the correlation functions over ranges $\eta_{i,\min} - \eta_{j,\max} \leq \Delta\eta_{ij} < \eta_{i,\max} - \eta_{j,\min}$. Numerical integration and averaging techniques are discussed in §11.5.

The interpretation of three-particle cumulants measured in heavy-ion collisions as functions $\Delta\phi_{12}$ and $\Delta\phi_{13}$ is greatly complicated by the presence of collective flow effects. Flow indeed contributes irreducible terms of the form $\langle v_p(1)v_m(2)v_n(3)\rangle\delta_{p,m+n}cos(p\phi_1 - m\phi_2 - n\phi_3)$ to the cumulant that are functions of the product of three flow coefficients known to be dependent on the particle momentum and subject to fluctuations as well as correlations. A discussion of these terms is beyond the scope of this textbook but may be found in ref. [160].

11.3 Integral Correlators

11.3.1 Introduction

Measurements of fluctuations in the relative production yields of two specific particle species are commonly performed to study the underlying particle production mechanisms and to investigate the nature of the system produced in elementary particle or nucleus–nucleus collisions. For example, it was predicted that event-by-event fluctuations of the ratio of the number of positively and negatively charged particles would be suppressed if a quark gluon plasma, as opposed to a hadron gas, were produced in high-energy nucleus-nucleus collisions. Similarly, fluctuations of the ratio of the yield of kaons to the yield of pions were also expected to be modified if the nuclear matter produced in collisions lies near a phase boundary or the critical point of nuclear matter. There also have been predictions and expectations for fluctuations of other types of particles or particle species.

While a measurement of event-by-event fluctuations of the ratio, R, of the yields of two particle types α and β observed in a specific fiducial momentum range, hereafter denoted N_α and N_β, may sound straightforward, complications may occur that make such measurements nontrivial. For instance, if the particle yields are small, there is a finite probability that the yield N_β may vanish in a given event, thereby leading to a divergent ratio

$R = N_\alpha/N_\beta$. Additionally, since particle detection efficiencies are typically functions of several kinematical and collision parameters, correcting the measured ratios for such fluctuations may become a rather complex task. Fortunately, a measurement of fluctuations of the ratio R may be replaced by an essentially equivalent measurement of two-particle integral correlators. Measurements of integral correlators are usually preferred relative to measurements of event-by-event fluctuations of particle yield ratios because they are not subject to divergences and can usually be corrected for detection efficiencies relatively easily. Integral correlators also present the advantage of being theoretically well defined and directly calculable. Their use for studies of relative yield fluctuations is consequently recommended over measurements of ratios.

11.3.2 Equivalence between Particle Yield Ratio Fluctuations and Integral Correlators

Let us first demonstrate the equivalence between measurements of ratio fluctuations and that of integral correlation functions. Let N_α and N_β be the yields of two particle types (e.g., positively and negatively charged particles) measured in a given event within a fiducial momentum range. We are interested in measuring the variance $\langle \Delta R^2 \rangle$ of the ratio $R = N_\alpha/N_\beta$ of these two yields. Let $\langle N_\alpha \rangle$ and $\langle N_\beta \rangle$ represent event averages, that is, the expectation values of the yields N_α and N_β. Let us further denote as ΔN_α and ΔN_β the deviations of N_α and N_β from their respective means, and define $\langle R \rangle$ as the ratio of these averages. The ratio R may thus be written

$$\begin{aligned} R &= \frac{\langle N_\alpha \rangle + \Delta N_\alpha}{\langle N_\beta \rangle + \Delta N_\beta} \\ &= \langle R \rangle \left(\frac{1 + \Delta N_\alpha / \langle N_\alpha \rangle}{1 + \Delta N_\beta / \langle N_\beta \rangle} \right). \end{aligned} \tag{11.63}$$

If the magnitude of the fluctuations ΔN_i are small relative to the means $\langle N_i \rangle$, one can write

$$R = \langle R \rangle \left(1 + \frac{\Delta N_\alpha}{\langle N_\alpha \rangle} - \frac{\Delta N_\beta}{\langle N_\beta \rangle} + O(1/N^2) \right). \tag{11.64}$$

The variance of the ratio normalized to the mean is thus

$$\frac{\langle \Delta R^2 \rangle}{\langle R \rangle^2} = \frac{\langle \Delta N_\alpha^2 \rangle}{\langle N_\alpha \rangle^2} + \frac{\langle \Delta N_\beta^2 \rangle}{\langle N_\beta \rangle^2} - 2 \frac{\langle \Delta N_\alpha \Delta N_\beta \rangle}{\langle N_\alpha \rangle \langle N_\beta \rangle}. \tag{11.65}$$

The preceding quantity is commonly denoted ν in the literature [161]. It may also be written

$$\nu \equiv \frac{\langle \Delta R^2 \rangle}{\langle R \rangle^2} = \left\langle \left(\frac{N_\alpha}{\langle N_\alpha \rangle} - \frac{N_\beta}{\langle N_\beta \rangle} \right)^2 \right\rangle. \tag{11.66}$$

In the limit of independent particle production (Poisson statistics), one expects that $\langle \Delta N_i^2 \rangle = \langle N_i \rangle$ and the correlator $\langle \Delta N_\alpha \Delta N_\beta \rangle$ vanishes. The so-called statistical limit of ν, more aptly called the independent particle production limit, may thus be written

$$\nu_{\text{stat}} = \frac{1}{\langle N_\alpha \rangle} + \frac{1}{\langle N_\beta \rangle}, \tag{11.67}$$

from which we find that the variance of the ratio fluctuations, $\langle \Delta R^2 \rangle$, becomes

$$\langle \Delta R^2 \rangle_{\text{stat}} = \nu_{\text{stat}} \langle R \rangle^2. \tag{11.68}$$

Such statistical fluctuations are typically of limited interest because they are predominantly determined by the magnitude of the yields $\langle N_\alpha \rangle$ and $\langle N_\beta \rangle$. Of greater interest is the deviation of ν from the statistical limit ν_{stat}. One consequently introduces the difference $\nu - \nu_{\text{stat}}$, known in the recent literature as a measure of dynamical fluctuations, noted ν_{dyn}. It is straightforward (Problem 11.1) to verify that

$$\nu_{\text{dyn}} = \frac{\langle N_\alpha(N_\alpha - 1)\rangle}{\langle N_\alpha \rangle^2} + \frac{\langle N_\beta(N_\beta - 1)\rangle}{\langle N_\beta \rangle^2} - 2\frac{\langle N_\alpha N_\beta \rangle}{\langle N_\alpha \rangle \langle N_\beta \rangle}. \tag{11.69}$$

This measure of fluctuations is of particular interest because the three terms it comprises are related to integral correlators $R_{\alpha\beta}$, as follows:

$$R_{\alpha\alpha} = \frac{\langle N_\alpha(N_\alpha - 1)\rangle}{\langle N_\alpha \rangle^2} - 1, \tag{11.70}$$

$$R_{\alpha\beta} = \frac{\langle N_\alpha N_\beta \rangle}{\langle N_\alpha \rangle \langle N_\beta \rangle} - 1.$$

The number of particle pairs $\langle N_\alpha(N_\alpha - 1)\rangle$ and $\langle N_\alpha N_\beta \rangle$ are given by integrals of the particle production cross sections

$$\langle N_\alpha(N_\alpha - 1)\rangle = \int_\Omega \rho_{\alpha\alpha}\, d\eta_1 d\phi_1 dp_{T,1} d\eta_2 d\phi_2 dp_{T,2}, \tag{11.71}$$

$$\langle N_\alpha N_\beta \rangle = \int_\Omega \rho_{\alpha\beta}\, d\eta_\alpha d\phi_\alpha dp_{T,\alpha} d\eta_\beta d\phi_\beta dp_{T,\beta}, \tag{11.72}$$

and the average single particle yields are given by

$$\langle N_\alpha \rangle = \int_\Omega \rho_\alpha\, d\eta_\alpha d\phi_\alpha dp_{T,\alpha}. \tag{11.73}$$

The quantities ρ_α and $\rho_{\alpha\beta}$ are single- and two-particle densities, respectively:

$$\rho_\alpha(\eta_\alpha, \phi_\alpha, p_{T,\alpha}) = \frac{d^3N}{d\eta_\alpha d\phi_\alpha dp_{T,\alpha}}, \tag{11.74}$$

$$\rho_{\alpha\beta}(\eta_\alpha, \phi_\alpha, p_{T,\alpha}, \eta_\beta, \phi_\beta, p_{T,\beta}) = \frac{d^6N}{d\eta_\alpha d\phi_\alpha dp_{T,\alpha} d\eta_\beta d\phi_\beta dp_{T,\beta}}. \tag{11.75}$$

The integrals may be taken over the entire acceptance of the detector or across specific narrow ranges in rapidity, production azimuth, and transverse momentum deemed suitable for the study of specific particle production mechanisms (e.g., emphasis on low- or high-p_T particles).

Clearly, the integral correlators, $R_{\alpha\beta}$, are in all ways similar to the differential correlation functions introduced earlier in this chapter, and consequently sharing the same attributes and properties. They scale as the inverse of the multiplicities N_α, and particle production by m distinct and independent sources (or mechanisms) should satisfy

$$R^{(m)}_{\alpha\beta} = \frac{1}{m} R^{(1)}_{\alpha\beta}, \tag{11.76}$$

where $R^{(1)}_{\alpha\beta}$ and $R^{(m)}_{\alpha\beta}$ are the correlators for a single process and a superposition of m identical such processes, respectively (see Problem 11.2). Integral correlators are also robust observables, i.e., observables independent of detection and measurement efficiencies – at least to first-order approximation, as we shall discuss in detail in §12.4. This implies, by construction, that the observable $\nu_{\rm dyn}$ also shares these characteristics. One expects in particular that $\nu_{\rm dyn}$ should scale as $1/m$ for nuclear collisions consisting of superpositions of m independent proton-proton (or perhaps parton–parton) processes (see Problem 11.3):

$$\nu^{(m)}_{\rm dyn} = \frac{1}{m}\nu^{(1)}_{\rm dyn}. \tag{11.77}$$

Note, on the other hand, that the observable $\nu = \langle \Delta R^2\rangle/\langle R\rangle^2$ is not a robust observable given its explicit dependence on $\langle N_\alpha\rangle$ and $\langle N_\beta\rangle$ in the independent-particle production limit. Measurements of yield fluctuations are thus best conducted in terms of the robust variable $\nu_{\rm dyn}$, which provides a simple and explicit connection to two-particle densities and is as such easily interpreted.

Obviously, Eq. (11.64) is an approximation strictly valid only for small deviations ΔN_i relative to the averages $\langle N_\alpha\rangle$. The correlator method discussed earlier is thus not an exact substitute for measurements of fluctuations of the ratios of particle species yields. Nonetheless, it remains the preferred observable given that (1) it does not suffer from the pathological behavior (divergence) associated with a ratio of numbers that may vanish, and (2) its interpretation in terms of integral correlators provides a strong and clear foundation for the interpretation of data.

In the following subsections, we discuss specific implementations of the $\nu_{\rm dyn}$ observable for the study of net charge fluctuations, and fluctuations of particle production.

11.3.3 Net Charge Fluctuations

Although electric charge is a conserved quantity, particle production in elementary particle and nuclear collisions is subject to net charge fluctuations. The net charge of particles produced in a given region of momentum space is expected to fluctuate collision by collision. The size of the fluctuations should be in part determined by the magnitude of the charge of the produced particles. In a quark gluon plasma (QGP), the charge carriers, the quarks, have fractional charges ($\pm 1/3$, $\pm 2/3$), and therefore fluctuations of net charge should be suppressed relative to particle production in a hadron gas where charge carriers have integer charges (± 1). Several theoretical works published in the 1990s in fact predicted that a signature of the production of QGP phase in relativistic heavy-ion collisions could be a substantial reduction of net charge fluctuations relative to that observed in lower energy collisions systems where no QGP is expected to be formed [46, 120, 121]. Subsequently, in the mid-2000s, several measurements were undertaken to find evidence for the predicted suppression of net charge fluctuations by SPS and RHIC experiments. Measurements of net charge fluctuations have also been conducted more recently at the LHC. While the results were somewhat inconclusive, the correlation functions developed to carry out the measurements have merits that extend beyond the search for explicit manifestation of the quark gluon plasma, and as such remain of general interest for the study of particle production dynamics in high-energy nuclear collisions.

The question arises as to what constitutes a reliable and significant measure of fluctuations of the net charge, $Q = N_+ - N_-$ (for a review, see [157]). Clearly, the size of the fluctuations must depend on the actual produced particle multiplicity, the magnitude of the individual charges, as well as the efficiency of the counting and detection processes. A measurement of the variance of the produced multiplicity would therefore be incomplete and inconclusive. A measurement of the ratio N_+/N_-, however, would obviously be sensitive to fluctuations of the net charge. Alternatively, one might also consider the variance of Q relative to the average total number of charge particles $\langle N_{\rm ch}\rangle = \langle N_+\rangle + \langle N_-\rangle$:

$$\omega_Q \equiv \frac{\langle \Delta Q^2\rangle}{\langle N_+\rangle + \langle N_-\rangle} = \frac{\langle Q^2\rangle - \langle Q\rangle^2}{\langle N_{\rm ch}\rangle}. \tag{11.78}$$

It is straightforward (see Problem 11.4) to show that one expects $\omega_Q = 1$, for a strictly Poissonian system. Particle production is, however, not a perfect Poisson process. For instance, Koch et al.[120] estimated that the production of resonances, such as the ρ-meson, in a hadron gas would reduce the fluctuations to $\omega_Q = 0.8$. They further predicted that a drastic reduction to $\omega_Q = 0.2$ would take place in the presence of a quickly expanding QGP [46, 121].

Inspection of the expressions for ω_Q reveals that this observable depends linearly on the efficiency, ϵ, of the detection and particle counting process. Because the efficiency may depend on the particle species considered as well as various other factors such as the detector occupancy, environmental features, defective detector components, and so on, the normalized variance constitutes a nonrobust observable, that is, one which requires a detailed calculation of the detection efficiency involving the various characteristics (flaws) and cuts used in the analysis. It is thus of interest to seek observables that are sensitive to net charge fluctuations but remain robust under practical experimental conditions. An obvious choice is the dynamic fluctuation observable, $\nu_{\rm dyn}$, introduced in the previous section. Let $N_\alpha = N_+$ and $N_\beta = N_-$, one gets

$$\nu_{+-,\rm dyn} = \frac{\langle N_+(N_+ - 1)\rangle}{\langle N_+\rangle^2} + \frac{\langle N_-(N_- - 1)\rangle}{\langle N_-\rangle^2} - 2\frac{\langle N_+N_-\rangle}{\langle N_+\rangle\langle N_-\rangle}. \tag{11.79}$$

The quantities $\langle N_+(N_+ - 1)\rangle$, $\langle N_-(N_- - 1)\rangle$, and $\langle N_+N_-\rangle$ are the average number of positively charged, negatively charged, and unlike-sign pairs, respectively, measured within a fiducial momentum volume over an ensemble of events. $\langle N_+\rangle$ and $\langle N_-\rangle$ are the average yields of positive and negative particles averaged over the same fiducial volume and event ensemble. $\nu_{+-,\rm dyn}$ shares all attributes and properties of $\nu_{\rm dyn}$ correlation functions. It is determined by the integral correlators R_{++}, R_{--}, and R_{+-} and is as such a robust observable.

$\nu_{+-,\rm dyn}$ may be measured as a function of global event observables such as the total transverse energy or the charged particle multiplicity produced in a selected part of the experimental acceptance. However, care must be taken to correct for finite bin width effects as discussed in §12.4.3.

Measurements of correlation functions in elementary particle collisions at the ISR, the Tevatron (FNAL), the SPS (CERN), and RHIC (BNL) have shown that $R_{++} \approx R_{--} < R_{+-}/2$ [85, 193]. The dynamic fluctuations $\nu_{+-,\rm dyn}$ are consequently found to be negative in such elementary processes. They are also found to be negative in nucleus–nucleus collisions, where $\nu_{+-,\text{dyn}}$ roughly scales inversely as the produced particle multiplicity [3, 159].

Charge conservation fixes the total charge produced in an elementary particle (or nuclear) collision. A measurement encompassing all particles produced by a collision (i.e., over 4π acceptance and perfect detection efficiency) would thus not be subject to net charge fluctuations. Particle production at SPS, RHIC, and LHC, however, spans several units of rapidity, and it is typically not possible to measure the full range of produced particles. Net charge fluctuations do take place on the scale of few units of rapidity. The strength of such fluctuations should in fact be sensitive to the degrees of freedom (i.e., hadronic vs. partonic) of the matter produced in high-energy collisions. Charge conservation does impact the measured fluctuations. One finds [157] that charge conservation yields a contribution to $\nu_{+-,\text{dyn}}$ on the order of

$$\nu^{cc}_{+-,\text{dyn}} = -\frac{4}{\langle N \rangle_{4\pi}}, \tag{11.80}$$

where the superscript cc indicates the "charge conservation" limit, and $\langle N \rangle_{4\pi}$ is the average total charged particle multiplicity produced by a collision system at a given impact parameter in nucleus–nucleus collisions (see also [46]).

Several other fluctuation observables have been proposed and used to measure net charge fluctuations. Of particular note is the Φ_q observable, specifically designed to identify dynamic fluctuations of the net charge [92] produced in nucleus–nucleus collisions. By construction, Φ_q yields a constant value for nucleus–nucleus collisions that could be reduced to a superposition of independent proton–proton interactions. It thus enables, at least in principle, identification of not only dynamic fluctuations, but fluctuations that might vary nontrivially as a function of collision system size, or collision centrality relative to proton–proton interactions. However, one can show that Φ_q can be expressed in terms of $\nu_{+-,\text{dyn}}$ [157], as follows:

$$\Phi_q \approx \frac{\langle N_+ \rangle^{3/2} \langle N_- \rangle^{3/2}}{\langle N \rangle^2} \nu_{+-,\text{dyn}}. \tag{11.81}$$

Given its explicit dependence on particle multiplicities, one concludes this observable is nonrobust against particle detection efficiencies. Use of the observable $\nu_{+-,\text{dyn}}$ is consequently recommended for practical measurements of net charge fluctuations. Changes in the collision dynamics may then be identified by scaling $\nu_{+-,\text{dyn}}$ with the produced charged particle multiplicity (corrected for detection efficiencies and charge conservation effects).

11.3.4 Kaon vs. Pion Yield Fluctuations

Anomalous fluctuations in the yield of kaons, relative to the yield of (charged) pions, were suggested as a potential signature of the formation of a quark gluon plasma in high-energy heavy-ion collisions [49]. Several measurements of such fluctuations have been conducted at the SPS (CERN) in terms of the variance, ΔR^2, of the ratio of the yield of charged kaons and charged pions measured in fixed rapidity and momentum ranges,

$$R = \frac{N_K}{N_\pi}, \tag{11.82}$$

relative to the mean ratio

$$\langle R \rangle = \frac{\langle N_K \rangle}{\langle N_\pi \rangle}. \tag{11.83}$$

Given that ΔR^2 is sensitive to trivial statistical fluctuations, as well as actual correlations in the particle production of kaons and pions, measurements in terms of this variable typically rely on a comparison of the variance measured in actual events (siblings) relative to those found in mixed events. Such a comparison is complicated, however, by various detector effects, and most particularly, by detection efficiencies.

In light of the discussion in §11.3.2, which establishes an equivalence between the dynamical fluctuations measured with $\nu_{\rm dyn}$ and deviations of the measured variance ΔR^2 from the independent particle limit, it is advisable to carry measurements of relative kaon and pion yields in terms of $\nu_{\rm dyn}$, which is by construction a robust quantity. Such measurements have been conducted at RHIC [4] in terms of the $\nu_{{\rm K}\pi,{\rm dyn}}$ observable, defined as:

$$\nu_{{\rm K}\pi,{\rm dyn}} = \frac{\langle N_K(N_K - 1)\rangle}{\langle N_K\rangle^2} + \frac{\langle N_\pi(N_\pi - 1)\rangle}{\langle N_\pi\rangle^2} - 2\frac{\langle N_K N_\pi\rangle}{\langle N_K\rangle\langle N_\pi\rangle}, \tag{11.84}$$

where $\langle N_K(N_K - 1)\rangle$ and $\langle N_\pi(N_\pi - 1)\rangle$ are second-order factorial moments of the produced kaon and pion yields, respectively, $\langle N_K N_\pi\rangle$ represents the average number of kaon + pion pairs observed event-by-event in the fiducial acceptance, whereas $\langle N_K\rangle$ and $\langle N_\pi\rangle$ are the mean kaon and pion yields, respectively. This observable is by construction robust against particle loses associated with detection efficiencies. It is thus suitable to measure the rather small dynamical fluctuations observed in large multiplicity A–A collisions.

11.3.5 Transverse Momentum Fluctuations

Introduction

Studies of event-by-event average transverse momentum fluctuations in heavy-ion collisions were initially undertaken to search for evidence of critical phenomena predicted to take place near the hadron–parton phase boundary and identify the formation of a quark gluon plasma [8, 9]. Although no conclusive evidence for the formation of a quark gluon plasma arose from early studies in Au–Au collisions, it was clear that transverse momentum fluctuations constituted a useful technique to study the collective dynamics of nucleus–nucleus collisions, most particularly radial flow effects. Studies were thus carried out by several groups for multiple colliding systems and at several colliding energies [6, 9, 12, 22, 104].

The event-wise mean transverse momentum is often defined as

$$\langle p_T \rangle \equiv \frac{1}{N}\sum_{i=1}^{N} p_{T,i}, \tag{11.85}$$

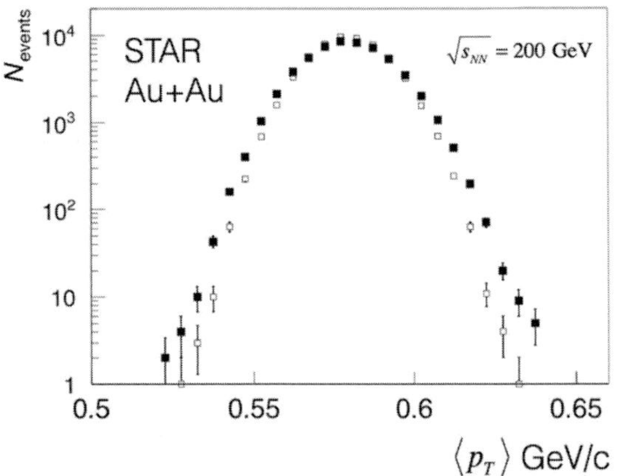

Fig. 11.9 Distribution of $\langle p_T \rangle$ per event (solid squares) measured for 5% most central Au+Au collision at $\sqrt{s_{NN}} = 200$ GeV, compared to a mixed event average p_T distribution (open squares). Data from STAR collaboration. (Adapted from J. Adams et al. Incident energy dependence of pt correlations at RHIC. *Physical Review C* 72:044902, 2005.)

where N is the total number of particles per event detected in the kinematical range of interest and $p_{T,i}$ represents their transverse momenta. The kinematical range may be selected in rapidity, azimuthal angle, and transverse momentum. The sum can in principle be carried on all detected particles, neutral or charged particles only, or some specific set of particle species. Figure 11.9 displays a measurement of the event-wise mean transverse momentum (Black squares) measured by the STAR collaboration in Au–Au collisions at $\sqrt{S_{NN}} = 200$ GeV [9], compared to a mean p_T distribution (open squares) obtained with mixed events. Mixed events, defined in §12.4.5, are obtained by mixing particles from different events and thus carry no intrinsic particle–particle correlation. They thus constitute a sensible reference to establish whether particles produced in a given event do exhibit correlations. One finds that the mean p_T distribution of actual events is slightly wider than the reference histogram, which means that dynamic mean p_T fluctuations take place in Au–Au collisions, that is, fluctuations exceed the statistical fluctuations expected for a stochastic system consisting of independently produced particles.

An intuitive and quantitative measure of dynamical fluctuations is the excess variance, δp_t^2, of the data relative to the reference:

$$\delta p_t^2 = \sigma_{\text{data}}^2 - \sigma_{\text{ref}}^2 \tag{11.86}$$

where σ_{data}^2 and σ_{ref}^2 are the variance of the real and reference mean p_T distributions, respectively. However, in view of limited particle detection efficiencies, the measured mean p_T distributions (real and mixed events) may be subject to artificial broadening associated with purely instrumental effects. The δp_t^2 observable also obscures the origins of the dynamical fluctuations, which are found in two- and multiparticle correlations. It is thus of greater interest to use observables that render this connection more explicit. Unfortunately, there exists several distinct ways to define and measure mean p_T observables that make this connection more explicit. Although qualitatively similar, the results obtained with these different definitions are found to be quantitatively distinct.

$\langle \Delta p_T \Delta p_T \rangle$ Correlation Functions

In this section, we introduce the so-called **event-wise** (EW) and **inclusive** (I) $\Delta p_T \Delta p_T$ observables. We show in the next section, that these observables can be both approximately related to the excess p_T variance defined by Eq. (11.86).

The event-wise average p_T per event is defined as

$$\langle\langle p_T \rangle\rangle_{EW} = \frac{1}{N_{\rm ev}} \sum_{\alpha=1}^{N_{\rm ev}} \frac{S_\alpha^{p_T}}{N_\alpha}, \tag{11.87}$$

where

$$S_\alpha^{p_T} = \sum_{k=1}^{N_\alpha} p_{T,k}, \tag{11.88}$$

which is the sum of the transverse momenta of the N_α particles in event α, and $N_{\rm ev}$ is the number of events considered.

The inclusive average p_T is obtained by taking the event averages of $S_\alpha^{p_T}$ and N_α separately:

$$\langle\langle p_T \rangle\rangle_I = \frac{\frac{1}{N_{\rm ev}} \sum_{\alpha=1}^{N_{\rm ev}} S_\alpha^{p_T}}{\frac{1}{N_{\rm ev}} \sum_{\alpha=1}^{N_{\rm ev}} N_\alpha} = \frac{\sum_{\alpha=1}^{N_{\rm ev}} S_\alpha^{p_T}}{\sum_{\alpha=1}^{N_{\rm ev}} N_\alpha} \equiv \frac{\langle S^{p_T} \rangle}{\langle N \rangle}. \tag{11.89}$$

While the definition of $\langle\langle p_T \rangle\rangle_{EW}$ may seem more "intuitive," the inclusive definition has the advantage of being directly related to the particle production cross section, and is thus recommended. Noting that $\sum S_\alpha^{p_T}$ is the sum of transverse momenta of all the particles measured in the data sample of interest, and that $\sum N_\alpha$ is the total number of such particles, one can write

$$\langle\langle p_T \rangle\rangle_I = \frac{\sum_{\rm all} p_T}{N_{\rm particles}}. \tag{11.90}$$

This expression is indeed just the average p_T of all the particles measured, and it can be written in terms of the inclusive single-particle cross section:

$$\langle\langle p_T \rangle\rangle_I = \frac{\int_\Omega \rho_1(\eta, \phi, p_T) p_T^2 \, d\eta d\phi dp_T}{\int_\Omega \rho_1(\eta, \phi, p_T) p_T \, d\eta d\phi dp_T}, \tag{11.91}$$

where the integration is taken over a selected subset Ω of the detector's acceptance, and

$$\rho_1 = \frac{d^3 N}{p_T \, d\eta d\phi dp_T}. \tag{11.92}$$

Writing a similar formula for $\langle\langle p_T \rangle\rangle_{EW}$ is possible but yields a rather complicated expression in terms of conditional cross-sections (see, e.g., [173] and Problem 11.6). One however expects that the two definitions converge in the large N particle production limit.

The event-wise and inclusive $\langle \Delta p_T \Delta p_T \rangle$ observables are defined as covariances relative to mean p_T averages, respectively. The event-wise $\langle \Delta p_T \Delta p_T \rangle_{EW}$ is given by

$$\langle \Delta p_T \Delta p_T \rangle_{EW} = \frac{1}{N_{\rm ev}} \sum_{\alpha=1}^{N_{\rm ev}} \frac{S_\alpha^{\Delta p_T \Delta p_T}}{N_\alpha (N_\alpha - 1)} \tag{11.93}$$

with

$$S_\alpha^{\Delta p_T \Delta p_T} = \sum_{i=1}^{N_\alpha} \sum_{j=1, j\neq i}^{N_\alpha} (p_{T,i} - \langle\langle p_T \rangle\rangle_{EW}) (p_{T,i} - \langle\langle p_T \rangle\rangle_{EW}). \tag{11.94}$$

The inclusive $\langle \Delta p_T \Delta p_T \rangle_I$ is similarly given by

$$\langle \Delta p_T \Delta p_T \rangle_I = \frac{\sum_{\alpha=1}^{N_{\rm ev}} S'^{\Delta p_T \Delta p_T}_\alpha}{\sum_{\alpha=1}^{N_{\rm ev}} N_\alpha (N_\alpha - 1)} = \frac{\langle S'^{\Delta p_T \Delta p_T} \rangle}{\langle N(N-1) \rangle} \tag{11.95}$$

with

$$S'^{\Delta p_T \Delta p_T}_\alpha = \sum_{i=1}^{N_\alpha} \sum_{j=1, j\neq i}^{N_\alpha} (p_{T,i} - \langle\langle p_T \rangle\rangle_I) \left(p_{T,j} - \langle\langle p_T \rangle\rangle_I\right). \tag{11.96}$$

As for $\langle p_T \rangle_{EW}$, the covariance $\langle \Delta p_T \Delta p_T \rangle_{EW}$ may seem more intuitive because it involves an average of $S_\alpha^{\Delta p_T \Delta p_T}$ calculated per pair of particles. The covariance $\langle \Delta p_T \Delta p_T \rangle_I$ is, however, of greater interest because it can be expressed easily as an integral of the two-particle cross section:

$$\langle \Delta p_T \Delta p_T \rangle_I = \frac{\int_{\rm accept} \rho_2 \Delta p_{T,i} \Delta p_{T,j}\, d\eta_1 d\phi_1 p_{T,1} dp_{T,1} d\eta_2 p_{T,2} d\phi_2 dp_{T,2}}{\int_{\rm accept} \rho_2\, d\eta_1 d\phi_1 p_{T,1} dp_{T,1} d\eta_2 d\phi_2 p_{T,2} dp_{T,2}}, \tag{11.97}$$

with

$$\Delta p_{T,i} = p_{T,i} - \langle\langle p_T \rangle\rangle_I, \tag{11.98}$$

and

$$\rho_2 = \frac{d^6 N}{p_{T,1} p_{T,2}\, d\eta_1 d\phi_1 dp_{T,1} d\eta_2 d\phi_2 dp_{T,2}} \tag{11.99}$$

Defined as a covariance, $\langle \Delta p_T \Delta p_T \rangle_I$ features the same properties as regular correlation functions. It is a robust variable against particle losses due to detection efficiencies and it scales inversely as the produced particle multiplicity as well as the number of independent particle sources (see Problem 11.8).

Relation between $\langle \Delta p_T \Delta p_T \rangle$ and δp_t^2

We show that the excess variance, δp_t^2, defined by Eq. (11.86) can be expressed in terms of the correlator $\langle \Delta p_T \Delta p_T \rangle$ by calculating the first and second moments of the sum S^{p_T} defined by Eq. (11.88).

For events with a fixed number of particles N, one can write

$$\begin{aligned}
\langle S^{p_T}\rangle_N &= \frac{1}{N}\left\langle\sum_{i=1}^{N} p_{T,i}\right\rangle = \frac{1}{N}\sum_{i=1}^{N}\langle p_{T,i}\rangle = \langle p_T\rangle \\
\langle S^{p_T}\rangle_N^2 &= \frac{1}{N^2}\sum_{i=1}^{N}\langle p_{T,i}\rangle^2 + \frac{1}{N^2}\sum_{i\neq j=1}^{N}\langle p_{T,i}\rangle\langle p_{T,j}\rangle \\
\langle (S^{p_T})^2\rangle_N &= \frac{1}{N^2}\left\langle\left(\sum_{i=1}^{N} p_{T,i}\right)\left(\sum_{j=1}^{N} p_{T,j}\right)\right\rangle \\
&= \frac{1}{N^2}\sum_{i=1}^{N}\langle p_T^2\rangle + \frac{1}{N^2}\sum_{i\neq j=1}^{N}\langle p_{T,i}p_{T,j}\rangle.
\end{aligned} \tag{11.100}$$

The variance of S^{p_T}, for fixed N, is thus

$$\begin{aligned}
\text{Var}\,[S^{p_T}] &= \langle (S^{p_T})^2\rangle_N - \langle S^{p_T}\rangle_N^2 \\
&= \frac{1}{N}\left(\langle p_T^2\rangle - \langle p_T\rangle^2\right) + \frac{1}{N^2}\langle S'^{\Delta p_T\Delta p_T}_{\alpha}\rangle.
\end{aligned} \tag{11.101}$$

First, note that the difference $\langle p_T^2\rangle - \langle p_T\rangle^2$ corresponds to the variance $\sigma_{p_T}^2$ of the inclusive p_T distribution, in other words, that obtained by plotting a histogram of all measured particles p_T values. Second, note that Var $[S^{p_T}]$ is actually the variance σ_{data}^2 of the histogram (data) of S^{p_T} discussed in §11.3.5. Clearly, the variance of an histogram of S^{p_T} accumulated with mixed events should have the same exact structure except for the fact that none of the particles composing a mixed-event are correlated. This implies that the quantity $\langle S'^{\Delta p_T\Delta p_T}_{\alpha}\rangle$ should be null for mixed events. The variance Var $[S^{p_T}]$ of the mixed event spectrum is thus simply equal to $\sigma_{p_T}^2$. We conclude that the excess variance δp_t^2 representing the difference between the same and mixed-event variance is equal to $\frac{1}{N^2}\langle S'^{\Delta p_T\Delta p_T}_{\alpha}\rangle$. One thus obtains the result

$$\delta p_t^2 = \frac{\langle N(N-1)\rangle}{N^2}\langle \Delta p_T\Delta p_T\rangle_I, \tag{11.102}$$

which tells us that the excess variance is approximately equal to the integral correlator $\langle \Delta p_T\Delta p_T\rangle_I$. It should be noted, however, that while the quantity $\langle \Delta p_T\Delta p_T\rangle_I$ is by construction robust against particle loss due to detector inefficiencies, the quantities δp_t^2 and $\sigma_{p_T}^2$ are not. It is consequently recommended to conduct p_T fluctuations studies on the basis of $\langle \Delta p_T\Delta p_T\rangle_I$ rather than using a mixed-event technique and a combination of the observables Var $[S^{p_T}]$, and δp_t^2 and $\sigma_{p_T}^2$. The use of $\langle \Delta p_T\Delta p_T\rangle_I$ rather than $\langle \Delta p_T\Delta p_T\rangle_{EW}$ is also deemed preferable because $\langle \Delta p_T\Delta p_T\rangle_I$ maps straightforwardly onto the well-defined correlation integral (11.97), while $\langle \Delta p_T\Delta p_T\rangle_{EW}$ yields a more convoluted expression in terms of conditional cross sections, expressed for fixed values of the event multiplicity.

11.4 Flow Measurements

Measurements of the momentum anisotropy of the particles produced in heavy-ion collisions constitute a central component of the RHIC and LHC heavy-ion programs because they provide tremendous insight into the nature and properties of the matter formed in these collisions. Theoretical studies indicate that the momentum anisotropy of produced particles finds its origin in the initial asymmetry of the geometry of the matter produced in heavy-ion collisions. The development of collective motion, or flow, results in part from anisotropic pressure gradients, momentum transport, and differential particle energy loss through the medium. From a practical standpoint, it is convenient to distinguish the radial and azimuthal components of the collective motion. The radial component, known as **radial flow**, may be estimated based of momentum spectra and momentum correlations, while measurements of the azimuthal component, known as **anisotropic flow** or simply **flow**, are achieved via two- and multiparticle azimuthal correlation functions. Since the spatial anisotropies vanish rapidly as collision systems expand and evolve, anisotropic flow is generally considered to be self-quenching and thus expected to originate mostly from the early phases of collisions, essentially during the first few fm/c (i.e., $\sim 3 \times 10^{-24}$) of heavy-ion collisions. Flow measurements are thus particularly sensitive to the early phases of collisions and the nature of the high-temperature and high-density matter produced in relativistic heavy-ion collisions. Flow measurements are also increasingly carried out on simpler colliding systems, such as p–Pb or even p–p to find out whether the energy densities and gradients produced in these systems are sufficient to produce radial and anisotropic flow.

11.4.1 Definition of Flow and Nonflow

Since the development of anisotropic flow is dependent on the initial geometry of collision systems, it is particularly important to clearly define the geometry and coordinates used to quantify measurements of flow.

Figure 11.10 presents a schematic illustration of the transverse profile of colliding nuclei and the participant region created by p–p interactions (which can be viewed as parton–parton or proton–proton interactions). Note that the use of a classical description (i.e., classical particle trajectories) of the nuclei profile is justified by the very high momentum of the colliding particles, which endows them with very short De Broglie wavelengths. The reaction plane is defined by the beam direction and the impact parameter vector consisting of a line joining the geometrical centers of the colliding nuclei. Particularly relevant for flow measurements is the azimuthal angle, Ψ_{RP}, of the reaction plane relative to some laboratory reference direction (not shown in the figure). The angle Ψ_{RP} cannot be dialed macroscopically nor observed directly and must consequently be inferred from the distribution of produced particles. All values of Ψ_{RP} are a priori equally probable; the probability of measuring a given value of Ψ_{RP} can thus be described as a uniform PDF in the range $[0, 2\pi]$. The initial anisotropic spatial geometry of the system causes anisotropic

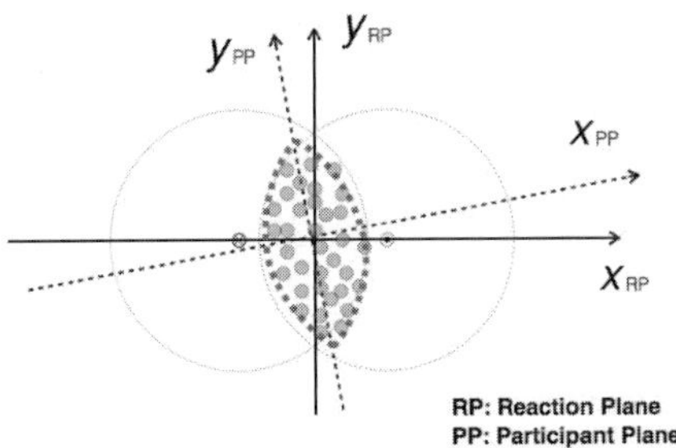

Fig. 11.10 The beam axis (extending out of page) and a line passing through the nuclei centers (x_{RP}) define the nominal **reaction plane** (RP). Event-by-event fluctuations in the location of p–p (i.e., nucleon–nucleon or parton–parton) interactions define the participant region with axes of symmetry x_{PP} and y_{PP}. The beam and x_{PP} axes define the **participant plane** (PP).

pressure gradients and differential energy losses. As illustrated in Figure 11.6, the energy and pressure gradient are maximum along the x-axis, and one thus expects the largest particle production of low momentum particles subjected to flow gradients along this axis. High-p_T particles produced during the earliest stages of a collision might not be driven by pressure gradients, but must nonetheless penetrate through the spatially anisotropic slower medium produced by the intersecting nuclei. Interactions with the medium are expected to produce energy losses. And since path lengths are shorter along the x-axis than along the y-axis, high-p_T particles are expected to suffer differential energy losses. They too should exhibit azimuthal anisotropies. The particle production cross section is thus expected to depend on the azimuthal angle ϕ relative to the orientation of the reaction plane Ψ. One may express this cross-section as a Fourier series:

$$E\frac{d^3N}{dp^3} = \frac{1}{2\pi}\frac{d^2N}{p_T\,dp_T dy}\left\{1 + 2\sum_{n=1}^{\infty} v_n \cos\left[n\left(\phi - \Psi_{\rm RP}\right)\right]\right\}, \tag{11.103}$$

where the coefficients, v_n, are known as **flow coefficients** or simply as **harmonic coefficients**. The introduction of a factor of 2 in the preceding decomposition will be justified in the discussion that follows.

As in prior sections, it is convenient to define shorthand notations $\rho_1(\phi, y, p_T)$ and $\rho_1(y, p_T)$ for the single particle density,

$$\rho_1(\phi, y, p_T) = \frac{1}{2\pi}\rho_1(y, p_T)\left\{1 + 2\sum_{n=1}^{\infty} v_n \cos\left[n\left(\phi - \Psi_{\rm RP}\right)\right]\right\} \tag{11.104}$$

The flow coefficients, v_n, may be obtained as the expectation value of $\cos\left[n\left(\phi - \Psi_{\rm RP}\right)\right]$,

$$v_n \equiv \left\langle \cos\left[n\left(\phi - \Psi_{\rm RP}\right)\right]\right\rangle \tag{11.105}$$

$$= \frac{\int_0^{2\pi} \rho_1(\phi, y, p_T)\cos\left[n\left(\phi - \Psi_{\rm RP}\right)\right]d\phi}{\int_0^{2\pi} \rho_1(\phi, y, p_T)\,d\phi}. \tag{11.106}$$

The integral in the denominator of Eq. (11.106) yields $\rho_1(y, p_T)$. To carry out the integral in the numerator, with $\rho_1(\phi, y, p_T)$ given by Eq. (11.104), recall the orthogonality relations of cosine and sine functions,

$$\int_0^{2\pi} \cos(n\phi)\cos(m\phi)\,d\phi = \pi\,\delta_{nm}, \tag{11.107}$$

$$\int_0^{2\pi} \sin(n\phi)\cos(m\phi)\,d\phi = 0, \tag{11.108}$$

defined for integer values of $n, m > 0$, and with the Kronecker delta function $\delta_{nm} = 1$ for $n = m$, and null otherwise. One can thus verify easily that the integral in the numerator yields $v_n \times \rho(y, p_T)$. Note that the factor of 2 originally inserted in the definition of the Fourier decomposition multiplies the factor π obtained from the integral $\int_0^{2\pi} \cos(n\phi)\cos(m\phi)\,d\phi$. This consequently yields a factor 2π that cancels the 2π normalization factor of the invariant cross section. One concludes that the expectation value $\langle \cos\left[n\left(\phi - \Psi_{\rm RP}\right)\right]\rangle$ indeed yields the flow coefficients v_n. Further note that Eq. (11.105) produces flow coefficients, known as **differential flow coefficients**, that are dependent on the rapidity, y, and transverse momentum, p_T, of the particles. **Integrated flow coefficients**, also known as average flow coefficients, may be obtained by further integrating the particle density over the fiducial rapidity and transverse momentum acceptance of the measurements. Flow coefficients vanish at null p_T and grow approximately linearly at low p_T while the density $\rho_1(y, p_T)$ is a steeply decreasing function at large p_T. Systematic errors associated with an integrated v_n measurement based on a finite p_T range can thus be controlled relatively easily.

Experimentally, the integrals in Eq. (11.105) are replaced by sums over all (or selected) particles measured in the fiducial acceptance. Given an estimate of the reaction plane angle, $\hat{\Psi}_{\rm RP}$, one obtains flow coefficients with

$$v_n = \frac{\left\langle \sum_{i=1}^{N_p} \cos[n(\phi_i - \hat{\Psi}_{\rm RP})]\right\rangle}{\langle N_p\rangle}, \tag{11.109}$$

where ϕ_i, $i = 1, N_p$, are the azimuthal angles of measured particles and the sum runs over all N_p measured particles in any given event, and the brackets stand for an average over events. The v_n may be obtained in bins of p_T and y or integrated over the entire fiducial range of the measurement.

The Fourier decomposition used in Eq. (11.103) is obviously incomplete. Sine terms were omitted because, on average, there should be an equal number of particles produced below and above the reaction plane. Sine terms being odd in $\phi - \Psi_{\rm RP}$ are thus expected to vanish, on average, and are consequently not required for a description of the averaged particle emission relative to the reaction plane. Fluctuations in the number of particles, however, take place on an event-by-event basis. One might thus be tempted to introduce sine terms in the Fourier decomposition (11.103). It is, however, usually deemed more convenient and physically meaningful to describe collisions on event-by-event using the notion of **participant plane** illustrated in Figure 11.10.

The first four harmonic coefficients, v_1, v_2, v_3, and v_4 are commonly known as **directed flow**, **elliptic flow**, **triangular flow**, and **quadrangular flow**, respectively, given their

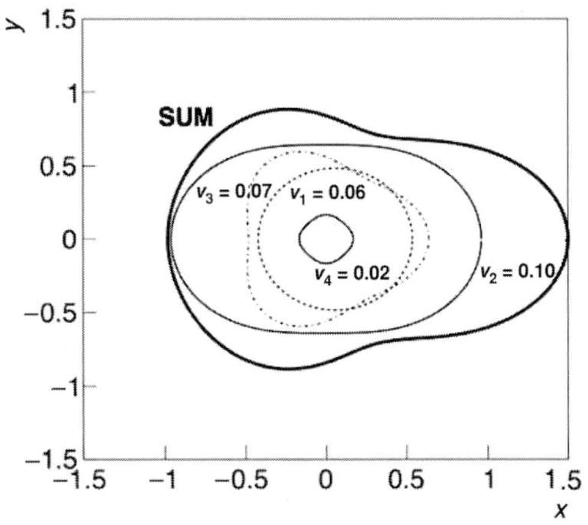

Fig. 11.11 Illustration of the geometrical interpretation of flow coefficients; the sizes of the radial profiles are plotted with $r_n(\phi) = k_n(1 + 2v_n \cos(n\phi))$, for given v_n values. The scaling coefficients k_n are set arbitrarily to facilitate the visualization of the different flow terms. The thick solid line represents a sum of the four terms.

obvious geometrical interpretation (Figure 11.11). Appellations in terms of dipole, quadrupole, and so on, should be frowned on since these names are usually reserved for the multipole expansion of three-dimensional charge or mass distributions, not the harmonic decomposition in azimuth of a flow field.

The reaction plane angle Ψ_{RP} is not readily accessible macroscopically but may be estimated based on the distribution of produced particles. Several of the techniques commonly used to estimate this angle and measure the flow coefficients based on two- and multiparticle correlation functions will be described in following sections and their relative merits discussed in §11.4.3. At this stage, it is important to point out that contributions to these correlation functions involving two- or n-particle correlations, resulting, for instance, from resonance decays or the production of jets, and having nothing to do with collective flow, may enter in the determination of flow coefficients. These contributions are commonly referred to as **nonflow**. They are noted δ_n and can in principle be estimated from two-particle correlations based on the following expression:

$$\langle \cos[n(\phi_i - \phi_j)] \rangle = \langle v_n^2 \rangle + \delta_n. \tag{11.110}$$

Measurements of flow coefficients are further complicated by fluctuations of the magnitude of the flow on an event-by-event basis, all other collision properties being equal. The variations are known as **flow fluctuations**, σ_{vn}^2, and formally defined as the variance of the flow coefficients:

$$\sigma_{vn}^2 = \langle v_n^2 \rangle - \langle v_n \rangle^2. \tag{11.111}$$

From Eqs. (11.110) and (11.111), one gets:

$$\langle \cos[n(\phi_i - \phi_j)] \rangle = \langle v_n \rangle^2 + \sigma_{vn}^2 + \delta_n, \tag{11.112}$$

which implies that measurements of $\langle\cos[n(\phi_i - \phi_j)]\rangle$ are determined by the square of the magnitude of the (average) flow, the variance of the flow, as well as nonflow effects. It initially appeared that flow fluctuations might be inextricably linked to nonflow effects. Recent developments, however, suggest flow fluctuations may largely be determined by fluctuations in the overlapping nuclei geometry, known as the participant nucleons region. Theoretical considerations further suggest that the principal axes of the participant region may in fact deviate substantially from the nominal average overlap region, as schematically illustrated in Figure 11.10. The flow coefficients measured according to the participant plane are always larger than those obtained relative to the nominal reaction plane. This leads to important contributions to the flow fluctuations.

11.4.2 Measurement Methods

In this section, we describe some of the many techniques developed over the years to estimate the flow coefficients v_n. These techniques vary in applicability based on the size of the data sample, ease of use, and their capacity to suppress or control nonflow effects.

Standard Event Plane Method

The **standard event plane** method, also known simply as **event plane** (EP) method, is the most basic of all techniques used to determine flow coefficients. Defined by Eq. (11.105), it requires the estimation of the reaction plane angle, $\Psi_{\rm RP}$, from the event plane computed on the basis of selected, or all, measured particles. The event plane determination proceeds on the basis of a 2D vector, denoted $\vec{Q}_n$, and known as the **event plane vector** of order n. It is calculated in the transverse plane, event-by-event, on the basis of the following expressions:

$$\vec{Q}_{n,x} = \sum_{i=1}^{N_p} w_i \cos(n\phi_i) \tag{11.113}$$

$$\vec{Q}_{n,y} = \sum_{i=1}^{N_p} w_i \sin(n\phi_i). \tag{11.114}$$

The sums run over the N_p measured particles. The ϕ_i, with $i = 1, N_p$, are the azimuthal angles of the measured particles in the laboratory reference frame. The coefficients w_i are weights assigned to each particle and designed to yield an optimal estimation of the event plane vector $\vec{Q}_n$, discussed in the text that follows. The event-plane angle, Ψ_n, and the modulus of $\vec{Q}_n$ are obtained with

$$\hat{\Psi}_n = \frac{1}{n}\arctan(Q_n, y/Q_n, x) \tag{11.115}$$

$$|\hat{Q}_n| = \sqrt{Q_n, x^2 + Q_n, y^2}. \tag{11.116}$$

The weights, w_i, are positive definite for even values of n, but must satisfy $w_i(-y) = -w_i(y)$ for odd harmonics. An optimal determination of the angle Ψ_n is achieved if the weights are set equal to the flow coefficients, $w_i = v_n(p_T, y)$ [182]. This is rather inconvenient because the v_n are not known a priori. It is, however, common and legitimate practice to use the p_T of the particles as weight, since the flow coefficients are typically proportional to the transverse momentum of the particles at low p_T. Estimates of the flow coefficients, noted v_n^{obs}, are obtained by replacing $\hat{\Psi}_{\text{RP}}$ by $\hat{\Psi}_n$ in Eq. (11.109),

$$v_n^{\text{obs}} = \left\langle \frac{1}{N_p} \sum_{i=1}^{N_p} \cos[n(\phi_i - \hat{\Psi}_n)] \right\rangle. \tag{11.117}$$

Note that the orientation of the $\vec{Q}$ vector is obviously influenced by the direction of all particles included in its calculation. As an extreme case, consider that if $\vec{Q}$ was determined on the basis of a single particle, this particle's momentum vector would be perfectly aligned with it. This leads to an autocorrelation effect that tends to skew the $\vec{Q}$ vector along the direction of high-p_T particles (most particularly if a weight proportional to p_T is used), and as a result inappropriately increases the value of the flow coefficients. To avoid this auto-correlation bias, one must recalculate the vector $\vec{Q}$, for each particle included in the v_n calculation, to exclude the contribution of this particle to the flow vector. This may be written

$$v_n^{\text{obs}} = \left\langle \frac{1}{N_p} \sum_{i=1}^{N_p} \cos[n(\phi_i - \hat{\Psi}'_n)] \right\rangle, \tag{11.118}$$

where $\hat{\Psi}'_n$ is obtained, for each particle, from

$$\hat{\Psi}'_n = \frac{1}{n} \arctan(Q'_{n,y}/Q'_{n,x}), \tag{11.119}$$

with

$$\vec{Q}'_{n,x} = Q_{n,x} - w_i \cos(n\phi_i) \tag{11.120}$$

$$\vec{Q}'_{n,y} = Q_{n,y} - w_i \sin(n\phi_i). \tag{11.121}$$

It is important to notice that for v_2, the definition (11.117) does not specify whether the anisotropy is in- or out-of-plane. Additional information, such as a measurement of the spectator plane using forward detectors, was required in practice to establish that the elliptic flow observed in heavy-ion collisions at RHIC and LHC energies is actually in-plane. It also worth noting that, mathematically, there is no intrinsic or a priori relationship between the different angles Ψ_n. However, the geometry and dynamics of nuclei–nuclei collisions may impart a finite degree of correlation between these angles. Measurements of their covariance $\text{Cov}[\Psi_m, \Psi_n]$ or average $\cos(\Psi_m - \Psi_n)$ are thus of interest and provide valuable insight about the collision geometry and dynamics.

While the aforementioned procedure corrects for autocorrelation effects, it does not account for fluctuations associated with the finite number of particles. Indeed, one can show that Eq. (11.118) produces a biased estimator of the flow coefficient v_n because the event plane vector, $\hat{Q}_n$, randomly deviates from the actual reaction vector due to the finite particle multiplicity. For large multiplicities, the fluctuations of $\hat{Q}_n$ can be shown to be Gaussian in the Q_x–Q_y plane, and thus the measured and true Fourier coefficients may be related by the following expression: [151]

$$v_n^{\text{obs}} \equiv \langle \cos [n(\phi - \Psi_{\text{RP}})] \rangle \tag{11.122}$$
$$= \langle \cos [n(\phi - \Psi_n)] \rangle \times \langle \cos [n(\Psi_n - \Psi_{\text{RP}})] \rangle \tag{11.123}$$
$$= v_n \times \langle \cos [n(\Psi_n - \Psi_{\text{RP}})] \rangle, \tag{11.124}$$

from which we conclude that a correction for **event plane resolution** may be achieved with

$$v_n = \frac{v_n^{\text{obs}}}{R_n}, \tag{11.125}$$

where the **event plane resolution** R_n is defined by

$$R_n = \langle \cos [n(\Psi_n - \Psi_{\text{RP}})] \rangle. \tag{11.126}$$

The event plane resolution coefficients R_n must be calculated for each harmonic n and obtained as an average taken over a large ensemble of events. Their magnitude depends on the strength of the flow v_n and the particle multiplicity M. They may be evaluated analytically if the $\hat{Q}_n$ fluctuations are Gaussian according to [151]

$$R_n(\chi) = \frac{\sqrt{\pi}}{2}\chi e^{-\chi^2/2}\left[I_{(n-1)/2}(\chi^2/2) + I_{(n+1)/2}(\chi^2/2)\right], \tag{11.127}$$

where $\chi = v_n\sqrt{M}$, and I_k are modified Bessel functions of order k. The functions $R_n(\chi)$ are plotted in Figure 11.12 for $k = 1$, $k = 2$, and so on. Eq. (11.127) is useful to model the behavior of the event plane resolution relative to the strength of the flow coefficients and the particle multiplicity, but, it is not readily sufficient for the experimental determination of the event plane resolution.

The event plane resolution can be estimated directly from the data using the **subevent method**, which consists of randomly subdividing the particles of each measured event into two subevents A and B of (approximately) equal size $M/2$. One may then calculate the flow vectors $\hat{Q}_n^A$ and $\hat{Q}_n^B$ of the two subevents and determine the correlation $\langle \cos[n(\hat{\Psi}_n^A - \hat{\Psi}_n^B)]\rangle$. For Gaussian fluctuations and sufficiently large multiplicities, the two angles $\hat{\Psi}_n^A$ and $\hat{\Psi}_n^A$ are statistically independent, and one may consequently write

$$\left\langle \cos\left[n\left(\hat{\Psi}_n^A - \hat{\Psi}_n^B\right)\right]\right\rangle = \left\langle \cos\left[n\left(\hat{\Psi}_n^A - \hat{\Psi}_{\text{RP}}\right)\right]\right\rangle \times \left\langle \cos\left[n\left(\hat{\Psi}_n^B - \hat{\Psi}_{\text{RP}}\right)\right]\right\rangle \tag{11.128}$$
$$= \left\langle \cos\left[n\left(\hat{\Psi}_n^A - \hat{\Psi}_{\text{RP}}\right)\right]\right\rangle^2.$$

The event plane resolution of subevents A (or B) is thus

$$R_{n,\text{sub}} = \sqrt{\left\langle \cos\left[n\left(\hat{\Psi}_n^A - \hat{\Psi}_n^B\right)\right]\right\rangle}. \tag{11.129}$$

The resolution for a full event may then be estimated from

$$R_{n,\text{full}} = R_n(\sqrt{2}\chi_{\text{sub}}). \tag{11.130}$$

The use of subevents is a powerful technique that enables a wide range of studies. One can produce subevents by random selection of particles of actual events, or on the basis of

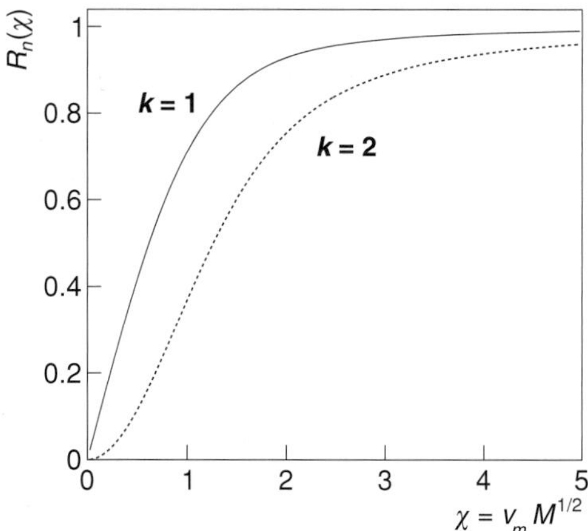

Fig. 11.12 Event plane resolution, R_n, as a function of $\chi = v_n\sqrt{M}$, where v_n is actual flow, and M is the number of particles involved in the estimation of the event plane resolution. The harmonic number of the correlation n is an integer k times the harmonic number m of the event plane.

the (pseudo)rapidity, charge, or any combination of these criteria. The use of a pseudorapidity gap, in particular, constitutes a straightforward method for suppressing short-range correlations and consequently suppressing nonflow effects.

Three techniques, known as ϕ weighting, recentering, and shifting, are commonly used to correct for detector artifacts in the evaluation of the event plane angle Ψ_n. They are described in §12.4.4.

Two and Multiparticle Correlation Methods

As discussed in §11.1.2, two-particle densities measured as function of the relative azimuthal angle between two produced particles are sensitive to collective flow. One indeed finds

$$\frac{dN_{\text{pairs}}}{d\Delta\phi} \propto 1 + \sum_{n=1}^{\infty} 2v_n^2 \cos(n\Delta\phi), \tag{11.131}$$

where all pairs of a particular momentum and rapidity range are selected for the calculation of this pair spectrum. It is therefore possible to obtain the **square** of the flow coefficients, v_n, from fits of pair azimuthal distributions. This technique is known as the **pairwise correlation method**. Experimentally, this can be readily accomplished by taking measurements of the normalized two-particle density $r_2(\Delta\phi)$ or the normalized cumulant $R_2(\Delta\phi)$ since, by construction, these observables yield measurements that are corrected for detection efficiencies. The properly normalized pair spectrum is thus

$$\frac{dN_{\text{pairs}}}{d\Delta\phi} = \frac{\langle n\rangle^2}{\langle n(n-1)\rangle} r_2(\Delta\phi), \tag{11.132}$$

where $\langle n\rangle$ and $\langle n(n-1)\rangle$ are the average number of particles and average number of pairs detected in the nominal momentum and rapidity range of the measurement of $r_2(\Delta\phi)$, respectively. The factor $\langle n\rangle^2/\langle n(n-1)\rangle$ is introduced to account for the fact that the numerator of $r_2(\Delta\phi)$ is proportional to the number of measured pairs while its denominator has an n^2 dependence on the number of particles.

The **two-particle cumulant method** is conceptually identical to the pairwise correlation method. However, instead of carrying out a fit of measured pair spectrum, one evaluates the average of the square of the flow coefficients directly by a measurement of $\langle\cos[n(\phi_1 - \phi_2)]\rangle$. Coefficients obtained with this method are denoted

$$v_n\{2\}^2 \equiv \langle\cos[n(\phi_1 - \phi_2)]\rangle, \tag{11.133}$$

where the average is calculated for all (selected) pairs of all (selected) events. It is important to realize that the coefficients $v_n\{2\}^2$ are actually a measure of the average of the square of the flow coefficient $\langle v_n^2\rangle$. They are thus sensitive to fluctuations as well as the magnitude of the flow coefficients and nonflow effects, as per Eq. (11.112).

It is useful and convenient to introduce the particle unit flow vector $u_{n,i}$, defined in the complex plane as

$$u_{n,i} \equiv e^{in\phi_i} = \cos n\phi_i + i\sin n\phi_i, \tag{11.134}$$

where ϕ_i, as in prior sections, denotes the azimuthal angle of particle i. Considering that the number of particles produced below and above the reaction are equal on average, one may write (see Problem 11.10)[3]:

$$\langle u_{n,1}u_{n,2}^*\rangle = \langle\cos n\phi_1 \cos n\phi_2\rangle = v_2\{2\}^2 \tag{11.135}$$

The unit vector $u_{n,i}$ consequently provides for an elegant technique to study differential flow (i.e., the flow dependence on transverse momentum and rapidity), known as the **scalar product method**:

$$v_n(p_T, y) = \frac{\langle Q'_{n,i}u_{n,i}^*(p_T, y)\rangle}{2\sqrt{Q_n^a Q_n^b}}, \tag{11.136}$$

where the average $\langle Q'_{n,i}u_{n,i}^*(p_T, y)\rangle$ is calculated for all particles of interest and averaged over all events. The event plane vector, hereafter expressed in the complex plane, is calculated as

$$Q'_{n,i} = \sum_{j\neq i} w_j u_{n,j}, \tag{11.137}$$

where the coefficients w_i are weights chosen to optimize the event plane determination. The sum over all particles j excludes particle i in order to avoid autocorrelations. Computationally, one can avoid the recalculation of $Q'_{n,i}$ for all particles by first computing $Q_n = \sum_j w_j u_{n,j}$ and using $Q'_{n,i} = Q_n - u_{n,i}$. The factor $2\sqrt{Q_n^a Q_n^b}$, in the denominator of Eq. (11.136), corrects for the event plane resolution and can be obtained as $\langle\cos[n(\Psi_a - \Psi_b)]\rangle$.

[3] The product $u_{n,1}u_{n,2}^*$ yields a term proportional to $\sin n\phi_1 \sin n\phi_2$, which is strictly null only in the absence of nonflow correlations.

The scalar method is shown in [10] to yield statistical errors that are slightly smaller than those achieved with the standard event plane method. Indeed, given that the event plane vector Q_n is in principle based on an arbitrary selection of particles, it can be obtained from single particles. The scalar product then reduces to the event plane method, but with poorer resolution. Additionally, note that division by the event resolution corrects the mean v_n values obtained but not their statistical fluctuations.

The two-particle correlators defined by Eqs. (11.125,11.133,11.136) are sensitive to collective flow as well as two- and few-particle correlations commonly known as nonflow. They thus tend to overestimate the flow coefficients v_n. To suppress nonflow contributions, it is thus desirable to utilize correlation functions that are by construction insensitive, or suppress the effects of two-, or few-particle correlations. We saw in §10.2 that cumulants of order n, noted C_n, are by construction insensitive to correlations of order $m < n$. Suppression of nonflow, dominated by two-particle correlations, in the determination of flow can thus be accomplished with higher order cumulants. However, note that since the correlators are obtained by averaging over azimuthal angles in the range $[0, 2\pi[$, odd order cumulants vanish by construction and are thus of little interest in the context of flow measurements. Measurements of flow, with suppressed nonflow, are thereby achieved with even-order cumulants exclusively. We here restrict our discussion to fourth-order cumulants, but extensions to higher orders are possible and well documented in the literature [182].

The fourth-order cumulant C_4 was first introduced in §10.2.2 in the context of generic particle correlation functions. In the context of azimuthal correlations, it may be written

$$\begin{aligned} C_4(1,2,3,4) = \rho_4(1,2,3,4) &- \sum_{(4)} \rho_1(1)\rho_3(2,3,4) \\ &- \sum_{(3)} \rho_2(1,2)\rho_2(3,4) + 2\sum_{(6)} \rho_1(1)\rho_1(2)\rho_2(3,4) \\ &- 6\rho_1(1)\rho_1(2)\rho_1(3)\rho_1(4). \end{aligned} \tag{11.138}$$

The indices 1, 2, 3, and 4 are here used as shorthand notations for the azimuthal angle ϕ_i, $i = 1, 2, 3, 4$, and the sums are carried over terms consisting of permutations of these indices. For flow measurements involving a sum over all 4-tuplets of particles, terms that contain odd-order densities must vanish when calculated for a large event ensemble since the average cosine of unpaired angles vanishes when average over $[0, 2\pi]$. The averaged four-cumulant consequently reduces two terms: one involving the four-particle density $\rho_4(1,2,3,4)$ and the other dependent on the product $\rho_2(1,2)\rho_2(3,4)$. Here again, it is convenient to use particle flow vectors $u_{n,i}$ defined in the complex plane. The fourth-order cumulant may thus be written

$$\langle\langle u_{n,1}u_{n,2}u^*_{n,3}u^*_{n,4}\rangle\rangle = \langle u_{n,1}u_{n,2}u^*_{n,3}u^*_{n,4}\rangle - 2\langle u_{n,1}u^*_{n,2}\rangle\langle u_{n,3}u^*_{n,4}\rangle \tag{11.139}$$

where the double brackets $\langle\langle\rangle\rangle$ indicate the correlator is a cumulant. One can verify (see Problem 11.14) that in the absence of flow fluctuations, the four-cumulant is equal to $-v_n^4$. One thus defines the fourth-order flow cumulant flow coefficients as

$$v_n\{4\} = \left(-\langle\langle u_{n,1}u_{n,2}u^*_{n,3}u^*_{n,4}\rangle\rangle\right)^{(1/4)}. \tag{11.140}$$

The $v_n\{4\}$ notation is universally used to identify flow coefficients determined on the basis of the fourth-order cumulant (11.139). Flow coefficients obtained based on higher order cumulants are likewise noted $v_n\{6\}$, $v_n\{8\}$, and so forth. However, note that because higher order cumulants involve the combination (subtraction) of several terms, they may yield negative values resulting from either fluctuations, or limited statistics. Their interpretation may consequently be somewhat challenging, and measurements of these quantities typically require substantially larger data samples to achieve the same statistical significance as that obtained with second-order cumulants.

Cumulants may be determined with generating functions, as discussed in §§2.13 and 10.2.2, or by direct calculation.

Other multiparticle cumulants are also of interest in the context of mixed harmonic studies. An important example of such studies involves the measurement of the three-particle correlator

$$\langle u_{n,1} u_{n,2} u^*_{2n,3} \rangle = v_n^2 v_{2n}, \tag{11.141}$$

where the particle flow vectors of particles 1 and 2 are calculated at order n while the flow vector of the third particle is obtained at order $2n$. This correlator was used successfully at RHIC to suppress nonflow effects in the study of v_1 and v_4 flow coefficients. Mixed harmonics are also extremely useful in studies (searches) of the **chiral magnetic effect** (CME) [125, 124, 183].

Q-Distribution Method

In previous sections, we showed how the event plane vector $\vec{Q}$ may be used to infer the orientation of the reaction plane of colliding nuclei. But $\vec{Q}$ also provides information about the magnitude of the flow itself. The flow vector $\vec{Q}$ may be determined based on a subset of or all measured particles. Its magnitude and direction are thus effectively determined by a random walk process, that which consists of the sum of all particle transverse momentum vectors. In the absence of flow and other forms of particle correlations, the random walk yields a vector $\vec{Q}$ whose magnitude grows in proportion to the square root of the number of particles M involved in its calculation. But in the presence of flow, that is, for finite flow coefficients, v_n, the magnitude of the vector grows proportionally to Mv_n. It is thus convenient to introduce a normalized flow vector

$$\vec{q}_n = \frac{\vec{Q}}{\sqrt{M}} \tag{11.142}$$

whose magnitude should be order unity in the absence of flow, and that should scale proportionally to $\sqrt{M}v_n$ in the presence of flow. Based on the discussion presented in §2.12.2, we conclude that $|\vec{q}_n| = v_m\sqrt{M}\langle p_T\rangle$, where $\langle p_T\rangle$ is the average transverse momentum of the particles selected for the calculation of $\vec{Q}$ (corresponding to the average step size in the language of a random walk used in §2.12.2). The magnitude of the flow vector $\vec{q}_n$ then has the following probability distribution in the large M limit:

$$\frac{1}{N}\frac{dN}{dq_n} = \frac{q_n}{\sigma_n^2} \exp\left(-\frac{v_n^2 M + q^2}{2\sigma_n^2}\right) I_o\left(\frac{q_n v_n \sqrt{M}}{\sigma_n^2}\right). \tag{11.143}$$

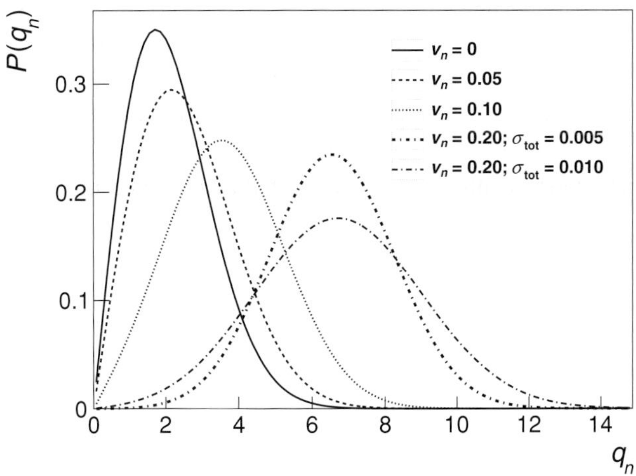

Fig. 11.13 Probability density function of the modulus of the normalized flow vector $\vec{q}_n$ for selected values of the flow coefficient v_n and the fluctuation parameter σ_{tot}.

where I_o is a modified Bessel function, v_n are flow coefficients, and σ_n is a measure of fluctuations.

$$\sigma_n^2 = \frac{1}{2}\left(1 + M\sigma_{\text{tot}}^2\right), \tag{11.144}$$

where

$$\sigma_{\text{tot}}^2 = \delta_n + 2\sigma_{vn}^2 \tag{11.145}$$

As illustrated in Figure 11.13, the presence of flow shifts the distribution toward larger q_n values, while increased fluctuations, determined by σ_{tot}, broaden the distribution. The **Q-distribution method** thus consists in estimating v_n and σ_n based on a fit of a measured q_n distribution with Eq. (11.143). The method enables the determination of σ_{tot} but cannot discern the effects of nonflow, δ_n, and flow fluctuations σ_{vn} separately.

Lee–Yang Zeros Method

The Lee–Yang zeros method is based on a technique developed in 1952 by Lee and Yang to detect a liquid–gas phase transition [2, 38, 53]. The technique involves the second-harmonic flow vector $\vec{Q}_2$ projected onto an arbitrary laboratory direction specified by an angle θ:

$$Q_2^\theta = \sum_{i=1}^{M} w_i \cos\left[2(\phi_i - \theta)\right]. \tag{11.146}$$

The sum is carried out over selected or all particles i with azimuthal angle ϕ_i and weight w_i. Typically, the projection is evaluated for five arbitrary but equally spaced values of θ in order to suppress detector acceptance effects. The method then entails finding the zero(s) of a complex generating function of the form

$$G_2^\theta(ir) = \left|\left\langle e^{irQ_2^\theta}\right\rangle\right|, \tag{11.147}$$

where r is variable along the imaginary axis of the complex plane and the average $\langle\rangle$ is taken over all events of interest. One uses the square of the modulus to determine the location of the first minimum, r_o^θ, corresponding to an angle θ, determined by the integrated flow

$$V_2^\theta = j_{01}/r_o^\theta, \tag{11.148}$$

$$v_2 = \left\langle V_2^\theta\right\rangle_\theta /M, \tag{11.149}$$

in which $j_{01} = 2.405$ is the first root of the Bessel function J_0 and M is the multiplicity of the event. The average $\langle\rangle_\theta$ is taken over the lab angles θ and yields the flow coefficient v_2 relative to the reaction plane axis. For implementation details and variants of the method, see, e.g., ref. [182] and references therein.

Fourier and Bessel Transforms Method

Let $f_o(Q_{x,n})$ denote the PDF of the x component of the Q_n flow vector in the absence of flow (i.e., $v_n = 0$) but finite nonflow correlations. Assume that the presence of a flow field does not otherwise influence the nonflow correlations. By virtue of the central limit theorem, the distribution of Q_n in the presence of flow can then be obtained simply by shifting the argument of f_o by an amount that depends on the reaction plane angle Ψ. Averaging over all values of this angle, one obtains

$$f(Q_{x,n}) \equiv \frac{1}{N}\frac{dN}{dQ_{x,n}} = \int \frac{d\Psi}{2\pi} f_o(Q_{x,n} - v_n M\cos(n\Psi)). \tag{11.150}$$

Next calculate the Fourier transform of this function:

$$\begin{aligned}\tilde{f}(k) &= \left\langle e^{ikQ_{n,x}}\right\rangle \\ &= \int \frac{d\Psi}{2\pi}\int dQ_{x,n} e^{ikQ_{n,x}} f_o(Q_{x,n} - v_n M\cos(n\Psi))\end{aligned} \tag{11.151}$$

Defining $t = Q_{x,n} - v_n M cos(n\Psi)$, Eq. (11.151) may then be written

$$\begin{aligned}\tilde{f}(k) &= \int \frac{d\Psi}{2\pi} e^{ikv_n M\cos(n\Psi)} \int dt e^{ikt} f_o(t) \\ &= J_0(kv_n M)\tilde{f}_o(k),\end{aligned} \tag{11.152}$$

in which one finds that the flow and nonflow contributions to the Fourier transform $\tilde{f}(k)$ are factorized: the transform $\tilde{f}_o(k)$ characterizes the nonflow correlations whereas $J_0(kv_nM)$ expresses the dependence on the flow magnitude v_n. The zeros of the Fourier transform are determined by the zeros of the Bessel function $J_0(kv_nM)$. One may then get an estimate of the flow coefficient v_n based on the first zero, k_1, of the Fourier transform

$$v_n = \frac{j_{01}}{k_1 M} \tag{11.153}$$

where, as in the previous section, j_{01} corresponds to the first zero of the Bessel function $J_0(z)$, and is equal to $j_{01} = 2.405$. This technique, called **Fourier and Bessel transforms method**, is equivalent to the Lee–Yang zeros method covered in the previous section.

A similar reasoning can be applied to the two-dimensional Fourier transform of $dN/dQ_{n,x}dQ_{n,y}$. One gets (see Problem 11.15)

$$\begin{aligned} \tilde{f}(k) &= \int dQ_{n,x} e^{ik_x Q_{n,x}} dQ_{n,y} e^{ik_y Q_{n,y}} \frac{d^2N}{dQ_{n,x}dQ_{n,y}} \\ &= dQ_n J_0(kQ_n) \frac{dN}{dQ_n} \sim J_0(kv_n M). \end{aligned} \tag{11.154}$$

The flow contribution is decoupled from all other correlation contributions. The Bessel transform, Lee–Yang zeros, and Q-distribution methods are thus similar if not totally equivalent. See [182] and references therein for a more in-depth discussion of this and related topics.

11.4.3 Pros and Cons of the Various Flow Methods

The methods presented in the previous sections may be compared based on their statistical accuracy (for equal data samples) and in their capacity to suppress or disentangle nonflow effects from flow. For instance, two-particle correlations obtained in a relatively narrow pseudorapidity range, with a single harmonic, measures flow in the participant plane, but the use of mixed harmonics, for instance the first harmonic from spectator neutrons and second harmonic for particles produced at central rapidities, should provide elliptic flow in the reaction plane.

Nonflow contributions, noted δ_n, are defined by Eq. (11.112) as the excess correlation from two- or few-particle correlations arising from particle production dynamics not related to collective effects (collective behaviors resulting from pressure gradients or differential attenuation determined by the collision geometry) and thus the collectivity of particles produced. Nonflow contributions should therefore be more or less independent of the particles' direction relative to the reaction plane. Sources of nonflow correlations include hadronization in jets, decays of short-lived particles, short-range correlations such as the Hanbury-Brown Twiss (HBT) effect, and energy/momentum conservation. Nonflow correlations tend to be stronger for particles emitted near one another in momentum space, particularly rapidity. It is thus possible to reduce the effects of nonflow in the evaluation of flow coefficients by considering particles emitted in distinct and well separated ranges of rapidity (pseudorapidity) and transverse momentum, or particle pairs with different charge combinations. The scalar product method, in particular, lends itself well to such measurements with a large η gap.

Nonflow effects are dominated by two-, three-, and few-particle correlations. As such, δ_n primarily scales as the inverse of the produced multiplicity (see §10.2.3) and thus lead to a nearly constant contribution of $M\langle u^*\rangle$ on collision centrality while flow's contribution rises and fall from peripheral to central collisions. Note, however, that the contribution of nonflow effects is likely to exhibit a small dependence on collision centrality, in A +A collisions, as a result of the changing relative probability of rare or high-p_T process with collision impact parameter. Be that as it may, nonflow contributions to the correlator $\langle uQ^*\rangle$ may be subtracted using the so-called **AA–pp method** given that nonflow contributions to this correlator should be nearly independent of collision centrality. It is worth noting that the AA–pp method can be used for any n-particle correlation measurement of harmonic coefficients.

Another technique commonly used to reduce nonflow effects in the determination of flow coefficients is the use of multiparticle cumulants, which by construction suppress by $\sim 1/M$ for each particle added to a correlator. Indeed, measurements taken at RHIC indicate that four-particle cumulants remove nonflow effects almost completely [182].

An additional uncertainty in the determination of flow coefficients arises from flow fluctuations. For a given collision system, beam energy, and impact parameter, one expects flow to reach an expectation value determined largely by the geometry of the collision system. By the very nature of microscopic systems, fluctuations in the initial geometry or the collision dynamics may occur. One thus expects the flow magnitude to exhibit event-by-event fluctuations. These fluctuations, however, affect the various flow measurement methods quite differently. The effects of flow fluctuations can be expressed formally as follows for two-, four-, and six-particle cumulants:

$$v_n\{2\} = \sqrt{\langle v_n^2\rangle} = \left(\langle v_n\rangle^2 + \sigma_v^2\right)^{1/2} \tag{11.155}$$

$$v_n\{4\} = \left(2\langle v_n^2\rangle^2 - \langle v_n^4\rangle\right)^{1/4} \tag{11.156}$$

$$v_n\{6\} = \left[(1/4)(\langle v_n^6\rangle - 9\langle v_n^4\rangle\langle v_n^2\rangle + 12\langle v_n^2\rangle^3)\right]^{1/6} \tag{11.157}$$

and so on, for higher-order cumulants. Clearly, while $v\{2\}$ is directly sensitive to $\langle v\rangle$ and σ_v, higher cumulants require knowledge of higher order moments of the distribution in v. Models can, however, be invoked to estimate the relations between these moments. For instance, in the limit of a Gaussian flow distribution, one finds [185]

$$v_n\{2\} = \left(\langle v_n\rangle^2 + \sigma_v^2\right)^{1/2} \approx \langle v_n\rangle + \sigma_v^2/(2\langle v\rangle) \tag{11.158}$$

$$v_n\{4\} = \left(\langle v_n\rangle^4 - 2\sigma_v^2\langle v_n\rangle^2 - \sigma_v^4\right)^{1/4} \approx \langle v_n\rangle - \sigma_v^2/(2\langle v\rangle) \tag{11.159}$$

$$v_n\{6\} = \left(\langle v_n\rangle^6 - 3\sigma_v^2\langle v_n\rangle^4\right)^{1/6} \approx \langle v_n\rangle - \sigma_v^2/(2\langle v_n\rangle) \tag{11.160}$$

We thus conclude that while flow fluctuations increase the magnitude of the coefficients estimated from two-particle correlations ($v\{2\}$), they reduce by an approximately equal amount the values obtained with four- and six-particle cumulants. The preceding relations strictly hold only if $\sigma_v \ll \langle v\rangle$. However, the flow vector distribution is, as we saw in §11.4.2, not perfectly Gaussian. Including effects of flow fluctuations, one gets [185]:

$$\frac{1}{N}\frac{dN}{v_n dv_n} = \frac{1}{\sigma_n^2}\exp\left(-\frac{v_n^2 + v_{n,o}^2}{2\sigma_n^2}\right) I_o\left(\frac{v_n v_{n,o}}{\sigma_n^2}\right). \tag{11.161}$$

where $v_{n,o}$ is the nominal value of the flow coefficient (i.e., its expectation value, $v_{n,o} = \langle v_n\rangle$). The v_2 cumulants can then be shown to be

$$v_2\{2\} = v_{2,o}^2 + 2\sigma_v^2, \tag{11.162}$$

$$v_2\{n\} = v_{2,o} \quad \text{for } n \geq 4. \tag{11.163}$$

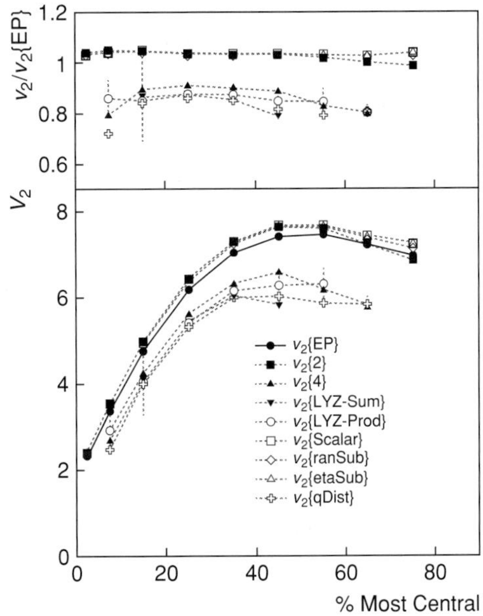

Fig. 11.14 Ratio of the elliptical flow coefficients v_2 obtained by several methods to those obtained with the event plane method in the analysis of charged hadrons measured in Au + Au collisions at $\sqrt{s_{NN}} = 200$ GeV [2, 7]. Ratios are shown for the random subevents, pseudorapidity subevents, scalar product, two-particle cumulants, four-particle cumulants, Q-distribution, and Lee–Yang zeros sum generating and product generating functions. (Data from STAR collaboration [182].)

Effects of nonflow on the event plane method are more complicated to evaluate but have been shown to range from $v_2\{\text{EP}\} = v_2\{2\} = \langle v^2\rangle^{1/2}$ to $v_2\{\text{EP}\} = \langle v\rangle$, depending on the reaction plane resolution [19]. The dependence of fluctuations of the Lee–Yang zeros method and its derivatives are nonlinear [184]. For Bessel–Gaussian distributions in v, these methods yield the same results as higher cumulants, namely $\langle v\rangle = v_o$ of the Bessel–Gaussian distribution. Consequently, if the v_2 distribution is Bessel–Gaussian, all multiparticle methods should yield the same result: $\langle v\rangle = v_o = v_{2,RP}$ [37].

To summarize, we note that the event plane method is a special case with results ranging between $v_2\{2\}$ and $v_2\{4\}$ depending on the reaction plane resolution. Higher-order cumulants $v_2\{6\}$ and Lee–Yang zeros results, however, tend to agree quantitatively with $v_2\{4\}$, as illustrated in Figure 11.14.

11.5 Appendix 1: Numerical Techniques Used in the Study of Correlation Functions

The calculation of products, such as $\rho_1 \otimes \rho_1(\Delta\phi)$ and $\rho_1 \otimes \rho_1 \otimes \rho_1(\Delta\phi_{12}, \Delta\phi_{12})$ required in the measurements of two-particle cumulant $C_2(\Delta\phi)$ and three-particle cumulant

$C_2(\Delta\phi_{12}, \Delta\phi_{12})$, is in principle based on integrals over coordinates ϕ_1, ϕ_2, and ϕ_3 (see, e.g., Eq. (11.10)). In practice, the densities $\rho_1(\phi_i)$ are estimated on the basis of histograms with a finite number of bins. One must thus replace the integrals over continuous variables by sums running over the histograms bins:

$$\begin{aligned}
\rho_1\rho_1(\Delta\phi_{12}) &\equiv \rho_1\rho_1(m) \\
&= \sum_{i,j=1}^{n} \rho_1(i)\rho_1(j)\delta_{m,j-i} \\
\rho_1\rho_1\rho_1(\Delta\phi_{12}, \Delta\phi_{13}) &\equiv \rho_1\rho_1\rho_1(m, p) \\
&= \sum_{i,j,k=1}^{n} \rho_1(i)\rho_1(j)\rho_1(k)\delta_{m,j-i}\delta_{p,k-i}.
\end{aligned} \tag{11.164}$$

The indices i, j, k are used to specify the ϕ bins of particles 1, 2, and 3, respectively, while the integer variables m and p correspond to bins in $\Delta\phi_{12}$ and $\Delta\phi_{13}$, respectively. The delta functions ensure the proper match between $\Delta\phi_{ij}$ and the difference $\phi_i - \phi_j$. This procedure assumes a one-to-one bin mapping. With n_ϕ bins in ϕ for each particle, this would require twice as many bins for the angle difference. However, given the periodicity of the ϕ and $\Delta\phi$ variable, one may transform the integers m and p according to

$$\text{if } m = i - j < 0 \text{ then, replace } m \text{ by } m + n_\phi. \tag{11.165}$$

and similarly for p. The number of bins in $\Delta\phi$, is consequently $n_{\Delta\phi} = n_\phi$. The aforementioned technique may similarly be applied to the determination of the terms $\rho_2 \otimes \rho_1(\Delta\phi_{ij}, \Delta\phi_{ik})$ of the C_3 cumulant. One gets

$$\begin{aligned}
\rho_2\rho_1(m, p)_{123} &= \sum_{i,j,k=1}^{n} \rho_2(i_1, j_2)\rho_1(k_3)\delta_{m,i_1+j_2}\delta_{p,i_1+k_3} \\
\rho_2\rho_1(m, p)_{231} &= \sum_{i,j,k=1}^{n} \rho_2(i_2, j_3)\rho_1(k_1)\delta_{m,i_1+j_2}\delta_{p,i_1+k_3} \\
\rho_2\rho_1(m, p)_{132} &= \sum_{i,j,k=1}^{n} \rho_2(i_1, j_2)\rho_1(k_3)\delta_{m,i_1+j_2}\delta_{p,i_1+k_3},
\end{aligned} \tag{11.166}$$

where the indices i_l, j_l, and j_l refer to bins in ϕ for particles l and the variables m and p correspond to bins in $\Delta\phi_{ij}$ and $\Delta\phi_{ik}$, respectively, as specified by the delta functions, and meant to satisfy the periodic boundary condition expressed by Eq. (11.165).

The aforementioned numerical technique is defined for finite width bins in ϕ and $\Delta\phi$. This implies a given bin in $\Delta\phi$ spans a range wider than its nominal width $2\pi/n_{\Delta\phi}$. For instance, for $i = j = 0$ one has ϕ_i and ϕ_j span the range $2\pi/n_\phi$. The difference $\phi_i - \phi_j$ consequently spans the range $[-2\pi/n_\phi, 2\pi/n_\phi]$ which is obviously wider than $[0, 2\pi/n_{\Delta\phi}]$ for $n_{\Delta\phi} = n_\phi$. The signal from a specific $\Delta\phi$ bin is thus effectively smeared across three bins in the numerical integrations of Eqs. (11.164) and (11.166). This effect, commonly known as **bin aliasing**, cannot be avoided but may be suppressed by use of a rebinning technique: first, calculate the integrals with $n_{\Delta\phi} = n_\phi$, and subsequently rebin the integrated signals to have $n_{\Delta\phi}/m$ bins using m as an integer multiple of 2. A value of $m \gg 2$ will greatly reduce the effects of the bin sharing. However, this may demand a large amount of statistics if the rebinning is applied on R_2 or R_3 to account for instrumental or detection inefficiencies.

The integration techniques expressed in Eqs. (11.164) and (11.166) can readily be applied to studies of correlations in $\Delta\eta$ rather than $\Delta\phi$. However, note that since no periodic boundary can be assumed in $\Delta\eta$, one must map the differences $\eta_1 - \eta_2$ obtained from n_η bins onto $n_{\Delta\eta} = 2n_\eta - 1$ bins in $\Delta\eta$. This leads to the mapping

$$m = i - j \text{ replaced by } m + n_{\Delta\eta} + 1, \tag{11.167}$$

where i, j are in the range $[1, n_\eta]$, and m lies in the range $[1, n_{\Delta\eta}]$, with $n_{\Delta\eta} = 2n_\eta - 1$. As for integrations (11.164) and (11.166), this mapping leads to smearing of the signals across bins. This effect can be suppressed by rebinding, as for measurements in $\Delta\phi$, but since $n_{\Delta\eta}$ is an odd number by construction one must effect a rebinning by an integer multiple of 3. Averaging in $\bar{\eta}$ over a square range of η_1, η_2 values yields $\Delta\eta$ values with different probabilities. One can simply count the relative probabilities of values of m by looping on all values of i and j and filling a relative probability histogram $H_m(m)$, one can use to obtain the integrals $\rho_1 \otimes \rho_1(\Delta\eta_{12})$, $\rho_1 \otimes \rho_1 \otimes \rho_1(\Delta\eta_{12}, \Delta\eta_{13})$, or $\rho_2 \otimes \rho_1(\Delta\eta_{ij}, \Delta\eta_{ik})$. For instance, for $\rho_1 \otimes \rho_1(\Delta\eta_{12})$, one gets

$$\rho_1 \otimes \rho_1(m) = \frac{1}{H_m(m)} \sum_{i,j=1}^{n} \rho_1(i)\rho_1(j)\delta(m - i + j + n_{\Delta\eta} + 1) \tag{11.168}$$

and with similar expressions for $\rho_1 \otimes \rho_1 \otimes \rho_1(m, p)$ or $\rho_2 \otimes \rho_1(m, p)$ (see Problem 11.9).

Exercises

11.1 Verify that the difference $\nu - \nu_{\text{stat}}$ yields Eq. (11.69).

11.2 Verify the scaling property expressed by Eq. (11.76).

11.3 Verify the scaling property expressed by Eq. (11.77).

11.4 Show that the quantity ω_Q defined by Eq. (11.78) tends to unity in the independent particle production limit.

11.5 Derive the equations (11.80) for the charge conservation limit in elementary or nuclear collisions that produce particles over a range of rapidity spanning several units.

11.6 Find an expression similar to Eq. (11.91) for the event-wise average $\langle p_T \rangle$ defined by Eq. (11.89). Hint: Express $\rho_1(\eta, \phi, p_T)$ as a sum of conditional cross sections $\rho_1(\eta, \phi, p_T|m)$, defined for a fixed event multiplicity m and each with probability $P(m)$, i.e. $\rho_1(\eta, \phi, p_T) = \sum_m P(m)\rho_1(\eta, \phi, p_T|m)$.

11.7 Find an expression similar to Eq. (11.97) for the event-wise average $\langle \Delta p_T \Delta p_T \rangle$ defined by Eq. (11.93). Hint: Decompose $\rho_2(\eta_1, \phi_1, p_{T,1}, \eta_2, \phi_2, p_{T,2})$ as a sum of conditional cross sections $\rho_2(\eta_1, \phi_1, p_{T,1}, \eta_2, \phi_2, p_{T,2}|m)$ similar to the decomposition of $\rho_1(\eta, \phi, p_T)$ used in Problem 11.6.

11.8 Show that the correlation $\langle \Delta p_T \Delta p_T \rangle_I^{(m)}$ applied to a colliding consisting of a superposition of m independent but identical processes, each with covariance $\langle \Delta p_T \Delta p_T \rangle_I^{(1)}$, scales as

$$\langle \Delta p_T \Delta p_T \rangle_I^{(m)} = \frac{1}{m} \langle \Delta p_T \Delta p_T \rangle_I^{(1)}. \tag{11.169}$$

11.9 Find expressions equivalent to Eq. (11.168) for the averages $\rho_1 \otimes \rho_1 \otimes \rho_1(\Delta\eta_{12}, \Delta\eta_{13})$ and $\rho_2 \otimes \rho_1(\Delta\eta_{ij}, \Delta\eta_{ik})$.

11.10 Verify that the product $\langle u_{n,1} u^*_{n,2} \rangle$ averaged over an ensemble events has a vanishing contribution from sine terms $\sin n\phi_1 \sin n\phi_2$.

11.11 Define a "triggered" correlation function $K_T(\Delta\phi)$ in terms of a two-particle cumulant C_2.

11.12 One expects the strength of the $\nu_{+-,\mathrm{dyn}}$ correlation function to be largely determined by the charge production mechanism. For instance, the production and decay of neutral resonances, such as the ρ-meson, should have a large impact on net charge fluctuation measured values of $\nu_{+-,\mathrm{dyn}}$. This can be illustrated with a simple model that includes the production of three particle types, π^+, π^-, and ρ^o. Assume the three species are produced independently and with relative fractions p_1, p_2, and p_3 respectively. Ignore effects associated with Bose statistics and assume the probability of producing n_1 π^+, n_2 π^-, and n_3 ρ^o may be expressed with a multinomial distribution:

$$P(n_1, n_2, n_3; N) = \frac{N!}{n_1! n_2! n_3!} p_1^{n_1} p_2^{n_2} p_3^{n_3}. \tag{11.170}$$

Further assume that all ρ^os decay into a pair π^+ and π^- and calculate the magnitude of dynamic fluctuations $\nu_{+-,dyn}$.

11.13 Verify that Eq. (11.50) is correct in cases where $\langle N_+ \rangle = \langle N_- \rangle$.

11.14 Verify Eq. (11.139) and show that the fourth-order cumulant equals $-v_n^4$ in the absence of flow fluctuations.

11.15 Demonstrate Eq. (11.154).

12 Data Correction Methods

No measurement is ever perfect. Measurement errors and uncertainties are indeed an intrinsic part of the scientific process. Skilled scientists, however, can devise techniques to minimize errors and correct for biases. In this chapter, we discuss the notions of accuracy, precision, and biases, and examine various sources and types of errors in §12.1. We then present, in §12.2, a discussion of specific sources of uncertainties arising in the nuclear sciences. Techniques to unfold detection efficiencies and resolution effects in the measurement of spectra and elementary cross sections are presented in §12.3 while correction techniques relevant for correlation and fluctuation observables are discussed in §12.4.

12.1 Experimental Errors

An experimental error may be defined as the difference between a measured (observed) value and the true value. This of course assumes the observable of interest is meaningfully defined and in fact has a true value. It should be clear, however, that there can be difficulties with this idea. Certain physical quantities such as temperature are defined only in the context of a large number limit. Indeed, while it is meaningful to speak of the temperature of gas consisting of a very large number of molecules, the notion of temperature becomes meaningless in the presence of only one or two particles. Limitations may also arise because of the quantum nature of phenomena. For instance, there is an intrinsic limitation in simultaneously measuring the instantaneous position and momentum of an elementary particle, although it is perfectly sensible to consider the expectation values (i.e., average) of these two observables.

The true value of a physics observable is, of course, unknown a priori otherwise there would be no reason to conduct an experiment. Knowing the error of a measurement is obviously also impossible, and one must then use the language of probability and statistics to estimate both the true value and the error. Properly conducted measurements, with a sound statistical analysis, are thus expected to yield values close to the true value but without guarantee of ever reaching it with infinite precision or accuracy.

The methods of probability and statistics covered in Part I of this book provide meaningful ways to estimate both the true value and error of any measurement. Most modern physical observables of interest, however, are quantities derived from measurements of one or several observable elementary quantities and calculations. Modern experiments (e.g., experiments at the Large Hadron Collider) are rather complex, and the estimation of errors may at first seem a daunting task. Fortunately, here again, the methods of statistics, and methods for the propagation of errors covered in Part I enable an estimate of the true values as well as errors on the derived quantities. It is thus useful to examine the various types of errors that can arise in an experiment, identify errors that can be reduced or suppressed by careful design of a measurement, and isolate those that are irreducible or depend on the amount of statistics gathered by an experiment.

Measurement errors and uncertainties can occur for a wide variety of reasons including simple mistakes, calculation errors, instrumentation reading error, instrumental or background processes or process noise that interfere with the measurement, as well as the basic stochastic nature of the measurement process. Improper "reading" of a gauge or the output of a measuring device can be greatly reduced by taking humans out of the measurement process and replacing them with robust and reliable sensors whose measurements are read out by automated computing systems. Errors and uncertainties remain nonetheless an intrinsic part of the experimental process. Indeed, even if human error can be avoided, there remain intrinsic uncertainties associated with physical processes and sensor technologies used to carry out measurements.

Consider, for instance, the identification of particle species in a magnetic spectrometer based on the energy loss (and momentum loss) particles undergo as they traverse sensing devices (see §8.3.3). By its very nature, the collisional processes that leads to energy deposition in a finite thickness sensor is stochastic. Measurement of position and energy loss (and deposition) consequently yield random values according to some PDF determined by the physics of the process (e.g., collisions) and the sensing device. As another example, consider measurements of particle time-of-flight with photosensitive devices, as described in §8.4.2. Particles passing through a material collide with electrons of the material and produce random molecular excitations that result in fluorescence. Timing measurements are obtained when the light collected produces an electronic signal that exceeds a "firing" threshold. The stochastic nature of the photon production and collection processes lead to irreducible fluctuations of the time when the firing threshold is reached and thus result in intrinsic timing uncertainties. The collection of finitely many ions or electrons in energy-sensing devices (such as calorimeters, pad chambers, and so on) similarly result in intrinsic uncertainties. In essence, all techniques relying on the counting of electrons, photons, or other types of particles are fraught with intrinsic uncertainties associated with the stochastic nature of the counting process. Such uncertainties may be mitigated by careful optimization of the detector design to yield a large number of counted particles, better collection, low noise amplification, and so forth, but they can never be eliminated. These sources of stochastic noise are commonly known as **measurement noise**.

Additional uncertainties also arise due to the physics of the detection or measurement process. For instance, measurements of particle momenta in a spectrometer rely on a precise determination of the radius of curvature of charged particle trajectories. The curvature may, however, be stochastically modified when particles pass through the various measuring devices used to establish their trajectory through the spectrometer. Limitations may also occur due to the dual nature of particles and light. The angular resolution of a telescope is, for instance, intrinsically limited by the diffraction of light and the counting of photons in the telescope's focal plane. Although diffraction effects can be suppressed by building increasingly large mirrors, they can never be completely eliminated. Likewise, the number of photons emitted by a star is finite, and ever increasing aperture sizes can reduce, but never eliminate, the stochastic noise of the photon collection process. In essence, the process used to measure a physical observable also limits the precision and accuracy that can be achieved in any given experiment. Uncertainties associated with the measurement process are usually referred to as **process noise**.

While measurement and process noises cannot be eliminated, careful experimental design can suppress their impact on the outcome of measurements and their uncertainties. It is also possible to handle series of measurements to partially filter out noise effects and obtain optimal treatment of process and measurement noises using the technique known as Kalman filter, already introduced in §5.6.

While measurement errors can be associated with a wide variety of causes and find an irreducible origin in measurement and process noises, one can in general categorize errors and uncertainties into two distinct classes. Errors associated with the randomness of the physical and measurement process are known as **statistical errors**, while those associated with uncertainties in the measurement protocol or method are referred to as **systematic errors**.

Statistical errors refer to the notion that repeated measurements of a given physical quantity yield different values with a "seemingly" random pattern. The randomness arises from the stochastic nature of the measurement and process noise discussed earlier. The random character of the measurement process can usually be described with a probability density function expressing the likelihood of measuring specific values or range of values. Statistical errors can be evaluated, characterized, and reported using the statistical methods discussed in Part I of this book. It is usually safe to assume that measured values are distributed with Poisson or Gaussian probability distributions about their true value, but important deviations from these distributions are also known to occur. Indeed, in practice, errors may "not" alway be perfectly Gaussian.

Systematic errors refer to the notion that a measurement may deviate from the true value because of the measurement method or protocol used to carry out the measurement or the statistical estimator used to in the determination of the observable. The error arises from the system or structure of the measurement technique or apparatus and is thus said to be **systematic**. Systematic errors may arise for various reasons such as instrumentation "reading error," flawed protocol, incorrect experimental model, improper correction methods, or a biased estimator.

Since the true value of an observable is not known, one cannot, by definition, determine the true error. One is thus reduced to estimating error(s) on the basis of our understanding of the measurement (measurement model), prior measurements, fundamental theories (although strictly speaking, measurements are tests of the theories, not the converse), or statistical estimates obtained by comparing several distinct measurements. It is useful to distinguish notions of **accuracy** and **precision**. The former is generally used to describe how close the result of an experiment approaches the true value of an observable, while the latter is used to ascribe how exact a result is obtained without particular reference to the unknown true value. A measurement is generally considered precise if repeated instances of the measurement yield tightly clustered values, that is, values with a small variance. A precise measurement may, however, suffer from biases and systematically deviate from the true value. It is then considered inaccurate. Conversely, a measurement protocol can be accurate in the sense that it yields results with zero bias but nonetheless remains imprecise because repeated measurements have a large variance. One thus wishes for measurements that are both precise and accurate, that is, with minimal variance and zero bias.

Biases can be introduced either by the **statistical estimators** used in the analysis of data (defined in §4.3) or from the experimental procedure used to carry out the measurement. For instance, a raw measurement of cross section is intrinsically biased since particle detection can never be accomplished with perfect efficiency. It is thus necessary to correct for such inefficiencies. As we shall see in §12.3, while correction methods used to account for limited detection efficiency may in principle be unbiased, practical considerations often limit the estimation of biases and errors, and one must estimate systematic uncertainties. It is important to note that estimation of systematic uncertainties is by far the most challenging component of error estimation. While there is no single or unique method to estimate systematic uncertainties, it is nonetheless possible to formulate generic principles and guidelines, which we discuss in §12.5.

It is also useful to distinguish between absolute and relative uncertainties or errors. Absolute error refers to the actual size of errors expressed in the same units as the actual measurement. For instance, the measurement of the length of a 1-meter rod could have an absolute uncertainty of 0.001 m. That is, the length is known up to a precision of 0.001 m. Repeated measurements of length are expected to cluster around the expectation value of 1 m with a standard deviation of 0.001 m. A relative error indicates the size of an error relative to the magnitude of the observable and is typically expressed as a percent error obtained by multiplying by 100 the ratio of the uncertainty by the expectation value of the observable (magnitude). The notion of relative error is particularly useful for the description of elementary particle production cross sections. Indeed, since particle production cross sections may have very large ranges of values (from femtobarns to barns), stating the error on a cross section in barns (or any subunits) is thus not particularly informative. Errors on cross sections are thus often reported in terms of relative errors.

The precision of a measurement is obviously manifest if the error is reported. But it is also evident in the number of **significant figures** used to report the measured value. **Significant figures** (also called significant digits) are those digits that carry meaning contributing to the precision of a number. Significant figures exclude all leading zeros or trailing zeros used as placeholders to indicate the scale of a number, as well as spurious digits introduced, for example, by calculations carried out to greater precision than that of the original data, or measurements reported to a greater precision than the equipment supports. By convention, if there is no digital point, the rightmost nonzero digit is the least significant digit. However, if there is a decimal point the rightmost digit is considered as the least significant digit even if zero. All digits between the least and most significant digits are counted as significant (see examples in Figure 12.1).

Use of the **normalized scientific notation** is strongly recommended to report all scientific results and avoid ambiguities on the number of significant figures. In the scientific notation, all numbers are written in the form $a \times 10^b$, where the exponent b is an integer and the coefficient a a real number. The exponent b is chosen so that the absolute value

3.142	0.3142	0.003142
2.718×10^{-5}	$2.718 \times 10^{+7}$	2718000
5.000	500.0	5000.

Fig. 12.1 All numbers shown have four **significant figures**.

of a lies in the range $1 \leq |a| \leq 10$. Thus 15600 is written as 1.56×10^4 while 230000. is denoted 2.30000×10^5. This form allows easy identification of the number of significant figures and comparison of numbers, because the exponent b gives the number's order of magnitude. The exponent b is negative for a number with absolute value between 0 and 1. For instance, 0.50 should be written as 5.0×10^{-1}. The 10 and its exponent are usually omitted when $b = 0$.

When quoting results, the number of significant figures should be approximately one more than that dictated by the experimental precision. This is useful mainly for calculation purposes, since it avoids illegitimate loss of precision in calculations. Additionally, when insignificant digits are dropped from a number, the last digit should be rounded off for best accuracy. One should truncate and treat excess digits as decimal fractions. The least significant digit should be incremented by one unit if the fraction is greater than 1/2, or if the fraction equals 1/2 and the least significant digit is odd.

As discussed earlier, statistical and systematic uncertainties have quite different origins and make different statements about the reliability of a measured value. It is thus common practice to report the two uncertainties explicitly and separately. For instance, if the measured cross section of a process is 240 mB with a statistical precision of 24 mB and a systematic uncertainty of 30 mB, one shall report the measurement as

$$240 \pm 24(stat) \pm 30(sys)\ \text{mB}, \tag{12.1}$$

in which the labels "stat" and "sys," which stand for *statistical* and *systematic*, respectively, are often omitted. As we already discussed in Part I of this book, the usual practice is to report one sigma (1σ) statistical uncertainties (whether based on the frequentist or Bayesian paradigms) but greater significance levels are occasionally used and identified as such.

The aforementioned notation implies the errors are symmetric, in other words, that it is equally probable that the true value of an observable be found below or above the quoted value and in the given range. It is often the case, however, that the confidence interval is in fact not symmetric about the quoted value. This may occur, for instance, when an observable A is obtained from a one parameter fit to some data. In this case, a 1σ interval should be determined by the range of the fitted parameter which, relative to the parameter value a at the minimum $\chi^2_{\min}$, yields a χ^2 increase of one unit. As schematically illustrated in Figure 12.2, if the χ^2 function, plotted vs. the fit parameter, is not symmetric relative to its minimum, the low-side range, $\Delta a_<$, and high-side range, $\Delta a_>$, required to produce a χ^2 increase of one unit are not equal. The uncertainty interval is thus asymmetric and is usually reported according to the low/high notation. One would, for instance, write

$$240^{+20}_{-6}\ \text{mB}. \tag{12.2}$$

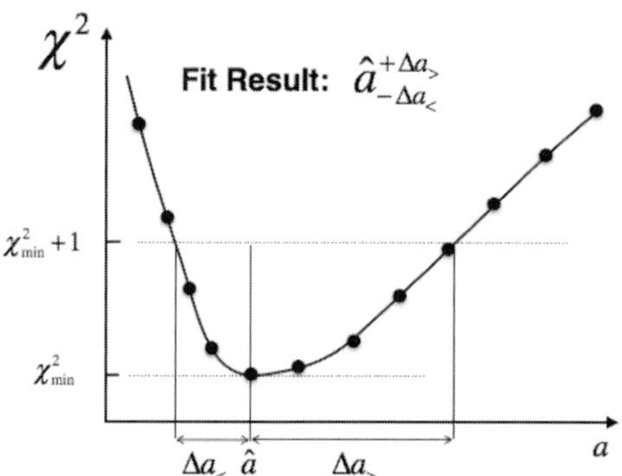

Fig. 12.2 Schematic illustration of conditions under which asymmetric error intervals are obtained from $\mathbb{LS}$ fits. The χ^2 function is plotted, for a given dataset, as a function of a model parameter "a." A one sigma error corresponds to the range in which the χ^2 increases by one unit relative to its minimum value. Given the dependency of χ^2 on a is asymmetric, one must quote the ranges $\Delta a_<$ and $\Delta a_>$ as low- and high-side errors, respectively, as shown.

Asymmetric error intervals may also occur in multiparameter fits or when the measured value is close to a physical boundary (see, e.g., §6.1.8). Asymmetric error intervals are also often reported for systematic errors. Indeed, it is quite possible that the measurement protocol might have a "one-side" effect on an observable, or that the effect is more prominent on the low- or high-side of the reported value. It is thus common to report low- and high-side intervals for both statistical and systematic uncertainties as follows

$$240^{+20}_{-6}(\text{stat})^{+33}_{-22}(\text{sys})\ \text{mB}. \tag{12.3}$$

There is no universally adopted standard notation or method to report errors in graphs. However, in physics, most particularly in high-energy physics, it has essentially become the norm to report statistical and systematic errors independently. Statistical errors are usually indicated with a simple vertical line (for the error on the dependent variable) passing through the data point it qualifies and the length of which corresponds to the size of the confidence interval. Some authors like to terminate the vertical line at both extremities with a short horizontal segment, but this practice has become rather infrequent in the last decade or so. Systematic errors are reported using a wide variety of notations and techniques depending on the circumstances and complexity of the data and errors. Some of these notations are schematically illustrated in Figure 12.3.

12.2 Experimental Considerations

All measurements of physical observables are subjected to limitations, constraints, and alterations associated with the nature of the measurement process as well as the properties

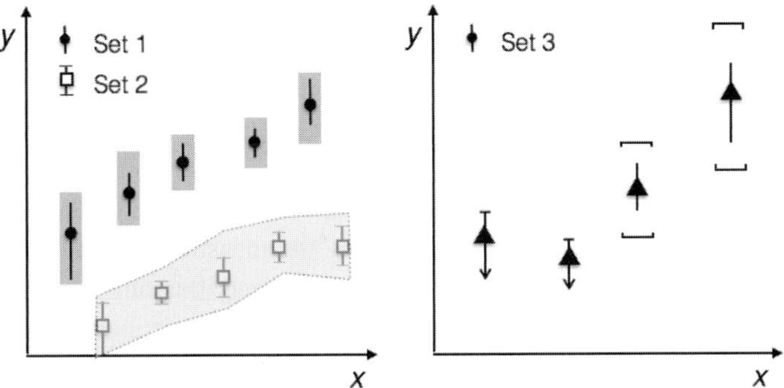

Fig. 12.3 Schematic illustration of notations and techniques commonly used for reporting errors in graphs displaying experimental results. Statistical error bars are shown with solid vertical lines. Error bars with downward pointing arrows are used to represent one-sided confidence levels with an upper limit. Systematic errors are shown with gray boxes for set 1, a shaded range for set 2, and brackets for set 3.

of the apparatus or devices used to carry them. These effects can be broadly categorized in terms of the following:

1. Experimental acceptance
2. Detection efficiency
3. Smearing and resolution
4. Ghost and background signals

Let us briefly discuss each of these effects.

12.2.1 Acceptance

The **acceptance** of a measurement corresponds to the range in which an observable of interest can be measured. The determination of the acceptance of a measurement is often a trivial matter, but it can also be rather complicated. At a basic level, the acceptance is determined by the device carrying out the measurement and the range in which an elementary observable is acquired. For instance, when measuring the time-of-flight (TOF) of particles, one is typically limited by the physical scale of the apparatus and the specific time-to-digital converter (TDC) used to carry out the time-to-digital conversion. The acceptance may also be determined by the measurement process itself. Consider, for instance, the measurement of the momentum of charged particles in a magnetic spectrometer. The momentum is proportional to the radius of the charged particle trajectories and the magnetic field. If the momentum of the particles is too small, their trajectory might not be measurable by the sensing devices and they will simply be lost, that is, not accepted in the measurement. The acceptance associated with certain measurements can be rather complex and depend both on the physical process involved and the measurement itself. This is particularly the case when the measurement process involves a cascade of different phenomena. For instance, the detection of a Higgs boson decaying into jets requires that one detects the jets, which in turn requires that one detects the charged and neutral particles of these jets. The acceptance of the Higgs is then dependent on several measurement processes and devices.

Given the range of conditions that may constrain the acceptance of a specific measurement, the actual determination of the acceptance may become an arduous task that is typically best carried out by modeling the measurement process and instruments by Monte Carlo simulations. We will consider selected illustrative examples in Chapter 14.

The acceptance of a measurement effectively defines the range of its applicability. Measured observables can be genuinely reported only if they are within the acceptance of the measurement. They thus cannot be corrected, strictly speaking, for unmeasured ranges of values. However, if there are observables whose behavior can be predicted with good certainty outside a measurement of interest (e.g., production cross section of particles vanishes at zero and infinite momentum), it may then be possible to carry out reliable extrapolations of the measurements, sometimes referred as **acceptance corrections**. It should be clear that such "corrections" are extrapolations whose validity and accuracy depend on the specificities of the observables and the measurement itself.

12.2.2 Efficiency

Whether one conducts a measurement of an elementary physical quantity (i.e., the amplitude of a magnetic field) or a more elaborate observable (e.g., the production cross section of a specific type of particle), one is invariably faced with the fact that some "events" will be lost either because of the nature of the phenomenon and observable being measured, that is, the process, or because of the measurement device itself. Consider, for instance, the detection of charged pions in a magnetic spectrometer. Since charged pions decay with a mean lifetime of 2.6×10^{-8} s, a substantial fraction may be lost because they decay in flight through the apparatus and consequently do not produce any or sufficient signals in sensing devices. Pions may also be lost because of various effects arising in the detection process such as fluctuations in energy deposition, charge collection, electronic noise, detector component malfunction, and so on.

The **efficiency**, ϵ, of a measurement is generically defined as the ratio of the expectation value of the number of measured "events," $\langle N_M \rangle$, by the expectation value of the number of produced (or true) events, $\langle N_T \rangle$:

$$\epsilon = \frac{\langle N_M \rangle}{\langle N_T \rangle}. \tag{12.4}$$

The word *event* refers to instances or occurrences of a particular phenomenon of interest, such as supernova explosions, proton–proton collisions, or the production of charged particles.

Methods for efficiency correction of spectra and correlation functions are presented in §§12.3 and 12.4.3, respectively, whereas techniques to determine the efficiency of measurements are discussed in §§12.4.6 and 14.2.4.

12.2.3 Signal Smearing, Precision, and Resolution

Strictly speaking, **signal smearing** refers to the notion that the **precision** of a measurement of a physical observable is invariably finite, while the term **resolution**, often called **resolving power**, corresponds to the ability of a particular measurement to distinguish or separate two "close-by" signals, phenomena, or entities, e.g., two peaks in a frequency or mass spectrum. In practice, the term resolution is often misused to refer to both precision and separation of close-by signals. As for acceptance and efficiency, both physical processes and the properties of a device have an influence on the **precision** and **resolution** of the measurement.

Consider, for instance, the measurement of the momentum of charged particles with a magnetic spectrometer, where the momentum is determined based on the radius of curvature of the charged particle trajectories. As they traverse detector components, scatterings with electrons and nuclei of the medium change their momentum and direction randomly. The radii and directions of the tracks are consequently randomly altered and yield a smearing of the measured momenta, and thus a **loss of precision**. Signal smearing also arises from the technique used to extract the radius of curvature. One must be able to sample hits or points where particles traverse specific components of the apparatus. Devices used to measure these points have finite granularity and feature finite signal-to-noise characteristics. They thus yield position measurements of finite precision, which in turn produce smearing, or **loss of precision**, in the determination of the radius of curvature of the charged particle tracks. The momentum of measured charged particles is thus limited by both scattering processes and the precision of hit measurements. This is true of all observables in general.

Process noise and measurement noise also limit the **resolving power** of a measurement. Consider once again the measurement of charge particles in a detector. Charged particles passing through a sensing device lose energy as they scatter with electrons and nuclei of the material. This lost energy is used to produce a signal and detect the passage of the particle. But, based on the need to collect the lost energy to produce a signal (often in terms of the amount of ionization produced by the particle), the sensing device must have a finite area. There can also be energy or signal sharing between nearby sensing units. This implies that the signals produced by the passage of two particles through a given detection unit may not be separable as two distinct signals and that they may then be confused as a single signal. The term resolution thus typically refers to the minimum signal separation (in the observable's units) required to reliably separate two distinct signals.

Process and measurement smearing, or noise, may be described, typically, in terms of a response function, $r(x_m|x_t)$, expressing the probability of obtaining a specific measured value, x_m, given an actual or true value, x_t. This response can in principle be expressed in terms of a well-defined PDF, with unit normalization:

$$\int_{\text{accept}} r(x_m|x_t)\, dx_m = 1. \tag{12.5}$$

The accuracy and precision of a measurement can be reported either in terms of its response function, if it is explicitly known, or in terms of its (estimated) moments. The first and second moments of $r(x_m|x_t)$ determine the bias and the accuracy of the measurement. The bias is obtain from the first moment

$$B_x \equiv \langle x_m \rangle - x_t = \int_{\text{accept}} (x_m - x_t)\, r(x_m|x_t)\, dx_m \tag{12.6}$$

whereas the accuracy is appraised on the basis of its mean square error (MSE).[1]

$$\text{MSE} = \left(\int_{\text{accept}} (x_m - x_t)^2\, r(x_m|x_t)\, dx_m \right)^{1/2}. \tag{12.7}$$

As already discussed in §4.4, the MSE depends on the bias B_x and the variance σ_x^2 of the measured value x_m (an estimator of x_t which in the context of §4.4 would be denoted $\hat{x}$) relative to its expectation value, $\langle x_m \rangle = E[x_m]$. The accuracy of the measurement may then be written

$$\text{MSE} = B_x^2 + \sigma_x^2, \tag{12.8}$$

where $\sigma_x^2 = E\left[(x_m - E\left[x_m\right])^2\right]$. Effectively, MSE provides an assessment of the overall accuracy of the measurement influenced in part by the bias B_x and the standard deviation σ_x of the measured value one would observe if the same experiment could be repeated several times.

The response function $r(x_m|x_t)$ can often be approximated with a Gaussian distribution. This is "guaranteed" by the central limit theorem, in particular, when the measurement outcome is a sum or superposition of several subprocesses. Examples of such processes include measurements of the energy of elementary particles with calorimeters, measurements of the curvature of tracks in a magnetic spectrometer, and the determination of particles' TOF with TOF detectors. However, $r(x_m|x_t)$ may also drastically deviate from a Gaussian distribution and feature long tails or various substructures. This is the case, for instance, of measurements of the energy loss of charged particles traversing a time projection chamber where scatterings can lead to large energy losses, or in the detection and measurement of the energy of jets where loss of particles (because of detection efficiencies) can lead to large amounts of missing jet energy.

The response $r(x_m|x_t)$ may be a rather complex function of intrinsic and extrinsic conditions. For instance, the transverse momentum resolution of a charged particle track typically depends on the actual momentum of the particle, its angle of emission (production), and its rapidity. It may also depend on detector-wide or event conditions. For instance, since the occupancy of a detector may influence the noise generated within the detector and the pattern recognition involved in finding and obtaining the momentum of tracks, the resolution or response function may then be dependent on such factors as the beam luminosity, the data acquisition rate, and so on.

There are a variety of techniques to determine the precision/resolution of measurements of physical observables. It is often possible to calibrate and measure the resolution of a device by measurements of well-known standards or calibrated inputs, but measurements with complex detectors typically rely on detailed simulations of the physical processes

[1] Technically, it is the MSE of the estimator of the observable of interest that one reports.

involved, as well as the performance of the detector components. Selected techniques used in the estimation of observable resolution and to account (i.e., correct) for finite resolution are discussed in §12.3 of this chapter, whereas basic simulation techniques used to simulate smearing effects are covered in §14.2.

12.2.4 Ghost Signals and Background Processes

Ghost signals refer to noise that may be construed as a signal either because it exceeds a specific measurement threshold or because it is explicitly reconstructed as such. Electronic noise and process noise produce signal fluctuations that occasionally have large amplitude and may thus fake actual signals. This is particularly the case when considering "elementary" observables measured by basic sensor units or detector components. More elaborate ghost signals may also occur as a result of such basic noise signal. For instance, fake tracks may be reconstructed in a spectrometer as a result of noisy electronics, correlated electronic signals, component malfunctions, or by incorrect association of true particle hits.

Background processes may also impede measurements of a particular observable. For instance, combinatorial pairs of particles may contribute a strong background in the reconstruction of the mass spectrum of decaying particles, and secondary particles (resulting from decays within the apparatus) contribute an undesirable background in measurements of particle production cross section and correlation functions.

Techniques to evaluate signal noise and background processes are as varied as the observables measured by scientists. A detailed discussion of such techniques is beyond the scope of this textbook.

12.3 Signal Correction and Unfolding

The problem of unfolding is formally introduced in §12.3.1. Given the intricacies of the unfolding problem, many scientists opt to compare their data to folded theoretical data rather than unfolding their measurements for instrumental effects. The relative merits of this alternative approach and unfolding of the detector effects are briefly discussed in §12.3.3. The correction coefficient method, commonly known as the bin-by-bin technique, and its limitations, are presented in §12.3.4. Actual unfolding techniques, including regularization, are described in §§12.3.6, 12.3.7, and 12.3.8.

12.3.1 Problem Definition

As discussed already in §12.2 of this chapter, measurements of physical observables involve a variety of instrumental effects that must be properly accounted for to obtain robust scientific results. These include effects associated with limited acceptance, detection efficiency, measurement resolution, possible nonlinearities, and the presence of noise and backgrounds. While extrapolation outside of the acceptance of a specific measurement is not usually warranted, corrections for efficiency, finite resolution, nonlinearities, and background processes are usually possible to perform in a well-controlled manner. Indeed, if an experiment and the observable X it measures are well defined, and if the experimental performance is reasonably stable, it is possible to account for efficiency and resolution effects in a deterministic fashion.

Let us consider a measurement of a specific observable X by a specific apparatus. We will represent by x_t the true value of the observable X, and by x_m the value observed with the apparatus. We will carry out our discussion assuming X is a simple observable (e.g., the transverse momentum, p_T), but the method discussed in the following is trivially extended to multidimensional observables (e.g., $\vec{p} = (p_T, \eta, \phi)$).

Let us first introduce the PDFs of the true, measured, and background distributions. Let $f_t(x_t)$ represent the probability (density) of observing true values x_t, with normalization:

$$\int_{\Omega_t} f_t(x_t)\, dx_t = 1, \tag{12.9}$$

where the integral is taken over the relevant domain Ω_t of the observable X. Likewise, let us denote the PDF of the measured distributions by $f_m(x_m)$, with similar normalization.

$$\int_{\Omega_m} f_m(x_m)\, dx_m = 1. \tag{12.10}$$

In general, the measured distribution f_m may involve a superposition of the true but smeared signal and some background. We will denote the PDF of this background as $f_b(x_m)$ and use the normalization

$$\int_{\Omega_m} f_b(x_m)\, dx_m = 1. \tag{12.11}$$

We will assume that it is possible, experimentally, to determine both the shape of this background PDF as well as the relative rate of background events.

If several background processes and sources of noises hamper the measurement, one can include them by estimating their respective distributions. Let us define p_α as the relative probability of a specific background process α and assume one can describe its distribution of measured values x_m in terms of a PDF, $f_{b,\alpha}(x_m)$, which represents the probability of measuring x_m given the background process α. The probability of observing values x_m produced by N_α background processes may then be written

$$f_b(x_m) = \sum_{\alpha=1}^{N_\alpha} p_\alpha f_{b,\alpha}(x_m), \tag{12.12}$$

where we require, by definition,

$$\int_{\Omega_m} f_{b,\alpha}(x_m)\, dx_m = 1, \tag{12.13}$$

$$\sum_{\alpha=1}^{N_\alpha} p_\alpha = 1.$$

We define the experimental response function, noted $R(x_m|x_t)$, as the probability (density) to observe a value x_m when the "event" has value x_t. This response function has two components:

$$R(x_m|x_t) = r(x_m|x_t) \times \epsilon(x_t). \tag{12.14}$$

The coefficient $r(x_m|x_t)$ is the smearing function of the measurement. It is determined by process noises and measurement noises, as briefly discussed in §12.2.4. It is a PDF representing the probability of finding a value x_m when the true value is x_t. By construction, one has the normalization

$$\int_{\Omega_m} r(x_m|x_t)\, dx_m = 1, \tag{12.15}$$

where Ω_m represents, here again, the domain of the measured values x_m.

The function $\epsilon(x_t)$ represents the efficiency of the measurement and thus corresponds to the probability of actually observing an "event" with an observable value x_t. Similarly to $r(x_m|x_t)$, the efficiency $\epsilon(x_t)$ may depend both on properties of the measurement process (e.g., decays typically reduce the efficiency of observing particles) and characteristics or performance of the apparatus. This could be, for instance, the probability of detecting charged particles within a magnetic spectrometer as a function of their transverse momentum, or the probability of finding and reconstructing a given type of collision at a proton collider.

In general, the functions $r(x_m|x_t)$ and $\epsilon(x_t)$ may have dependencies on other variables such as external or global conditions of the apparatus and collisions being measured. For example, both the efficiency and the smearing function of charged particle measurement usually depend on their direction (rapidity, azimuth angle), as well as the complexity of the events of which they are part of, the detector occupancy, the data acquisition rate, and so on.

Experimentally, given finite statistics and resolution, it is convenient to discretize the true, measured, and background distributions using histograms $\vec{\mu} = (\mu_1, \mu_2, \ldots, \mu_M)$, $\vec{\nu} = (\nu_1, \nu_2, \ldots, \nu_N)$, $\vec{\beta} = (\beta_1, \beta_2, \ldots, \beta_N)$. In general, the true and measured histograms may have different number of bins, M and N, as well as different ranges. We will see later in this section how these differences may impact the determination of the true distribution in practice.

Let us denote the number of true, measured, and background events observed in a given experiment by m_{tot}, n_{tot}, and b_{tot}, respectively. Their expectation values are

$$\mu_{\text{tot}} = \mathrm{E}\left[m_{\text{tot}}\right], \tag{12.16}$$

$$\nu_{\text{tot}} = \mathrm{E}\left[n_{\text{tot}}\right], \tag{12.17}$$

$$\beta_{\text{tot}} = \mathrm{E}\left[b_{\text{tot}}\right]. \tag{12.18}$$

We will show in the following that the number of measured events shall be equal to

$$\nu_{\text{tot}} = \langle\epsilon\rangle\mu_{\text{tot}} + \beta_{\text{tot}}, \tag{12.19}$$

where $\langle\epsilon\rangle$ represents the average efficiency of the measurement. We will assume it is possible to determine β_{tot} unambiguously, although this may prove to be challenging in practice.

The probability, p_j, of x_t being "produced" in bin j is the integral of the PDF $f_t(x_t)$ across that bin:

$$p_j = \int_{\text{bin } j} dx_t f_t(x_t). \tag{12.20}$$

The expectation value of the number of entries in bin j of the true value histogram is thus

$$\mu_j = \mu_{\text{tot}} \times p_j = \mu_{\text{tot}} \int_{\text{bin } j} dx_t f_t(x_t). \tag{12.21}$$

Likewise, the expectation value of the number of entries in bin i of the measured and background distributions shall be

$$\nu_i = \nu_{\text{tot}} \int_{\text{bin } i} dx_m f_m(x_m), \tag{12.22}$$

$$\beta_i = \beta_{\text{tot}} \int_{\text{bin } i} dx_m f_b(x_m). \tag{12.23}$$

Evidently, the number of entries ν_i is a superposition of the smeared true signal, including the detection efficiency and fake signals from background processes. First ignoring backgrounds, one can write

$$\nu_i = \mu_{\text{tot}} \times \text{Probability(event in bin } i). \tag{12.24}$$

The probability of having an event in bin i is determined by the product of the smearing function $R(x_m|x_t)$ and the PDF $f_t(x_t)$. One can thus write

$$\nu_i = \mu_{\text{tot}} \int_{\Omega_t} dx_t \text{Prob}(x_m \text{ in } i|\text{true } x_t, \text{detected}), \tag{12.25}$$

$$\times \text{Prob(detect } x_t) \times \text{Prob(produce } x_t),$$

$$= \mu_{\text{tot}} \int_{\text{bin } i} dx_m \int_{\Omega_t} dx_t r(x_m|x_t)\epsilon(x_t) f_t(x_t). \tag{12.26}$$

The integral over x_t may be partitioned into M bins j by including a sum over all such bins:

$$\nu_i = \mu_{\text{tot}} \int_{\text{bin } i} dx_m \sum_{j=1}^{M} \int_{\text{bin } j} dx_t r(x_m|x_t)\epsilon(x_t) f_t(x_t). \tag{12.27}$$

We seek to express ν_i in terms of the true histogram values μ_j. Multiplying the argument of the integral by $\mu_j/\mu_j = 1$ and rearranging, we get

$$\nu_i = \sum_{j=1}^{M} \int_{\text{bin } i} dx_m \int_{\text{bin } j} dx_t \frac{r(x_m|x_t)\epsilon(x_t) f_t(x_t)}{\mu_j/\mu_{\text{tot}}} \mu_j. \tag{12.28}$$

This may be written

$$\nu_i = \sum_{j=1}^{M} R_{ij}\mu_j, \tag{12.29}$$

with the introduction of the response matrix R_{ij}, defined as

$$R_{ij} = \frac{\int_{\text{bin } i} dx_m \int_{\text{bin } j} dx_t r(x_m|x_t)\epsilon(x_t)f_t(x_t)}{\int_{\text{bin } j} dx_t' f_t(x_t')}, \tag{12.30}$$

where we used Eq. (12.21) to express the ratio μ_j/μ_{tot}. The numerator of Eq. (12.30) corresponds to the joint probability of measuring a signal in x_m bin i when the true value is in bin j, while the denominator amounts to the probability of finding the true signal in bin j. This ratio is thus the conditional probability of getting a measured signal in bin i when the true signal is in bin j:

$$R_{ij} = \text{Prob(observed in bin } i|\text{true in bin } j). \tag{12.31}$$

Summing over all observed bins i, one gets

$$\sum_{i=1}^{N} R_{ij} = \epsilon_j, \tag{12.32}$$

where ϵ_j represent the average detection efficiency in bin j:

$$\epsilon_j = \frac{\int_{\text{bin } j} dx_t \epsilon(x_t) f_t(x_t)}{\int_{\text{bin } j} dx_t f_t(x_t)}. \tag{12.33}$$

The expression Eq. (12.29) of the expectation value of the measured signal ν_i is incomplete since we have so far neglected background processes. Assuming the density $f_b(x_m)$ and the number of background events can be reliably estimated, one can write

$$\nu_i = \sum_{j=1}^{M} R_{ij}\mu_j + \beta_i, \tag{12.34}$$

where β_j represents the expectation value of the background processes obtained with Eq. (12.23). This expression can be conveniently represented in terms of a matrix equation as follows:

$$\vec{\nu} = R\vec{\mu} + \vec{\beta}, \tag{12.35}$$

where $\vec{\nu}$ and $\vec{\beta}$ represent $N \times 1$ column vectors, $\vec{\mu}$ is an $M \times 1$ column vector, and R is an $N \times M$ matrix.

The histogram $\vec{\nu}$ represents the expectation value of the measured distribution. In practice, an experiment shall obtain a specific histogram, hereafter noted $\vec{n}$, exhibiting statistical deviations from $\vec{\nu}$. However, repeated measurements of $\vec{n}$ should, on average, converge toward $\vec{\nu}$. Indeed, one should have

$$E[\vec{n}] = \vec{\nu}. \tag{12.36}$$

It thus seems legitimate to replace $\vec{\nu}$ by $\vec{n}$ in Eq. (12.35), and seek an inverse of the matrix R in order to obtain an estimate of $\vec{\mu}$ according to

$$\vec{m} \equiv \hat{\mu} = R^{-1}\left(\vec{n} - \vec{\beta}\right). \tag{12.37}$$

It is relatively straightforward to show that the estimator $\vec{m}$ corresponds to the solution of the least square problem with

$$\chi^2(\vec{\mu}) = \sum_{i,j=1}^{N} (\nu_i - n_i) \left(V^{-1}\right)_{ij} (\nu_i - n_i), \tag{12.38}$$

where ν_i are treated as functions of fit parameters μ_j defined by Eq. (12.34), and the matrix V^{-1} is the inverse of the covariance matrix:

$$V_{ij} = \text{Cov}\left[n_i, n_j\right]. \tag{12.39}$$

Assuming the inverse R^{-1} exists, one finds (see Problem 12.4) that the minimal $\chi^2(\vec{\mu})$ obtained by simultaneously solving

$$\frac{\partial \chi^2(\vec{\mu})}{\partial \mu_k} = 0 \tag{12.40}$$

indeed corresponds to Eq. (12.37). It is also possible to demonstrate that the estimator $\vec{m}$, given by Eq. (12.37), corresponds to the solution obtained by maximization of the log-likelihood function (see Problem 12.7):

$$\log L(\mu) = \sum_{i=1}^{N} \log\left(P(n_i|\nu_i)\right), \tag{12.41}$$

where $P(n_i|\nu_i)$ is a Poisson distribution (or binomial distribution).

The estimator $\vec{m}$, defined by Eq. (12.37), is in fact an unbiased estimator of $\vec{\mu}$:

$$\begin{aligned} \text{E}\left[m_j\right] &= \sum_{i=1}^{N} \left(R^{-1}\right)_{ji} \left(\text{E}\left[n_i\right] - \beta_i\right), \\ &= \sum_{i=1}^{N} \left(R^{-1}\right)_{ji} \left(\nu_i - \beta_i\right), \\ &= \mu_j, \end{aligned} \tag{12.42}$$

insofar as the noise spectrum $\vec{\beta}$ is properly estimated.

The covariance between distinct bins of the estimator is calculated similarly and amounts to

$$\text{Cov}\left[m_i, m_j\right] = \sum_{k,k'=1}^{N} \left(R^{-1}\right)_{ik} \left(R^{-1}\right)_{jk'} V_{ij}, \tag{12.43}$$

where V_{ij} is the covariance matrix defined in Eq. (12.39). Defining matrix elements $U_{ij} = \text{Cov}\left[m_i, m_j\right]$, the preceding may be written in the compact form:

$$U = R^{-1} V \left(R^{-1}\right)^T, \tag{12.44}$$

where the label T indicates the transpose of the matrix R^{-1}. If the measured n_i obey Poisson statistics, one has $V_{kk'} = \delta_{kk'} n_k$. The expectation of the covariance matrix U is thus

$$U_{ij} \equiv \mathrm{Cov}\left[m_i, m_j\right] = \sum_{k=1}^{N} \left(R^{-1}\right)_{ik} \left(R^{-1}\right)_{jk} \nu_k, \tag{12.45}$$

from which we conclude that if R is diagonal or nearly diagonal, so should the matrix U.

One can additionally verify that the variance of each component m_j corresponds to the RCF bound obtained with the maximum log-likelihood $\log L(\mu)$. The estimator $\vec{m}$ defined by Eq. (12.37) has zero bias and finite covariance. It should thus provide us, in principle, with a solution to the inverse problem. It should then be possible to correct a measurement for instrumental effects, provided one can invert the response function R. In practice, life is a little more complicated. There are, in fact, several issues with this inverse problem, which we discuss in details in the following section.

12.3.2 Issues with the Inverse Problem...

The first issue encountered with the inverse problem is that the calculation of R requires knowledge of $f_t(x_t)$. This is a problem because this quantity is precisely the answer we seek with our solution of the inverse problem. Indeed, the calculation of the integrals in Eq. (12.30) does require knowledge of f_t. In practice, however, this is only a minor issue because one usually carries out a calculation of the response function $R(x_m|x_t)$ using a MC procedure based on a model PDF, f_{MC}, chosen to be a close approximation of the data. One can then verify by explicit calculations, or by iterative procedures, whether the model f_{MC} enables a reliable estimation of the response matrix R_{ij}. Additionally, it is easy to verify that if $f_t(x_t)$ is approximately constant throughout any given bin j, that it can be factorized out of the integral in x_t and thus cancels out:

$$R_{ij} \approx \frac{f_t(x_{t,j}) \int_{\mathrm{bin}\ i} dx_m \int_{\mathrm{bin}\ j} dx_t r(x_m|x_t)\epsilon(x_t)}{f_t(x_{t,j}) \int_{\mathrm{bin}\ j} dx_t'}, \tag{12.46}$$

$$= \frac{1}{\Delta x_{t,j}} \int_{\mathrm{bin}\ i} dx_m \int_{\mathrm{bin}\ j} dx_t r(x_m|x_t), \epsilon(x_t), \tag{12.47}$$

where $\Delta x_{t,j}$ represents the width of bin j.

A far more serious issue with our solution to the inverse problem is that it typically produces wildly oscillating solutions that are not usable in practice. To illustrate this problem, consider the true signal (thick solid line) in Figure 12.4 and the response/smearing function plotted in Figure 14.8. Application (folding) of the smearing function produces the measured expectation value shown in Figure 12.4 as a thin solid line. In an actual experiment, the measured data has finite statistics and exhibit bin-to-bin fluctuations, shown as a dash-line histogram. Unfolding of this measured histogram with Eq. (12.37) unfortunately yields a corrected distribution with large oscillations as illustrated in Figure 12.4 by the solid dots. In spite of all good intents and purposes, this method simply does not work. The unfolded distribution is totally different from the original distribution.

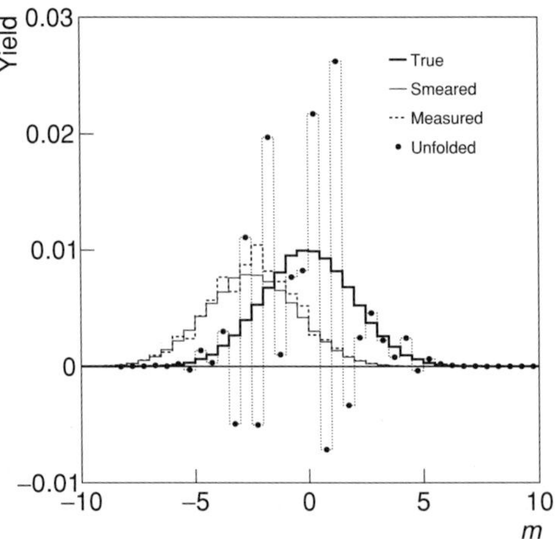

Fig. 12.4 Unregularized unfolding is inherently unstable because the procedure attempts to unsmear peaks and consequently yield spiky structures for spectra with finite statistics.

The failure is no mere numerical instability or coding error. It is an intrinsic feature of the unfolding procedure. Indeed, consider that the effect of the response function is to smear a sharp peak into a wider distribution. The inverse problem thence has a tendency to take wide peaks and transform them into narrow peaks. Statistical fluctuations produce small peaks, which the inversion transforms into large peaks! The inverse problem is intrinsically unstable and thus seems doomed to failure.

Fortunately, instabilities can be tamed provided one is willing to introduce somewhat of a bias in the inversion procedure. This is a general feature of all estimators. The introduction of external constraints may reduce the variance of an estimator but results in a biased estimator. The result may nonetheless be acceptable if the bias is kept under control and reasonably small.

12.3.3 To Fold or Unfold, That Is the Question!

Unfolding of instrumental effects is challenging and prone to many numerical instabilities, as is painfully obvious from the discussion in the previous section. However, unfolding of instrumental effects is not always required. It may be sufficient to compare "raw" experimental data (i.e., not unfolded) with model calculations that account for instrumental effects. It is in fact much simpler to run model predictions through an experimental filter than to unfold experimental data. All that is needed is a sufficiently detailed model of the experimental response. In nuclear and particle physics, this is best accomplished with the GEANT package and a performance simulator specific to the apparatus. The idea is to simulate collisions and the production of particles according to the model of interest. The GEANT package is used to simulate the propagation of produced particles through the apparatus. It is possible to account for particle interactions within the detector, decays, and other forms of backgrounds as well as a detailed response of the apparatus. It is even possible to simulate varying or faulty performance of the instruments according to a detailed history of the performance of the apparatus. Model data filtered according to such a detailed simulation of the detector performance can then be meaningfully compared to experimental data. It is therefore possible to establish whether the model properly account for the observed data.

Comparisons of filtered theoretical models to experimental data are relatively common and quite useful in challenging the predictions of such theoretical models. The problem, however, is that in the absence of unfolded data, all theoretical models need to be propagated through the same experimental filter which accounts for all specificities and idiosyncrasies of the experiment. While it is in principle possible to publish a detailed description of the filter or even a computer code that replicates the filter, it is not always possible, in practice, because the filter may be too complex, or too fastidious to replicate and use. The complexity of the problem is further exacerbated if the experimental setup, particularly the trigger, is evolving through time. It must then be made clear which dataset and results correspond to which filter. It is also the case that the burden of proof, so to speak, is transferred to all scientists proposing a new model of the experimental data. It makes the comparison between models far less transparent and more prone to misrepresentations. How indeed can theorists be sure they are using the right experimental filter? This seems hardly fair. As a result, scientists tend to think that it is far more appropriate for experimenters to correct, that is, unfold their data for instrumental effects. It is also the case that the filtering of model predictions can grossly suppress features of interest and intuitive or explicit judgment of the validity of a model can be severely compromised.

Finally, and perhaps most importantly, detectors typically feature drastically different acceptance, resolution, and overall responses. Their effect on a measured observable may then be quite different. Any comparisons of uncorrected observables are consequently not possible or meaningful. Most scientists thus tend to regard the use of experimental filters to treat theoretical model predictions for comparison with experimental data as a last resort to be used exclusively when unfolding is too difficult or fraught with too many uncertainties.

12.3.4 Bin-by-Bin Method

The bin-by-bin method is commonly used in particle and nuclear physics to correct measurements of cross section and various other observables. The method is based on the assumption that bin-to-bin sharing is modest or negligible and that one can obtain a reasonable correction of measured data using a single multiplicative correction factor for each bin as follows:

$$\mu_i = C_i \left(n_i - \beta_i \right), \tag{12.48}$$

where β_i are background estimates for bin $i = 1, \ldots, m$, and C_i are coefficients designed to effect an appropriate correction of the measured distribution n_i to obtain a corrected distribution μ_i.

The coefficients C_i can generally be determined from MC simulations according to

$$C_i = \frac{\mu_i^{\mathrm{MC}}}{\nu_i^{\mathrm{MC}}}, \tag{12.49}$$

where μ_i^{MC} represents the true signal generated by the MC model and ν_i^{MC} is the simulated experimental spectrum accounting for all relevant instrumental effects. Given sufficient computing power and resources, the coefficients C_i can in principle be determined with arbitrarily small statistical errors. If computing resources are scarce and statistical errors large, the usual methods of error propagation may be used to determine statistical errors on the corrected spectrum μ_i. However, it should be noted that strong correlations exist, by construction, between μ_i^{MC} and ν_i^{MC}. Care must thus be exercised in the estimation of errors on the coefficients C_i. A sensible determination of the errors may be accomplished with the unified approach discussed in §6.2.2 (frequentist approach) or by calculation of a credible region (Bayesian approach) based on the posterior probability $p(\mu_i^{MC}|\nu_i^{MC})$ as discussed in §7.2.4.

The bin-by-bin method is obviously much simpler in application than the unfolding procedure outlined in §12.3.1. But is it precise enough or susceptible of skewing the corrected distribution μ_i? The answer to this question, of course, depends on the response function R defined in §12.3.1. If R is essentially diagonal, one can write

$$R_{ij} \approx \delta_{ij}\varepsilon_j. \tag{12.50}$$

In this context, the background subtracted signal is simply

$$\nu_i^{\text{signal}} = n_i - \beta_i = \varepsilon_i\mu_i. \tag{12.51}$$

The coefficients C_i are consequently the multiplicative inverse of the efficiencies ε_i. The bin-by-bin method thus amounts to an efficiency correction only. But when R is nondiagonal, the factors $1/C_i$ may become arbitrarily large, even larger than unity. The estimator $\hat{\mu}_i$ defined by (12.48) may thus be arbitrarily biased. Indeed, the expectation value of $\hat{\mu}_i$ is given by

$$\text{E}[\hat{\mu}_i] = C_i\text{E}[n_i - \beta_i] = C_i(\nu_i - \beta_i). \tag{12.52}$$

Inserting the values given by Eq. (12.49) for C_i, one finds

$$\text{E}[\hat{\mu}_i] = \frac{\mu_i^{MC}}{\nu_i^{MC}}\nu_i^{\text{sig}} \tag{12.53}$$

$$= \left(\frac{\mu_i^{MC}}{\nu_i^{MC}} - \frac{\mu_i}{\nu_i^{\text{sig}}}\right)\nu_i^{\text{sig}} + \mu_i, \tag{12.54}$$

in which the first term represents the estimator's bias. The covariance matrix U_{ij} of the corrected signals is evaluated in a similar fashion (see Problem 12.5).

$$U_{ij} = \text{Cov}\left[\hat{\mu}_i, \hat{\mu}_j\right] = C_i^2\text{Cov}\left[n_i, n_j\right], \tag{12.55}$$

$$= C_i^2\nu_i^{\text{sig}}\delta_{ij}, \tag{12.56}$$

where, in the second line, we assumed the measured n_i are uncorrelated and determined by Poisson statistics.

The bias term B_i may be written

$$B_i = \left(\frac{1}{\varepsilon_i^{MC}} - \frac{1}{\varepsilon_i^{\text{true}}}\right)\nu_i^{\text{sig}}. \tag{12.57}$$

It is null only if the detection efficiencies determined by MC simulation, $\varepsilon_i^{\text{MC}}$, are identically equal to the true (unknown) efficiency, $\varepsilon_i^{\text{true}}$. However, this may be strictly possible only if the model properly reproduces the data perfectly and the simulation accounts for all instrumental effects appropriately. Since the MC model used to simulate the data is unlikely, in general, to perfectly reproduce the true signal, the estimator is by construction biased, and the bin-by-bin correction method shall lead to finite systematic errors. To the extent that the simulation provides a proper account of experimental effects, though, it is possible to use the bin-by-bin method iteratively to improve the correction coefficients. First, an arbitrary MC model is used to generate a first-order correction of the measured data. Then the corrected spectrum is used to produce a new set of MC data and evaluate tuned correction factors C_i', which can be used to produce a new updated corrected spectrum. The convergence of this iterative method can be tested with a **closure test** (see §12.4.7).

12.3.5 Correcting for Finite Bin Widths

Experimental distributions μ_i corrected for efficiency are commonly obtained with histograms of finite bin widths and must consequently be also be corrected for averaging effects through the bins. Recall from §12.3.1 that the values μ_i correspond to integrals (12.21) of the true distribution $f_t(x)$:

$$\mu_j = \mu_{\text{tot}} \times p_j = \mu_{\text{tot}} \int_{\text{bin } j} dx_t f_t(x_t).$$

The value of the function $f_t(x)$ may thus be simply evaluated in each bin j as

$$f_t(x_j) = \frac{\mu_j}{\Delta x_j \mu_{\text{tot}}}, \tag{12.58}$$

where $\Delta x_j = x_{\max,j} - x_{\min,j}$ corresponds to the width of bin j. It might be tempting to report this function by plotting the values $f_t(x_j)$ for each bin center position, $x_c = (x_{\max,j} + x_{\min,j})/2$. This would constitute a formally valid plot only if the function $f_t(x_j)$ varies linearly through all bins, however. Indeed, remember from our discussion in §4.6 that one can write

$$\int_{x_{\min,i}}^{x_{\max,i}} f(x)\,dx = f(x_i)\left(x_{\max,i} - x_{\min,i}\right) = f(x_{o,i})\Delta x_i,$$

where $x_{o,i}$ is a value of x within the interval $[x_{\min,i}, x_{\max,i}]$ but not the center of the bin in general. It is straightforward to show that the value $x_{o,i}$ corresponds to the center of the bin i only if the function $f(x_i)$ varies linearly though the bin. If $f(x)$ is nonlinear, one must use the dependence of the function in the vicinity of the bin i to estimate the appropriate value $x_{o,i}$. This can be accomplished by fitting a suitable model function $f_m(x; \vec{a})$ to the spectrum to obtain a parameterization $\vec{a}$ that enables calculation of the preceding integral for each bin:

$$I_i = \int_{x_{\min,i}}^{x_{\max,i}} f_m(x;\vec{a})\,dx = f_m(x_{o,i})\Delta x_i.$$

The values $x_{o,i}$ are then obtained by inversion of the function f_m, as follows:

$$x_{o,i} = f_m^{-1}(I_i/\Delta x_i). \tag{12.59}$$

As an example of application of this method, consider the determination of transverse momentum values $p_{T,i}$ that should be used to properly represent a transverse momentum spectrum $\frac{1}{p_T}\frac{dN}{dp_T}$. Since transverse momentum spectra measured in high-energy collisions are typically steeply falling, one can use a power law function

$$f_m(p_T) = A\left(1 + \frac{p_T}{B}\right)^{-n} \tag{12.60}$$

to fit a spectrum $\frac{1}{p_T}\frac{dN}{dp_T}$. One writes

$$\frac{dN}{dp_T} = p_T f(_m(p_T) \tag{12.61}$$

$$I_i = \int_{\text{bin}-\text{i}} \frac{dN}{dp_T} dp_T = f(p_{T,i})\Delta p_i \tag{12.62}$$

$$= \int_{\text{bin}-\text{i}} p_T f_m(p_T)\,dp_T = \langle p_T\rangle_i \int_{\text{bin}-\text{i}} f_m(p_T)\,dp_T \tag{12.63}$$

where $\langle p_T\rangle_i$ is the mean p_T in bin i. Once the fit parameters A and B are known, one can thus get the values $p_{T,i}$ according to

$$p_{T,i} = \frac{\int_{\text{bin}-\text{i}} p_T f(p_T)\,dp_T}{\int_{\text{bin},i} f(p_T)\,dp_T}. \tag{12.64}$$

12.3.6 Unfolding with Regularization Methods

As we argued in §12.3.3, it is usually deemed preferable to correct raw data for instrumental effects rather than apply experimental filters on theoretical predictions to test theoretical models, and it is essential for proper comparison of results from different experiments. Since the bin-by-bin method is applicable only when the response matrix R_{ij} is diagonal or near diagonal, one is compelled to tackle the unfolding problem directly. Alas, a simple inversion of the response matrix typically leads to disastrous results: inverted spectra are numerically unstable and tend to feature wild oscillatory structures.

Meaningful unfolding is nonetheless possible. The general idea is to introduce a small bias in order to suppress fluctuations in the inverted instrumental response. This is accomplished by artificially imposing a smoothness constraint on the inverted solution using a smoothness function, $S(\vec{\mu})$, called the **regularization function**. The concept of regularization is rather general and is applied to machine learning and inverse problems. It involves the addition of possibly ad hoc information in order to solve an ill-posed problem or to prevent overfitting. It can be viewed in a Bayesian approach as imposing prior conditions on model parameters.

The use of a regularization function implies the obtained solution $\vec{\mu}$ will have a sub-optimal χ^2:

$$\chi^2 = \chi^2_{\min} + \Delta\chi^2, \tag{12.65}$$

or likelihood:

$$\log L(\vec{\mu}) \sim \log L_{\max} - \Delta \log L. \tag{12.66}$$

The key is to decide what deviation from the optimum, $\Delta\chi^2$ or $\Delta \log L$, can be meaningfully tolerated.

A wide range of regularization techniques and unfolding methods are documented in the scientific literature and commonly used in data analyses [13, 23, 48, 68, 91, 106]. A discussion of all these methods is well beyond the scope of this text. We focus our discussion on the singular value decomposition (SVD) method in §12.3.7 and the Bayesian unfolding method §12.3.8.

12.3.7 SVD Unfolding Method

We begin this section with a short summary of the notion of singular value decomposition in §12.3.7. An implementation of regularization in the context of the SVD method is presented in §§12.3.7 and 12.3.7. An example of application is presented in §12.3.7.

SVD of a Matrix

The SVD of a real $m \times n$ matrix $\mathbf{A}$ amounts to a factorization of the form

$$\mathbf{A} = \mathbf{U}\mathbf{S}\mathbf{V}^T, \tag{12.67}$$

where $\mathbf{U}$ and $\mathbf{V}$ are orthogonal matrices of size $m \times m$ and $n \times n$, respectively, while $\mathbf{S}$ is an $m \times n$ diagonal matrix with nonnegative diagonal elements. These conditions are expressed as follows

$$\mathbf{U}\mathbf{U}^T = \mathbf{U}^T\mathbf{U} = \mathbf{I}, \tag{12.68}$$

$$\mathbf{V}\mathbf{V}^T = \mathbf{V}^T\mathbf{V} = \mathbf{I}, \tag{12.69}$$

$$\mathbf{S}_{ij} = 0 \text{ for } i \neq j, \tag{12.70}$$

$$\mathbf{S}_{ii} = s_i \geq 0, \tag{12.71}$$

where the s_i are called **singular values** of the matrix $\mathbf{A}$, while the columns of $\mathbf{U}$ and $\mathbf{V}$ are known as the **left** and **right singular vectors**. One can verify that if $\mathbf{A}$ is itself orthogonal, all its singular values are equal to unity (see Problem 12.6), whereas a degenerate matrix features at least one null singular value.

As it happens, a singular value decomposition is very useful toward the solution of linear equation systems of the type encountered in the unfolding problem

$$\mathbf{A}\vec{x} = \vec{b}, \tag{12.72}$$

where $\vec{x}$ and $\vec{b}$ are $n \times 1$ and $m \times 1$ column vectors, respectively. With $m \geq n$, the folded distribution $\vec{b}$ has more bins than $\vec{x}$ and thus constitutes an over constrained system. Formally, the solution of Eq. (12.72) may be written

$$\vec{x} = \mathbf{A}^{-1}\vec{b} = \left(\mathbf{USV}^T\right)^{-1}\vec{b} = \mathbf{VS}^{-1}\mathbf{U}^T\vec{b}, \tag{12.73}$$

and is exact provided the matrix $\mathbf{A}$ and the measured vector $\vec{b}$ have no errors. In practice, one finds that finite errors imply that diagonal elements s_i may be small and very imprecise for values i exceeding some threshold k. Hocker and Kartvelishvili [106] have shown that these components of the decomposition lead to solutions with fast numerical oscillations which render the technique ineffective unless a proper regularization technique is applied. We outline, in the next section, the procedure they developed to apply such a regularization.

SVD with Regularization

The unfolding of the overdetermined linear equation system (12.72) may be viewed as a solution of the least-squares problem

$$\chi^2 = \sum_{i=1}^{m} \left(\sum_{j=1}^{n} A_{ij}x_j - b_i \right)^2, \tag{12.74}$$

and is appropriate if the equations are exact or if the errors in $\vec{b}$ are all equal. But this is usually not the case, and it is thus better to consider a weighted least-squares system of equations

$$\chi^2 = \sum_{i=1}^{m} \left(\frac{\sum_{j=1}^{n} A_{ij}x_j - b_i}{\Delta b_i} \right)^2, \tag{12.75}$$

where Δb_i represent errors on b_i. With the inclusion of errors Δb_i, all terms i contribute more or less equally. If the errors on b_i and b_j are correlated, one must introduce a covariance matrix $B_{ij} = \text{Cov}[b_i, b_j]$, and the preceding equation may be written in matrix form as

$$\chi^2 = \left(\mathbf{A}\vec{x} - \vec{b}\right)^T \mathbf{B}^{-1} \left(\mathbf{A}\vec{x} - \vec{b}\right). \tag{12.76}$$

The matrix $\mathbf{A}$ represents the probability that events in a bin i might end up in bin j. Since it is usually generated on the basis of Monte Carlo simulations (see §14.2.3), one typically ends up with a situation where rare i bins are not strongly populated and the corresponding coefficients A_{ij} may be artificially large (and even equal to unity) and rather imprecise thereby contributing a source of numerical oscillations. Hocker and Kartvelishvili [106] showed this problem can be partially mitigated by scaling the values A_{ij} and x_j

according to

$$A_{ij} \to A'_{ij} = \lambda_j A_{ij} \quad \text{for } i = 1, m, \tag{12.77}$$
$$x_j \to x'_j = x_j/\lambda_j,$$

since these scalings do not formally change Eq. (12.72), but may be chosen to produce a matrix $\mathbf{A}'$ where statistically strong terms are large and weaker terms are small. This is achieved by introducing $\omega_j = x_j/x_j^{\text{mc}}$ where x_j^{mc} are the MC yields used to generate the matrix $\mathbf{A}$. One then writes

$$\sum_{j=1}^{n} A'_{ij}\omega_j = b_i, \tag{12.78}$$

which must be solved for ω_j. At the end of the unfolding procedure, one shall simply multiply these by x_j^{mc} to obtain the desired solutions x_j.

$$x_j = \omega_j \times x_j^{\text{mc}} \quad \text{for } j = 1, \ldots, n. \tag{12.79}$$

Typically, experimentalists aim to use MC models that provide a reasonable representation of the data. The values ω_j are thus approximately close to unity and feature only small fluctuations, thus requiring fewer terms in the SVD decomposition. The use of the scaled A'_{ij} also permit the statistics to be better represented with large terms, where statistical accuracy is large, and smaller terms otherwise. One can then proceed to obtain a formal solution which includes both scaled A'_{ij} coefficients and errors on b_i.

In general, the covariance matrix $\mathbf{B}$ may not be diagonal. But given it should be symmetric and positive definite, its SVD may be written

$$\mathbf{B} = \mathbf{Q}\mathbf{R}\mathbf{Q}^T, \tag{12.80}$$

where $\mathbf{R}$ is a strictly diagonal matrix with $R_{ii} = r_i^2 \neq 0$. Its inverse $\mathbf{B}^{-1}$

$$\mathbf{B}^{-1} = \mathbf{Q}\mathbf{R}^{-1}\mathbf{Q}^T \tag{12.81}$$

can then be substituted in (12.76) with scaled $\mathbf{A}'$ and $\vec{\omega}$:

$$\chi^2 = \left(\mathbf{A}'\vec{\omega} - \vec{b}\right)^T \mathbf{Q}\mathbf{R}^{-1}\mathbf{Q}^T \left(\mathbf{A}'\vec{\omega} - \vec{b}\right). \tag{12.82}$$

Introducing

$$\tilde{A}_{ij} = \frac{1}{r_i}\sum_m Q_{im}A'_{mj}, \tag{12.83}$$
$$\tilde{b}_i = \frac{1}{r_i}\sum_m Q_{im}b_m,$$

one can write

$$\chi^2 = \left(\tilde{\mathbf{A}}\vec{\omega} - \tilde{b}\right)^T \left(\tilde{\mathbf{A}}\vec{\omega} - \tilde{b}\right). \tag{12.84}$$

Minimization of this χ^2 yields

$$\sum_j \tilde{A}_{ij}\omega_j = \tilde{b}_i, \tag{12.85}$$

where $\tilde{b}_i$ now has a covariance equal to the unit matrix $\mathbf{I}$ and all equations i have the same weight. The problem remains, unfortunately, that an exact solution of this system of equations still features a rapidly oscillating distribution. But, it is now in a form such that spurious oscillatory components can be suppressed by a judicious choice of regularization. This is achieved in practice by writing,

$$\left(\tilde{\mathbf{A}}\vec{\omega} - \tilde{b}\right)^T \left(\tilde{\mathbf{A}}\vec{\omega} - \tilde{b}\right) + \tau \left(\mathbf{C}\vec{\omega}\right)^T \left(\mathbf{C}\vec{\omega}\right) = \min . \tag{12.86}$$

The matrix $\mathbf{C}$ defines the a priori condition on the solution of the equations and enables the implementation of a suitable regularization controlled by the regularization parameter τ. Given the solution is expected to be "smooth," it is reasonable to seek to minimize the local curvature of the distribution. This may be written

$$\sum_i \left[(\omega_{i+1} - \omega_i) - (\omega_i - \omega_{i-1})\right]^2 . \tag{12.87}$$

One then writes $\mathbf{C}$ as

$$\mathbf{C} = \begin{pmatrix} -1 & 1 & 0 & 0 & \cdots & 0 \\ 1 & -2 & 1 & 0 & \cdots & 0 \\ 0 & 1 & -2 & 1 & \cdots & 0 \\ 0 & & \cdots & & & \\ 0 & & \cdots & 1 & -2 & 1 \\ 0 & & \cdots & 0 & 1 & -1 \end{pmatrix} . \tag{12.88}$$

Minimization of Eq. (12.86) then leads to overdetermined system of equations

$$\begin{bmatrix} \tilde{\mathbf{A}} \\ \sqrt{\tau}\mathbf{C} \end{bmatrix} \omega = \begin{bmatrix} \tilde{b} \\ 0 \end{bmatrix}, \tag{12.89}$$

which can be solved by SVD. To avoid solving iteratively for several values of τ, Hocker and Kartvelishvili [106] found Eq. (12.89) may be written

$$\begin{bmatrix} \tilde{\mathbf{A}}\mathbf{C}^{-1} \\ \sqrt{\tau}\mathbf{I} \end{bmatrix} \mathbf{C}\omega = \begin{bmatrix} \tilde{b} \\ 0 \end{bmatrix}, \tag{12.90}$$

and solved for $\tau = 0$ provided one modifies the matrix $\mathbf{C}$ such that its inverse exists. One also needs to assume that the value s_i can be sorted in a nonincreasing sequence by appropriately swapping pairs of singular values. This can be accomplished by adding a small value ξ of order 10^{-3} or 10^{-4} to all terms of the diagonal:

$$\mathbf{C} = \begin{pmatrix} -1+\xi & 1 & 0 & 0 & \cdots & 0 \\ 1 & -2+\xi & 1 & 0 & \cdots & 0 \\ 0 & 1 & -2+\xi & 1 & \cdots & 0 \\ 0 & & \cdots & & & \\ 0 & & \cdots & 1 & -2+\xi & 1 \\ 0 & & \cdots & 0 & 1 & -1+\xi \end{pmatrix} . \tag{12.91}$$

For $\tau = 0$, a decomposition of $\tilde{\mathbf{A}}\mathbf{C}^{-1}$ may be expressed as

$$\tilde{\mathbf{A}}\mathbf{C}^{-1} = \mathbf{U}\mathbf{S}\mathbf{V}^T . \tag{12.92}$$

Writing $\vec{d} = \mathbf{U}^T\tilde{b}$ and $\vec{z} = \mathbf{V}^T\mathbf{C}\vec{\omega}$, one then has

$$s_i \times z_i = d_i, \quad \text{for } i = 1, \ldots, n. \tag{12.93}$$

The orthogonality of $\mathbf{U}$ and the fact that the covariance matrix of $\tilde{b}$ is equal to unity insure that $\vec{d}$ also has unit covariance. Solution of (12.93) yields (for $\tau = 0$):

$$\begin{aligned} z_i^{(0)} &= \frac{d_i}{s_i}, \\ \vec{\omega}^{(0)} &= \mathbf{C}^{-1}\mathbf{V}\vec{z}^{(0)}. \end{aligned} \tag{12.94}$$

With $\tau = 0$, there is no regularization, so this solution is effectively useless. However, introducing finite (i.e., nonzero) values of τ, one can show that values d_i may be obtained according to

$$d_i^{(\tau)} = d_i \frac{s_i^2}{s_i^2 + \tau}, \tag{12.95}$$

which finally yield the sought for solution

$$\begin{aligned} z_i^{(\tau)} &= \frac{d_i s_i}{s_i^2 + \tau}, \\ \vec{\omega}^{(\tau)} &= \mathbf{C}^{-1}\mathbf{V}\vec{z}^{(\tau)}. \end{aligned} \tag{12.96}$$

A finite value of τ thus avoids numerical instabilities by effectively providing a cutoff of high-frequency terms (fast oscillations) of the decomposition.

The covariance matrices $\mathbf{Z}$ and $\mathbf{W}$ are found to be

$$\mathbf{Z}_{ik}^{(\tau)} = \frac{s_i^2}{\left(s_i^2 + \tau\right)^2}\delta_{ik}, \tag{12.97}$$

$$\mathbf{W}^{(\tau)} = \mathbf{W}^{-1}\mathbf{V}\mathbf{Z}^{(\tau)}\mathbf{C}^{-1}, \tag{12.98}$$

and the solution of the inverse problem thus become

$$x_i^{(\tau)} = x_i^{\text{mc}}\omega_i^{(\tau)}, \tag{12.99}$$

$$X_{ik}^{(\tau)} = x_i^{\text{mc}}\mathbf{W}_{ik}^{(\tau)}x_k^{\text{mc}}. \tag{12.100}$$

Choice of the Regularization Parameter τ and Errors

Plotting the values d_i (or $\ln |d_i|$) vs. i, one finds, typically, that the coefficients decrease with increasing i, until a value $i = k$ beyond which they oscillate or remain more or less constant because they are not statistically well determined. The critical value $i = k$ beyond which the coefficients d_i amount to Gaussian noise, can thus be used to set the regularization parameter

$$\tau = s_k^2. \tag{12.101}$$

The coefficients $s_i^2/(s_i^2 + \tau)$ in (12.95) suppress contributions $i > k$, and one obtains solutions with only modest fluctuations.

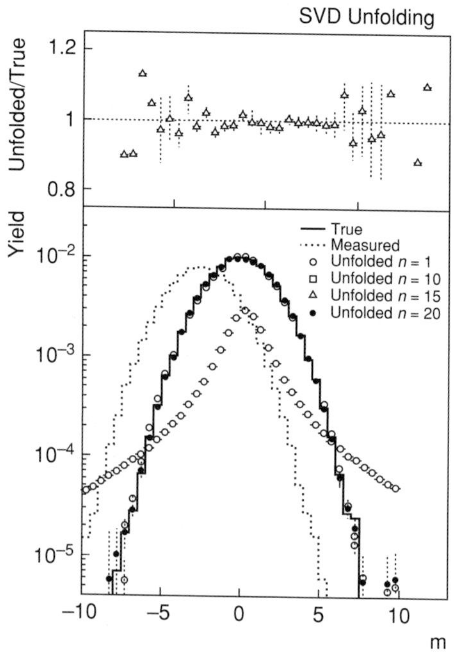

Fig. 12.5 Example of application of the SVD unfolding method. (Bottom panel) Simulated true (Gaussian) and measured distributions obtained with Gaussian smearing as described in the text; open and closed symbols represent unfolded distributions obtained after 1, 10, 15, and 20 applications of the SVD unfolding procedures. (Top panel) Ratio of the unfolded distribution obtained with 15 iterations to the true distribution. (Figure courtesy of S. Prasad.)

An alternative and effective technique to determine the optimal regularization parameter τ consists in generating a test spectrum $\vec{x}^{\text{test}}$ close but statistically distinct from the spectrum $\vec{x}^{\text{mc}}$ used to generate the unfolding matrix $\mathbf{A}$. One can then apply the folding and unfolding procedure onto $\vec{x}^{\text{mc}}$ and determine which value of τ yields a minimum χ^2, that is, such that the difference between the original test spectrum and its unfolded value best agree.

Hocker and Kartvelishvili [106] provide a detailed algorithm of the procedure as well as detailed examples. Functions to calculate SVD are available in a variety of software packages, including ROOT [59] with the class TDecompSVD.

Example of SVD Unfolding

We present an example of SVD unfolding accomplished with the RooUnfold framework [13] for a measurement of particle multiplicity. The generation of the response matrix R_{ij} is discussed in §14.2.3. We assume, arbitrarily, that the parent distribution has a Breit–Wigner shape with a mean of $\mu_t = 0.3$ and a width $\Gamma = 2.5$, while the response is described with a Gaussian distribution with a mean -2.5 and a standard deviation of 0.2. The true multiplicity distribution $P(m_t)$, shown with a solid line in Figure 12.5, is based on 100,000 events sampled from a Gaussian distribution centered at $m_t = 0$ and a standard deviation of 2.0. It was sampled and smeared with the preceding response function to generate the measured distribution, represented in Figure 12.5 with a dotted curve.

The SVD algorithm was used iteratively on the measured distribution, with the simulated response curve, to obtain the unfolded distributions shown in Figure 12.5 with open and closed points. One finds that a simple application of the unfolding method yields a distribution centered at the proper multiplicity but with a shape heavily reminiscent of the Breit-Wigner distribution used in the generation of the response matrix. However, the unfolded distribution obtained with 15 iterations of the procedure reproduces the true distribution extremely well and yields a ratio of the unfolded to true distribution which is within statistical errors, consistent with unity (top panel of the figure). In fact, a detailed study shows that the method yields numerically stable results, with unfolded to true distribution ratio equal to unity, within statistical errors, from 10 to 20 iterations. The SVD method is thus producing unfolding distributions that are robust and rather insensitive to the number of iterations, once convergence is achieved.

12.3.8 Bayesian Unfolding

As discussed in details in Chapter 7, **Bayesian inference** encompasses a wide range of techniques to learn about physical quantities (or scientific hypotheses) based on experimental data. Formally, it first requires the construction of deterministic and probabilistic models of the observable of interest and its dependencies on measurements and other assumed knowledge. In essence, this entails the determination of the probability of an event A might follow from an event B, noted $P(A|B)$, and the converse $P(B|A)$. In the context of an unfolding problem, B might represent the true spectrum whereas A would correspond to the measured data. The task is then to use a model of the measurement $P(A|B)$ based on some prior hypothesis for B, and rely on Bayes' theorem to inverse the problem and use the measured A to determine the true B. The technique is known as **Bayesian unfolding**, of which we present a brief outline largely following the seminal work of G. D'Agostini [73].

Inference models may be represented graphically as networks, called **Bayesian networks**, or **belief networks** [73]. Figure 12.6 illustrates a fit model used to establish the value of a parameter θ, which determines the relation between a dependent variable y and an independent variable x. The network explicitly identifies the components of the problem, their relations, and the assumptions used in the solution. The unfolding problem requires a slightly more elaborate network we introduce in the next section.

Bayesian Network

A Bayesian network can be regarded as a set of connections between nodes representing causes and effects. In the context of the unfolding problem, the causes correspond to the true values μ_i sought for, whereas the effects are the measured values n_i determined by the measurement protocol, the performance of the instruments (uncertainties and errors), and the actual values of the observables. A simple cause-and-effect network, sufficient for the description of the unfolding problem, is illustrated in Figure 12.7. The n nodes C_i represent the causes μ_i to be determined by unfolding whereas the m nodes E_j are effect

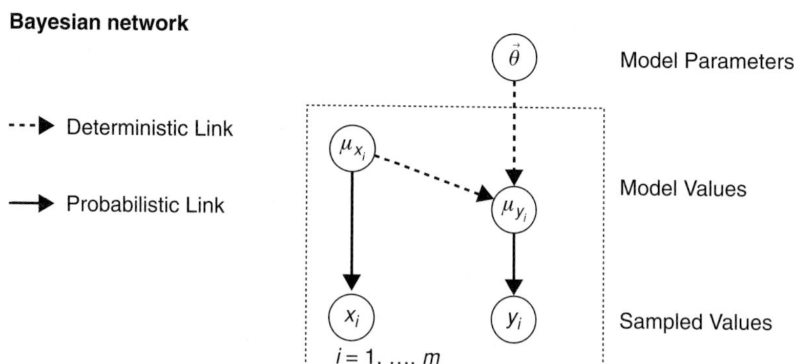

Fig. 12.6 Belief network used for the formulation of a model and fit of an experimental data distribution. The model parameters $\vec{\theta}$ are believed to lead to dependent values μ_{y_i} given independent values μ_{x_i} but the stochastic nature of the measurement process yields sampled values x_i and y_i for the independent and dependent values, respectively. One could enrich the network by including explicit assumptions concerning correlations between the variables and/or the width and bias of the fluctuations. (Adapted with permission from G. D'Agostini. Fits, and especially linear fits, with errors on both axes, extra variance of the data points and other complications. *ArXiv Physics e-prints*, November 2005.)

nodes, corresponding to measured n_j. The trash node T is used to represent detection inefficiencies: a certain fraction of each cause node C_i may remain undetected and ends up in the trash node.

The goal of unfolding is to find values $\vec{\mu} = (\mu_1(C_1), \ldots, \mu_n(C_n))$, given a set of measured effects $\vec{n} = (n_1(E_1), \ldots, n_m(E_m))$. The cause-to-effect links illustrated in Figure 12.7 have a probabilistic nature. Each cause C_i has a probability $P(E_j|C_i)$ of leading to an effect E_j. The inverse problem is thus statistical and probabilistic in nature also. Effectively, one must evaluate the probability $P(\vec{\mu}|\vec{n}, R, I)$ of the causes to be given by specific values $\vec{\mu}$ based on measured values $\vec{n}$, a response function R, and possibly some other system information I. As in previous sections of this chapter, the quantities n_j represent the number of

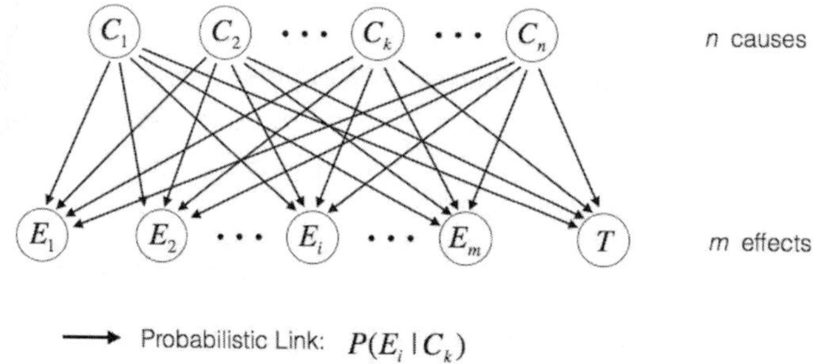

Fig. 12.7 Cause–effect network. The probabilistic links indicate the probability $P(E_i|C_k)$ a cause C_k may lead to an effect E_i. T is a trash node corresponding to "unmeasured events." (Adapted with permission from G. D'Agostini. Improved iterative Bayesian unfolding. *ArXiv e-prints*, October 2010.)

"events" measured in bin j of the observable of interest x_m, while μ_i stands for the number of events of type C_i. By construction, R expresses the probability of observing a certain effect E_j given a specific cause C_i:

$$R_{ji} \equiv P(E_j|C_i, I) \tag{12.102}$$

where I once again represents what is known about the system (measurement protocol, instrument, etc.). The conditional probabilities are usually determined from MC simulations and consequently subjected to finite statistical uncertainties of their own. Given the probabilistic nature of the measurement process, the unfolding process is equally probabilistic, not deterministic. One can thus argue that the unfolding problem is a priori not amenable to a deterministic operation such as a matrix inversion, which also fails, as we saw, for purely technical reasons. This implies the need for a probabilistic solution of the problem instead of a simple matrix inversion, which effectively corresponds to a deterministic "rotation" in a space of n-dimensions.

One finds that the inversion "works" in the large number of events limit, such that stochastic effects are negligible and the observations $\vec{n}$ correspond to their expectation values ν. Indeed, the expectation values of n_j given the cause C_i can be written

$$\mathrm{E}\left[n_j|\mu_i\right] = P(E_j|C_i, I)\mu_i = R_{ji}\mu_i. \tag{12.103}$$

The expectation value of n_j given a set of causes $\vec{\mu}$ is thus

$$\mathrm{E}\left[n_j|\vec{\mu}\right] = \sum_{i=1}^{n} R_{ji}\mu_i, \tag{12.104}$$

or in matrix notation

$$\vec{n} = R\vec{\mu}, \tag{12.105}$$

which yields the known inverse problem

$$\mathrm{E}\left[\vec{\mu}\right] = R^{-1}\mathrm{E}\left[\vec{n}\right]. \tag{12.106}$$

where we explicitly include expectation values to emphasize that this result holds exclusively in the large number limit, that is, for the expectation values of $\vec{n}$, and not necessarily for a specific instance or measurement. Bayesian purists argue that the idea of inverting R is logically flawed because it is strictly valid for the expectation value of the measurements $\vec{n}$ not the actual measurements. However, as we saw, if the background is well known, then the preceding expression does yield, at least in principle, an unbiased estimator of $\vec{\mu}$.

Practical Considerations

By virtue of Bayes' theorem (Eq. 2.17) and the law of total probability (Eq. 2.19), one can write

$$P(\vec{\mu}|\vec{n}, R, I) = \frac{P(\vec{n}|\vec{\mu}, R, I)P(\vec{\mu}|I)}{\sum_{\vec{\mu}'} P(\vec{n}|\vec{\mu}', R, I)P(\vec{\mu}'|I)}, \tag{12.107}$$

where the sum, in the denominator, is taken over all possible sets of events $\vec{\mu}'$ and essentially amounts to a normalization factor. The interesting components of the theorem reside in its numerator. The probability $P(\vec{n}|\vec{\mu}, R, I)$ is known as the likelihood of $\vec{n}$ given $\vec{\mu}$, the experimental response, and any other prior information I available about the system. It expresses the probability of obtaining a specific set of yields $\vec{n}$ given a set of values $\vec{\mu}$. The probabilities $P(\vec{\mu}|I)$ and $P(\vec{\mu}|\vec{n}, R, I)$ are known as **prior**, and **posterior**, respectively (For a detailed discussion of these concepts, see §§7.2.2 and 7.2.3). The prior expresses the probability of values $\vec{\mu}$ before a measurement is conducted (or perhaps based on prior measurements), while the posterior gives the probability of the values $\vec{\mu}$ after a measurement of $\vec{n}$ has been completed. Effectively, the preceding expression provides updated knowledge of the probability of a certain set of values $\vec{\mu}$ given a measurement $\vec{n}$. It is important to note that, nominally, the law of total probability does not yield a specific value $\vec{\mu}$ but rather the probability (or density of probability) of a specific set of values $\vec{\mu}$.

Clearly, the presence of a prior is logically required to obtain a posterior from the probability inversion provided by the law of total probability. But what prior should one use, since the choice should affect the outcome of the inversion? An obvious positive aspect of the notion of prior is that it enables one to insert in the solution of the inverse problem any knowledge that might be available ab initio about the system before the measurement is even attempted. That said, prior information is typically rather vague and may have a negligible impact on the inference process which then ends up being dominated by the likelihood $P(\vec{n}|\vec{\mu}, R, I)$. As we saw in §7.3, vagueness can be expressed by writing

$$P(\vec{\mu}|I) = \text{constant}. \tag{12.108}$$

The inference may then be performed according to the likelihood only:

$$P(\vec{\mu}|\vec{n}, R, I) \propto P(\vec{n}|\vec{\mu}, R, I). \tag{12.109}$$

The most probable spectrum $\vec{\mu}$ is thus the one that maximizes the likelihood in the absence of any contradictory information. The Bayesian approach effectively recovers the maximum likelihood estimators of the frequentist approach discussed in prior sections.

Given a full knowledge of the response function R, one could examine all values of $\vec{\mu}$, find their respective probabilities, obtain the most probable (or the expectation value), and consequently assess the uncertainty of the result. This is, however, not possible in practice with the usual knowledge one has of the likelihood function $P(\vec{n}|\vec{\mu}, R, I)$. To see why this is the case, consider the distribution of the cause events in bin C_i into all effect bins E_j. This can be described by a multinomial distribution

$$P(\vec{n}|\mu_i, R, I) = \frac{\mu_i!}{\prod_{k=1}^{m+1} n_k!} \prod_{j=1}^{m+1} R_{ji}^{\,n_j}. \tag{12.110}$$

The probability $P(\vec{n}|\vec{\mu}, R, I)$ is thus a sum of multinomial distributions – for which no closed formula exists.

The elements of R are typically obtained by MC simulations. A large number of "events" is to be generated for each bin μ_i and one must count where they end up in n_j bins. Intuitively, one can write

$$R_{ji} = \frac{n_j^{\rm MC}}{\mu_j^{\rm MC}}. \tag{12.111}$$

One should, however, be aware of the fact that finite statistics in the generation of MC events leads to statistical errors on the coefficients. Bayes' theorem may be applied toward the determination of the coefficients R_{ji}. Defining column vectors R_i as $\vec{R}_i = (R_{1i}, R_{1i}, \ldots, R_{n+1,i})$, one can write

$$P(R_i|\vec{n}^{\rm MC}, \mu_i, I) \propto P(\vec{n}^{\rm MC}|\mu_i, R_i, I) \times P(R_i|I), \tag{12.112}$$

which provides a technique to improve our knowledge of the coefficient R_{ji}. Since $P(\vec{n}|\mu_i, R, I)$ is a multinomial distribution, it is convenient to choose a prior distribution in the form of a Dirichlet distribution (defined in §3.8):

$$P(\vec{\mu}^{\rm prior}|I) \propto p_1^{\alpha_1-1} p_2^{\alpha_2-1} \cdots p_m^{\alpha_m-1}, \tag{12.113}$$

where the values p_i express the probabilities of each cause i and the values α_i are specific values of the actual yield in each cause bin i. The multinomial and Dirichlet distributions are conjugate distributions (as defined in §7.3.3). Indeed, the product of a multinomial distribution by a Dirichlet distribution is itself a Dirichlet distribution with new values α_i'. It is thus possible to recalculate the coefficients based on these new values α_i'. This may also come in handy when estimating errors on the coefficients.

Basic Algorithm for Bayesian Unfolding

Bayes' theorem can be applied to the cause-and-effects bins to infer the (unknown) value of the cause bins. However, instead of using Eq. (12.107), one writes

$$P(C_i|E_i, I) = \frac{P(E_j|C_i, I) \times P(C_i|I)}{\sum_{k=1}^{m} P(E_j|C_k, I) \times P(C_k|I)}. \tag{12.114}$$

The probabilities $P(E_j|C_i, I) = R_{ij}$ are the likelihood, given by Eq. (12.102), of causes C_i yielding events in effect bins E_j, while the probabilities $P(C_i|I)$ represent the prior knowledge of the probability of the cause bins C_i. The probabilities $P(C_i|E_i, I)$ thus represent the posteriors or updated knowledge about the probability of the cause bins C_i. It is convenient to define coefficients θ_{ij} according to

$$\theta_{ij} = P(C_i|E_i, I) = \frac{R_{ji} P(C_i|I)}{\sum_{k=1}^{m} R_{jk} P(C_k|I)}. \tag{12.115}$$

In this formulation, the coefficients $P(C_i|I)$ assign a specific (prior) probability to the cause bins C_i, not the whole set of cause bins $\vec{\mu}$. If no prior information is available, one can set all bins to be equally probable. Note that this is different from assuming that all spectra $\vec{\mu}$

are equally probable. Indeed, assuming a flat spectrum is far more restrictive than assuming that all spectra are equally probable. In practice, many implementations of the algorithm use the measured spectrum as the prior.

Given the probabilities θ_{ij}, one can share the counts (yields) of the measured bin E_j and estimate the causes according to

$$\mu_i|_{n_j} = P(C_i|E_j, I)n_j = \theta_{ij}n_j. \tag{12.116}$$

Summing over all observations, we get

$$\mu_i|_{\vec{n}} = \sum_{j=1}^{N} \theta_{ij}n_j. \tag{12.117}$$

We must also correct for efficiencies

$$\mu_i|_{\vec{n}} = \frac{1}{\varepsilon_i}\sum_{j=1}^{N} \theta_{ij}n_j, \tag{12.118}$$

where by definition

$$\varepsilon_i = \sum_{j=1}^{N} P(E_j|C_i, I) = \sum_{j=1}^{N} R_{ji} = 1 - R_{N+1,i}, \tag{12.119}$$

with coefficients R_{ij} estimated from MC studies with an appropriate simulation of the measurement according to

$$R_{ji} \approx \frac{n_j^{\rm MC}}{\mu_i^{\rm MC}}. \tag{12.120}$$

As already mentioned in the previous section, the accuracy of the coefficients R_{ij} is intrinsically limited by the statistical accuracy achieved in the determination of the yields $n_j^{\rm MC}$ and $\mu_i^{\rm MC}$. Since these two yields are correlated a priori by the MC generation process, it is prudent to avoid using the usual error propagation technique and instead make use of a sampling method.

By construction, the unfolded spectrum tends to "remember" the prior spectrum, and thus, it is convenient to iterate the preceding algorithm. Regularization and smoothing may also be added to the algorithm.

Bayesian Unfolding Example

Figure 12.8 displays an example of the application of Bayesian unfolding obtained with the response matrix, Figure 14.8, already used in §12.3.7 to illustrate SVD unfolding. Unfolded distributions are shown with open and closed symbols for 1, 3, 4, and 5 iterations of the algorithm. The top panel presents ratios of the unfolded distributions to the true distribution (solid line). One observes that the Bayesian method yields an unfolded distribution in excellent agreement with the true distribution, within statistical errors, and which is stable when repeated iteratively several times.

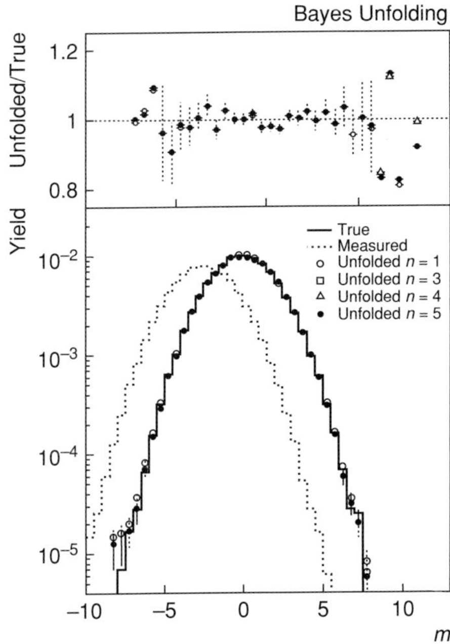

Fig. 12.8 Example of application of the Bayes unfolding method. (Bottom panel) Simulated true (Gaussian) and measured distributions obtained with Gaussian smearing as described in the text; open and closed symbols represent unfolded distributions obtained after 1, 3, 4, and 5 applications of the Bayesian unfolding procedures. Top panel: ratio of the unfolded to true distributions obtained with 4 and 5 iterations. (Figure courtesy of S. Prasad.)

12.4 Correcting Measurements of Correlation Functions

Integral and differential correlation functions are defined in terms of combinations of single-, two-particle, and more generally multiparticle yields (or cross sections). As such they are subjected to the same instrumental limitations and distortions as the yields on which they are based. The impact of instrumental effects may be exacerbated by the fact that the determination of cumulants involves subtraction of two or more terms of similar magnitude. So while instrumental effects on cross sections (or production yields) may be relatively modest, they can have disastrous consequences on correlation functions, most particularly when correlations are weak because of dilution effects associated with large numbers of correlation sources. Fortunately, correlation functions expressed in terms of ratios of cumulants to products of single-particle yields are approximately robust, that is, independent of detection efficiencies. They thus lend themselves to a variety of simple correction techniques that are relatively easy to implement.

We begin with a discussion of issues involved in the correction of measurement of variance, which leads us naturally into a discussion of the robustness of ratios of factorial moments and corrections of differential correlation functions. Irreducible instrumental effects associated with variations in the instrumental response are discussed in the latter part of this section.

12.4.1 Correction Of Integrated Yield Measurements

Measurements of integrated particle production yields over a finite kinematic range are of interest as estimators of collision hardness (e.g., in proton–proton interactions) or collision centrality (e.g., in heavy-ion collisions) and for measurements of yield fluctuations. Insofar as the kinematic volume of integration considered is much larger than the resolution of the kinematical parameters of interest, smearing effects may be neglected and one is chiefly concerned with correcting for detector efficiencies.

From a theoretical standpoint, the average integrated yield, $\langle N \rangle$, over a specific kinematic domain, Ω, is determined by the particle production cross section (see §10.2):

$$\langle N \rangle = \int_{\Omega} \frac{d^3 N}{dp^3} dp^3. \tag{12.121}$$

Experimentally, however, one observes that the produced number of particles fluctuates collision by collision owing to the stochastic nature of the particle production process. The fluctuations may be described by a probability function, $P_{\text{prod}}(N)$, determined by the dynamics and correlations involved in the particle production process. The average multiplicity is thus

$$\langle N \rangle = \int_0^\infty P_{\text{prod}}(N) N dN. \tag{12.122}$$

Measurements of particle production are usually subject to losses and a variety of other instrumental effects. For large detectors, one can usually assume that the probability of detecting one particle is independent of the probability of detecting others. One can then model the detection of a single-particle with a Bernoulli distribution (defined in §3.1). Let ε represent the probability to measure any produced particle in the kinematic domain Ω:

$$\begin{aligned} P_{\text{single}}(n|\varepsilon) &= 1 - \varepsilon \quad \text{probability of not observing}, \ n = 0 \\ &= \varepsilon \qquad\quad \text{probability of observing}, \ n = 1. \end{aligned} \tag{12.123}$$

As we saw in §3.1, the probability of obtaining n successes out of N trials determined by the same Bernoulli distribution amounts to a binomial distribution. We can then express the probability of simultaneously detecting n particles in the domain Ω as a binomial distribution with success probability ε:

$$P_{\text{det}}(n|N, \varepsilon) = \frac{N!}{n!(N-n)!} \varepsilon^N (1-\varepsilon)^{N-n}. \tag{12.124}$$

For fixed N, the distribution has a mean (see Table 3.1)

$$\langle n \rangle_{\text{N}} = \text{E}[n] = \int P_{\text{det}}(n|N, \varepsilon) n\, dn = \varepsilon N, \tag{12.125}$$

and a variance

$$\langle (n - \langle n \rangle)^2 \rangle_N = \int P_{\text{det}}(n|N, \varepsilon)\,(n - \langle n \rangle)^2\, dn \tag{12.126}$$
$$= N\varepsilon(1 - \varepsilon).$$

One must account, however, for fluctuations of N according to $P_{\text{prod}}(N)$. The probability of observing n particles may then be written

$$P_{\text{meas}}(n|\varepsilon) = \int dN P_{\text{det}}(n|N, \varepsilon) P_{\text{prod}}(N). \tag{12.127}$$

The mean measured multiplicity $\langle n \rangle$ can be calculated in terms of $\langle N \rangle$ even if $P_{\text{prod}}(N)$ is unknown:

$$\langle n \rangle = \int dn n P_{\text{meas}}(n|\varepsilon), \tag{12.128}$$

$$= \int dn n \int dN P_{\text{det}}(n|N, \varepsilon) P_{\text{prod}}(N). \tag{12.129}$$

Indeed, changing the order of integration, we get

$$\langle n \rangle = \int dN P_{\text{prod}}(N) \int dn n P_{\text{det}}(n|N, \varepsilon), \tag{12.130}$$

$$= \varepsilon \int dN P_{\text{prod}}(N) N, \tag{12.131}$$

$$= \varepsilon \langle N \rangle. \tag{12.132}$$

We thus formally obtain the intuitively obvious result that one can correct the measured average multiplicity by simply dividing by the efficiency of the measurement:

$$\langle N \rangle = \frac{\langle n \rangle}{\varepsilon}. \tag{12.133}$$

This result is correct even in cases where the efficiency varies over time. To demonstrate this, let us assume, for simplicity's sake, that an experiment can be divided into two time periods featuring particle detection efficiencies ε_1 and ε_2. Let us also assume that the probability of observing the events during the two time periods is unmodified by this change, and let us denote the number of events detected in the two periods as N_1^{ev} and N_2^{ev}.

The average multiplicities measured during the two periods are noted $\langle n_i \rangle$ with $i = 1, 2$. Given the definition of efficiency and the preceding result, one can write

$$\langle n_i \rangle = \varepsilon_i \langle N \rangle. \tag{12.134}$$

The average efficiency is calculated as a weighted average of the efficiencies of the two periods

$$\varepsilon_{\text{avg}} = \frac{N_1^{\text{ev}} \varepsilon_1 + N_2^{\text{ev}} \varepsilon_2}{N_1^{\text{ev}} + N_2^{\text{ev}}}. \tag{12.135}$$

The multiplicity measured across the two periods is:

$$\langle n \rangle = \varepsilon_{\text{avg}} \langle N \rangle. \tag{12.136}$$

Extraction of the true mean multiplicity $\langle N\rangle$ can thus be obtained for either time periods

$$\langle N\rangle = \frac{\langle n\rangle_1}{\varepsilon_1} = \frac{\langle n\rangle_2}{\varepsilon_2}, \tag{12.137}$$

or globally from the average of the two periods,

$$\langle N\rangle = \frac{\langle n\rangle}{\varepsilon_{\text{avg}}}, \tag{12.138}$$

since

$$\langle n\rangle_{\text{avg}} = \frac{N_1^{\text{ev}}\langle n\rangle_1 + N_2^{\text{ev}}\langle n\rangle_2}{N_1^{\text{ev}} + N_2^{\text{ev}}} \tag{12.139}$$

$$= \frac{N_1^{\text{ev}}\varepsilon_1\langle N\rangle + N_2^{\text{ev}}\varepsilon_2\langle N\rangle}{N_1^{\text{ev}} + N_2^{\text{ev}}} \tag{12.140}$$

$$= \frac{N_1^{\text{ev}}\varepsilon_1 + N_2^{\text{ev}}\varepsilon_2}{N_1^{\text{ev}} + N_2^{\text{ev}}}\langle N\rangle \tag{12.141}$$

$$= \varepsilon_{\text{avg}}\langle N\rangle. \tag{12.142}$$

This conclusion can clearly be generalized to multiple time periods when the detection efficiency might have taken different values. For measurements of average particle production yields, it thus does not matter that the experimental response changes over time as long as one can track these changes and estimate the detection efficiency during each period independently or globally for the entire data-taking run. This means that insofar as it possible to obtain an average efficiency, it is not necessary to carry the analysis of each time period separately; and one can use Eq. (12.138) to determine the average produced particle yield. However, we will see in the next section that such simple treatment is not warranted for measurements of fluctuations and correlation functions.

The detection efficiency is rarely uniform across the acceptance of a measurement. One must thus also consider whether the average efficiency correction procedure outlined earlier is robust when the detection efficiency is a function of kinematical variables. For simplicity's sake, let us partition the measurement acceptance into two parts of size Ω_1 and Ω_2 with respective efficiencies ε_1 and ε_2. The average multiplicity measured in these two regions may be written

$$\langle n_i\rangle = \varepsilon_i\langle N_i\rangle, \tag{12.143}$$

where

$$\langle N_i\rangle = \int_{\Omega_i} \frac{d^3N}{dp^3}dp^3, \tag{12.144}$$

with the full kinematical range Ω corresponding to

$$\Omega = \sum_{i=1}^{2}\Omega_i. \tag{12.145}$$

The average number of produced particles can be properly determined by summing corrected yields in parts 1 and 2 individually:

$$\langle N \rangle = \sum_{i=1}^{2} \langle N_i \rangle = \sum_{i=1}^{2} \frac{\langle n_i \rangle}{\varepsilon_i}. \tag{12.146}$$

If the fractions $f_i = \langle N_i \rangle / \langle N \rangle$ of the total yield produced in the two parts of the acceptance are known a priori, one can write an average efficiency (as in the case of the time-varying efficiency discussed earlier):

$$\varepsilon_{\text{avg}} = \frac{f_1 \varepsilon_1 + f_2 \varepsilon_2}{f_1 + f_2} = f_1 \varepsilon_1 + f_2 \varepsilon_2, \tag{12.147}$$

since $f_1 + f_2 = 1$ by definition. Unfortunately, the fractions f_i are in general not known a priori, and it is thus not possible to formally define a model independent average efficiency across the full acceptance Ω. However, in cases where the production cross section is nearly constant within the experimental acceptance, one can write

$$f_i = \frac{\int_{\Omega_i} \frac{d^3N}{dp^3} dp^3}{\int_{\Omega} \frac{d^3N}{dp^3} dp^3} \approx \frac{\int_{\Omega_i} dp^3}{\int_{\Omega} dp^3} = \frac{\Omega_i}{\Omega}, \tag{12.148}$$

which satisfies $\sum_i f_i = 1$. The size of the fractions f_i is then fixed by the relative sizes of the Ω_i, and a model independent average efficiency can be formulated. The efficiency correction, defined by Eq. (12.138), is thus applicable in spite of the fact the efficiency may vary through the acceptance of the measurement.

12.4.2 The Unfriendly Variance

Studies of the variance (as well as higher moments) of physical observables are often of interest to probe the dynamics of physical systems. We have discussed in §11.3.3, for instance, that the study of fluctuations of the net charge of produced particles or ratios of species integrated yields are particularly useful to probe the collision dynamics of large nuclei at relativistic energies. Unfortunately, while the variance of an observable is in principle a good measure of its fluctuations, correcting a measurement of variance for instrumental effects is anything but trivial. For illustrative purposes, let us continue the discussion initiated in the previous section and examine how the detection efficiency affects the variance of the measured integrated yield within a specific acceptance Ω.

We use Eq. (12.127) to calculate the variance of the measured multiplicity:

$$\text{Var}[n] = \langle (n - \langle n \rangle)^2 \rangle = \langle n^2 \rangle - \langle n \rangle^2. \tag{12.149}$$

The mean $\langle n \rangle$ is known from Eq. (12.130). The calculation of the second moment $\langle n^2 \rangle$ proceeds as in (12.130):

$$\begin{aligned} \langle n^2 \rangle &= \int dN P_{\text{prod}}(N) \int dn n^2 P_{\text{det}}(n|N, \varepsilon), \\ &= \int dN P_{\text{prod}}(N) N\varepsilon(1 - \varepsilon + N\varepsilon), \\ &= \varepsilon(1 - \varepsilon)\langle N \rangle + \varepsilon^2 \langle N^2 \rangle, \end{aligned} \tag{12.150}$$

where in the second line we used the second moment of the binomial distribution (see Table 3.1) and in the third line the definition of the average $\langle N \rangle$ and $\langle N^2 \rangle$. The variance of the measured distribution is thus

$$\text{Var}[n] = \varepsilon^2 \text{Var}[N] + \varepsilon(1 - \varepsilon)\langle N \rangle. \tag{12.151}$$

This expression contains terms linear and quadratic in ε. Correcting the measured variance of the multiplicity thus cannot be achieved by dividing the measured variance by the square of the efficiency as one might intuitively be inclined to do. Since one is usually interested in detecting small deviations of the variance from values expected for a Poisson distribution, it is thus imperative that the efficiency ε be known very accurately. Unfortunately, achieving the required level of efficiency is often rather challenging, and it may not be possible to reliably correct the measured variance.

All is not lost, however, since two alternative approaches are possible. The first approach involves a measurement of factorial moments while the second is based on a comparison of the measured variance with that obtained with mixed events (defined in §12.4.5). Both approaches are commonly used in practical applications.

The variance of integrated particle yields is determined by the production dynamics of the particle and the extent to which the produced particles are correlated. It can thus be regarded as having a purely "statistical" or Poisson component, σ^2_{stat}, and a "dynamical" component, $\Delta\sigma^2_{\text{dyn}}$. Instrumental effects may unfortunately modify the size of the variance and introduce a "shift" $\Delta\sigma^2_{\text{inst}}$. The measured variance may thus be written

$$\sigma^2_{\text{meas}} = \sigma^2_{\text{stat}} + \Delta\sigma^2_{\text{dyn}} + \Delta\sigma^2_{\text{inst}}. \tag{12.152}$$

The Poisson component is trivial and of limited interest. Measurements thus typically aim at the identification of the dynamical component. Unfortunately, this component is often much smaller than the size of the correction implied by Eq. (12.151). A precise assessment of the efficiency correction is therefore essential to achieve a meaningful measurement.

Rather than attempting a precise evaluation of the efficiency and other instrumental effects based on some simulation model of an experiment, it is possible to use the data itself to estimate the efficiency based on the mixed-event technique. As its name suggests, this technique, consists in synthesizing artificial events based on actual data by mixing particles from different actual events. Particles produced in different events are de facto noncorrelated by the production process but bear the effects of the instrumentation. The variance or correlation function of mixed events should thus be determined solely by the stochastic nature of the production process and instrumental effects. Since particles from different events are uncorrelated, the variance of integrated yields should be determined by Poisson statistics as well as correlation induced by detector effects. Under ideal conditions (discussed later in this chapter) one can hypothesize that the instrumental effects are essentially identical for real and mixed events. One can thus write

$$\sigma^2_{\text{mixed}} = \sigma^2_{\text{stat}} + \Delta\sigma^2_{\text{inst}}. \tag{12.153}$$

Subtraction of the variance of mixed events from that of real events therefore enables, in principle, the extraction of the dynamical component of the variance:

$$\Delta\sigma^2_{\text{dyn}} = \sigma^2_{\text{meas}} - \sigma^2_{\text{mixed}}. \tag{12.154}$$

It is important to realize that the dynamical correlations may either increase or decrease the variance of the integrated yield. The dynamical component $\Delta\sigma^2_{\text{dyn}}$ may be positive or negative, respectively. For negative values of $\Delta\sigma^2_{\text{dyn}}$, it is obviously not meaningful to extract a square root and quantify dynamical effects as an imaginary value. The fact of the matter is that the value $\Delta\sigma_{\text{dyn}}$, that is, without the square, is not particularly significant. This shift is really the result of correlation effects involving two or more particles. It can indeed be positive, negative, or even null. The temptation to extract a square root is thus simply an artifact of the poor historical choice of notation used to denote dynamical fluctuations.

Correcting for detector effects and the determination of the dynamical correlation may also be carried out based on ratios of factorial moments and corrected measurements of integrated multiplicity. The variance of the measured integrated yield, $\langle (n - \langle n\rangle)^2\rangle$, may be written

$$\langle (n - \langle n\rangle)^2\rangle = \langle n^2\rangle - \langle n\rangle^2. \tag{12.155}$$

Inserting $-\langle n\rangle + \langle n\rangle$ yields

$$\langle (n - \langle n\rangle)^2\rangle = \langle n^2\rangle - \langle n\rangle + \langle n\rangle - \langle n\rangle^2, \tag{12.156}$$

$$= \langle n(n-1)\rangle + \langle n\rangle - \langle n\rangle^2. \tag{12.157}$$

We thus find that the variance of n may be expressed in terms of the factorial moment $\langle n(n-1)\rangle$ and two terms that depend on the mean multiplicity $\langle n\rangle$. Recall from §10.2 that the difference $\langle n(n-1)\rangle - \langle n\rangle^2$ corresponds to a factorial cumulant C_2, which constitutes a proper measure of particle correlations. One can thus write

$$\langle (n - \langle n\rangle)^2\rangle = C_2 + \langle n\rangle. \tag{12.158}$$

Since the intent of a measurement of variance is to identify the strength of particle correlations, the preceding expression suggests that rather than measuring the variance of n, one should measure the cumulant C_2 directly. We will see in the next section that while the cumulant C_2 is itself dependent on the detection efficiency, the correlation function R_2 obtained by dividing C_2 by the square of the integrated average multiplicity is in fact independent of detection efficiencies and thus constitutes a robust observable. Measurements of C_2 or R_2 should thus be favored over measurements of variances insofar as the intent is to identify correlations induced by particle production dynamics.

12.4.3 Correcting Differential Correlation Functions and Factorial Cumulants

Robustness of Normalized Cumulants R_2

We demonstrated in §10.2 that cumulants and factorial cumulants constitute genuine measures of particle correlations, while n-particle densities suffer from trivial "combinatorial backgrounds" and thus do not provide a reliable measure of correlations. We will now proceed to show that ratio of differential cumulants and factorial cumulants to products of single-particle yields are robust variables. We begin our demonstration with two-particle correlation functions and show that the robust ratios can be trivially extended to higher order cumulants.

Measurements of single- and two-particle differential cross sections (e.g., $dN/d\eta$) are usually carried out using histograms with bins of finite width. Recall from §4.6 that one can estimate functions by dividing the histogram bin content (already scaled by the number of integrated events to obtain a per event average) by the width(s) of the bins:

$$\hat{f}(x_i) = \frac{h_i}{\Delta x_i}, \tag{12.159}$$

$$\hat{f}(x_i, y_j) = \frac{h_{i,j}}{\Delta x_i \Delta y_j}. \tag{12.160}$$

Estimates of the single-particle differential cross section, $\rho_1(x) \equiv dN/dx$, and two-particle differential cross section, $\rho_2(x_1, x_2) \equiv dN/dx_1 dx_2$ expressed as a function of some kinematical variable x (e.g., the transverse momentum, the pseudo-rapidity, etc.) may then be written

$$\hat{\rho}_1(x_i) = \frac{\langle N(x_i)\rangle}{\Delta x_i}, \tag{12.161}$$

$$\hat{\rho}_2(x_{1,i}, x_{2,j}) = \frac{\langle N(x_{1,i})N(x_{2,j})\rangle}{\Delta x_{1,i}\Delta x_{2,j}}, \tag{12.162}$$

where $\langle N(x_i)\rangle$ and $\langle N(x_i)N(x_j)\rangle$ are the average number of particles in bin i and the average number of pairs of particles in bins i and j, respectively.

$$\langle N(x_i)\rangle = \int_{\text{bin,i}} \frac{dN}{dx} dx, \tag{12.163}$$

$$\langle N(x_{1,i})N(x_{2,j})\rangle = \int_{\text{bin,i}} dx_1 \int_{\text{bin,j}} dx_2 \frac{d^2N}{dx_1 dx_2}. \tag{12.164}$$

Experimentally, instrumental losses may occur. Ignoring smearing effects, the measured average number of particles in bin i and pairs of particles in bins i and j are thus

$$\langle n(x_i)\rangle = \varepsilon_1(x_i)\langle N(x_i)\rangle, \tag{12.165}$$

$$\langle n(x_{1,i})n(x_{2,j})\rangle = \varepsilon_2(x_{1,i}, x_{2,j})\langle N(x_{1,i})N(x_{2,j})\rangle. \tag{12.166}$$

One can then define estimators of the observed single-particle density and two-particle densities as follows:

$$\hat{\nu}_1(x_i) = \langle n(x_i)\rangle/\Delta x_i, \tag{12.167}$$

$$\hat{\nu}_2(x_{1,i}, x_{2,j}) = \langle n(x_{1,i})n(x_{2,j})\rangle/\Delta x_i \Delta x_j. \tag{12.168}$$

Based on the preceding expressions, we can write

$$\hat{\nu}_1(x_i) = \varepsilon_1(x_i)\hat{\rho}(x_i), \tag{12.169}$$

$$\hat{\nu}_2(x_{1,i}, x_{2,j}) = \varepsilon_2(x_{1,i}, x_{2,j})\hat{\rho}_2(x_{1,i}, x_{2,j}), \tag{12.170}$$

where $\varepsilon_1(x_i)$ is the efficiency for detecting particles in bin x_i and $\varepsilon_2(x_{1,i}, x_{2,j})$ is the joint efficiency corresponding to the probability of simultaneously observing particles in bins $x_{1,i}$ and $x_{2,j}$.

In general, for particles emitted at different momenta and directions, it is reasonable to assume that $\varepsilon_2(x_{1,i}, x_{2,j})$ may be expressed as a simple product of the single-particle efficiencies, that is the efficiencies for independently observing each particle in bins x_i and x_j:

$$\varepsilon_2(x_{1,i}, x_{2,j}) = \varepsilon_1(x_{1,i})\varepsilon_1(x_{2,j}). \tag{12.171}$$

One can then readily verify that the ratio r_2 and the correlation function R_2 defined in §11.1.1 by Eqs. (10.36) and (10.37) are robust variables. Indeed, calculating r_2 in terms of measured densities $\hat{\nu}_1(x_i)$ and $\hat{\nu}_2(x_{1,i}, x_{2,j})$, we have

$$\hat{r}_2^{\text{meas}}(x_{1,i}, x_{2,j}) = \frac{\hat{\nu}_2(x_{1,i}, x_{2,j})}{\hat{\nu}_1(x_{1,i})\hat{\nu}_1(x_{2,j})}. \tag{12.172}$$

Inserting the expressions (12.170) and (12.169) for $\hat{\nu}_2$ and $\hat{\nu}_1$, respectively, we get

$$\hat{r}_2^{\text{meas}}(x_{1,i}, x_{2,j}) = \frac{\varepsilon_1(x_{1,i})\varepsilon_1(x_{2,j})\hat{\rho}_2(x_{1,i}, x_{2,j})}{\varepsilon_1(x_{1,i})\hat{\rho}_1(x_{1,i})\varepsilon_1(x_{2,j})\hat{\rho}_1(x_{2,j})}, \tag{12.173}$$

$$= \frac{\hat{\rho}_2(x_{1,i}, x_{2,j})}{\hat{\rho}_1(x_{1,i})\hat{\rho}_1(x_{2,j})}, \tag{12.174}$$

$$= \hat{r}_2^{\text{prod}}, \tag{12.175}$$

which is the definition of the normalized cross section r_2 expressed in terms of the produced densities $\hat{\rho}_1(x_{1,i})$, $\hat{\rho}_1(x_{2,j})$, and $\hat{\rho}_2(x_{1,i}, x_{2,j})$. Since $R_2 = r_2 - 1$, we conclude that both r_2 and R_2 are independent of detection efficiency and thus robust observables.

However, note that whenever the bins x_i and x_j are identical, the number of pairs must be calculated as $\langle n(x_i)(n(x_i) - 1)\rangle$. The preceding formulas are otherwise unchanged. Our discussion also holds irrespective of the size of the bins used to carry out the analysis as long as smearing effects are negligible. We thus conclude that our statement of robustness holds for both differential and integral correlation functions (factorial cumulants).

Rather than measuring $R_2(x_1, x_2)$, it is often sufficient or more appropriate to consider the dependence of the cumulant on the difference $\Delta x = x_1 - x_2$ (e.g., $\Delta\varphi = \varphi_1 - \varphi_2$, or $\Delta\eta = \eta_1 - \eta_2$ and report $R_2(\Delta x)$. As we discussed in §11.1.2, this can be viewed as an average over the domain spanned by the variable $\bar{x} = (x_1 + x_2)/2$ determined by the acceptance for x_1 and x_2:

$$\hat{R}_2(\Delta x) = \frac{1}{\Omega(\Delta x)} \int_{\Omega(\Delta x)} d\bar{x}\hat{R}_2(\Delta x, \bar{x}). \tag{12.176}$$

Since $R_2(\Delta x, \bar{x})$ may be obtained directly from $R_2(x_1, x_2)$ by means of a change of variables $x_1, x_2 \to \Delta x, \bar{x}$, it is by construction also independent of detection efficiencies and consequently robust. However, recall from §11.1.1 that two distinct methods may be used to estimate the function $\hat{R}_2(\Delta x)$. Method 2 yields the preceding formula (12.176) and is consequently robust by construction, while Method 1 is based on a ratio of averages:

$$\hat{R}_2(\Delta x) = \frac{\hat{\nu}_2(\Delta x)}{\nu_1 \otimes \nu_1(\Delta x)}, \tag{12.177}$$

where the product $\nu_1 \otimes \nu_1(\Delta x)$ is typically determined based on a mixed-event method:

$$\nu_1 \otimes \nu_1(\Delta x) = \hat{\nu}_2^{\text{Mixed}}(\Delta x). \tag{12.178}$$

Formally, the density $\hat{\nu}_2(\Delta x)$ may be calculated as

$$\hat{\nu}_2(\Delta x) = \frac{1}{\Omega(\Delta x)} \int_{\Omega(\Delta x)} d\bar{x} \hat{\nu}_2(\Delta x, \bar{x}), \tag{12.179}$$

while the product $\nu_1 \otimes \nu_1(\Delta x)$ may be similarly written

$$\hat{\nu}_1 \otimes \nu_1(\Delta x) = \frac{1}{\Omega(\Delta x)} \int_{\Omega(\Delta x)} d\bar{x} \hat{\nu}_1(x_1) \hat{\nu}(x_2). \tag{12.180}$$

Substituting the expressions for $\hat{\nu}_2(\Delta x, \bar{x})$ and $\nu_1(x_i)$ in Eq. (12.177), we get

$$\hat{R}_2^{\text{meas}}(\Delta x) = \frac{\int_{\Omega(\Delta x)} d\bar{x} \varepsilon_1(x_{1,i}) \varepsilon_1(x_{2,j}) \hat{\rho}_2(x_1, x_2)}{\int_{\Omega(\Delta x)} d\bar{x} \varepsilon_1(x_{1,i}) \rho_1 \varepsilon_1(x_{2,j}) \rho_1(x_2)}. \tag{12.181}$$

Clearly, if the efficiencies $\varepsilon_1(x_1)$ and $\varepsilon_1(x_2)$ have explicit dependencies on x_1 or x_2, the preceding expression shall yield a ratio R_2 that may differ appreciably from the result obtained with Eq. (12.176). We thus conclude that Method 1 cannot be considered robust in general. However, if the efficiencies are constant throughout the acceptance of the measurement, than they can be factored out of the integral and hence cancel out of Eq. (12.181). Method 1 can thus be considered approximately robust. Noting that $\nu_2(\Delta x, \bar{x})$ may be written $\nu_1(x_1)\nu_1(x_2)R_2(\Delta x, \bar{x})$, the preceding expression can be rewritten as

$$\hat{R}_2^{\text{M1}}(\Delta x) = \frac{\int_{\Omega(\Delta x)} d\bar{x} \varepsilon_1(x_1) \rho_1 \varepsilon_1(x_2) \rho_1(x_2) R_2(\Delta x, \bar{x})}{\int_{\Omega(\Delta x)} d\bar{x} \varepsilon_1(x_1) \rho_1 \varepsilon_1(x_2) \rho_1(x_2)}. \tag{12.182}$$

We thus conclude that the ratio (12.181) shall also be robust if $\hat{R}_2(\Delta x, \bar{x})$ is approximately independent of $\bar{x}$:

$$\hat{R}_2^{\text{M1}}(\Delta x) \approx \hat{R}_2(\Delta x, \bar{x}) \frac{\int_{\Omega(\Delta x)} d\bar{x} \varepsilon_1(x_1) \rho_1 \varepsilon_1(x_2) \rho_1(x_2)}{\int_{\Omega(\Delta x)} d\bar{x} \varepsilon_1(x_1) \rho_1 \varepsilon_1(x_2) \rho_1(x_2)}, \tag{12.183}$$

$$= \hat{R}_2(\Delta x, \bar{x}). \tag{12.184}$$

So, although R_2 may depend on Δx, Method 1 shall remain robust as long as it is constant for any given value of Δx, in other words, $R_2(\Delta x, \bar{x}) = K(\Delta x)$, where $K(\Delta x)$ is some smooth function of Δx but independent of $\bar{x}$.

One can also show that Method 1 yields a robust result for azimuthal correlations (e.g., functions of $\Delta\varphi$) with periodic boundary conditions (see Problem 12.1).

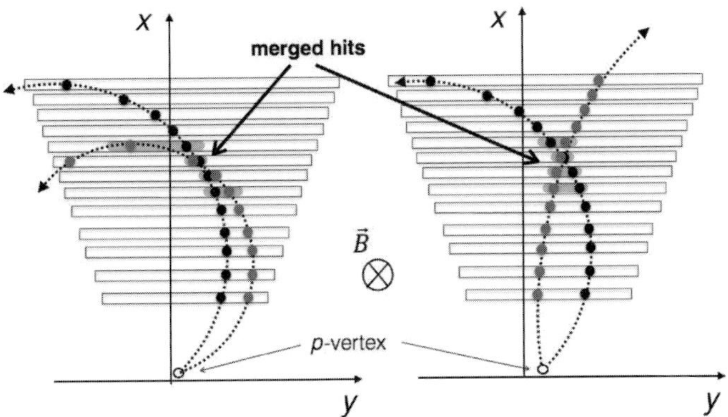

Fig. 12.9 Schematic illustration of track merging between equal sign and unequal sign charged particle tracks. The rectangular boxes indicate detection planes while dark circles represent the track energy deposition and detected hits. Wherever two tracks pass in close proximity the hits they produced may end up merging. The disturbed shape and centroid of these hits hinder the track reconstruction quality and efficiency.

Basic Instrumental Effects

Neither Method 1 nor Method 2 is robust for particle pairs featuring a small separation $\Delta\eta$, $\Delta\varphi$, and Δp_T. As illustrated in Figure 12.9, like-sign charged particles with similar momenta traverse the detector in close proximity of another and may thus not be well resolved. Such two tracks may then appear to partially or totally merge with one another. In the same way, unlike-sign particles with similar $\Delta\eta$ and $\Delta\varphi$ may cross the detection planes at nearly identical positions and become unresolved, thereby also resulting in track pair losses. **Track merging** and **track crossing** thus create an irreducible loss of particle pairs. Effectively, this implies the joint efficiency ε_2 does not factorize for particle pairs with $\Delta\eta \approx 0$, $\Delta\varphi \approx 0$, and $\Delta p_T \approx 0$. The observables r_2 and R_2 are consequently not robust for particles with small track separation.

In complex detectors consisting of multiple segments or components, it is also possible for tracks to become broken, as shown in Figure 12.10. There is typically a finite probability that a single track might be reconstructed as two or more tracks, resulting in a phenomenon known as **track splitting**. This produces extra tracks which artificially increase the average particle yield $\langle n \rangle$ and create artificial particle correlations. Splitting produces like-sign tracks with very similar parameters (e.g., η, φ, and p_T) and may thus produce artificially enhanced correlations for pairs with $\Delta\eta \approx 0$, $\Delta\varphi \approx 0$, and $\Delta p_T \approx 0$. In data analyses, particularly for measurements of correlation functions, effects associated with track splitting may be easily suppressed by using exclusively tracks that feature a number of hits well in excess of half of the number of possible hits.

The notion of robustness also breaks down when detectors are operated in occupancy conditions that exceed their design specifications. Detector occupancy refers to the fraction of detector sensors being "hit" at any given time. In low track or hit multiplicity environments, the probability of track or hit merging is small, and ambiguities resulting in track merging or splitting are typically limited. However, high-occupancy conditions may

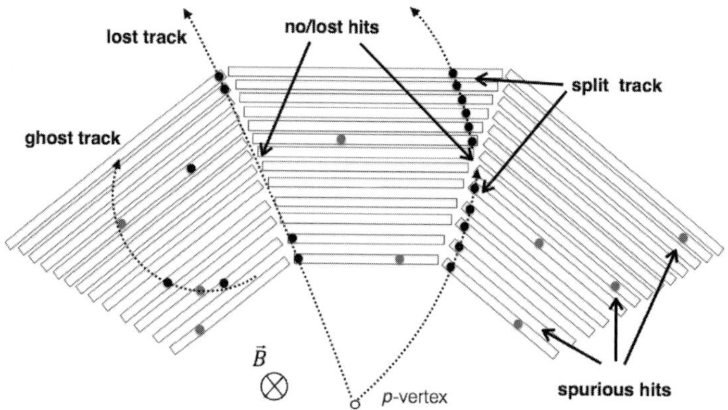

Fig. 12.10 Schematic illustration of split tracks, ghost tracks, as well as lost tracks.

induce ambiguities in hit and track reconstruction, as illustrated in Figure 12.10, which result in the reconstruction of false tracks, also known as **ghost tracks**. Admixtures of ghost tracks may impair the reconstruction efficiency of proper tracks and induce instrumental correlations if they are limited to specific regions of a complex detector. The reconstruction of uncorrelated ghost tracks may also change the magnitude of the normalized cumulant R_2 because the artificially inflated single-track density increases the size of its denominator but not its numerator. A similar effect may be produced by the decay of long-lived particles such as K_s^0 or Λ^0.

The robustness of correlation functions may be impaired by a variety of other instrumental effects, some of which we discuss in the next section.

Instrumentally Induced Correlations

We have seen in §12.4.3 that the normalized density r_2 and normalized cumulant R_2 are robust when measured according to Method 2 and to a lesser extent when estimated with Method 1, even when the efficiency varies throughout the acceptance of the measurement. However, the robustness is lost if the detection efficiency varies with time or event-by-event relative to some global event parameter. Let us illustrate this statement for a detection system featuring two performance states, that is, with efficiency of detection globally taking two distinct values for two classes of events, either separated in time, run, or some other "external" parameter.

For simplicity's sake, let us assume a correlation measurement between identical particles in the same kinematic range. The detection of both particles of the pair thus has the same efficiency dependence on the measured coordinates. However, let us also assume that two distinct classes of events of knowable size (i.e., how many events belong to each class) exist and feature different detection efficiencies we shall denote as ε_i, with $i = 1, 2$. The approach discussed here will hold for higher numbers of performance classes. We saw earlier in this chapter that is possible to define an effective or average efficiency ε_{avg}, provided the relative sizes of the two event classes are known. This average efficiency may then be

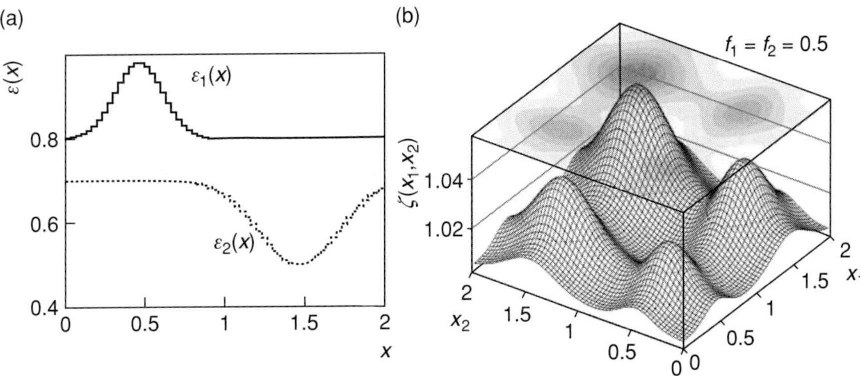

Fig. 12.11 (b) Robustness function $\xi(x_1, x_2)$ for a two-performance state experiment calculated for the efficiencies shown in (a), and assuming the same amount of data were acquired with the two efficiency curves.

used to correct average single-particle densities according to Eq. (12.133). Such correction is unfortunately not possible for correlation functions if the two sets of events are mixed. To demonstrate this statement, let us define the fraction of events reconstructed with efficiencies ε_i as f_i, and satisfying $\sum_i f_i = 1$. Let us also assume the two-particle efficiency factorizes (i.e., $\varepsilon_i(x_1, x_2) = \varepsilon_i(x_1)\varepsilon_i(x_2)$). If the data are analyzed indiscriminately of event classes, the measured single-particle and two-particle densities may be written

$$\hat{n}_1(x) = [f_1\varepsilon_1(x) + f_2\varepsilon_2(x)]\,\rho_1(x), \tag{12.185}$$

$$\hat{n}_2(x_1, x_2) = [f_1\varepsilon_1(x_1)\varepsilon_1(x_2) + f_2\varepsilon_2(x_1)\varepsilon_2(x_2)]\,\rho_2(x_1, x_2). \tag{12.186}$$

It is thus easy to verify that the ratio r_2 is not robust under such circumstances:

$$\hat{r}_2^{\text{meas}}(x_1, x_2) = \xi(x_1, x_2)\hat{r}_2(x_1, x_2),$$

with

$$\hat{r}_2(x_1, x_2) \equiv \frac{\rho_2(x_1, x_2)}{\rho_1(x_1)\rho_1(x_2)}, \tag{12.187}$$

and

$$\xi(x_1, x_2) = \frac{f_1\varepsilon_1(x_1)\varepsilon_1(x_2) + f_2\varepsilon_2(x_1)\varepsilon_2(x_2)}{[f_1\varepsilon_1(x_1) + f_2\varepsilon_2(x_1)]\,[f_1\varepsilon_1(x_2) + f_2\varepsilon_2(x_2)]}. \tag{12.188}$$

which is manifestly different from unity in general. The function $\xi(x_1, x_2)$ measures the extent to which the robustness of the observable r_2 is broken by having different efficiencies for two classes of events. This is illustrated in Figure 12.11, which displays the function $\xi(x_1, x_2)$ for two arbitrary efficiency curves $\varepsilon_1(x)$ and $\varepsilon_2(x)$ and $f_1 = f_2 = 0.5$ corresponding to a situation where an equal amount of data is acquired with the efficiency curves. All this said, it is quite remarkable that while large and arbitrary nonuniformities (of order 10–20%) have been assumed in the efficiency distributions shown in Figure 12.11, their "average" leads to relatively small deviations of less than 5%. This maximum difference essentially amounts to half the difference between the two spatial averages of the efficiencies. The R_2 observable thus remains relatively robust in spite of the time dependence of the efficiency. The necessity to correct for such effects is then determined by the level of accuracy sought after in the measurement, as well as the size of these variable efficiency features.

If the efficiencies $\varepsilon_i(x)$ are actually independent of x but depend on time, one finds that the factor $\xi(x_1, x_2)$ merely produces a shift in the amplitude of the correlation function. However, in cases where $\varepsilon_i(x)$ varies appreciably with x, particularly near the edges of the acceptance, variations of $\varepsilon_i(x)$ with external conditions or parameters such as the detector occupancy or the collision vertex position may induce sizable artificial structures throughout a correlation function [163]. Such effects may also be induced at the boundaries between detector components with performances that vary in time or with external conditions such as the detector occupancy.

Robustness of the r_2 and R_2 observables may be recovered, at least partially, if the analysis may be carried according to event classes, that is, by measuring these observables for each event class independently and combining the measurements into a global average.

$$r_2^{\text{class}}(x_1, x_2) = f_1 \frac{\varepsilon_1(x_1)\varepsilon_1(x_2)\rho_2(x_1, x_2)}{\varepsilon_1(x_1)\rho_1(x_1)\varepsilon_1(x_2)\rho_1(x_2)} + f_2 \frac{\varepsilon_2(x_1)\varepsilon_2(x_2)\rho_2(x_1, x_2)}{\varepsilon_2(x_1)\rho_1(x_1)\varepsilon_2(x_2)\rho_1(x_2)}. \tag{12.189}$$

The efficiencies cancel out and one gets

$$r_2^{\text{class}}(x_1, x_2) = f_1 \frac{\rho_2(x_1, x_2)}{\rho_1(x_1)\rho_1(x_2)} + f_2 \frac{\rho_2(x_1, x_2)}{\rho_1(x_1)\rho_1(x_2)}, \tag{12.190}$$

$$= r_2(x_1, x_2), \tag{12.191}$$

since, by definition, $f_1 + f_2 = 1$. This form of computation enables full-use of datasets even though the efficiency may vary with time or experimental conditions. The technique is applicable, in particular, for measurements of correlation functions where the acceptance and efficiency are dependent on the position of the collision vertex. It then suffices to bin the analysis of two- and single-particle densities in terms of the position of the vertex. With n bins, corrected yields and correlations may thus be obtained with

$$r_2^{\text{class}}(x_1, x_2) = \sum_{i=1}^{n} f_i \frac{\nu_{2,i}(x_1, x_2)}{\nu_{1,i}(x_1)\nu_{1,i}(x_2)}, \tag{12.192}$$

where the coefficients f_i represent the relative frequency of events in each z-vertex bin and satisfy $\sum_{i=1}^{n} f_i = 1$.

Collision Centrality Averaging in Heavy-Ion Collisions

In studies of high-energy heavy-ion collisions, it is useful and convenient to study the strength of single-particle yields and correlation functions as a function of collision centrality estimated on the basis of global event observables, such as the total transverse energy, zero degree energy, or integrated multiplicity measured in a selected portion of the detector acceptance. Two types of experimental effects must be accounted for in principle. The first effect arises because cumulants and single-particle yields scale differently with the number of sources of correlated particle production, while the second effect has to do with possible efficiency dependencies on the event multiplicity (i.e., number of particles produced in an event) or the detector occupancy.

First, consider that given sufficient statistics and computing resources, it should be possible to determine the normalized two-particle density r_2 as a function of some collision impact measure, b, with arbitrarily fine granularity (resolution) using Method 2:

$$r_2(x_1, x_2|b) = \frac{\rho_2(x_1, x_2|b)}{\rho_1(x_1|b)\rho_1(x_2|b)}. \tag{12.193}$$

An average can then be taken over collision centralities in a specific range $[b_{\min}, b_{\max}]$:

$$r_2(x_1, x_2|[b_{\min}, b_{\max}]) = \int_{b_{\min}}^{b_{\max}} r_2(x_1, x_2|b)P(b)\, db, \tag{12.194}$$

where $P(b)$ is a PDF expressing the probability of collisions of centrality b. In practice, it may not be possible to measure the single and pair densities with fine granularity. One then gets densities as averages over collision centrality between the limits, $b_{\min}$, and $b_{\max}$.

$$\rho_1(x|[b_{\min}, b_{\max}]) = \int_{b_{\min}}^{b_{\max}} \rho_1(x|b)P(b)\, db, \tag{12.195}$$

$$\rho_2(x_1, \eta_2|[b_{\min}, b_{\max}]) = \int_{b_{\min}}^{b_{\max}} \rho_2(x_1, x_2|b)P(b)\, db. \tag{12.196}$$

An estimate of the correlation function (Eq. 12.194) can then be obtained from the ratio of the averages.

$$r_{2,\text{est}}(x_1, x_2|[b_{\min}, b_{\max}]) = \frac{\rho_2(x_1, x_2|[b_{\min}, b_{\max}])}{\rho_1(x_1; [b_{\min}, b_{\max}])\rho_1(x_2|[b_{\min}, b_{\max}])} - 1. \tag{12.197}$$

This estimate is biased, however, by the finite centrality bin width used in the measurement. To demonstrate this bias, let us approximate the densities as

$$\rho_1(x_1|b) = n(b)h_1(x_1), \tag{12.198}$$

$$\rho_2(x_1, x_2|b) = n(b)(n(b) - 1)h_2(x_1, x_2), \tag{12.199}$$

where $n(b)$ and $n(b)(n(b) - 1)$ denote the average single and pair densities evaluated as function of the centrality parameter b, while the functions $h_1(x_1)$ and $h_2(x_1, x_2)$ denote the dependence of the number of singles and pairs on the position x_1 and x_2, and are here assumed, for simplicity's sake, to be independent of collision centrality. The preceding estimate of the correlation function then becomes

$$r_{2,\text{est}}(x_1, x_2|[b_{\min}, b_{\max}]) = Q(b_{\min}, b_{\max})\frac{h_2(x_1, x_2)}{h_1(x_1)h_1(x_2)}, \tag{12.200}$$

with

$$Q(b_{\min}, b_{\max}) = \frac{\int_{b_{\min}}^{b_{\max}} n(b)\,(n(b) - 1)\,P(b)\, db}{\left(\int_{b_{\min}}^{b_{\max}} n(b)P(b)\, db\right)^2}, \tag{12.201}$$

which is clearly different than the result obtained with Eq. 12.194

$$\int_{b_{\min}}^{b_{\max}} \frac{n(b)\,(n(b)-1)}{n(b)n(b)} P(b)\,db, \tag{12.202}$$

which tends to unity in the large $n(b)$ limit. The estimate Eq. (12.200) is consequently more and more biased with increasing bin width. This effect may, however, be suppressed by dividing $R_{2,\text{est}}$ explicitly by $Q(b_{\min}, b_{\max})$. Such correction could not compensate, however, for possible dependencies of the shape of the functions $h_1(x_1)$ and $h_2(x_1, x_2)$ with the centrality parameter b. If the data sample is sufficiently large, the best approach is thus to carry the analysis in fine bins of b, which one can then subsequently combine, by weighted average, into wider centrality bins.

The estimate Eq. (12.200) is not only biased as a result of the width of the centrality bin but also intrinsically nonrobust against detection efficiencies. Indeed, since detection efficiencies typically decrease with higher detector occupancy, one expects particle tracks produced in high-multiplicity collisions (central collisions) to be reconstructed with smaller efficiencies than those produced in peripheral or low-multiplicity collisions. This dependence effectively produces different classes of events, each with a different efficiency response through the detector acceptance. Average of correlation function across finite centrality bins may thus be subjected to instrumentally induced changes in amplitude or shape. To illustrate this, let us express the efficiency as a function $\varepsilon(x|M)$ of collision centrality measured by some global observable M. The robustness measure $\xi(x_1, x_2)$ may then be written (see Problem 12.2)

$$\xi(x_1, x_2) = \frac{\int dMP(M)\varepsilon(x_1|M)\varepsilon(x_2|M)}{\int dMP(M)\varepsilon(x_1|M)\int dM'P(M')\varepsilon(x_2|M')}, \tag{12.203}$$

where $P(M)$ corresponds to the probability of events with centrality M, with $\int dMP(M) = 1$ within a specific centrality bin. It is quite obvious that unless $P(M)$ is constant across a bin, or the efficiency independent of M, the factor $\xi(x_1, x_2)$ shall in general deviate from unity and contribute a bias in the estimation of correlation functions. This issue can be avoided, in principle, if it is possible to use fine centrality bins and calculate the correlation function according to Eq. (12.192). This calculation technique yields properly corrected and robust correlation functions in the limit of very narrow centrality bins. Its feasibility may, however, be limited by the size of the data sample. If too fine a binning in b is attempted, the sampled single-particle yield $\rho_1(x|b)$ may be null in one or several x bins. The ratio $R_2(x_1, x_2|b)$ would then diverge in those x bins, and the method would consequently fail. This implies that, in practice, it is necessary to systematically test the correction method and verify its convergence while changing the impact parameter bin size used to carry out centrality bin-width corrections.

Weighing Techniques

We saw in §12.4.3 that whenever the detection efficiency exhibits dependencies on global event observables such as the vertex position, detector occupancy, or beam instantaneous luminosity, one may need to partition the data analysis into a large number of different bins in order to obtain conditions under which detection efficiencies are approximately constant within one bin. Correlation studies typically require large-size histograms and may thus consume a large amount of computer memory. Having to replicate the same histograms several times to account for efficiency dependencies may become a cumbersome and memory expensive proposition. Fortunately, a simple solution exists in the form of a weighing technique.

Weighing techniques are rather general and may be used toward the study of R_2 as well as other types of correlation functions such as $\langle \Delta p_T \Delta p_T(\Delta\eta, \Delta\phi)\rangle$ discussed in §11.1.5. The weights are designed to be inversely proportional to the detection efficiency and as such equalize the detector response. A technique to obtain them is discussed below. Once the weights are available, one can proceed to carry out the correlation analysis using either Method 1 or Method 2, and by incrementing histograms with weights $\omega(\eta_1, \phi_1, p_{T,1})\times \omega(\eta_2, \phi_2, p_{T,2})$ rather than unity, for R_2 analyses, and $\omega(\eta_1, \phi_1, p_{T,1}) \times \omega(\eta_2, \phi_2, p_{T,2}) \times \Delta p_{T,1} \times \Delta p_{T,2}$ for $\langle \Delta p_T \Delta p_T\rangle$ analyses.

In the context of correlation studies carried out as a function of $\Delta\eta$ vs. $\Delta\phi$, one must account for the fact that detection efficiencies are complicated functions of ϕ, η, and p_T that may evolve with the position, z, of the collision vertex. One must consequently obtain weights, $\omega(\eta, \phi, p_T, z)$, that depend simultaneously on all four of these variables. The weights thus acquire a dual function: they account for the z dependence as well as the p_T versus ϕ dependencies simultaneously. The purpose of the weights is to equalize the response in p_T across all values of ϕ, η, and for all z. They can therefore be calculated as

$$\omega(\eta, \phi, p_T, z) = \frac{\int d\phi \int d\eta \int_{z_{\min}}^{z_{\max}} dz \langle n(\eta, \phi, p_T, z)\rangle}{\langle n(\eta, \phi, p_T, z)\rangle}, \tag{12.204}$$

where the integration on ϕ and η covers the fiducial acceptance of interest. By construction, the p_T spectra and $\langle p_T\rangle$ become independent of ϕ as well as η. Independence relative to η is likely acceptable at LHC and RHIC in the context of narrow η acceptance detectors such as STAR and ALICE, but better η dependent treatment may be required for wider acceptances.

Realistic detector performance simulations can be used to estimate absolute detection efficiencies as a function of track parameters as well as global event conditions (e.g., centrality, detector occupancy, etc.) and calculate weights, $\omega(\eta, \phi, p_T)$, in terms of their multiplicative inverse.

$$\omega(\eta, \phi, p_T) = \frac{1}{\varepsilon(\eta, \phi, p_T)}. \tag{12.205}$$

Alternatively, detection efficiencies may also be estimated using embedding techniques (§12.4.6).

12.4.4 Flow Measurement Corrections

Measurements of flow correlation functions such as those defined in §11.4 require the determination of the flow plane angle Ψ and nominally assume this angle is uniformly distributed in the $[0, 2\pi[$ range. In practice, this is rarely the case because of various instrumental effects. Fortunately, there exist several simple techniques to remedy this problem.

The first technique, known as **Phi weighing**, gives each *particle* a weight $\omega(\phi)$ proportional to the inverse of the azimuthal distribution $h(\phi)$ of the particles averaged over a large ensemble of events.

$$\omega(\phi) = \left(\frac{h(\phi)}{\frac{1}{n_\phi} \sum_\phi h(\phi)} \right)^{-1}, \tag{12.206}$$

where n_ϕ is the number of bins in ϕ.

The second technique, called **recentering**, uses a modified event **Q**-vector obtained by subtracting an event averaged **Q**-vector from each event's nominal **Q**-vector.

$$\vec{Q}'_n = \vec{Q}_n - \langle \vec{Q}_n \rangle. \tag{12.207}$$

The third technique, commonly referred to as **shifting**, gives a weight $\omega(\Psi)$ to *events* based on the event ensemble distribution of reaction plane angles $h(\Psi)$.

$$\omega(\Psi) = \left(\frac{h(\Psi)}{\frac{1}{n_\Psi} \sum_\Psi h(\Psi)} \right)^{-1}, \tag{12.208}$$

where n_Ψ is the number of bins in Ψ.

Other correction techniques are also documented in the literature [182].

12.4.5 Mixed-Event Correction Technique

Event mixing is a technique commonly used to decorrelate observables that can be combined into one quantity, such as the invariance mass of particle pairs, two-, or multiparticle correlation functions. The technique was initially developed in the context of analyses of intensity interferometry of pions to obtain uncorrelated particle pairs and determine effects of acceptance and detection efficiency directly from data [131]. The method was extended later on for the generation of background distributions for a wide range and variety of correlation functions, invariant mass distributions, and many other applications.

Event mixing involves the generation of "artificial" events consisting of uncorrelated objects (hits, charged particle tracks, jets, etc.) based on actual data. Objects are sampled from different events and are as such physically uncorrelated, that is, from the point of view of production processes. Their detection is nonetheless subject to essentially all instrumentation effects (e.g., detector resolution, acceptance, efficiency, etc.). The analysis of mixed events enables correlation analyses of ab initio uncorrelated objects and thus provides a baseline for actual correlation measurements.

As a practical example, let us consider a measurement of the normalized cumulant $R_2(\varphi_1, \varphi_2)$ for particles produced in nucleus–nucleus collisions. Recall from §10.2.4 that

R_2 is defined as a ratio of the cumulant $C_2(\varphi_1, \varphi_2)$ by the product of measured single particle densities $\nu_1(\varphi_1)\nu_1(\varphi_2)$:

$$R_2(\varphi_1, \varphi_2) = \frac{C_2(\varphi_1, \varphi_2)}{\nu_1(\varphi_1)\nu_1(\varphi_2)}, \tag{12.209}$$

$$= \frac{\nu_2(\varphi_1, \varphi_2)}{\nu_1(\varphi_1)\nu_1(\varphi_2)} - 1. \tag{12.210}$$

In a background-free measurement, the measured densities are determined by the production cross section and detection efficiencies. Rather than trying to determine the efficiencies explicitly, one can mobilize the fact that R_2 is by construction a robust observable and utilize mixed events to estimate the denominator of the correlation function. The basic idea of the event-mixing technique is to "fabricate" artificial events consisting of an equal number of particles (tracks) as actual events, but with tracks selected randomly from a large pool of events of similar characteristics. One can then proceed to analyze these mixed events as if they were real events and obtain mixed two-particle densities $\nu_2^{\text{mix}}(\varphi_1, \varphi_2)$. Since the tracks composing a mixed event are by construction uncorrelated, the two-particle density is then approximately equivalent to the product of single particles $\nu_1(\varphi_1)\nu_1(\varphi_2)$:

$$\nu_2^{\text{mix}}(\varphi_1, \varphi_2) \approx \nu_1(\varphi_1)\nu_1(\varphi_2). \tag{12.211}$$

This technique accounts for instrumentation effects but does not contain correlations associated with the production process. One can then obtain R_2 from the ratio

$$R_2^{\text{meas}}(\varphi_1, \varphi_2) = \frac{\nu_2(\varphi_1, \varphi_2)}{\nu_2^{\text{mix}}(\varphi_1, \varphi_2)} - 1, \tag{12.212}$$

which accounts for detection efficiencies. Indeed, given that the measured densities are

$$\nu_2(\varphi_1, \varphi_2) = \varepsilon_1\varepsilon_2\rho_2(\varphi_1, \varphi_2), \tag{12.213}$$

$$\nu_2^{\text{mix}}(\varphi_1, \varphi_2) = \varepsilon_1\varepsilon_2\rho_1(\varphi_1)\rho_1(\varphi_2), \tag{12.214}$$

one obtains the desired result

$$R_2^{\text{meas}}(\varphi_1, \varphi_2) = \frac{\rho_2(\varphi_1, \varphi_2)}{\rho_1(\varphi_1)\rho_1(\varphi_2)} - 1. \tag{12.215}$$

Similar constructions can be done for correlation analyses involving three or more particles, the study of invariant mass spectra, and essentially all types of analyses requiring a baseline or reference consisting of uncorrelated objects.

While the concept of mixed-event analysis is simple, certain precautions must be taken to avoid biases and various technical issues. For instance, in heavy-ion experiments the produced particle multiplicities are a steep function of the collision impact parameter. Considering that most instruments have an efficiency that monotonically decreases with rising detector occupancy, one must be careful to only mix events of approximately the same multiplicity and trigger type, otherwise the event mixing may result in associating incorrect efficiencies or trigger biases and thus lead to a biased measurement of cross section, or correlation function.

Care must also be taken to mix events without carrying actual particle correlations into the synthesized mixed events. For instance, if mixed events are assembled by picking several particles from a small number of distinct events, one is effectively transposing some of the correlations that might be present in those events into the mixed events. Although the strength of these correlations is clearly diluted by event mixing, it is safest to carry out event-mixing using a single-particle per event. One must also pay attention to the fact that the number of mixed events that can be synthesized can be deceptively large. Indeed, given appropriate computing resources, it might be tempting to generate a number of mixed events far larger than the actual number of real events. Such **oversampling** of mixed events should definitely be avoided because it leads to the false impression that the efficiency correction accomplished, for instance, by calculating the normalized cumulant R_2 based on mixed events could receive no statistical error contribution from the mixed-event sample. Alas, mixed events are built from actual events. A finite number of actual events implies that single- and two-particle yields have finite statistical fluctuations. It is consequently impossible to eliminate these statistical errors and achieve mixed events with better precision than that obtained with actual events.

As a specific example, consider the generation of mixed events with a multiplicity of M particles from a pool of N events of same multiplicity M. Let us count the number of distinct mixed events that can be synthesized by randomly picking one particle per event out of M events. There are N ways to pick the first event, $N - 1$ ways to pick up the second, and so on. The numbers of event combinations, $N_{\rm EC}$, is thus $N_{\rm EC} = N!/(N - M)!M!$. Since M events are sampled, there are thus M^M ways to pick up one particle per event. The resulting number of possible mixed event permutations, $N_{\rm EP}$, is thus: $N_{\rm EP} = M^M N!/[(N - M)!M!]$. For $M = 50$, and $N = 1000$, one can use the Stirling approximation to evaluate the preceding expression. One gets $N_{\rm EP} \approx 10^{150}$ which is indeed a huge number of permutations. The number can be even larger if $q > 1$ particles are picked from each event.

While the number of permutations achievable with mixed events is very large, one must realize that fluctuations of $\nu_2^{\rm mix}$ are also unusually large because each particle included in a mixed event is reused several times in the formation of pairs. These fluctuations were first evaluated in ref. [180] using a pair counting technique. A more direct calculation is however possible based on the realization that the synthesis of mixed events amounts to multinomial sampling. For a fixed event size M, the bin content n_i of a histogram of the density $\rho(x)$ shall have expectation values p_i, with $\sum_i^m p_i = 1$, where m is the number of bin in the histogram. The values p_i represent the probability of getting counts in each bin. Since all particles are independent because they actually originate from distinct events, the bin contents are also uncorrelated. The number of particles n_i obtain in each bin i is thus determined by a multinomial distribution.

$$P_M(n_1, \ldots, n_m|M; p_1, \ldots, p_m) = \frac{M!}{\prod_{i=1}^{m} n_i!} \prod_{i=1}^{m} p_i^{n_i}, \tag{12.216}$$

with $\sum_{i=1}^m n_i = M$. Recall from §2.10 that moments of a PDF can be calculated based on derivatives of its characteristic function. The characteristic function of a multinomial distribution (§3.2) is

$$\phi(t_1, t_2, \ldots, t_m) = \left(\sum_{j=1}^{m} p_j e^{it_j} \right)^n. \tag{12.217}$$

One can then readily verify (Problem 12.3) that the lower-order moments of n_i are

$$\langle n_i \rangle = \mathrm{E}[n_i] = M p_i \tag{12.218}$$
$$\langle n_i^2 \rangle = \mathrm{E}[n_i^2] = M(M-1)p_i^2 + M p_i \tag{12.219}$$
$$\langle n_i n_j \rangle = \mathrm{E}[n_i n_j] = M(M-1)p_i p_j \tag{12.220}$$
$$\begin{aligned} \langle n_i^2 n_j^2 \rangle &= \mathrm{E}[n_i^2 n_j^2] \\ &= M(M-1)(M-2)(M-3)p_i^2 p_j \\ &\quad + M(M-1)(M-2)\left[p_i^2 p_j + p_i p_j^2\right] + M(M-1)p_i p_j. \end{aligned} \tag{12.221}$$

The variance of the pair yield $n_i n_j$ expressed in terms of the averages $\langle n_i \rangle$ is thus

$$\begin{aligned} \mathrm{Var}\left[n_i n_j\right] &= \left(-\frac{4}{M} + \frac{10}{M^2} - \frac{6}{M^3}\right) \langle n_i \rangle^2 \langle n_j \rangle^2 \\ &+ \left(1 - \frac{3}{M} + \frac{2}{M^2}\right) \left[\langle n_i \rangle^2 \langle n_j \rangle + \langle n_i \rangle \langle n_j \rangle^2\right] \\ &+ \left(1 - \frac{1}{M}\right) \langle n_i \rangle^2 \langle n_j \rangle. \end{aligned} \tag{12.222}$$

For large M, the variance of fluctuations of $n_i n_j$ thus approximately scale as $\langle n \rangle^3$. One can then write

$$\delta \langle n_i n_j \rangle \approx \langle n_i \rangle^{3/2} \approx \langle n_i n_j \rangle^{3/4}. \tag{12.223}$$

By contrast, errors on $\langle n_i \rangle$ scale as $\langle n_i \rangle^{1/2}$. We thus conclude that errors on mixed events grow faster than errors on the mean number of entries in single particle spectra. Large datasets are thus required to obtained equivalent size errors on mixed events.

Alternatively, one may also consider the number of pairs, say at specific φ_1, φ_2 values to be of the order of the product of the number of singles $n(\varphi_1)n(\varphi_2)$. Based on a full data sample, the relative error on this product, of order $n_1^{-1} + n_2^{-1}$, effects a minimum bound on the error on the number of pairs and thus two-particle densities that might be derived from them. The redeeming factor is that one is seldom interested in the density $\rho_2(\varphi_1, \varphi_2)$ directly but instead compute an average over $\bar{\varphi}$ to obtain $\rho_2(\Delta\varphi)$ which combines from one to m bins of $\rho_2(\varphi_1, \varphi_2)$. The error on the number of pairs is thus on the order or smaller than $n_1^{-1} + n_2^{-1}$. There is thus no point in generating a number of mixed events that would attempt to improve the statistical accuracy much below this bound.

12.4.6 Embedding Techniques

Embedding is a technique commonly used in experimental nuclear and particle physics to evaluate the instrumental efficiency in finding and reconstructing objects (e.g., hits, tracks, jets, etc.) of interest. The technique involves the insertion of tagged objects generated via MC simulations into actual events before they are reconstructed. One measures the objects' reconstruction efficiency as the degree to which they are reconstructed when embedded into actual events. This efficiency corresponds to the ratio of the number of objects N_F found by the number embedded N_E:

$$\varepsilon = \frac{N_F}{N_E}. \tag{12.224}$$

Embedding may be carried out with different levels of refinement and detail. For instance, for the evaluation of track reconstruction efficiency, one can embed tracks at the point or hit level or at the detector level, that is, in terms of simulated analog-to-digital converter (ADC) signals. While embedding points enables rapid testing of track reconstruction algorithms, it does not usually properly account for issues such as hit loss or hit overlap/merging. Full embedding simulations, for example the simulation of signal generation and propagation through the detector, is usually accomplished with the computer code GEANT with add-ons specific to each experiment. While considerably slower, full simulations make it possible to account for signal losses, interferences, merging, and so on. They thus permit as realistic as possible a simulation of the instrumental performance.

The evaluation of the error on the estimated efficiency ε requires some particular attention. Given ε is obtained from a ratio, one might at first be inclined to simply apply the quotient rule for the error on a ratio of two quantities. That would be incorrect, however, because the two quantities are not independent. In fact, by construction, the number of observed tracks is necessarily a subset of the produced or embedded tracks and one has $N_F \leq N_E$ by definition (unless there are ghost tracks). Clearly, the efficiency cannot exceed unity. One might then be inclined to use a lower limit only if the ratio is near unity, or a two sided error interval if the estimated efficiency is considerably smaller than unity. This amounts to flip-flopping, as discussed in §6.1.8. Fortunately, the unified approach introduced in §6.1.8 alleviates this problem and enables the definition of error intervals without flip-flopping or empty intervals, and no undercoverage. Its application for binomial sampling, relevant for efficiency estimates by embedding and related techniques, is discussed in §6.2.2.

12.4.7 Closure Test

Data correction methods can get pretty involved and complicated. But just because sophisticated unfolding techniques promise properly corrected data does not mean they deliver precise and unbiased results. It is thus useful to carry out a **closure test**.

The basic idea of a closure test is quite simple: (1) define a parent distribution f_P that mimics or might closely resemble the actual distribution; (2) produce a large number of events on the basis of this distribution and process them through a performance simulator of the experiment, its event reconstruction software, and the data analysis to obtain a simulated raw distribution f_{raw}; (3) proceed to unfold or correct the raw data to obtain a corrected (unfolded) distribution f_{corr}; (4) compare the unfolded distribution f_{corr} with the original distribution f_P to verify whether the unfolding procedure yields an unfolded distribution statistically compatible with the parent distribution.

Comparison of the f_{corr} and f_P distributions may begin with a simple visual inspection. It should be possible to instantly recognize improperly unfolded data if the two distributions deviate significantly from one another. However, given finite computing resources, the distributions shall have finite statistical errors, and it may not be visually obvious whether the distributions are in fact in mutual agreement, or not. It is then necessary to make use of a statistical test (§6.4). The null hypothesis of this test shall be that the unfolded distribution f_{corr} is undistinguishable from the parent distribution f_P. The test may then be implemented as a histogram compatibility test (§6.6.6) or Kolmogorov–Smirnov test (§6.6.7) using a predetermined significance level β. The significance level shall be chosen, based on the desired level of experimental accuracy, to minimize the risk of an error of the second kind, that is, to falsely accept the null hypothesis and consider the corrections are properly carried out when in fact they are not.

12.5 Systematic Errors

12.5.1 Definition

Systematic errors are measurements errors associated with the "system," that is, the measurement technique, the experimental protocol, or the analysis method used to carry out a measurement. Unlike statistical errors, which can be arbitrarily reduced in magnitude by repeating measurements of a specific phenomenon, statistical errors are an intrinsic flaw of the "system" and thus cannot be reduced or eliminated by repeating a measurement. Indeed, nothing is gained by repeating a flawed procedure several times. Systematic errors are due to biases introduced by the measurement technique or protocol. Since a bias does not disappear when a measurement is repeated, there is, in fact, no point in repeating the same measurement many times. The bias will persist, and the repeated measurements will have small statistical errors and thus might give the illusion that the results are accurate and precise. However, while increased statistics improve the precision of a measurement, the measurement remains inaccurate because it is systematically biased. Measured values tend to be consistently either to small or too large and having a zillion data points cannot fix that basic problem. This said, having a large data sample may enable a careful and detailed analysis which might unravel the presence of biases. But at the end of the day, understanding biases and systematic errors requires one has an in-depth understanding of all aspects of a particular measurement.

12.5.2 The Challenge

Systematic errors often affect (but not always) all data points of a particular measurement in a similar way: all points tend to be either too low or too large. For instance, consider a measurement of distances made with a stretched out tape measure. If the fabric or material

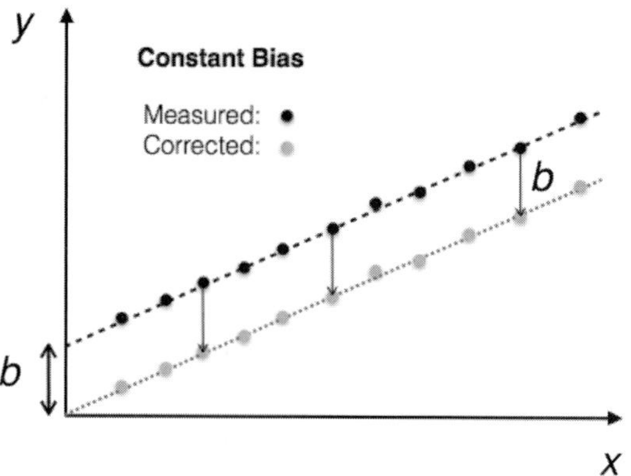

Fig. 12.12 Correcting for constant bias is possible if a signal can be measured, or extrapolated as shown, at a point where the magnitude is null or known to have a specific value by definition or construction. Here, the signal is expected to be null at $x = 0$; one can then use the observed offset b as an estimate of the measurement bias to correct all data points.

of the tape measure is worn out and stretched due to repeated usage or improper handling, distances or sizes measured will appear consistently smaller than they really are no matter what the size or distance is. The measured distances will thus be systematically too small. However, when measuring functional dependencies, for instance, $y \equiv y(x)$, one may find that the bias on y may itself be a function of x and y. The bias might indeed tend to be larger (or smaller) for a certain range of values of either x or y, and it might in fact be an arbitrarily complicated function of either variables.

Constant biases are least problematic. Indeed, if a bias δy_{bias} can be identified at specific value of x, and one has good reasons to believe it is constant for all values of x, than measured values $y(x)$ can be corrected for by subtracting the identified bias from all measured values:

$$y_{\text{corrected}} = y_{\text{measured}} - \delta y_{\text{bias}}. \tag{12.225}$$

This situation arises when the signal y is expected to be null, by construction or by definition, for a specific value of x. For instance, if one expect y to be strictly proportional to x and null for $x = 0$, one can detect the presence of a bias, as illustrated in Figure 12.12, by extrapolating measured data points to the origin, $x = 0$. Unfortunately, systematic errors are not always obvious or constant, and a lot of "digging" may be required to identify their presence and approximate magnitude. Failure to identify and correct for systematic biases may have consequences ranging from mild inconvenience to catastrophic failure and erroneous scientific conclusions.

The biggest problem with systematic biases is that it is usually not possible to precisely determine the magnitude or the functional dependence of the bias. Indeed, it is one thing to identify the presence of a source of bias, but it is usually a bigger challenge to determine the magnitude of the effect. One is thus required to make "educated" guesses as to what the magnitude of the biases might be and how they could impact the measured data. In the end, one is forced to acknowledge that the "final" data might be over- or undercorrected for biases. One must then estimate the range of one's ignorance (minimum and maximum corrections possible and consistent with the experimental method and the measured data) on the proper magnitude of corrections and determine how such uncertainties impact the corrected data. The range of applicable corrections then determines a range of measured values that one can quote as systematic errors when reporting the measured data.

As a specific example, let us consider the fixed bias correction illustrated in Figure 12.12. Here, the presence of a bias is readily identified because the signal y is expected to vanish at $x = 0$ but manifestly does not. One uses the extrapolated signal at $x = 0$ to estimate the bias b and correct all data points (gray dots). While straightforward, this correction procedure entails several issues. First, the extrapolation to the origin is only as good the linear fit to the data points. If the statistic is poor, the extrapolated value $y(0)$ may be arbitrarily large or small, and the corrected data shall correspondingly be too low or too high. Second, this procedure assumes the data is strictly governed by a linear dependence. What if, in fact, the phenomenon has a quadratic (or higher order) dependence on x? The extrapolation would then most likely be wrong and so would the corrected data. And third, it is also possible that while the linear model is correct, various instrumental effects may cause an artificial higher-order polynomial dependence of the signal y on x (in other words, the bias depends on x). Use of a linear extrapolation would once again produce an incorrect (or at the very least incomplete) correction. The task of the experimentalist is thus to evaluate, to the best of their ability, the magnitude of such effects. The error on b (which is subsequently subtracted from all data points) obtained from a linear would constitute a first estimate of systematic errors on the data points. One could also use higher-order (nonlinear) fits to extrapolate to the origin and possibly enlarge the range of the offset b consistent with the data, and which is subtracted to obtain corrected data.

The preceding example is rather simplistic. In practice, the presence of biases may not be obvious and evaluating the magnitude of known systematic effects is not a trivial matter. In fact, it is often the case that far more time and efforts are spent understanding and evaluating systematic errors than carrying out the measurement in the first place. It is also the case that the challenge of the task typically increases in proportion to the accuracy one wishes to achieve in a particular measurement.

12.5.3 Reporting and Improving Errors

Statistical and systematic errors have distinct causes and origins. It has thus become common practice, in most scientific disciplines, to report measurements with statistical and systematic error bars independently. This enables an explicit acknowledgment of the precision and accuracy achieved in the reported measurement. Assuming the statistical estimator used to determine the reported parameters are unbiased (or asymptotically unbiased but with a very large dataset), the statistical error bars provide an indication of the level of precision achieved in the measurement, while the systematic errors indicate the estimated level of accuracy. It is then possible to evaluate, at a glance, the merits of an experiment and identify its weaknesses. Experiments are said to be **statistics limited** if the statistical error bars are larger, for the most part, than systematic errors. Conversely, if the latter predominate, the experiment is described as **systematics limited**.

Achieving a better statistical error is conceptually simple, it suffices to significantly increase the size of the data sample. In particle and nuclear physics experiments, this can be achieved by increasing the duration of the experiment, the intensity of the beam and target thickness (the luminosity of the machine), the acceptance, or detection efficiency of the apparatus. Improving on a systematic limited experiment can be significantly harder. One must indeed achieve better control over the various biases and improve on the techniques used to suppress or correct for these biases. Such improvements typically require novel experimental techniques or better ways to analyze the data.

12.5.4 Brief Survey of Sources of Systematic Errors

Sources of systematic errors are as varied as the observables and measurements techniques used in experimental physics. It is consequently not possible to provide an exhaustive list of all types of systematic errors that may arise in practice. At the outset, it is quite important to have a good understanding of the experimental apparatus, the measurement protocol, as well as the properties and limitations of the observables under study. Indeed, only a careful examination of the properties and attributes of the detector components and analysis techniques used in a measurement of interest may reveal sources of biases and systematic errors. Nevertheless, one may formulate a few generic remarks about the process of understanding systematic errors and their most common sources.

As we have discussed at great length in this and earlier chapters, scientific measurements involve a large number of steps and operations, each of which is susceptible of introducing some kind of bias or systematic error. Two broad classes of systematic errors are worth considering. The first class involves sources of errors associated with the measurement itself whereas the second is concerned with "physical backgrounds."

Measurement Biases

The reconstruction of events and measurements of cross sections in nuclear (or elementary particle) scatterings involve several steps, including signal calibration, correction for detection efficiencies and experimental acceptance, as well as signal smearing (resolution effects). Each of these may indeed contribute a source of bias and systematic errors.

The calibration of energy and timing signals typically involves the subtraction of an ADC (TDC) pedestal and multiplication by an appropriate gain factor. Pedestals and gains are determined by experimental procedures and protocols which may be statistic or systematic limited (§9.1.3). They are thus known with finite precision and accuracy. They may also change over time for a variety of reasons. The pedestals and gains used in an analysis may then be slightly off, and induce biases in the energy (amplitude) or time scales of the measurement. This is particularly the case, for energy measurements, when the reference used to obtain the gain calibration lies at much lower energy than typical energy signals of interest in the physical measurement. The signal gain calibration is then effectively based on an extrapolation, which may result in gross under- or overestimation of the measured amplitudes, energies, or timing signals. Additional issues may arise if nonlinearities are involved in the energy or charge collection process. Together, these effects introduce a systematic uncertainty on the energy (charge, amplitude, or time) scale of the measurement.

Most measurements carried out in particle and nuclear physics involve some form of counting. For instance, one associates and counts hits to form tracks, tracks to form events, which may, in turn, also be counted. All these counting steps are predicated on some quality control cuts: not all produced objects (e.g., hits, charged particle track, energy shower, collision event, etc.) are duly counted and an arbitrary fraction may be missed. It is the purpose of efficiency (and acceptance) corrections to account for these losses. Obviously, although many distinct techniques might be used to estimate detection efficiencies, there remains the possibility that efficiencies may be, as for gains, under- or overestimated. Such uncertainties may arise for a variety of reasons, including the finite size of data samples, simulation samples, as well as improper characterization or modeling of the measurement process. Together, these effects lead to systematic uncertainties on the magnitude of cross sections.

Measurement errors affecting both the energy scale and the magnitude of the cross section may also arise due to finite resolution or signal smearing, whether unfolded or not. For instance, momentum or energy smearing of a steeply falling cross section spectrum (i.e., with increasing momentum) has a tendency to spread cross sections from low to high momenta, thereby disturbing the functional shape of the cross section dependence on energy or momentum.

Systematic uncertainties on cross sections may also arise because of faked (often called ghost) objects, most particularly tracks. For example, faked hits may be produced by noisy detector signals and lead to random associations of hits into faked or improperly measured tracks. Faked or ghost tracks may then end up being counted as real tracks and artificially inflate the measured cross sections. Ghost tracks are most common at low momenta because spurious hits are more easily associated to large curvature tracks than straight tracks.

Various other forms of **pile-up** may occur and lead to biases. An obvious example of pile-up involves the detection of tracks from several distinct collisions in slow detectors such as time projection chambers. Tracks from different collisions may appear unresolved and result in apparent high-multiplicity events, thereby shifting the production cross section from low to high multiplicity.

Biases may also occur because of the physical processes involved in the detection and measurement of signals. For instance, charged particles lose energy as they penetrate through the various materials composing a detector. Energy losses imply a reduction in momentum, which in turn results in an increase of curvature for charged particle tracks. The track energy loss thus introduces a small downward bias in the moment of particles, which cannot, typically, be compensated for equally well for all particle masses or species. Since energy losses are largest in the $1/\beta^2$ regime, this results in a slight distortion of the energy scale, which typically affects low momenta the most.

Event triggering (i.e., the selection of events with particular features), whether performed with actual triggering detectors, or in software during the data analysis may also induce various forms of biases, most particularly on the efficiency and cross section of processes. For instance, requiring that a scattering event contains one or two high-energy jets tends to skew the event multiplicity distribution toward high-multiplicity values. It obviously also affects the measured momentum distribution. Scattering measurements are often recorded with **minimum-bias** triggers. These are typically based on large detector arrays capable of detecting a sizable fraction of the produced particles. They are usually designed to collect all inelastic collisions with uniform efficiency and irrespective of the produced particle multiplicity. In practice, the finite size and limited efficiency of these detectors results in cross section biases, most particularly for low-multiplicity events. There is thus no such a thing as a truly unbiased minimum-bias trigger.

Physical Backgrounds

Physical backgrounds may also constitute sources of biases, particularly in measurements of cross sections. They can be broadly be divided into two classes. The first class consists of physical processes taking place within the detector because of electromagnetic or hadronic interactions of produced particles (i.e., particles produced by the scattering under study) with materials of the detector, while the second involves processes produced directly by the collision under study (i.e., irrespective of the presence of a detector).

At the outset of this discussion, recall the notions of primary, secondary, and tertiary tracks that we introduced in §8.3.1. Primary tracks are those tracks that are produced at the primary vertex of an interaction whereas secondary tracks originate from decays of primary particles as they traverse the detection apparatus. We have called tertiary tracks those tracks that are produced by interaction (electromagnetic or hadronic) of primary and secondary particles with materials of the detector. Experimentally, the segregation of primary, secondary, and tertiary tracks (particles) is largely based on their distance of closest approach to the primary vertex of an interaction. However, by virtue of the stochastic nature of decays and interaction with materials, secondary and tertiary particles may occur at arbitrarily small distances from the primary vertex and thus be indistinguishable, in practice, from primary particles. Secondary (and tertiary) particles produced near the vertex thus constitute a background in the determination of primary particle production cross section. Conversely, because finite resolution effects smear the DCA of tracks, primary tracks may appear as secondary tracks and thus contribute a sizable combinatorial background in the identification of decaying particles via topological cuts or invariant mass reconstruction (§8.5.1). Similarly, tertiary tracks may also be mis-identified as either primary or secondary tracks and thus constitute a source of background (and cross section bias) for these particle types. Cross contamination of primary and secondary (as well as tertiary) is typically based on distance of closest approach (DCA) cuts. Tight (short) maximum DCA cuts are required to select primary and reject secondary tracks whereas wide minimum DCA cuts are used to select secondary particles. No matter what cut values are used, however, some secondary tracks end up being selected as primary particles, and some primary particles are categorized as secondary. This unavoidable misidentification of the two types introduces backgrounds and thus biases in cross section measurements of both types of particles, and although these biases can be suppressed to some degree by judicial DCA cuts, they can never be eliminated. It is thus a common feature of all cross section analyses to study particle yields as a function of DCA cut values in order to estimate the relative contributions of the two types of particles. Unfortunately, such studies cannot be performed with perfect accuracy and one is thus required to include systematic errors on particle yields (or cross sections) based on trends and extrapolations.

Contamination from secondary particles is particularly insidious for measurements of correlation functions. Secondary particles produced by two- or three-body decays are de facto correlated by virtue of energy-momentum conservation. They thus naturally produce spurious correlation features in two- (or three-) particle correlation functions. As for measurements of single particle cross sections, one can use DCA cuts to reduce or suppress the impact of these spurious correlation features but they cannot be completely eliminated. One must then resort to systematic studies of the strength and shape of the correlation functions when DCA cuts are changed. It then becomes possible to report systematic errors on the amplitude of the correlation functions.

Various other physical and detector induced processes produce backgrounds in single particle cross section measurements and spurious features in two- or three-particle correlation functions. Particularly important in measurements of correlation functions are processes of pair creation by photons as they interact electromagnetically with high atomic number materials of a detector. Pair creation produces a forward going pair of positron and electron which result in a sharp peak at the origin of $\Delta\eta - \Delta\phi$ correlation functions of unlike-sign charged particle pairs. This type of contamination may be suppressed by rejecting electron-like particle signals (e.g., based on dE/dx, TOF, or Cerenkov radiation) and a minimum cut on the invariant mass of the pair (since pair conversion tend to produce a very low mass peak). Contamination may persist and it is thus necessary to assess systematic effects on correlation functions associated with these conversions.

Biases in measurements of single particle correlation and correlation functions may also arise because of track splitting and track merging (defined in §12.4.3). Track splitting leads to an artificial increase of the cross section and spurious correlation features at small $\Delta\eta$ and $\Delta\phi$ values. It can usually be suppressed significantly by requiring "long" tracks, that is, tracks that contain at least 50% of possible hits. Splitting contamination may remain, however, in spite of cuts, and it is necessary to estimate track splitting effects, for instance, by varying the minimum number of hits requirement.

Track merging poses a more serious difficulty. Indeed, the crossing of tracks with similar pseudorapidity and azimuth of production yields merged or unresolved hits which, in turn, lead to the reconstruction of fewer track pairs. One is then faced with a difficult loss of single and particle-pair to evaluate. These result in a reduction of the single particle cross section and a loss of particle pairs at small $\Delta\eta - \Delta\phi$ values. Such losses can in principle be modeled with Monte Carlo simulations. In practice, the accuracy of this type of simulations is severely limited by complicated and changing response features of particle detectors. One is then required, once again, to estimate systematic effects associated with the loss of particles, and most particularly pairs of particles produced at small relative pseudorapidities and azimuthal angles.

The aforementioned effects are the most common and salient effects encountered in measurements. However, many other effects may need to be considered in practical situations, such as loss of signals or spurious noise associated with various types of detector malfunction.

Exercises

12.1 Use an arbitrary Fourier decomposition of the efficiencies $\varepsilon(x_i)$ and the correlation function R_2 to demonstrate that Eq. (12.181) yields a robust result, i.e., is independent of efficiencies. Verify that is not the case for coordinates without periodic boundary conditions (e.g. $\Delta\eta$).

12.2 Verify Eq. (12.203) expressing the robustness measure $\xi(x_1, x_2)$.

12.3 Use the characteristic function $\phi(t_1, t_2, \ldots, t_m)$ of the multinomial distribution given by Eq. (12.217) to derive the lowest order moments of the distribution listed in Eqs. (12.218–12.221).

12.4 Show that $\frac{\partial \chi^2(\vec{\mu})}{\partial \mu_k} = 0$ is satisfied by Eq. (12.37).

12.5 Derive the expression of the covariance matrix U_{ij} given by Eq. (12.55).

12.6 Show that all singular values of an orthogonal matrix $\mathbf{A}$ are equal to unity.

12.7 Verify that the estimator (12.37) corresponds to the solution obtained by maximization of the log-likelihood function and yield

$$\log L(\mu) = \sum_{i=1}^{N} \log\left(P(n_i|\nu_i)\right),$$

where $P(n_i|\nu_i)$ is a Poisson distribution (or binomial distribution).

PART III

SIMULATION TECHNIQUES

13 Monte Carlo Methods

Modern science deals with increasingly challenging and complex systems. While relatively simple mathematical models can often be formulated, their implementation for the description of real-world systems is typically not amenable to analytical solutions. For instance, one may wish to "evolve" the state of a system (e.g., in a microscopic model of nucleus-nucleus collisions) with time to study its many configurations or one may want to calculate the acceptance of a large detector for the study of particles produced by elementary particle collisions (proton–proton or heavy ion collisions at the Large Hadron Collider), and so on. The number of states, configuration, or variables may, however, be prohibitively large, or the boundary conditions of the problem very complex, so it is not possible to sample or integrate the full configuration of the phase space of a system by conventional integration techniques. In these instances, it is possible to resort to Monte Carlo methods to sample the many possible states of the system or to obtain the integral of arbitrary functions over the system's configuration or phase space.

We begin our description of Monte Carlo methods with a short discussion of the principles of the method in §13.1. We then show, in §13.2, that Monte Carlo simulations essentially amount to a form of integration that can be most effectively used to compute multidimensional integrals. The notion of pseudorandom numbers and basic techniques commonly used for the generation of pseudorandom numbers are then presented in §13.3. Selected examples of basic applications are introduced in §13.4. More elaborate techniques for applications in nuclear sciences are presented in Chapter 14.

13.1 Basic Principle of Monte Carlo Methods

Monte Carlo methods use "chance" or more properly said random numbers to sample all possible states of a system. The name *Monte Carlo* is inspired from games of chance played at the famous Monte Carlo casinos in Monaco, located south of France. The concept and first implementation of the method originates from Los Alamos National Laboratory, New Mexico, where mathematicians John von Neumann and Stanislaw Ulam, working on the Manhattan Project in the 1940s, proposed carrying out theoretical calculations of radiation shielding using experimental modeling on a computer based on chance. Given the secrecy of the project, von Neuman allegedly chose the name "Monte Carlo" in reference to the casinos where Ulam's uncle would borrow money to gamble.

Random methods of calculation, or stochastic simulations, may be traced back to the pioneers of probability theory, but more specifically to the preelectronic era. Monte Carlo simulations were essential for the development of the Manhattan Project, though, severely limited by the computational tools available at the time. It was only with the development of increasingly powerful computers that Monte Carlo methods were systematically explored and increasingly used for military purposes and later in physical sciences and operations research. The Rand Corporation and the U.S. Air Force were two of the major organizations responsible for funding and disseminating information on Monte Carlo methods during this time, and they began to find a wide application in many different fields. In fact, one of the early pseudorandom number generator function, rand, is named after the Rand Corporation.

Simulations, modeling, or Monte Carlo calculations require a reliable set of random numbers. Truly random numbers might in principle be obtained by observation of stochastic phenomena. One could, for instance, utilize the time interval between decays of radio isotopes or the time interval between the passages of cosmic rays through a particle detector. One could also use the electronic noise intrinsically present in any electronic circuit. By their nature, stochastic phenomena are such that given one "event" it is impossible to predict when the next will occur. Additionally, one does not expect correlations between the occurrence of the events; a sequence of events is intrinsically unpredictable. The generation of random numbers in this fashion would require a device to measure the natural phenomenon (e.g., electronic noise) and digitization of the signal. Both are often impractical and even problematic. For instance, use of electronic noise would require white noise, for example, signals whose Fourier transform yields a spectrum where all frequencies have equal probability (uniform probability) density over a wide range of frequencies. The digitization process is also challenging. A useful random generation must produce numbers with a fine granularity. It should have the capacity to produce billions of different numbers in order to permit simulation with high resolution and avoid coarse binning of the observables or variables being simulated. This would in practice require digitization with an exceeding large number of bits to enable production of random number sets with fine granularity that provide a reasonable approximation of a subset of the set of real numbers, $\mathbb{R}$. Although these requirements can be met in principle, their realization is impractical for most computing applications, and generally considered not worth doing. Mathematicians, scientists, and engineers alike thus commonly use pseudorandom numbers generated algorithmically by a computer program. This said, note that there exists providers of true random numbers based, for instance, on atmospheric noise [100].

The production and study of pseudorandom numbers is an area of research of its own. We here focus on the essential concepts commonly used in the generation of random numbers but include references to more advanced or specialized works.

13.2 Monte Carlo Integration

It is convenient to first introduce the Monte Carlo method as a technique to carry out integrals of single and multidimensional functions. This is of particular interest because, in essence, all applications of the Monte Carlo technique may be reduced to the calculation of some integral.

In order to motivate the Monte Carlo integration method, we first review a few theorems of basic calculus (without demonstration).

Mean Value Theorem for Integrals

If a function, $f(x)$, is continuous over a closed interval $[a, b]$, then there exists a number c, with $a < c < b$, such that

$$f(c) = \frac{1}{b-a} \int_a^b f(x)\,dx. \tag{13.1}$$

Composite Midpoint Rule

The area under a curve, $f(x)$, in a given interval $[a, b]$, can be calculated by partitioning the interval in n subintervals $\{[x_{k-1}, x_k]\}_{k=1}^n$ of equal width, h. As per the mean value theorem stated previously, there are values c'_k for each subinterval such that

$$\int_{x_{k-1}}^{x_k} f(x)\,dx = (x_k - x_{k-1})\,f(c'_k) = hf(c_k'). \tag{13.2}$$

The integral of $f(x)$ in the interval $[a, b]$ may then be written

$$\int_a^b f(x)\,dx = \sum_{k=1}^n \int_{x_{k-1}}^{x_k} f(x)\,dx = h \sum_{k=1}^n f(c'_k). \tag{13.3}$$

If instead of using the values c'_k satisfying Eq. (13.2), one uses the midpoints $c_k = a + (k - \frac{1}{2})h$, for $k = 1, 2, \ldots, n$, then one can write:

$$\int_a^b f(x)\,dx = M(f, h) + E_M(f, h), \tag{13.4}$$

where

$$M(f, h) = \frac{b-a}{n} \sum_{k=1}^n f(c_k) \tag{13.5}$$

is known as the **composite midpoint rule** and constitutes an approximation of the integral in Eq. (13.4) with error $E_M(f, h)$. One can show that there exists a value c with $a < c < b$ such that the error $E_M(f, h)$ is given by

$$E_M(f, h) = \frac{1}{24}(b-a)\,h^2 f^2(c). \tag{13.6}$$

As such, the error $E_M(f, h)$ is proportional to the square of the bin size h while the area is linear in h. This implies it is possible to obtain an arbitrarily suitable approximation of the integral by choosing a sufficiently small value of the bin width h.

Monte Carlo Integration Rule

The Monte Carlo method is based on the composite midpoint rule approximation. However, rather than evaluating the functions at the midpoint of bins of equal width, one chooses n randomly distributed points x_1, x_2, x_n in the interval $[a, b]$ and evaluates the average of the function at these points:

$$\hat{f} = \frac{1}{n}\sum_{i=1}^{n} f(x_i). \tag{13.7}$$

An approximation of the integral is then obtained by multiplying this average by the width of the interval $[a, b]$:

$$\int_a^b f(x)\,dx = (b-a)\hat{f} + E_{MC}(f, n). \tag{13.8}$$

The error on the integral, $E_{\rm MC}(f, n)$, can be estimated from

$$E_{\rm MC}(f, n) = (b-a)\sqrt{\frac{\widehat{f^2} - (\hat{f})^2}{n}}, \tag{13.9}$$

where

$$\widehat{f^2} = \frac{1}{n}\sum_{i=1}^{n} f^2(x_i). \tag{13.10}$$

The error is thus proportional to the variance of $f(x_i)$ and inversely proportional to the square root of the number of points (samples) included in the MC calculation. As such, one concludes that the estimate converges slowly to zero with $n \to \infty$ and is of order $n^{-1/2}$ for finite n.

For single-variable functions, the Monte Carlo integration converges slower than conventional techniques such as the trapezoid method, which converges as n^{-2}, or the m-point Gauss method, which converges as n^{-2m+1}. A numerical integral of a single variable function is thus most effectively obtained with the Gauss method. But integration over multidimensional space using the trapezoid or Gauss methods (or similar methods) can quickly become prohibitive. For instance, a system with m particles has as many as $3 \times m$ degrees of freedom (excluding rotational degrees of freedom): integration using say 100 bins in each dimension would require $100^{3\times m}$ bins and is thus clearly impractical for large values of m, even with modern computers. However, one can show that the convergence of a conventional integration technique in d dimensions scales as $n^{-2/d}$ and is thus slower than the MC convergence ($n^{-1/2}$) for problems involving $d \geq 5$ dimensions. The use of Monte Carlo techniques for problems of five or more dimensions (degrees of freedom) is thus clearly advantageous and commonly used in practical applications.

13.3 Pseudorandom Number Generation

13.3.1 Definition and Basic Criteria

It is usually preferable to use pseudorandom numbers rather than attempting the production of true random numbers. Pseudorandom numbers are generated by use of computer algorithms designed to produce a sequence of number of apparently uncorrelated numbers that are uniformly distributed over a predetermined range. Modern algorithms provide very long sequences of random numbers that fulfill typical requirements of simulations carried out in physics as well as in the simulation of engineered systems.[1] Pseudorandom generators also provide the possibility of repeating the same exact experiment (simulation) many times, thereby making debugging of complex simulation programs considerably easier. In addition, modern random number generators are portable, thereby enabling the production of identical number sequences on different computers and even different operating systems, compilers, and computer architectures. A program can thus be used and tested on different computers with predictably identical results.

Random number generators are typically required to meet the following basic criteria:

1. They must be uniformly distributed within a specific range.
2. They must satisfy statistical tests of randomness, lack predictability, and there should be no correlations between neighboring numbers of a sequence.
3. An algorithm should produce a large sequence of different numbers before repeating a cycle.
4. The calculation should be very fast.

Let us consider each of these points in some detail. (1) Uniformly distributed numbers are required to ensure no region of the parameter space used in a simulation is more probable than others. (2) Randomness and unpredictability are obviously required as the basis of Monte Carlo simulations. Absence of correlations between nearby numbers of a sequence is required to ensure there are no unintended autocorrelations or intercorrelations between the generated simulation variables. (3) Given their algorithmic nature, pseudorandom numbers based on a finite number of bits architecture may feature two undesirable traits: Poor designs lead to algorithms getting stuck on a specific value, for instance zero. The presence of such a fixed end-point implies a sequence is usable only for a finite number of calls to the generator. Good designs feature sequences that do not get stuck on a specific value but eventually repeat: they are cyclical with a finite period. Obviously, cycles should be as long as possible to ensure truly uniform and nonclumpy distributions. (4) Lastly, speed is required to enable rapid and efficient simulations where the CPU's time is spent on the model calculations rather than the generation of random numbers.

13.3.2 Production of Uniform Deviates

There are numerous random number generators on the market (for a relatively recent review, see [114]). These may be included as part of software packages such as Mathematica®, MATLAB®, and ROOT, or as standalone codes provided with compiler libraries. Source codes are available from many vendors as well.

In essence, all generators use the same fundamental concept based on a multiplication method called **uniform deviates method**. The method requires an integer r_0 be chosen as a seed, which provides a starting value, and two integer constants a and m. Successive

[1] Actually, Monte Carlo simulations are now commonly used in all sciences.

pseudorandom numbers r_i are produced sequentially using the recursion relation

$$r_{i+1} = (a \times r_i) \quad \text{mod } m, \tag{13.11}$$

where mod stands for the "modulo" operation that yields the remainder of the left number division by the integer m. Studies have demonstrated that with a careful choice of the constants a and m, one can obtain finite sequences of numbers that appear to be randomly distributed between 1 and $m - 1$. It thus suffices to divide the numbers by m to obtain seemingly continuous random numbers in the range $[0, 1[$. The length of a sequence is determined by the choice of the constants and is limited by the computer word size, that is, the number of bits used in integer computation. For instance, values $m = 37$, and $a = 5$ lead to a sequence of 36 "random" numbers. Obviously, practical implementations involve much larger constant values, which produce much longer sequences. There are also several variations of this multiplication algorithm [114].

Detailed statistical studies must be made to ensure that untested random generators produce acceptable sequences of numbers. Failure to use generators abiding by the criteria listed in the previous section may lead to erroneous results and conclusions. In the 1970s, a number of commercial generators provided with FORTRAN compilers and operating systems featured unsuitable degrees of correlations. However, in recent years, better care has been taken by suppliers, and commercially available generators are typically quite reliable.

The numbers produced by the multiplication algorithm, Eq. (13.11), are not truly random, and one might be concerned about the existence of hidden correlations. A basic technique used to improve the generated sequences is to shuffle the numbers of a sequence. In effect, one uses two sequences of numbers simultaneously with distinct values of constant a and m: one sequence is stored in an array and a number from the second sequence is used as an index to select numbers from the first sequence. Obviously, this technique is limited by storage and capacity. Local shuffling within a block of numbers can be also used.

The ROOT environment provides four distinct uniform random generators named TRandom, and TRandom$X$, with $X = 1, 2, 3$. The TRandom3, based on the "Mersenne Twister generator" [141], features a period of about 10^{6000}, is quite fast, and is as such most likely the best choice for the generation of uniform deviates. Note that the availability of several distinct generators makes it possible to carry out calculations with more than one generator and verify that the results are reasonably generator independent.

A discussion of techniques to evaluate the degree of correlation produced by a given generator is beyond the scope of this textbook but may be found, for instance, in ref. [189].

13.3.3 Inversion (Transformation) Method

Consider the problem of generating random numbers according to a nonnegative function, $f(x)$, in some allowed range $x_{\min} \le x < x_{\max}$. We wish to produce numbers x randomly in such a way that the probability of such generated numbers to be within the interval $[x, x + dx]$ is proportional to $f(x)\,dx$. The function $f(x)$ does not need be normalized so that its integral over the chosen applicability range equals unity. In fact, as discussed in the previous section, a Monte Carlo technique can actually be used to evaluate that integral.

The **inversion method** is applicable to a function $f(x)$ whenever it is possible to find a primitive function $F(x)$ and calculate its inverse $F^{-1}(x)$. Given that the goal is to produce random numbers in the range $x_{\min} \le x < x_{\max}$ with probability $f(x)\,dx$, consider the integral

$$\int_{x_{\min}}^{x} f(x')\,dx' = \int_{x_{\min}}^{x} dF(x') = F(x) - F(x_{\min}) \tag{13.12}$$

The weight associated to the integral in dF is unity. Effectively, this corresponds to having uniformly distributed random numbers, that is, the distribution of F is uniformly distributed. One can then use a uniform random generator to select a value of F and use its inverse F^{-1} to determine the corresponding value x. The integral of $f(x)$ over the range $x_{\min} \le x' < x$ is a linear fraction R of the integral over the range $x_{\min} \le x' < x_{\max}$ which one may write

$$\int_{x_{\min}}^{x} f(x')dx' = F(x) - F(x_{\min}) = R\left(F(x_{\max}) - F(x_{\min})\right). \tag{13.13}$$

Solving the preceding for $F(x)$,

$$F(x) = R\left(F(x_{\max}) - F(x_{\min})\right) + F(x_{\min}), \tag{13.14}$$

and inverting the function to obtain the value of x, one gets

$$x = F^{-1}\left(R\left(F(x_{\max}) - F(x_{\min})\right) + F(x_{\min})\right). \tag{13.15}$$

The random generation of numbers distributed according to a function $f(x)$ is thus relatively straightforward and may be obtained according to the following basic algorithm. First, use a uniform random generator bound between 0 and 1 to determine a value of R, and next insert this value in Eq. (13.15) to produce a value of x randomly distributed according to $f(x)$.

Before we proceed with the implementation of the inversion method in two examples below, note that, although simple and efficient, the method is unfortunately applicable only in a very limited number of cases: those for which the function $f(x)$ has an analytical primitive (i.e., indefinite integral) $F(x)$ easily invertible to provide an analytical solution. If the function $F(x)$ exists but cannot be inverted, Eq. (13.15) cannot be applied directly, but it is nonetheless possible to obtain a generator by using a numerical inversion technique. For cases where no primitive function $F(x)$ exists, one must consider alternative approaches, such as those presented in §§13.3.4 and 13.3.5.

Example 1: Linearly Distributed Random Numbers

Problem

Build a random generator that produces numbers linearly distributed, that is, according to $f(x) = ax$, in an arbitrary range $[x_{\min}, x_{\max}]$, with $x_{\max} > 0$.

Solution

The function $f(x)$ is integrable, which means it is possible to use the inversion method. The integral of $f(x)$ is

$$y = F(x) = \frac{1}{2}ax^2. \tag{13.16}$$

Given a value y, one solves for x according to

$$x = \sqrt{\frac{2y}{a}}. \tag{13.17}$$

The inverse function F^{-1} may then be written

$$F^{-1}(y) = \sqrt{\frac{2y}{a}}. \tag{13.18}$$

Now insert this expression into Eq. (13.15) to obtain a linearly distributed value of x:

$$x = \sqrt{Rx_{\max}^2 + (1 - R)x_{\min}^2}. \tag{13.19}$$

Example 2: Exponentially Distributed Random Numbers

Problem

Build a random generator that produces numbers distributed according to an exponential distribution:

$$f(x) = ke^{-x/\lambda}. \tag{13.20}$$

Solution

The indefinite integral of the preceding exponential is $F(x) = -k\lambda e^{-x/\lambda}$. Its inverse is

$$F^{-1}(y) = -\lambda \ln\left(-\frac{y}{k\lambda}\right). \tag{13.21}$$

Application of Eq. (13.15) yields

$$x = -\lambda \ln\left(R \exp(-x_{\max}/\lambda) + (1 - R)\exp(-x_{\min}/\lambda)\right), \tag{13.22}$$

where R is a uniformly distributed random number in the range [0, 1].

13.3.4 Acceptance/Rejection Method

Often times, the transformation method is not applicable because the analytical integral of a function does not exist, or because, if it does exist, it is impossible to invert it. One must then explore alternative methods. We first consider the **acceptance/rejection method**, which is widely applicable provided the maximum $f_{\max}$ of the function $f(x)$ is known, that is, if the function satisfies $f(x) \leq f_{\max}$ in the range of interest.

To apply this method, consider an arbitrary value of x in the range $[x_{\min}, x_{\max}]$ has been selected. The quantity $f(x)dx$ is proportional to the probability of having this particular

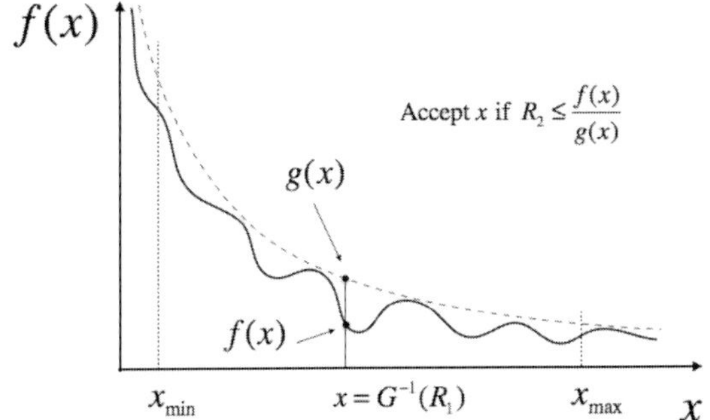

Fig. 13.1 Schematic illustration of the principle of the inversion + acceptance hybrid method.

value of x. Given the function maximum $f_{\max}$ is known, one can express this probability as $f(x)/f_{\max}$, and one can then use a uniformly generated random number in the range [0, 1] to decide whether the selected value of x is accepted, that is, one accepts the value of x provided the random number R is smaller than the relative probability $f(x)/f_{\max}$.

The acceptance/rejection method requires the use of two random numbers: the first is used to select a specific value of x and the second to determine whether this value is accepted. It may then be implemented according to the following algorithm:

1. Generate two uniformly distributed random numbers r_1 and r_2 in the range [0, 1].
2. Use the first number r_1 to select a value x with a uniform probability in the allowed range $[x_{\min}, x_{\max}]$:

$$x = x_{\min} + r_1 (x_{\max} - x_{\min}). \tag{13.23}$$

3. Compare r_2 with the ratio $f(x)/f_{\max}$; if $r_2 > f(x)/f_{\max}$, and then reject this instance of x and start over at step 1 for a new try. Otherwise retain the value of x as the final answer.

The acceptance/rejection method works for any function $f(x)$ with an upper-bound $f_{\max}$, but it may become highly inefficient (and slow) if the function $f(x)$ has large spikes or varies dramatically over the range of interest $[x_{\min}, x_{\max}]$. The efficiency of a pseudorandom generator method is defined as the probability that a value of x will be retained at any given try. In the context of the acceptance/rejection method, it amounts to the ratio of the integral of the function over the range $[x_{\min}, x_{\max}]$ by the area $f_{\max}(x_{\max} - x_{\min})$. The efficiency can thus be low if the integral of $f(x)$ is much smaller than the area, that is, if $f(x)$ is in general much smaller than $f_{\max}$.

There are also cases when an upper bound cannot be defined. This is the case, for instance, for functions of the form $1/x^{\alpha}$, with $\alpha > 0$, evaluated with a lower bound at zero. Variable transformations may then be used to make the function smoother. For instance, using $y = \ln x$ would enable a roughly constant function over a narrow interval and therefore yield a fairly large efficiency.

In view of the limited efficiency of the acceptance/rejection method, it is often convenient to combine it with the inversion method discussed in the previous section. This is accomplished, as illustrated in Figure 13.1, by using an *envelope* function $g(x)$, which satisfies the two following conditions: (1) The value of the function $g(x)$ is such that

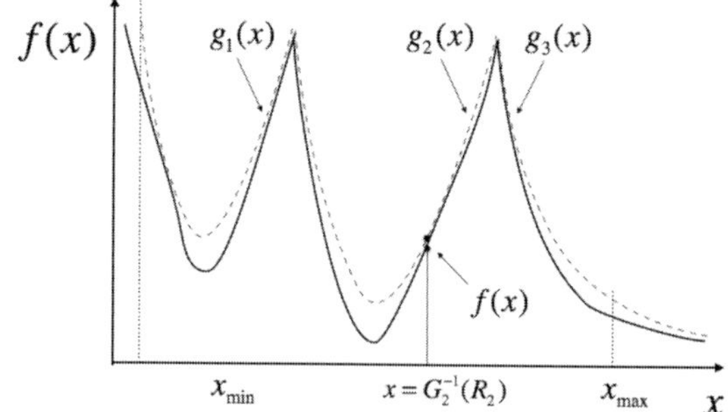

Fig. 13.2 Schematic illustration of the principle of the composite inversion + acceptance hybrid method.

$f(x) \le g(x)$ everywhere over the interval of interest and (2) the function $g(x)$ must have an easily calculable primitive function $G(x)$, which is invertible, that is, with a known analytical solution for $G^{-1}(y)$. A trial value x is generated according to $x = G^{-1}(R_1)$, with uniform deviate $R_1 \in [0, 1]$, and a second uniform deviate $R_2 \in [0, 1]$ is used to decide whether the selected value of x shall be accepted and returned. The value is accepted if R_2 is smaller than the ratio $f(x)/g(x)$. This hybrid method operates similarly as the acceptance/rejection method but offers better efficiency because rather than choosing x values uniformly over the interval of interest, one uses a function $g(x)$ to **presample** x. Ideally, a function $g(x)$ mimics the overall behavior of the function $f(x)$, thereby providing an increased efficiency. The efficiency of the method is determined by the ratio of the integral of the two functions over the interval. And, provided $g(x) < f_{max}$ over the interval $[x_{min}, x_{max}]$ of interest, one finds the efficiency of this method is greater than that of the basic acceptance/rejection method. The efficiency of the method is

$$\epsilon = \frac{\int_{x_{min}}^{x_{max}} f(x)\,dx}{\int_{x_{min}}^{x_{max}} g(x)\,dx} > \frac{\int_{x_{min}}^{x_{max}} f(x)\,dx}{f_{max}(x_{max} - x_{min})}. \tag{13.24}$$

The algorithm of this hybrid method can be summarized as follows:

1. Generate a random number R_1 with uniform distribution in the range $[0, 1]$, and use the inverse function $G^{-1}(R_1)$ to obtain a trial version of x.
2. Generate a random number R_2 with uniform distribution in the range $[0, 1]$.
3. If $R_2 > f(x)/g(x)$, reject this value of x value and start over. Otherwise retain the value of x as the final answer.

The method first selects x according to the probability $g(x)\,dx = dG(x)$. Then the acceptance/rejection part of the algorithm accepts x with a probability $f(x)/g(x)$. The overall probability of getting a specific value of x is thus $g(x)\,dx \times f(x)/g(x) = f(x)\,dx$ as needed to properly sample the function $f(x)$.

Functions used in theoretical calculations often feature several spikes, as illustrated in Figure 13.2. It may then be difficult to find a "simple" function with similar behavior and each spike may have to be handled separately. This is achieved by partitioning the range of interest in n subdomains where one can apply n sampling functions $g_k(x)$, with $k = 1, \ldots, n$. Clearly, each of the functions must satisfy the condition $f(x) \le g_k(x)$ in their respective range of validity and have an invertible primitive function $G_k^{-1}(y)$. One can thus define a composite function $g(x)$ as

$$g(x) = \begin{cases} g_1(x) \text{ for } x_{\text{min}} \leq x < x_1 \\ \dots \quad \dots \\ g_k(x) \text{ for } x_{k-1} \leq x < x_k \\ \dots \quad \dots \\ g_n(x) \text{ for } x_{n-1} \leq x < x_{\text{max}} \end{cases} \tag{13.25}$$

such that the functions $g_k(x)$ have primitive functions, and can be inverted. Generation of pseudorandom numbers may then be achieved with the following algorithm:

1. Generate a uniform deviate $R_1 \in [0, 1[$ to select an integer value k satisfying $P_{k-1} \leq R_1 < P_K$ at random in the range $[1, n]$, based on the cumulative probabilities $P_k = \sum_{i=1}^{k} p_i$ for $1 \leq k \leq n$ and $P_0 = 0$, where the coefficients p_i represent integrals of the functions $g_i(x)$ in their respective ranges of applicability
$$p_i = \int_{x_{min,i}}^{x_{max,i}} g_i(x)\, dx. \tag{13.26}$$
2. Given the function g_k selected, generate a random number R_2 with uniform distribution in the range $[0, 1[$ and use the inverse method to generate a value x according to
$$x = G_k^{-1}\left(G_k(x_{\text{min,k}}) + R_2\left[G_k(x_{\text{max,k}}) - G_k(x_{\text{min,k}})\right]\right) \tag{13.27}$$
3. Generate a random number R_3 and compare it to the ratio $f(x)/g_k(x)$; if larger, then reject the value and return to step 1 for a new try. Otherwise, retain and use the most recent value of x as the final answer.

13.3.5 Integrated Histogram Method

The **Integrated Histogram method** is a numerical technique based on the inversion method discussed in §13.3.3. It is commonly used either as a practical technique to handle non-integrable functions or for the generation of random numbers based on histograms measured experimentally, particularly in the context of analyses of the performance of a measurement or detection system.

Recall that the inversion method requires inversion of the indefinite integral, $F(x)$, of a function $f(x)$ one wishes to sample. The integrated histogram method replaces the function $f(x)$ by a histogram $h(x)$ meant to be a reasonable representation or approximation of the function. Additionally, it replaces the analytical inversion of the function $F(x)$ by a numerical inversion based on a binary search across H_x, a histogram consisting of a running sum of h_x, as schematically illustrated in Figure 13.3. Let us examine the method in more detail.

Consider the generation of random numbers in the range $[x_{\text{min}}, x_{\text{max}}]$ of interest. The basic idea of the method is to replace the function $f(x)$ by a histogram h_x with sufficiently many bins to enable a reasonable representation of the function. If $f(x)$ is known exactly, one has the latitude to decide the precision with which the histogram must be binned.

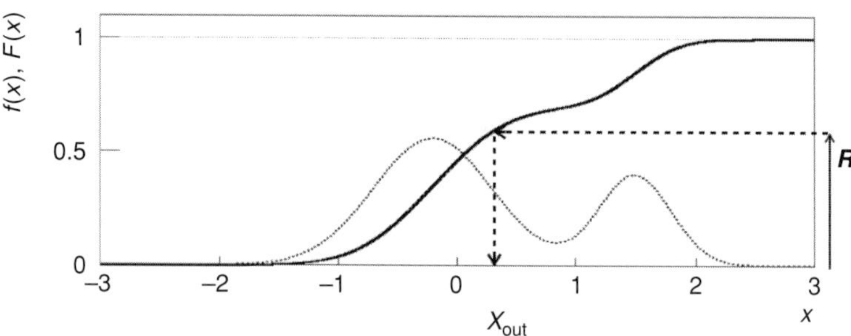

Fig. 13.3 Illustration of the integrated histogram method. The solid line represents the cumulative distribution function, $F(x)$, of the PDF $f(x)$, shown as a dotted line. The random number R is produced with a uniform PDF in the range [0, 1]. The method outputs x_{out}, which corresponds to the middle of bin m satisfying the condition (13.30). The functions $f(x)$ and $F(x)$ are here plotted with 1,000 bins to give the illusion of a continuous function.

If the histogram is obtained from a measurement, one can only hope that the statistical accuracy and the binning are both sufficient to obtain a reasonable representation of $f(x)$. To simplify, let us assume the histogram is partitioned in n bins of equal width Δx in the range $[x_{\min}, x_{\max}]$. Formally, the content of each bin, $h_x(i)$, with $i = 1, \ldots, n$, is given by the integral of the function $f(x)$ over the interval spanned by the bin:

$$h_x(i) = \int_{x_{\min,i}}^{x_{\max,i}} f(x)\,dx, \tag{13.28}$$

where $x_{\min,i} = x_{\min} + (i-1) \times \Delta x$ and $x_{\max,i} = x_{\min} + i \times \Delta x$. In practice, if the bins are sufficiently narrow, one may replace the preceding integral by $f(x_i)\Delta x$ where x_i is the midpoint of each bin. It may also be sufficient to use the (average) value of the function in the bin rather than the area under the curve. The definite integral $F(x) - F(x_{\min})$ may then be replaced by the sum

$$H_x(m) \equiv \sum_{i=1}^{m} h_x(i). \tag{13.29}$$

In the original inversion method, one uses a linear fraction $F(x) = R(F(x_{\max}) - F(x_{\min}))$ determined by a random number R to obtain a value x. This linear fraction is here replaced by $R \times H_x(n)$ or simply R if the sum $H_x(n)$ is normalized to unity. One must then find the value m satisfying

$$H_x(m-1) \le R \times H_x(n) < H_x(m). \tag{13.30}$$

The generated value x may then be chosen has the middle of bin m. Alternatively, one can carry out a linear or quadratic interpolation based on nearby bins $H_x(m \pm 1)$. For sufficiently narrow bins, the method produces random deviates that mimic the function $f(x)$ (or the histogram representing the function).

The search for the value m that satisfies the preceding condition is a weakness of the method. If the number of bins, n, is large, the task of finding which value m satisfies the condition can be rather CPU intensive. A linear search, that is, proceeding by systematic inspection of all values of m starting from 1, takes an average of $n/2$ steps. However, a binary tree search requires an average number of steps that scales as $\ln n$ and is thus considerably faster for histograms with a large number of bins.

13.3.6 Multidimensional Pseudorandom Generation

The simultaneous generation of random numbers for a system involving n variables may be handled as the random selection or production of an n-dimension vector $\vec{x}$ in an n-dimensional space. There are two basic methods, each with several variations, to handle the generation of vectors $\vec{x}$ in an n-dimensional space.

The first method assumes the generation of numbers can be factorized (at least approximately) whereas the second method is applicable whenever the range of the components exhibit a simple hierarchical relation.

Factorization Method

Production of random numbers distributed according to a function $f(\vec{x})$ with the **factorization method** is possible provided there exists a function $g(\vec{x})$ everywhere equal or larger than $f(\vec{x})$ and which can be factorized into a product of n distinct functions:

$$g(\vec{x}) = g_1(x_1) \times g_2(x_2) \times \cdots \times g_n(x_n) = \prod_{i=1}^{n} g_i(x_i), \tag{13.31}$$

where x_i, $i = 1, 2, n$ are the components of the n-dimensional vector $\vec{x}$. Each of the functions $g_i(x_i)$ may be simple or composite. The generation of vectors proceeds as follows:

1. First generate n x_i values independently using each of the g_i functions.
2. Generate a random number R in the range $[0, 1]$ and compare it to the ratio $f(x)/g(x)$; if R is smaller than the ratio, retain the point, otherwise repeat step 1.

If the function $f(x)$ is itself factorable as a product of n functions (one for each dimension), it then suffices to set $g(x) = f(x)$ and the method reduces to step 1.

Hierarchical Boundary Conditions Method

The generation of random numbers can become an arbitrarily complicated problem if the ranges of the n components of the vector $\vec{x}$ are interdependent. In the simplest cases, the ranges of the n components are independent of each other and the generation of $\vec{x}$ then takes place in an n-dimensional hypercube. But there can be systems such that the generation of the different components of $\vec{x}$ have intricate interdependencies. This would be the case, for instance, if the range of x_3 depends on x_2 whose range itself depends on the value of x_1 or other values, etc.

The hierarchical boundary conditions method is useful if the boundaries of allowed regions can be written in a form where the range of x_1 is known, the allowed range of x_2 depends only on x_1, that of x_3 only on x_1 and x_2, and so on, until x_n whose range may depend on all the preceding variables. Under such conditions, it may be possible to find a function $g(x)$ that can be integrated over x_2 through x_n to yield a simple function of x_1, according to which x_1 is selected. Having done that, x_2 is selected according to a distribution which now depends on x_1, but with x_3 through x_n integrated over. The procedure is continued until x_n is reached. In the end, the ratio $f(x)/g(x)$ is used to determine whether to retain the point.

13.3.7 Gaussian Generator

By virtue of the central limit theorem, a superposition of n random variables (i.e., a sum) leads to a Gaussian distribution in the limit $n \to \infty$. The Gaussian PDF is thus of special interest since it may be used to represent the fluctuations of a wide variety of processes and systems. One indeed finds that many random observables and systems can be described to a very good approximation with a Gaussian PDF. However, the generation of Gaussian deviates is not a trivial task. Indeed, first note that the Gaussian distribution cannot be integrated analytically. This implies one cannot use the inversion method and must resort to alternative techniques. Obviously, the rejection method is readily applicable for a Gaussian. However, it is quite inefficient and slow because the function has tails extending to infinity. Alternatively, one might consider using the central limit theorem itself since the sum of n numbers generated according to a uniform random generator in the range $[0, 1]$ has a Gaussian distribution in the large n limit. But this method is also rather inefficient and CPU-intensive given the production of a single Gaussian deviate as a sum requires the generation of a large number of uniform deviates. A much more efficient technique is based on the simple mathematical construction discussed in the text that follows.

A Gaussian PDF depends on two parameters: a centroid μ and a width σ. One might thus have the impression that an infinite number of different generators might be required to match arbitrary values of μ and σ. This is not the case, however, since any Gaussian can be mapped onto the standard normal distribution $\phi(z)$ defined such that the probability of finding z in the interval $[z, z + dz]$ is given by

$$\phi(z)dz = \frac{1}{\sqrt{2\pi}} \exp\left(-\frac{z^2}{2}\right) dz. \tag{13.32}$$

A simple change of variable $x = \sigma z + \mu$ thus provides for a Gaussian PDF with arbitrary mean μ and standard deviation σ. Indeed, since $z = (x - \mu)/\sigma$, and $dz = dx/\sigma$, one gets

$$p_G(x)dx = \frac{1}{\sqrt{2\pi}\,\sigma} \exp\left(-\frac{(x-\mu)^2}{2\sigma^2}\right) dx. \tag{13.33}$$

Gaussian deviates with arbitrary mean μ and standard deviation σ may thus be obtained by scaling random numbers z generated with a standard normal distribution according to $x = \sigma z + \mu$.

Normally distributed random numbers can be efficiently produced with an algorithm devised by Box and Muller [57]. Although a one-dimensional Gaussian PDF is not integrable, a simple transformation enables the integration of a two-dimensional Gaussian. It is thus possible to efficiently produce pairs of Gaussian deviates. Indeed, consider the two-dimensional (standard) normal distribution defined as follows:

$$\phi_2(x, y) = \frac{1}{2\pi} \exp\left(-\frac{x^2 + y^2}{2}\right) = \phi(x)\phi(y). \tag{13.34}$$

The technique introduced by Box and Muller is based on the transformation

$$x = \sqrt{-2\ln r} \cos 2\pi\phi, \tag{13.35}$$

$$y = \sqrt{-2\ln r} \sin 2\pi\phi, \tag{13.36}$$

where r and ϕ are two uniformly distributed deviates in the semi-open range $]0, 1]$. The applicability of the method is readily verified since by definition of the variables x and y one gets

$$x^2 + y^2 = -2\ln r, \tag{13.37}$$

$$\exp\left(-\frac{x^2 + y^2}{2}\right) = \exp(\ln r) = r. \tag{13.38}$$

Values of r and ϕ uniformly distributed in the range $]0, 1]$ thus indeed map onto Gaussian deviates x and y. It is consequently possible to write a computer function that produces two Gaussian deviates per call to the function. The method is very CPU-efficient because it requires the generation of (only) two uniform deviates to produce two Gaussian deviates.

13.4 Selected Examples

We consider a few practical examples of random number generation commonly encountered in nuclear and particle physics.

13.4.1 Generation of Random Two-Dimensional Vectors

Let us begin by considering the generation of random vectors $\vec{r}$ in a two-dimensional configuration space. Obviously, two random numbers will be required for each random vector. However, one has a choice of representation: should one express the vectors in Cartesian coordinates or in polar coordinates? In the first case, one would have to generate two numbers r_x and r_y corresponding to positions along axes x and y. In the second case, one needs to produce a radius, r, and a polar angle θ.

The choice of representation shall depend on the specificities of the problem considered, as illustrated in Figure 13.4. For rectangular geometry involving, for instance, the simulation of particles entering detectors of rectangular cross section, it is obviously simplest to generate numbers x and y in Cartesian coordinates corresponding to the position of the particle's entry point along the face of the detector. On the other hand, if the vector $\vec{r}$ represents a displacement bound in some range $r_{\min} \le |\vec{r}| < r_{\max}$, it might be more convenient to generate vectors in terms of radii r and polar angles θ.

To generate a random position uniformly in a rectangle defined by two corners $(x_{\min}, y_{\min})$ and $(x_{\max}, y_{\max})$, it is simplest to use rectangular coordinates as illustrated in

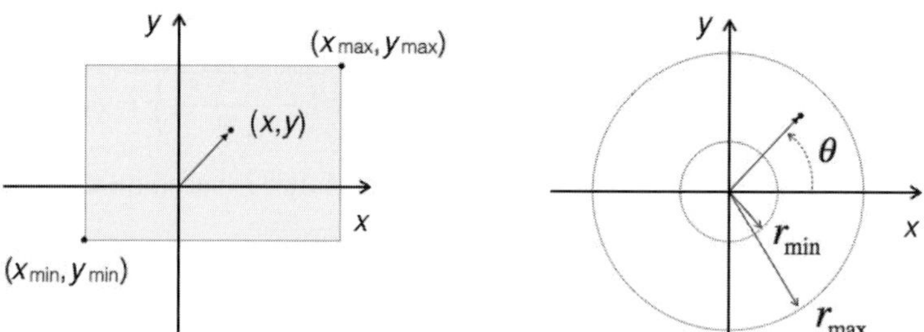

Fig. 13.4 Illustration of two geometries used for the generation of 2D random vectors. (a) Rectangular geometry, (b) circular geometry.

Figure 13.4a:

$$x = x_{\min} + r_1 \times (x_{\max} - x_{\min}), \tag{13.39}$$

$$y = y_{\min} + r_2 \times (y_{\max} - y_{\min}), \tag{13.40}$$

where r_1 and r_2 are random numbers uniformly distributed in the range [0, 1].

The generation of random displacements expressed in polar coordinates constitutes a more interesting case. If all directions are uniformly probable, the direction may be randomly produced with $\theta = 2\pi r_1$ in which r_1 is a uniform deviate in the range [0, 1]. For displacement magnitudes uniformly distributed in the range $r \leq r_{\max}$, one might initially be tempted to use a linear expression such as $r_2 \times r_{\max}$. Note, however, that in polar coordinates, the element of area is $rd\theta\, dr$. Consequently, uniform sampling in r would lead to a decreasing number of "hits" per unit of area for increasing values of radius r. However, given the element of area can also be expressed as $d\theta d\left(r^2\right)$, the sampling should be uniform in r^2 rather than r and one writes

$$r^2 = r_{\min}^2 + r_2 \times \left(r_{\max}^2 - r_{\min}^2\right), \tag{13.41}$$

where r_2 is a uniform deviate in the range [0, 1]. Formally, this can be derived as follows: since by assumption the displacement density is set to be uniform per unit of area, the displacement density can be written

$$\frac{dN}{dxdy} = \frac{1}{A_{\mathrm{TOT}}}, \tag{13.42}$$

where A_{TOT} is a constant denoting the total area under consideration, that is, the fiducial area in which displacements can be generated. Transforming the (x, y) coordinates in polar coordinates (r, θ), one gets

$$\frac{dN}{drd\theta} = \frac{dN}{dxdy}\left|\frac{\partial(x,y)}{\partial(r,\theta)}\right| = \frac{r}{A_{\mathrm{TOT}}}. \tag{13.43}$$

The PDF determining the r distribution is thus

$$f(r) = \frac{dN}{dr} = \frac{2\pi r}{A_{\mathrm{TOT}}}. \tag{13.44}$$

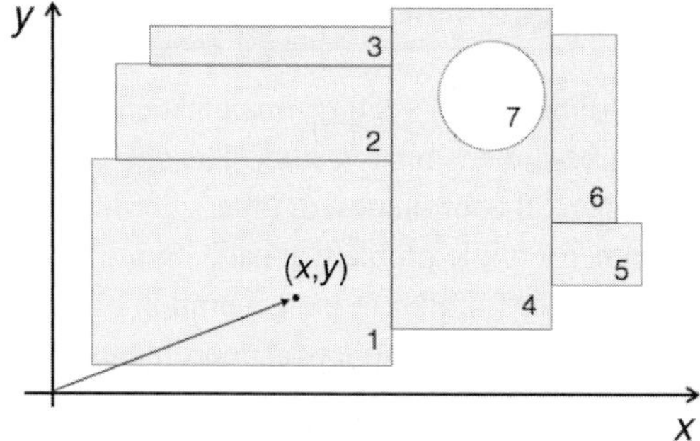

Fig. 13.5 Illustration of random number generation with complex masks featuring a combination of inclusion (gray) and exclusion regions (white).

Assuming minimum and maximum radii $r_{\min}$ and $r_{\max}$, let us proceed to derive a formula for the generation of r values using the inversion method. Integration of $f(r)$ yields $F(r)$ as follows:

$$\int_{r_{\min}}^{r} f(r)dr = F(r) - F(r_{\min}) = \frac{\pi}{A_{\text{TOT}}}\left(r^2 - r_{\min}^2\right). \tag{13.45}$$

Finally, inversion of $F(r)$ produces the anticipated result:

$$r = \sqrt{r_2\left(r_{\max}^2 - r_{\min}^2\right) + r_{\min}^2}, \tag{13.46}$$

where r_2 is a random uniform deviate in the range [0, 1]. The displacement vector is thus $\overrightarrow{\Delta r} = (r\cos\theta, r\sin\theta)$.

We have assumed that the radial displacements would generate a uniform density per unit of area. This is unlikely to be the case in general. A more appropriate probability density $f(r)$ that matches the specificities of a given system has to be identified and the size r of the displacement calculated accordingly. For instance, if $f(r)$ is constant in the range $[r_{\min}, r_{\max}]$, than radii may be generated according to $r = r_2\left(r_{\max} - r_{\min}\right) + r_{\min}$. If instead, $f(r) = 1/r$, then one gets $r = \exp\left[r_2\left(\ln r_{\max} - \ln r_{\min}\right) + \ln r_{\min}\right]$.

Perfectly rectangular or circular geometries are obviously uncommon in real-world situations. And it is in fact where Monte Carlo methods find their most interesting and powerful uses. In two-dimensional geometries involving nonrectangular boundaries (i.e., curved boundaries) or forbidden regions, one must exercise judgment and creativity as to what shall constitute the most CPU-effective generation process. As illustrated in Figure 13.5, one might, for instance, use a large rectangle that encloses the fiducial region (i.e., the region of interest) and use “masks” to dismiss points produced in forbidden regions.

In the case illustrated in Figure 13.5, this amounts to having mathematical formulation of the boundaries, that is, minimum and maximum value expressions for both the x and y. This can be accomplished with one or multiple functions. Alternatively, one can partition the sample space and use a two-step approach: for instance, if all positions are uniformly probable (in the allowed regions), one first randomly selects in which partition the position should be generated according to the relative size of the area of the various partitions, and subsequently generate a specific position within the selected partition.

13.4.2 Generation of Random Three-Dimensional Vectors

As with two-dimensional vectors, one must choose a coordinate system for the generation of random three-dimensional vectors. The choice of Cartesian coordinates, cylindrical coordinates, spherical coordinates, or other coordinate systems is determined by convenience and the symmetry of the problem at hand. Since the generation in Cartesian and cylindrical coordinates is rather similar to the generation of two-dimensional vectors, we focus on the generation of vectors using spherical coordinates, which requires the production of a radius r, a polar angle θ, and an azimuthal angle ϕ.

Let us consider all positions within a sphere of a radius $r_{\max}$ to be equally probable. The radius r is to be generated in the range $[0, r_{\max}]$, while the polar and azimuthal angles are in the ranges $0 \le \theta \le \pi$ and $0 \le \phi < 2\pi$, respectively. Since all positions are equally probable, the number of points per unit of volume $r^2 \sin\theta d\theta d\phi dr$ must be generated uniformly. Rewriting the element of volume $d\phi d(\cos\theta) d(r^3)/3$, we conclude that equally probable positions, within a sphere of radius $r_{\max}$, may be achieved by generating values for ϕ, $\cos\theta$, and r^3 using three random numbers r_i, $i = 1, 2, 3$ in the range $[0, 1]$ as follows:

$$\phi = 2\pi r_1, \tag{13.47}$$

$$\cos\theta = 2r_2 - 1, \tag{13.48}$$

$$r^3 = r_3 r_{\max}^3. \tag{13.49}$$

Note that by generating $\cos\theta$ in the range $[-1, 1]$, one ensures that all polar angles in the range $[0, \pi]$ are produced, and that all elements of volumes, being proportional to $d\cos\theta$, are equally probable. Likewise, the generation of r^3 with a uniform PDF, rather than r, insures that the point density remains constant throughout the volume of the sphere.

The generation of three-dimensional vectors whose probability are functions of the coordinates r, θ, and ϕ should be treated according to the specificities of the system to be simulated. Section 14.1.1 discusses the generation of momentum vectors according to a Maxwell–Boltzmann distribution as an example of nontrivial dependence on the length (radius) of the vector. Dependencies on θ and ϕ are obviously also possible. For instance, anisotropic flow production in heavy ion collisions may be simulated with a simple function of the azimuthal angle ϕ presented in §14.1.5.

13.4.3 Weighing Method

The generation of one or several random variables according to a specific PDF is not always practical or even desirable because the PDF might be too complicated or because the geometry and boundary conditions at hand are too intricate. It remains, nonetheless, that the relative frequency of random variables may have to obey specific probability density in order to represent the particular phenomenon being modeled. The **weighing method** may then provide a simple solution.

Let us consider the generation of "events" consisting of n random numbers $\vec{x} = (x_1, x_2, \ldots, x_n)$ in some domain D. Rather than generating each $\vec{x}$ according to a complicated procedure, one may elect to produce each of the values x_i, $i = 1, \ldots, n$ according

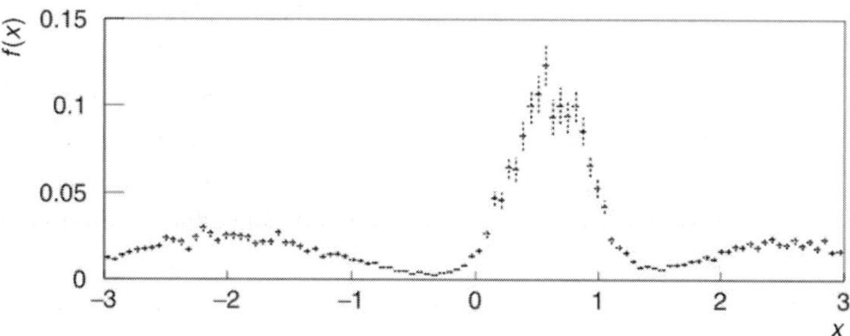

Fig. 13.6 Random triple Gaussian deviates generated with the weighing method according to Eq. (13.50).

to uniform generators, each in some appropriate or convenient range. The technique then involves the assignment of a weight to each $\vec{x}$ produced according to some weight function $W(\vec{x})$. The variable or variables required are first generated according to uniform distributions in their respective ranges of interest. Once the variables are generated, one verifies whether the event lies within the domain of interest. If it does not, the event is simply ignored and the procedure repeated. If the event is deemed of interest, one calculates its weight according $W(\vec{x})$. Then, rather than counting each event as a unit, one increments histograms of interest in proportion to the weight $W(\vec{x})$ of the event.

Consider, as a simple example, the production of events consisting of a single random value x to be distributed according to a triple Gaussian distribution in the range $-3 < x \le 3$ for specific values of the means μ_i, widths σ_i, and normalizations f_i for $i = 1, \ldots, 3$. Let us ignore the fact that there exists several techniques to obtain Gaussian deviates and generate values x according to a uniform distribution in the range $-3 < x \le 3$ and calculate for each value of x, the weight W according to

$$W(x) = \sum_{i=1}^{3} f_i \frac{1}{\sqrt{2\pi}\sigma_i} \exp\left[-\frac{(x-\mu_i)^2}{2\sigma_i^2}\right]. \tag{13.50}$$

Figure 13.6 displays a triple Gaussian distribution obtained with 10,000 events and $\mu_1 = -2$, $\sigma_1 = 0.75$; $\mu_2 = 0.6$, $\sigma_2 = 0.30$; $\mu_3 = 2.5$, $\sigma_3 = 0.60$.

The weight technique is particularly useful for the simulation of events involving several variables that are correlated but difficult to generate ab initio. It also presents the advantage that given a set of existing events, one can apply weights a posteriori and study, for instance, the effect of different weight functions on measured distributions.

Exercises

13.1 Use the inversion method to create a generator producing random numbers according to a parabolic PDF, $f(x) = ax^2$. First find the inversion expression and then implement it in a computer program. Use your program to generate a histogram of the PDF in the range $-10 \le x < 10$, and inspect the histogram for 100, 500, 1,000, 10,000 or more events.

13.2 Use the acceptance/rejection method to create a generator producing numbers according to $f(x) \propto x \exp(-x^2/2\sigma^2)$ in the range [0, 10].

13.3 Use the acceptance/rejection method to create a generator producing numbers according to $f(x, y) \propto x^2y^2 \exp\left(-\left(x^2 + y^2\right)/2\sigma^2\right)$ in the domain $-4 \le x < 4$, $-5 \le y < 5$, for $\sigma = 1.5$.

13.4 Use the MC method defined by Eq. (13.7) to calculate the area of a circle and determine the value of π.

13.5 Write a function that produces Gaussian deviates based on the Box and Muller method.

13.6 Write a function that produces Gaussian deviates based on the central limit theorem and uniform deviates in the range [0, 1]. Study the precision of the method by varying the number of deviates used in the sum.

13.7 Write a function that produces Gaussian random numbers based on the integrated histogram inversion method. Hint: Select a domain of applicability and produce a table or histogram containing a running integral of the Gaussian distribution.

13.8 Find a method to generate Gaussian random numbers in the x–y plane with expectation values $E[x] = \mu_x$, $E[y] = \mu_y$, and variances $\text{Var}[x] = \pi y$, and $\text{Var}[y] = \pi x$.

13.9 Using the generator written in Problem 13.1, study the precision of the mean, variance, skewness of the distributions obtained with 100, 500, 1,000, 10,000 or more events.

14 Collision and Detector Modeling

In particle and nuclear physics, Monte Carlo simulations are used for a wide variety of purposes, including prediction and interpretation of results, studies of the performance of detectors and their components, as well as corrections of measured data to obtain accurate outcomes that are independent of experimental or instrumental effects. A detailed coverage of all the types of simulations used by particle and nuclear physicists is thus well beyond the scope of this text. We focus our discussion, in §14.1, on basic techniques used in the simulation of simple physics observables, such as momentum spectra or correlation functions, while in §14.2, we present simple techniques used in the simulation of detector performance and data correction for smearing and efficiency effects.

14.1 Event Generators

Simulations of particle production by elementary particles or nuclei collisions are based on phenomenological and theoretical models of varying complexity. A large variety of models predicated on a wide range of assumptions are described in the literature and used as practical event generators by theoreticians and experimentalists to describe and interpret measured results or analyze the performance of particle detectors. Commonly used event generators for the simulation of proton–proton collisions include JETSET [174], PYTHIA [175], HERWIG [66], PHOJET [52], EvtGen [133], and several others. There is an even greater variety of models for the generation and simulation of heavy-ion collisions. Early models used in the simulation of AGS, SPS, and RHIC collisions include HIJING [188], RQMD [176], URQMD [29], AMPT [137], VENUS [191], and EPOS [192]. More recent developments for the study of the interaction of jets within the medium produced in heavy-ion collisions include Martini [169], Jewel [198], Cujet [61], YaJEM [164], and many others. As noted earlier, we restrict the scope of our discussion to a small selection of phenomenological and theoretically motivated models, which illustrate how more complex generators are built and can be used to carry out simple simulations and analyses.

14.1.1 Basic Particle Generators

Basic simulations of particle production in elementary particle collisions often assume that produced particles are uncorrelated and have a kinematical distribution determined by a specific PDF. The choice of PDF used in a particular simulation is dictated by the goals of the study and the needs for realistic reproduction of the attributes of collisions of interest. We focus our discussion on a few illustrative examples.

Isotropic Exponential Distribution

We first consider the simulation of isotropic particle production according to an exponential momentum distribution

$$\frac{dN}{dp} = \frac{1}{\lambda} e^{-p/\lambda} \tag{14.1}$$

over the momentum range $[p_{\min}, p_{\max}]$.

The generation of random momentum vectors requires three random numbers, r_1, r_2, and r_3, in the range $[0, 1]$. Isotropic emission is achieved by generating polar and azimuth angles according to

$$\begin{aligned} \phi &= 2\pi r_1 \\ \cos\theta &= 2r_2 - 1, \end{aligned} \tag{14.2}$$

whereas the generation of particle momenta according to the exponential distribution, Eq. (14.1), can be accomplished using the inversion technique, Eq. (13.22), introduced in §13.3.3:

$$p = -\lambda \ln\left(r_3 \exp(-p_{\max}/\lambda) + (1 - r_3)\exp(-p_{\min}/\lambda)\right), \tag{14.3}$$

where $p_{\min}$ and $p_{\max}$ define the range over which the momentum spectrum is to be generated.

Power-Law Distribution

Particle production, particularly at high momentum, can often be described as a power law over a restricted momentum range.

Strictly speaking, a power-law distribution is a function characterized by a scale invariance. For instance, a function of the type $f(x) = ax^k$ is invariant in form under scaling of the argument x by a constant factor c:

$$f(cx) = a\,(cx)^k = c^k f(x) \propto f(x). \tag{14.4}$$

Indeed, scaling the argument by a factor c only multiplies the amplitude of the function by a constant value. This behavior is in principle readily identified graphically, with data, by plotting the logarithm of the dependent variable f as a function of the logarithm of the independent variable x. The end result is a straight line with a slope equal to the exponent of the power law:

$$\ln(f) = \ln(ax^k) = k\ln(x) + \ln(a). \tag{14.5}$$

Approximate power-law behaviors have been identified in several areas of science (e.g., scale of earthquakes, sizes of moon craters, wealth of individuals, etc.). In particle physics, one observes that the production of high-momentum particles can be described with a steeply decreasing power law originating from the approximate self-similar behavior associated with parton fragmentation.

Power-law probability distributions are often extended by adding a slowly varying coefficient $S(x)$:

$$f(x) \propto S(x)x^{-\alpha}, \tag{14.6}$$

where $\alpha > 1$, and $S(x)$ is required to satisfy $\lim_{x\to\infty} S(tx)/S(x) = 1$ for an arbitrary constant $t > 0$. Obviously, for $S(x) = c$, with c a constant, the power law holds for all positive values of x. If $S(x)$ is not exactly constant but varies very slowly, it may possible to find a minimum value $x_{\min}$ beyond which the power law takes the form

$$p_{\text{pl}}(x) = \frac{\alpha - 1}{x_{\min}} \left(\frac{x}{x_{\min}}\right)^{-\alpha}, \tag{14.7}$$

where the coefficient $(\alpha - 1)/x_{\min}$ defines the PDF's normalization.

Moments μ_n of the power-law function $p_{\text{pl}}(x)$ above the cutoff value $x_{\min}$ are obtained by direct integration. One finds (see Problem 14.1):

$$\mu_n = \int_{x_{\min}}^{\infty} x^n p_{\text{pl}}(x)\, dx = \frac{\alpha - 1}{\alpha - 1 - n} x_{\min}^n, \tag{14.8}$$

which is strictly defined only for $n < \alpha - 1$. One readily verifies that moments $n \geq \alpha - 1$ indeed diverge. For instance, for $\alpha < 2$, the mean and all higher moments are infinite while for $2 < \alpha < 3$, the mean exists but all other moments (including the variance) do not.

Power-law behaviors are typically observed over a finite range of the independent variable x beyond which they break down. For instance, in particle physics a strict power-law behavior as a function of the momentum (or transverse momentum) of the produced particles would imply that particles with arbitrarily large momenta can be produced. Such a behavior is clearly impossible because it would violate conservation of energy: given a system with a specific collision energy, there is an energy bound beyond which individual particles cannot be produced. This problem can be solved or at the very least suppressed with the introduction of a power law with exponential cutoff:

$$f_{\text{plc}}(x) \propto \left(\frac{x}{x_{\min}}\right)^{-\alpha} e^{-\lambda x}, \tag{14.9}$$

where the exponential $e^{-\lambda x}$ dominates the power law at very large values of x. Strictly speaking, this function does not globally scale as and is not asymptotically a power law, but it does exhibit approximate scaling behavior between the minimum $x_{\min}$ and an upper cutoff determined by the exponential.

Generation of random numbers according to the power law Eq. (14.7) is readily achieved with the transformation method (§13.3.3) since the function $f_{\text{pl}}(x)$ is integrable. One finds (see Problem 14.2) that random continuous values of x distributed according to $f_{\text{pl}}(x)$ can be obtained with

$$x = x_{\min}\,(1 - r)^{\frac{1}{1-\alpha}}, \tag{14.10}$$

where r is a uniformly distributed random number in the range [0, 1]. Note that for application in high energy physics, one may substitute either the momentum p or transverse momentum p_T of the particle for x, as we discuss in the next section.

Flat Rapidity Distribution

The production of particles in high-energy collisions, either proton–proton or nucleus–nucleus, is hardly isotropic and is typically better described with an approximately flat rapidity distribution – particularly in the central rapidity region, i.e., $y \approx (y_{\text{beam}} + y_{\text{target}})/2$. The invariant cross section Eq. (8.73) may then be modeled according to a function of the form

$$E\frac{d^3\sigma}{dp^3} = \frac{d^2\sigma}{\pi dp_T^2 dy} = kf(p_T), \tag{14.11}$$

where $k = d\sigma/dy$ is assumed to be constant in the kinematical range of interest, and $f(p_T)$ is a suitably chosen model for the transverse momentum spectrum of produced particles. Depending on the p_T range of interest, $f(p_T)$ can be chosen to be an exponential function (as in §14.1.1), a power-law distribution (§14.1.1), a Maxwell–Boltzmann momentum distribution (§14.1.1) or other appropriate functions. For illustrative purposes, let us consider the generation of particles with a flat rapidity distribution and a power-law distribution in p_T, which should be suitable (approximately) for the simulation of high p_T particles or jets.

The azimuth angle is assumed to be distributed uniformly and is generated according to

$$\phi = 2\pi r_1, \tag{14.12}$$

where r_1 is a uniformly distributed random number in the range [0, 1]. The rapidity distribution is assumed to be flat in a range $[y_{\min}, y_{\max}]$ and is thus generated with

$$y = y_{\min} + r_2 (y_{\max} - y_{\min}), \tag{14.13}$$

where $r_2 \in [0, 1]$. Finally, the p_T of the particle is produced according to

$$p_T = p_{T,\min} (1 - r_3)^{\frac{1}{1-\alpha}}, \tag{14.14}$$

with $r_3 \in [0, 1]$, a preset minimum transverse momentum $p_{T,\min}$, and a suitably chosen value of α. The energy and moment components of the particle may then be obtained according to Eqs. (8.39 and 8.41):

$$E = m_T \cosh y, \tag{14.15}$$

$$p_x = p_T \cos\phi, \tag{14.16}$$

$$p_y = p_T \sin\phi, \tag{14.17}$$

$$p_z = m_T \sinh y, \tag{14.18}$$

with $m_T = \sqrt{p_T^2 + m^2}$, where m is the mass of the particle.

In certain situations, it might be desirable to produce particles according to a flat pseudorapidity, η, distribution. One then generates η according to

$$\eta = \eta_{\min} + r_2 \left(\eta_{\max} - \eta_{\min}\right). \tag{14.19}$$

The polar angle θ of the particle is obtained by inverting Eq. (8.31)

$$\theta = 2\tan^{-1}\left[\exp(-\eta)\right], \tag{14.20}$$

and the momentum components are

$$p_x = p_T \cos\phi \sin\theta, \tag{14.21}$$

$$p_y = p_T \sin\phi \sin\theta, \tag{14.22}$$

$$p_z = p_T \cos\theta, \tag{14.23}$$

while the energy is

$$E = \sqrt{p_x^2 + p_y^2 + p_z^2 + m^2}. \tag{14.24}$$

Maxwell–Boltzmann Momentum Distribution

The Maxwell–Boltzmann distribution, introduced in §3.16, describes the momentum distribution of molecules of nonrelativistic systems in (near) thermodynamic equilibrium. It is often used to model thermalized particle production in heavy-ion collisions.

It is convenient to express the Maxwell–Boltzmann distribution as

$$f_{\rm MB}(\vec{p}) = \left(\frac{1}{2\pi mkT}\right)^{3/2} \exp\left(-\frac{p_x^2 + p_y^2 + p_z^2}{2mkT}\right), \tag{14.25}$$

which may also be written

$$f_{\rm MB}(\vec{p}) = \frac{1}{N}\frac{d^3N}{p^2 dp d\cos\theta d\phi} = f_{\rm 1D}(p_x|T) f_{\rm 1D}(p_y|T) f_{\rm 1D}(p_z|T), \tag{14.26}$$

with

$$f_{\rm 1D}(p_i|T) = \left(\frac{1}{2\pi mkT}\right)^{1/2} \exp\left(-\frac{p_i^2}{2mkT}\right). \tag{14.27}$$

One can then generate momenta distributed according to a Maxwell–Boltzmann distribution using three random Gaussian deviates p_x, p_y, and p_z determined by

$$P(p_i) \propto \exp\left(-\frac{p_i^2}{2\sigma^2}\right), \tag{14.28}$$

with $\sigma = mkT$. The generation of p_x, p_y, and p_z as Gaussian deviates produces an isotropic distribution. In order to obtain particles with a flat rapidity distribution and a Maxwell–Boltzmann transverse profile, it suffices to generate p_x and p_y according to the preceding PDF and the rapidity according to Eq. (14.13). The p_z component of the momentum and the energy are then obtained with Eqs. (14.18) and (14.15), respectively.

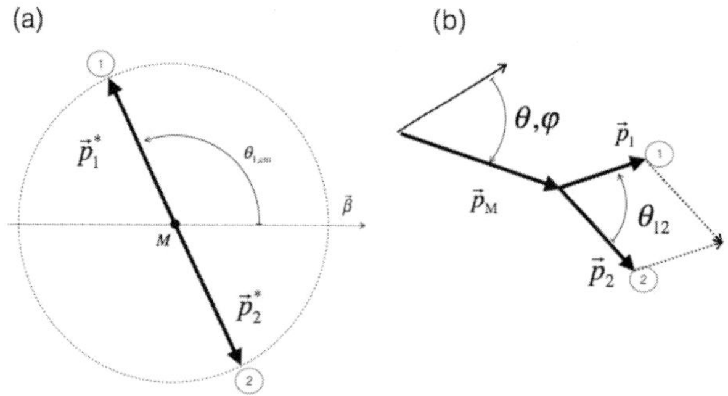

Center of mass reference frame   Laboratory reference frame

Fig. 14.1 Definition of kinematic variables required for the simulation of two-body decays. (a) Rest frame of the parent particle. (b) Laboratory frame.

14.1.2 Simulation of Resonance Decays

Corrections for detection efficiencies, acceptance, and other instrumental effects require that one can accurately model the decay of short-lived particles and take into account both kinematical and instrumental effects. We first consider the decay of narrow-width resonances, which can be represented by a fixed mass value. Decay simulations involving finite widths are considered next, as are examples of three-body decay that may be represented as a succession of two-body decays.

Fixed-Mass Two-Body Decays

The generation of two-body decays involves (1) the generation of the two daughter particles in the rest frame of the parent, (2) generation of the kinematic parameters of the parent in the laboratory reference frame, and (3) boost of the two daughter particles according to the speed and direction of the parent. The relevant variables are defined in Figure 14.1.

In the rest frame of the parent, the two daughter particles have momenta of equal magnitude, p^*, but in opposite directions. It is thus sufficient to generate the momentum vector of the first particle $\vec{p}_1^{\,*}$, and the momentum of the second particle is simply $\vec{p}_2^{\,*} = -\vec{p}_1^{\,*}$. The magnitude of the momentum is determined by the mass M of the parent, as well as the masses m_1 and m_2 of the daughter particles according to Eq. (8.165):

$$p^* = \frac{1}{2M}\{[M^2 - (m_1 - m_2)^2][M^2 - (m_1 + m_2)^2]\}^{1/2}. \tag{14.29}$$

The generation of the polar and azimuth angles of particle 1 are carried out according to

$$\phi_1^* = 2\pi r_1, \tag{14.30}$$

$$\cos\theta_1^* = 1 - 2r_2, \tag{14.31}$$

where r_1 and r_2 represent random numbers in the range [0, 1] generated for each decay. The *CM* momentum $\vec{p}_1^{\,*}$ and energy E_1^* of particle 1 is thus

$$p_{1,x}^* = p^* \sin\theta_1^* \cos\phi_1^*, \tag{14.32}$$

$$p_{1,y}^* = p^* \sin\theta_1^* \sin\phi_1^*, \tag{14.33}$$

$$p_{1,z}^* = p^* \cos\theta_1^*, \tag{14.34}$$

$$E_1^* = \sqrt{(p^*)^2 + m_1^2}, \tag{14.35}$$

and for particle 2, one has

$$p_{2,x}^* = -p_{1,x}^*, \tag{14.36}$$

$$p_{2,y}^* = -p_{1,y}^*, \tag{14.37}$$

$$p_{2,z}^* = -p_{1,z}^*, \tag{14.38}$$

$$E_2^* = \sqrt{(p^*)^2 + m_2^2}. \tag{14.39}$$

Boosting the daughter particles in the laboratory frame requires knowledge of the direction and speed of the parent particle. In turn, generation of the momentum vector of the parent $\vec{p}_P$ requires a specific production scenario or model such as those discussed in prior sections. Let us here assume that the momentum of the parent $\vec{p}_P$ is known. Its velocity vector may then be calculated according to

$$\vec{\beta}_P = \frac{\vec{p}_P}{E_P}. \tag{14.40}$$

The daughter particles $i = 1, 2$ can thus be boosted in the laboratory frame (Problem 14.3) according to

$$\vec{p}_i^{\,\text{lab}} = \vec{p}_i^{\,*} + [(\gamma_P - 1)\vec{p}_i^{\,*} \cdot \hat{\beta}_P + \gamma_P E_i^*]\hat{\beta}_P, \tag{14.41}$$

where $\gamma = (1 - \beta_P^2)^{-1/2}$.

Finite-Width Resonance Two-Body Decays

While energy-momentum conservation constrains the energy and momenta of particles produced in elementary particle collisions, it does not uniquely specify the mass of short-lived particles, which can then nominally be produced event by event with values determined by a Breit–Wigner distribution (§3.15). Simulations of the decays of short-lived (i.e., finite width) resonances thus requires generation of a mass value M for the parent according to

$$f_{BW}(M|M_0, \Gamma) = \frac{1}{\pi} \frac{\Gamma/2}{\Gamma^2/4 + (M - M_0)^2}, \tag{14.42}$$

where M_0 and Γ are the nominal mass and width of the parent particle, respectively. Note that since the Breit–Wigner distribution has no a priori bounds, imposing an artificial upper-value cut to account for kinematical limitations encountered in finite beam energy experiments may be required. A lower cut must de facto be imposed to account for the

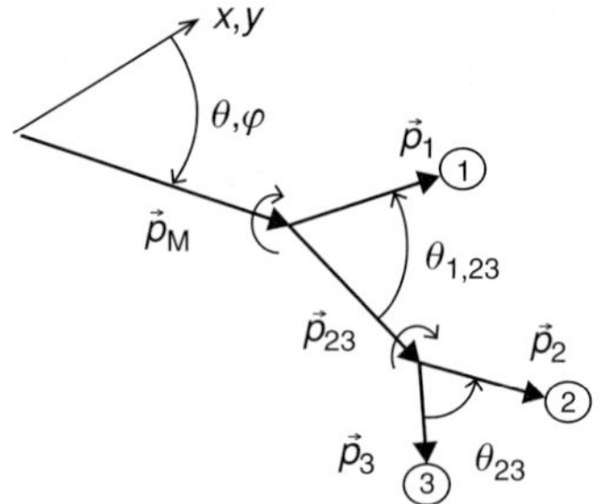

Fig. 14.2 Three-body decays as sequential two-body decays. In general, the two decays do not occur in the same plane.

finite masses of the particles and the need for $|\vec{p}^*| > 0$. With the mass in hand, the generation of daughter particles proceeds in the same manner as for a fixed mass covered in the previous paragraph.

Three-Body Decays Involving Sequential Two-Body Decays

A number of three-body decays may be described as a sequence of two-body decays, as illustrated in Figure 14.2.

Simulations of the decay of a parent particle in terms of a sequence of two-body decays proceeds quite similarly as the simpler case of a single two-body decay. Given the mass M of the parent particle, one generates the first decay into particles of mass m_1 and m_{23} according to the two-body decay algorithm. If the mass m_{23} is not fixed, one must first determine its value according to an appropriate PDF such as a Breit–Wigner distribution. Once the momentum $\vec{p}_{23}$ is determined, one can then proceed to decay the particle into particles 2 and 3 using the two-body decay algorithm.

14.1.3 Simple Event Generators

The generation of collision events with large numbers of particles of different species requires specific assumptions be made concerning the relative probabilities of emission of each type of particle and their momentum distributions. In this section, we first describe a rudimentary technique that neglects energy and momentum conservation, and in the next section, we show how one can modify the technique to achieve particle generation that conserves energy and momentum event by event.

For illustrative purposes, let us assume the number m of particles in an event can be modeled ab initio with a PDF $P_m(m)$ which we choose to be uniform in the range $[m_{\min}, m_{\max}]$, while the production of n different particle species of mass m_i, charge q_i, are specified by probabilities p_i, with $i = 1, \ldots, n$ such that

$$\sum_{i=1}^{n} p_i = 1. \tag{14.43}$$

Let us further assume that all particle species can be characterized by a flat rapidity distribution in the range $[y_{\min}, y_{\max}]$ and Maxwell–Boltzmann distributions in the transverse plane, with temperatures T_i.

Before one begins with the generation of particles, it is convenient to define the cumulative sum of probabilities P_i, for $i = 0, \ldots, n$, according to

$$P_0 = 0 \qquad \text{for } i = 0, \tag{14.44}$$

$$P_i = \sum_{k=1}^{i} p_k \qquad \text{for } 1 \leq i \leq n, \tag{14.45}$$

with $P_1 = p_1, P_2 = p_1 + p_2$, and so on, up to $P_n = 1$ by virtue of Eq. (14.43).

The generation of events can then be accomplished according to the following algorithm:

1. For each event, ...
   a. Determine the multiplicity (number of particles) of the event on the basis of uniform PDF $P_m(m)$:

   $$m = int\left[m_{\min} + r_1\left(m_{\max} - m_{\min}\right)\right], \tag{14.46}$$

   where the notation $int\,[O]$ indicates one needs to round the number O to the nearest integer.
   b. For each particle,
      i. Generate a random number r_2 in the range $[0, 1]$ to determine the species (mass m_i, charge q_i, and emission temperature T_i) of the particle. The species of the particle is specified by the integer i that satisfies $P_{i-1} \leq r_2 < P_i$.
      ii. Generate the momentum vector of the particle assuming a mass m_i and temperature T_i with a flat rapidity distribution and transverse Maxwell–Boltzmann distribution based on the techniques presented in earlier sections.

Note that if some of the produced particles are short-lived resonances, one can apply the techniques presented in §14.1.2 to simulate their decay and the generation of daughter particles.

The aforementioned event-generating technique is quite simple but somewhat ad hoc. It provides a simple method for fast generation of particles and simple modeling of the production of particles in elementary collisions, but given it is based on simplistic and purely phenomenological models, it should not be expected to provide a very accurate account of particle production cross sections and correlations. Although it can be readily modified to use a more realistic multiplicity distribution or particle momentum distributions, it features an intrinsic flaw that cannot be fixed by substitutions of more realistic PDFs: the events produced do not have a specific total energy or net momentum and, as such, do not satisfy laws of energy–momentum conservation. This may be appropriate as long as the particles generated are meant to represent only a fraction of all observable particles produced by actual collisions. But if the generated particles are meant to be representative of whole events, one must use a slightly modified technique to ensure that the total momentum and the total energy of the particles are produced with predefined values; in other words, the particle generation must obey laws of energy–momentum conservation. It is in principle also necessary to account for the conservation of other conserved quantities (quantum numbers) such as electric charge, strangeness, baryon number, and so on. A particle generation technique that accounts for global energy and momentum conservation is presented in the next section. Better modeling of collisions still can be achieved with computer codes based on comprehensive theoretical frameworks of particle interaction and particle production, such as those already cited [29, 52, 61, 66, 133, 137, 169, 174, 175, 176, 188, 191, 192, 198].

14.1.4 Multiparticle Generation with Energy–Momentum Conservation

We saw in §8.2.2 that the cross section of a process producing n particles is determined by the square of the transition amplitude $|\mathfrak{M}|$ and an n-body phase factor dR_n (8.79). It is thus convenient to define the n-body phase integral R_n as follows:

$$R_n = \int_{4n} \delta^{(4)}\left(P - \sum_{j=1}^{n} p_j\right) \prod_{i=1}^{n} \delta\left(p_i^2 - m_i^2\right) d^4 p_i, \tag{14.47}$$

where P is the total 4-momentum of the n-body system, while p_i and m_i are the 4-momenta and masses of the produced particles, respectively. The differential cross section in terms of a certain kinematical parameter α (particle momentum, emission angle, etc.) may then be written

$$\frac{d\sigma}{d\alpha} = \frac{d}{d\alpha}\left(|\mathfrak{M}|^2 R_n\right). \tag{14.48}$$

One can equally sample the entire n-particle phase space by assuming that $|\mathfrak{M}| = 1$. This shall produce differential cross sections (spectra) that are determined exclusively by the phase-space of the n outgoing particles, that is, cross sections corresponding to a uniform n-body phase space. One notes that by virtue of the factor $\delta^{(4)}(P - \sum_{j=1}^{n} p_j)$, generation according to R_n shall automatically conserve both energy and momentum. Additionally, given that

$$\delta\left(p_i^2 - m_i^2\right) d^4 p_i = \frac{p_i^2}{E_i} dp_i d\cos\theta_i d\phi_i, \tag{14.49}$$

it should also produce particles in proportion to the density of states: events with a higher density of states should be more probable. It is also worth mentioning that R_n leads to an intrinsic n-body particle correlation determined solely by energy–momentum conservation. This type of correlation is to be distinguished from those implied by the production process embodied in the amplitude $|\mathfrak{M}|$. Separating these two sources of correlation experimentally is unfortunately a rather nontrivial task. Finally, consider that if a flat spectrum in n-body phase-space is not appropriate, one can always assign a weight to generated events based on the desired transition amplitude $|\mathfrak{M}|$ after an n-body event has been produced according to R_n.

The notion of using a Monte Carlo method to calculate the R_n integral was first discussed by Kopylov [130], while Srivastava and Sudarshan [177] derived the covariant form,

Eq. (14.47), which lends itself to a recursive calculation of the integral. We here follow the M-generator algorithm presented by James [113].

The integral Eq. (14.47) may be written

$$R_n(P; m_1, \ldots, m_n) = \int_{4n} \left[\delta^{(4)} \left(P - p_n - \sum_{j=1}^{n-1} p_j \right) \prod_{i=1}^{n} \delta \left(p_i^2 - m_i^2 \right) d^4 p_i \right] \times \delta \left(p_n^2 - m_n^2 \right) d^4 p_n, \tag{14.50}$$

where the expression in square brackets is $R_{n-1}(P - p_n; m_1, \ldots, m_{n-1})$. One can thus indeed calculate $R_n(P; m_1, \ldots, m_n)$ recursively

$$R_n(P; m_1, \ldots, m_n) = \int_{4n} R_{n-1}(P - p_n; m_1, \ldots, m_{n-1}) \delta \left(p_n^2 - m_n^2 \right) d^4 p_n. \tag{14.51}$$

Making use of Eq. (14.49), Eq. (14.51) may be written

$$R_n(P; m_1, \ldots, m_n) = \int_{4n} R_{n-1}(P - p_n; m_1, \ldots, m_{n-1}) \frac{d^3 p_n}{2E_n}, \tag{14.52}$$

which in principle provides a basis for the calculation of R_n since it is expressed in terms of a Lorentz invariant and can thus be calculated recursively in arbitrary frames of reference. The problem resides in the choice of efficient bounds of the momentum p_n. It turns out to be more efficient and practical to modify the preceding recursion formula to obtain an expression in terms of particle masses. This is accomplished by noting that

$$\delta^{(4)} \left(P - \sum_{j=1}^{n} p_j \right) = \int \delta^{(4)} \left(P - P_l - \sum_{j=l+1}^{n} p_j \right) \delta^{(4)} \left(P_l - \sum_{k=1}^{l} p_k \right) d^4 P_l, \tag{14.53}$$

which enables us to write

$$R_n \left(P; m_1, \ldots, m_n \right) = \int \delta^{(4)} \left(P - P_l - \sum_{j=l+1}^{n} p_j \right) \prod_{j=l+1}^{n} \delta \left(p_j^2 - m_j^2 \right) d^4 p_j \times \int \delta^{(4)} \left(P_l - \sum_{k=1}^{l} p_k \right) \prod_{j=1}^{l} \delta \left(p_j^2 - m_j^2 \right) d^4 p_j d^4 P_l. \tag{14.54}$$

Noting that

$$1 = \int_0^\infty \delta \left(P_l^2 - M_l^2 \right) dM_l^2, \tag{14.55}$$

one can then express R_n as

$$R_n(P; m_1, \ldots, m_n) = \int_0^\infty \left[\int \delta^{(4)}\left(P - P_l - \sum_{j=l+1}^{n} p_j\right) \right. \times \prod_{j=l+1}^{n} \delta(p_j^2 - m_j^2)\delta\left(P_l^2 - M_l^2\right) d^4p_j d^4P_l \times \left. \int \delta^{(4)}\left(P_l - \sum_{k=1}^{l} p_k\right) \prod_{j=1}^{l} \delta(p_j^2 - m_j^2) d^4p_j \right] dM_l^2, \tag{14.56}$$

which gives us

$$R_n(P; m_1, \ldots, m_n) = \int_0^\infty R_{n-l+1}(P; M_l, m_{l+1}, \ldots, m_n) \times R_l(P_l; m_1, \ldots, m_l)\, dM_l^2. \tag{14.57}$$

One can show that repeated applications of this "splitting" relation, starting with $l = 2$, yields the recurrence relation

$$R_n = \int dM_{n-1}^2 \cdots \int dM_2^2 \prod_{i=1}^{n-1} R_2(M_{i+1}; m_i, m_{i+1}), \tag{14.58}$$

where

$$R_2(M_{i+1}; M_i, m_{i+1}) = \frac{2\pi}{M_{i+1}} \sqrt{M_{i+1}^2 + \left(\frac{M_i^2 - m_{i+1}^2}{M_{i+1}}\right)^2 - 2\left(M_{i+1}^2 + m_{i+1}^2\right)}. \tag{14.59}$$

Transforming the integrals dM^2 into $2MdM$, one obtains

$$R_2 = \frac{1}{2m_1} \int \int \prod_{i=1}^{n-1} 2M_i R_2(M_{i+1}; M_i, m_{i+1})\, dM_{n-1} \cdots dM_2, \tag{14.60}$$

which is pictorially represented in Figure 8.12.

One can then proceed to the generation of n-body processes as if they were a succession of two-body decays. One only needs to apply a two-body phase-space factor for each "decay vertex." The boundaries of integration are a tricky issue. Nominally, one might be tempted to write

$$M_{j-1} + m_j < M_j < M_{j+1} - m_{j+1}. \tag{14.61}$$

But this implies the boundaries of jth integral depend on other integrals, which leads to incorrect sampling of the phase space in a Monte Carlo integration. The masses can, however, be chosen according to the less restrictive condition

$$\sum_{i=1}^{j} m_i < M_j < M_n - \sum_{i=j+1}^{n} m_i, \tag{14.62}$$

using the generating technique

$$M_j = r_j \left(M_n - \sum_{i=j+1}^{n} m_i \right) + \sum_{i=1}^{j} m_j, \tag{14.63}$$

where r_j are random numbers in the range [0, 1]. One verifies that the M_j generated with the preceding expression will satisfy the boundaries Eq. (14.61) provided

$$0 < r_1 < \cdots < r_j < \cdots < r_{n-2} < 1. \tag{14.64}$$

For the generation of an event with n particles, it thus suffices to generate $n - 2$ random numbers and sort them in ascending order for the calculation of the masses using Eq. (14.63). However, one more step is required at each decay vertex. Indeed the aforementioned condition specifies the mass M_i but it does not dictate the direction of the produced pair. One must then randomly choose, for each decay, a direction for the mass M_j.

$$\phi_j = 2\pi r_{j'}, \tag{14.65}$$

$$d\cos\theta_j = -1 + 2 \times r_{j''}, \tag{14.66}$$

where $r_{j'}$ and $r_{j''}$ are random numbers in the range [0, 1]. It is important to note that isotropic emission is only expected in the rest frame of the mass M_j. To obtain the required Lorentz invariance, one must generate two-body decays successively in the rest frame of each mass M_j. This means one must successively Lorentz-transform each momentum into the group of particles preceding it.

This algorithm was first made available as a program named GenBod in CERN software libraries but is now available within the ROOT framework [59] as class TGenPhaseSpace.

14.1.5 Correlated Particles Generators

A variety of techniques may be used to generate correlated particles. At the outset, note that the production of daughter particles resulting from a two- or three-body decay and boosted in the laboratory frame according to the speed and direction of their parent particle produce correlated particles: for a high-velocity parent particle, the daughter particles tend to be separated by a small angle that decreases for increasing velocity (momentum) of the parent. Similarly, the production of a finite number m of particles with the multiparticle generation technique presented in §14.1.4 also results in net or global correlations between all particles produced [54, 55]. Correlations shall also result from conservation laws such as (electric) charge, strangeness, baryon number conservation, and so on.

Kinematically correlated particles may be generated if they can be produced by means of hierarchical processes, such as sequential two-body decays or successive particle generation accounting individually for energy and momentum conservation. Correlations may also be achieved, for instance, by shifting (§14.1.5) or boosting the momentum of groups of correlated particles [162]. In general, it may be cumbersome or technically difficulty to impart elaborated correlations between generated particles. Fortunately, it is always possible to generate particles independently and assign events or particle n-tuplets a weight according to a correlation function ansatz (§14.1.5).

Anisotropic Particle Generation

Anisotropic flow, relative to an event plane, may be represented according to a Fourier decomposition

$$\frac{1}{N}\frac{dN}{d\phi} \propto 1 + 2\sum_{n=1}^{n_{\max}} v_n \cos\left(n(\phi - \Psi_n)\right) \tag{14.67}$$

where Ψ_n represent the orientation of the event plane of order n (see §11.1.1).

Generation of random angles according to the preceding expression with the inversion method is not possible because its integral cannot be inverted, while the use of the acceptance/rejection method is somewhat inefficient. On the other hand, the integrated histogram method is simple and reasonably efficient. It suffices to produce a finely binned histogram of the distribution Eq. (14.67) and apply the histogram method presented in §13.3.5.

In some situations, one may wish to introduce flow artificially after the fact, that is, after particles have been generated by some third-party event generator. Use of the acceptance/rejection method would be a bad choice in this case because it would change the integrated particle production cross section (i.e., the integral of the momentum distribution). While the aforementioned Fourier decomposition cannot be achieved from *scratch,* a reasonable approximation may be obtained by shifting particles in the transverse plane according to

$$p_x{}' = p_x \times \left(1 + 2\sum_{n=1}^{n_{\max}} v_n \cos(n\Psi_n)\right), \tag{14.68}$$

$$p_y{}' = p_y \times \left(1 + 2\sum_{n=1}^{n_{\max}} v_n \sin(n\Psi_n)\right),$$

where the event plane angles Ψ_n are chosen randomly *event by event* in the range $[0, 2\pi]$. The flow coefficients v_n can be arbitrary constants or even functions of the transverse momentum of the particles. The above "shift" of the pair (p_x, p_y) does not change the integral of the momentum distribution, provided that the angles Ψ_n are generated as uniform deviates in the range $[0, 2\pi]$, since integrals of the sine and cosine functions in this range vanish. The shift also conserves momentum in the transverse plane. Indeed, if the original events were generated with momentum conservation, that is, such that

$$\sum_{i=1}^{m} p_{x,i} = 0, \tag{14.69}$$

$$\sum_{i=1}^{m} p_{y,i} = 0, \tag{14.70}$$

where m is the number of particles in the event, then a random shift of all particles of an event according to a specific set of angles Ψ_n and fixed v_n coefficients (i.e., common to all particles of an event) yields

$$\sum_{i=1}^{m} \vec{p}'_{x,i} = \left(\sum_{i=1}^{m} p_{x,i}\right) \times \left(1 + 2\sum_{n=1}^{n_{\max}} v_n \cos(n\Psi_n)\right) = 0, \tag{14.71}$$

$$\sum_{i=1}^{m} \vec{p}'_{y,i} = \left(\sum_{i=1}^{m} p_{y,i}\right) \times \left(1 + 2\sum_{n=1}^{n_{\max}} v_n \sin(n\Psi_n)\right) = 0, \tag{14.72}$$

and thus conserves momentum in the transverse plane. Note, however, that this simple factorization breaks down if the flow coefficients are functions of the momentum of the particles. The introduction of flow after the fact, with this technique, thus does not strictly conserve momentum if the coefficients v_n are functions of p_T and/or rapidity. It may prove adequate, nonetheless, for simulations of the performance of large detectors that include particle losses and resolution smearing.

Simulation of Correlations with Weights

Simulations of correlated particles production can be achieved by applying an ad hoc weight to particle pairs, n-tuplets, or events after their production based on some function of the particle momenta and energies. This enables the simulation of correlated particles based on the generation of particles using simple algorithms that nominally produce independent and uncorrelated particles.

Let us consider, as an example, the generation of particles with a peaked correlation in relative azimuth and rapidity. One can generate events with uncorrelated particles as in §14.1.3. Correlation may, however, be simulated with the use of weights. For instance, in order to simulate a two-particle correlations of the form

$$w(\Delta\eta, \Delta\phi) = 1 + A\frac{1}{2\pi\sigma_\eta\sigma_\phi} \exp\left(-\frac{\Delta\eta^2}{2\sigma_\eta^2}\right) \exp\left(-\frac{\Delta\phi^2}{2\sigma_\phi^2}\right), \tag{14.73}$$

it suffices to generate independent particles with, say, flat rapidity distributions and assign each pair a weight $w(\Delta\eta, \Delta\phi)$. To simulate a two-particle cumulant, one may generate two sequences of events with identical multiplicity distributions. Pairs of the first sequence are given the weight $w(\Delta\eta, \Delta\phi)$ to simulate "real" events, $\rho_2(\Delta\eta, \Delta\phi)$, while pair of the second sequence are given a unit weight and simulate "mixed" events, $\rho_1 \otimes \rho_1(\Delta\eta, \Delta\phi)$. One next calculates the ratio, bin by bin, of the $\rho_2(\Delta\eta, \Delta\phi)$ and $\rho_1 \otimes \rho_1(\Delta\eta, \Delta\phi)$ histograms, and subtract one, to obtain a normalized cumulant, as illustrated in Figure 14.3.

Similarly, Hanbury-Brown Twiss (HBT) type correlations can be simulated with a weight of the form

$$w(\vec{p}_1, \vec{p}_2) \propto \exp\left(-\frac{q^2}{2\sigma_{\text{HBT}}^2}\right), \tag{14.74}$$

with $q^2 = (\vec{p}_1 - \vec{p}_2)^2$ for pairs of identical particles (e.g., π^+ or π^-) with momenta $\vec{p}_1$ and $\vec{p}_2$. The width σ_{HBT} of the correlation functions is known to be inversely proportional to the size of the emitting source size r.

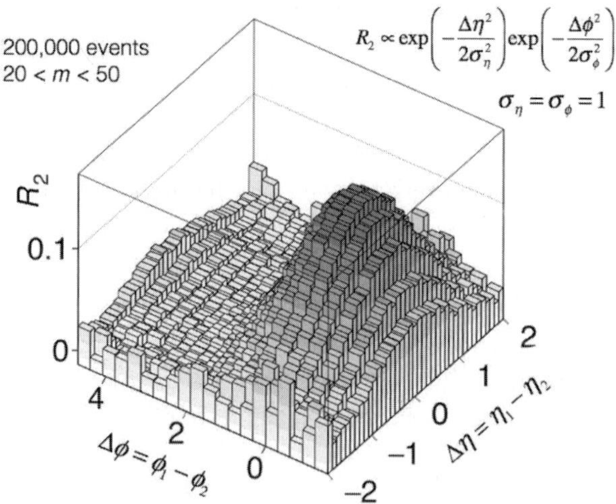

Fig. 14.3 Simulation of a two-particle correlation function using the weight method. Particle pairs were given the weight shown for real events and unit weight for mixed events. The simulation included 200,000 events with a multiplicity of particles in the range $20 < m \leq 50$. The cumulant is calculated as $\rho_2(\Delta\eta, \Delta\phi)/\rho_1 \otimes \rho_1(\Delta\eta, \Delta\phi) - 1$.

Quite obviously, arbitrarily complex and more sophisticated correlation functions can be modeled using this weight technique.

14.2 Detector Simulation

Detector simulations are commonly used to (1) understand the performance of a detection system, (2) determine the acceptance, efficiency, and resolutions of specific measurements, and (3) obtain response matrices that can be used toward the correction of simulated data. Such tasks can be best accomplished with detailed and comprehensive studies of the detector performance carried out within detector simulation environments such as that provided by the computer code GEANT, introduced in §14.2.4. But since the use of GEANT requires quite a bit of setup and computer coding, and considering that GEANT simulations can be rather CPU-intensive, performance studies are often conducted with **fast simulators** involving simple and sometimes rather primitive models of a detector's response.

We introduce basic exemplary techniques for the simulation of the effects of finite efficiency, detector resolution, and acceptance determination in §14.2.1, and provide a simple example of a calculation of a response matrix for jet measurements in §14.2.3.

14.2.1 Efficiency, Resolution, and Acceptance Simulators

Efficiency Simulation

The efficiency of a detection system towards measurement of a specific type of particle (or range of particle species) may be obtained by means of detailed simulations based on a realistic particle production model and a detailed simulation of the detector performance. Alternatively, one can also embed simulated tracks (or other relevant objects) into actual events and determine what fraction of the embedded tracks are actually recovered by the track reconstruction software.

Let us here assume that the detection efficiency for a specific particle species is known and can be parameterized with a model $\epsilon(\eta, \phi, p_T)$. In some cases, it might be possible to factorize the efficiency into a product of three functions $\epsilon(\eta, \phi, p_T) = \epsilon(\eta)\epsilon(\phi)\epsilon(p_T)$. It may even be possible to assume that dependencies on rapidity and azimuth are negligible and treat $\epsilon(\eta)$ and $\epsilon(\phi)$ as constants. One can then use this model to carry out fast simulations of the effects of the limited detection efficiency on momentum spectra, invariance mass spectra, correlation functions, or more complex objects such as jets.

While the effect of detection can in principle be trivially determined by multiplying the efficiency by the production cross section, it is often more convenient to carry out a simulation involving the production model considered and a model $\epsilon(\eta, \phi, p_T)$ of the efficiency. This has the added benefit of providing a track-by-track account of detector resolution effects. In such simulations, it can often be assumed that the detection efficiency for two or more particles factorizes and that one can apply the function $\epsilon(\eta, \phi, p_T)$ for each particle independently. A simulation may be based on a simple home-brewed particle generator or more sophisticated and theoretically motivated event generators whose events are either generated on the fly (event-by-event) or read from stored files. A typical simulation thus proceeds as follows:

1. Generate or read an event from file.
2. Efficiency/smearing: For each particle of an event
    a. Decide whether the particle should be accepted.
    b. Smear kinematical parameters of the particle (if needed/desired).
    c. Store smeared parameters.
3. Analysis: Carry out the required analysis of the generated event based on accepted (and smeared) particles. Optionally carry out the analysis on all generated and unsmeared particles to obtain a "perfect detection" reference.

The decision whether to accept a particle is based on three steps: (1) given the (unsmeared) kinematical parameters of the particle (η, ϕ, p_T), calculate the efficiency of detection $\epsilon(\eta, \phi, p_T)$; (2) generate a random number r with a uniform distribution in the range $[0, 1]$; and (3) accept the particle if $r \le \epsilon(\eta, \phi, p_T)$.

Figure 14.4 presents an example of the application of an efficiency function $\epsilon(p_T)$ on a parent distribution $f(p_T) = \lambda^{-1}\exp(-p_T/\lambda)$. It shows, in particular, that the measured transverse momentum mean $\langle p_T\rangle$ can be significantly altered by the detector's efficiency dependence on transverse momentum, $\epsilon(p_T)$.

Resolution Smearing

Smearing of kinematical parameters is based on models of the detector response. For instance, to simulate instrumental effects (smearing) on a transverse momentum measurement, one requires a PDF $f(p_T|\vec{\theta})$ describing fluctuations of the measured p_T determined

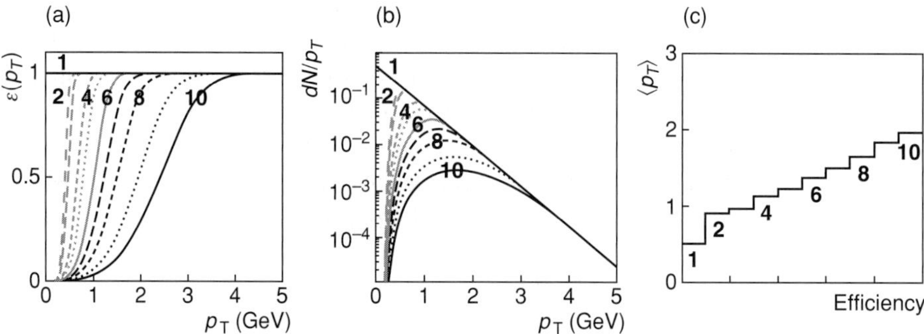

Fig. 14.4 Illustration of the effect of efficiency function $\epsilon(p_T)$ on a p_T spectrum and measured mean $\langle p_T \rangle$. (a) Selected efficiency response curves. (b) Product of the efficiency curves by the input spectrum ($f(p_T) = \lambda^{-1} \exp(-p_T/\lambda)$, with $\lambda = 0.5$). (c) Average p_T obtained with the selected efficiency curves.

by some set of parameters $\vec{\theta}$. The parameters $\vec{\theta}$ are here assumed to be known a priori. In practice, they could be obtained from either detailed simulations of the detector performance (§14.2.4) or an explicit measurement. Often times, but not always, the smearing response of the detection system may be assumed to be (approximately) Gaussian. Specification of the smearing function $f(p_T|\vec{\theta})$ thus reduces to the expression of the mean and root mean square (rms) of a Gaussian as a function of the true p_T of the particles. If the mean is null, the measurement can be considered without bias, and only fluctuations need considering. However, note that given particles lose energy (and thus momentum) as they traverse a detector, the track reconstruction software must compensate for such losses (see §9.2). The problem arises that at a given momentum, energy losses depend on the mass (or particle identification, PID) of the particle, which is unknown a priori. Corrections are thus typically made assuming the PID of the most abundantly produced particle species (e.g., pions). This invariably leads to a bias (i.e., a systematic shift) in the reconstruction of the momentum of particles. Such a shift should ideally be accounted for in simulations of the performance of detection system. Thus for each simulated species production, not only should the rms be provided as a function of the transverse momentum of the particles but also a function estimating the momentum bias achieved in a typical event reconstruction. Additionally, note that the reconstruction of charged particle tracks yield Gaussian fluctuations in the track curvature (C), which is inversely proportional to the p_T of the particle. Fluctuations in p_T are thus generally non-Gaussian. However, for simplicity's sake, we here illustrate a smearing simulation procedure assuming Gaussian fluctuations.

We assume the fluctuations in momentum measurements are determined by a Gaussian PDF with mean $\mu \equiv \mu(p_{T,0})$ and rms $\sigma \equiv \sigma(p_{T,0})$ where $p_{T,0}$ is the true momentum of the particles. The bias $\mu \equiv \mu(p_{T,0})$ is typically smallest for minimum ionizing particles and grows approximately in inverse relation of the momentum at small momenta and proportionally to the momentum at higher momenta. It could thus be modeled according to

$$\mu(p_{T,0}) = a_{-1}p_{T,0}^{-1} + a_0 + a_1 p_{T,0}, \tag{14.75}$$

where the coefficients a_n must be suitably fitted to represent the actual measurement bias of each particle species. Energy loss fluctuations typically scale with the average energy loss. Since energy losses have a $1/\beta^2$ dependence at low momentum and a logarithmic rise at higher momenta, one may model the momentum dependency of the fluctuations accordingly. However, note that the p_T resolution becomes exceedingly poor for very small curvature because the finite granularity of the hit detectors intrinsically limits the position resolution. One can thus in general model the rms with a low-order power series, such as

$$\sigma(p_{T,0}) = b_{-2}p_{T,0}^{-2} + b_{-1}p_{T,0}^{-1} + b_0 + b_1 p_{T,0} + b_2 p_{T,0}^2, \tag{14.76}$$

where the coefficients b_n must also be suitably fitted to represent the actual fluctuations for each measured particle species.

Assuming the coefficients a_n and b_n are known, either from a detailed simulation or actual measurements, one can thus carry out (fast) simulations of the detector response (smearing) according to the algorithm presented earlier in this section. Smearing requires calculation track by track of both $\mu(p_{T,0})$ and $\sigma(p_{T,0})$ with Eqs. (14.75, 14.76). Smeared track momenta are then obtained based on

$$p_T = p_{T,0} + r_G[\mu(p_{T,0}), \sigma(p_{T,0})], \tag{14.77}$$

where $r_G[\mu(p_{T,0}), \sigma(p_{T,0})]$ represents Gaussian deviates with mean $\mu(p_{T,0})$ and rms $\sigma(p_{T,0})$ calculable according to the algorithm presented in §13.3.7.

A fast simulator including smearing may, for instance, be used to study the effects of smearing on momentum spectra (Figure 14.5), the reconstruction of short-lived decaying particles based on the invariant mass technique, or jet measurements.

14.2.2 Kinematic Acceptance of Two-Body Decays

Monte Carlo simulations based on a simple particle generator provide a quick and easy method to determine the acceptance of a detector toward measurements of momentum spectra, particularly those of short-lived resonances. Let us consider, as an example, simulations of the decay of kaons K_s^0 into a pair $\pi^+ + \pi^-$ detected by invariant mass reconstruction (§8.5.1).

We will use the algorithm presented in sections §§14.1.1, 14.1.3, and 14.2.1 to generate K_s^0 with a Maxwell–Boltzmann distribution in transverse momentum and a uniform pseudorapidity in the range $|\eta| < 3$. Kaons are decayed into $\pi^+ + \pi^-$ pairs exclusively. Pions are assumed detectable with a 100% efficiency if in the ranges $0.2 < p_T \le 1.5$ and $|\eta| < 2$. However, their momenta are smeared according to Eq. (14.76) using resolution parameters of curve 2 in Figure 14.5.

The simulated invariant mass spectra and K_s^0 detection efficiency obtained with these detection conditions are shown in Figure 14.6. One finds that the pion acceptance dramatically shapes the K_s^0 detection efficiency and acceptance.

More generally, one can use k_B functions $B_i^{\min/\max}(\eta, \phi, p_T)$, $i = 1, \ldots, k_B$ to define the boundaries within which a specific particle species is considered detectable. For instance,

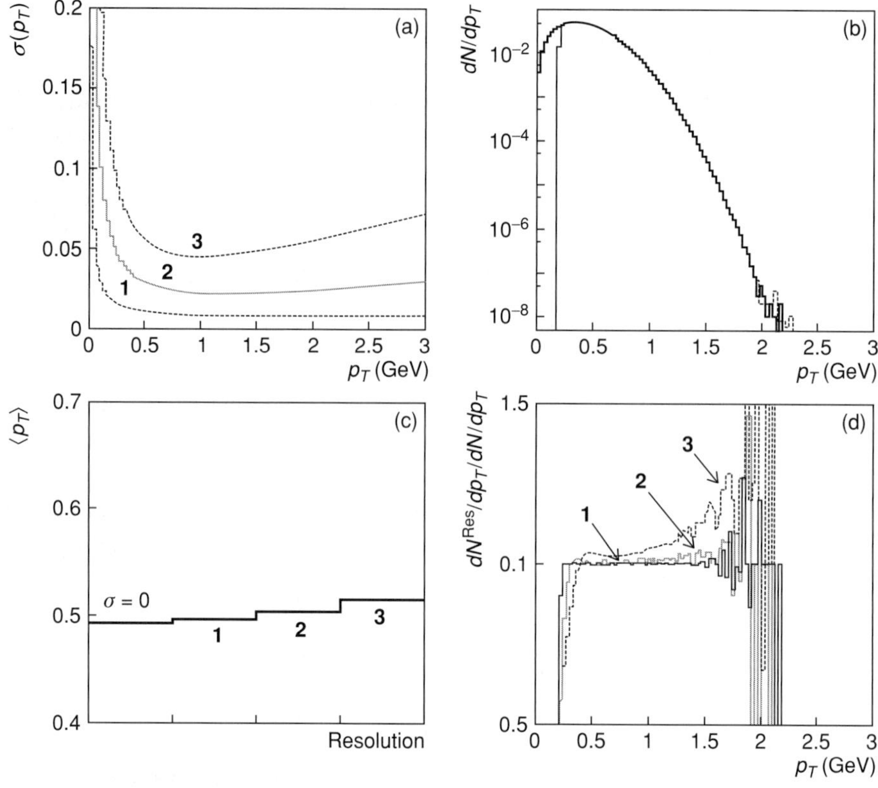

Fig. 14.5 Illustration of the effect of resolution smearing on p_T spectra. (a) Resolution responses used in the simulations. (b) Simulation of p_T spectrum using a Maxwell–Boltzmann distribution and smearing parameters shown in (a). (c) Average p_T obtained with perfect and finite resolution. (d) Ratio of the smeared spectra to the original spectrum. Note that a p_T threshold of 0.2 GeV has been used to illustrate the joint effects of finite resolution and acceptance.

as illustrated in Figure 14.7, a set of functions $p_{T,i}^{\min/\max}(\eta)$ can express the minimum and maximum p_T values detectable by an apparatus. A particle with momentum vector $\vec{p} = (\eta, \phi, p_T)$ would be considered detectable if and only if it satisfies the conditions

$$p_{T,i}^{\min}(\eta) \le p_T \le p_{T,i}^{\max}(\eta), \quad \text{for all } i = 1, \ldots, k \tag{14.78}$$

14.2.3 Response Matrix Determination

We saw in §12.3 that the unfolding of measured distributions requires knowledge of a response matrix describing the efficiency and smearing imparted on a measured signal by instrumental effects. The determination of a response matrix typically requires simulation of the effects of the instrumentation of the measured particle kinematical parameters, and their impact on other quantities such as detected particle multiplicities, resonance decays, or measurements of jets and their properties. Such studies are best conducted based on detailed simulations of the detector performance (§14.2.4) but can often also be accomplished

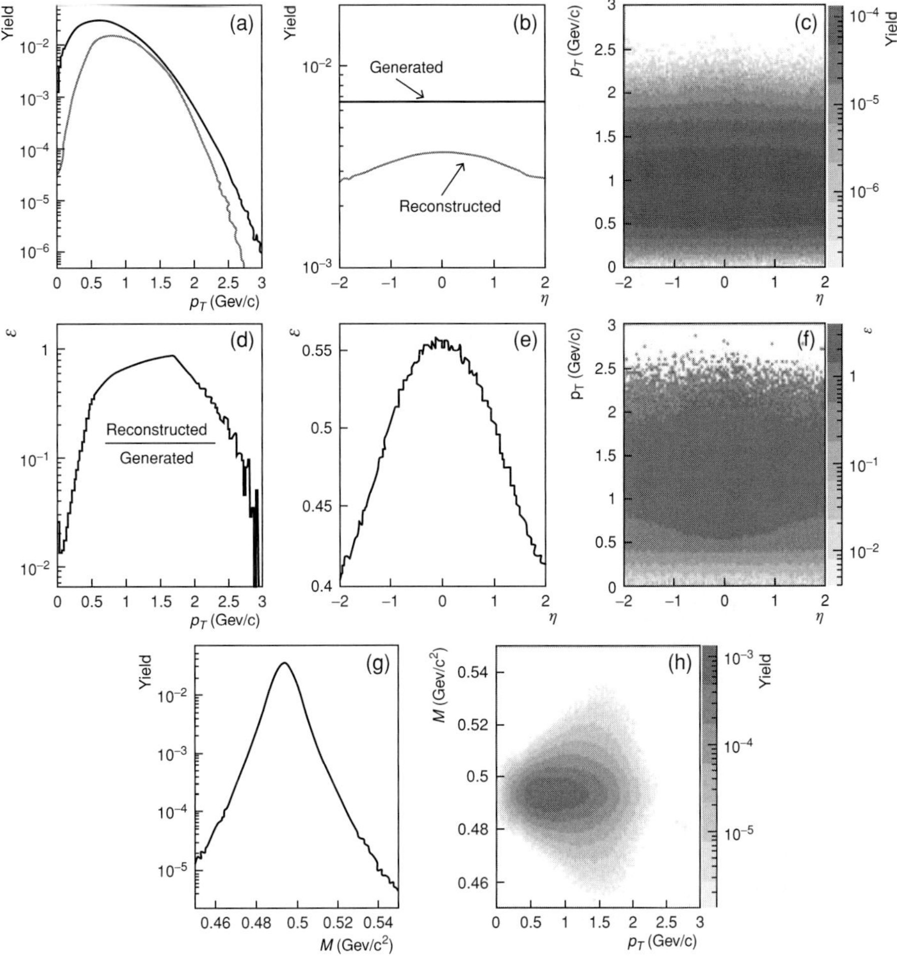

Fig. 14.6 Simulation of the decay and reconstruction of $K_s^0 \to \pi^+ + \pi^-$ with the invariant mass technique. (a) Generated (solid line) and reconstructed (dashed line) Maxwell–Boltzmann transverse momentum (p_T) spectra. (b) Generated and reconstructed pseudorapidity distributions. (c) Reconstructed distribution in p_T vs. η. (d) Acceptance and efficiency vs. p_T. (d) Efficiency vs. η; efficiency vs. p_T and η; (g) K_s^0 reconstructed mass spectrum. (h) Mass vs. reconstructed p_T. Pions are assumed detectable with a 100% efficiency if in the ranges $0.2 < p_T \leq 1.5$ and $|\eta| < 2$. Pion momentum resolution as in curve 2 of Figure 14.5.

using fast simulators such as those already discussed in previous sections. We here illustrate the determination of a response matrix toward corrections of measured charged particle multiplicity spectra.

For the sake of simplicity, we assume the particle multiplicity amounts to an average value $\langle m \rangle$, which we subtract ab initio from the simulated produced and measured distributions. The apparatus is assumed to have an efficiency such that measured multiplicities are, on average, 2.5 units lower than the actual values, and smeared with a resolution of 0.2

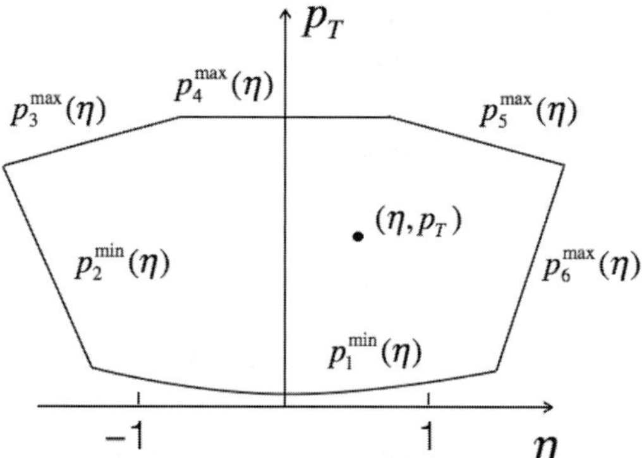

Fig. 14.7 Acceptance definition with k_B boundary functions $p_{T,i}^{\min/\max}(\eta, \phi), i = 1, \ldots, 6$.

units. The response is simulated with a Gaussian distribution

$$P(m_m|m_t) = \frac{1}{\sqrt{2\pi}\sigma_m} \exp\left(-\frac{[m_m - m_t - \mu_m]^2}{2\sigma_m^2}\right), \tag{14.79}$$

with $\mu_m = 2.5$, $\sigma_m = 0.2$, and where m_t and m_m represent the *true* and *measured* multiplicities, respectively. One million events were generated to obtain the response matrix displayed in Figure 14.8, which was used in §§12.3.7 and 12.3.8 to present examples of unfolding with the SVD and Bayesian methods, respectively.

More generally, one might wish to account for specificities of the apparatus resolution of particle momenta as well as a more detailed description of its acceptance. The simulation could then proceed as follows:

1. Create a response matrix histogram to store event values of produced and measured multiplicities noted N and n, respectively.

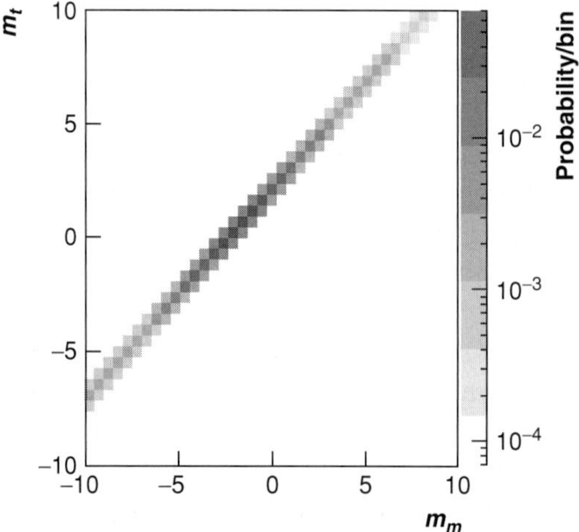

Fig. 14.8 Response matrix used for examples of SVD and Bayesian discussed in §§12.3.7 and 12.3.8.

2. Select a large number of events N_{ev} to be produced.
3. Define or select the function $\epsilon(\eta, \phi, p_T)$ to be used in the simulation of detection efficiency.
4. Define or select the conditions $B_i(\eta, \phi, p_T), i = 1, k$ to be used in the simulation of the experimental acceptance.
5. For each event,
   a. Determine the total multiplicity M of the event randomly according to an appropriate distribution in the multiplicity range or interest (e.g., $10 \leq M < 1{,}000$).
   b. Set particle counters N and n to zero,
   c. Generate an event with M particles according to appropriate rapidity and transverse momentum distributions (e.g., flat rapidity and Maxwell–Boltzmann p_T spectrum).
      i. Generate particles according to chosen production model,
      ii. For each generated particle,
         (a) Determine if the particle is within the rapidity, azimuthal angle, and p_T boundaries of the experiment.
         (b) If it is, increment true detected multiplicity N by one unit.
         (c) Use the efficiency function $\epsilon(\eta, \phi, p_T)$ to determine if the particle is detected.
         (d) If it is, smear the momentum and rapidity of the particle and determine whether the smeared momentum and rapidity fall within the kinematical boundaries of the experiment.
         (e) If it does, increment the measured multiplicity n by one unit.
   d. Increment the response function histogram according to the event's (N, n).
6. Normalize the response function histogram

14.2.4 Detailed Detector Simulations Using GEANT

The de facto standard in particle and nuclear physics for comprehensive simulations of detector performance is the program GEANT [14]. Specialized computer codes such as EGS 5 (Electron Gamma Shower)[105] and FLUKA [32] are also available to simulate the development of electromagnetic and hadronic showers within materials.

GEANT provides a comprehensive computing environment for simulating the passage of particles through matter. It includes components for modeling detector geometry and material properties, the propagation of particles through electric or magnetic fields, as well as the modeling of particle energy loss and deposition in detector materials. GEANT can thus be used to simulate the detection of arbitrarily complex collisions and study detector performance attributes such as detection acceptance, efficiency, momentum resolution, and much more. GEANT, now in its fourth version, provides tools to simulate with high precision a vast array of particle interactions within detector materials and sensors, including electromagnetic, hadronic and optical processes, as well as the decay of long-lived particles. This is accomplished through modeling of electromagnetic and hadronic processes over a wide energy range starting, in some cases, from 250 eV and extending in others to the TeV energy range. Written in C++, GEANT was designed and constructed to expose the physics models utilized, handle complex geometries, and enable its easy adaptation for optimal use in different applications. GEANT is the result of a worldwide collaboration of physicists and software engineers. It has a wide community of users and finds applications in particle physics, nuclear physics, accelerator design, space engineering, and medical physics. A detailed discussion of the Monte Carlo techniques and methods used in GEANT is well beyond the scope of this textbook but can be found in various publications (see ref. [14] and references therein).

While GEANT enables accurate and detailed simulations of detector performance, its application for certain tasks such as the description of hadronic showers in calorimeter can be rather fastidious and slow. Various groups have thus designed **fast simulators** to achieve the same goals as GEANT but requiring only a small fraction of its computing time, albeit with perhaps slightly less accuracy [17, 98].

Exercises

14.1 Derive the expression (14.8) of the moments of the power-law distribution.

14.2 Verify the expression (14.10) for the generation of random numbers according to a power law of the form (14.7).

14.3 Derive the expression (14.41) for the boost of a vector $\vec{p}^*$ in an arbitrary direction and velocity $\vec{\beta}_P$. Hint: Decompose the vector $\vec{p}^*$ in terms of components parallel and perpendicular to the boost direction and use the boost formula, Eq. (8.17), to carry the boost in the parallel direction.

14.4 Write a Monte Carlo program to calculate the acceptance and detection efficiency of K_s^0 decaying into $\pi^+ + \pi^-$. Assume the K_s^0 are produced with a Maxwell–Boltzmann distribution with $T = 0.40$ GeV. Consider the acceptance for a detector capable of identifying pions ($\pi^\pm$) in the momentum range $0.2 \le p \le 1.5$ GeV. Plot the acceptance and efficiency as a function of transverse momentum p_T and pseudorapidity η.

14.5 Write a Monte Carlo program using the weight method to simulate the production of correlated pions according to the HBT effect assuming source sizes of 2, 5, and 15 fm. Additionally assume the pions are produced with a Maxwell–Boltzmann distribution with $T = 0.40$ GeV.

References

[1] K. Aamodt, et al. (ALICE Collaboration). Higher harmonic anisotropic flow measurements of charged particles in pb-pb collisions at $\sqrt{s_{NN}} = 2.76$ TeV. *Physical Review Letters*, 107:032301, 2011.

[2] B. I. Abelev, et al. (STAR Collaboration). Centrality dependence of charged hadron and strange hadron elliptic flow from s(NN)**(1/2) = 200-GeV Au + Au collisions. *Physical Review C*, 77:054901, 2008.

[3] B. I. et al. (STAR Collaboration). Beam-energy and system-size dependence of dynamical net charge fluctuations. *Physical Review C*, 79:024906, 2009.

[4] B. I. Abelev, M. M. Aggarwal, Z. Ahammed, et al. K/π fluctuations at relativistic energies. *Physical Review Letters*, 103(9):092301, 2009.

[5] M. Abramowitz and I. A. Stegun. *Handbook of mathematical functions*. New York: Dover Publications, 1965.

[6] D. Adamova, et al. (CERES Collaboration). Event-by-event fluctuations of the mean transverse momentum in 40, 80, and 158 agev/c Pb + Au collisions. *Nuclear Physics A*, 727(1):97–119, 2003.

[7] J. Adams, et al. (STAR Collaboration). Azimuthal anisotropy in Au+Au collisions at s(NN)**(1/2) = 200-GeV. *Physical Review C*, 72:014904, 2005.

[8] J. Adams, et al. (STAR Collaboration). Event by event pT fluctuations in Au – Au collisions at $\sqrt{s_{NN}} = 130$ GeV. *Physical Review C*, 71:064906, 2005.

[9] J. Adams, et al. (STAR Collaboration). Incident energy dependence of pt correlations at RHIC. *Physical Review C*, 72:044902, 2005.

[10] C. Adler, et al. (STAR Collaboration). Elliptic flow from two- and four-particle correlations in Au+Au collisions at $\sqrt{s_{NN}} = 130$ GeV. *Physical Review C*, 66:034904, 2002.

[11] C. Adler, et al. (STAR Collaboration). Disappearance of back-to-back high-p_T hadron correlations in central Au+Au collisions at $\sqrt{s_{NN}} = 200$ GeV. *Physical Review Letters*, 90:082302, 2003.

[12] S. S. Adler, et al. (PHENIX Collaboration). Measurement of nonrandom event-by-event fluctuations of average transverse momentum in $\sqrt{s_{NN}} = 200$ GeV – Au and $p-p$ collisions. *Physical Review Letters*, 93(9):092301, 2004.

[13] T. Adye. Unfolding algorithms and tests using RooUnfold. *ArXiv e-prints*, May 2011.

[14] S. Agostinelli, J. Allison, K. Amako, et al. Geant4, a simulation toolkit. *Nuclear Instruments and Methods in Physics Research Section A: Accelerators, Spectrometers, Detectors and Associated Equipment*, 506(3):250–303, 2003.

[15] Murray A Aitkin, Brian Francis, and John Hinde. *Statistical modelling in GLIM 4*, Volume 32. Oxford: Oxford University Press, 2005.

[16] N. N. Ajitanand, J. M. Alexander, P. Chung, et al. Decomposition of harmonic and jet contributions to particle-pair correlations at ultrarelativistic energies. *Physical Review C*, 72:011902, Jul 2005.

[17] M. Albrow and R. Raja. In *Proceedings, Hadronic Shower Simulation Workshop*, Batavia, September 6–8, 2006. *AIP Conference Proceedings*, 896:1–254, 2007.

[18] T. Alexopoulos, M. Bachtis, E. Gazis, and G. Tsipolitis. Implementation of the legendre transform for track segment reconstruction in drift tube chambers. *Nuclear Instruments and Methods in Physics Research Section A: Accelerators, Spectrometers, Detectors and Associated Equipment*, 592(3):456–462, 2008.

[19] B. Alver, B. B. Back, M. D. Baker, et al. Importance of correlations and fluctuations on the initial source eccentricity in high-energy nucleus-nucleus collisions. *Physical Review C*, 77:014906, 2008.

[20] C. Amsler et al. (Particle Data Group). Review of particle physics. *Physics Letters B*, 667(1):1–6, 2008.

[21] A. Andronic and J. P. Wessels. Transition radiation detectors. *Nuclear Instruments and Methods in Physics Research A*, 666:130–147, 2012.

[22] T. Anticic, B. Baatar, D. Barna, et al. Transverse momentum fluctuations in nuclear collisions at 158A gev. *Physical Review C*, 70(3):034902, 2004.

[23] V. B. Anykeyev, A. A. Spiridonov, and V. P. Zhigunov. Comparative investigation of unfolding methods. *Nuclear Instruments and Methods in Physics Research Section A: Accelerators, Spectrometers, Detectors and Associated Equipment*, 303(2):350–369, 1991.

[24] J. P. Archambault, A. Artamonov, M. Cadabeschi, et al. Energy calibration of the atlas liquid argon forward calorimeter. *Journal of Instrumentation*, 3(2):P02002, 2008.

[25] Paul Avery. Applied fitting theory vi: Formulas for kinematic fitting. *CLEO Note CBX*, 98–37, 1998.

[26] Roger J. Barlow. *Statistics: A guide to the use of statistical methods in the physical sciences*, volume 29. New York: John Wiley & Sons, 1989.

[27] V. E. Barnes, P. L. Connolly, D. J. Crennell, et al. Observation of a hyperon with strangeness minus three. *Physical Review Letters*, 12:204–206, 1964.

[28] R. M. Barnett, C. D. Carone, D. E. Groom, et al. Review of particle physics. *Physical Review D*, 54:1–708, Jul 1996.

[29] S. A. Bass, M. Bleicher, M. Brandstetter, C. Ernst, L. Gerland, et al. Urqmd: A new molecular dynamics model from ganil to CERN energies. In *Structure of vacuum and elementary matter: Proceedings, International Conference on Nuclear Physics at the Turn of the Millennium*, Wilderness, South Africa, March 10–16, 1996, pp. 399–405.

[30] Steffen A. Bass, Pawel Danielewicz, and Scott Pratt. Clocking hadronization in relativistic heavy-ion collisions with balance functions. *Physical Review Letters*, 85:2689–2692, 2000.

[31] Steffen A. Bass, Pawel Danielewicz, and Scott Pratt. Clocking hadronization in relativistic heavy ion collisions with balance functions. *Physical Review Letters*, 85:2689–2692, 2000.

[32] Giuseppe Battistoni, F. Cerutti, A. Fasso, A. Ferrari, S. Muraro, J. Ranft, S. Roesler, and P. R. Sala. The fluka code: Description and benchmarking. In *Hadronic Shower Simulation Workshop (AIP Conference Proceedings*, Volume 896), pp. 31–49, 2007.

[33] S. S. Baturin and A. D. Kanareykin. Cherenkov radiation from short relativistic bunches: General approach. *Physical Review Letters*, 113:214801, 2014.

[34] Florian Beaudette. The CMS particle flow algorithm. arXiv:1401.8155 [hep-ex].

[35] J. Beringer, et al. (Particle Data Group). Review of particle physics. *Physical Review D*, 86:324–325, 2012.

[36] Didier H. Besset. *Object-oriented implementation of numerical methods: An introduction with Java and Smalltalk*. Burlington, MA: Morgan Kaufmann, 2001.

[37] Rajeev S. Bhalerao, Jean-Yves Ollitrault, and Subrata Pal. Characterizing flow fluctuations with moments. *Physics Letters*, B742:94–98, 2015.

[38] R. S. Bhalerao, N. Borghini, and J. Y. Ollitrault. Analysis of anisotropic flow with Lee-Yang zeroes. *Nuclear Physics A*, 727:373–426, 2003.

[39] Hans Bichsel. A method to improve tracking and particle identification in tpcs and silicon detectors. *Nuclear Instruments and Methods in Physics Research Section A: Accelerators, Spectrometers, Detectors and Associated Equipment*, 562(1):154–197, 2006.

[40] Pierre Billoir and Sijin Qian. Simultaneous pattern recognition and track filtering by the Kalman filtering method. *Nuclear Instruments and Methods in Physics Research A*, 294(1–2):219–228, 1990.

[41] Pierre Billoir. Further test for the simultaneous pattern recognition and track filtering by the Kalman filtering method. *Nuclear Instruments and Methods in Physics Research A*, 295(3):492–500, 1990.

[42] Pierre Billoir. Track fitting with multiple scattering: A new method. *Nuclear Instruments and Methods in Physics Research*, 225(2):352–366, 1984.

[43] Pierre Billoir. Progressive track recognition with a Kalman-like fitting procedure. *Computer Physics Communications*, 57(1):390–394, 1989.

[44] Pierre Billoir, R. Frühwirth, and Meinhard Regler. Track element merging strategy and vertex fitting in complex modular detectors. *Nuclear Instruments and Methods in Physics Research Section A: Accelerators, Spectrometers, Detectors and Associated Equipment*, 241(1):115–131, 1985.

[45] Pierre Billoir and Sijin Qian. Fast vertex fitting with a local parametrization of tracks. *Nuclear Instruments and Methods in Physics Research Section A: Accelerators, Spectrometers, Detectors and Associated Equipment*, 311(1):139–150, 1992.

[46] M. Bleicher, S. Jeon, and V. Koch. Event-by-event fluctuations of the charged particle ratio from nonequilibrium transport theory. *Physical Review C*, 62:061902, 2000.

[47] V. Blobel. Alignment algorithms. In S. Blusk, O. Buchmuller, A. Jacholkowski, T. Ruf, J. Schieck, and S. Viret (eds.), *Proceedings, First LHC Detector Alignment Workshop*, CERN, Geneva, Switzerland, September 4–6, 2006, p. 5.

[48] Volker Blobel. An unfolding method for high energy physics experiments. *arXiv preprint hep-ex/0208022*, 2002.

[49] Christoph Blume. Search for the critical point and the onset of deconfinement. *Central European Journal of Physics*, 10(6):1245–1253, 2012.

[50] Rudolf K. Bock and Angela Vasilescu. *The particle detector briefbook*. Berlin and Heidelberg: Springer-Verlag 1998.

[51] G. Bonvicini. Private communication, 2009.

[52] Fritz W. Bopp, R. Engel, and J. Ranft. Rapidity gaps and the PHOJET Monte Carlo. pp. 729–741, 1998.

[53] N. Borghini, R. S. Bhalerao, and J.-Y. Ollitrault. Anisotropic flow from lee yang zeros: A practical guide. *Journal of Physics G: Nuclear and Particle Physics*, 30(8):S1213, 2004.

[54] Nicolas Borghini. Multiparticle correlations from momentum conservation. *European Physical Journal*, C30:381–385, 2003.

[55] Nicolas Borghini. Momentum conservation and correlation analyses in heavy-ion collisions at ultrarelativistic energies. *Physical Review C*, 75:021904, 2007.

[56] Nicolas Borghini, Phuong Mai Dinh, and Jean-Yves Ollitrault. Flow analysis from cumulants: A practical guide. In *International Workshop on the Physics of the Quark Gluon Plasma Palaiseau*, France, September 4–7, 2001.

[57] G. E. P. Box and Mervin E. Muller. A note on the generation of random normal deviates. *Annals of Mathematical Statistics*, 29(2):610–611, 1958.

[58] Lawrence D. Brown, T. Tony Cai, and Anirban DasGupta. Interval estimation for a binomial proportion. *Statistical Science*, 16(2):101–133, 2001.

[59] Rene Brun and Fons Rademakers. ROOT, an object oriented data analysis framework. *Nuclear Instruments and Methods in Physics Research Section A: Accelerators, Spectrometers, Detectors and Associated Equipment*, 389(1):81–86, 1997.

[60] Lars Bugge and Jan Myrheim. Tracking and track fitting. *Nuclear Instruments and Methods*, 179(2):365–381, 1981.

[61] Alessandro Buzzatti and Miklos Gyulassy. An overview of the CUJET model: Jet Flavor Tomography applied at RHIC and LHC. *Nuclear Physics A*, 910–911:490–493, 2013.

[62] Vladimir Cerny. Thermodynamical approach to the traveling salesman problem: An efficient simulation algorithm. *Journal of Optimization Theory and Applications*, 45(1):41–51, 1985.

[63] Zhe Chen. Bayesian filtering: From kalman filters to particle filters, and beyond. *Statistics*, 182(1):1–69, 2003.

[64] Sen Cheng, Silvio Petriconi, Scott Pratt, Michael Skoby, Charles Gale, et al. Statistical and dynamic models of charge balance functions. *Physical Review C*, 69:054906, 2004.

[65] C. J. Clopper and E. S. Pearson. The use of confidence or fiducial limits illustrated in the case of the binomial. *Biometrika*, 26(4):404–413, 1934.

[66] G. Corcella, I. G. Knowles, G. Marchesini, S. Moretti, K. Odagiri, et al. HERWIG 6: An event generator for hadron emission reactions with interfering gluons (including supersymmetric processes). *JHEP*, 0101:010, 2001.

[67] Glen Cowan. *Statistical data analysis*. Oxford: Oxford University Press, 1998.

[68] Glen Cowan. A survey of unfolding methods for particle physics. In *Prepared for Conference on Advanced Statistical Techniques in Particle Physics*, Durham, England, pp. 18–22, 2002.

[69] Richard T. Cox. Probability, frequency and reasonable expectation. *American Journal of Physics*, 14(1):1–13, 1946.

[70] J. L. Crassidis and J. L. Junkins. *Optimal estimation of dynamic systems*. Boca Raton, FL: CRC Press, 2004.

[71] S. Cucciarelli. The performance of the cms pixel detector and the primary-vertex finding. *Nuclear Instruments and Methods in Physics Research Section A: Accelerators, Spectrometers, Detectors and Associated Equipment*, 549(1):49–54, 2005.

[72] G. D'Agostini. Fits, and especially linear fits, with errors on both axes, extra variance of the data points and other complications. *ArXiv Physics e-prints*, November 2005.

[73] G. D'Agostini. Improved iterative Bayesian unfolding. *ArXiv e-prints*, October 2010.

[74] R. H. Dalitz. Cxii. on the analysis of τ-meson data and the nature of the τ-meson. *Philosophical Magazine*, 44(357):1068–1080, 1953.

[75] R. H. Dalitz. Decay of τ mesons of known charge. *Physical Review*, 94(4):1046, 1954.

[76] J. Damgov, V. Genchev, and S. C. Mavrodiev. Nonlinear energy calibration of CMS calorimeters for single pions. 2001.

[77] S. R. Das. On a new approach for finding all the modified cut-sets in an incompatibility graph. *IEEE Transactions on Computers*, 100(2):187–193, 1973.

[78] F. W. Dyson, A. S. Eddington, and C. Davidson. A determination of the deflection of light by the sun's gravitational field, from observations made at the total eclipse of May 29, 1919. *Philosophical Transactions of the Royal Society of London A: Mathematical, Physical and Engineering Sciences*, 220(571–581):291–333, 1920.

[79] Christian W. Fabjan and T. Ludlam. Calorimetry in high-energy physics. *Annual Review of Nuclear and Particle Science*, 32(1):335–389, 1982.

[80] Gary J. Feldman and Robert D. Cousins. Unified approach to the classical statistical analysis of small signals. *Physical Review D*, 57(7):3873, 1998.

[81] Thomas Ferbel. *Experimental techniques in high-energy nuclear and particle physics*. Singapore: World Scientific, 1991.

[82] Daniel Fink. A compendium of conjugate priors. See www.people.cornell.edu/pages/df36/CONJINTRnew%20TEX.pdf, p. 46, 1997.

[83] Y. Fisyak. Private communication, 2015.

[84] R. Fletcher. *Practical methods of optimization: Unconstrained optimization*. New York: John Wiley & Sons, 1980.

[85] L. Foa. Inclusive study of high-energy multiparticle production and two-body correlations. *Physics Reports*, 22:1–56, 1975.

[86] R. Frühwirth. Application of Kalman filtering to track and vertex fitting. *Nuclear Instruments and Methods in Physics Research Section A: Accelerators, Spectrometers, Detectors and Associated Equipment*, 262(2–3):444–450, 1987.

[87] R. Frühwirth, Pascal Vanlaer, and Wolfgang Waltenberger. Adaptive vertex fitting. Technical report, CERN-CMS-NOTE-2007-008, 2007.

[88] E. Widl and Rudolph Frühwirth. Application of the Kalman alignment algorithm to the CMS tracker. *Journal of Physics: Conference Series*, 291(3):32–65, 2010.

[89] Rudolf Frühwirth and Meinhard Regler. *Data analysis techniques for high-energy physics*, Vol. 11. Cambridge: Cambridge University Press, 2000.

[90] Daihong Fu, Kenneth C Dyer, Stephen H. Lewis, and Paul J. Hurst. A digital background calibration technique for time-interleaved analog-to-digital converters. *IEEE Journal of Solid-State Circuits*, 33(12):1904–1911, 1998.

[91] N. D. Gagunashvili. Densities mixture unfolding for data obtained from detectors with finite resolution and limited acceptance. *Nuclear Instruments and Methods in Physics Research Section A: Accelerators, Spectrometers, Detectors and Associated Equipment*, 778:92–101, 2015.

[92] M. Gazdzicki and S. Mrowczynski. A Method to study 'equilibration' in nucleus-nucleus collisions. *Zeitschrift für Physik C*, 54:127–132, 1992.

[93] A. Gelb. *Applied optimal estimation*. Cambridge, MA: MIT Press, 2004.

[94] A. Glazov, I. Kisel, E. Konotopskaya, and G. Ososkov. Filtering tracks in discrete detectors using a cellular automaton. *Nuclear Instruments and Methods in Physics Research Section A: Accelerators, Spectrometers, Detectors and Associated Equipment*, 329(1):262–268, 1993.

[95] I. A. Golutvin, Yu T. Kiryushin, S. A. Movchan, G. A. Ososkov, V. V. Pal'chik, and E. A. Tikhonenko. Robust optimal estimates of the parameters of muon track segments in cathode strip chambers for cms experiments. *Instruments and Experimental Techniques*, 45(6):735–741, 2002.

[96] T. G. Gregoire. Design-based and model-based inference in survey sampling: Appreciating the difference. *Canadian Journal of Forest Research*, 28(10):1429–1447, 1998.

[97] Phil Gregory. *Bayesian logical data analysis for the physical sciences: A comparative approach with mathematica support*. Cambridge: Cambridge University Press, 2005.

[98] Guenter Grindhammer, M. Rudowicz, and S. Peters. The fast simulation of electromagnetic and hadronic showers. *Nuclear Instruments and Methods*, 290:469, 1990.

[99] Claus Grupen and Boris Schwartz. Particle detectors. http://ajbell.web.cern.ch/ajbell/Documents/eBooks/Particle_Detectors_Grupen.pdf 2008.

[100] M. Haahr. True random number service. www.random.org/.

[101] F. R. Hampel, E. Ronchetti, P. Rousseeuw, and W. Stahel. *Robust statistics: The approach based on influence functions*. New York: John Wiley & Sons, 1986.

[102] M. Hansroul, H. Jeremie, and D. Savard. Fast circle fit with the conformal mapping method. *Nuclear Instruments and Methods in Physics Research Section A: Accelerators, Spectrometers, Detectors and Associated Equipment*, 270(2):498–501, 1988.

[103] H. L. Harney. *Bayesian inference: Parameter estimation and decisions*. Advanced Texts in Physics. Berlin and Heidelberg: Springer, 2014.

[104] Stefan Heckel, et al. (ALICE Collaboration). Event-by-event mean pt fluctuations in pp and PPb collisions measured by the alice experiment at the lhc. *Journal of Physics G: Nuclear and Particle Physics*, 38(12):124095, 2011.

[105] H. Hirayama, Y. Namito, A. F. Bielajew, S. J. Wilderman, and W. R. Nelson. The EGS5 code system. 2005.

[106] Andreas Hocker and Vakhtang Kartvelishvili. SVD approach to data unfolding. *Nuclear Instruments and Methods A*, 372:469–481, 1996.

[107] John H. Holland. *Adaptation in natural and artificial systems*. Cambridge, MA: MIT Press, 1992.

[108] C. Hoppner, S. Neubert, B. Ketzer, and S. Paul. A novel generic framework for track fitting in complex detector systems. *Nuclear Instruments and Methods in Physics Research Section A: Accelerators, Spectrometers, Detectors and Associated Equipment*, 620(2–3):518–525, 2010.

[109] P. V. C. Hough. Machine analysis of bubble chamber pictures. *Conference Proceedings*, C590914:554–558, 1959.

[110] Kerson Huang. *Introduction to statistical mechanics*. Hoboken, NJ: John Wiley & Sons, 2001.

[111] W. D. Hulsbergen. The global covariance matrix of tracks fitted with a Kalman filter and an application in detector alignment. *Nuclear Instruments and Methods in Physics Research Section A: Accelerators, Spectrometers, Detectors and Associated Equipment*, 600(2):471–477, 2009.

[112] David J. Jackson. A topological vertex reconstruction algorithm for hadronic jets. *Nuclear Instruments and Methods in Physics Research Section A: Accelerators, Spectrometers, Detectors and Associated Equipment*, 388(1):247–253, 1997.

[113] F. James. Monte Carlo phase space. *CERN 68-15*, 1968.

[114] F. James. A review of pseudorandom number generators. *Computer Physics Communications*, 60(3):329–344, 1990.

[115] F. James and M. Roos. Minuit: A system for function minimization and analysis of the parameter errors and correlations. *Computer Physics Communications*, 10:343–367, 1975.

[116] Frederick James. *Statistical methods in experimental physics*, Vol. 7. Singapore: World Scientific, 2006.

[117] Edwin T. Jaynes. *Probability theory: The logic of science*. Cambridge: Cambridge University Press, 2003.

[118] Harold Jeffreys. *Scientific inference*. Cambridge: Cambridge University Press, 1973.

[119] Harold Jeffreys. *The theory of probability*. Oxford: Oxford University Press, 1998.

[120] S. Jeon and V. Koch. Fluctuations of particle ratios and the abundance of hadronic resonances. *Physical Review Letters*, 83:5435–5438, 1999.

[121] S. Jeon and V. Koch. Charged particle ratio fluctuation as a signal for QGP. *Physical Review Letters*, 85:2076–2079, 2000.

[122] Rudolph Emil Kalman. A new approach to linear filtering and prediction problems. *Journal of Fluids Engineering*, 82(1):35–45, 1960.

[123] M. G. Kendall and A. Stuart. *The advanced theory of statistics*, Vol. 1. London: C. Griffin and Co., 1969.

[124] Dmitri E. Kharzeev. The chiral magnetic effect and anomaly-induced transport. *Progress in Particle and Nuclear Physics*, 75(0):133–151, 2014.

[125] Dmitri E. Kharzeev, Larry D. McLerran, and Harmen J. Warringa. The effects of topological charge change in heavy ion collisions: "Event by event p. and cp violation." *Nuclear Physics A*, 803(3):227–253, 2008.

[126] Scott Kirkpatrick, C. Daniel Gelatt, Mario P. Vecchi, et al. Optimization by simulated annealing. *Science*, 220(4598):671–680, 1983.

[127] Ivan Kisel. Event reconstruction in the cbm experiment. *Nuclear Instruments and Methods in Physics Research Section A: Accelerators, Spectrometers, Detectors and Associated Equipment*, 566(1):85–88, 2006.

[128] W. Kittel and E. A. De Wolf. *Soft Multihadron Dynamics*. Singapore: World Scientific, 2005.

[129] Andrej N. Kolmogorov. *Sulla determinazione empirica di una legge di distribuzione*. na, 1933.

[130] G. Kopylov. *JETP*, 8:996, 1959.

[131] G. I. Kopylov. Like particle correlations as a tool to study the multiple production mechanism. *Physics Letters B*, 50(4):472–474, 1974.

[132] Thomas S. Kuhn. *The structure of scientific revolutions*. Chicago: University of Chicago Press, 2012.

[133] D. J. Lange. The EvtGen particle decay simulation package. *Nuclear Instruments and Methods*, A462:152–155, 2001.

[134] Bin Le, T. W. Rondeau, J. H. Reed, and C. W. Bostian. Analog-to-digital converters. *IEEE Signal Processing Magazine*, 22(6):69–77, Nov 2005.

[135] William R. Leo. *Techniques for nuclear and particle physics experiments: A how-to approach*. Heidelberg and New York: Springer-Verlag, 1994.

[136] Kenneth Levenberg. A method for the solution of certain non–linear problems in least squares. 1944.

[137] Zi-wei Lin, Subrata Pal, C.M. Ko, Bao-An Li, and Bin Zhang. Multiphase transport model for heavy ion collisions at RHIC. *Nuclear Physics A*, 698:375–378, 2002.

[138] Louis Lyons. Open statistical issues in particle physics. *Annals of Applied Statistics*, 887–915, 2008.

[139] Rainer Mankel. A concurrent track evolution algorithm for pattern recognition in the hera-b main tracking system. *Nuclear Instruments and Methods in Physics Research Section A: Accelerators, Spectrometers, Detectors and Associated Equipment*, 395(2):169–184, 1997.

[140] Donald W Marquardt. An algorithm for least-squares estimation of nonlinear parameters. *Journal of the Society for Industrial and Applied Mathematics*, 11(2):431–441, 1963.

[141] Makoto Matsumoto and Takuji Nishimura. Mersenne twister: A 623-dimensionally equidistributed uniform pseudo-random number generator. *ACM Transactions on Modeling and Computer Simulation*, 8(1):3–30, January 1998.

[142] S. Mertens, T. Lasserre, S. Groh, et al. Sensitivity of next-generation tritium beta-decay experiments for keV-scale sterile neutrinos. *JCAP*, 1502(02):20, 2015.

[143] Nicholas Metropolis, Arianna W. Rosenbluth, Marshall N. Rosenbluth, Augusta H. Teller, and Edward Teller. Equation of state calculations by fast computing machines. *The Journal of Chemical Physics*, 21(6):1087–1092, 1953.

[144] Zbigniew Michalewicz. *Genetic Algorithms + Data Structures = Evolution Programs* (2nd, extended ed.). New York: Springer-Verlag, 1994.

[145] Melanie Mitchell. *An introduction to genetic algorithms*. Cambridge, MA: MIT Press, 1996.

[146] Stanislaw Mrowczynski. Transverse momentum and energy correlations in the equilibrium system from high-energy nuclear collisions. *Physics Letters*, B439:6–11, 1998.

[147] John A. Nelder and Roger Mead. A simplex method for function minimization. *Computer Journal*, 7(4):308–313, 1965.

[148] Robert G. Newcombe. Two-sided confidence intervals for the single proportion: comparison of seven methods. *Statistics in Medicine*, 17(8):857–872, 1998.

[149] Jerzy Neyman. Outline of a theory of statistical estimation based on the classical theory of probability. *Philosophical Transactions of the Royal Society of London. Series A, Mathematical and Physical Sciences*, 236(767):333–380, 1937.

[150] K. A. Olive, et al. (Particle Data Group). Review of particle physics. *Chinese Physics*, C38:090001, 2014.

[151] Jean-Yves Ollitrault. On the measurement of azimuthal anisotropies in nucleus-nucleus collisions. 1997.

[152] W. Pinganaud. Contributions to the development of the silicon microstrip detector for the star experiment at rhic. 2000. Thesis, Universite de Nantes.

[153] J. Podolanski and R. Armenteros. Iii. analysis of v-events. *Philosophical Magazine*, 45(360):13–30, 1954.

[154] M. J. D. Powell. An efficient method for finding the minimum of a function of several variables without calculating derivatives. *The Computer Journal*, 7(2):155–162, 1964.

[155] Scott Pratt. General charge balance functions: A tool for studying the chemical evolution of the Quark-Gluon Plasma. *Physical Review C*, 85:014904, 2012.

[156] William H Press. *Numerical recipes, 3rd ed.: The art of scientific computing*. Cambridge: Cambridge University Press, 2007.

[157] C. Pruneau, S. Gavin, and S. Voloshin. Methods for the study of particle production fluctuations. *Physical Review C*, 66:044904, 2002.

[158] C. A. Pruneau. Methods for jet studies with three-particle correlations. *Physical Review C*, 74:064910, 2006.

[159] C. A. Pruneau. Net charge fluctuations at RHIC. *Acta Physica Hungarica*, A25:401–408, 2006.

[160] C. A. Pruneau. Three-particle cumulant study of conical emission. *Physical Review C*, 79:044907, 2009.

[161] C. A. Pruneau, S. Gavin, and S. Voloshin. Net charge dynamic fluctuations. *Nuclear Physics A*, 715:661–664, 2003.

[162] Claude A. Pruneau, Sean Gavin, and Sergei A. Voloshin. Transverse radial flow effects on two- and three-particle angular correlations. *Nuclear Physics A*, 802(14):107–121, 2008.

[163] Shantam Ravan, Prabhat Pujahari, Sidharth Prasad, and Claude A. Pruneau. Correcting Correlation Function Measurements. *Physical Review C*, 89(2):024906, 2014.

[164] Thorsten Renk. YaJEM: A Monte Carlo code for in-medium shower evolution. *International Journal of Modern Physics E*, 20:1594–1599, 2011.

[165] F. Ronchetti and the ALICE collaboration. The alice electromagnetic calorimeter project. *Journal of Physics: Conference Series*, 160(1):012012, 2009.

[166] Peter J. Rousseeuw and Annick M. Leroy. *Robust regression and outlier detection*, Vol. 589. Hoboken, NJ: John Wiley & Sons, 2005.

[167] Richard Royall. *Statistical evidence: A likelihood paradigm*, Vol. 71. Boca Raton, FL: CRC press, 1997.

[168] Andrew P. Sage and James L. Melsa. *Estimation theory with applications to communications and control*. New York: McGraw-Hill, 1971.

[169] Bjoern Schenke, Charles Gale, and Sangyong Jeon. MARTINI: An Event generator for relativistic heavy-ion collisions. *Physical Review C*, 80:054913, 2009.

[170] Ekkard Schnedermann, Josef Sollfrank, and Ulrich Heinz. Thermal phenomenology of hadrons from 200 A gev s+s collisions. *Physical Review C*, 48:2462–2475, Nov 1993.

[171] Michael Schulz, Daniel Fischer, Thomas Ferger, Robert Moshammer, and Joachim Ullrich. Four-particle dalitz plots to visualize atomic break-up processes. *Journal of Physics B: Atomic, Molecular and Optical Physics*, 40(15):3091, 2007.

[172] Claude E. Shannon and Warren Weaver. *The mathematical theory of communication*. University of Illinois Press, 2015.

[173] M. Sharma and C. A. Pruneau. Methods for the study of transverse momentum differential correlations. *Physical Review C*, 79:024905, 2009.

[174] Torbjorn Sjostrand and Mats Bengtsson. The Lund Monte Carlo for jet fragmentation and e+ e- physics. Jetset Version 6.3: An update. *Computer Physics Communications*, 43:367, 1987.

[175] Torbjorn Sjostrand, Stephen Mrenna, and Peter Z. Skands. A brief introduction to PYTHIA 8.1. *Computer Physics Communications*, 178:852–867, 2008.

[176] H. Sorge, R. Mattiello, A. Jahns, Horst Stoecker, and W. Greiner. Meson production in P-A and A-A collisions at AGS energies. *Physics Letters*, B271:37–42, 1991.

[177] P. Srivastava and G. Sudarshan. Multiple production of pions in nuclear collisions. *Physical Review*, 110:765–766, 1958.

[178] Are Strandlie and Rudolf Frühwirth. Track and vertex reconstruction: From classical to adaptive methods. *Reviews of Modern Physics*, 82(2):1419, 2010.

[179] S. Viret, C. Parkes, and M. Gersabeck. Alignment procedure of the "LHCb" vertex detector. *Nuclear Instruments and Methods in Physics Research Section A: Accelerators, Spectrometers, Detectors and Associated Equipment*, 596(2):157–163, 2008.

[180] Sergei Voloshin. Fluctuations in the mixed event technique. 1994.

[181] Sergei Voloshin. Transverse radial expansion in nuclear collisions and two particle correlations. *Physics Letters B*, 632(4):490–494, 2006.

[182] Sergei Voloshin, Arthur M. Poskanzer, and Raimond Snellings. Collective phenomena in non-central nuclear collisions. 2008.

[183] Sergei A. Voloshin. Discussing the possibility of observation of parity violation in heavy ion collisions. *Physical Review C*, 62:044901, Aug 2000.

[184] Sergei A. Voloshin. Toward the energy and the system size dependence of elliptic flow: Working on flow fluctuations. 2006.

[185] Sergei A Voloshin, Arthur M Poskanzer, Aihong Tang, and Gang Wang. Elliptic flow in the gaussian model of eccentricity fluctuations. *Physics Letters B*, 659(3):537–541, 2008.

[186] R. H. Walden. Analog-to-digital converter survey and analysis. *IEEE Journal on Selected Areas in Communications*, 17(4):539–550, Apr 1999.

[187] Wolfgang Waltenberger. Development of vertex finding and vertex fitting algorithms for CMS.

[188] Xin-Nian Wang and Miklos Gyulassy. HIJING: A Monte Carlo model for multiple jet production in p p, p A and A A collisions. *Physical Review D*, 44:3501–3516, 1991.

[189] Yongge Wang and Tony Nicol. *Statistical Properties of Pseudo Random Sequences and Experiments with PHP and Debian OpenSSL*, Vol. 8712 of *Lecture Notes in Computer Science*. Cham, Switzerland: Springer International Publishing, 2014.

[190] Hans-Peter Wellisch, J. P. Kubenka, H. Oberlack, and P. Schacht. *Hadronic calibration of the H1 LAr calorimeter using software weighting techniques*. Max-Planck-Inst. für Physik, 1994.

[191] K. Werner. Analysis of proton – nucleus and nucleus-nucleus scattering at 200-agev by the multistring model Venus. *Physics Letters*, B208:520, 1988.

[192] Klaus Werner. The hadronic interaction model EPOS. *Nuclear Physics Proceedings*, 175–176:81–87, 2008.

[193] J. Whitmore. Multiparticle Production in the Fermilab Bubble Chambers. *Physics Reports*, 27:187–273, 1976.

[194] Richard Wigmans. Advances in hadron calorimetry. *Annual Review of Nuclear and Particle Science*, 41(1):133–185, 1991.

[195] M. C. S. Williams. Particle identification using time of flight. *Journal of Physics G*, 39:123001, 2012.

[196] Edwin B. Wilson. Probable inference, the law of succession, and statistical inference. *Journal of the American Statistical Association*, 22(158):209–212, 1927.

[197] William A Zajc. kno scaling isn't what it used to be. *Physics Letters B*, 175(2):219–222, 1986.

[198] Korinna C. Zapp, Frank Krauss, and Urs A. Wiedemann. A perturbative framework for jet quenching. *JHEP*, 1303:080, 2013.

[199] P. Zarchan and H. Musoff. *Fundamentals of Kalman filtering: A practical approach*. American Institute of Aeronautics and Astronautics, 2000.

Index